T0140478

Advances in Intelligent Systems and Computing

Volume 711

Series editor

Janusz Kacprzyk, Polish Academy of Sciences, Warsaw, Poland
e-mail: kacprzyk@ibspan.waw.pl

The series "Advances in Intelligent Systems and Computing" contains publications on theory, applications, and design methods of Intelligent Systems and Intelligent Computing. Virtually all disciplines such as engineering, natural sciences, computer and information science, ICT, economics, business, e-commerce, environment, healthcare, life science are covered. The list of topics spans all the areas of modern intelligent systems and computing such as: computational intelligence, soft computing including neural networks, fuzzy systems, evolutionary computing and the fusion of these paradigms, social intelligence, ambient intelligence, computational neuroscience, artificial life, virtual worlds and society, cognitive science and systems, Perception and Vision, DNA and immune based systems, self-organizing and adaptive systems, e-Learning and teaching, human-centered and human-centric computing, recommender systems, intelligent control, robotics and mechatronics including human-machine teaming, knowledge-based paradigms, learning paradigms, machine ethics, intelligent data analysis, knowledge management, intelligent agents, intelligent decision making and support, intelligent network security, trust management, interactive entertainment, Web intelligence and multimedia.

The publications within "Advances in Intelligent Systems and Computing" are primarily proceedings of important conferences, symposia and congresses. They cover significant recent developments in the field, both of a foundational and applicable character. An important characteristic feature of the series is the short publication time and world-wide distribution. This permits a rapid and broad dissemination of research results.

More information about this series at http://www.springer.com/series/11156

Himansu Sekhar Behera
Janmenjoy Nayak · Bighnaraj Naik
Ajith Abraham
Editors

Computational Intelligence in Data Mining

Proceedings of the International Conference on CIDM 2017

 Springer

Editors
Himansu Sekhar Behera
Department of Computer Science and
    Engineering & Information Technology
Veer Surendra Sai University of Technology
Sambalpur, Odisha
India

Janmenjoy Nayak
Department of Computer Science and
    Engineering
Sri Sivani College of Engineering (SSCE)
Srikakulam, Andhra Pradesh
India

Bighnaraj Naik
Department of Computer Application
Veer Surendra Sai University of Technology
Sambalpur, Odisha
India

Ajith Abraham
Machine Intelligence Research (MIR) Lab
Auburn, WA
USA

and

Technical University of Ostrava
Ostrava
Czech Republic

ISSN 2194-5357   ISSN 2194-5365   (electronic)
Advances in Intelligent Systems and Computing
ISBN 978-981-10-8054-8   ISBN 978-981-10-8055-5   (eBook)
https://doi.org/10.1007/978-981-10-8055-5

Library of Congress Control Number: 2017964255

Printed on acid-free paper

This Springer imprint is published by the registered company Springer Nature Singapore Pte Ltd.
The registered company address is: 152 Beach Road, #21-01/04 Gateway East, Singapore 189721, Singapore

Preface

In the next decade, the growth of data both structured and unstructured will present challenges as well as opportunities for industries and academia. The present scenario of storage of the amount of data is quite huge in the modern database due to the availability and popularity of the Internet. Thus, the information needs to be summarized and structured in order to maintain effective decision-making. With the explosive growth of data volumes, it is essential that real-time information that is of use to the business can be extracted to deliver better insights to decision-makers, understand complex patterns, etc. When the quantity of data, dimensionality and complexity of the relations in the database are beyond human capacities, there is a requirement for intelligent data analysis techniques, which could discover useful knowledge from data. While data mining evolves with innovative learning algorithms and knowledge discovery techniques, computational intelligence harnesses the results of data mining for becoming more intelligent than ever. In the present scenario of computing, computational intelligence tools offer adaptive mechanisms that enable the understanding of data in complex and changing environments.

The Fourth International Conference on "Computational Intelligence in Data Mining (ICCIDM 2017)" is organized by Veer Surendra Sai University of Technology (VSSUT), Burla, Sambalpur, Odisha, India, during 11–12 November 2017. ICCIDM is an international forum for representation of research and developments in the fields of data mining and computational intelligence. More than 250 prospective authors submitted their research papers to the conference. After a thorough double-blind peer review process, editors have selected 78 papers. The proceedings of ICCIDM is a mix of papers from some latest findings and research of the authors. It is being a great honour for us to edit the proceedings. We have enjoyed considerably working in cooperation with the international advisory, programme and technical committee to call for papers, review papers and finalize papers to be included in the proceedings.

This international conference on CIDM aims at encompassing new breed of engineers, technologists making it a crest of global success. All the papers are focused on the thematic presentation areas of the conference, and they have provided ample opportunity for presentation in different sessions. Research in data mining has its own history. But, there is no doubt about the tips and further advancements in the data mining areas will be the main focus of the conference. This year's programme includes exciting collections of contributions resulting from a successful call for papers. Apart from those, two special sessions named "Computational Intelligence in Data Analytics" and "Applications of Computational Intelligence in Power and Energy Systems" have been proposed for more discussions on the theme-related areas. The selected papers have been divided into thematic areas including both review and research papers and highlight the current focus on computational intelligence techniques in data mining.

We hope the author's own research and opinions add value to it. First and foremost are the authors of papers, columns and editorials whose works have made the conference a great success. We had a great time putting together this proceedings. The ICCIDM conference and proceedings are a credit to a large group of people, and everyone should be congratulated for the outcome. We extend our deep sense of gratitude to all those for their warm encouragement, inspiration and continuous support for making it possible.

We hope all of us will appreciate the good contributions made and justify our efforts.

Sambalpur, India	Himansu Sekhar Behera
Srikakulam, India	Janmenjoy Nayak
Sambalpur, India	Bighnaraj Naik
Auburn, USA/Ostrava, Czech Republic	Ajith Abraham

Conference Committee

Chief Patron and President

Prof. E. Saibaba Reddy, Vice Chancellor, VSSUT, B.Tech., M.E. (Hons.) (Roorkee), Ph.D. (Nottingham, UK), Postdoc (Halifax, Canada), Postdoc (Birmingham, UK)

Honorary Advisory Chair

Prof. S. K. Pal, Sr., Member IEEE, LFIEEE, FIAPR, FIFSA, FNA, FASc, FNASc, FNAE, Distinguished Scientist and Former Director, Indian Statistical Institute, India
Prof. V. E. Balas, Sr., Member IEEE, Aurel Vlaicu University, Romania

Honorary General Chair

Prof. Rajib Mall, Indian Institute of Technology (IIT) Kharagpur, India
Prof. P. K. Hota, Dean, CDCE, VSSUT, India

General Chair

Prof. Ashish Ghosh, Indian Statistical Institute, Kolkata, India
Prof. B. K. Panigrahi, Indian Institute of Technology (IIT) Delhi, India

Programme Chair

Dr. H. S. Behera, Veer Surendra Sai University of Technology (VSSUT), Burla, Odisha, India

Chairman, Organizing Committee

Prof. Amiya Ku. Rath, HOD, Department of CSE and IT, Veer Surendra Sai University of Technology (VSSUT), Burla, Odisha, India

Vice-Chairman, Organizing Committee

Dr. S. K. Padhy, HOD, Department of Computer Application, Veer Surendra Sai University of Technology (VSSUT), Burla, Odisha, India

Convenor

Dr. Bighnaraj Naik, Department of Computer Application, Veer Surendra Sai University of Technology (VSSUT), Burla, Odisha, India

International Advisory Committee

Prof. A. Abraham, Machine Intelligence Research Labs, USA
Prof. Dungki Min, Konkuk University, Republic of Korea
Prof. Francesco Marcelloni, University of Pisa, Italy
Prof. Francisco Herrera, University of Granada, Spain
Prof. A. Adamatzky, Unconventional Computing Centre, UWE, UK
Prof. H. P. Proença, University of Beira Interior, Portugal
Prof. P. Mohapatra, University of California
Prof. S. Naik, University of Waterloo, Canada
Prof. George A. Tsihrintzis, University of Piraeus, Greece
Prof. Richard Le, La Trobe University, Australia
Prof. Khalid Saeed, AUST, Poland
Prof. Yew-Soon Ong, Singapore
Prof. Andrey V. Savchenko, NRU HSE, Russia
Prof. P. Mitra, P.S. University, USA
Prof. D. Sharma, University of Canberra, Australia
Prof. Istvan Erlich, University of Duisburg-Essen, Germany
Prof. Michele Nappi, University of Salerno, Italy
Prof. Somesh Jha, University of Wisconsin, USA
Prof. Sushil Jajodia, George Mason University, USA
Prof. S. Auephanwiriyakul, Chiang Mai University, Thailand
Prof. Carlos A. Coello Coello, Mexico
Prof. M. Crochemore, University de Marne-la-Vallée, France
Prof. T. Erlebach, University of Leicester, Leicester, UK
Prof. T. Baeck, Universiteit Leiden, Leiden, The Netherlands

Prof. J. Biamonte, ISI Foundation, Torino, Italy
Prof. C. S. Calude, University of Auckland, New Zealand
Prof. P. Degano, Università di Pisa, Pisa, Italy
Prof. Raouf Boutaba, University of Waterloo, Canada
Prof. Kenji Suzuki, University of Chicago
Prof. Raj Jain, WU, USA
Prof. D. Al-Jumeily, Liverpool J. Moores University, UK
Prof. M. S. Obaidat, Monmouth University, USA
Prof. P. N. Suganthan, NTU, Singapore
Prof. Biju Issac, Teesside University, UK
Prof. Brijesh Verma, CQU, Australia
Prof. Ouri E. Wolfson, University of Illinois, USA
Prof. Klaus David, University of Kassel, Germany
Prof. M. Dash, NTU, Singapore
Prof. L. Kari, Western University, London, Canada
Prof. A. S. M. Sajeev, Australia
Prof. Tony Clark, MSU, UK
Prof. Sanjib ku. Panda, NUS, Singapore
Prof. R. C. Hansdah, IISC Bangalore
Prof. G. Chakraborty, Iwate Prefectural University, Japan
Prof. Atul Prakash, University of Michigan, USA
Prof. Sara Foresti, University of degli Studi di Milano, Italy
Prof. Pascal Lorenz, University of Haute Alsace, France
Prof. G. Ausiello, University di Roma "La Sapienza", Italy
Prof. X. Deng, University of Liverpool, England, UK
Prof. Z. Esik, University of Szeged, Szeged, Hungary
Prof. A. G. Barto, University of Massachusetts, USA
Prof. G. Brassard, University de Montréal, Montréal, Canada
Prof. L. Cardelli, Microsoft Research, England, UK
Prof. A. E. Eiben, VU University, The Netherlands
Prof. Patrick Siarry, Université de Paris, Paris
Prof. R. Herrera Lara, EEQ, Ecuador
Prof. M. Murugappan, University of Malaysia

National Advisory Committee

Prof. P. K. Pradhan, Registrar, VSSUT, Burla
Prof. R. P. Panda, VSSUT, Burla
Prof. A. N. Nayak, Dean, SRIC, VSSUT, Burla
Prof. D. Mishra, Dean, Students' Welfare, VSSUT, Burla
Prof. P. K. Kar, Dean, Faculty & Planning, VSSUT, Burla

Prof. P. K. Das, Dean, Academic Affairs, VSSUT, Burla
Prof. S. K. Swain, Dean, PGS&R, VSSUT, Burla
Prof. S. Panda, Coordinator TEQIP, VSSUT, Burla
Prof. D. K. Pratihar, IIT Kharagpur
Prof. K. Chandrasekaran, NIT Karnataka
Prof. S. G. Sanjeevi, NIT Warangal
Prof. G. Saniel, NIT Durgapur
Prof. B. B. Amberker, NIT Warangal
Prof. R. K. Agrawal, JNU, New Delhi
Prof. U. Maulik, Jadavpur University
Prof. Sonajharia Minz, JNU, New Delhi
Prof. K. K. Shukla, IIT, BHU
Prof. A. V. Reddy, JNTU, Hyderabad
Prof. A. Damodaram, Sri Venkateswara University
Prof. C. R. Tripathy, SU, Odisha
Prof. P. Sanyal, WBUT, Kolkata
Prof. G. Panda, IIT, BBSR
Prof. B. B. Choudhury, ISI Kolkata
Prof. G. C. Nandy, IIIT Allahabad
Prof. R. C. Hansdah, IISC Bangalore
Prof. S. K. Basu, BHU, India
Prof. J. V. R. Murthy, JNTU, Kakinada
Prof. D. V. L. N. Somayajulu, NIT Warangal
Prof. G. K. Nayak, IIIT, BBSR
Prof. P. P. Choudhury, ISI Kolkata
Prof. D. Vijaya Kumar, AITAM, Srikakulam
Prof. Sipra Das Bit, IIEST, Kolkata
Prof. S. Bhattacharjee, NIT Surat

Technical Committee Members

Dr. Adel M. Alimi, REGIM-Lab, ENIS, University of Sfax, Tunisia
Dr. Chaomin Luo, University of Detroit Mercy Detroit, Michigan, USA
Dr. Istvan Erlich, Department of EE & IT, University of Duisburg-Essen, Germany
Dr. Tzyh Jong Tarn, Washington University in St. Louis, USA
Dr. Simon X. Yang, University of Guelph, Canada
Dr. Raffaele Di Gregorio, University of Ferrara, Italy
Dr. Kun Ma, Shandong Provincial Key Laboratory of Network Based Intelligent Computing, University of Jinan, China
Dr. Azah Kamilah Muda, Faculty of ICT, Universiti Teknikal Malaysia Melaka, Malaysia
Dr. Biju Issac, Teesside University, Middlesbrough, England, UK
Dr. Bijan Shirinzadeh, Monash University, Australia

Dr. Enver Tatlicioglu, Izmir Institute of Technology, Turkey

Dr. Hajime Asama, The University of Tokyo, Japan

Dr. N. P. Padhy, Department of EE, IIT Roorkee, India

Dr. Ch. Satyanarayana, Department of Computer Science and Engineering, JNTU Kakinada, India

Dr. B. Majhi, Department of Computer Science and Engineering, NIT Rourkela, India

Dr. M. Murugappan, School of Mechatronic Engineering, University Malaysia Perlis, Perlis, Malaysia

Dr. Kashif Munir, King Fahd University of Petroleum and Minerals, Hafr Al-Batin Campus, Kingdom of Saudi Arabia

Dr. L. Sumalatha, Department of Computer Science and Engineering, JNTU Kakinada, India

Dr. K. N. Rao, Department of Computer Science and Engineering, Andhra University, Visakhapatnam, India

Dr. S. Das, Indian Statistical Institute, Kolkata, India

Dr. D. P. Mohaptra, National Institute of Technology (NIT) Rourkela, India

Dr. A. K. Turuk, Head, Department of CSE, NIT RKL, India

Dr. M. P. Singh, Department of CSE, NIT Patna, India

Dr. R. Behera, Department of EE, IIT Patna, India

Dr. P. Kumar, Department of CSE, NIT Patna, India

Dr. A. Das, Department of CSE, IIEST, WB, India

Dr. J. P. Singh, Department of CSE, NIT Patna, India

Dr. M. Patra, Berhampur University, Odisha, India

Dr. A. Deepak, Department of CSE, NIT Patna, India

Dr. D. Dash, Department of CSE, NIT Patna, India

Finance Committee

Prof. Amiya Ku. Rath, Department of CSE & IT, Veer Surendra Sai University of Technology (VSSUT), Burla, Odisha, India

Prof. U. R. Jena, COF, Veer Surendra Sai University of Technology (VSSUT), Burla, Odisha, India

Dr. H. S. Behera, Department of CSE & IT, Veer Surendra Sai University of Technology (VSSUT), Burla, Odisha, India

Dr. S. K. Padhy, Department of Computer Application, Veer Surendra Sai University of Technology (VSSUT), Burla, Odisha, India

Dr. Bighnaraj Naik, Department of Computer Application, Veer Surendra Sai University of Technology (VSSUT), Burla, Odisha, India

Publication Chair

Dr. Janmenjoy Nayak, Sri Sivani College of Engineering, Srikakulam, Andhra Pradesh, India

Special Session Chairs

Special Session on "Computational Intelligence in Data Analysis":

Dr. Asit Kumar Das, Indian Institute of Engineering Science and Technology (IIEST), IIIT Shibpur, WB, India
Dr. Imon Mukherjee, International Institute of Information Technology (IIIT) Kalyani, WB, India

Special Session on "Computational Intelligence in Power & Energy Systems"

Prof. (Dr.) P. K. Hota, Veer Surendra Sai University of Technology (VSSUT), Odisha, India
Dr. Sasmita Behera, Veer Surendra Sai University of Technology (VSSUT), Odisha, India

Publicity Chair

Prof. P. C. Swain, VSSUT, Burla
Dr. Santosh Kumar Majhi, VSSUT, Burla
Mrs. E. Oram, VSSUT, Burla

Registration Chair

Dr. Sucheta Panda, VSSUT, Burla
Mrs. Sasmita Acharya, VSSUT, Burla
Mrs. Sasmita Behera, VSSUT, Burla
Ms. Gargi Bhattacharjee, VSSUT, Burla
Mrs. Santi Behera, VSSUT, Burla

Sponsorship Chair

Dr. Satyabrata Das, VSSUT, Burla
Mr. Sanjaya Kumar Panda, VSSUT, Burla
Mr. Sujaya Kumar Sathua, VSSUT, Burla
Mr. Gyanaranjan Shial, VSSUT, Burla

Web Chair

Mr. D. C. Rao, VSSUT, Burla
Mr. Kishore Kumar Sahu, VSSUT, Burla
Mr. Suresh Kumar Srichandan, VSSUT, Burla

Organizing Committee Members

Dr. Manas Ranjan Kabat, VSSUT, Burla
Dr. Rakesh Mohanty, VSSUT, Burla
Dr. Suvasini Panigrahi, VSSUT, Burla
Dr. Manas Ranjan Senapati, VSSUT, Burla
Dr. P. K. Sahu, VSSUT, Burla
Mr. Satya Prakash Sahoo, VSSUT, Burla
Mr. Sanjib Nayak, VSSUT, Burla
Ms. Sumitra Kisan, VSSUT, Burla
Mr. Pradipta Kumar Das, VSSUT, Burla
Mr. Atul Vikas Lakra, VSSUT, Burla
Ms. Alina Mishra, VSSUT, Burla
Ms. Alina Dash, VSSUT, Burla
Dr. M. K. Patel, VSSUT, Burla
Mr. D. P. Kanungo, VSSUT, Burla
Mr. D. Mishra, VSSUT, Burla
Mr. S. Mohapatra, VSSUT, Burla

International Reviewer Committee

Arun Agarwal, Siksha 'O' Anusandhan University, Odisha, India
M. Marimuthu, Coimbatore Institute of Technology, Coimbatore, India
Sripada Rama Sree, Aditya Engineering College (AEC), Surampalem, Andhra Pradesh, India
Harihar Kalia, Seemanta Engineering College, Odisha, India
A. S. Aneeshkumar, Alpha Arts and Science College, Chennai, Tamil Nadu, India
Kauser Ahmed P., VIT University, Vellore, Tamil Nadu, India
Ajanta Das, Birla Institute of Technology, Mesra, India
Manuj Darbari, BBD University, Lucknow, India
Manoj Kumar Patel, Veer Surendra Sai University of Technology, Odisha, India
Nagaraj V. Dharwadkar, Rajarambapu Institute of Technology, Maharashtra, India
Chandan Jyoti Kumar, Department of Computer Science and Information Technology, Cotton College State University, Assam, India
Biswapratap Singh Sahoo, Department of Electrical Engineering, National Taiwan University, Taipei, Taiwan.

S. P. Tripathy, NIT Durgapur, West Bengal, India

Suma V., Dean, Dayananda Sagar College of Engineering, Odisha, India

R. Patel, BVM Engineering College, Gujarat, India.

P. Sivakumar, SKP Engineering College, Tamil Nadu, India

Partha Garai, Kalyani Government Engineering College, Kalyani, West Bengal, India

Anuranjan Misra, Noida International University, India

K. G. Srinivasagan, National Engineering College, Tamil Nadu, India

Jyotismita Chaki, Jadavpur University, West Bengal, India

Sawon Pratiher, IIT Kharagpur, India

R. V. S. Lalitha, Aditya College of Engineering and Technology, Surampalem, India

G. Rosline Nesa Kumari, Saveetha University, Chennai, Tamil Nadu, India

Narayan Joshi, Parul University, Vadodara, India

S. Vaithyasubramanian, Sathyabama University, Chennai, Tamil Nadu, India

Abhishek Kumar, Lovely Professional University, Punjab, India

Prateek Agrawal, Lovely Professional University, Punjab, India

Sumanta Panda, Veer Surendra Sai University of Technology, Odisha, India

Jayakishan Meher, Centurion University of Technology and Management, Odisha, India

Balakrushna Tripathy, VIT University, Tamil Nadu, India

Rajan Patel, Sankalchand Patel College of Engineering, NG, India

Narender Singh, Chhaju Ram Memorial Jat College, Haryana, India

P. M. K. Prasad, GMR Institute of Technology, Andhra Pradesh, India

Chhabi Rani Panigrahi, Central University Rajasthan, Gujarat, India

Srinivas Sethi, Indira Gandhi Institute of Technology, Odisha, India

Mahmood Ali Mirza, DMS SVH College of Engineering, Andhra Pradesh, India

P. Kumar, Zeal College of Engineering and Research, Pune, India

B. K. Sarkar, BIT Mesra, Ranchi, India

Trilochan Panigrahi, National Institute of Technology, Goa, India

Deepak D. Kshirsagar, College of Engineering Pune (COEP), Pune, India

Rahul Paul, University of South Florida, USA

Maya V. Karki, Professor, MSRIT, Bangalore, India

A. Anny Leema, B. S. Abdur Rahman University, Chennai, India

S. Logeswari, Bannari Amman Institute of Technology, Sathyamangalam, Tamil Nadu, India

R. Gomathi, Bannari Amman Institute of Technology, Sathyamangalam, Tamil Nadu, India

R. Kavitha, Sastra University, Thanjavur, Tamil Nadu, India

S. Das, Veer Surendra Sai University of Technology, Odisha, India

S. K. Majhi, Veer Surendra Sai University of Technology, Odisha, India

Sucheta Panda, Veer Surendra Sai University of Technology, Odisha, India

P. K. Hota, Veer Surendra Sai University of Technology, Odisha, India

G. T. Chandrasekhar, Sri Sivani College of Engineering, Srikakulam, Andhra Pradesh, India

Janmenjoy Nayak, Sri Sivani College of Engineering, Srikakulam, Andhra Pradesh, India
Bighnaraj Naik, Veer Surendra Sai University of Technology, Odisha, India

Acknowledgements

The theme and relevance of ICCIDM attracted more than 250 researchers/academicians around the globe which enabled us to select good quality papers and serve to demonstrate the popularity of the ICCIDM conference for sharing ideas and research findings with truly national and international communities. Thanks to all those who have contributed in producing such a comprehensive conference proceedings of ICCIDM.

The organizing committee believes and trusts that we have been true to the spirit of collegiality that members of ICCIDM value even as also maintaining an elevated standard as we have reviewed papers, provided feedback and presented a strong body of published work in this collection of proceedings. Thanks to all the members of the organizing committee for their heartfelt support and cooperation.

After the three successful versions of ICCIDM, it has indeed been an honour for us to edit the proceedings of this fourth series of ICCIDM. We have been fortunate enough to work in cooperation with a brilliant international as well as national advisory board, reviewers, and programme and technical committee consisting of eminent academicians to call for papers, review papers and finalize papers to be included in the proceedings.

We would like to express our heartfelt gratitude and obligations to the benign reviewers for sparing their valuable time, putting an effort to review the papers in a stipulated time and providing their valuable suggestions and appreciation in improvising the presentation, quality and content of this proceedings. The eminence of these papers is an accolade not only to the authors but also to the reviewers who have guided towards perfection.

Last but not least, the editorial members of Springer Publishing deserve a special mention and our sincere thanks to them not only for making our dream come true in the shape of this proceedings, but also for its hassle-free and in-time publication in the reputed Advances in Intelligent Systems and Computing, Springer.

The ICCIDM conference and proceedings are a credit to a large group of people, and everyone should be proud of the outcome.

Contents

About the Editors

Prof. Himansu Sekhar Behera is working as Head of the Department and Associate Professor in the Department of Information Technology, Veer Surendra Sai University of Technology (VSSUT) (a unitary technical university, established by Government of Odisha), Burla, Odisha. He has received his M.Tech. degree in Computer Science and Engineering from NIT Rourkela (formerly REC, Rourkela) and Doctor of Philosophy in Engineering (Ph.D.) from Biju Patnaik University of Technology (BPUT), Rourkela, Government of Odisha, respectively. He has published more than 80 research papers in various international journals and conferences and edited 11 books and is acting as a member of the editorial/reviewer board of various international journals. He is proficient in the field of computer science engineering, served the capacity of programme chair and tutorial chair and acts as advisory member of committees of many national and international conferences. His research interests include data mining and intelligent computing. He is associated with various educational and research societies like OITS, ISTE, IE, ISTD, CSI, OMS, AIAER, SMIAENG, SMCSTA. He is currently guiding seven Ph.D. scholars, and three scholars have been awarded under his guidance.

Dr. Janmenjoy Nayak is working as an Associate Professor at Sri Sivani College of Engineering, Srikakulam, Andhra Pradesh, India. He has been awarded INSPIRE Research Fellowship from the Department of Science and Technology, Government of India (both as JRF and as SRF levels) for doing his Doctoral Research in the Department of Computer Science and Engineering & Information Technology, Veer Surendra Sai University of Technology, Burla, Odisha, India. He completed his M.Tech. degree (gold medallist and topper of the batch) in Computer Science from Fakir Mohan University, Balasore, Odisha, India, and M.Sc. degree (gold medallist and topper of the batch) in Computer Science from Ravenshaw University, Cuttack, Odisha, India. He has published more than 50 research papers in various reputed peer-reviewed international conferences, refereed journals and chapters. He has also published one textbook on "Formal Languages and Automata Theory" in Vikash Publishing House Pvt. Ltd., which has been widely acclaimed throughout the country and abroad by the students of all levels. He is the recipient of "Young Faculty in Engineering" award from Centre of Advance Research and Design, VIFA-2017, Chennai, India, for exceptional academic records and research excellence in the area of computer science engineering. He has been serving as an active member of reviewer committee of various reputed peer-reviewed journals such as *IET Intelligent Transport Systems, Journal of Classification, Springer, International Journal of Computational System Engineering, International Journal of Swarm Intelligence, International Journal of Computational Science and Engineering, International Journal of Data Science*. He is the life member of some of the reputed societies like CSI, India, OITS, OMS, IAENG (Hong Kong). His areas of interest include data mining, nature-inspired algorithms and soft computing.

Dr. Bighnaraj Naik is an Assistant Professor in the Department of Computer Applications, Veer Surendra Sai University of Technology, Burla, Odisha, India. He received his doctoral degree from the Department of Computer Science and Engineering & Information Technology, Veer Surendra Sai University of Technology, Burla, Odisha, India; master's degree from the Institute of Technical Education and Research, SOA University, Bhubaneswar, Odisha, India; and bachelor's degree from the National Institute of Science and Technology, Berhampur, Odisha, India. He has published more than 50 research papers in various reputed peer-reviewed international conferences, refereed journals and chapters. He has more than 8 years of teaching experience in the fields of computer science and information technology. He is the life member of International Association of Engineers (Hong Kong). His areas of interest include data mining, soft computing. He is the recipient of "Young Faculty in Engineering" award for 2017 from Centre of Advance Research and Design, VIFA-2017, Chennai, India, for exceptional academic records and research excellence in the area of computer science engineering. He has been serving as an active member of reviewer committee of various reputed peer-reviewed journals such as *Swarm and Evolutionary Computation*, Elsevier; *Journal of King Saud University*, Elsevier; *International Journal of Computational System Engineering*, Inderscience; *International Journal of Swarm Intelligence*, Inderscience; *International Journal of Computational Science and Engineering*, Inderscience; *International Journal of Data Science*, Inderscience. Currently, He is serving as editor of the book entitled "Information Security in Biomedical Signal Processing", IGI-Global (publisher), USA. Also, he is the Guest Editor of *International Journal of Computational Intelligence Studies*, Inderscience Publication, and *International Journal of Data Science and Analytics*, Springer.

Prof. Ajith Abraham received M.S. degree from Nanyang Technological University, Singapore, and Ph.D. degree in Computer Science from Monash University, Melbourne, Australia. He is currently a Research Professor at the Technical University of Ostrava, Czech Republic. He is also the Director of Machine Intelligence Research Labs (MIR Labs), Scientific Network for Innovation and Research Excellence, which has members from more than 75 countries. He serves/has served the editorial board of over 50 international journals and has also guest-edited 40 special issues on various topics. He is an author/co-author of more than 700 publications, and some of the works have also won the best paper awards at international conferences and also received several citations. Some articles are available in the ScienceDirect Top 25 hottest articles. He serves the IEEE Computer Society's Technical Committee on Scalable Computing and was the General Chair of the 8th International Conference on Pervasive Intelligence and Computing (PICOM 2009) and 2nd International Conference on Multimedia Information Networking and Security (MINES 2010). He is also the General Co-chair of MINES 2011. Since 2008, he is the Chair of IEEE Systems Man and Cybernetics Society Technical Committee on Soft Computing. He is a Senior Member of IEEE, the IEEE Computer Society, the Institution of Engineering and Technology (UK) and the Institution of Engineers Australia (Australia). He is the founder of several IEEE-sponsored annual conferences, which are now annual events—Hybrid Intelligent Systems (HIS—11 years); Intelligent Systems Design and Applications (ISDA—11 years); Information Assurance and Security (IAS—7 years); Next Generation Web Services Practices (NWeSP—7 years), Computational Aspects of Social Networks (CASoN—3 years), Soft Computing and Pattern Recognition (SoCPaR—3 years), Nature and Biologically Inspired Computing (NaBIC—3 years) are some examples.

BER Performance Analysis of Image Transmission Using OFDM Technique in Different Channel Conditions Using Various Modulation Techniques

Arun Agarwal, Binayak Satish Kumar and Kabita Agarwal

Abstract It can be clearly seen in the modern world of digital era that we depend a lot on mobile/smartphones which in return depends on data/information to proof its worthy. To meet this increased demand of data rate, new technology is needed as for old technology like GPRS and EDGE which is beyond their capacity. So orthogonal frequency division multiplexing (OFDM) takes this place as it increases the data rate in the same bandwidth, as it is also fixed. In this paper, we present a comparison between data handling using different modulations, which in turn can be assumed for bit error rate (BER) and signal-to-noise ratio (SNR) curve, of non-OFDM and OFDM channel under different fading environments.

Keywords Data rate · BER · Fading channels · SNR · Convolutional coding
High-speed Internet · OFDM · Modulation

1 Introduction

In modern world for data need of every individual person, the technologies like GPRS and EDGE are not sufficient enough as they have far less data throughput than the need is. In this place comes orthogonal frequency division multiplexing

A. Agarwal
Department of EIE, ITER, Siksha 'O' Anusandhan Deemed to be University, Khandagiri Square, Bhubaneswar 751030, Odisha, India
e-mail: arunagrawal@soa.ac.in

B. S. Kumar (✉)
Department of ECE, ITER, Siksha 'O' Anusandhan Deemed to be University, Khandagiri Square, Bhubaneswar 751030, Odisha, India
e-mail: binayakdsip008@gmail.com

K. Agarwal
Department of ETC, C. V. Raman College of Engineering, Bhubaneswar 752054, Odisha, India
e-mail: akkavita22@gmail.com

© Springer Nature Singapore Pte Ltd. 2019
H. S. Behera et al. (eds.), *Computational Intelligence in Data Mining*, Advances in Intelligent Systems and Computing 711, https://doi.org/10.1007/978-981-10-8055-5_1

(OFDM) which is a high data throughput technology where a single large data stream transmission takes place over a number of low-rate sub-carriers [1]. These multiple sub-carriers are maintained orthogonal to each other in OFDM, by the addition of cyclic prefix/guard interval. So, this helps in utilizing lesser bandwidth, i.e., more data in same bandwidth.

OFDM nowadays finds tremendous applications in variety of wireless and wired digital communication channels for broadband data such as Wi-fi. Due to use of cyclic prefix in OFDM, it efficiently mitigate the effect of multi-path fading channel and hence resistant to intersymbol interference (ISI) [1–8]. OFDM also allows the implementation of single-frequency networks (SFN), whereby network transmitters can cover large area with the same transmission frequency. Such advantage allows OFDM to find application in high data rate digital services such as digital video broadcasting (DVB) and digital audio broadcasting (DAB).

In this work, we evaluate the BER performance with various digital modulation techniques with OFDM that support high data throughput [9, 10]. The performance of QPSK, 16-QAM, and 64-QAM was compared with BPSK which is the most common type of digital modulation with lowest BER-to-SNR ratio for achieving higher data rate with minimal transmitter power in Rayleigh fading environment.

The contribution of this paper has been divided into three sections. Section 1 explained about introduction. Section 2 will brief about fading channels. Finally, in Sect. 3 simulation results are presented and we compare BER for image transmission using OFDM with that of non-OFDM transmission technique.

2 Fading Environment

If it would have been only the condition of sending and receiving the information through communication channel, then there would have been no problem, but in real world it's not. In real world, the communication channel wired or wireless is prone to introduction of noises which is due to reflection of signal, refraction of signal, interference from electronic appliances, etc., and this causes in fading of signal strength.

2.1 AWGN Channel

Additive white Gaussian noise (AWGN) channel is the most common type channel model that adds white Gaussian noise to the signal that passes through it. Here, a white noise with fixed spectral density (watts/hertz of bandwidth) is simply added with amplitude that follows a Gaussian distribution. This channel does not have fading component. It has a very simple mathematical model which is useful for deeper system analysis.

The AWGN channel adds white Gaussian noise to a real or complex input signal. When the input signal is real, this channel adds real Gaussian noise and produces a real output signal. When the input signal is complex, this channel adds complex Gaussian noise and produces a complex output signal.

The AWGN channel model is widely used in LOS microwave links, satellite and deep space communication links. AWGN is frequently used model for most terrestrial path modeling which evaluates the effect of background additive noise of the channel.

2.2 Rician Channel

The model for Rician fading channel is same as Rayleigh fading except that in Rician channels in which we have a strong line of sight component along with reflected waves. A Rician fading channel can be characterized by two parameters, K and Ω. K represents a ratio between the direct path power and the power in the scattered paths. Ω is defined to be the total power from both paths and acts as a scaling factor to the distribution.

3 Proposed Work

An image is transmitted using BPSK, QPSK, 16-QAM, and 64-QAM digital modulation scheme using OFDM technique in AWGN, Rayleigh, and Rician channel. Then, the BER graph of the same is compared with non-OFDM technique to justify/verify that OFDM is a better technique for transmission in fading channels.

4 Simulation Results and Discussion

In this section, all results of the simulation are presented and analyzed. Simulated BER plots are shown for different modulation techniques and with and without OFDM. Performance analysis was carried out in three channels, namely AWGN, Rayleigh, and Rician to test suitability of the communication system model proposed. Figure 1 below presents BER performance curve of non-OFDM in AWGN channel.

From the Table 1 of SNR versus BER of AWGN channel, it can be seen that what it looks like a very small improvement around 0.2 dB in SNR value in OFDM in comparison with non-OFDM method.

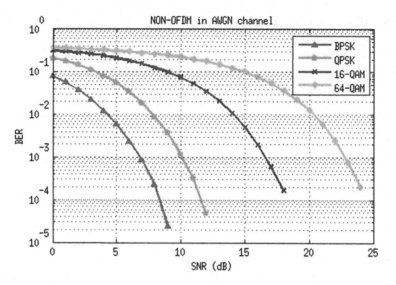

Fig. 1 BER performance curve of non-OFDM in AWGN channel

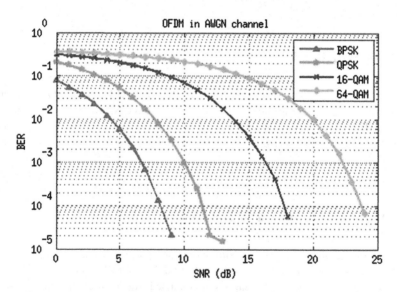

Fig. 2 BER performance curve of OFDM in AWGN channel

From the Table 2 of SNR versus BER of Rician channel, it can be seen that now notable improvement of about 3.5 dB in SNR value in OFDM in comparison with non-OFDM method.

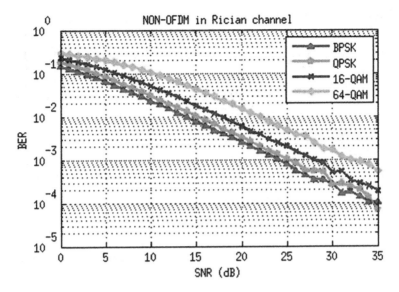

Fig. 3 BER performance curve of non-OFDM in Rician channel

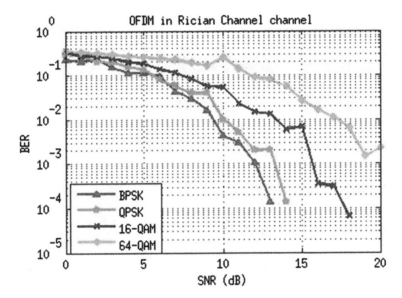

Fig. 4 BER performance curve of OFDM in Rician channel

From the Table 3 of SNR versus BER of Rayleigh channel, it can be seen an improvement of about 7 dB that can be worth looking for in SNR value in OFDM in comparison with non-OFDM method.

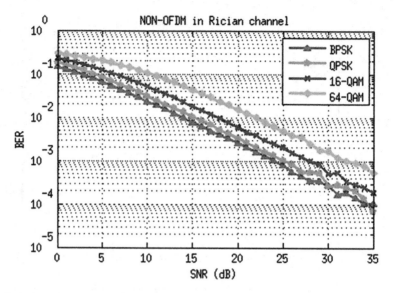

Fig. 5 BER performance curve of non-OFDM in Rayleigh channel

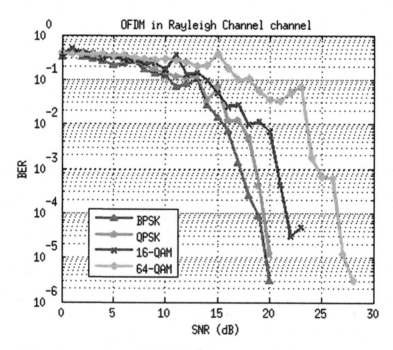

Fig. 6 BER performance curve of OFDM in Rayleigh channel

Table 1 SNR for BER at 10^{-3} in AWGN channel for Figs. 1 and 2

Modulation method	Modulation scheme			
	BPSK (dB)	QPSK (dB)	16QAM (dB)	64QAM (dB)
Non-OFDM	7	10.5	16.8	23
OFDM	6.8	10	16.3	22.8

Table 2 SNR for BER at 10^{-3} in Rician channel for Figs. 3 and 4

Modulation method	Modulation scheme			
	BPSK (dB)	QPSK (dB)	16QAM (dB)	64QAM (dB)
Non-OFDM	17	18	21	26
OFDM	12	13.5	17.8	23.5

Table 3 SNR for BER at 10^{-3} in Rayleigh channel for Figs. 5 and 6

Modulation method	Modulation scheme			
	BPSK (dB)	QPSK (dB)	16-QAM (dB)	64-QAM (dB)
Non-OFDM	24	25	28	32
OFDM	15.5	17.5	19	25

5 Conclusion

As, we are seeing a 0.2–0.5 dB improvement in SNR in AWGN channel, then a 2.5–5.0 dB improvement in SNR in Rician channel and finally a 7–9 dB improvement in SNR in Rayleigh channel. Therefore it confirms that proposed model is suitable for Rayleigh fading channel.

It can be concluded from the result of our simulation that in Rayleigh fading environment, with AWGN noise, the BER curve showed an improvement in SNR with OFDM. We carried out comparisons between OFDM technology and non-OFDM technology for transmission of image. The performance of BER curve is better with OFDM technology in all channels. Finally to conclude OFDM is a promising technology for upcomiong 5G networks with MIMO.

References

1. Gangadharappa, Mandlem, Rajiv Kapoor, and Hirdesh Dixit. "An efficient hierarchical 16-QAM dynamic constellation to obtain high PSNR reconstructed images under varying channel conditions." IET Communications 10.2 (2016): pp. 139–147.
2. Song, Jie, and K.j.r. Liu. "Robust Progressive Image Transmission over OFDM Systems Using Space-time Block Code." IEEE Transactions on Multimedia 4.3 (2002): pp. 394–406.

3. Alok, and Davinder S. Saini. "Performance analysis of coded-OFDM with RS-CC and Turbo codes in various fading environment." Information Technology and Multimedia (ICIM), 2011 International Conference on. IEEE, 2011: pp. 1–6.

4. Arun Agarwal, S.K. Patra "Performance prediction of OFDM based DAB system using Block coding techniques", presented in proceedings of the IEEE International Conference on Emerging Trends in Electrical and Computer Technology (ICETECT-2011), Mar 23rd–24th, 2011 at Nagarcoil, Kanyakumari, India.

5. Arun Agarwal, Saurabh N. Mehta, "Combined Effect of Block interleaving and FEC on BER Performance of OFDM based WiMAX (IEEE 802.16d) System", American Journal of Electrical and Electronic Engineering, © Science and Education Publishing, USA, Vol. 3, No. 1, pp 4–12, March 2015, https://doi.org/10.12691/ajeee-3-1-2.

6. Arun Agarwal, Kabita Agarwal, "Performance Prediction of WiMAX (IEEE 802.16d) Using Different Modulation and Coding Profiles in Different Channels", pp. 2221–2234, Volume 10, Number 2, February 2015 in International Journal of Applied Engineering Research (IJAER).

7. Arun Agarwal, Saurabh N. Mehta, "Design and Performance Analysis of MIMO-OFDM System Using Different Antenna configurations", in conference Proceedings of the IEEE International Conference on Electrical, Electronics, and Optimization Techniques (IEEE-ICEEOT-2016), Chennai, Tamil Nadu, India, 3rd to 5th March 2016, pp 1373–1377.

8. Arun Agarwal, Kabita Agarwal, "Design and Simulation of COFDM for high speed wireless communication and Performance analysis", in IJCA - International Journal of Computer Applications, pp 22–28, Vol-2, Oct-2011., ISBN: 978-93-80865-49-3.

9. Nyirongo, Nyembezi, Wasim Q. Malik, and David J. Edwards. "Concatenated RS-convolutional codes for ultrawideband multiband-OFDM." 2006 IEEE International Conference on Ultra-Wideband. IEEE, 2006: pp. 137–142.

10. Haque, Md Dulal, Shaikh Enayet Ullah, and Md Razu Ahmed. "Performance evaluation of a wireless orthogonal frequency division multiplexing system under various concatenated FEC channel-coding schemes." Computer and Information Technology, 2008. ICCIT 2008. 11th International Conference on. IEEE, 2008; pp. 94–97.

On Understanding the Release Patterns of Open Source Java Projects

Arvinder Kaur and Vidhi Vig

Abstract Release length is of great significance to companies as well as to researchers as it provides a deeper insight into the rules and practices followed by the applications. It has been observed that many Open Source projects follow agile practices of parallel development and Rapid Releases (RR) but, very few studies till date, have analyzed release patterns of these Open Source projects. This paper analyzes ten Open Source Java projects (Apache Server Foundation) comprising 718 releases to study the evolution of release lengths. The results of the study show that: (1) eight out of ten datasets followed RR models. (2) None of these datasets followed RR models since their first release. (3) The average release length was found to be four months for major versions and one month for minor versions (exceptions removed). (4) There exists a negative correlation between number of contributors and release length.

Keywords Release cycle · Frequent versions · Evolution · Open Source Repositories

1 Introduction

Agile methodologies like XP [1] instituted the notion of *faster or Rapid* Release cycles and advocate the benefits of using them for both companies and customers. Soaring market competition has forced many software companies to exercise shorter release cycles and release their products within a span of weeks or days [2]. Mozilla Firefox migrated to Rapid Release concept after facing huge competition from Google Chrome and shifted from its Traditional Release model of one year for a major release to 6 weeks from version 5.0 [3, 4].

A. Kaur · V. Vig (✉)
University School of Information, Communication and Technology, Guru Gobind
Singh Indraprastha University, Sec 16-C, Dwarka, New Delhi 110078, India
e-mail: vidhi.ipu@gmail.com

A. Kaur
e-mail: arvinderkaurtakkar@yahoo.com

© Springer Nature Singapore Pte Ltd. 2019
H. S. Behera et al. (eds.), *Computational Intelligence in Data Mining*,
Advances in Intelligent Systems and Computing 711,
https://doi.org/10.1007/978-981-10-8055-5_2

These shorter cycles allow faster customer feedback thereby enabling companies to schedule their succeeding releases more easily and do not pressurize the developer to complete the entire feature at once. The features can be published in incremental releases, and meanwhile, the developer can pay attention on quality assurance [5] resulting in faster bug detection and correction [4]. Shorter release cycles have been adapted in many software and embedded domains. A recent study [6] analyzed release patterns in mobile domain and found that frequently updated mobile applications (i.e., shorter release cycles) on Google Play Store were highly ranked by the users, irrespective of their high update frequency [6]. In fact, updated versions were embraced more quickly by the users worldwide [5]. Still, there exist certain disadvantages to this methodology in terms of lesser time to fix bugs, lack of stabilized platforms [7], increased user cost due to continuous updates [8], and inability to test all configuration of released product [9].

From testing perspective, some recent studies [10, 11] advocate using RR models since it allows testing to be more concentrated and allowed extended investigation of features pertaining to highest risks. Alternatively, releasing versions faster makes testing more continuous and deadline oriented. However, some studies found that testing is hampered in such models as there is lesser time to test [12].

This study is an endeavor to analyze the characteristics of datasets and their release patterns in Open Source. The ease of replication is the prime USP (Unique Selling Point) of OSS. Not only it offers pliability of replication, but also enables confirmation or revision to a larger research group, unlike closed source and industrial data. Since Open Source is the breeding ground of many software and researches, it must be explored for every new technique and domain.

The paper presents the data collection methodology in Sect. 2, followed by analyses in Sect. 3, discusses the exceptions in Sect. 3.3, and concludes the result in Sect. 4.

2 Data Collection Methodology

Apache Server Foundation (ASF) [13] hosts numerous projects in multiple languages like Java, Python, Ruby, C, and C++. Out of all, Java holds the maximum portion, thereby making it an ideal choice for study. The required software artifacts were obtained from the Git [14] and Jira [15] repository. While selection, the following points were kept in mind:

- The projects must be in "Active" state, i.e., last developmental activity not later than six months
- Projects labeled as "Retired" by ASF were ignored
- Projects must have a development life cycle of more than five years

Firstly, Apache Server Foundation was explored, and projects written in Java language were kept in pool for consideration. Then, the above-mentioned *selection criteria* were applied, and projects fulfilling the criteria were separated from others. These projects were now subjected to Simple Random Sampling (since it enhances

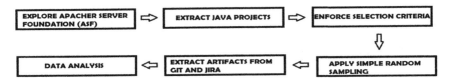

Fig. 1 Data collection process

generalizability), and ten datasets were finally selected for the analysis. The artifacts of the selected datasets were obtained from Git and Jira (which are source and bug repositories, respectively). The process of data collection is explained graphically in Fig. 1 given below, and the characteristics of the studied projects are presented in Table 1.

Note: The word Release and Version are used synonymously in the study.

3 Analysis

This section of the paper presents the outcome and possible interpretation of the results obtained. At first, the study investigates the release trends and patterns of the datasets selected, then the major and minor versions of the datasets are investigated, and finally, exceptions are identified and examined.

3.1 Datasets and Their Entire Release History

Rapid Release (RR) model encourages releases to be published in days or weeks in comparison with the Traditional Release (TR) model where the releases are not so deadline oriented. The terms "Rapid Release" and "Traditional Releases" are used antonyms by the research papers published in this domain [4, 6, 10, 12]. For analysis, the study first collected the release dates of all the released versions and then calculated the release time between each version. Release time of major versions (T), calculated using Eq. 1, is the time between two subsequent major versions.

$$T(R) = Time(Version(i+1) - Version(i)) \tag{1}$$

where T(R) is the time in days, weeks, and months, Version (i) = Release date of Version (i), Version (i + 1) = Release date of next subsequent Version (i + 1).

Next, the mean time of these releases was calculated using Eq. 2, in order to see the average time taken by each dataset to release their version.

Table 1 Studied software projects

Projects	Avro	Camel	Cordova	Chukwa	Groovy	Hive	Jclouds	Pig	Rat	Zookeeper
First release	14-Jul-09	2-Jul-07	23-Dec-09	8-Nov-09	13-Dec-03	30-Apr-09	6-Oct-10	29-Oct-07	10-Apr-09	13-Nov-07
Last release	7-Nov-16	22-Jan-17	4-Jan-17	16-Jul-16	10-Jan-17	13-Jan-17	18-Nov-16	7-Jun-16	14-Jun-16	3-Sep-16
Total release	67	93	94	6	150	86	99	53	8	62
No. of major releases	9	26	36	6	10	18	10	17	8	10
No. of minor releases	58	67	58	0	140	68	89	36	0	52

Note Data was last collected on 18 January 2017. Changes beyond this date don't contribute to the study

Table 2 Arithmetic Mean of releases of all the datasets (in days, weeks and months)

Projects	Avro	Camel	Cordova	Chukwa	Groovy	Hive	Jclouds	Pig	Rat	Zookeeper
Days	41	38	27	407	32	33	23	59	328	54
Weeks	6	5	4	58	5	5	3	8	47	8
Months	1	1	1	14	1	1	1	2	11	2

$$AM = \frac{1}{n} \sum_{i=1}^{n} T(R)i \qquad (2)$$

where AM = Arithmetic Mean, n = number of datasets.

This Arithmetic Mean of the releases gives the average time (days, weeks, months) taken by each dataset to air their release and further enable the study to identify the release models followed by these datasets. However, the study assumes datasets with release time in days or weeks under RR model and release time in months under TR models. The result of the analysis is presented in Table 2.

It can be observed that the average release length of eight out of ten selected datasets is one month while two datasets, Chukwa and Rat, take an average of a year to publish their release. On further investigation, it was found that these two datasets had the minimum number of releases (six for Chukwa and eight for Rat) in their entire life span of seven years. This clearly indicates that these two datasets do not follow the RR model and fall under the percentage of projects that follow the TR models. It was found that more than 54% of the datasets follow the RR model in Open Source repositories [16], and our results not only comply with their results but also provide stronger proof of RR trend in OSS. Figure 2 given below presents the release patterns of Chukwa and Rat.

The average of these AM, i.e., the Grand Mean (GM) was calculated using Eq. 3, to check average release time of the datasets together.

$$GM = \sum AM / N \qquad (3)$$

where, GM = Grand Mean, AM = Arithmetic Mean, N = total number of datasets.

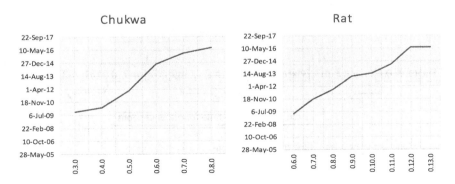

Fig. 2 Release cycle of Chukwa and Rat

It was discovered that the GM of all the datasets taken together yielded an average time of three months for a release while the GM of the eight datasets that follow the RR model was one month. Since most of the datasets followed the RR model, it can be concluded that a RR model was favored over TR model by the datasets.

3.2 Major and Minor Versions of the Datasets

During data analysis of these projects, following assumptions were made:

- All the release candidates and maintenance versions would be counted as well and would be conceived as minor versions
- Release naming convention usually follows as *.*.* Releases with last bit nonzero were usually bug fixes or maintenance releases. Since no new enhancement or improvement followed such releases, therefore, all such releases will be perceived as minor releases only

The release time of the major and minor versions was calculated as follows:

Release time of major versions (T) is the time between two subsequent major versions which was calculated using Eq. 4 given below

$$T(M) = Time(Major(i+1) - Major(i)) \tag{4}$$

where T(M) is the time in days, weeks, and months, Major (i) = Release date of major version (i), Major (i + 1) = Release date of next subsequent major version.

The release time of the minor versions was the time taken to publish the minor releases between two major releases. This can be better understood by Fig. 3.

Since Chukwa and Rat don't follow the RR model, therefore it will not be considered for further analysis in this section, though it will be discussed in Sect. 3.3 later in the study. Consequently, the Arithmetic Mean (AM) and the Grand Mean (GM) of both major and minor versions of eight datasets were

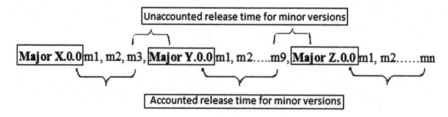

Fig. 3 Calculation of release time of minor versions

Table 3 Arithmetic and Grand Mean of major and minor versions

Analysis of AM of major versions								
Datasets	Avro	Camel	Cordova	Groovy	Hive	Jclouds	Pig	Zookeeper	GM (Major)
Days	266	96	90	294	156	199	137	262	187
Weeks	38	14	13	42	22	28	20	37	27
Months	9	3	3	10	5	7	5	9	6
Analysis of AM of minor versions								GM (Minor)
Days	42	29	28	33	36	23	65	56	39
Weeks	6	4	4	5	5	3	9	8	6
Months	1	1	1	1	1	1	2	2	1

calculated (Eqs. 2, 3). The value "N" changes for number of major and minor accordingly. The output of the datasets is presented in Table 3.

The simple analysis of Grand Mean of Table 3 shows that on an average, the average time taken by datasets to release a major version is six months and minor version is one month. It was further be observed from Table 3 that both Avro and Groovy have worst release time for major versions but best release time for minor versions. However, on deeper analysis, it was discovered that both these datasets released numerous minor versions between major versions which resulted in long gaps between major versions and short gaps in minor versions.

3.3 The Exceptions

3.3.1 Datasets Following the TR Models

The study discovered that two of the datasets (Chukwa, Rat) out of ten didn't follow RR models (Sect. 3.1). These datasets continued to produce their releases in years, and huge gaps between their subsequent releases were not a result of inactivity in the datasets. As a matter of fact, both these datasets were found active in terms of development, testing, and customer feedback.

The study observed one strikingly different feature in datasets following the RR model in OSS, which was "Number of contributors." The number of contributors in the datasets following the RR model was significantly higher than the datasets following the TR model. Figure 4 given below represents the number of contributor in each dataset.

It can be interpreted from Fig. 4 that datasets having maximum number of contributors had the maximum number of commits. While datasets Camel, Cordova, Groovy, Hive, and Jclouds had more than hundred contributors and thousand commits, Rat and Chukwa had only five and eleven contributors, respectively, resulting in lesser number of commits.

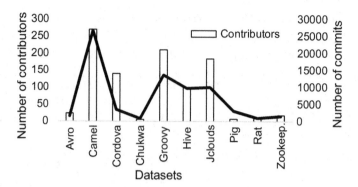

Fig. 4 Commits and contributors in each dataset

The study also found that dataset Pig had only seven contributors (less than rat) but had more than three thousand commits (commits nearly equal to Cordova that had more than hundred contributors). In fact, Pig had the least number of releases among all the datasets following the RR model. This indicates that the larger number of contributors doesn't necessarily imply better development effort, and datasets Pig (7 contributors, 3031 commits) and Cordova (139 contributors, 3374 commits) are very good examples of this scenario.

Finally, the study found negative correlation between number of contributors and release length and positive correlation between number of contributors and number of releases, i.e., if the contributors are more, the release length will be shorter and number of releases will be more. However, correlation between number of contributors and number of commits could not be established uniformly in the study.

3.3.2 Datasets Following RR Models at Later Stages

It was discovered that all of the selected datasets started following the RR release model few years post the onset of development phase and finally aggregated to shorter release cycles. Datasets Cordova, Hive switched to RR model 2 years post their first delivery while Groovy, Pig, Camel switched 4 years post their first release. Zookeeper migrated the fastest, i.e., within half year, to RR model. It is interesting to observe that this trend was adopted by all the datasets that claimed RR model by averaging out their release cycle less than a month.

It was also interesting to know that most of these datasets switched their model in the year 2011 irrespective of when they initiated. Can this be the phase of revolution for RR model? Further studies are required to conform the results.

3.3.3 Datasets Following the RR Models with Few Exceptional Cases

Datasets Avro and Zookeeper, somewhere in the middle of their life cycle, deviated from the shorter release cycles and released their major releases in a gap of 1+ years. Avro released their minor version 1.7.3 on 8-12-12 and published their next minor version in 27-02-14 and major version 1.8.0 on 29-1-16. Zookeeper on the other hand published their minor version 3.4.6 on 12-3-14 and their next release 3.5.0 made its way in public on 6-4-15.

This gap pushed the study to investigate if the dataset went dormant during that phase or not. It was found that both Git [14] and Jira [15] logged bug reporting and resolving activity during the year 2013 as well, for Avro when no release was published. Zookeeper too showed the same trend on both the repositories in the year 2014. Availability of stabilized platform can be one of the reasons for not publishing a new version, but the study suggests more research before concluding the results.

4 Conclusions

The soaring market thrust has compelled many software companies to revamp their strategies and embrace new software development models to gain user's confidence and participation. The concept of Rapid Release or faster versions has been thoroughly explored for Open Sources like Firefox. This paper explores the Open Source repositories for release patterns. The paper analyzed and examined ten Open Source Apache projects to identify their release patterns and found that most have them have already shifted their models to Rapid Release with production of major versions in four months and minor versions in one month. The study also found correlation between release lengths, number of contributors, and number of releases.

Number of contributors, incapability to cope with deadline pressures and personal choices, might influence the choice of models, but Open Source community seems to favor RR over TR and this study gives ample evidence of that. However, the study suggests a deeper analysis on larger datasets, before generalizing the results. Future scope of the study includes the analysis of RR models on software bugs, quality, test effort estimation, and prediction.

References

1. Beck, K. and Andres, C. Extreme Programming Explained: Embrace Change (2nd Edition). Addison-Wesley, 2004. and I. N. Sneddon, "On certain integrals of Lipschitz-Hankel type involving products of Bessel functions," Phil. Trans. Roy. Soc. London, vol. A247 (1955) 529–551

2. Shorten release cycles by bringing developers to application lifecycle management. HP Applications Handbook, Retrieved on February 08, 2012. [Online]. Available: http://bit.ly/x5PdXl

3. Mozilla puts out firefox 5.0 web browser which carries over 1,000 improvements in just about 3 months of development. InvestmentWatch on June 25th, 2011. Retrieved on January 12, 2012. [Online]. Available: http://bit.ly/aecRrLB

4. F. Khomh, B. Adams, T. Dhaliwal and Y. Zou., "Understanding the impact of rapid releases on software quality" Empirical Software Engineering, vol, 20, pp 336–373. 2015

5. F. Khomh, T. Dhaliwal, Y. Zou and B. Adams, "Do faster releases improve software quality? an empirical case study of mozilla firefox." in Proceedings of the 9th working conference on mining software repositories (MSR). pp. 179–188, 2012

6. S. McIlroy, N. Ali and A.E. Hassan, "Fresh apps: an empirical study of frequently-updated mobile apps in the Google play store," Empirical Software Engineering, vol 21, pp. 1346–1370, 2016

7. Shankland, S.: Google ethos speeds up chrome release cycle. (2010) [Online]. Available: http://cnet.co/wlS24U

8. Kaply M. (2011) Why do companies stay on old technology? Retrieved on January 12, 2012

9. Porter, A., Yilmaz, C., Memon, A.M., Krishna, A.S., Schmidt, D.C., Gokhale, A. Techniques and processes for improving the quality and performance of open-source software. Software Process: Improvement and Practice, 11(2) (2006) 163–176

10. Baysal, O., Davis, I., Godfrey, M.W. A tale of two browsers. In Proc. of the 8th Working Conf. on Mining Software Repositories (MSR), (2011) 238–241

11. K. Petersen and C. Wohlin, "A comparison of issues and advantages in agile and incremental development between state of the art and an industrial case," Journal of System and Software, vol 82, pp. 1479–1490, 2009

12. M.V. Mäntylä, B. Adams, F. Khomh, E. Engström, and K. Petersen, "On rapid releases and software testing: a case study and a semi-systematic literature review." Empirical Software Engineering, pp 1384–1425, 2015

13. Apache Server Foundation(ASF): https://www.apache.org/: Last accessed in January 2017

14. Github: https://www.github.com Last accessed in January 2017

15. Jira: https://issues.apache.org/jira/ Last accessed in January 2017

16. Otte, T., Moreton, R., Knoell, H.D. Applied quality assurance methods under the open source development model. In Proc. of the 32nd Annual IEEE Intl. Computer Software and Applications Conf. (COMPSAC) (2008) 1247–1252

A Study of High-Dimensional Data Imputation Using Additive LASSO Regression Model

K. Lavanya, L. S. S. Reddy and B. Eswara Reddy

Abstract With the rapid growth of computational domains, bioinformatics finance, engineering, biometrics, and neuroimaging emphasize the necessity for analyzing high-dimensional data. Many real-world datasets may contain hundreds or thousands of features. The common problem in most of the knowledge-based classification problems is quality and quantity of data. In general, the common problem with many high-dimensional data samples is that it contains missing or unknown attribute values, incomplete feature vectors, and uncertain or vague data which have to be handled carefully. Due to the presence of a large segment of missing values in the datasets, refined multiple imputation methods are required to estimate the missing values so that a fair and more consistent analysis can be achieved. In this paper, three imputation (MI) methods, mean, imputations predictive mean, and imputations by additive LASSO, are employed in cloud. Results show that imputations by additive LASSO are the preferred multiple imputation (MI) method.

Keywords High-dimensional data · Multiple imputations · Regression
Missing data

K. Lavanya (✉)
Department of Computer Science and Engineering, JNTUA, Anantapur 515822,
Andhra Pradesh, India
e-mail: lavanya.kk2005@gmail.com

L. S. S. Reddy
Department of Computer Science and Engineering, KLU, Guntur 522502,
Andhra Pradesh, India
e-mail: drlssreddy@kluniversity.in

B. Eswara Reddy
Department of Computer Science, JNTUA, Anantapur 517234, Andhra Pradesh, India
e-mail: eswarcsejntua@gmail.com

© Springer Nature Singapore Pte Ltd. 2019
H. S. Behera et al. (eds.), *Computational Intelligence in Data Mining*,
Advances in Intelligent Systems and Computing 711,
https://doi.org/10.1007/978-981-10-8055-5_3

1   Introduction

Extracting the knowledge and analyzing high-dimensional data are one the key challenges in which variety and veracity are the two distinct characteristics. In this concern, every one needs to be aware of the design of high-performance proposals with the analysis of high-dimensional data [1]. As a result of knowledge analysis, people can easily find desired data. In the field of bioinformatics, extensive analysis is more essential for processing data which resides in wide range of expression profiling studies.

1.1   Origins of Missing Data

Missing data are of two types, namely item non-response and unit non-response.

1.1.1   Unit Non-response

Unit non-response occurs when a respondent cannot be contacted to complete a survey or does not return a questionnaire [2, 3]. Consequently, respondent is excluded from the study. Methods for managing unit non-response are weighting methods.

1.1.2   Item Non-response

At most often, item non-response is derived at all throughout self-administered questionnaires; once the applicant finishes his working survey, however, some cases missed few options to fill or neglect to complete the portion of the document. The missing values may occur due to a participant refuse or forget to answer a survey question, and an applicant also forgot to come back when a question is skipped. However, in addition to that applicant feels extra barrier to answer long questionnaires, it might be case near to due date. It might often item non-response in some other studies, which includes missing data values, data lost in collection and process. In such a cases, where it is also not possible to collect data due to device failure or a participant simply forgot to raise the answer for a question. The extensive survey of handling item non-response results good solutions which are methods based on weighting factor, few of imputation techniques, and some methods on basis of maximum likelihood which includes EM algorithm. The study of this work completely handles only item non-response. Type of missingness can be defined mathematically as the three ways. These procedures have been comprehensively defined by Little and Rubin [2–7]: (1) missing completely at random (MCAR), (2) missing at random (MAR), and (3) missing not at random (MNAR).

1.1.3 Missing Completely At Random (MCAR)

MCAR occurs when the missing value does not depend on outcome and covariates, i.e., the probability that an observation x_i is missing is not related to the value of x_i or to the value of any other variables. For example, in survey research, the participant may unintentionally turn over two pages instead of just one and thus cause a whole page of missing observations. Little's MCAR test is the most familiar test for missing data which is of the type missing completely at random. If the p value for Little's MCAR test is not significant, then the data may be assumed to be MCAR and missingness is believed not to issue for the analysis [2, 7]. Listwise deletion of values is appropriate, provided the number of missing values is not very large.

1.1.4 Missing At Random (MAR)

In MAR if missingness is related to other measured variables in the analysis model, but not to the underlying values of the incomplete variable i.e., if the data meet the necessity that missingness does not depend on the value of x_i after controlling for another variable [8, 9]. In MAR, missingness can well predicted from observed variables, and then, multiple imputation (MI) is appropriate.

1.1.5 Missing Not at Random (MNAR)

When missing values are neither MCAR nor MAR, then they are classed as missing not at random (MNAR). MNAR, also known as non-ignorable, occurs when missing values depend on non-observed data even after taming of all the observed data. One approach to MNAR (non-ignorable missingness) is to impute values based on external research design as the variable with missing data is not sufficiently correlated with other variables in the dataset.

1.2 High-Dimensional Data

Extracting the knowledge and analyzing high-dimensional data are one the key challenges in which variety and veracity are the two distinct characteristics. Developing a high-performance application that can effectively analyze high-dimensional data and discovering useful patterns from that data are a challenging task. High-dimensional data can be defined as follows: Let $Q = (x_1, c_1), \ldots, (x_n, c_n)$ be labeled data, where

$x_i \in \mathbb{R}^d$ denotes the ith feature vector and c_i is the corresponding class label. One of the prominent problems in high-dimensional data is missingness [1].

2 Research Background

2.1 Multiple Imputations

Multiple imputations [10–15] may be a statistically high-principled methodology which has been a good solution to missing data issues. The technique is very simple, and it produces a result balance to original data. In multiple imputations, missing values for any variable are predicted using existing values from other variables [16]. The predicted values, referred to as imputes, are substituted for the missing values leading to a full dataset called an imputed dataset. This method is performed multiple times, producing multiple imputed datasets. The standard statistical analysis is carried out on each imputed dataset, producing multiple analysis results. Multiple imputations account for missing data by restoring not solely the natural variability within the missing data, however additionally by incorporating the uncertainty caused by estimating missing data [7, 8, 17, 18]. Maintaining the original variability of the missing data is completed by building imputed values which are based on variables related to the missing data and causes of missingness. Uncertainty is accounted by creating completely different versions of missing data sets and watching the variability between imputed datasets and original data sets [9]. Thus, while doing multiple imputations, a researcher is fascinated by conserving vital characteristics of the dataset as a whole (e.g., means, variances, regression parameters). The degree to which these techniques are appropriate depends on how the missing data relates to the other variables. It is, therefore, desirable to combine a multiple imputation method with additive least absolute shrinkage and selection operator (LASSO) for imputation data into the dataset. House Price dataset with various missing data is used, and simulated results are shown. This study concludes that the combined multiple imputation and additive LASSO model performed well in the experimental results.

2.2 Additive LOSSO Model

The linear regression assumes that vector of responses $n \times 1$ in Q, q_i and X with $n \times xp$ matrix of predictors whose having ith subject values with row vector x_i, whereas Lasso Regression assumes that either of the observations which are independent or they can have q_i which are provisionally independent given with the values x_{ij} where they x_{ij} have been consistent so that $\sum_i x_{ij}/n = 0$ and $\sum_i x_{ij}^2/n = 1$ In addition, the value q_i has been inclined to center to have sample mean 0 under these considerations, and the Lasso estimates are given by

$$\widehat{\beta} = \arg\min_{\beta} \sum_{i=1}^{n} (q_i - \beta'_j x_i)^2 \tag{1}$$

subject to $\sum_{j=1}^{p} |\beta_i| \leq t$ where $t \geq 0$ is a tuning value which will control the amount of shrinkage applied to the estimated parameters and, therefore, the degree of variable selection

2.2.1 Coefficient Estimates Can Be Calculated

Let A be "active set" of covariates where most correlated with the "current" residual.

Initially, $A = \{x_{j1}\}$ for some covariate x_{j1}

Take the largest possible step in the direction of x_{j1} until another covariate x_{j2} enters A

Continue in the direction equiangular between x_{j1} and x_{j2} until third covariate x_{j3} enters A

Continue in the direction equiangular between x_{j1}, x_{j2}, x_{j3} until a fourth covariate x_{j4} enters A

This procedure continues until all covariates are added to the missing point.

$$\widehat{\beta} = \begin{cases} c_j + \lambda/a_j & c_j \leftarrow \lambda \\ 0 & c_j \in [-\lambda, \lambda] \\ (c_j - \lambda)/a_j & c_j > \lambda \end{cases} \tag{2}$$

3 Related Work and Result Analysis

3.1 Data Exploration

3.1.1 Histogram and Plot

The frequency of item is calculated with reference to area of each bar in plot. A total of 81 variables with 1460 observations are in the given dataset. The results shown in Fig. 1a and b provide frequency distribution of data in terms of variables including LotFrontAge, GarageType, GarageFinish, and GarageYrBlt with the usage of histogram and gplot. Moreover, distribution of missing in each variable along with missingness is explored in gplot.

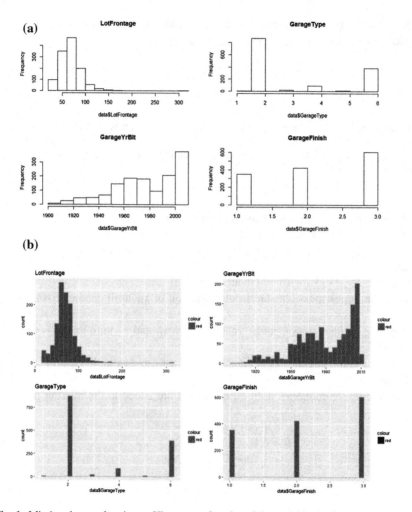

Fig. 1 Missing data exploration. **a** Histograms, **b** gplot of the variables in dataset

3.1.2 Aggregation Plot

Aggregation Plot explores more on geographical positions of missing samples and their frequencies. In Fig. 2, the information includes results of missing data exploration is on absolute proportion of each variable in dataset. However, the plot in left side describes observations with missingness. At right side, results are projected with a combination of red and yellow boxes and represent variables missingness. However, plots also derive proportion and frequency of missingness in each variable combination and are shown in a separate box.

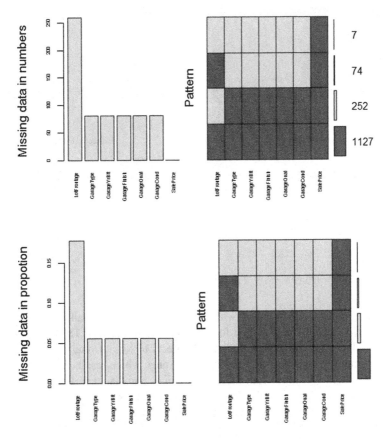

Fig. 2 Aggregation plot of the house dataset. Missing data distribution in proportion and frequencies

3.2 Imputation Techniques

3.2.1 Deletion Technique

The deletion method shown in this work is recommended for all categories of missing data. But deletion method results in partial estimated parameters to data if it is missing completely at random. Moreover, the derived imputed data are small in size when compared to original data and also result in larger error rate. With the above observations, it is not good for complete case analysis and also affects biased results for further analysis. However, some of the existing techniques are multiple imputation and direct maximum likelihood work for data with MAR. Similarly, data with MNAR retained statistical models. In the case of complete case analysis, the original data were reduced and it can also result to reduce test of power made by data.

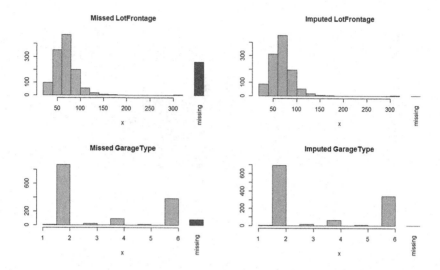

Fig. 3 Data exploration after imputation with deletion technique

Referring to Fig. 3, the model with complete data set containing 1460 observations and 81 variables is applied for complete case analysis. The complete case
analysis reduces the variables to 18 which effects to loss of data. The imputation
data exploration with different plots and histograms, which results no missing data,
is on LotFrontAge, GarageType, and GarageFinish, and GarageYrBlt variables. But
the SalesPrice variable is used to predict decisions. If the listwise deletion is used,
some of the parameters were deleted and are required to estimate predicting success.

3.2.2 Predictive Mean

Predictive mean produces best results where imputing continuous variables are not
normally distributed. The method imputed missing values with very closure to real
values, and this cannot be possible with exiting methods based on linear regression.
The range and skewness of both original and imputed are very close in this method.
In this method, missing values are replaced with its nearby predicated mean.
Moreover, the distribution in this approach follows linear regression which results
in higher variability in imputed data.

The results in Fig. 4a show plot which results in original and imputed values of
three variables LotFrontAge, GarageType, and GarageYrBlt. The plot describes
distribution of both imputed and Observed values. However, the figure shapes both
points are looking close and matching shape almost describes that original and
imputed values are significantly similar. Figure 4b shows another plot which
describes density of the imputed data in color magenta and observed data is in blue.
However, sometimes it is recommended to see distribution on each individual

(a)   **(b)**

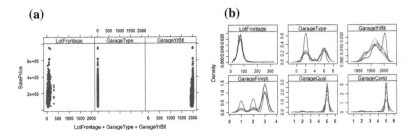

Fig. 4 **a** Data comparison with XY plot. **b** Density plot of imputed data

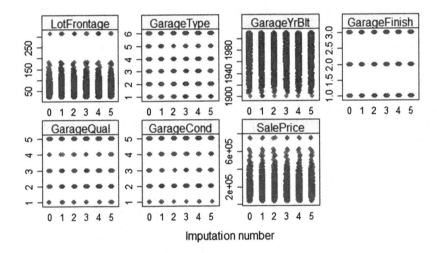

Fig. 5 Striplot of imputed data

variable like LotFrontAge, GarageType, GarageFinish, and GarageYrBlt with the usage of histogram and gplot in dataset, and it can be possible with striplot; results are shown in Fig. 5.

Table 1 shows analysis result of predicted mean with estimates for the interception, LotFrontAge, GarageType, GarageFinish, and GarageYrBlt. The study in

Table 1 Predictive mean imputation analysis of dataset

Observation	Estimate	Standard_Error_Rate	t_value	Pr > |t|
Interception	−1829448.914	161523.382	−11.326	0.0e+00
LotFrontage	807.1172	86.869	9.291	3.3e-10
GarageType	−2405.713	1166.793	−2.061	4.1e-02
GarageYrBlt	1031.187	81.099	12.715	0.0e+00
GarageFinish	−26734.811	2625.464	−10.183	0.0e+00
GarageCond	5889.830	3624.750	1.6249	1.0e-01
GarageQual	−9809.785	3134.182	−3.129	1.79e-03

which still some of the observations are still producing relatively larger errors includes GarageCond and GarageQual. It is observed that some of the observations which include LotFrontAge, GarageType, and GarageQual are significantly reduced error rate with this method.

3.2.3 Additive LASSO Model

Figure 6 shows a set of plots, where variables are imputed with LASSO predicators and the total plot displays with single red color. The scatter plots shows the imputation results of two variables i.e., GarageType and LotFrontage of Houseprice Data. From Eqs. 1 and 2, the following variables are imputed, i.e., LotFrontage, log-transformed LotFrontage, GarageType, GarageCond, GarageYrBlt, and GarageFinal. It is clearly visible that LotFrontage, GarageType, and GarageFinal are recommended to include predicated list for further analysis. The frequency of missingness describes completely zero and indicates with red numbers. Moreover, distribution of imputed data is shown in blue and red boxes. The result describes combination of histogram and bar plot and null distribution of missingness. Similarly, plots in Fig. 6 results strong association over two variables include LotFrontage and GarageType. The frequency of missingness describes completely zero

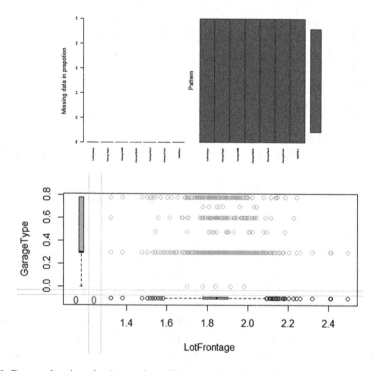

Fig. 6 Data exploration after imputation with proposed method

Table 2 Additive LASSO model imputation analysis of dataset

Observation	Estimate	Standard_Error_Rate	t_value	Pr > |t|	F_value	Pr(>F)
LotFrontage	880.027	68.796	12.792	1.4e-35	328.996	<2e-16
GarageType	−2928.253	1036.944	−2.824	4.8e-03	233.270	<2e-16
GarageYrBlt	1042.229	80.976	12.871	5.6e-36	387.957	<2e-16
GarageFinish	27877.938	2508.624	−11.113	1.4e-27	125.372	<2e-16
GarageCond	3789.322	3558.773	1.065	2.9e-01	1.134	0.2872
GarageQual	−8062.642	3004.422	−2.684	7.4e-03	6.084	0.0138

and indicates with red numbers. Moreover, distribution of imputed data shown in blue and red boxes.

The Table 2 shows analysis result of proposed Additive LASSO imputation technique with estimates better results in terms of LotFrontAge, GarageType, GarageFinish and GarageYrBlt. The results shows less standard errors and estimates proposed variables to significant. Proposed result analysis helps to estimates future vector for further analysis like classification and prediction. After imputation model is fitted into linear regression and shows that less error rate compared to the predictive mean imputation technique.

4 Conclusion

Three imputation (MI) methods, mean, imputations predictive mean, and imputations by additive LASSO, are employed. Results show that imputations by additive LASSO are the preferred multiple imputation (MI) method.

References

1. Fanyu Bu, Zhikui Chen, Qingchen Zhang Laurence T. Yang," Incomplete high-dimensional data imputation algorithm using feature selection and clustering analysis on cloud, J Supercomput, (2016) 72:2977–2990.
2. Rubin, D.B.: Multiple imputation for nonresponse in surveys, 1st ed., New York: John Wiley and Sons, Inc., (1987). 258 pages.
3. Schafer, J.L.: Multiple imputation: a primer, Statistical Methods in Medical Research, 8, (1999). 3–15.
4. Little, R.J.A. and Rubin, D.B.: Statistical analysis with missing data, 2nd ed., New York: John Wiley and Sons, Inc., (2002). 381 pages.
5. Little, R.J.A.: A test of missing completely at random for multivariate data with missing values, Journal of American Statistical Association, 83, (1988). 1198–1202.
6. Little, R.: Calibrated Bayes, for Statistics in general, and missing data in particular, Statistical Science, 26, (2011). 162–174.

 7. Rubin, D.B. and Schemer, N.: Multiple imputation in health-care databases: An overview and some applications, Statistics in Medicine, 10, (1991). 585–598.
 8. Schafer, J.L. and Olsen, M.K.: Multiple imputation for multivariate missing-data problems: A data analyst's perspective, Multivariate Behavioral Research, 33, (1998). 545–571.
 9. Schneider, T.: Analysis of incomplete climate data: Estimation of mean values and covariance matrices and imputation of missing values, Journal of Climate, 14, (2001). 853–871.
10. Jolani, S., Debray, T., Koffijberg, H., van Buuren, S., and Moons, K.: Imputation of systematically missing predictors in an individual participant data meta-analysis: a generalized approach using MICE. Statistics in Medicine, 34(11): (2015). 1841–1863.
11. Kropko, J., Goodrich, B., Gelman, A., and Hill, J.: Multiple imputation for continuous and categorical data: Comparing joint multivariate normal and conditional approaches. Political Analysis, 22(4): (2014). 497–519.
12. Langan, D., Higgins, J., and Simmonds, M.: Comparative performance of heterogeneity variance estimators in meta-analysis: a review of simulation studies. Research Synthesis Methods. To appear. (2016).
13. Lassus, J., Gayat, E., Mueller, C., Peacock, W., Spinar, J., Harjola, V., van Kimmenade, R., Pathak, A., Mueller, T., and et al. (2013). Incremental value of biomarkers to clinical variables for mortality prediction in acutely decompensated heart failure: the Multinational Observational Cohort on Acute Heart Failure (MOCA) study. International Journal of Cardiology, 168 (3):2186–2194.
14. Quartagno, M. and Carpenter, J.: Multiple imputation for IPD meta-analysis: allowing for heterogeneity and studies with missing covariates. Statistics in Medicine, 35(17): (2016). 2938–2954.
15. Yucel, R.: Random-covariances and mixed-effects models for imputing multivariate multilevel continuous data. Statistical modelling, 11(4): (2011). 351–370.
16. Erler, N., Rizopoulos, D., van Rosmalen, J., Jaddoe, V., Franco, O., and Lesaffre, E.: Dealing with missing covariates in epidemiologic studies: A comparison between multiple imputation and a full Bayesian approach. StatMed. (2016).
17. van Buuren, S.: Flexible Imputation of Missing Data (Chapman & Hall/CRC Interdisciplinary Statistics). Chapman and Hall/CRC. (2016).
18. Vink, G., Lazendic, G., and van Buuren, S.: Partitioned predictive mean matching as a multilevel imputation technique. Psychological Test and Assessment Modeling, 57(4): (2015). 577–594.

An Efficient Multi-keyword Text Search Over Outsourced Encrypted Cloud Data with Ranked Results

Prabhat Keshari Samantaray, Navjeet Kaur Randhawa
and Swarna Lata Pati

Abstract Cloud computing offers efficient deployment options that motivate large enterprises to outsource the data to the cloud. However, outsourcing sensitive information may compromise the privacy of the data. To enable keyword-based search over encrypted data, we proposed a multi-keyword search scheme on a tree-based encrypted index data structure to retrieve information from encrypted cloud data. In this model, the document collection is clustered using a hierarchical k-means method. A vector space model was used to create an encrypted index and query vectors, and a depth-first search algorithm is proposed for efficient search mechanism. The results were ranked based on relevance score between the encrypted index and query vectors. Rigorous experiments show the performance and efficiency of the proposed methods.

Keywords Cloud computing · Symmetric searchable encryption
Document clustering · Multi-keyword ranked search

1 Introduction

Symmetric searchable encryption (SSE) provides an efficient way of implementing keyword-based search scheme on outsourced sensitive encrypted cloud data. First SSE scheme was proposed by Song et al. [1]. Several improved SSE schemes include single keyword-based search [2–7], single keyword ranked search [8, 9]

P. K. Samantaray (✉) · N. K. Randhawa · S. L. Pati
Department of Computer Science and Engineering, College of Engineering
and Technology, Ghatikia, Bhubaneswar 751003, India
e-mail: prabhat.samantray@gmail.com

N. K. Randhawa
e-mail: navjeetkaur30@gmail.com

S. L. Pati
e-mail: spati@cet.edu.in

© Springer Nature Singapore Pte Ltd. 2019
H. S. Behera et al. (eds.), *Computational Intelligence in Data Mining*,
Advances in Intelligent Systems and Computing 711,
https://doi.org/10.1007/978-981-10-8055-5_4

and multi-keyword ranked search [10–15]. The multi-keyword ranked search scheme provided by SE has got much attention in recent years.

In this paper, we proposed a multi-keyword ranked search scheme based on tree-based index constructed from clustered document collection. We will use vector space model to construct the index and search query and calculate the relevance between a document and query using term frequency (TF) X inverse document frequency (IDF) rule. The vectors are encrypted using secure kNN algorithm [16], the document clustering is done using bisecting k-means clustering method [17], and search operation is done using an efficient depth-first search algorithm. Two encrypted schemes are proposed: basic multi-keyword ranked search on clustered documents (BMRSCD) to address ciphertext threat model and enhanced multi-keyword ranked search on clustered documents (EMRSCD) to address known background threat model.

2 Background

2.1 System Model

The system model shown in Fig. 1 comprises of the data owner, the data user and the cloud server. The data owner partitions the documents collection D into z clusters, builds an encrypted searchable index tree I, creates encrypted document

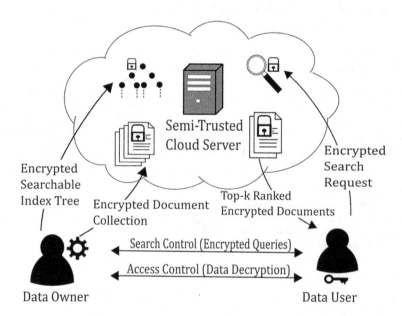

Fig. 1 System architecture of ranked search over outsourced encrypted cloud data

collection E and finally uploads I and E to the cloud server. The data user generates an encrypted query called trapdoor (*TDR*), using secret key shared by data owner, and sends a search request to the cloud server. The cloud server stores the encrypted index tree I and encrypted document collection E, and it performs search operation on I to fetch the relevance documents and send it back to the data user.

2.2   Threat Models

Since the cloud system is considered as a semi-trusted system, it can perform an attack by using the encrypted index tree I and trapdoor *TDR* in known ciphertext model or it may perform a statistical attack by using additional background information in the known background model.

2.3   Preliminaries

Vector Space Model: Vector space model is text mining technique that uses $TF \times IDF$ rule [18] to evaluate the relevance between documents and query. The term frequency is defined by the frequency of a keyword present in the document, whereas the inverse document frequency is the ratio of a number of documents in the collection to the number of documents that contains the keyword. Each document represented by a vector containing normalized *TF* and each query represented by a vector containing normalized *IDF*. The relevance between the document and query can be computed by taking the dot product of the document vector and query vector.

Document Clustering: The document collection is clustered using bisecting k-means clustering [17] method. It is a hierarchical k-means clustering method in which it inherits the feature of high-quality clustering technique from hierarchical clustering method and the feature of efficient clustering technique from k-means clustering method. The algorithm of bisecting k-means clustering method is given in Algorithm 1.

Algorithm. 1. *ClusterDocumentCollection(D, z)*
1   Create initial cluster having all the documents in the collection.
2   Insert the initial cluster to *ClusterSet*.
3   if size of *ClusterSet* is not equal to z then
4   Choose a cluster, C from *ClusterSet* to split.
5   Find two sub-clusters $\{C_1, C_2\}$ using k-means algorithm (Bisection Step).
6   Repeat step 5, the bisection step until the split produces highest overall similarity of the sub-clusters C_1, C_2.
7   Remove Cluster C from *ClusterSet*
8   Insert the sub-clusters C_1, C_2 into *ClusterSet*.
9   Return *ClusterSet*

Index Tree Construction: The searchable index tree in this model is adapted from keyword balanced binary tree (KBB) from Xia et al. [15]. Similar to KBB, the searchable index tree in this model stores vector V at each node i and we denote it as V_i. A node i in the index tree can be defined as 5-tuple:

$$i = \langle ID, V, Lch, Rch, DocID \rangle$$

where *ID* stores the unique identity value for the node i, *Lch* and *Rch* stores the reference of the left child and right child node, respectively. If the node i is an internal node, then *Doc_ID* is set to null and the elements of the index vector V_i are assigned to the values computed from its child nodes as given in Eq. (1).

$$V[r] = max\{i.Lch \rightarrow D[r], i.Rch \rightarrow D[r]\} \tag{1}$$

for each $r = 1, 2, \ldots m$.

The leaf nodes are represented by clusters, and *DocID* field contains the list of documents in the cluster. Figure 2 shows index tree construction from the clusters of documents. The cluster vector CV_j of cluster j that represents the leaf node of the tree is calculated as given in Eq. (2).

$$CV_j[r] = max\{d_1[r], d_2[r], \ldots d_s[r]\} \tag{2}$$

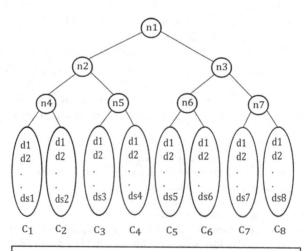

C₁ to C₈ represents the document clusters
s1,s2,.....,sj represents number documents in cluster Cj
d1,d2,.....,dsj represents the documents in cluster Cj

Fig. 2 Index tree construction from the cluster of documents

Cluster C_j

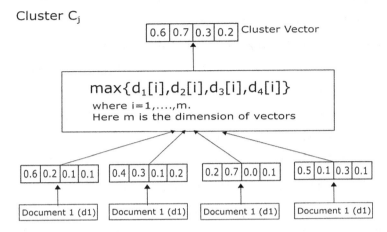

Fig. 3 Calculation of the cluster vector CV_j for the cluster Cj

for each $r = 1, 2 \ldots$ m and s = size of the cluster or the number of documents in cluster j.

The calculation of the cluster vector CV_j for a cluster j is illustrated in Fig. 3.

Finally, the index tree is built similar to *KBB* from the cluster vectors which are treated at leaf nodes. Algorithm 2 shows the index tree construction process.

Algorithm 2. *CreateIndexTree (D, z)*

Inputs: The document collection $D = \{d_1, d_2, \ldots . d_n\}$. A parameter z is provided which denote the number of clusters to be created.

Output: An unencrypted searchable index tree T.

1:   *ClusterList = ClusterDocumentCollection(D, z)*;

2:   For each cluster j in *ClusterList* Create leaf node for cluster vector CV_j calculated using Eq (2).

3:   Create the Index Tree based on KBB defined in Xia et al. [15]

Search Process in Index Tree: The proposed search algorithm on the index tree is a recursive search algorithm based on greedy depth-first search (GDFS) defined in [15]. The search algorithm is given in Algorithm 3.

Algorithm.3. *SearchIndexTree(TreeNode i)*

Input: The root node of the index tree.

Output: The result list *Result_List*, which contains top-k ranked documents with corresponding relevance score.

1:	if the node i is not a leaf node then
2:	if *Relevance_Score(V_i, Q) > MinScore* then
3:	*SearchIndexTree(i. high_relevance_child)*;
4:	*SearchIndexTree(i. low_relevance_child)*;
5:	Else Return;
7:	End if
8:	Else
9:	If *Relevance_Score(V_i, Q) > MinScore* then
10:	For each document d in the cluster i
11:	If *Relevance_Score(d.V, Q) > MinScore* then
12:	Insert *<Relevance_Score(d.V, Q),DocID>* into ResultList
13:	End for
14:	Sort the elements in *Result_List*
15:	Keep the top-k elements in *Result_List* and remove the rest;
16:	End If
17:	Return
18:	End If

3 Encrypted Schemes

3.1 The BMRSCD Scheme

In this scheme, a basic multi-keyword ranked search with clustered document collection is proposed in known ciphertext model and used secure kNN algorithm [16] to encrypt the vectors. The functions are defined as follows.

$C \leftarrow CreateDocumentClusters (D, z)$—The document collection is clustered using bisecting k-means method to generate a list of document clusters $C = \{c_1, c_2, \ldots c_z\}$.

$SK \leftarrow GenSecretKeys ()$—The data owner generates the secret key SK which includes (1) a binary m-bit secret vector S where m denotes the dictionary size and (2) two $(m \times m)$ Gaussian random invertible matrices M_1, M_2. The secret key is denoted as $SK = \{S, M_1, M_2\}$.

$I \leftarrow GenEncryptedIndex (C, SK)$—Initially, an unencrypted index tree T is created using function $T \leftarrow CreateIndexTree (D, z)$. The index vector V_i is encrypted by splitting it into vectors $\{V_i', V_i''\}$, such that if $S[r] = 0$, then $V_i'[r]$ and $V_i''[r]$ are set to $V_i[r]$, and if $S[r] = 1$, then $V_i'[r]$ and $V_i''[r]$ are set to two random values whose sum is equal to $V_i[r]$. The encrypted index vector EV_i for the node i in the index tree I is calculated as $EV_i = \{M_1^T V_i', M_2^T V_i''\}$.

$TDR \leftarrow GenTrapdoor (K_q, SK)$—The unencrypted query vector Q containing normalized IDF values of K_q with dimension m. The query vector is partitioned into two random vectors Q', Q'' such that if $S[r] = 0$, then $Q'[r]$ and $Q''[r]$ are set to two random values whose sum is equal to $Q[r]$ else $Q'[r]$ and $Q''[r]$ are set to $Q[r]$. Finally, the trapdoor is created as $TDR = \{M_1^{-1}Q', M_2^{-1}Q''\}$.

Result_List ← *Search (TDR, I, k)*—The server executes depth-first search *SearchIndexTree(TreeNode i)* on index tree *I* and returns the top-*k* ranked most relevant encrypted documents.

Relevance Score calculation: The relevance score calculated from encrypted index and trapdoor is equal to that of unencrypted vectors. It can be shown in Eq. (3).

$$
\begin{aligned}
EV_i \cdot TDR &= \left(M_1^T V_i'\right)\left(M_1^{-1} Q'\right) + \left(M_2^T V_i''\right)\left(M_2^{-1} Q''\right) \\
&= \left(M_1^T V_i'\right)^T \left(M_1^{-1} Q'\right) + \left(M_2^T V_i''\right)^T \left(M_2^{-1} Q''\right) \\
&= V_i' Q' + V_i'' Q'' = V_i \cdot Q = Rel\_Score(V_i, Q)
\end{aligned}
\tag{3}
$$

3.2 The EMRSCD Scheme

Although the vectors are well protected in known ciphertext model, BMRSCD cannot ensure privacy in known background model because the keywords can be identified by the server exploiting statistical information [9, 11, 19]. The main reason is that the relevance score calculated between EV_i and TDR is same as that between V_i and Q. To address this issue, a tunable randomness is added to the relevance score evaluation. The functions in EMRSCD scheme are basically similar to the BMRSCD scheme except that the dimension of the vectors is increased to $(m + m')$.

The random invertible matrices will have dimension $(m + m') \times (m + m')$. For the index vectors, the extended term in $V_i[m + u], u = 1, \ldots m'$ is set to a random number ε_u. The extended elements m' in Q contain binary values such that a number of m'' elements set to 1 and the rest are set to 0.

The relevance score evaluation between EV_i and TDR is calculated as given in Eq. (4).

$$
Rel\_Score(EV_i, TDR) = V_i \cdot Q + \sum \varepsilon_w
\tag{4}
$$

where, $w \in \{u | Q[m + u] = 1\}$.

4 Performance Analysis

The performance of the schemes is evaluated using the dataset built with text documents from BBC news archives [20]. In this section, we test the efficiency of index tree construction and search process.

4.1 Index Tree Construction

In index tree construction process, the clustering of document collection takes $O(nzlog_2(z))$ time. The unencrypted index tree generates O(z) nodes, and the encryption of the index vector during splitting process takes O(m) time. The two $(m \times m)$ matrix multiplications with m-dimensional vector take $O(m^2)$ time. The total time taken to construct the index tree is $O(z(nlog_2(z) + m^2))$. Figure 4 shows the comparisons of index tree construction time of proposed models.

4.2 Search Efficiency

The search operation is performed in depth-first until the leaf node and linear search within the cluster. Let θ be the number of clusters that contains the resultant documents, k be the number of search result, z be the number of clusters in the index tree, n be the cardinality of document collection and s be the size of the cluster, and we assume $\theta < k$ and $\theta << n$. The running time of relevance score evaluation is $O(m)$, and as a result, the complexity of search is $O(\theta m \, log(z) + s)$. Figure 5 shows the comparison of search efficiency of proposed model.

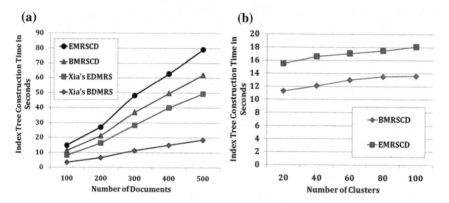

Fig. 4 Comparison of index tree construction time: **a** BDMRS, BMRSCD, EDMRS and EMRSCD with number of clusters 30 with dictionary size of 1000 keywords, **b** with different number of clusters with dictionary size of 1000 keywords

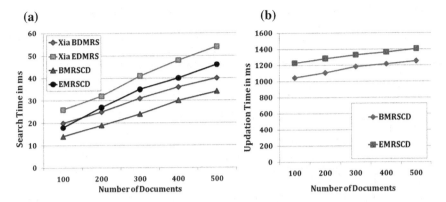

Fig. 5 Comparison of search operation: **a** BDMRS, BMRSCD, EDMRS and EMRSCD with number of keywords in query 25, **b** the time to update a document in BMRSCD and EMRSCD with dictionary size 1000 and number of clusters 20

5 Conclusion and Future Work

In this paper, we proposed a tree-based secure index constructed using vector space model on clustered document collection, and a depth-first search algorithm is used for the efficient search process. The index tree and search queries are encrypted using the secure kNN algorithm to prevent the cloud server from determining the keywords. The relevance score between a document and a query for document ranking is calculated by using $TF \times IDF$ rule. Two secure schemes are defined to address various privacy requirements in two threat models. Finally, through experimental analysis, the efficiency of the proposed schemes is provided.

Symmetric SE is an open research area, and still, there are a lot of issues needed to be addressed. In the proposed scheme, the ranking information of the retrieved documents is not hidden from the cloud server due to efficiency reason which may help the cloud server to gather some information about underlying encrypted index tree. In this model, we assume the data users are trustworthy but it may be the case the data user shares the secret keys with their relatives or some users having the wrong intention of using extracted information.

References

1. Song, D.X., D. Wagner, and A. Perrig,: Practical techniques for searches on encrypted data. In Proceedings IEEE Symposium on Security & Privacy, 2000, pp. 44–55.
2. Goh, E.-J.: Secure Indexes. IACR Cryptology ePrint Archive, Vol. 2003, p. 216, 2003.
3. Curtmola, R., Garay, J., Kamara, S., Ostrovsky, R.: Searchable Symmetric Encryption: Improved definitions and efficient constructions. In Proc. 13th ACM Conf. Computer and Communications Security, 2006, pp. 79–88.

4. Li, J., Wang, Q., Wang, C., Cao, N., Ren, R., Lou, W.: Fuzzy keyword search over encrypted data in cloud computing. In IEEE Proceedings INFOCOM, 2010, pp. 1–5.
5. Kuzu, M., Islam, M. S., Kantarcioglu, M.: Efficient similarity search over encrypted data. In Proceedings IEEE 28th International Conference on Data Engineering, 2012, pp. 1156–1167.
6. Wang, C., Ren, K., Yu, S., Urs, K. M. R.: Achieving usable and privacy-assured similarity search over outsourced cloud data. In Proceedings IEEE INFOCOM, 2012, pp. 451–459.
7. Wang, B., Yu, S., Lou, W., Hou, Y. T.: Privacy-preserving multi-keyword fuzzy search over encrypted data in the cloud. In Proceedings IEEE INFOCOM, 2014, pp. 2112–2120.
8. Wang, C., Cao, N., Li, J., Ren, K., Lou, W.: Secure Ranked Keyword Search over Encrypted Cloud Data. In Proceedings IEEE 30th International Conference Distributed Computing Systems (ICDCS 2010), 2010, pp. 253–262.
9. Wang, C., Cao, N., Ren, K., Lou, W.: Enabling Secure and Efficient Ranked Keyword Search over Outsourced Cloud Data. IEEE Transaction on Parallel and Distributed Systems, vol. 23, no. 8, pp. 1467–1479, 2012.
10. Cao, N., Wang, C., Li, M., Ren, K., Lou, W.: Privacy-preserving multi-keyword ranked search over encrypted cloud data. In Proceedings IEEE INFOCOM, 2011, pp. 829–837.
11. Sun, W., Wang, B., Cao, N., Li, M., Lou, W., Hou, Y. T., Li, H.: Privacy-preserving multi-keyword text search in the cloud supporting similarity-based ranking. In Proceedings 8th ACM SIGSAC Symposium on Information, Computer and Communications Security, 2013, pp. 71–82.
12. Orencik, C., Kantarcioglu, M., Savas, E.: A practical and secure multi-keyword search method over encrypted cloud data. In Proceedings IEEE 6th International Conference on Cloud Computing, 2013, pp. 390–397.
13. Zhang, W., Xiao, S., Lin, Y., Zhou, T., Zhou, S.: Secure ranked multi-keyword search for multiple data owners in cloud computing. In Dependable Systems and Networks (DSN), IEEE 44th Annual IEEE/IFIP International Conference, 2014, pp. 276–286.
14. Pang, H., Shen, J., Krishnan, R.: Privacy-preserving similarity-based text retrieval. ACM Transactions on Internet Technology, Vol. 10, no. 1, p. 4, 2010.
15. Xia, Z., Wang, X., Sun, X., Wang, Q.: A Secure and Dynamic Multi-Keyword Ranked Search Scheme over Encrypted Cloud Data. IEEE Transactions on Parallel and Distributed Systems, vol. 27, no. 2, pp. 340–352, February 2016.
16. Wong, W.K., Cheung, D.W.-L., Kao, B., Mamoulis, N.: Secure KNN computation on encrypted databases. In Proceedings of ACM SIGMOD International Conference on Management of Data, pp. 139–152, June 2009.
17. Steinbach, M., Karypis, G., Kumar, V.: A Comparison of Document Clustering Techniques. In KDD Workshop on Text Mining, vol. 400, no. 1, pp. 525–526, August 2000.
18. Manning, C.D., Raghavan, P., Schutze, H.: Introduction to Information Retrieval. Cambridge, UK, Cambridge University Press, 2008.
19. Zerr, S., Olmedilla, D., Nejdl, W., Siberski, W.: Zerber + r: Top-k retrieval from a confidential index. In Proceedings of the 12th International Conference on Extending Database Technology: Advances in Database Technology, pp. 439–449, 2009.
20. BBC Datasets [Online]. Available: http://mlg.ucd.ie/datasets/bbc.html.

Robust Estimation of IIR System's Parameter Using Adaptive Particle Swarm Optimization Algorithm

Meera Dash, Trilochan Panigrahi and Renu Sharma

Abstract This paper introduces a novel method of robust parameter estimation of IIR system. When training signal contains strong outliers, the conventional squared error-based cost function fails to provide desired performance. Thus, a computationally efficient robust Hubers cost function is used here. As we know that the IIR system falls in local minima, gradient-based algorithm cannot be used. Therefore, the parameters of the IIR system are estimated using adaptive particle swarm optimization algorithm with Hubers cost function. The simulation results show that the proposed algorithm provides better performance than Wilcoxon norm-based robust algorithm and conventional error squared based PSO algorithm.

Keywords IIR system · Impulsive noise · Robust estimation · Wilcoxon norm
Hubers cost function · Adaptive particle swarm optimization

1 Introduction

Parameter estimation for finite impulse response (FIR) system has been discussed in literature. It is because, the FIR system is simple and inherently stable [1, 2]. On the other hand, most of the natural systems are modeled as infinite impulse response (IIR) or autoregressive and moving average (ARMA) systems. For example in applications such as target tracking, industrial control, fast routing, data

M. Dash (✉)
Department of ECE, ITER Siksha 'O' Anusandhan University,
Bhubaneswar 751030, Odisha, India
e-mail: meeranayak@soauniversity.ac.in

T. Panigrahi
Department of ECE, National Institute of Technology Goa, Ponda 403401, India
e-mail: tpanigrahi@nitgoa.ac.in

R. Sharma
Department of EE, ITER Siksha 'O' Anusandhan University,
Bhubaneswar 751030, Odisha, India
e-mail: renusharma@soauniversity.ac.in

© Springer Nature Singapore Pte Ltd. 2019   41
H. S. Behera et al. (eds.), *Computational Intelligence in Data Mining*,
Advances in Intelligent Systems and Computing 711,
https://doi.org/10.1007/978-981-10-8055-5_5

reduction and data aggregation. The estimation of IIR system is difficult due to instability caused by poles and multimodal cost function [3].

Local optimization techniques such as LMS algorithm which work well for FIR systems are not suitable here because they are likely to be trapped in the local minima [4]. During the learning process when the poles move outside the unit circle, the IIR systems become unstable. We also know that the adaptive IIR system converges slowly which needs further attention.

In recent days, in order to overcome the local minima problem in the squared error space, a stochastic search algorithm is proposed as an alternative to the gradient descent-based technique. Since the gradient is not calculated, these stochastic algorithms are structure independent. The structure of the adaptive system does not directly influence the parameter updates during the learning process. In literature, for design and identification of IIR system, many evolutionary algorithms are used. These are genetic algorithm (GA) [5], particle swarm optimization (PSO) and its variants [6], harmonic search (HS) algorithm [7]. But these algorithms are not robust to the outliers as all have used conventional error squared cost function. The presence of such outliers severely degrades the estimation performance of the IIR system [8]. Thus, there is a need for the development of robust estimation evolutionary algorithms to alleviate the effect of outliers in the training data.

In this paper, the modified PSO known as adaptive PSO algorithm is used to estimate the IIR system parameter. Robust Huber cost function is used to make the optimization algorithm robust against outliers. The proposed robust cost function is computationally efficient than the statistical Wilcoxon norm. The results are compared with Wilcoxon norm-based robust PSO algorithm [9] and conventional error squared based PSO. The simulation result shows that the proposed robust algorithm gives faster convergence and better estimation accuracy.

2 Problem Formulation

Consider an IIR system with N and M number of poles and zeros, respectively. The noisy output (measured data) of the IIR system $d(i)$ at ith instant is related to the input vector by its constant coefficient difference equation, given in (1). The difference equation is as follows [4]

$$d(i) = \sum_{n=1}^{N} a_n d(i-n) + \sum_{m=0}^{M} b_m x(i-m) + v(i) \qquad (1)$$

where $\{a_n\}_{n=1}^{N}$ and $\{b_m\}_{m=0}^{M}$ are the feedback and feed-forward parameters of the IIR system to be estimated and $v(i)$ denotes the noise at ith instant. In general, the noise is considered as normalized Gaussian noise with zero mean and variance σ_v^2. In fact, in most of the applications, the data is mixed with impulsive noise. The input data vector $\mathbf{x}_i = [x(i-1), \ldots, x(i-M)]^T$ at time instant i. The current estimate output depends upon N number of previous outputs. The corresponding transfer function of the IIR system (1) is given as

$$H(z) = \frac{B(z)}{A(z)} = \frac{\sum_{m=0}^{M} b_m z^{-m}}{1 - \sum_{n=1}^{N} a_n z^{-n}} \tag{2}$$

where $A(z) = 1 - \sum_{n=1}^{N} a_n z^{-n}$ and $B(z) = \sum_{m=0}^{M} b_m z^{-m}$ are feedback and feed-forward coefficient polynomials in Z-domain of IIR system, respectively. During the estimation, the unknown parameters of the IIR system having transfer function $H(z)$ given in (2) is to be estimated by identifying new adaptive system $\hat{H}(z)$ in such a way that the outputs from both the systems matches closely to the same given input. The adaptive system is modeled as IIR filter. The output of the estimated system to the input $x(i)$ is given in (3) as

$$y(i) = \sum_{n=1}^{N} \hat{a}_n(i)y(i-1) + \sum_{m=0}^{M} \hat{b}_m(i)x(i-1) \tag{3}$$

where $y(i)$ is the adaptive IIR system's output. The transfer function of the adaptive IIR system whose parameters to be estimated is given as

$$\hat{H}(z) = \frac{\hat{B}(z)}{\hat{A}(z)} = \frac{\sum_{m=0}^{M} \hat{b}_m z^{-m}}{1 - \sum_{n=1}^{N} \hat{a}_n z^{-n}} \tag{4}$$

where $\hat{A}(z) = 1 - \sum_{n=1}^{N} \hat{a}_n z^{-n}$ and $\hat{B}(z) = \sum_{m=0}^{M} \hat{b}_m z^{-m}$ are feedback and feed-forward coefficient polynomials in Z-domain of adaptive system.

In the conventional estimation problem, mean square error (MSE) which is given in (5) is used as the objective function. This is considered as error square fitness function and given as

$$MSE(\mathbf{x}, \mathbf{w}) = \frac{1}{N_s} \sum_{i=1}^{N_s} e^2(i) \tag{5}$$

where N_s is the number of input samples and the error $e(i)$ is given as

$$e(i) = d(i) - y(i) \tag{6}$$

The vector $\mathbf{w}$ is the IIR system parameters to be determined. The main job of the parameter estimation algorithm is to update the adaptive IIR system's parameter of (4) iteratively using appropriate optimization algorithms in such a way that the output of the adaptive system matches to the output of the unknown IIR system for the same given input. The basic block diagram for the robust parameter estimation of IIR system using adaptive IIR system is shown in Fig. 1.

Fig. 1 Robust IIR system parameter estimation

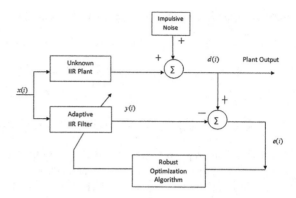

3 Robust Estimation of IIR System's Parameter

The estimation and identification algorithms and their performance analysis for IIR system are available in the literature. In most of the cases, the noise is simply considered as additive white Gaussian noise. In a real-world problems, outliers is encountered. The outliers are modeled as two independent components of a Gaussian mixture [10] which has wide applications in communication and signal processing. The outliers may be generated using the following Eq. (7).

$$v(i) = v_g(i) + b(i)v_{im}(i) \tag{7}$$

where $v_g(i)$ and $v_{im}(i)$ are independent zero mean normalized Gaussian random variable with different variances σ_g^2 and σ_{im}^2, respectively; $b(i)$ is modeled as Bernoulli random process. The probability of occurrence $P(b(i) = 1) = p$ and $P(b(i) = 0) = 1 - p$. High variance of $v_{im}(i)$ is chosen than that of $v_g(i)$ so that when outliers occur that is $b(i) = 1$, a large magnitude of noise is experienced in $v(i)$. The probability density function of $v(i)$ is given in Eq. (8) as

$$f_v(x) = \frac{1 - p}{\sqrt{2\pi}\sigma_g} \exp\left(\frac{-x^2}{2\sigma_g^2}\right) + \frac{p}{\sqrt{2\pi}\sigma_T} \exp\left(\frac{-x^2}{2\sigma_T^2}\right) \tag{8}$$

where $\sigma_T^2 = \sigma_g^2 + \sigma_{im}^2$.

3.1 Robust Fitness Function

In a conventional optimization problem, the least mean square objective function such as $MSE(\mathbf{x}, \mathbf{w})$ is used in the optimization algorithm to estimate the parameter of IIR system. The error squared is sensitive to outliers, because the value increases when it occurs. Therefore, to make the optimization algorithm robust, the least-squared objective function is replaced with a robust cost function, $h(\mathbf{x}, \mathbf{w})$ [11]. To make the classification algorithm robust, different distance-based learning algorithms are used in literature [12–14].

The robust function h is designed in such a way that it gives less weight to data points which deviate greatly from the desired parameter (error is high when impulsive noise occur), $\mathbf{w}$. The cost function (5) is modified for a robust estimation and given as

$$f_{robust}(\mathbf{w}) = \frac{1}{N_s} \sum_{i=1}^{N_s} h(\mathbf{x}_i, \mathbf{w}) \tag{9}$$

In literature, different robust functions are available to define robust cost function (9). Statistical norm-based robust function known as Wilcoxon norm is used in robust identification and prediction problem using PSO algorithm is proposed in [9]. But there is no control on the convergence speed and steady-state performance. One of the standard robust cost functions is the Huber loss function [15] which is defined as

$$h(x; \boldsymbol{\theta}) = \begin{cases} e^2(i)/2, & \text{for } e(i)| \leq \gamma \\ \gamma e(i) - \gamma^2/2, & \text{for } e(i) > \gamma \end{cases} \tag{10}$$

Huber function given in (10) is convex, differentiable, and also robust to outliers. The choice of threshold γ is crucial in estimation problem. The steady-state performance may poor if the γ value is large, but will achieve faster convergence, whereas for small γ value the steady-state performance is better, but the convergence will be slow.

4 Modified PSO for Robust Parameter Estimation of IIR System

Let us consider P number of particles in the swarm. In the search space, every particle is characterized by two factors known as position and velocity which are given for jth particle at ith time instant as $\mathbf{w}_j^i = [\hat{a}_{j1}, \ldots, \hat{a}_{jN}, \hat{b}_{j0}, \ldots, \hat{b}_{jM}]$, where $\hat{a}_j$ and $\hat{b}_j$ represent the feedback and feed-forward parameters of the IIR system and $\mathbf{v}_j^i = [v_{j1}^i, v_{j2}^i, \ldots, v_{jD}^i]^T$, respectively. Each particle flies through the D-dimensional $(D = N + M + 1)$ hyperspace with respect to two reference points known as *pbest* and *gbest*. For optimization, the velocity and position of every particle's update as per the following equations [16] are given as:

$$\mathbf{v}_j^{i+1} = \omega^i \mathbf{v}_j^i + c_1 \mathbf{r}_1^i \odot (\mathbf{p}_j^i - \boldsymbol{\phi}_j^i) + c_2 \mathbf{r}_2^i \odot (\mathbf{p}_g^i - \boldsymbol{\phi}_j^i) \tag{11}$$

$$\mathbf{w}_j^{i+1} = \mathbf{w}_j^i + \mathbf{v}_j^{i+1} \tag{12}$$

where $\odot$ represents the element-wise multiplication, $j = 1, 2, \ldots, P$ is the particle index, $i = 1, 2, \ldots$, is iteration number, and ω is the inertial weight. The constants c_1 and c_2 are called acceleration coefficients. Two D-dimensional independent and uniformly distributed random vectors $\mathbf{r}_1$ and $\mathbf{r}_2$ in the range [0 1] used to vary the relative pull of *pbest* and *gbest* stochastically. This makes the PSO algorithm heuristic in nature [16].

In PSO, the velocity update in Eq. (11) is a key in the optimization process. In literature, different methods are available to change the ω value [17] for fast convergence. In one approach, ω decreases during the optimization process for a certain number of iterations and then remains minimum. This helps the algorithm to achieve better global search ability during the initial run and more local search ability toward the end which is very common in literature [9].

Since it changes with iteration only, the search algorithm is not verified during a course of run to choose the value of ω. In [18], the authors used PSO for direction of arrival estimation where the particle position is adjusted in such a way that the best-fitted particle moves slowly in comparison to the least fitted particle. In a particular iteration, this can be achieved by choosing ω value in accordance with their rank, given as:

$$w_i = w_{min} + \frac{(w_{max} - w_{min}) \times \text{Rank}_i}{P} \tag{13}$$

In Eq. (13), the first rank is assigned to the highly fitted particle and the lowest fitted particle assigned last rank. Here, the modified adaptive PSO algorithm is used to estimate the IIR system parameters. Each particle evaluates their fitness by calculating the Hubers cost function (10).

5 Simulation Results and Discussions

In this section, simulation study is carried out here to show the robustness of the proposed algorithm for the estimation of standard benchmark IIR system against outliers. The performance of proposed Hubers cost function-based robust adaptive PSO algorithm (HUAPSO) is compared with two other algorithms. One is the conventional error squared based PSO algorithm (ESPSO) and another algorithm is the Wilcoxon norm-based robust PSO algorithm (WNPSO) [9]. The simulation results show that the new algorithm outperforms over the existing algorithms. The standard temporally correlated data model given in (14) is used here to generate the input data [10].

$$x(i) = \alpha \cdot x(i - 1) + \beta \cdot n(i), \qquad i > -\infty \tag{14}$$

Here, $\alpha \in (0, 1)$ is the temporal correlation index, $n(i)$ is a spatially independent normalized Gaussian noise having unit variance, and $\beta = \sqrt{\sigma_x^2 \cdot (1 - \alpha^2)}$.

The following two examples given in (15) and (17) are used to demonstrate the results. These are

System 1: First consider a second-order IIR system at each sensor. The difference equation is given as

$$d(i) = \sum_{n=1}^{2} a_n d(i-n) + \sum_{m=0}^{1} b_m x(i-m) + v(i) \tag{15}$$

where the feedback and feed-forward parameters are $\mathbf{a} = [1.1314, -0.25]$ and $\mathbf{b} = [0.05, -0.4]$. The second-order adaptive IIR model (16) is used to estimate the second-order IIR system parameters which is

$$y(i) = \sum_{n=1}^{2} \hat{a}_n y(i-n) + \sum_{m=0}^{1} \hat{b}_m x(i-m) \tag{16}$$

System 2: A fourth-order IIR system (17) is considered in System 2 [6] whose transfer function is

$$H(z) = \frac{1 - 0.9z^{-1} + 0.81z^{-2} - 0.729z^{-3}}{1 + 0.04z^{-1} + 0.2775z^{-2} - 0.2101z^{-3} + 0.14z^{-4}} \tag{17}$$

Fourth-order adaptive IIR system (18) is used as the model to estimate the parameters of the fourth-order IIR system

$$\hat{H}(z) = \frac{\hat{b}_0 + \hat{b}_1 z^{-1} + \hat{b}_2 z^{-2} + \hat{b}_3 z^{-3}}{1 - \hat{a}_1 z^{-1} - \hat{a}_2 z^{-2} - \hat{a}_3 z^{-3} - \hat{a}_4 z^{-4}} \tag{18}$$

For the simulation purpose, we chose 50 number of input signals, number of particles is 30, Vmax = 0.5, $c_1 = c_2 = 1.49$. The results are generated by running 100 independent experiments and then plotted the averaged values. The number of iterations is chosen 2000. The percentage of outliers is 10%. The threshold value $\gamma = 0.1$ for Huber cost function. The variance of the Gaussian noise is 0.001, whereas the variance of impulsive noise is 10000 times the variance of normal Gaussian noise.

In Fig. 2a, b, the MSE performances for Systems 1 and 2 are plotted. It has been seen from the figures that the proposed HUAPSO algorithms outperforms over conventional ESPSO and robust WNPSO algorithms. The conventional error squared based ESPSO algorithm complete fails to estimate the parameters when outliers are present in the training data. It is because when outliers occur, the error is very high and that leads to the deviation in the estimation process, whereas in robust algorithm the error is controlled by different mechanism.

Although both the robust algorithms are robust against the outliers, but the HUAPSO algorithm takes less number of iterations to achieve the minimum MSE value compared to that of WNPSO. The convergence of the algorithm further can control by choosing different value of γ. The steady-state performance of the

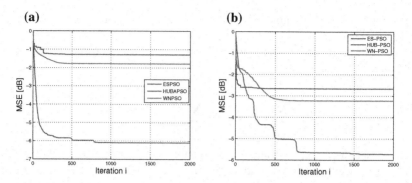

Fig. 2 MSE performances of ESPSO, WNPSO, and HUAPSO algorithms for second- and fourth-order IIR systems. **a** MSE performance of system 1 with 10% outliers. **b** MSE performance of system 2 with 10% outliers

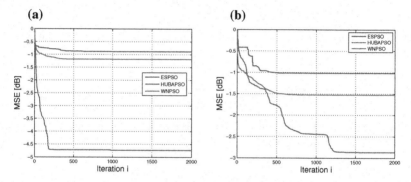

Fig. 3 MSE performances of ESPSO, WNPSO, and HUAPSO algorithms for second- and fourth-order IIR systems. **a** MSE performance of system 1 with 20% outliers. **b** MSE performance of system 2 with 20% outliers

proposed algorithm is nearly 4 dB and 3 dB better than existing WNPSO algorithm for System 1 and 2, respectively.

The robustness of the proposed algorithm is tested in presence of 20% of outliers in Fig. 3a, b. From the figures, it has been seen that the MSE performance of all the algorithm degrades when the percentage of outliers increased. But the proposed Huber function-based adaptive PSO (HUBAPSO) algorithm provides much better performance compared to the other algorithms.

6 Conclusion

A novel method of robust estimation of IIR system's parameter in presence of outliers is presented here. A robust Huber function is used here to alleviate the effect of outliers instead of conventional squared error as a fitness function in adaptive particle swarm optimization algorithm. The robust cost function is computationally efficient than Wilcoxon norm and has a option to control the convergence speed of the optimization algorithm. The performance of new algorithm is compared with Wilcoxon norm-based robust algorithm and conventional PSO algorithm. The simulation results show that the proposed robust adaptive PSO algorithm is more robust, converges fast, and provides better estimation accuracy compared to other existing algorithms.

References

1. J. I. Ababneh and M. H. Bataineh, "Linear phase fir filter design using particle swarm optimization and genetic algorithms," Digital Signal Processing, vol. 18, no. 4, pp. 657–668, 2008.
2. M. Nayak, T. Panigrahi, and R. Sharma, "Distributed estimation using multi-hop adaptive diffusion in sparse wireless sensor networks," in International Conference on Microwave, Optical and Communication Engineering (ICMOCE), Dec 2015, pp. 318–321.
3. J. J. Shynk, "Adaptive IIR filtering," IEEE ASSP Magazine, vol. 6, no. 2, pp. 4–21, April 1989.
4. B. Widrow and S. D. Strearns, Adaptive Signal Processing. Englewood Cliffs, NJ:Prentice-Hall, 1985.
5. L. Xue, Z. Rongchun, and W. Qing, "Optimizing the design of IIR filter via genetic algorithm," in Neural Networks and Signal Processing, 2003. Proceedings of the 2003 International Conference on, vol. 1, Dec. 2003, pp. 476–479 Vol. 1.
6. G. Panda, P. M. Pradhan, and B. Majhi, "IIR system identification using cat swarm optimization," Expert Systems with Applications, vol. 38, no. 10, pp. 12671–12683, 2011.
7. S. Saha, R. Kar, D. Mandal, and S. Ghoshal, "Harmony search algorithm for infinite impulse response system identification," Computers and Electrical Engineering, vol. 40, no. 4, pp. 1265–1285, 2014.
8. S. R. Kim and A. Efron, "Adaptive robust impulsive noise filtering," IEEE Transactions on Signal Processing, vol. 43, no. 8, pp. 1855–1866, Aug. 1995.
9. B. Majhi, G. Panda, and B. Mulgrew, "Robust identification and prediction using wilcoxon norm and particle swarm optimization," in 17th European Signal Processing Conference, Aug 2009, pp. 1695–1699.
10. T. Panigrahi, G. Panda, and B. Mulgrew, "Error saturation nonlinearities for robust incremental LMS over wireless sensor networks," ACM Trans. on Sensor Network, vol. 11, no. 2, pp. 27:1–27:20, Dec. 2014.
11. T. Panigrahi, B. Mulgrew, and B. Majhi, "Robust distributed linear parameter estimation in wireless sensor network," in 2011 International Conference on Energy, Automation and Signal, Dec 2011, pp. 1–5.
12. Y. Zhong, Y. Deng, and A. K. Jain, "Keystroke dynamics for user authentication," in 2012 IEEE Computer Society Conference on Computer Vision and Pattern Recognition Workshops, June 2012, pp. 117–123.
13. K. Q. Weinberger and L. K. Saul, "Distance metric learning for large margin nearest neighbor classification," Journal of Machine Learning Research, vol. 10, pp. 207–244, Jun. 2009.
14. M. Kstinger, M. Hirzer, P. Wohlhart, P. M. Roth, and H. Bischof, "Large scale metric learning from equivalence constraints," in 2012 IEEE Conference on Computer Vision and Pattern Recognition, June 2012, pp. 2288–2295.
15. T. Panigrahi, M. Panda, and G. Panda, "Fault tolerant distributed estimation in wireless sensor networks," Journal of Network and Computer Applications, vol. 69, pp. 27–39, 2016.
16. J. Robinson and Y. Rahmat-Samii, "Particle swarm optimization in electromagnetics," IEEE Transactions on Antennas and Propagation, vol. 52, no. 2, pp. 397–407, Feb. 2004.

17. A. Nickabadi, M. M. Ebadzadeh, and R. Safabakhsh, "A novel particle swarm optimization algorithm with adaptive inertia weight," Applied Soft Computing, vol. 11, no. 4, pp. 3658–3670, 2011.

18. T. Panigrahi, D. H. Rao, G. Panda, B. Mulgrew, and B. Majhi, "Maximum likelihood DOA estimation in distributed wireless sensor network using adaptive particle swarm optimization," in *the Proc. of ACM International Conference on Communication, Computing and Security (ICCCS2011)*, Feb. 2011, pp. 134–136.

A Shallow Parser-based Hindi to Odia Machine Translation System

Jyotirmayee Rautaray, Asutosh Hota and Sai Sankar Gochhayat

Abstract This paper describes a Hindi to Odia machine translation system developed using a popular open-source platform called Apertium. With population of over 1.27 billion, 18 officially recognized languages, 30 regional languages, and over 2000 dialects, the multilingual society of India needs well-developed ICT tools for the citizens to exchange and share information and knowledge between them easily. Though Hindi is the national language of India, still a lot of people of Odisha are unable to understand the information written in Hindi. In this scenario, a suitable Hindi to Odia machine translation system will help the people to understand and use Hindi in a more productive way. For development of such a machine translation system, we decided to use the Apertium platform due to several reasons. It is well suited for building machine translation systems between closely related language pairs, such as Hindi and Odia due to its shallow parser level transfer modules. The use of FST in all the modules makes this much faster as compared to other shallow parser-based platforms. Also, it is available in GPL license under free open-source software. In this paper, we have also demonstrated the linguistic and computational challenges in building linguistic resources for both Hindi and Odia languages. Specifically, the use of TAM (Tense, Aspect, and Modality) concept in transfer module is a unique approach for building transfer rules between Hindi and Odia in Apertium platform. This work can be easily extended to develop MT systems for other Indian language pairs easily.

Keywords Apertium · Hindi · Odia · TAM · Anusaaraka · Transfer rules
Bilingual dictionaries

J. Rautaray (✉) · A. Hota · S. S. Gochhayat
Department of Computer Science and Engineering, College of Engineering and Technology,
Bhubaneswar, Odisha, India
e-mail: jyotirmayee.1990@gmail.com

A. Hota
e-mail: asutosh.hota@gmail.com

S. S. Gochhayat
e-mail: saisankargochhayat@gmail.com

© Springer Nature Singapore Pte Ltd. 2019
H. S. Behera et al. (eds.), *Computational Intelligence in Data Mining*,
Advances in Intelligent Systems and Computing 711,
https://doi.org/10.1007/978-981-10-8055-5_6

1   Introduction

Machine translation is one of the subfields of artificial intelligence and is a process of translating from one language known as source language to another language known as target language [1]. As we know that language is a medium of communication or we can say that through language information can be exchanged. Languages have information at various levels such as morphological, syntactic, semantic, and pragmatic [1]. In a sentence, one word can have multiple meaning, e.g., *bank* (here meaning of bank can be money bank, blood bank, etc.). Hence the extracting exact information needs extralinguistic information of receiving person. Though machines have high computational power and huge storage capacity, it is pragmatically difficult for computers to have a human understanding of the day-to-day world. It is thus a major problem for the machines to govern on the correct sense of ambiguity on the same words which often lead to various different contextual implications. MT was one of the first conceptualized applications of computers back in the 1950s. Even after more than 60 years of research demand for MT is growing steadily and still is an open problem. The system provides (FAH-QUT) Fully Automated High Quality Translation of Unrestricted Text is still a distant dream [1]. However, a number of machine translation systems have been developed such as Bing translator, Baidu, IBM Watson, [2, 3] and Google Translate. Machine translation systems are divided into four types such as rule-based, statistical, example-based and hybrid and neural net-based. This work focuses on establishing a rule-based machine translation system where outputs are displayed layer by layer, representing various stages of translation where each incremental layer is closer to translation [1, 4].

2   Theoretical Foundation

The major contribution of this paper is to establish a machine translation system that would translate from Hindi to Odia language. The translation would equip the population in remote areas to have better understanding of the national language. The proposed work uses the Apertium machine translation library which comes under the FOSS (Free open-source software) [5–8]. As the name suggests the whole system is driven by the set of linguistic rules. This platform provides an engine and toolbox to write rules specific for your language pair to build the machine translation system between those pairs. Basically, it is a shallow-transfer machine translation system which processes the input in several stages. In the initial layers, the analysis of source text is done which includes morph analysis, part of speech (PoS) tagging and chunking. In the further layers, the transfer rules are applied in 3 successive layers and finally the generator layer produces the target language as end output [4, 9]. Figure 1 shows the Apertium pipeline and its modules.

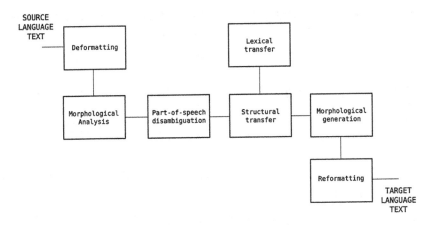

Fig. 1 Apertium system architecture

2.1 De-formatter

The separation of the text is to be translated from the format information, i.e., markup language (RTF and HTML tags, white space, etc.) [1]. The rest of the modules would be treated as blanks between the words as the format information is encapsulated. A system for annotating a document in a way that is syntactically distinguishable from the text is known as the markup language.

2.2 Morphological Analyzer

It tokenizes the text in surface forms and delivers, for each surface form, one or more lexical forms consisting of lemma, lexical category, and morphological inflection information.

2.3 Part-of-Speech (PoS)

A PoS tagger [10] which chooses, using a first-order hidden Markov model (Cutting et al., 1992) (HMM), one of the lexical forms corresponding to an ambiguous surface form. The process of assigning an unambiguous grammatical category to words in context is known as part-of-speech. The surface forms of words can be often assigned to multiple parts-of-speech by morphological analysis which leads to ambiguity. For instance, the word "trap" can be either a singular noun ("It's a trap!") or a verb ("I'll trap it") in the English language. A hidden Markov model (HMM) can often be expressed in the form of two matrices each representing transition and emission probabilities as well as a vector representing the foremost probabilities. This can be suitably expressed as:

$$M = (A, B, \pi)$$

Here, A is the matrix of transition probabilities, M is the model, B is the matrix of emission probabilities, and π is the vector of initial probabilities. These probabilities are calculated between the tag set and the ambiguity classes from a training set. This is referred to as parameter estimation.

2.4   Transfer Modules

Apertium consists of data for many language pairs and hence, these linguistic datasets include monolingual as well as bilingual dictionaries. Each SL lexical form is read by a lexical transfer and the corresponding TL lexical form is delivered by referring to a bilingual dictionary. A finite-state chunker is used by the structural transfer module to detect patterns of lexical forms which need to be processed for word reordering, agreement, etc. and then performs several operations. The structural transfer rules are employed to perform grammatical and other language-based transformations, whereas the lexical data transfer is employed for the part-of-speech tagger [10], which is in charge of the disambiguation of the source language text.

2.5   Generators and Re-formatters

After suitable inflection by a morphological generator, the TL surface helps to generate a TL lexical form. A post-generator performs orthographic operations such as contractions (e.g., Spanish del = de + el) and apostrophations (e.g., Catalan linstitut = el + institut) [11]. The translated text is achieved by removal of encapsulation sequences used to protect character in SL text and restoration of format information from the de-formatter. Apertium also provides a utility tool to see the intermediate outputs of each module. Figure 2 shows the screenshot of the tool called Apertium viewer.

3   Problem Formulation

In this paper, we are going to describe our research work for developing a machine translation system from Hindi to Odia, where Hindi is our source language and Odia is our target language. For developing such a machine translation system, we decided to work on Apertium platform. This platform provides us with the transfer

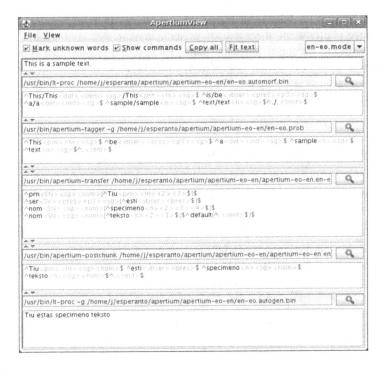

Fig. 2 Apertium viewer

engine and toolbox which allows us to build machine translation systems between any two language pairs [12]. We have to build the required data and linguistic rule to run the MT engine. We have used the following linguistic resources.

- Morphological dictionary for Hindi language.
- Morphological dictionary for Odia language.
- Bilingual dictionary between Hindi and Odia language.
- Transfer rule for Hindi to Odia structural change.

4 Motivation

India is the largest democratic country in the world with more than 30 languages and approximately 2000 dialects used for communication by the Indian people [13, 14]. Out of these languages, English and Hindi are often used for official work. Though Hindi is recognized as national language of India, still many people exist in Odisha who neither speak nor understand Hindi. For the larger benefit of such people, we need to develop an automatic machine translation system for various ICT-based applications. Through these not only the information exchange will be easier but a lot of knowledge sharing is also possible. Apertium is a very popular open-source platform and there already exists many successfully built MT system between various European language pairs. However, till date there is no such acceptable machine translation system from Hindi to Odia. So this motivated us for doing research in developing machine translation system between Hindi and Odia language.

5 Proposed System and Performance Analysis

The major work that has to be done in developing such a MT system in Apertium platform is to develop the various required linguistic resources [12]. As mentioned earlier, the various linguistic resources that we use in this MT system are

- Morphological dictionary for Hindi language.
- Morphological dictionary for Odia language.
- Bilingual dictionary between Hindi and Odia language.
- Transfer rule for Hindi to Odia structural change.

In the further subsections, we will briefly describe the details about the file structures of this linguistic data. No prior work has been done taking into consideration Hindi to Odia machine translations.

5.1 Hindi Morphological Dictionary

It is used to get the morphological analysis of the source language, i.e., Hindi. The dictionary is an XML file. It uses several XML tags for writing the linguistic data. Below are the descriptions of several tags used in this dictionary file [6].

- Alphabet: It defines the set of letters that may be used in the dictionary
- sdef: Defines symbols. In the context of Apertium symbol refers to a grammatical symbol label
- n: for noun
- pl: for plural

  Other examples of symbol are

- sg: for singular
- p1: first person
- pri: present indicative

  Paradigms are defined in pardef tag.

- e: for entry
- p: for pair

- l: for left and it is used for analysis
- r: for right and it is used for generation.

5.2 Hindi-Odia Bilingual Dictionary

Bilingual dictionary is also called as translation dictionary which is used to translate words or phrases from one language to another. Here, we have developed a standard dictionary called Shabdanjali consisting of approx. 30,000 words which is available as an open-source dictionary between English and Hindi. We created the parallel Hindi to Odia dictionary using the words available in English-Hindi pair [7, 9].

5.3 Transfer Rule for Hindi-Odia Structural Change

This module is responsible for doing the structural changes from Hindi language to Odia. Normally rules are transferred from source language to target language in 3 stages known as intra-chunk, inter-chunk, and post-chunk. Intra-chunk module is responsible for doing the structural changes inside a single chunk element [12]. Inter-chunk module is responsible for doing the structural changes among the various chunks present in the input sentence and also for modifying the syntactic information associated with each chunk. Post-chunk is responsible to modify the output of the inter-chunk module and to reformat it in chunk formats accepted by the generator module [7–9].

In fig. 3, shows an example transfer rule. The symbols used in the rule are

- Pattern: Rule is applied if the given pattern is found in the sentence. For each pattern in the rule, there is an associated action, which produces an associated output called out [6].
- lu: The output as a lexical unit.
- Clip: This tag allows a user to select and manipulate attributes of each captured word through the pattern. Information from both source language (side sl) and target language (side tl) lexical forms can be accessed.
- pos: It's an index of the captured words through pattern [1].
- side: It specifies if the selected clip is from source language or target language.
- part: It indicates which part of the lexical form is processed. Its value is already defined in ¡section-def-attrs¿. It can also take 4 predefined values like.
- lem: It refers to lemma of the lexical form.
- lemh: It refers to the first part of a split lemma.
- lemq: It refers to the queue of a split lemma.
- whole: It represents the whole lemma.
- lit: It specifies the value of a string.

- lit-tag: It specifies the value of grammatical symbols or tags by means of attribute.
- when: It describes a conditional option. It contains condition to be tested with test and one block of zero or more instruction of the kind choose, let, out, modify-case, call-macro, or append which will be executed if the above condition is met.
- test: It contains a conjunction "and," a disjunction "or" and a negation "not."
- equal: It is used if two strings are equal.
- begins-with: It is used in case of string beginning.
- ends-with: It is used in case of string ending.
- contains-substring: It is used if string contains substring.
- in: It specifies the inclusion in a set.
- append: If the form is found then append it to the list. It is similar to concatenating string.
- let: It indicates value assignment.

5.4   Tense Aspect Modality (TAM)

As we translate a sentence from Hindi to Odia, the major structural changes occur at the verb phrase level. For e.g., in a Hindi sentence "ladakA jA rahA hE" its equivalent Odia translation will be "bALaka jauci." If we analyze the translation, we can mark that the verb "jA rahA hE" is grouped to one word "jAuCi" in Odia. This is due to the agglutinative nature of the Odia language. In an example sentence: In Hindi—"ladaki jA rahi hE" and the equivalent Odia translation "bALikA jAuci." We can see that while translating from Hindi to Odia, a change is gender has no effect on the verb. From verb part if we extract the root part, then the remaining part is called as TAM. Now by analyzing TAM of the verb phrase "piwA hE" will be "wA hE" and its corresponding Odia translation will be "piuCi" where the TAM will be "uCi." For these changes in TAM, we write rules in the transfer module. For specific pattern, there exist several TAMs, e.g., "rAma kAma kiyA" for which Odia translation will be "rAma kAma karilA." This is an example of a sentence with a single verb. So, we need to write rule for a single verb pattern of TAM "yA." Under one specific pattern several TAMs can be found. Similarly, we have rules for verb having two words in the phrase, e.g., (KAwA hE) and three words in the phrase, e.g., (KAwA rahA hE).

In the rule, TAM part of source language will be compared with the equivalent TAM of bilingual dictionary. Lemma part of bilingual dictionary will be checked and if it is found then that lemma along with the TAM part will be in output of that source language which is our required target language. Similarly, we also have written rules for double verb pattern (i.e., verb verb) e.g.: in Hindi—"rAma kAma karawA WA" which Odia equivalent will be "rAma kAma karuWilA," verb-verb-verb pattern e.g.: In Hindi—"rAjA kAma kara rahA hogA" which Odia meaning will be "rAjA kAma karuWiba," verb-prawyaya-verb pattern e.g.: In Hindi—"rAma kAma karane vAlA WA" which Odia meaning will be "rAma kAma karibAra WilA," verb-verb-verb-verb pattern e.g.: in Hindi—"rAjA se kAma karAyA jA sakawA hE" which Odia meaning will be "rAjA xbArA kAma kar-AjAipAriba," verb-prsg pattern e.g.: in Hindi—"rAma Kane se UT gayA" which Odia meaning will be "rAma KAibATAru uTigalA," prsg-verb pattern e.g.: In Hindi—"rAma kAma nahIM karegA" which Odia meaning will be "rAma kAma kariba nAhIM," prsg-verb-verb-verb pattern e.g.: In Hindi—"rAma kAma nahIM kara sakawA WA" which Odia meaning will be "rAma kAma karipArinaWAnwA" and prsg-verb-verb-verb-verb pattern e.g.: In Hindi—"rAma se kAma nahIM kar-AyA jA sakawA WA" which Odia meaning will be "rAma xbArA kAmA karAjAipArinaWAnwA."

5.5 Generation

The generation module carries out the generation of the final words taking into consideration their characteristics and attributes. The morphological analyzer present in Apertium is similar to the generator, however, they variate in the orientation of processing the input string. Hence, the Odia morphological analyzer is essentially used in the reverse direction for generation of the final translated Odia sentences [12, 15].

By giving source language in Apertium viewer it will generate its analysis layer by layer and give its proper output. Where Apertium viewer is a utility program to view and edit output at various stages of Apertium system. In this section, Figs. 3, 4, 5 and 6 show the various generative outputs from the Apertium platform.

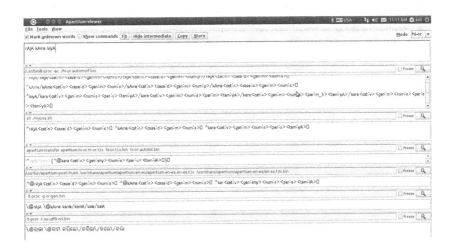

Fig. 3 Output representing a single verb pattern

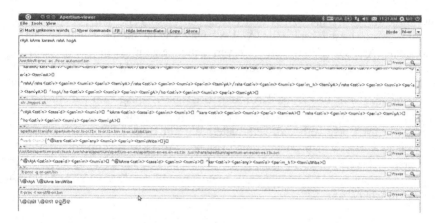

Fig. 4 Output of sentence of verb-verb-verb pattern

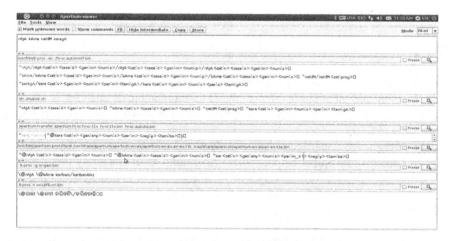

Fig. 5 Results for the sentence of negation verb type

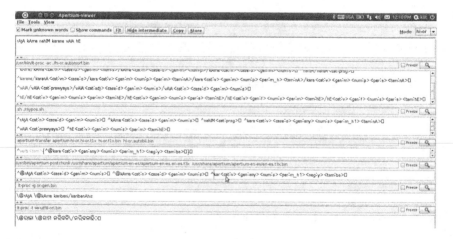

Fig. 6 Results for sentence of prsg verb prawyaya verb pattern

6 Conclusion and Future Work

In this paper, we have tried to propose how machine translation system from Hindi to Odia language has been developed. We described the role of TAM in Hindi and Odia languages, respectively. This open-source model provides a great opportunity for the users to develop various applications on the top of it and improve the system as well. Further, a suitable word sense disambiguation (WSD) module can be attached to further tune the system. As we know that Apertium is a shallow-transfer-based machine translation system [6], so we will try to improve the output by using the output from a deep parser for handling various complex phenomena like word sense disambiguation, co-reference resolution.

Acknowledgements The authors would like to thank Prof. Vineet Chaitanya for giving the basic insights to develop the system. We are thankful to Mr. Sriram Chaudhury (Asst. Prof., KIIT University) for his continuous support and guidance. We are also thankful to IIIT-Hyderabad, Hyderabad Central University and the Apertium group for facilitating us with useful tools and linguistic resources for the successful development of this system.

References

1. Sriram Chaudhury et al., "Anusaaraka: An Expert system based MT System," in the proceedings of IEEE conference on Natural language processing and knowledge management (IEEE-NLP KE 2010), Beijing, China.
2. IBM Watson, http://www.ibm.com/watson/services/language-translator/.
3. Bing Translator, http://www.bing.com/translator/.
4. Akshar Bharati et al., "Natural Language Processing: A Paninian Perspective," Prentice-Hall of India, New Delhi, 1995.
5. Mikel L. Forcada, "Apertium: free/open-source rule-based machine translation", Presentation at Fourth Machine Translation Marathon "Open Source Tools for Machine Translation," Dublin, Ireland, 29 Jan. 2009.
6. Apertium Org, https://www.apertium.org/index.eng.html?dir=nld-afr#translation.
7. Francis M. Tyers et al., "Free/open-source resources in the Apertium platform for machine translation research and development," The Prague Bulletin of Mathematical Linguistics, No. 93, 2010, pp. 67–76.
8. Amba P. Kulkarni, "Design and Architecture of 'Anusaaraka'—An Approach to Machine Translation," Satyam Technical Review, vol 3, Oct. 2003.
9. Mikel L. Forcada et al., "Documentation of the Open-Source Shallow-Transfer Machine Translation Platform Apertium," Departament de Llenguatges i Sistemes Inform`atics, Universitat d'Alacant, Alicante, Spain, Technical report, Mar. 10, 2009.
10. D. Cutting et al., "A practical part-of-speech tagger," in Proceedings of Third Conference on Applied Natural Language Processing, Association for Computational Linguistics, Trento, Italy, 1992, pp. 133–140.
11. Computational Processing of the Portuguese Language: 7th International Workshop, PROPOR 2006, Itatiaia, Brazil, May 13–17, 2006, Proceedings.
12. Felipe Sánchez-Martínez et al., "Integrating corpus-based and rule-based approaches in an open-source machine translation system," E-03071, Department de Lenguatges i Sistemes Informatics, Universitat d'Alacant, Alacant, Spain.

13. Mall, Shachi, and Umesh Chandra Jaiswal. "Developing a system for machine translation from Hindi language to English language", ICCCT, 2013.
14. Akshar Bharati et al., "Anusaaraka: Overcoming the Language Barrier in India," appeared in "Anuvad", Sage Publishers, New Delhi, 2002.
15. Akshar Bharati et al., "LERIL: Collaborative Effort for Creating Lexical Resources," in Proc. of Workshop on Language Resources in Asian Languages, together with 6th NLP Pacific Rim Symposium, Tokyo, Nov. 30, 2001.

Coherent Method for Determining the Initial Cluster Center

Bikram Keshari Mishra and Amiya Kumar Rath

Abstract Several aspects of research works are now carried out on clustering of objects where the main focus is on finding the near-optimal cluster centers and obtaining the best possible clusters into which the objects fall into so that the desired expectations are met. This is because a bad selection of cluster center may result in dragging a data very far away from its actual cluster resulting in deficient clustering. Hence, we have accentuated on determining the near-optimal cluster centers and also position the data in their real clusters. We have explored three kinds of clustering techniques, viz. K-Means, FEKM-, and TLBO-based clusterings applied on quite a few data sets. Analysis was made considering two factors, namely cluster validation and average quantization error. Dunn's index, Davies–Bouldin index, silhouette coefficient, and C index were used for quantitative evaluation of the clustering results. As per our anticipation, almost all validity indices provide promising outcome for both FEKM- and TLBO-based clusterings than K-Means inferring superior cluster formation. Further tests support that FEKM- and TLBO-based clustering has smaller value of quantization error than K-Means.

Keywords Optimal centroid · Cluster validation · K-Means
FEKM- and TLBO-based clustering

B. K. Mishra (✉) · A. K. Rath
Department of Computer Science and Engineering, Veer Surendra Sai University
of Technology, Burla 768018, India
e-mail: bikrammishra2012@gmail.com

A. K. Rath
e-mail: amiyaamiya@rediffmail.com

© Springer Nature Singapore Pte Ltd. 2019
H. S. Behera et al. (eds.), *Computational Intelligence in Data Mining*,
Advances in Intelligent Systems and Computing 711,
https://doi.org/10.1007/978-981-10-8055-5_7

1 Introduction

Clustering [1, 2] is a vital tool for data mining since it can establish key patterns without any previous supervisory information. It is used to partition a data set into various groups according to different significant factors and thus determine valuable patterns from data. Clustering is done using the similarity criteria with the purpose of maximizing the similarity within a cluster and minimizing the similarity between clusters. Clustering classifies the raw data persuasively and discovers the hidden patterns that may exist in the data sets [3].

Consequently, researchers are now focusing on different techniques with an aim to improve the formation of good clusters [4, 5], reduce noisy data [6], find optimum number of cluster centers [7], and introduce cluster technique in various application fields.

There are certain critical issues like determination of cluster centers [7–9] which is important to avoid malicious clustering, and to obtain good clusters, we were motivated to develop a technique which gives a solution to the above issues. In addition, we have used the fundamental concept of TLBO [10] for achieving near-optimal centers.

After the creation of desired clusters, it is quite important to determine how good the partitions are formed. This can be achieved by using several conventional clustering validation indexes [11]. In this work, we have used some extensive validity indices and enforce them on numerous data sets to obtain more precise clusters.

We evaluate FEKM- [12] and TLBO-based clustering [13] with the clustering results obtained from K-Means [14]. The comparison is carried out based on (i) validity indices and (ii) average quantization errors obtained from clusters. All these techniques are compared taking numerous data sets from UCI repository [15]. As per our expectations, the quality of clustering produced by FEKM- and TLBO-based clustering is much acceptable than the other considered.

The offerings of this paper are as follows:

(a) Highlighting on much improved selection of initial cluster centers.
(b) Obtaining the efficient cluster groups.
(c) Testing the quality of clustering.
(d) Executing the methods on variety of data sets.

1.1   Organization of the Paper

The paper is organized as follows: The concept of cluster validation and some broadly employed validity index for judging the formation of groups that efficiently fits the elemental data is presented in Sect. 2. We presented in Sect. 3 the skillful and innovative exploration in this related realm made by various researchers.

Several techniques used for clustering with their pros and cons are mentioned in Sect. 4. The results obtained from various simulations are given in Sect. 5. Last but not least, Sect. 6 concludes the paper citing some future scenarios of this work.

2   Cluster Validation

Cluster analysis is a group of multivariate approaches with an objective to group objects within the clusters in a closer proximity, while the separation between clusters will be far more apart. Each cluster C^i can be recognized by its cluster center. The centroid c^i can be determined [16] as:

$$c^i = \frac{1}{|C^i|} \sum_{x \in C^i} x \tag{1}$$

where $|C^i|$ is the cardinality of cluster C^i.

Cluster validation [11, 17] is meant for evaluating the end result of a clustering algorithm. The group of data with similar characteristics are positioned closer to each other than those present outside the cluster. Here, we discuss few *internal validity indices* that have been used for probing the accuracy of clusters.

2.1   Dunn's Index (DI)

Dunn's index [18] discovers the *'compactness and well separated clusters.'* DI is defined as follows:

$$DI(c) = \min_{i \in c} \left\{ \min_{j \in c, j \neq i} \left\{ \frac{\delta(A_i, A_j)}{\max_{k \in c} \{\Delta(A_k)\}} \right\} \right\} \tag{2}$$

where

$$\delta(A_i, A_j) = \min\{d(\underline{x_i}, \underline{x_j}) | \underline{x_i} \in A_i, \underline{x_j} \in A_j\} \tag{3}$$

$$\Delta(A_k) = \max\{d(\underline{x_i}, \underline{x_j}) | \underline{x_i}, \underline{x_j} \in A_i\} \tag{4}$$

d is a function for calculating the distance between cluster center and data point, and A_j is a set of elements whose data are assigned to ith cluster. A *larger* DI value indicates the presence of compact and well-separated clusters.

2.2 Davies–Bouldin Index (DBI)

DBI [19] is ratio of sum of inside-cluster allocation to across-cluster separation. The inside ith cluster distribution is given by:

$$S_{i,\,q} = \left(\frac{1}{|A_i|} \sum_{\underline{x}\,\in\,A_i} \|\underline{x} - \underline{v}_i\|_2^q \right)^{1/q} \tag{5}$$

The across ith and jth division is given by:

$$d_{ij,\,t} = \left\{ \sum_{s=1}^{p} |v_{si} - v_{sj}|^t \right\}^{1/t} = \|\underline{v}_i - \underline{v}_j\|_t \tag{6}$$

where $\underline{v}_i$ is the ith cluster center, $(t, q) \geq 1$, and $|A_i|$ is quantity of elements in A_i. $R_{i,\,qt}$ is given by:

$$R_{i,\,qt} = \max_{j\,\in\,c,\,j\,\neq\,i} \left\{ \frac{S_{i,\,q} + S_{j,\,q}}{d_{ij,\,t}} \right\} \tag{7}$$

Ultimately, DBI is:

$$DB(c) = \frac{1}{c} \sum_{i=1}^{c} R_{i,\,qt} \tag{8}$$

A *minimum* DBI value implies good clustering.

2.3 Silhouette Coefficient (SC)

In SC [20], for an individual data i, first the average distance from i to all points present in its cluster is calculated, which is a. Second, calculate b, which is the minimum of average distance of data i to all points present in another cluster.

Finally, SC for a data point is given by:

$$s = \begin{cases} 1 - a/b & \text{if } a < b \\ 0 & \text{if } a = b \\ b/a - 1 & \text{if } a > b \end{cases} \tag{9}$$

s is between 0 and 1. When s is closer to 1, it is taken as 'well classified.'

2.4 C Index

C index [21] is defined as follows:

$$C_{index} = \left(\frac{S - S_{min}}{S_{max} - S_{min}} \right) \tag{10}$$

where S is the sum of distances of all pairs of data within a cluster, S_{min} is the sum of n smallest distance from all data pairs, S_{max} is the sum of n largest distance from all data pairs, and n is the number of those pairs. The C index is limited to interval [0, 1] and should be minimized for better clustering.

3 Related Works

A number of relevant works can be seen in recent years about diversity in which clustering methods have been used. In this section, we have discussed a few that are most appropriate to the approach presented in this text.

The limitations of K-Means are discussed in [22], and a proficient way in which data are assigned to clusters is proposed. This technique lessens implementation time of K-Means. R. Xu et al. [23] surveyed a range of clustering methods along with their applications and discussed several proximity criteria and validity measures. A comprehensive form of K-Means performing accurate clustering with no pre-assignment of cluster numbers is proposed by Y. M. Cheung [24]. This method is valid to ellipse-shaped data clusters also.

C. S. Li [8] suggested the concept of nearest neighbor pair for determining the initial centroids for K-means. This method searches two nearest neighbor pairs that are largely unlike and present in different clusters. A novel approach for determining initial centers and assigning data points to clusters was suggested in [25] but with a limitation that initial number of cluster has to be given as input. Since in K-Means algorithm initial cluster centers are chosen randomly hence, [9] suggested a method in which cohesion degree of the neighborhood of an object and the coupling degree amid neighborhoods of objects are defined based on a model. In addition, a new initialization method was proposed in this novel work.

A new k-medoids method can be seen in [17] for obtaining initial medoids. The distance matrix is calculated once and used for finding new medoids at each iterative step. Experimental results show this method has better performance than K-means.

A population-based TLBO method proposed by [26] is used to solve the clustering problem. Result shows that this method provides the optimum value and small standard deviation when compared with SA, PSO, ACO, and K-means.

While clustering using fuzzy c-means, finding the initial center is important in its ultimate result. TLBO suggested in [27] addresses this problem. Initially, TLBO explores to determine the near-optimal centers. Then, these are treated as the pre-liminary cluster center for c-mean. Results show TLBO minimizes the difficulty in selecting the premium centers for the c-means.

4 Clustering Techniques

Now, we elaborate few clustering means we have used and compared in this work.

4.1 K-Means

This is simplest unsupervised learning [28] algorithms used for clustering. Steps of K-Means are as follows:

(1) Randomly select the initial cluster centers.
(2) Assign each data to the cluster that is nearest to the cluster center.
(3) Recalculate the center by taking the mean of all data present in the cluster.
(4) Repeat steps 2 and 3 until centers no longer change.

But, the outcome of K-Means heavily relies on the randomly selected initial centers.

4.2 Far Efficient K-Means Algorithm (FEKM)

Keeping the limitations of K-Means in mind, we came up with FEKM [12], for efficiently selecting the initial cluster centers. Initially, the distance between each pair of data is calculated and two farthest away data d_1 and d_2 are selected as two initial cluster centers. Data closest to d_1 are included in its cluster and are eliminated from the set till amount of data in d_1 attain a threshold value. Then, mean of d_1 cluster determines its new centroid c_1. Similar process is adapted to d_2 to get its centroid c_2. For the third center, choose a data d_i in such a way that:

$$max(min(distance(\{d_i, c_1\}, \{d_i, c_2\}))).$$

Once the third center is found, data to that cluster are allotted till a threshold is reached, and data in this new cluster are removed from data set. The mean of d_i determines its new centroid c_i. This process is continued for K number of clusters.

Experimental result shows that FEKM is an improved means of clustering than K-Means by effectively determining the initial cluster centers rather than randomly selecting them.

4.3 Teaching–Learning-Based Optimization (TLBO)

Rao et al. [10, 29] suggested two vital forms of learning:

(i) Learning through the teacher (*teacher phase*),
(ii) Learning through interaction (*learner phase*).

(A) Teacher Phase

Teacher is the best learner. Teacher T_1 aims to bring mean learning intensity of class M_1 toward his/her own level so that learner's level rises to a new mean M_2. But, the entire class gains knowledge according to the value of teaching delivered and the quality of students in the class which is the mean value of population P_i in the class. T_1 raises the student quality from M_1 to M_2, and thus, they dub for a new teacher T_2 of greater quality. Let M_i be mean learning value of class and T_i be a teacher so, T_i shift M_i to its own rank, and hence, we get a new mean M_{new}. Therefore,

$$\text{Difference\_mean}_i = r_i(M_{new} - T_f M_i) \tag{11}$$

where T_f is a teaching factor whose value can be either 1 or 2, and r_i is a random number between [0,1]. T_f is given randomly as:

$$T_f = \text{round}[1 + \text{rand}(0, 1)*(2 - 1)] \tag{12}$$

This difference transforms the existing result which is:

$$X_{new,\, i} = X_{old,\, i} + \text{Difference\_mean}_i \tag{13}$$

(B) Learner Phase

Learners raise their learning level by two ways:

(i) Effort from teacher.

(ii) By interaction among themselves.

Learner modification is given by:
```
for i=1 to Pₙ
{
  Pick two learners Xᵢ and Xⱼ randomly such that i ≠ j
  if (f(Xᵢ) < f(Xⱼ))
        Xₙₑw,ᵢ = Xₒₗd,ᵢ + rᵢ(Xᵢ −Xⱼ)
  else
        Xₙₑw,ᵢ = Xₒₗd,ᵢ + rᵢ(Xⱼ −Xᵢ)
  end if
}
  Select Xₙₑw if it produces an improved function value.
```

4.4   TLBO Means of Clustering

TLBO means of clustering suggested by [13] is done using two stages:

(i) With TLBO for getting initial cluster centers.
(ii) Using enhanced clustering approach for performing clustering.

Phase I: *Deciding optimal cluster centers using TLBO*:

To achieve good clusters, TLBO has been fused with our enhanced clustering approach [14]. TLBO proposed by [10] is used to choose the primary centroids. Centers with min. quantization error values given by S.C Satpathy and A. Naik [30] are preferred as desired centers. The suitability of learner considered as the quantization error is obtained by:

$$
J_e = \frac{\sum_{j=1}^{N_c} \left[\sum_{\forall Z_p \in C_{ij}} d\left(Z_p, m_{ij}\right) / \left| C_{ij} \right| \right]}{N_c}
\tag{14}
$$

where N_c is no. of cluster centroid vector, m_{ij} is jth cluster center vector of the ith particle in cluster C_{ij}, z_p is pth data vector, and $d(Z_p, m_{ij})$ is distance matrix to all C_{ij}. Then, using the method of Sect. 4.3, we optimally get desired cluster centers.

Phase II: *Enhanced Clustering Methodology for clustering*:

This involves clustering of data using enhanced clustering methodology [14]. Once TLBO attains stopping condition, the resultant vector obtained from learner group with min. quantization error is considered as initial centroids. Then, every data point is allotted to its nearby center, and a record of its cluster index regarding its current cluster position alongside its distance from its center is recorded by using two matrices. The centers are again determined by taking mean of the data present in the cluster. Next, inside every group each datum with its newly obtained center distance is calculated. If new distance is greater than old distance, then the data do not stay in that cluster. Otherwise, distance of that data with each left over center is recalculated and is assigned to the cluster which is nearby to its center. This is repeated until convergence.

5   Experimental Results

We evaluated the results of K-Means, FEKM, and TLBO means of clustering on diverse data sets [15] which are shown in Table 1.

The clustering qualities of mentioned algorithms are assessed using DI, DBI, silhouette coefficient, and C index. After comparing their clustering results, it was

Table 1 Features of a few data sets used

Data set used	Attributes of each data	No. of classes	Instances present
Iris	4	3	150
Wine	13	3	178
Seed	7	3	210
Balance	4	3	625
Mushroom	22	2	8124
Abalone	8	3	4177
Glass	11	2	214

found that almost all values of different validity indices show encouraging results for both FEKM and TLBO means of clustering as per our expectation. When K is kept 3 and number of iterations 20, Table 2a, it was examined that majority values of DI acquired for every data set for FEKM- and TLBO-based clustering are larger than K-Means implying better clustering. Likewise, majority of DBI values of FEKM- and TLBO-based clustering are smaller than those of K-Means. As expected, SC values of FEKM- and TLBO-based clustering are nearer to one indicating good clustering. Finally, the values for C index are minimum for FEKM- and TLBO-based clustering in comparison with K-Means. These are signs of superior cluster formation. Similar sorts of outcome were produced when number of iterations were chosen as 40 and 80. This can be viewed from Table 2b, c. A comparison of all the referred algorithms with 20 iterations and K chosen as 3 for DI, DBI, silhouette coefficient, and C index is shown in Fig. 1a–d. Experiments were furthermore conducted with different iterations and number of clusters, and only few results are shown. The average quantization error showed calculated values of FEKM- and TLBO-based clustering that are less than K-Means, as shown in Table 3 (Fig. 2).

6   Observations and Conclusion

The selection of initial centers is a decisive factor for clustering for the reason that incorrectly selected centroids may possibly affect the final outcome of clustering. Unlike K-Means where initial centers are obtained randomly, FEKM- and TLBO-based clustering is an approach in the direction of determining the near-optimal cluster centers.

In this manuscript, three kinds of clustering techniques, viz. K-Means, FEKM-, and TLBO-based clustering, applied on quite a few data sets are explored. When assessment was made between them considering two factors namely cluster validity and average quantization error, it was found that almost every value of DI, DBI, silhouette coefficient, and C index provides promising outcome for both FEKM- and TLBO-based clusterings than K-Means. Hence, our purpose for determining near about optimal initial cluster centers was fulfilled to some extent. The quantization error parameter also produces smaller value for FEKM- and TLBO-based

Table 2 Analysis of K-Means, FEKM-, and TLBO-based clustering taking into account DI, DBI, SC, and C index

	K-Means				FEKM				TLBO-based clustering			
	DI	DBI	SC	C index	DI	DBI	SC	C index	DI	DBI	SC	C index
(a) when K = 3 and no. of iterations is 20												
Iris	0.038	0.577	0.709	0.367	0.109	0.031	0.791	0.347	0.056	0.680	0.690	0.461
Wine	0.047	0.519	0.934	0.510	0.072	0.031	0.902	0.466	0.063	0.596	0.832	0.603
Abalone	0.030	0.760	0.936	0.650	0.038	0.049	0.893	0.830	0.093	0.756	0.883	0.820
Glass	0.070	0.011	0.734	0.520	0.094	0.020	0.799	0.613	0.085	0.024	0.807	0.487
Mushroom	0.094	0.097	0.892	0.839	0.097	0.051	0.936	0.708	0.087	0.062	0.971	0.731
Seed	0.083	0.307	0.800	0.567	0.090	0.033	0.887	0.456	0.098	0.038	0.700	0.636
Balance	0.084	0.248	0.687	0.504	0.104	0.005	0.610	0.519	0.094	0.047	0.726	0.413
(b) when K = 3 and no. of iterations is 40												
Iris	0.085	0.575	0.821	0.517	0.130	0.049	0.884	0.529	0.117	0.489	0.787	0.541
Wine	0.098	0.598	0.843	0.585	0.113	0.048	0.773	0.499	0.123	0.508	0.873	0.565
Abalone	0.094	0.749	0.968	0.705	0.139	0.060	0.834	0.615	0.153	0.665	0.960	0.560
Glass	0.140	0.328	0.785	0.506	0.146	0.127	0.940	0.561	0.136	0.222	0.807	0.506
Mushroom	0.137	0.504	0.762	0.736	0.151	0.068	0.877	0.717	0.135	0.663	0.781	0.613
Seed	0.135	0.444	0.826	0.647	0.136	0.060	0.770	0.596	0.152	0.345	0.704	0.652
Balance	0.147	0.520	0.777	0.514	0.154	0.230	0.836	0.426	0.139	0.428	0.786	0.469
(c) when K = 3 and no. of iterations is 80												
Iris	0.095	0.568	0.790	0.036	0.126	0.023	0.883	0.031	0.134	0.361	0.782	0.060
Wine	0.107	0.591	0.796	0.063	0.128	0.018	0.789	0.140	0.140	0.380	0.786	0.074
Abalone	0.090	0.740	0.581	0.202	0.143	0.029	0.657	0.255	0.170	0.536	0.616	0.171
Glass	0.168	0.169	0.890	0.051	0.255	0.249	0.835	0.026	0.253	0.195	0.849	0.059
Mushroom	0.450	0.356	0.683	0.265	0.354	0.429	0.690	0.262	0.452	0.531	0.588	0.167
Seed	0.348	0.334	0.810	0.028	0.362	0.337	0.746	0.047	0.319	0.201	0.839	0.056
Balance	0.355	0.267	0.678	0.293	0.381	0.344	0.760	0.122	0.349	0.234	0.697	0.196

Table 3 Analysis of K-Means, FEKM-, and TLBO-based clustering taking into account average quantization error

	K-Means	FEKM	TLBO-based clustering
Iris	1.2546	0.9195	1.1020
Wine	1.3517	0.7019	0.8671
Abalone	1.9245	1.7362	1.7928
Glass	1.7454	0.8804	1.2602
Mushroom	2.1601	1.8925	2.0145
Seed	1.3895	0.7016	0.8624
Balance	1.8582	1.6797	1.8101

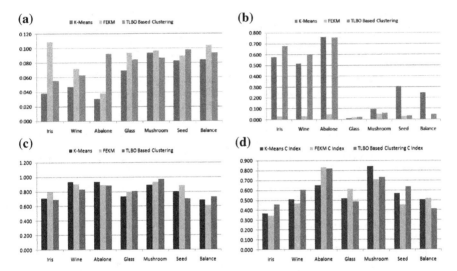

Fig. 1 **a** Performance based on DI values with K = 3 and iterations = 20. **b** Performance based on DBI values with K = 3 and iterations = 20. **c** Performance based on SC values with K = 3 and iteration = 20. **d** Performance based on C index values with K = 3 and iteration = 20

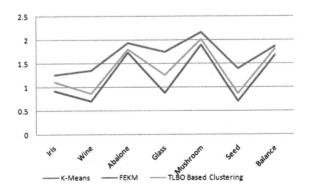

Fig. 2 Performance analysis based on average quantization error on different data sets

clustering. These two factors considered indicates the formation of suitable clusters in which data are present in a closer proximity within a cluster and the interval of data from one cluster to another is quite far.

Acknowledgements We are extremely thankful to Sagarika Swain who provided expertise that greatly assisted the work. The authors also express gratitude to the editors and the anonymous referees for any productive suggestions on the paper.

References

1. Jain, A.K., Topchy, A., Law, M.H.C., and Buhmann J.M., "Landscape of clustering algorithms", *'in Proc. IAPR International conference on pattern recognition, Cambridge, UK'*, pp. 260–263, 2004.
2. L. Kaufman, P.J. Rousseeuw, "Finding Groups in Data: An Introduction to Cluster Analysis", John Wiley & Sons, 1990.
3. Huang Z, "Extensions to the k-means algorithm for clustering large data sets with categorical values," *Data Mining and Knowledge Discovery*", Vol. 2, pp. 283–304, 1998.
4. A.K. Jain, M.N. Murty, P.J. Flynn, "Data Clustering: A Review", *ACM Computing Surveys*, Vol. 31, No. 3, pp 264–323, September, 1999.
5. Vladimir Estivill Castro, "Why so many clustering algorithms—A Position Paper", *'SIGKDD Explorations'*, vol. 4, issue 1, pp 65–75,2002.
6. H. Xiong, G. Pandey, M. Steinbach and V. Kumar, "Enhancing Data Analysis with Noise Removal", *"IEEE Transactions on Knowledge and Data Engineering"*, volume: 18, Issue: 3, pp. 304–319, 2006.
7. M Erisoglu, N Calis, S Sakallioglu, "A new algorithm for initial cluster centers in k-means algorithm", *"Pattern Recognition Letters"*, volume 32, Issue 14, Pages 1701–1705, 2011.
8. C.S. Li, "Cluster Center Initialization Method for K-means Algorithm Over Data Sets with Two Clusters", *"2011 International Conference on Advances in Engineering, Elsevier"*, pp. 324–328, vol. 24, 2011.
9. Fuyuan Cao, Jiye Liang, Guang Jiang, "An initialization method for the K-Means algorithm using neighborhood model", *'Computers and Mathematics with Applications'*, pp. 474–483, 2009.
10. R.V. Rao, V. J. Savsani and D.P. Vakharia, 'Teaching–learning-based optimization: A novel method for constrained mechanical design optimization problems'. *Computer-Aided Design* 43, pp. 303–315, 2011.
11. M. Halkidi, Y. Batistakis, M. Vazirgiannis, Clustering validity checking methods: Part ii, SIGMOD Record 31 (3) 2002, pp. 19–27.
12. B.K. Mishra, N.R. Nayak, A.K. Rath and S. Swain, "Far Efficient K-Means Clustering Algorithm", *"Proceedings of the International Conference on Advances in Computing, Communications and Informatics"*, ACM, pp. 106–110, 2012.
13. B.K. Mishra, N.R. Nayak and A.K. Rath, Assessment of basic clustering techniques using teaching-learning-based optimization, Int. J. Knowledge Engineering and Soft Data Paradigms, Vol. 5, No. 2, pp. 106–122, 2016.
14. B.K. Mishra, N.R. Nayak, A.K. Rath and S. Swain, "Improving the Efficiency of Clustering by Using an Enhanced Clustering Methodology", *"International Journal of Advances in Engineering & Technology"*, Vol. 4, Issue 2, pp. 415–424, 2012.
15. C. Merz and P. Murphy, UCI Repository of Machine Learning Databases, Available: http://ftp.ics.uci.edu/pub/machine-learning-databases.
16. Bagirov, A.M and Yearwood, J, "A new non-smooth optimization algorithm for minimum sum-of-squares clustering problems", *EJOR 170*, 2 (2006), pp. 578–596.

17. H.S. Park and C.H. Jun,"A simple and fast algorithm for K-medoids clustering", *"Expert System with Applications"*, pp. 3336–3341, 2009.
18. J. C. Dunn, 'A Fuzzy Relative of the ISODATA Process and Its Use in Detecting Compact Well-Separated Clusters'. *'J. Cybernetics'*, vol. 3, pp. 32–57, 1973.
19. D. L. Davies and D. W. Bouldin, 'A Cluster Separation Measure', *'IEEE Trans Pattern Analysis & Machine Intelligence'*, vol. 1, pp 224–227, 1979.
20. P. J. Rousseeuw, "Silhouettes: a graphical aid to the interpretation and validation of cluster analysis", *"Journal of Computational and Applied Mathematics"*, vol. 20, pp. 53–65, 1987.
21. L. J. Hubert and J. R. Levin, "A general statistical framework for accessing categorical clustering in free recall", *"Psychological Bulletin 83"*, pp. 1072–1080, 1976.
22. Shi Na, L. Xumin and G. Yong, "Research on K-Means clustering algorithm-An Improved K-Means Clustering Algorithm". *"IEEE 3rd International Symposium on Intelligent Information Technology and Security Informatics"*, pp. 63–67, 2010.
23. R. Xu and D. Wunsch, 'Survey of Clustering Algorithms', *"IEEE Transactions on Neural networks"*, vol. 16, no. 3, 2005.
24. Y. M. Cheung, 'A New Generalized K-Means Clustering Algorithm'. *'Pattern Recognition Letters'*, vol. 24, issue 15, pp. 2883–2893. 2003.
25. K. A. Abdul Nazeer, M. P. Sebastian, "Improving the Accuracy and Efficiency of the k-means Clustering Algorithm", *"Proceedings of the World Congress on Engineering"*, Vol I, 2009.
26. B. Amiri, (2012). 'Application of Teaching-Learning-Based Optimization Algorithm on Cluster Analysis'. *Journal of Basic and Applied Scientific Research*, 2(11), pp. 11795–11802.
27. A. Naik. S. C Satpathy and K. Parvathi, 'Improvement of initial cluster centre of c-means using Teaching learning based optimization'. *'2nd International Conference on Communication, Computing & Security'*, pp. 428–435, 2012.
28. J. Mac Queen, "Some methods for classification and analysis of multivariate observations", *"Fifth Berkeley Symposium on Mathematics, Statistics and Probability"*, pp. 281–297, University of California Press, 1967.
29. R.V. Rao and V. Patel, 'An elitist teaching-learning-based optimization algorithm for solving complex constrained optimization problems'. *International Journal of Industrial Engineering Computations*, pp. 535–560, 2012.
30. S. C Satpathy and A. Naik, 'Data Clustering Based on Teaching-Learning-Based Optimization'. *SEMCCO, LNCS 7077*, pp. 148–156, 2011.

Digital Image Watermarking Using (2, 2) Visual Cryptography with DWT-SVD Based Watermarking

Kamal Nayan Kaur, Divya, Ishu Gupta and Ashutosh Kumar Singh

Abstract It has become earnest necessity to protect the digital multimedia content for the possessors of the documents and service providers. Watermarking is such a technique which helps us to attain copyright protection. The concerned literature includes various methods which help to embed information into various multimedia elements like images, audio, and video. In the given paper, we have reviewed DWT-SVD watermarking technique for image watermarking. We have projected a new algorithm for image watermarking using visual cryptography that generates two shares with DWT-SVD. The scheme is highly protected and vital to the image processing assault. The following material also gives an insight into implementation of the contained algorithm step-by-step and shows the future prospects.

Keywords DFT · DCT · Discrete wavelet transform (DWT)
Image watermarking · Singular-value decomposition (SVD)

1 Introduction

All the data which are managed on Internet and multimedia network system are processed on data machine, and in today's era, the copying of such data without much loss in quality is not that tough to achieve. Thus, digitization has come at the cost of many disadvantages. Internet plays a vital role in the distribution of digital

K. N. Kaur (✉) · Divya · I. Gupta · A. K. Singh
National Institute of Technology, Kurukshetra 136119, Haryana, India
e-mail: kamal.nayan1910@gmail.com

Divya
e-mail: divya.dulyan7@gmail.com

I. Gupta
e-mail: ishugupta23@gmail.com

A. K. Singh
e-mail: ashutosh@nitkkr.ac.in

© Springer Nature Singapore Pte Ltd. 2019
H. S. Behera et al. (eds.), *Computational Intelligence in Data Mining*,
Advances in Intelligent Systems and Computing 711,
https://doi.org/10.1007/978-981-10-8055-5_8

content which is illegal and is not authorized [1]. This leads to such an exposure that results in breach of owner's right and hampers the genuineness of a digital content. We can safeguard the digital content by implanting extra information called watermark into it. This watermarking (data hiding) technique is the one in which we can add some multimedia image, audio, or video [2]. Nowadays, watermarking is being used in each and every field, either the data are saved on cloud, screaming on television, printed on any piece of paper or the data are saved anywhere in any kind of form [3–5].

This technique basically involves embedding secret information into the digital image which needs to be protected in the form of watermark such that the resulting image is resistant and robust to numerous standard data processing techniques such as filtering, re-sizing, and cropping to name a few. But with the advancement in technology, the integrity of schemes is also being compromised as attackers have developed techniques to discard the watermark to make it protected. A watermark thus should be added in a way that it remnants irremovable and perceptible as long as its enduring feature of the digital data and information remains standardized.

2 Related Work

A watermark system is usually divided into two distinct steps (According to [6]): (a) Embedding (b) Extraction.

- *Embedding*: In embedding, an algorithm secures the prominent data which is to be embedded and a watermark signal is created.
- *Extraction*: Extraction involves further detection and identification. Detection is the process in which it is conceived that if the watermark is embedded in the received content. The effectiveness can be checked only through Type I and Type II errors. In Type I error, it gives the false image of watermark, whereas watermark is not present. In Type II error, the watermark is not detected although it is present in the data. Whereas identification helps to decode the watermark. The errors in identification can further be classified as "open set" or "closed set." Open set defines that there is possibility of the existence of one of N or no watermark, whereas closed set defines the problem where one of N possible watermarks is known to be in the received data and the detector has to choose the most one having the highest possibility [7, 8].

In earlier times, in spatial domain, the pixel values of host image were changed to put the watermark. Under spatial domain, we have least significant bit (LSB) in which we modify the LSB's of both host and marked images which are stationed on the theory that the bits of LSB are normally imperceptible. No doubt that the spatial domain watermarking is too fragile but at the same time it is easy to implement in data processing. In pursuance of having more appropriate techniques, researches worked in the direction that the watermark image is never added to the intensities of

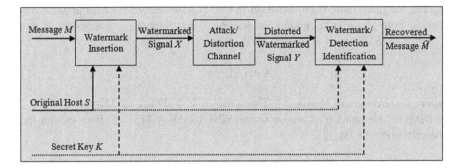

Fig. 1 Life cycle of watermarking

cover image but to the values of its transform coefficients. After that, the water-marked image is obtained by performing the inverse transformation.

As described in [1, 9–11], most of the transform-based algorithms of watermarking techniques are discrete cosine transform (DCT) and discrete wavelet transform (DWT). The wavelet transform creates a data structure that is recognized as scale-space representation. Discrete cosine transform (DCT) emblematizes data in terms of frequency space in place of an amplitude space. This is beneficial by reason of it can be comparable as humans notice light. So, the parts that are not anticipated and it can be determined and squandered. The watermarking techniques of DCT are more appropriate as compared to spatial domain techniques [12]. In today's world, discrete wavelet transform is commonly used in digital image processing, compression, watermarking etc. This change is based on small waves, called wavelet, of alternate frequency and limited duration. The wavelet transform crumbles the image into horizontal, vertical, and diagonal spatial directions [13]. In discrete Fourier transform (DFT), the function is changed by its frequency components. It is robust against geometric attacks like circumvolution, mounting, cropping, and adaptation. DFT shows invariant translation. Discrete wavelet transform (DWT) is generally appropriate in modernistic watermarking schemes. In a DWT-based scheme, the value of DWT diagonal coefficients is changed with the bits in the diagonal position that represents the watermark [14, 15]. Life cycle of watermarking scheme is shown in Fig. 1.

3 Watermarking Technique and Algorithms

The indispensable idea of proposing an image watermarking manner is to atone both indistinguishable and robustness requisite. This article states about the amalgamation of image watermarking scheme hinges on discrete wavelet transform (DWT) and singular-value decomposition (SVD). According to this technique, the watermark is not put directly on the wavelet coefficients but instead of it, it is put on the elements of singular values of the overlay image's DWT sub-bands. In linear algebra, singular-value decomposition (SVD) is a powerful tool with applications varying from image compression and watermarking. Suppose A is a matrix of size $n \times n$, then SVD of (A) is given in Eq. (1).

$$A = S * U * V'. \tag{1}$$

Where U: Orthogonal Matrix; V': Transpose of V; S: Diagonal Matrix.

Diagonal elements of S are known as SINGULAR VALUES. They exhibit the property given in Eq. (2).

$$S(1, 1) > S(2, 2) > S(3, 3) > \cdots > S(n, n). \tag{2}$$

By merging DWT with SVD, we discuss a basic watermarking algorithm as described in [13]. DWT splits the image into four frequency bands: LL, HL, LH, and HH band, where LL band: low frequency; HL, LH band: middle frequency; HH band: high frequency as represented in Fig. 2.

Approximate details of an image are exhibited by LL band. HL band represents horizontal details, LH exhibits vertical details and HH band focuses on diagonal details of the image. Generally, HH band is selected to put the watermark as it possesses the minutest details and its contribution to image energy is very little. Therefore, watermark embedding does not result in the attachment of cover image. Also, watermark added in HH band has the capability to bear certain image processing distortions, for example, noised in, intensity direction and constraint of the human visual system can be exploited by adding the given watermark into HH band. HVS is not able to identify the changes made to HH band. This scheme represents the idea of changing singular characters of the diagonal of the HH band with the singular characters of diagonal of the watermark. If a watermark is elected in such a way that the characters of singular diagonal lie between the given range then both the energies, i.e., energy of the singular characters of watermark and the energy of singular characters of HH band will be relatively comparable. Therefore, the change in the singular characters will not alter the trait of image and would not change the value of energy of high-frequency band.

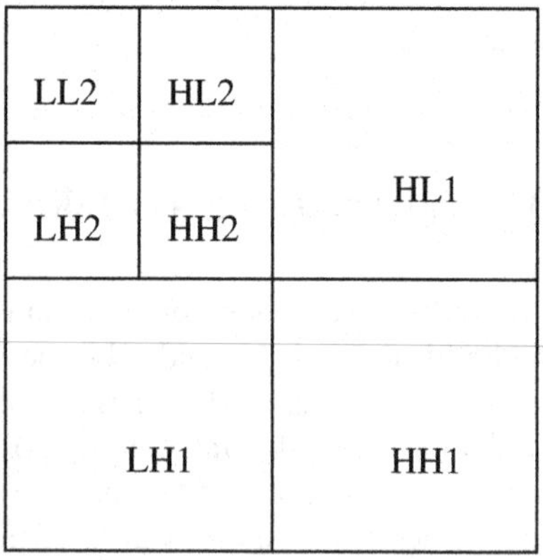

Fig. 2 DWT decomposition model

LL2	HL2	HL1
LH2	HH2	
LH1		HH1

3.1 Watermark Embedding Algorithm

1. Convert the cover image and watermark image into grayscale. Split the watermark image (the image to be hidden) into two shares using (2, 2) visual cryptography. This is a random share generation scheme.
2. One of the above generated shares is put into the cover image while other is provided to the rightful user for watermark generation. Using SVD, split the share to be embedded into the orthogonal matrices using Eq. (3)

$$\text{Share } 1 = Uw * Sw * Vw'. \tag{3}$$

3. Split the cover image (in which the image is to be embedded) into four sub-bands: LL, HL, LH, and HH by applying Haar wavelet.
4. Use SVD to HH band of the cover image using Eq. (4).

$$H = UH * SH * VH'. \tag{4}$$

5. Singular values of HH band of the cover image are replaced by singular values of the share to be embedded.
6. Implement inverse SVD to obtain the modified HH band by applying Eq. (5)

$$H = UH * Sw * VH'. \tag{5}$$

3.2 Watermark Extraction Algorithm

To produce watermarked cover image, inverse DWT is applied.

1. Split the noisy watermarked image into four sub-bands by applying the Haar wavelet in DWT transform.
2. Apply SVD on HH band using Eq. (6)

$$H = UH * SH * VH'. \tag{6}$$

3. Singular values from HH band are extracted.
4. Watermark is constructed by applying singular values. Orthogonal matrices Uw and Vw are obtained by making use of SVD of original watermark. Shares can be obtained by applying Eq. (7)

$$\text{Share}1 = Uw * SH * Vw. \tag{7}$$

4   Results

The proposed algorithm has been implemented using MATLAB 2016 on Intel i3 2.0 GHZ system running windows 8. First of all, we have read cover image represented in Fig. 3a and watermark logo is shown in Fig. 3b. Both the images are then converted into its respective black and white images. Watermark is embedded in the black and white converted image. After that, we have applied visual cryptography and generated two shares as exhibited in Fig. 4a and Fig. 4b, respectively.

First of all, the cover image is decomposed in one step decomposition as shown in Fig. 5a then using DWT function of MATLAB, it is decomposed in four steps decomposition represented in Fig. 5b. Now using the steps as discussed in the

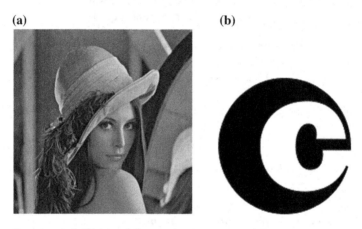

Fig. 3 a Cover image. **b** Watermark logo

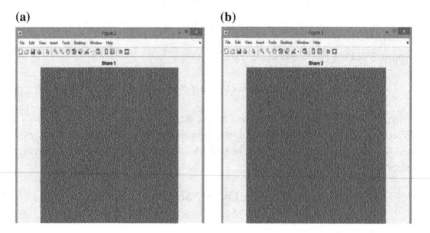

Fig. 4 a Share1. **b** Share2

(a)

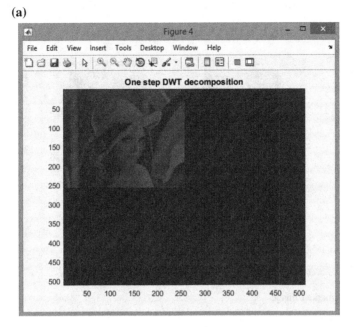

(b)

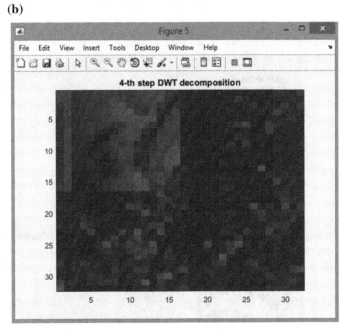

Fig. 5 a One step DWT decomposition. **b** 4th step DWT decomposition

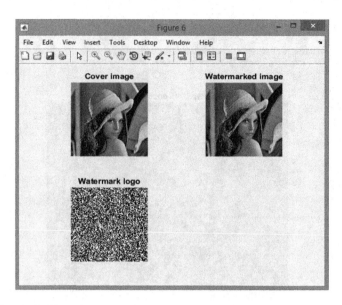

Fig. 6 Watermarked image

(a)

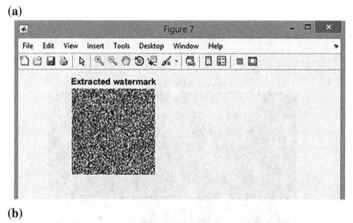

(b)

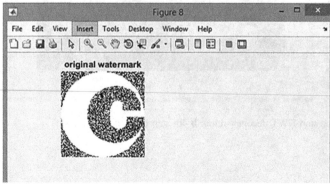

Fig. 7 **a** Extracted watermark. **b** Original watermark

above algorithm, we generate the signature, embedded it and thus, watermarked image is obtained in Fig. 6. The presence of watermark is checked, if its presence is there then it is authenticated else the authentication fails. Now using Haar wavelet and inverse DWT, we decompose the watermarked image and generated the signature and finally, we constructed the watermark as exhibited in Fig. 7a and b correspondingly.

5 Conclusion and Future Work

This analysis specifies a robust and efficient watermarking algorithm along with providing an insight into the watermarking process. The watermark will be embedded in the original image and will be extracted from it later on. Also, meanwhile the image will be processed in various ways. All of these result in making the data distributed over the network secure and safe.

References

1. V. M. Potdar, S. Han and E. Chang, "A survey of digital image watermarking techniques," *INDIN '05. 2005 3ʳᵈ IEEE International Conference on Industrial Informatics*, 2005, pp. 709–716.
2. H. Wei, M. Yuan, J. Zhao and Z. Kou, "Research and realization of digital watermark for picture protecting," *2009 IEEE First International Workshop on Education Technology and Computer Science (ETCS'09)*, China, vol. 1, 2009, pp. 968–970.
3. S. Chhabra and A. K. Singh, "Dynamic data leakage detection model based approach for MapReduce computational security in cloud," *2016 Fifth International Conference on Eco-friendly Computing and Communication Systems (ICECCS)*, Bhopal, 2016, pp. 13–19.
4. H. Taneja, Kapil and A. K. Singh, "Preserving Privacy of Patients based on Reidentification Risk," *Fourth International Conference on Eco-friendly Computing and Communication Systems (ICECCS)*, 2015, pp. 448–454.
5. J. Kumar and A. K. Singh, "Dynamic resource scaling in cloud using neural network and black hole algorithm," *2016 Fifth International Conference on Eco-friendly Computing and Communication Systems (ICECCS)*, Bhopal, 2016, pp. 63–67.
6. C. I. Podilchuk and E. J. Delp, "Digital watermarking: algorithms and applications," *IEEE Signal Processing Magazine*, vol. 18, no. 4, July 2001, pp. 33–46.
7. J. Kapur and A. J. Baregar, "Security using image processing." *International Journal of Managing Information Technology (IJMIT)*, vol. 5, no. 2, pp. 13–21, May 2013.
8. P. Singh and R. S. Chadha, "A survey of digital watermarking techniques, applications and attacks," *International Journal of Engineering and Innovative Technology (IJEIT)*, vol. 2, no. 9, pp. 165–175, March 2013.
9. Ma Bin, "Experimental research of image digital watermark based on DWT technology," *IEEE International conference on Uncertainty Reasoning and Knowledge Engineering (URKE)*, Bali, vol. 2, 2011, pp. 9–12.
10. Y. Qianli and C. Yanhong, "A digital image watermarking algorithm based on discrete wavelet transform and Discrete Cosine Transform," *2012 IEEE International Symposium on Information Technology in Medicine and Education (ITME)*, Japan, vol. 2, 2012, pp. 1102–1105.

11. M. Thapa, Dr. S. K. Sood and A. P. M. Sharma, "Digital image watermarking techniques based on different attacks," *International Journal of Advanced Computer Science and Applications (IJACSA)*, vol. 2, no. 4, pp. 14–19, 2011.
12. N. Divecha and Dr. N. N. Jani, "Implementation and performance analysis of DCT-DWT-SVD based watermarking algorithms for color images," *Proceeding in 2013 IEEE International Conference on Intelligent Systems and Signal Processing (ISSP)*, 2013, pp. 204–208.
13. D. V. S. Chandra, "Digital image watermarking using Singular Value Decomposition," *Proceeding of 45th IEEE symposium on circuits and systems*, vol. 3, pp. 264–267, 2002.
14. E. Ganic and A. M. Eskicioglu, "Robust DWT-SVD domain image watermarking: embedding data in all frequencies," *Proceedings of the 2004 ACM Workshop on Multimedia and Security*, Sept. 2004, pp. 166–174.
15. C. C. Lai and C. C. Tsai, "Digital Image Watermarking Using Discrete Wavelet Transform and Singular Value Decomposition," in *IEEE Transactions on Instrumentation and Measurement*, vol. 59, no. 11, pp. 3060–3063, Nov. 2010.

Modeling of Nexa-1.2kW Proton Exchange Membrane Fuel Cell Power Supply Using Swarm Intelligence

Tata Venkat Dixit, Anamika Yadav and Shubhrata Gupta

Abstract The heuristic approach of simulator design based on swarm intelligence of Nexa-1.2kW Ballard proton exchange membrane fuel cell (PEMFC) has been presented. The parameters of the Nexa-1.2kW PEMFC simulator are determined using particle swarm optimization (PSO) algorithm. The results of PEMFC simulator are experimentally verified. Further, the discrete PI controlled SEPIC converter has been used for interconnecting a fuel cell to a load. The fuel cell simulator, converter integration, and its control are implemented in MATLAB/SIMULINK environment. Finally, the effect of load variation and stack temperature on fuel cell power conditioning unit has been investigated. The rise in stack temperature results in slight reduction in cell current and considerable rise in terminal voltage of the fuel cell.

Keywords Nexa-1.2kW PEM fuel cell · SEPIC · PSO

1 Introduction

The major part of the energy demands of the world is fulfilled by fossil fuel which has significant adverse environmental impact. Also, the fossil fuels are costly and limited in availability. Nowadays, to meet the trend of increasing energy demand the world has shifted their focus on sustainable energy sources such as wind, solar and fuel cell. The unpredictability of output of wind and solar energy has made the fuel cell as important energy technology for reliable power generation. The fuel cell

T. V. Dixit (✉) · A. Yadav · S. Gupta
Department of Electrical Engineering, National Institute of Technology,
Raipur 492001, Chhattisgarh, India
e-mail: tvdixit@gmail.com

A. Yadav
e-mail: ayadav.ele@nitrr.ac.in

S. Gupta
e-mail: sgupta.ele@nitrr.ac.in

© Springer Nature Singapore Pte Ltd. 2019
H. S. Behera et al. (eds.), *Computational Intelligence in Data Mining*,
Advances in Intelligent Systems and Computing 711,
https://doi.org/10.1007/978-981-10-8055-5_9

is an electrochemical device that converts chemical energy into electrical energy. The fuel cell system is non-polluting, efficient, and performance is independent of geographical and metrological factors. Fuel cells are widely used in space, vehicular, military, and hybrid electric applications [1–3]. Among all fuel technologies, PEM fuel cell system is highly preferred due to its fast response, high efficiency, low operating temperature, and low corrosion [4].

To simulate the PEM, fuel cell based on electrochemical, fluid dynamics, and thermal phenomenon of many models have been reported in the literature [5–8]. In past few decades, stochastic population search techniques are getting the attention in developing efficient methods for solving optimization problem of various mathematical models. Nowadays, optimization techniques have attracted much attention for parameter identification of PEMFC such as simulated annealing [9], differential evolution [10, 11], particle swarm optimization (PSO) [12]. Tang et al. [13] have built a small sample nonlinear model for the methanol fuel cell (DMFC) using a support vector regression (SVR) approaches combined with PSO algorithm for its parameter optimization.

The use of actual Nexa-1.2kW PEM fuel cell power conditioning unit in design and testing for interconnecting source and load is expensive and any failure may lead to damage of the source and power conditioning unit. Hence, simple Nexa-1.2kW simulator has been developed and experimentally verified. In power conditioning unit, DC–DC converter is used to convert unregulated voltage into regulated output voltage at desired voltage under different operating conditions. The SEPIC converter has low switching loss, high efficiency better transient response, and output current characteristics due to inductor at input and output stage. The proper selection of input and output stage inductor minimizes the current ripple [14]. This feature of SEPIC converter has motivated for replacing Buck–Boost converter.

In this paper, we reported the application of PSO algorithm to determine the parameters of mechanistic model which simulates the actual Nexa-1.2kW PEM fuel cell with fairly good accuracy. Further, simulated PEMFC module connected to SEPIC converter for ensuring the regulated output load voltage. The output voltage of DC–DC converter is regulated through digital PI controller. Also, the effect of load variation on fuel cell power supply has been investigated.

2 Modeling of PEMFC Simulator

The PEMFC consists of a solid polymeric membrane electrolyte pressed between anode and cathode electrodes, and it consumes hydrogen to produce electrical energy. The various electrochemical reactions inside the PEMFC are as follows [9, 12]:

$$\text{At Anode: } H_2 \rightarrow 2H^+ + 2e^- \tag{1}$$

$$\text{At Cathode: } 4H^+ + O_2 + 4e^- \rightarrow H_2O \tag{2}$$

$$\text{Overall: } 2H_2 + O_2 \rightarrow 2H_2O \tag{3}$$

The electrode potential under thermodynamic balance (at no-load) is known as reversible cell potential or internal potential. It is to be noted that [9, 12]

$$E_{Nernst} = f\{T, P_{H_2}, P_{O_2}\} \tag{4}$$

where T is stack temperature (°K), Ed is the voltage drop during load transient caused by delay in hydrogen and oxygen flow under steady state and it is assumed to be zero, P_{H2} and P_{O2} are the partial pressure of hydrogen and oxygen, respectively, in mbar. The ideal standard potential of a fuel cell is 1.229 V/cell with water in the form of moisture as byproduct [6]. The Nernst voltage is always less than the standard potential of a cell due to irresistible losses in the cell as expressed in Eq. 5 [12].

$$E_{Nernst} = 1.229 - (8.5e - 4)(T - 298.15) + (4.308e - 5)T\left(\ln\left(P_{H_2}P_{O_2}^{0.5}\right)\right) - E_d \tag{5}$$

The activation loss in PEMFC is due to sluggishness in the chemical reactions that take place on the active surface of the electrode. On the basis of several experiments, an empirical formula to estimate the activation over-potential is given in Eq. 6 [9]:

$$V_{act} = \zeta_1 + \varphi T + \zeta_3 T \ln(C_{O_2}) + \zeta_4 T \ln I \tag{6}$$

$$\text{Where, } \varphi = \zeta_2 + 2 \times 10^{-4} \ln A + 4 \times 10^{-5} \ln(C_{H_2}) \tag{7}$$

where I is the stack current, ζ_1, ζ_2, ζ_3, and ζ_4 are the parametric coefficients for each cell. These coefficients are based on electrochemical, thermodynamics, fluid mechanics, and dissolved oxygen and hydrogen concentration at the cathode/membrane interface. The reactant concentration at the cathode/membrane interface is defined by Henry's law expression as follows [12]:

$$C_{O_2} = 1.97 \times 10^{-7} \exp(498/T)P_{O_2} \tag{8}$$

$$C_{H_2} = 9.17 \times 10^{-7} \exp(-77/T)P_{H_2} \tag{9}$$

The ohmic over-potential can be expressed as [5]

$$V_{ohm} = (R_M + R_t)I \tag{10}$$

$$R_M = \frac{\rho_M l}{A} \tag{11}$$

where I is the stack current, l is the thickness of membrane (cm), A is active area of membrane (cm^2), ρ_M is the specific resistivity of membrane for electron flow (Ω cm), R_M and R_t are the membrane and transfer equivalent resistance of the cell, respectively. The resistivity of Nafion series PEMFC can be estimated by Eq. 12 [12]

$$\rho_M = \frac{181.6\left[1 + 0.03\frac{I}{A} + 0.062\left(\frac{T}{303}\right)^2\left(\frac{I}{A}\right)^{2.5}\right]}{\left[\sigma - 0.634 - \frac{3I}{A}\right]\left[\exp\left(4.18\left(\frac{T-303}{T}\right)\right)\right]} \tag{12}$$

The adjustable fitting parameter sigma (σ) is affected by membrane manufacturing process. The ohmic losses can be minimized by reducing the thickness of the electrodes and using high conductivity electrodes and interconnect.

The concentration potential drop results from the concentration gradient of reactant as they are consumed in the reaction. The empirical equation to calculate concentration potential drop is expressed as [7]

$$V_{con} = \frac{RT}{nF}\ln\left(1 - \frac{J}{J_{max}}\right) \tag{13}$$

where J and J_{max} are the actual and maximum current density (A/cm^2), n = 2. Finally, Eq. 14 expresses the output stack voltage of fuel cell system [9].

$$V_{fc} = N_{cell} \times (E_{Nernst} + V_{act} - V_{ohm} + V_{con}) \tag{14}$$

3 Modeling of Nexa-1.2kW Fuel Cell Using Swarm Intelligence

This section addresses how to find the optimal values of six parameters of highly nonlinear fuel cell model which best fits a given real Nexa-1.2kW PEMFC system. Since a system is nonlinear, a heuristic optimization technique (PSO) has been adopted to determine optimal set of parameters for voltage model. In the PSO algorithm, each particle adjusts its position in search space by its own as well as other particle flying experience to reach the best global solution. The performance of this technique can be improved by defining the initial population range of each swarm such that each particle closely bounds to the expected domain of a feasible region. A real-coded PSO estimator is proposed to identify the ζ_1, ζ_2, ζ_3, ζ_4, R_t, and 'σ' parametric coefficients for each Nexa-1.2kW fuel cell.

The mathematical model of optimization problem can be described as follows:

$$\min or \max f\{X, V_{error}\} \tag{15}$$

where X is some equality and non-equality constraints, and V_{error} is the objective function to be minimized. The selected PSO input parameters are as follows:

$$X = \left[I_{st}, T, H_2^{con}, O_2^{con}, A, l \right] \tag{16}$$

3.1 Fitness Function and Imposed Parametric Constraints

The selection of proper objective function is the milestone of any optimization technique. Therefore, root mean square error between experimental data and parameter optimized simulator output voltage has been considered to form an objective function. It can be expressed as follows:

$$F_\Delta = \frac{1}{\sqrt{N}} \left\| \left(V_{\exp} - V_{fc}(I, \theta) \right) \right\| \tag{17}$$

where $V_{\exp}$ is the measured voltage, V_{fc} is the simulator output voltage at certain current, and $\theta = [\zeta_1, \zeta_2, \zeta_3, \zeta_4, \sigma]$ is set of identified parameters of the mathematical model of fuel cell. The N is the number of experimental data taken into consideration. Noteworthy, the goal of this optimization is to reduce the error between experimental and simulated data with respect to parameter values. As we know in the heuristics method of optimization, different run gives different optimal points due to random initialization of states. Therefore, averaging of the objective function is one of the ways to encounter the above state problem. The average objective function for 'κ' runs can be defined as:

$$F_{\Delta avg} = \sum_{i=1}^{\kappa} F_{\Delta i} / \kappa \tag{18}$$

During this study, Eqs. 19 and 20 show the imposed parametric constraints reported by Gong and Cai [11]:

$$\left. \begin{array}{l} -1.19969 \leq \zeta_1 \leq -0.8532; \quad 0.001 \leq \zeta_2 \leq 0.005; \\ 3.6e - 5 \leq \zeta_3 \leq 9.8e - 5; \; -2.6e - 4 \leq \zeta_4 \leq -9.54e - 5; \end{array} \right\} \tag{19}$$

$$10 \leq \sigma \leq 23; 0.0001 \leq R_t \leq 0.0008 \tag{20}$$

In this work, the performance of swarms is made better by putting parametric constraints (lower and upper bound constraints) as shown in (19–20). In PSO, each particle moves about the cost surface with a velocity. The equation to update velocity and position based on the global and local best solution of each particle and the condition of stability is demonstrated in [15, 16], and due to space limitation, these equations are omitted in this paper. The PSO parameters are as follows: number of variables to be optimized = 20; number of particles = 50; $C_1 = 2$; $C_2 = 2$; number of generations = 200; $\omega_{max} = 0.9$, $\omega_{min} = 0.4$ and boundary Constraint = 'Penalize.'

3.2 Detection of Model Parameters of PEMFC Using Swarm Intelligence

Fig. 1 shows the convergence behavior of PSO fitness value for PEMFC parameter identification. Presently, the PSO estimator is set for 200 generations, but PSO displayed a premature convergence, i.e., mean fitness value reaches to best score in less time. The mean of optimized fitness function reaches to best score 0.03259 at 167 generation. The optimized parameters of voltage model of Nexa-1.2kW PEM fuel cell are as reported in Table 1.

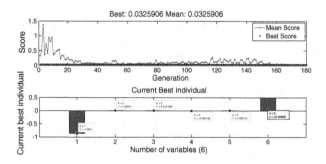

Fig. 1 Convergence process of PSO algorithm and best individual value of parameters

Table 1 Voltage model identified parameters using PSO

Parameters	Value	Parameters	Value	Parameters	Value
ζ_1	−0.864	ζ_2	0.00281	ζ_3	827e-5
ζ_4	−0.000148	R_t	0.0001	σ	22.999998

4 Experimental Setup of Nexa-1.2kW PEM Fuel Cell

The experimental setup of Nexa-1.2kW PEM fuel cell module in our institute has been shown in Fig. 2. It is a fully integrated safe system for indoor applications. In this stack, 36 cells are connected in the series which has a capacity to carry 60 A steady-state current. It can produce DC power with voltage ranging from 36 V at no-load to 18 V at the full load. The maximum steady-state current and operating temperature of the PEM Nexa-1.2kW fuel cell model is 60 A and 65 °C, respectively. The performance of fuel cell mainly depends on stack temperature, hydrogen and oxygen pressure, partial pressures. The NEXA-OSC software facilitates the data acquisition, monitoring, and controlling of the module through a PC interface. In this module, the active area of membrane (cm^2) is 120 cm^2, membrane length 0.02 cm, the hydrogen and oxygen pressure to the stack is normally maintained at 0.3 bar and 0.1 bar, respectively.

5 SIMULINK Model of Fuel Cell Power Supply

The use of DC–DC converter with fuel cell ensures efficient conversion of power from nonlinear source to load. The DC–DC converter converts unregulated output voltage of fuel cell into regulated output voltage at desired voltage level under various operating conditions. Here, PI controlled SEPIC converter is an integral part of fuel cell power conditioning unit, which ensures constant 60 V load voltage under load variation. A SIMULINK circuit diagram of fuel cell simulator of PI controlled SEPIC based on designed parameter given in Table 2 is shown in Fig. 3.

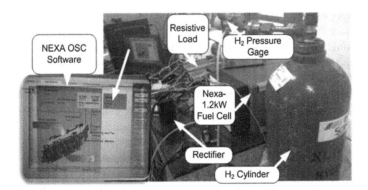

Fig. 2 Experimental setup of Nexa-1.2kW Ballard stack of 36 cells system

Table 2 Parameters of SEPIC and PI controller

L1	L2	C1	C2	RL	Kp	Ki
1.5 mH	1.5 mH	330 μF	2.2 mF	12 Ω	0.0002	0.78

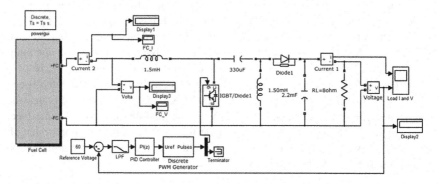

Fig. 3 SIMULINK diagram of Nexa-1.2kW PEMFC fed SEPIC converter power conditioning unit

6 Results and Discussion

In this section, the results of a PSO optimized simulator model with DC–DC converter are presented and discussed. The real-coded PSO algorithm has been implemented in MATLAB-7.10(R2010a) for parameter detection of the Nexa-1.2kW PEM fuel cell simulator. The identified parameters are listed in Table 1. The cell voltage versus stack current plot corresponding to these parameters is shown in Fig. 4a. In Fig. 4a, it can be noticed that some experimental data samples do not agree with mechanistic model results. It is due to the reduction of voltage drop with rise in stack temperature while other operating conditions remain same.

The effect of stack current and temperature on cell performance is shown in Fig. 5. Where the variation of stack current, voltage and temperature (represented in °C) with time could be observed. Due to the limitation of available load in the laboratory, test has been conducted till 49 A stack current. The effect of the adjustable fitting parameter (σ) on polarization characteristics is presented in Fig. 4b. It has been observed that increasing 'σ' shifts the polarization curve upwards, i.e., increased value of 'σ' results in higher cell voltage. In other words, the effect of ohmic voltage drop decreases at high value of 'σ.'

The load current and output voltage of SEPIC converter are shown in Fig. 6a. The selected values of the PI controller ensure the regulated output voltage of 60 V irrespective of a load disturbance introduced at t = 1 s. From Fig. 6a, it could be observed that at t = 1 s load is increased from 300 to 600 W; as a result, load current jumped from earlier steady-state value 5 A to new steady-state value 10 A but steady-state load voltage (~60 V) remains same. Further, the temperature rise in fuel

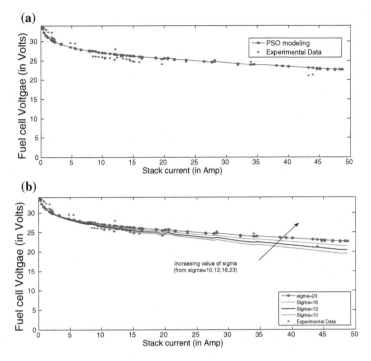

Fig. 4 **a** Output voltage versus stack current under different temperatures of PEMFC. **b** Effect of sigma ('σ') on polarization characteristics

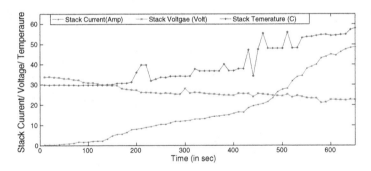

Fig. 5 Stack current, voltage, temperature versus time plot of PEMFC

cell at t = 2.5 s (from 300 to 315 °K) has been introduced, and it is observed that its effect on load voltage and current was insignificant. The effect of load and temperature disturbance on fuel cell performance has been presented in Fig. 6b. Where due to the load disturbance, considerable voltage drop at terminal voltage has been observed. On the other side, the rise in stack temperature causes the slight reduction in cell current and considerable voltage rise in terminal voltage of fuel cell.

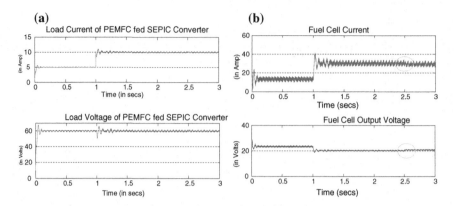

Fig. 6 a Load current and voltage versus time under load and stack temperature variation. **b** Fuel cell current and voltage versus time under load and stack temperature variation

7 Conclusion

In this paper, a simulator of Nexa-1.2kW fuel cell has been developed and validated with experimental data. The accuracy of simulator depends on how best variables and their interconnection in the process within the cell are formulated. The effect of adjustable parameter 'σ' on the polarization curve has been investigated, and it is concluded that at high value of 'σ' the ohmic drop decreases as a result output voltage has increased. The significant parameters of simulator are determined by the PSO optimization technique. The PEMFC simulator is integrated with SEPIC DC–DC converter to capture the transient performance of the cell under load variations. The load resistance is reduced from 12 to 6 Ω at t = 1 s. During this load disturbance, PI controller effectively regulated the load voltage at 60 V under steady state and introduced 12.5% overshoot during the load disturbance. The rise in stack temperature from 300 to 315 °K introduces 4.25% rise in stack voltage and 1.39% reduction in cell current, but this effect is insignificant at load side.

References

1. Kanhu Charan Bhuyan, 'Development of Controllers Using FPGA for Fuel Cells in Standalone and Utility Applications Kanhu Charan Bhuyan', National Institute of Technology Rourkela, 2014.
2. T. Lajnef, S. Abid, and A. Ammous, 'Modeling, control, and simulation of a solar hydrogen/fuel cell hybrid energy system for grid-connected applications', *Adv. Power Electron.*, vol. 2013, p. 352765 (9 pp.), 2013.
3. I. Soltani, 'An Intelligent, Fast and Robust Maximum Power Point Tracking for Proton Exchange Membrane Fuel Cell', *World Appl. Program.*, vol. 3, no. July, pp. 264–281, 2013.
4. C. Ziogou, E. N. Pistikopoulos, M. C. Georgiadis, S. Voutetakis, and S. Papadopoulou, 'Empowering the performance of advanced NMPC by multiparametric programming - An

application to a PEM fuel cell system', *Ind. Eng. Chem. Res.*, vol. 52, no. 13, pp. 4863–4873, 2013.

5. J. M. Corrêa, F. A. Farret, L. N. Canha, and M. G. Simoes, 'An electrochemical-based fuel-cell model suitable for electrical engineering automation approach', *IEEE Trans. Ind. Electron.*, vol. 51, no. 5, pp. 1103–1112, 2004.

6. R. I. Salim, H. Noura, M. Nabag, and A. Fardoun, 'Modeling and Temperature Analysis of the Nexa 1.2 kW Fuel Cell System', *J. Fuel Cell Sci. Technol.*, vol. 12, no. 6, pp. 1–9, 2015.

7. A. Gebregergis and P. Pillay, 'Implementation of fuel cell emulation on DSP and dSPACE controllers in the design of power electronic converters', *IEEE Trans. Ind. Appl.*, vol. 46, no. 1, pp. 285–294, 2010.

8. N. Benchouia and A. Hadjadj, 'Modeling and validation of fuel cell PEMFC', *Rev. des Energies ...*, vol. 16, pp. 365–377, 2013.

9. M. T. Outeiro, R. Chibante, A. S. Carvalho, and A. T. de Almeida, 'A parameter optimized model of a Proton Exchange Membrane fuel cell including temperature effects', *J. Power Sources*, vol. 185, no. 2, pp. 952–960, 2008.

10. U. K. Chakraborty, T. E. Abbott, and S. K. Das, 'PEM fuel cell modeling using differential evolution', *Energy*, vol. 40, no. 1, pp. 387–399, 2012.

11. W. Gong and Z. Cai, 'Accelerating parameter identification of proton exchange membrane fuel cell model with ranking-based differential evolution', *Energy*, vol. 59, pp. 356–364, 2013.

12. Q. Li, W. Chen, Y. Wang, S. Liu, and J. Jia, 'Parameter Identification for PEM Fuel-Cell Mechanism Model Based on Effective Informed Adaptive Particle Swarm Optimization', *IEEE Trans. Ind. Electron.*, vol. 58, no. 6, pp. 2410–2419, 2011.

13. J. L. Tang, C. Z. Cai, T. T. Xiao, and S. J. Huang, 'Support vector regression model for direct methanol fuel cell', *Int. J. Morden Phys. C*, vol. 23, no. 7, pp. 1–8, 2012.

14. E. Durán, M. B. Ferrera, J. M. Andújar, and M. S. Mesa, 'I-V and P-V Curves Measuring System for PV Modules based on DC-DC Converters and Portable Graphical Environment', in *Industrial Electronics (ISIE), 2010 IEEE International Symposium on*, 2010, pp. 3323–3328.

15. S. K. Sahu and D. D. Neema, 'A robust speed sensorless vector control of multilevel inverter fed induction motor using particle swarm optimization', *Int. J. Innov. Res. Electr. Electron. Instrum. Control Eng.*, vol. 3, no. 1, pp. 23–32, 2015.

16. C. C. Kuo, 'A novel coding scheme for practical economic dispatch by modified particle swarm approach', *IEEE Trans. Power Syst.*, vol. 23, no. 4, pp. 1825–1835, 2008.

Survey of Different Load Balancing Approach-Based Algorithms in Cloud Computing: A Comprehensive Review

Arunima Hota, Subasish Mohapatra and Subhadarshini Mohanty

Abstract The Internet has become the basic necessity of day-to-day activity. It has a greater impact in modernizing the digital world. Consequently, cloud computing is one of the promising technical advancements in recent days. It is widely adopted by the different community for its abundant opportunities. It provides services and resources on ad hoc basis. Still, it has numerous issues related to resource provisioning, security, real-time data access, event content dissemination, server consolidation, virtual machine migration. These issues are to be addressed and resolved to provide a better quality of service in this computing paradigm. Load balancing is one of the vexing issues in the cloud platform. It ensures reliability and availability in this computing environment. It increases the efficiency of the system by equally distributing the workload among competing processes. The primary goal of load balancing is to minimize response time, cost, and maximize throughput. In the past decades, researchers have proposed different methodologies in order to resolve this issue. However, different load balancing parameters are yet to be optimized. This survey paper presents a comprehensive and comparative study of various load balancing algorithms. The study also portrays the merits and demerits of all the state-of-the-art-schemes which may prompt the researchers for further improvement in load balancing algorithms.

Keywords Cloud computing · Load balancing · Virtual machine
CloudSim

A. Hota · S. Mohapatra (✉) · S. Mohanty
Department of Computer Science and Engineering, College of Engineering
and Technology, Bhubaneswar 751003, India
e-mail: smohapatra@cet.edu.in

A. Hota
e-mail: arunimahota123@gmail.com

S. Mohanty
e-mail: sdmohantycse@cet.edu.in

© Springer Nature Singapore Pte Ltd. 2019
H. S. Behera et al. (eds.), *Computational Intelligence in Data Mining*,
Advances in Intelligent Systems and Computing 711,
https://doi.org/10.1007/978-981-10-8055-5_10

1 Introduction

The word 'cloud' acts as a metaphor for the Internet [1]. It provides services on requirement basis. The services are in the form of hardware, software, application, or database. Cloud computing provides three categories of services such as software as a service (SaaS), platform as a service (PaaS), infrastructure as a service (IaaS). In SaaS, the providers store their application in the cloud and make it available for common users on payment basis. Similarly, in PaaS, different computing platforms like operating system, Web server, database, programming languages are provided for fascinating the services. IaaS provider makes an entire computing infrastructure available as a service. It facilitates accessing virtualized servers, hardware, and network as an Internet-based service [2]. It enables a homogeneous virtualized environment where specific software can be installed and executed. The major components of cloud systems are Clients: who request for services, broker: who acts as an intermediate between client and cloud provider. It keeps all the cost-effective services after negotiating with different cloud vendors, and cloud provider: It is a firm or company that provides resources of varying capabilities and different services with different pricing models. Besides these advantages, [3] there are some challenges in cloud computing such as security and privacy, resource scheduling, efficient load balancing, performance monitoring, scale and quality of service (QoS) management. Recently, IT companies are focusing on minimizing the execution time as well as the cost associated with virtual machine migration and data transfer.

In the cloud system, several requests may raise from different geographical channels concurrently. The requests are assigned randomly to various cloud providers which ensure unequal assignment of load (request) at a given node. For which, some of the nodes are heavily loaded and some of them are under loaded. Such situation can lead to poor performance of the system. In order to overcome this situation, proper load balancing is essential. In this paper, a comparative analysis of different approaches in the cloud environment is presented.

2 Load Balancing

Load balancer is a device, which efficiently distribute the workloads across a group of backend servers. Modern Web sites serve millions of concurrent requests from users and return the correct information as in the form of text, images, and video, all in a fast and reliable manner. To meet the requirements, cloud computing generally need more servers for facilitating the services which is cost effective [4]. Working of load balancer is described in Fig. 1. Load balancing performs the following functions:

- Distribute user requests effectively across multiple servers
- It ensures high availability and reliability by sending requests only to the servers that are online

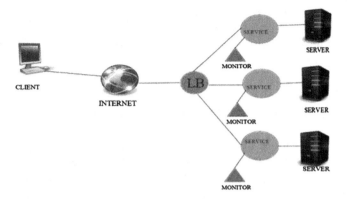

Fig. 1 Basic load balancing model

- On demand basis, it provides flexibility to add or subtract servers

Different types of load balancing approaches are present.

- Static: It is non-preemptive, i.e., once the load is allocated to the node, it cannot be transferred to another node. It focuses only on the information about the average behavior of the system, ignoring the current state of the system.
- Dynamic: It works on current state of node and distributes the load among nodes.

Dynamic approach is further divided into two types:

- Centralized: Here only a single node is responsible for managing and distribution of load within the system.
- Dynamic: It works on current state of node and distributes the load among nodes.

Different types of qualitative metrics that can enhance the performance of load balancer in cloud are as follows:

- Throughput: The tasks that have completed their execution within a stipulated time period. Tasks having high throughput perform better than the other system.
- Fault tolerance: The property of a system which continues its work even after the failure of some of its component. Every load balancing algorithm should possess good fault-tolerant approach.
- Migration time: It is the transferred time between the tasks that migrate from one machine to another in a system. It should be minimized for better performance.
- Response time: The amount of time taken by load balancing algorithm to response a task in a system. This parameter should be minimized for better performance of a system.
- Scalability: It works with a limited number of processor and machines. This parameter can be improved for better system performance.

- Resource Utilization: It gives the information within which the present resource is utilized. For efficient balancing the load, resource utilization should be maximized.

3 Survey of Different Load Balancing Algorithms

Cloud computing consists of distributed computing resources such as the processor, storage, and network. It deals with different fundamental technologies such as virtualization, interoperability, scalability. The objective of cloud computing is to make a better use of heterogeneous resources and improve the overall performance of the system. Load balancing algorithms can be categorized into three different types, namely heuristic, metaheuristic, and hybrid based on the type of algorithm used.

3.1 Heuristic Approach-Based Load Balancing Algorithms

Heuristic algorithms are of static or dynamic type. Heuristic is a group of constraints that intends to find the best possible solution for a particular problem. The constraints are highly problem dependent, and they are carefully designed to obtain a solution within a restricted time. In this paper, the different approach-based algorithms have various constraints; hence, they are designed to obtain a solution in a limited time. These types of algorithms are advantageous because they are capable of finding a satisfactory solution efficiently. Compared to metaheuristic algorithms, heuristic algorithms are easy to implement. Different heuristic-based load balancing algorithms with their merits and demerits are listed in Table 1.

3.2 Metaheuristic Approach-Based Load Balancing Algorithms

Metaheuristic algorithms are dynamic in nature. All evolutionary and swarm intelligence algorithms come under the metaheuristic algorithm. Metaheuristic algorithms differ from heuristic algorithms as it requires more time to run and settle upon the final solution. In most of the cases, the solution space might be considerably large in comparison to heuristic algorithms, metaheuristic algorithms need more time to run and find the final solution as its solution space can be quite large. Besides, metaheuristics are in generally random processes. Their convergent time and solution are highly dependent on the nature of the problem, initial configurations, and the way to search the solution. Different metaheuristic-based load balancing algorithms with their merits and demerits are listed in Table 2.

Table 1 Survey of different heuristic-based load balancing algorithms

Algorithm	References	Environment	Optimization measures	Benefits	Limitations
Improved Max-Min Algorithm	[5]	Cloud (CloudSim)	Overall completion time, makespan	It performs better than previous max-min algorithm	Comparisons are only made with RASA and max-min algorithm
Round Robin Algorithm	[6]	Cloud (Cloud Analyst)	Time slices	It works well with no of processes	Based on processing overhead, it is not enough for load balancing
Balance Reduce Algorithm	[7]	Cloud (Java)	Makespan	Works well under various network state	It is hard to obtain a near optimal solution when the cost changes frequently
Threshold Algorithm	[8]	Cloud	Reduce the interposes communication time	It improves the performance of the system	When all remote processors are over loaded, it causes disturbance in load balancing
FCFS Algorithm	[9]	Cloud (CloudSim)	Computing time	Perform better than RR	Difficult to reach an ideal effect of load balancing
Round Robin VM Load Balancing Algorithm	[10]	Cloud (Cloud Analyst)	Deadlock and server overflow	Take less response time and data center processing time over RR method	Special priority is not given for important task

Table 2 Survey of different metaheuristic-based load balancing algorithm

Algorithm	References	Environment	Optimization measures	Benefits	Limitations
Artificial Bee Colony Algorithm	[11]	Cloud	Makespan	Improves the max throughput	Lack of using secondary information, slow down when used in sequential process and the solution increases the computational cost
Honey Bee Algorithm	[12]	Cloud	Minimize execution time, waiting time, response time	Maximized the throughput	It does not consider the quality of service (QoS) factor when executing the tasks
	[13]	Cloud	Minimize the processing cost, time, execution capacity	It determines the resources according to their priorities	It adds more parameters for determining load
Min-min Scheduling Algorithm	[14]	Cloud	Minimize the time of completion, processing speed, task size, makespan	It provides better utilization ratio	Here limitations are not set properly. It does not consider the quality of service parameters when performing the task
Firefly Algorithm (FA)	[15]	Cloud	Minimizes the response time, CPU utilization rate, processing time	It provides better search space	Here results are dependent on distance as well as radiance factors
	[16]	Cloud	Reduce the computation time, cost, bandwidth	It provides better migration of tasks	Instability may arise due to complex nature

(continued)

Table 2 (continued)

Algorithm	References	Environment	Optimization measures	Benefits	Limitations
Particle Swarm Optimization Algorithm (PSO)	[17]	Cloud	Minimizes the fitness value, makespan	It provides better utilization ratio.	It is valid for equal-sized population
	[18]	Cloud (Cloudsim)	Execution cost with deadline constraint	For IaaS clouds	PSO performances are problem dependent. The swarms may prematurely converged
	[19]	Cloud (CloudSim)	Throughput, response time	PSO is used in online scheduling scenario	PSO performances are problem dependent. The swarms may prematurely converged
	[20]	Cloud (CloudSim, JSwarm)	Makespan	Load is transferred from an overloaded virtual machine to another physical machine	PSO performances are problem dependent. The swarms may prematurely converged
	[21]	Cloud (CloudSim)	Energy conservation	Considers Euclidean distance to calculate fitness function	PSO performances are problem dependent. The swarms may prematurely converged
ACO Algorithm	[22]	Cloud environment (CloudSim toolkit)	Makespan	Use of positive feedback mechanism	Overhead, searching for to a certain extent
	[23]	Cloud (CloudSim)	Load balancing, SLA violation, energy consumption	Efficient distribution of load among nodes, maximize the utilization of resource	Slower convergence than other heuristics
	[24]	Cloud (CloudSim)	Reduce the physical server	Effectively utilizes the	Slower convergence

(continued)

Table 2 (continued)

Algorithm	References	Environment	Optimization measures	Benefits	Limitations
			significantly, when the number of VM is large	resource when VM size is large	than other heuristics
	[25]	Cloud (Cloud Analyst)	Energy conservation, SLA constraints of throughput, response time	Flexible, it removes the cloned VMs from the nodes if they are underutilized or if the performance is more than required. It also removes the VMs from the nodes whose service agreement get expire	Slower convergence than other heuristics
	[26]	Cloud	Reduce the number of physical servers	It improves resource utilization	Slower convergence than other heuristics
GA Algorithm	[27]	Cloud (Cloud Analyst)	Reduce cost and response time	It provides better load balancing solution	Computational complexity is more
	[28]	Cloud (Cloud Analyst)	Makespan	It provides efficient utilization of resource in cloud	It does not provide same priority job
	[29]	Cloud (MATLAB)	Makespan	Enhances the overall performance of cloud computing	It provides poor efficiency, when search space is enlarged
	[30]	Cloud	Resource, energy consumption	Can be used in green data centers	Computational complexity is more due to different stages
	[31]	Cloud (CloudSim)	Minimizes the completion time, costs of task	Maximizes the resource utilization and Improve the performance system	Computational complexity is more due to different stages

(continued)

Table 2 (continued)

Algorithm	References	Environment	Optimization measures	Benefits	Limitations
League Championship Algorithm	[32]	Cloud (MATLAB)	Minimize makespan	It gives better result than first come first serve (FCFS), last job first (LCF), and best Effort First (BEF)	Self-adaptive strategy may be applied for controlling and tuning the parameter in LCA
BAT Algorithm	[33]	Cloud	Minimize makespan	It has high accuracy and efficiency than GA	Convergence very quickly at early stage, there is no mathematical analysis to link the parameters with convergence rates
	[34]	Cloud	Minimize processing cost	It performs better in terms of processing cost	Convergence very quickly at early stage, there is no mathematical analysis to link the parameters with convergence rates

3.3 Hybrid Approach-Based Load Balancing Algorithms

Hybrid algorithms are combination of different heuristic or metaheuristic algorithms, which reduces the computation time as well as the cost. Also, it provides an effective result than the other algorithms. Different hybrid-based load balancing algorithms with their merits and demerits are listed in Table 3.

Table 3 Survey of different hybrid-based load balancing algorithm

Algorithm	References	Environment	Optimization measures	Benefits	Limitations
GA algorithm hybridized with Tabu search	[35]	Cloud	Minimizes cost, size of data	It provides better computation for load balancing	This approach is limited to simple tasks
GA algorithm hybridized with multi-agent techniques	[36]	Cloud	Maximizes the resource utilization	It enhances convergence time	Computational complexity is high
GA algorithm hybridized with PSO	[37]	Cloud (Cloud Analyst)	Minimize execution time, communication cost, and data transfer cost	It transfers the task from heavy-loaded VM to lightly loaded VM	Better results in comparison to individual GA and PSO
PSO hybridized with cuckoo search	[38]	Cloud (CloudSim)	Minimize makespan	Perform better than PSO	Explore the algorithm with larger scale in cloud computing

4 Conclusion

Cloud computing system has widely been adopted by the industry and academic. However, there are many issues which exist in this environment like load balancing, migration of virtual machines, server constellation which has not been yet fully addressed. Beside this, load balancing is the most promising issue in the system. Various techniques and algorithms have been used to solve this problem for providing better quality of service and efficient utilization of the resources with reduced cost and complexity. In this paper, various load balancing algorithms have been presented with their advantage and disadvantage. The survey paper provides abundant scope for researchers to develop efficient load balancing algorithms for cloud environment.

References

1. Mell, Peter, and Grance, Tim.: "The NIST definition of cloud computing." (2011)
2. Alkhanak, Ehab Nabiel, et al.: "Cost optimization approaches for scientific workflow scheduling in cloud and grid computing: A review, classifications, and open issues." Journal of Systems and Software. 113 (2016) 1–26

3. Kaur, Rajwinder, and Luthra, Pawan.: "Load balancing in cloud computing." Second Symposium on Cloud computing. (2012)
4. Farrag, Aya, Salah, A., Mahmoud, Safia Abbas., and Sayed, M. El.: "Intelligent cloud algorithms for load balancing problems: A survey." Intelligent Computing and Information Systems (ICICIS), 2015 IEEE Seventh International Conference on. IEEE (2015)
5. Elzeki, O. M., Reshad, M. Z., and Elsoud, M. A.: "Improved max-min algorithm in cloud computing." International Journal of Computer Applications 50(12) (2012)
6. Samal, Pooja, and Mishra, Pranati.: "Analysis of variants in Round Robin Algorithms for load balancing in Cloud Computing." International Journal of computer science and Information Technologies 4(3) (2013) 416–419
7. Jin, Jiahui., Luo, Junzhou., Song, Aibo., Dong, Fang., Xiong, Runqun.: "BAR: An Efficient Data Locality Driven Task Scheduling Algorithm for Cloud Computing". IEEE (2011)
8. Sharma, Sandeep., Singh, Sarabjit., and Sharma, Meenakshi.: "Performance analysis of load balancing algorithms." World Academy of Science, Engineering and Technology. 38(3) (2008) 269–272
9. Agarwal, Dr, and Jain, Saloni.: "Efficient optimal algorithm of task scheduling in cloud computing environment." arXiv preprint arXiv:1404.2076 (2014)
10. Mahajan, Komal., Makroo, Ansuyia., and Dahiya, Deepak.: "Round robin with server affinity: a VM load balancing algorithm for cloud based infrastructure." Journal of information processing systems. 9.3(2013) 379–394
11. Karaboga, Dervis, and Bahriye Akay.: "A comparative study of artificial bee colony algorithm." Applied mathematics and computation . 214.1(2009) 108–132
12. Gupta, Harshit., Sahu, Kalicharan.: "Honey Bee Behaviour Based Load Balancing of Tasks in Cloud Computing," International journal of Science and Research. Vol. 3 Issue 6 June (2014)
13. Soni, Ashish., Vishwakarma, Gagan., Jain, Kumar, Yogendra.: "A Bee Colony based Multi-Objective Load Balancing Technique for Cloud Computing Environment," International Journal of Computer applications. Vol. 114, No.4, March (2015) 0975–8887
14. Chen, Huankai., Wang, Frank., Helian, Na., Akanmu, Gbola.: "User-Priority Guided Min-Min Scheduling Algorithm For Load Balancing in Cloud Computing"
15. Florence, Paulin A., Shanthi, V.: "A Load Balancing Model Using Firefly Algorithm In Cloud Computing," Journal Of Computer Science, 10 (7) (2014) 1156–1165
16. Susila, N., Chandramathi, S., Kishore, Rohit.:" A Fuzzy-based Firefly Algorithm for Dynamic Load Balancing in Cloud Computing Environment, "Journal Of Emerging Technologies In Web Intelligence, Vol. 6, No. 4, November (2014)
17. Kai Pan and Jiaqi Chen.: "Load Balancing In Cloud Computing Environment Based on An Improved Particle Swarm Optimization," 6th IEEE International Conference on Software Engineering and Service Science. IEEE (2015) 595–598
18. Rodriguez Sossa M, Buyya R.: Deadline based resource provisioning and scheduling algorithm for scientific workflows on clouds. IEEE Trans Cloud Comput vol: 2 (2014) 222–35
19. Beegom ASA, Rajasree MS.: A particle swarm optimization based pareto optimal task scheduling in cloud computing. In: Adv swarm intell notes comput sci. Springer (2014) 79–86
20. Ramezani F, Lu J, Hussain FK.: Task-based system loadbalancing in cloud computing using particle swarm optimization. Int J Parallel Program 42 (2014) 739–54
21. Xiong A, Xu C.: Energy efficient multiresource allocation of virtual machine based on PSO in cloud data center. Math Probl Eng (2014)
22. Dam, Santanu., Mandal, Gopa., Dasgupta, Kousik., Dutta, Paramartha.: "An Ant Colony Based Load Balancing Strategy in Cloud Computing," Advanced Computing, Networking and Informatics – Volume 2, Smart Innovation, Systems and Technologies 28, Springer International Publishing Switzerland (2014)
23. Pacini E, Mateos C, García C.: Balancing throughput and response time in online scientific clouds via ant colony optimization. Adv Eng Software [Elsevier] 84 (2015) 31–47

24. Khan S, Sharma N.: Effective scheduling algorithm for load balancing (SALB) using ant colony optimization in cloudcomputing. Int J Adv Res Computer Science Software Eng vol: 4 (2014) 966–73
25. Dam S, Mandal G, Dasgupta K, Dutta P.: An ant colony based load balancing strategy in cloud computing. Adv Comput Network Informatics vol: 2 (2014) 403–13
26. Liu X, Zhan Z, Du K, Chen W.: Energy aware virtual machine placement scheduling in cloud computing based on ant colony optimization. In: Proc conf genet evol comput. ACM (2014) 41–7
27. Joshi Garima, Verma S.K.: "Load Balancing Approach in Cloud Computing using Improvised Genetic Algorithm: A Soft Computing Approach," International Journal of Computer Applications Vol. 122 No. 9 July (2015) 0975–8887
28. Dasgupta K, Mandal B, Dutta P, Mandal JK, Dam S.: A Genetic Algorithm (GA) based load balancing strategy for cloud computing. Proc Technol vol: 10 (2013) 340–7
29. Wang T, Liu Z, Chen Y, Xu Y, Dai X.: Load balancing task scheduling based on genetic algorithm in cloud computing. In: IEEE 12th int conf dependable auton secur comput; (2014) 146–52
30. Joseph CT, Chandrasekaran K, Cyriac R.: A novel family genetic approach for virtual machine allocation. Proc Comput Sci 46 (2015) 558–65
31. Hamad, Safwat A., and Fatma A. Omara.: "Genetic-Based Task Scheduling Algorithm in Cloud Computing Environment." International Journal of Advanced Computer Science & Applications 1.7 (2016) 550–556
32. Sun J, Wang X, Li K, Wu C, Huang M, Wang X.: An auction and League Championship Algorithm based resource allocation mechanism for distributed cloud. Lect Notes Comput Sci (Including Subser Lect Notes Artif Intell Lect Notes Bioinformatics) vol: 8299 (2013) 334–46
33. Jacob L.: Bat algorithm for resource scheduling in cloudcomputing. Int J Res Appl Sci Eng Technol vol: 2 (2013) 53–7
34. Raghavan S, Marimuthu, C, Sarwesh, P, & Chandrasekaran K.: Bat algorithm for scheduling workflow applications in cloud. Int Conf Electron Des Comput Networks Autom Verif (EDCAV). IEEE (2015) 139–44
35. Adamuthe A.C., Thampi G.T., Bagane P.A.: "Genetic Algorithms and Tabu Search for Solving Workflow Scheduling Application in Cloud," ICCN, Elsevier; (2013) 216–223
36. Zhu, Kai., Song, Huaguang., Liu, Lijing., Gao, Jinzhu., Cheng, Guojian.: "Hybrid Genetic Algorithm for Cloud Computing Applications", IEEE, (2012)
37. Richhariya, Vineet, Dubey, Ratnesh., and Siddiqui, Rozina.: "Hybrid Approach for Load Balancing in Cloud Computing." (2015)
38. Al–maamari, A., and Omara, F.A.: "Task Scheduling using Hybrid Algorithm in Cloud Computing Environments," IOSR Journal of Computer Engineering, vol. 17 no. 3 (2015) 96–106

Analysis of Credit Card Fraud Detection Using Fusion Classifiers

Priyanka Kumari and Smita Prava Mishra

Abstract Credit card fraud detection is a critical problem that has been faced by online vendors at the finance marketplace every now and then. The rapid and fast growth of the modern technologies causes the fraud and heavy financial losses for many financial sectors. Different data mining and soft computing-based classification algorithms have been used by most of the researchers, and it plays an essential role in fraud detection. In this paper, we have analyzed some ensemble classifiers such as Bagging, Random Forest, Classification via Regression, Voting and compared them with some effective single classifiers like K-NN, Naïve Bayes, SVM, RBF Classifier, MLP, Decision Tree. The evaluation of these algorithms is carried out through three different datasets and treated with SMOTE, to deal with the class imbalance problem. The comparison is based on some evaluation metrics like accuracy, precision, true positive rate or recall, and false positive rate.

Keywords Credit card fraud · Fraud detection · SMOTE · Cross-validation
MLP · Bagging · Voting · Random forest · Classification via regression

1 Introduction

In today's environment, the e-commerce platform has become most enormous place for financial organizations; they choose e-commerce and Internet services to enhance their productivity and to minimize consumption of time and effort. The online marketplace is very dynamic and massive. The fraudsters always observe the

P. Kumari (✉)
Computer Science and Engineering, ITER, SOA University,
Bhubaneswar 751030, Odisha, India
e-mail: kumari.priyanka522@gmail.com

S. P. Mishra
Computer Science and Information Technology, ITER, SOA University,
Bhubaneswar 751030, Odisha, India
e-mail: smitaprava@yahoo.com

© Springer Nature Singapore Pte Ltd. 2019
H. S. Behera et al. (eds.), *Computational Intelligence in Data Mining*,
Advances in Intelligent Systems and Computing 711,
https://doi.org/10.1007/978-981-10-8055-5_11

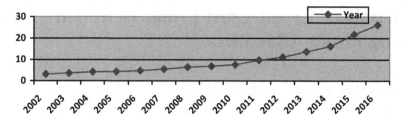

Fig. 1 Annual global fraud losses (credit card and debit card)

customer's behavior and dynamically adapt to it. As a result, identification of fraudulent cases becomes more challenging. The e-commerce system is used by both the authorized as well as unauthorized persons. There is always a risk of credit card fraud while using any card for online or offline transactions. However, wrong detection leads to dissatisfied users and no detection incurs heavy financial losses.

The credit card fraud is an unfaithful act which is performed by the fraudsters. Online transaction using a credit card may be either online purchase or transfer of money. The card-based transactions can be classified into two types: virtual card and physical card [1]. In the past few decades, credit card has become a rigorous target of the fraudsters. Credit card fraud can be done either online or offline. Online fraud is done by using sensitive information regarding card, and offline fraud is accomplished by using stolen physical cards [2]. Many financial sectors lose millions of dollars every year because of credit card fraud. Figure 1 shows the rate of annual global losses of previous fourteen years due to the credit or debit card fraud according to Nilson Report, October 2016.
The main objectives of the work may be summarized as follows:

1. To propose a fusion model by using ensemble classifiers based on different data mining techniques that are used for fraud detection.
2. To apply and analyze the effect of SMOTE technique on the datasets for dealing with imbalance nature of the datasets.
3. To focus on the new directions toward increase in performance and reliability of credit card fraud detection methods.

This paper is ordered as follows: Sect. 2 reviews previous fraud detection models, Sect. 3 discusses methodologies for developing data mining-based fraud detection models, Sect. 4 comprises of implementation and results discussion, and Sect. 5 concludes the observations and highlights future directions.

2 Literature Review

In the credit card fraud detection domain, many review works have been performed which report the various types of frauds and fraud detection methodologies [3–5]. C. Carter and J. Catlett [6] investigated machine learning approach such as ID3 of decision trees and probability trees for credit assessment to assess the costs of a good or bad credit risk. Hanagandi et al. [7] presented a fraud score and they used radial basis function network with density-based clustering and historical information on transactions for computing credit score. Ghosh and Reilly [8] proposed a fraud detection model based on feed-forward neural network using account transactions dataset of a particular customer. They found that more fraud detected through rule-based detection system and false positive rate is also decreased with this system. Dorronsoro et al. [9] used neural network and a nonlinear Fisher's discriminate to propose an online system for credit card fraud detection. K. R. Seeja et al. [10] proposed a novel credit card fraud detection model using frequent item set mining. S. Suganya et al. [11] developed a Firefly metaclassification method for credit card fraud detection. K. Hassibi [12] used feed-forward artificial neural network (ANN) and trained this network on different back-propagation training algorithms for detection of fraudulent transactions. Maes et al. [13] have implemented Bayesian belief networks and ANN for credit card fraud detection and they found that performance of Bayesian network was better than ANN for fraud detection. Some researchers used clustering technique, self-organizing map neural network for detection of credit card fraud based on customer behavior [14, 15]. Srivastava et al. [16] developed a model for credit card fraud detection using hidden Markov model. Moreover, Sanchez et al. [17] applied fuzzy association rule mining in extracting knowledge from transactional credit card database for fraud detection. Lu and Ju [18] designed a class-weighted support vector machine for classifying the imbalanced transaction data. Wong et al. [19] searched an application of credit card fraud detection based on artificial immune system.

Concisely, various machine learning, data mining, soft computing, adaptive pattern recognition, ANNs, fuzzy systems, and statistical modeling were used to develop several fraud detection models to provide a decision with some degree of certainty about whether a particular transaction is fraudulent and to minimize the detection of false transactions.

3 Methodologies

We explored the performance of some effective techniques for credit card fraud detection. In this section, we briefly described the classifiers involved in this evaluation.

Neural Network: We have taken multilayer perceptron (MLP) and radial basis function (RBF) classifier as neural networks in our evaluation. MLP is a feed-forward neural network with one or more hidden layers between input and output layer. In our analysis, for performance evaluation of MLP, the momentum value is set to 0.2 and learning rate is set to 0.3. Regarding the MLP, a complete investigation and testing has been reported by Hassoun [20] and by Żak [21].

RBF Classifier is an artificial neural network. It uses radial basis function as activation function, and the output of RBF network is radial basis function as a linear combination. RBF neural networks have one input layer, one hidden layer, and one output layer. The hidden layer calculates the norm of the input from the neuron. It passes the norm through a nonlinear activation function for obtaining the output.

Decision Tree: A Decision Tree is a tree structure which attempts to separate the given records into mutually exclusive subgroups. To perform this, it starts from the root node; each node is split into child nodes in a binary or a multisplit fashion. Each Decision Tree method uses its own splitting algorithms and splitting metrics. The correctly classified object by a Decision Tree is known as its accuracy. From the well-known Decision Tree algorithms, we include Simple CART for performance evaluation in this work.

Support Vector Machine: Support vector machine (SVM) was proposed by Vapnik and his group at AT&T Bell Laboratories [22]; it is a promising binary classification technique. SVM tries to separate the two classes by creating a hyper plane in between them with maximal marginal value for better performance results while minimizing the classification error [23]. SVM has been successfully applied to many areas such as telecommunication fraud detection [24], pattern recognition [23], and system intrusion detection [25]. In this paper, we used SVM by using a radial basis kernel function.

Naïve Bayes Classifier: The Naïve Bayes classifier is a probabilistic method based on two assumptions. Firstly, all attributes in an entry that require to be classified are contributing in the decision. Secondly, all attributes should be statistically independent, which means that value of one attribute does not give any information about the value of other attributes. The Bayes rule is used to classify the class of new instance. The new instance belongs to that class which has the higher posterior probability, in the fraud detection task.

K-Nearest Neighbor: K-nearest neighbor (K-NN) is a simple instance-based learning algorithm that compares each new incoming instance with the existing instances by calculating the distance from nearest neighbor instances, and the class of minimum distance instances is assigned to the new instance. The euclidian distance formula used for calculating the distance between new and existing instance in this work is as presented in Eq. 3.1.

$$\text{dists}(X_1, Y_1) = \text{sqrt}\left((X_2 - X_1)^2 + (Y_2 - Y_1)^2\right) \tag{3.1}$$

Ensemble Classifier: We used an ensemble or fusion of classifiers because that can be more accurate and reliable than single classifiers. Recently, in the area of credit card fraud detection, the concept of combining classifiers is found to be a new direction for improvement in the performance of individual classifiers in terms of precise and accurate results.

Fig. 2 Workflow model for credit card fraud detection

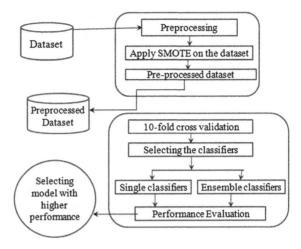

Ensemble is the concept of combining a series of models with the aim of creating an improved model. We used some of the popular ensemble methods such as Bagging, Voting, Classification via Regression, and Random Forest to analyze their performances for fraud detection. The accuracy is a critical factor of any proposed model for a proper classification of fraudulent or legal transactions [26]. Many researchers investigated that an ensemble of classifiers gives better results than any single classifier. Random Forests are very efficient for the classification and regression problems [27]. A Random Forest is a collection of decision trees. The reputation of Random Forest is due to its high performance compared with the other algorithms [28] (Fig. 2).

4 Implementation and Result Discussion

4.1 Datasets and Preprocessing

Datasets: In order to evaluate the performances of classifiers, three datasets have been taken. The first two dataset collected from UCI repository that is German credit dataset (*jg.arff*) and Australian credit approval dataset (http://tunedit.org/repo/UCI/australian.arff), and the third dataset is bank-data.csv (http://read.pudn.com/downloads151/sourcecode/asm/658135/bank-data.csv.arff__.htm). In the German credit dataset, each instance contains 7 numerical, 13 categorical attributes, and 1 class attribute (good or bad). In the Australian credit dataset, each instance contains 8 numeric, 6 nominal, and 1 class attribute (accepted or rejected). In the bank-data. csv, dataset contains 1 class attribute (yes or no), 3 numeric, and 9 nominal attributes. Table 1 represents the description of datasets in detail before applying SMOTE on it.

Table 1 Description of dataset before the SMOTE

Dataset	Instance	Attribute	Class1		Class2	
jg.arff	1300	21	Good	700	Bad	600
australian.arff	690	15	Accept	307	Reject	383
bank-data.csv.arff	1422	12	Yes	1096	No	326

Table 2 Description of dataset after the SMOTE

Dataset	Instance	Attribute	Class1		Class2	
jg.arff	1900	21	Good	700	Bad	1200
australian.arff	997	15	Accept	614	Reject	383
bank-data.csv.arff	1748	12	Yes	1096	No	652

Data Preprocessing: The datasets were preprocessed for the purpose of improving the performance. The all three data were preprocessed by using a supervised technique that is synthetic minority oversampling technique (SMOTE). SMOTE has been used for the purpose of handling imbalance nature of the datasets that are used for credit card fraud detection and increasing the accuracy of the classifiers. In this paper, the evaluation has been performed on the dataset with SMOTE and without using SMOTE and then a comparison chart of performance of all the classifiers is given. Table 2 presents the datasets after being treated with SMOTE.

4.2 Experimental Setup

WEKA Tool: In the presented paper, the experimentation has been done by using *WEKA* tool. *WEKA* is the abbreviation of Waikato Enviroment of Knowledge Analysis. It is an open-source, handy, and reliable tool for machine learning techniques [29]. It is a standardized tool for the existing methodologies. If we write the code for algorithm, then validity of the code as well as the outcomes are questionable. Hence, a stabilized tool is preferred for acceptance, reliability, and ease of use. It is used for several applications such as classification, clustering, feature selection, regression and association, and standard data mining problems.

Ten-Fold Cross-Validation: The validations are performed with *ten-fold cross-validation* test model. In this test model, the dataset is divided into 10 sets; one set is used as test samples and other nine as training samples. Test samples are classified after each training sample, and same procedure is repeated with all other samples. The accuracies and computational time for each classifier are observed.

4.3 Performance Metrics

The performance of fraud detection systems has been evaluated by using number of performance metrics which is suitable for credit card fraud detection. Here, four widely used evaluation metrics taken which helps to evaluate the performance of the proposed model are described as follows:

(a) **Accuracy**: It is the overall performance of the classifiers. It shows that how many of the total instances have been classified correctly by the chosen classifier [30].

$$Acurracy = \frac{TP + TN}{TP + FP + TN + FN} \tag{4.3.1}$$

(b) **Precision**: The precision metrics shows the reliability of the output of a classifier. It is also called confidence of a classifier.

$$Precision = \frac{TP}{TP + FP} \tag{4.3.2}$$

(c) **Recall or True Positive Rate (TPR)**: It finds the efficiency of the classifier in detecting the actual fraudulent data. In the fraud detection, it is known as fraud catching rate. It is also called as sensitivity. It shows the part of positive, which is classified as positive. The fraud catching rate for the proposed model should be high for a good detection system.

$$Recall = \frac{TP}{TP + FN} = \frac{TP}{P} \tag{4.3.3}$$

(d) **False Positive Rate (FPR)**: It denotes the part of fraud transaction which is classified as positive. In the fraud detection, it is known as false alarm rate, which should be lower for better performance of any model. It classifies fraud class as positive.

$$FPR = \frac{FP}{N} \tag{4.3.5}$$

However, the accuracy rate is more realistic evaluation metrics for fraud detection system. The other metrics are equally valuable for the fraud detection system. The FPR should be minimized to achieve better performance, and TPR should be maximized as it is defined as fraud catching rate. The ten-fold cross-validation is performed for each experiment.

Table 3 Performance evaluation of classifiers using German data

Algorithm		Experiment 1: Using German dataset with SMOTE				Experiment 1: Using German dataset without SMOTE			
		Acc (%)	Precision	Recall or TPR	FPR	Acc (%)	Precision	Recall or TPR	FPR
K-NN		93.15	0.932	0.932	0.091	90.30	0.903	0.903	0.904
NB		**84.31**	0.843	0.843	0.181	79.69	0.797	0.797	0.206
MLP		93.73	0.937	0.937	0.072	90.07	0.901	0.901	0.101
RBF		86.15	0.861	0.862	0.168	80.69	0.807	0.807	0.199
SVM		83.26	0.833	0.833	0.232	79.46	0.796	0.795	0.205
DT	Simple CART	89.47	0.895	0.895	0.122	85.38	0.854	0.854	0.149
	RF	94.94	0.950	0.949	0.054	**93.93**	0.929	0.929	0.073
Classification via Regression		**95.21**	0.952	0.952	0.050	92.76	0.848	0.848	0.157
Bagging	RF	94.78	0.948	0.948	0.057	92.15	0.955	0.955	0.082
	J48	92.31	0.923	0.923	0.086	87.30	0.873	0.873	0.132
	RT	93.78	0.938	0.938	0.067	90.69	0.908	0.907	0.099
Voting	Majority voting	93.05	0.930	0.931	0.086	90.84	0.908	0.908	0.094
	Average probability	90.46	0.905	0.905	0.097	90.61	0.906	0.906	0.095

4.4 Results and Discussion

In this paper, we performed three experiments on the three different datasets. All the experiments show the comparison between the performance of single classifiers such as K-NN, NB, MLP, RBF, SVM, Simple CART with the ensemble classifiers like Random Forest, Bagging, Voting, Classification via Regression and evaluated with applying SMOTE and without applying SMOTE on the datasets. The ten-fold cross-validation model has been used for validating all the classifiers in all three experiments.

Table 3 represents the results of Experiment 1, which shows that, when we applied SMOTE on German data, the ensemble method Classification via Regression outperforms all the classifiers in terms of accuracy, precision, recall or TPR, FPR. It gives highest 95.21% accuracy rate. The ensemble method Random Forest and Bagging with Random Forest also work well with the accuracy rates as 94.94%, and 94.78%, respectively. The Voting algorithm achieves 93.05% accuracy rate. Without applying SMOTE on this data, the best result can be obtained by Random Forest that is 93.93% accuracy rate. Here, the accuracy rate increased by 1.28% when SMOTE is applied on German dataset.

The results of Experiment 2 are described in Table 4. It shows that, when SMOTE is applied on Australian dataset, then the same Classification via Regression gives highest accuracy rate of 91.17% and also outperforms all the

Table 4 Performance evaluation of classifiers using Australian data

Algorithm		Experiment 2: Using Australian dataset with SMOTE				Experiment 2: Using Australian dataset without SMOTE			
		Acc (%)	Precision	Recall or TPR	FPR	Acc (%)	Precision	Recall or TPR	FPR
K-NN		87.46	0.874	0.875	0.142	82.02	0.820	0.820	0.187
NB		77.93	0.822	0.799	0.170	77.53	0.792	0.775	0.261
MLP		88.76	0.887	0.888	0.130	83.76	0.838	0.838	0.164
RBF		90.67	0.907	0.907	0.119	86.23	0.865	0.862	0.134
SVM		89.96	0.902	0.900	0.138	85.50	0.866	0.865	0.131
DT	Simple CART	89.86	0.900	0.899	0.133	84.92	0.859	0.849	0.141
	RF	90.87	0.908	0.909	0.111	86.23	0.862	0.862	0.143
Classification via Regression		**91.17**	0.911	0.912	0.106	85.50	0.856	0.855	0.145
Bagging	RF	90.97	0.910	0.910	0.111	86.37	0.864	0.864	0.139
	J48	90.67	0.907	0.907	0.118	86.37	0.864	0.864	0.138
	RT	90.37	0.904	0.904	0.122	84.92	0.850	0.849	0.152
Voting	Majority voting	90.47	0.905	0.905	0.119	87.24	0.873	0.872	0.129
	Average probability	90.47	0.905	0.905	0.119	**87.39**	0.874	0.874	0.128

classifiers in terms of accuracy, FPR, precision, and recall or TPR. The K-NN, Naïve Bayes, MLP, Simple CART performed poorly on this dataset. Without using SMOTE, the best result can be obtained by Average of Probability of Voting ensemble classifier, the accuracy rate is 87.39%. The accuracy rate increases by 3.78% when SMOTE is applied on Australian dataset.

The results of Experiment 3 are shown in Table 5 which shows that while SMOTE is applied on bank-data, the best result is achieved by the Naïve Bayes algorithm with the accuracy rate of 90.16%. The Naïve Bayes classifier performs well in terms of all metrics such as accuracy, FPR, recall or TPR, and precision. This dataset contains maximum categorical attribute values, as a result of which the accuracy rate decreases, and some classifiers do not perform well with the categorical data. It shows that SMOTE does not perform well with this type of dataset as well. The highest accuracy rate of 86.23% is obtained by Majority Voting, without applying SMOTE on Bank dataset. Here, the difference between accuracy rate of classifier with SMOTE and without SMOTE is 3.93%.

Table 5 Performance evaluation of classifiers using bank-data

Algorithm		Experiment 3: Using Bank dataset with SMOTE				Experiment 3: Using Bank dataset without SMOTE			
		Acc (%)	Precision	Recall or TPR	FPR	Acc (%)	Precision	Recall or TPR	FPR
K-NN		87.29	0.872	0.873	0.161	85.61	0.843	0.852	0.372
NB		**90.61**	0.906	0.906	0.117	84.95	0.841	0.850	0.366
MLP		77.46	0.771	0.775	0.129	78.41	0.812	0.784	0.722
RBF		84.03	0.849	0.840	0.238	77.04	0.594	0.770	0.771
SVM		73.56	0.752	0.736	0.406	77.07	0.594	0.771	0.771
DT	Simple CART	85.92	0.885	0.859	0.237	77.07	0.594	0.771	0.771
	RF	89.07	0.903	0.891	0.178	77.07	0.594	0.771	0.771
Classification via Regression		85.92	0.885	0.859	0.237	77.07	0.594	0.771	0.771
Bagging	RF	88.95	0.902	0.890	0.179	77.07	0.594	0.771	0.771
	J48	85.92	0.885	0.859	0.237	77.07	0.594	0.771	0.771
	RT	89.53	0.907	0.895	0.170	78.41	0.812	0.784	0.722
Voting	Majority voting	79.17	0.793	0.792	0.290	**86.23**	0.863	0.862	0.138
	Average probability	79.23	0.794	0.792	0.293	86.23	0.863	0.862	0.138

5 Conclusion

From the above experimentation and results, we conclude that there is no single classifier in data mining that can perform better than the ensemble classifiers. The Classification via Regression ensemble classification technique performs well on both German data with accuracy of 95.21% and Australian data with accuracy of 91.17%. This method performs well with large dataset with less categorical attribute. But it gives low accuracy while using bank-data because this dataset contains more categorical attributes which cause lower accuracy. So, this method can not perform well with categorical dataset. The idea of applying SMOTE technique on datasets gives reason to the better performance and better handling of class imbalance problem. We found that by applying SMOTE on all datasets, the accuracy is enhanced by 2–4%. The accuracy of Classification via Regression using Random Forest outperforms all the single classifiers on both the datasets when treated with SMOTE. Random Forest is the ensemble approach of the Decision Tree and also performs well with all the datasets. The accuracy of classifiers decreases when using bank-data due to the presence of categorical attributes in the dataset.

The accuracy of classifiers could be enhanced by developing a fraud detection model on some selected attributes of the dataset and by using the datasets which have less categorical attributes. It would decrease the computational time or time taken to build a model. In future, more analysis could be done using other combination of classifiers. We can take other combination of ensemble classifiers with different datasets of differents attributes.

Acknowledgements I would like to express my heartfelt thanks to Prof. Dr. Debahuti Mishra, Head of the Dept (Computer Science and Engineering) at ITER, for her encouragements and support throughout the work.

References

1. Ravindra P. S., Vijayalaxmi K., "Survey on Credit Card Fraud Detection Techniques", International Journal of Engineering and Computer Science, Vol. 4, No. 11, pp. 15010–15015, 2015.
2. Linda D., Hussein A., Pointon J., "Credit card fraud and detection techniques: a Review", Banks and Bank Systems, Vol. 4, No. 2, 2009.
3. Bolton R. J., Hand D. J., "Statistical fraud detection: a review", Statistical Science, Vol. 17, No. 3, pp. 235–255, 2002.
4. Bhattacharyya S., Jha S., Tharakunnel K., Westland J. C., "Data mining for credit card fraud: a comparative study", Decision Support Systems, Vol. 50, No. 3, pp. 602–613, 2011.
5. Ngai E. W. T., Hu Y., Wong Y. H., Chen Y., Sun X., "The application of data mining techniques in financial fraud detection: a classification framework and an academic review of literature", Decision Support Systems, Vol. 50, No. 3, pp. 559–569, 2011.
6. Carter C., Catlett J., "Assessing credit card applications using machine learning", IEEE Expert: intelligent systems and their applications, Vol. 2, pp. 71–79, 1987.
7. Hanagandi V., Dhar A., Buescher K., "Density-based clustering and radial basis function modeling to generate credit card fraud scores", Computational Intelligence for Financial Engineering, 1996.
8. Ghosh S., Reilly D. L., "Credit card fraud detection with a neural network", In Proceedings of the 27 th Hawaii International Conference on System Sciences, Vol. 3, pp. 621–630, 1994.
9. Dorronsoro J. R., Ginel F., Sanchez C., Cruz C. S., "Neural fraud detection in credit card operations", IEEE Transactions on Neural Networks, Vol. 8, pp. 827–834, 1997.
10. Seeja K. R., Masoumeh Z., "Fraud Miner: a novel credit card fraud detection model based on frequent itemset mining", The Scientific World Journal, Vol. 10, pp. 1–10, 2014.
11. Suganya S., Kamalra M., "Meta classification technique for improving credit card fraud detection", International Journal of Scientific and Technical Advancements, Vol. 2, No. 1, pp. 101–105, 2016.
12. Hassibi K., Lisboa P. J. G., Vellido A., Edisbury B., "Detecting payment card fraud with neural networks," Business application of Neural Networks, Singapore: World Scientific, 2000.
13. Maes S., Tuyls K., Vanschoenwinkel B., Manderick B., "Credit card fraud detection using Bayesian and neural networks," In Proceedings of the 1st International NAISO Congress on Neuro Fuzzy Technologies, pp. 261–270, 1993.
14. Zaslavsky V., Strizhak A., "Credit card fraud detection using self organizing maps", Information and Security, Vol. 18, pp. 48–63, 2006.
15. Quah J. T. C, Sriganesh M., "Real-time credit card fraud detection using computational intelligence", Expert Systems with Applications, Vol. 35, No. 4, pp. 1721–1732, 2008.
16. Srivastava L., Kundu A., Sural S., Majumdar A. K., "Credit card fraud detection using hidden Markov model", IEEE Transactions on Dependable and Secure Computing, Vol. 5, No. 1, pp. 37–48, 2008.

17. Sanchez D., Vila M. A., Cerda L., Serrano J. M., "Association rules applied to credit card fraud detection", Expert Systems with Applications, Vol. 36, No. 2, pp. 3630–3640, 2009.
18. Lu Q., Ju C., "Research on credit card fraud detection model based on class weighted support vector machine", Journal of Convergence Information Technology, Vol. 6, No. 1, pp. 62–68, 2011.
19. Wong N., Ray P., Stephens G., Lewis L., "Artificial immune systems for the detection of credit card fraud", Information Systems, Vol. 22, No. 1, pp. 53–76, 2012.
20. Zak S.H., (2003), "Systems and Control" NY: Oxford University Press.
21. Hassoun M.H., (1999), "Fundamentals of Artificial Neural Networks", Cambridge, MA: MIT press.
22. Cortes C., Vapnik V., "Support vector network. Machine Learning", Vol. 20, pp. 273–297, 1995.
23. Burges C. J. C., "A Tutorial on Support Vector Machines for Pattern Recognition, Data Mining and Knowledge Discovery", Vol. 2, No. 2, pp. 955–974, 1998.
24. Kim H. C., Pang S., Je H. M., Kim D., Bang S. Y., "Constructing support vector machine ensemble. Pattern Recognition", Vol. 36, pp. 2757–2767, 2003.
25. Chen R. C., Chen J., Chen T. S., Hsieh C. H., Chen T. Y., Wu K. Y., "Building an Intrusion Detection System Based on Support Vector Machine and Genetic Algorithm", Lecture Notes in Computer Science (LNCS), Vol. 3498, pp. 409–414, 2005.
26. Louzada F., Ara A., "Bagging k-dependence probabilistic networks: An alternative powerful fraud detection tool", Expert System Applications, Vol. 39, No. 14, pp. 11583–11592. 2012.
27. Elghazel H., Aussem A., Perraud F., "Trading-off diversity and accuracy for optimal ensemble tree selection in random forests", Ensembles in Machine Learning Applications Chennai, India, pp. 169–179, 2011.
28. Río S., "On the use of Map Reduce for imbalanced big data using random forest. Information Sciences", vol. 285, pp. 112–137, 2014.
29. Vaithiyanathan V., "Comparison of different classification techniques using different datasets", International Journal of Advances in Engineering & Technology, Vol. 6, No. 2, pp. 764, 2013.
30. Fadaei N. F., Moattar M., "Ensemble classification and extended feature selection for credit card fraud detection", Journal of AI and Data Mining, 2016.

An Efficient Swarm-Based Multicast Routing Technique—Review

Priyanka Kumari and Sudip Kumar Sahana

Abstract Multicast routing is emerging as a popular communication format for networks where a sender sends the same data packet to multiple nodes in the network simultaneously. To support this, it is important to construct a multicast tree having minimal cost for every communication session. But, because of dynamic and unpredictable environment of the network, multicast routing turns into a combinatorial issue to locate a best path connecting a source node and destination node having minimum distance, delay and congestion. To overcome this, various multicast conventions have been proposed. As of late, swarm and evolutionary techniques such as ant colony optimization (ACO), particle swarm optimization (PSO), artificial bee colony (ABC) and genetic algorithm (GA) have been adopted by the researchers for multicast routing. Out of these, ACO and GA are most popular. This paper shows an important review of existing multicast routing techniques along with their advantages and limitations.

Keywords Multicast routing · Ant colony optimization (ACO)
Particle swarm optimization (PSO) · Artificial bee colony (ABC)
Genetic algorithm (GA)

1 Introduction

Multicast routing is a bandwidth-conserving technology which sends similar data to various hubs in the network system by simultaneously reducing the traffic. The methodologies of multicast routing are source-based tree (establishes a turn-around way to the source node) and core-based tree (establishes an invert way to the core route).

P. Kumari (✉) · S. K. Sahana
Department of Computer Science & Engineering, Birla Institute of Technology,
Mesra, Ranchi 835215, Jharkhand, India
e-mail: phdcs10004.16@bitmesra.ac.in

S. K. Sahana
e-mail: sudipsahana@gmail.com

© Springer Nature Singapore Pte Ltd. 2019
H. S. Behera et al. (eds.), *Computational Intelligence in Data Mining*,
Advances in Intelligent Systems and Computing 711,
https://doi.org/10.1007/978-981-10-8055-5_12

123

The performances of network such as throughput, congestion control, delay and reliability directly depend on routing, that's why routing is a very important aspect of any network communication. With the rapid improvement of communication technology, multicast routing performs a vital role in most of the applications such as video conference, software delivery, emergency search, interactive multimedia games, Web-based learning. Due to technical evolutionary, multicast routing faces some challenges such as I. optimal route, II. congestion control and III. Quality of Service. The network topology changes dynamically and unpredictably as nodes move freely [1]. There is also no centralized control [2]. To support multicast directing, it is necessary to build a multicast tree having minimal cost. But it becomes a combinatorial problem which is a challenging task. Greedy method such as Dijkstra's algorithm gives the best path, but it is applicable only for a small number of nodes. It takes more computation time to find the best route for a large number of nodes.

Swarm intelligence is a computational technique for solving distributed problems inspired from biological social insects like bees, ants. Ant colony optimization (ACO) is an optimization algorithm for solving combinatorial problems. Also it has known to be a good and decent technique in support of routing algorithms in the communication network systems.

2 Classification of Multicast Routing Protocols

On the premise of directing information update mechanism, multicast routing protocols [3] are classified into three sorts: reactive algorithm, proactive algorithm and hybrid algorithm. Comparison between reactive, proactive and hybrid algorithms are shown in Table 1.

Table 1 Comparison between reactive, proactive and hybrid algorithms

Parameters	Reactive	Proactive	Hybrid
Routing information	Doesn't store?	Keep stored in a table	Depends on requirements
Route availability	Computed as per requirement	Always available	Depends on location of the target
Latency	High latency	Lower latency	Low latency
Loop-free	Yes	No	Yes
Routing overhead	Low overhead	More overhead	Lower overhead as compared to proactive
Route quality	No information about route quality	Maintain information about routes' quality	Maintain information about route's quality
Examples	DSDV [4], OLSR [5]	AODV [6], DSR [7]	ZRP [8]

3   Classification of Multicast Routing Techniques

Various techniques have been proposed to improve multicast routing techniques. These techniques can be categorized into three sorts—conventional method [9], evolutionary method [10] and swarm intelligence-based method [11]. Classification of multicast routing techniques is shown in Fig. 1.

3.1   Conventional Methods

The authors of [12] proposed a proactive protocol 'Destination-Sequenced Distance Vector (DSDV)' based on the design of traditional distance vector routing algorithm with specific changes, for example loop-free. Dynamic source routing (DSR) algorithm is proposed by the authors in [7] on the concept of source routing. It does not utilize periodic updates as it computes paths when there is need and after that maintains them. In research paper [13], the authors proposed wireless routing protocol (WRP) based on the design of distance vector routing algorithm similar to DSDV. This algorithm finds path having shortest distance with avoiding loop and count-to-infinity problem. The authors of [6] proposed ad hoc on-demand distance vector (AODV), which is based on the mixture of DSDV and DSR algorithms.

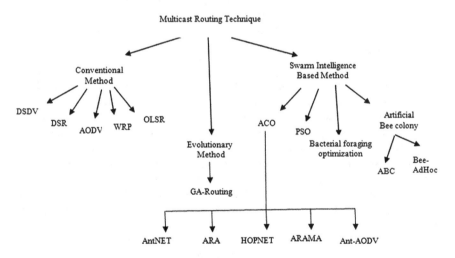

Fig. 1 Classification of multicast routing techniques

Optimized link state routing (OLSR) is proposed by the authors in [5]. Multipoint relays (MPRs) are the key idea of OLSR which diminishes control overhead by lessening the amount of broadcasts as compared to unadulterated flooding systems.

3.2   Evolutionary Methods

Genetic algorithm belongs to the category of evolutionary algorithms (EAs) which is described by Charles Darwin. It is a searching algorithm used for optimization [10]. The authors of [14] proposed a genetic-inspired multicast routing optimization algorithm with bandwidth and end-to-end delay constraints. In research paper [15], the authors proposed a well-formed hierarchical hybrid parallel genetic algorithm (PGA) based on crude-grained and well-grained algorithm.

3.3   Swarm-Based Multicast Routing Methods

Swarm intelligence (SI) [16] enquire about is enlivened by the aggregate conduct of some social living creatures, for example, ants, birds, bees and fishes, which was first presented by Gerardo Beni and Jing Wang in 1989. Swarm intelligence-based algorithms are used to get the optimal solutions in the real-world problems. Further classification of swarm intelligence-based algorithms is shown in Fig. 1.

3.3.1   Ant-Colony-Based Multicast Routing Algorithms

ACO is a famous swarm intelligence approach. It is based on the inspiration from the behaviour of real ants, which are wondering about their homes to scrounge for look for food [17]. The ants get the shortest way from home to food. The issue of finding shortest routes by ant is like the issue of routing in communication networks. The authors of [18] proposed mobile agents for adaptive routing (Ant-NET) algorithm based on ACO principle. This algorithm gives better performance, but it takes more delays in propagating routing algorithm. The authors in [19] proposed ant-colony-based routing algorithm (ARA), which works like AODV. This algorithm reduces routing overhead and gives better performance but in case of high mobility, it is not true. The authors in [20] proposed ant ad hoc on-demand distance vector (Ant-AODV), a hybrid of reactive routing algorithm AODV and proactive routing algorithm ACO. It reduces end-to-end delay and route discovery latency in dynamic networks. In research paper [21], the authors proposed ant routing algorithm for mobile ad hoc networks (ARAMA), which is an ant-based proactive routing algorithm. This algorithm reduces overheads of route discovery and route maintenance. The authors of [22] proposed HOPNET routing algorithm which combines the benefits of ACO algorithm and zone routing algorithm. HOPNET routing algorithm is more scalable for large networks and also gives efficient results. The authors of [23] have proposed a modified ant colony optimization (MACO) algorithm which depends on ant colony system (ACS) with some alteration in the arrangement of starting movement and in neighbouring updating method.

3.3.2 Particle Swarm Optimization-Based Multicast Routing Algorithms

Kennedy and Eberhart proposed PSO [24] in 1995. It is a populace-based optimization technique where the populace is called as a swarm. The authors of [25] proposed a PSO-based multicast routing algorithm. The running time of this algorithm is smaller than the previous algorithms, and it also produces better results in terms of quality. The authors in [26] proposed a hybrid PSO-GA routing algorithm which combines PSO with genetic operators to enhance the searching process in QoS multicast routing. The authors of [27] proposed the hybrid ACO/PSO (HACOPSO)-based algorithm using the concept of ACO and PSO. The cost of the multicast tree created by HACOPSO is less, and it also satisfies loss rate, delay jitter and bandwidth constraints.

3.3.3 Artificial Bee Colony (ABC)-Based Multicast Routing Algorithms

'Artificial bees' colony algorithm is a type of swarm intelligent optimization algorithm. Bees' group collects nectar with very high efficiency in the environment and it is dynamic. The author of [28] proposed human bees algorithm based on bee behaviour. The authors of [29] proposed BeeAdHoc which is a bee-inspired routing algorithm. This algorithm gives better execution in terms of packet delay and packet delivery ratio with smaller consumption. The authors of [30] proposed ABC multicast routing which optimizes the Steiner tree problem. This algorithm converges well and gives optimum solution quickly. Table 2 shows the comparison of multicast routing methods.

Table 2 Comparison of multicast routing methods

S. No.	Algorithms	Methods based on	Routing approach	Route metric	Remarks
1	DSDV [12]	Conventional method	Proactive	Shortest path	• Loop-free maintains shortest path for every node • High overhead and suffers from count-to-infinity problem • Does not fit in the large network
2	DSR [7]	Conventional method	Reactive	Shortest path or next available	• Loop-free maintains shortest paths • Low overhead and provides multiple paths • Large delays and scalability problems due to source routing and flooding • Requires more processing resources
3	WRP [13]	Conventional method	Proactive	Shortest path	• Loop-free but not instantaneous • Types of path created—single • Routing information is accurate • Eliminates count-to-infinity problem • Fast convergence • Memory overhead
4	AODV [6]	Conventional method	Reactive	Newest path	• Shortest routes and loop-free • Minimizing the network load • Supports for both unicast and multicast data packets transmission • Adaptable to highly dynamic topologies • Large delays and scalability problems • Does not provide any security
5	OLSR [5]	Conventional method	Proactive	Shortest path	• Loop-free and reduced control overhead by using multipoint relays • Does not require central administrative system to handle routing process • It provides paths which is not necessarily the shortest path

(continued)

Table 2 (continued)

S. No.	Algorithms	Methods based on	Routing approach	Route metric	Remarks
					• Suitable for large and dense ad hoc networks • Does not allow long delays in packets transmission
6	Ant-NET [18]	Ant colony optimization principle	Proactive	Pheromone table	• Operators-pheromone update and measures trail evaporation • Type of path created—single • Takes more time in propagating routing information and high overhead
7	ARA [19]	Ant-colony-based	Reactive	Pheromone table	• Route discovery approach —flooding • Type of ants in use—forward and backward ants • Types of path created—multipath • Technique for detecting route failure—backtracking • Low overhead in low network load • Does not adapt the changes in a dynamic network
8	Ant-AODV [20]	Conventional method AODV and ant colony optimization	Hybrid	Newer and shorter route	• Route discovery approach —flooding • Type of ants in use—forward and backward ants • Type of route created—single path • Technique for detecting route failure—local repair error message • Packet delivery fraction —high • Low end-to-end delay but high overhead • Link failure notification —Yes
9.	ARAMA [21]	Ant-colony-based method	Hybrid	Pheromone table and probability routing table	• Operators-Pheromone update and measures trail evaporation • Routing overheads—controlled

(continued)

Table 2 (continued)

S. No.	Algorithms	Methods based on	Routing approach	Route metric	Remarks
					• Provides a number of unessential paths between each source node and destination node • Initialization period of first ant to find the destination node is very long
10	BeeAdHoc [29]	Bee-inspired method	Reactive	Probability of optimal route	• Inspired by foraging principles of honey bees • Operators-reproduction, replacement of bee selection • Using broadcast approach • Consumes less energy and low delays • Packet delivery ratio—high • It uses the higher memory for storing every forager
11	GA routing [14]	Genetic algorithm	Reactive	Genetic algorithm based on multihop routing	• Operators—selection, crossover and mutation • Cost of multicast tree—minimum • Maximize network (small) lifetime by using minimum number of aggregation trees
12	PSO [25]	Particle swarm optimization	Reactive	Newer and best route	• Execution time is less • Cost of multicast tree—minimum • Execution time grows very slowly with the size of network • Execution time becomes about to equal to ACO when the number of nodes are more than 200
13	HOPNET [22]	ACO and zone routing framework	Hybrid	Inter-RT and Intra-RT	• Operators—pheromone update and measures trail evaporation • Multiple path and low end-to-end delay • High packet delivery ratio high overhead • More scale for large networks

(continued)

Table 2 (continued)

S. No.	Algorithms	Methods based on	Routing approach	Route metric	Remarks
14	PSO-GA [26]	PSO and GA	Hybrid	Shortest route	• Each gene in the particle encoding represents a possible route the source node to one of the destinations • Give better QoS performance • Fast convergence and avoids early convergence to local optimal
15	ABC [30]	Bee-inspired method	Hybrid	Shortest route	• Generating through simulation of the honey bees' behaviour • Operators—selection, replacement of bee, reproduction of bee • Fast convergence • When the number of bee is low, it is hard to find the best solution
16	PGA [15]	Genetic algorithm	Hybrid	Shortest path	• Operators—selection, crossover, mutation • Combination of well-grained and the crude-grained genetic algorithm • Takes small time to choose the shortest path and reduces communication overhead
17	HACOPSO [27]	ACO and PSO	Hybrid	Shortest path	• Construct the multicast tree patterns more sensibly such that the cost of tree is minimum • Satisfies the QoS constraints • Execution time is less
18	Modified ACO [23]	Ant colony system	Reactive	Optimal path	• Operators—pheromone update and measures trail evaporation • Convergence rate is faster • Gives optimal path • Cost of multicast tree—minimum • For small network size, it slightly takes less execution time than the conventional time. • It is not applicable for large networks

4 Discussion

Conventional methods such as DSDV, DSR, WRP, AODV and OLSR are used for multicast routing algorithms. DSDV and DSR are popular routing methods which are able to discover the shortest route for each node, but these are not suited for large networks. Similarly, AODV finds routes quickly and accurately, whereas its performance decreases with increasing the number of nodes. It gives better performance up to 100 nodes. DSDV devours more bandwidth as it broadcasts routing information at regular intervals, whereas in AODV there is no need to maintain route table, which results in less bandwidth consumption and in addition less overhead. OLSR and WRP are suitable for large networks, but it is not necessary that the path discovered by OLSR method will be shortest and WRP does not avoid loop instantaneously. All conventional methods take a long time to find the best paths. Ant-NET, ARA, Ant-AODV, ARAMA, MACO are ant-based multicast routing algorithms to find the shortest paths and take less time than all conventional methods. However, Ant-NET, Ant-AODV, ARAMA suffer from routing overhead, which affects their QoS performance. Like DSDV and DSR, MACO is also not applicable for large networks. BeeAdHoc and ABC are bee-inspired methods for multicast routing, which has very good optimization ability and also gives solution quickly. For large networks, these methods may not be suitable. Genetic-based multicast routing method PGA takes less time to choose the shortest path, but this method does not mention about QoS performance, for example jitter, end-to-end delay and bandwidth. The hybridization methods like hybrid PSO-GA and HACOPSO multicast routing methods give shorter routes and better performance in terms of QoS parameters in small networks. For large networks, these methods do not give better performance.

5 Conclusions

For supporting real-time and non-real-time applications, multicast routing should guarantee for optimal path, congestion control and QoS parameters. But it is a genuine dynamic issue in communication networks, and this issue is an NP-hard problem as it cannot be solved in a polynomial time. Various multicast routing techniques have been proposed which are based on conventional methods, evolutionary methods and swarm intelligence-based methods. Due to multicast routing is NP-hard problem, conventional methods take more time to locate the best path. Many evolutionary and swarm-based algorithms have been proposed to locate the best path and also improve the QoS parameters. It is noted that the biologically inspired techniques give more beneficial answers, but hybridization techniques exhibit even more honest solutions. Thus, our study suggests that different evolutionary and swarm-based algorithms are employed for solving multicast routing problem and for enhancing the effectiveness either by improving the technique and/or by using the hybridized approach. These approaches can also be employed in the implementation of other applications like Job Shop Scheduling Problem, Vehicle Routing problem and many more applications.

References

1. D. Sivakumar, B. Suseela, R.Varadharajan, "A Survey of Routing Algorithms for MANET", IEEE- International Conference On Advances In Engineering, Science And Management (ICAESM- 2012), March 30, 31, 2012, pp. 625–640.
2. P. M. Pardalos and C. A. S. Oliveira, "A survey of combinatorial optimization problems in multicast routing," Comput. Oper. Res., ELSEVIER, 2005, vol. 32, no. 8, p. 1953–1981 M. R. Macedonia, D. P. Brutzman, "MBone provides audio and video across the internet", Computer, IEEE, 1994, Volume: 27, p. 30–36.
3. N. Al-Karaki and A.E. Kamal, "Routing techniques in wireless sensor networks: A survey", IEEE Wireless Commun. Mag., Dec- 2004, vol.11, no.6, pp.-6–28.
4. Anuj K. Gupta, Harsh Saadawarti, and Anil K. Verma, "Review of various routing protocols for MANETs", International Journal of information and Electronics, IEEE, vol. 1, No. 3, Nov-2011, p. 251–259.
5. T. Clausen, P. Jacquet, A. Laouiti, P. Muhlethaler, A. Qayyum, and L. Viennot, "Optimized link state routing protocol for ad-hoc networks" in: Proceedings of IEEE INMIC, December 2001, pp. 62–68.
6. Charles E. Perkins, Elizabeth M. Royer, "Ad-hoc On-Demand Distance Vector Routing", In Proceedings of the 2nd IEEE Workshop on," Mobile Computing Systems and Applications", Feb-1999, pp. 90–100.
7. David A. Maltz, David B. Johnson, "Protocols for adaptive wireless and mobile computing." In IEEE Personal Communications, 3(1), February 1996.
8. Z. J. Haas, M. Perlman, "The performance of query control schemes of the zone routing protocol", IEEE/ACM Transactions on networking," vol. 9. 4, pp. 427–438, Aug-2001.
9. Mehtab Alam, Asif Hameed Khan, Ihtiram Raza Khan, "Intelligence in MANETs: A Survey", International Journal of Emerging Research in Management & Technology (IJERMT), May-2016, Vol-5, pp. 141–150.
10. Y. Yen, Y. Chan, H.Chao, J. H. park, "A genetic Algorithm for energy-efficient based Multicast Routing in MANETs", ELSEVIER., pp. 858–869, Mar.-2008.
11. E. Bonabeau, M. Dorigo, G. Theraulaz, FROM Natural to Artificial Systems, Oxford university press, New York, NY, 1999.
12. C.E. Perkins and P. Bhagwat, "Highly Dynamic Destination-Sequenced Distance-Vector (DSDV) for Mobile Computer", "Proc. ACM Conf. Communications Architectures and Protocols", London, UK, August 1994, pp. 234–244.
13. S. Murthy and J. J. Garcia-Luna-Aceves, "An efficient routing protocol for wireless networks", ACM Mobile Networks and App. J. Special Issue on Routing in Mobile Communication Networks, Oct. 1996, pp. 183–197.
14. Sanghoun Oh, Chang Wook Ahn and R.S. Ramakrishna, "A Genetic-Inspired Multicast Routing Optimization Algorithm with Bandwidth and End-to-End Delay Constraints", SPRINGER, Berlin, Heidelberg, ICONIP, Part III, LNCS 4234, pp. 807–816.
15. Neeraj, A. Kumar, "Efficient Hierarchical Hybrids Parallel Genetic Algorithm For Shortest Path Routing", Proceedings in 5th International Conference-Confluence the Next Generation Information Technology Summit (Confluence) IEEE, Sept.-2014, pp. 257–261.
16. E. Bonabeau, M. Dorigo, G. Theraulaz, Swarm Intelligence. FROM NATURAL to Artificial Systems, Oxford University press, New York, NY, 1999.

17. Marco Dorigo, Mauro Birattari, Thomas stutzle, Ant Colony Optimization", IEEE Computational Intelligence Magazine, 2006, vol. 1, p. 28–39.
18. Gianni Di Caro, Marco Dorigo, "Mobile Agents for Adaptive Routing", Proceedings of the Thirty-First Hawaii International Conference on System Sciences, IEEE, 1998, PP. 74–83.
19. Mesut, Gunes, Udo Sorges, Imed Bouazizi, "ARA-The Ant-Colony Based Routing Algorithm for MANETs", In Proceedings IEEE Computer Society ICPPW'02 Workshop, 2002.
20. Shivanjay Marwaha, Chen khong tham, Dipti Srinivasan, "A Novel Routing Protocol Using Mobile Agents and Reactive Route Discovery For AD-Hoc Wireless Networks", Global Telecommunications Conference, 2002. GLOBECOM '02. IEEE, 2002 pp. 163–167.
21. O. Hossein and T. Saadawi, "Ant routing algorithm for mobile adhoc networks (ARAMA)", Proceedings of the 22nd IEEE International Performance, Computing and Computing, and Communications Conference, Phoenix, Arizona, USA, April 2003, pp. 281–290.
22. J. Wang, E. Osagie, P. Thulasiraman, R. Thulasiram, "Hopnet: a hybrid ant colony optimization routing algorithm for mobile ad hoc network," Ad Hoc Networks 7 (4), ELSEVIER, 2009, pp. 690–705.
23. S. K. Sahana, Mohammad AL-Fayoumi, P. K. Mahanti,: "Application of Modified Ant Colony Optimization (MACO) for Multicast Routing Problem", I.J. Intelligent Systems and Applications, 2016, 4, p- 43–48.
24. R. C. Eberhart, J. Kennedy, "A new optimizer using particle swarm theory", Proceedings of the 6th Symposium Micro Machine and Human Science, IEEE Press, Los Alamitos, CA, October 1995, pp. 39–43.
25. Ziqiang Wang, Xia Sun, Dexian Zhang. "A PSO-Based Multicast Routing Algorithm", 3rd International Conference On Natural Computation (ICNC), Computer Society, IEEE, 2007, p. 664–667.
26. Rehab F. Abdel Kader, "Hybrid discrete PSO with GA operators for efficient QoS-multicast routing", Ain Shams Engineering Journal, Production and hosting by ELSEVIER, March 2011, pp. 21–31.
27. Manoj Kumar Patel *, Manas Ranjan Kabat, Chita Ranjan Tripathy, "A hybrid ACO/PSO based algorithm for QoS multicast routing problem", Ain Shams Engineering Journal, ELSEVIER, 2014, pp. 113–120.
28. Karaboga, D., "An idea based on honey bee swarm for numerical optimization", Tech. Rep. TR06, Erciyes Univ. Press, Erciyes, 2005.
29. Horst. F. Wedde and Muddassar, "The Wisdom Of The Hive Applied To Mobile Ad-Hoc Networks", Proceedings IEEE Swarm Intelligence Symposium, 2005, pp. 341–348.
30. Zhenhua Zheng, Hua Wang, Lin Yao, "An artificial bee colony optimization algorithm for multicast routing", 14th International Conference on advanced Communication Technology (ICACT), IEEE, 2012, pp. 168–172.

A New Howard–Crandall–Douglas Algorithm for the American Option Problem in Computational Finance

Nawdha Thakoor, Dhiren Kumar Behera, Désiré Yannick Tangman and Muddun Bhuruth

Abstract The unavailability of a closed-form formula for the American option price means that the price needs to be approximated by numerical techniques. The valuation problem can be formulated either as a linear complementarity problem or a free-boundary value problem. Both approaches require a discretisation of the associated partial differential equation, and it is common to employ standard second-order finite difference approximations. This work develops a new procedure for the linear complementarity formulation. Howard's algorithm is used to solve the discrete problem obtained through a higher-order Crandall–Douglas discretisation. Speed and error comparisons indicate that this approach is more efficient than the procedures for solving the free-boundary value problem.

Keywords Computational finance · American option · Policy iteration Howard's algorithm

1 Introduction

A put option is a financial contract which gives to its holder the right to buy an asset at an exercise price specified in the contract. The option is an American put if the holder can exercise his right at any time up to maturity date of the contract.

N. Thakoor · D. Y. Tangman · M. Bhuruth (✉)
Department of Mathematics, University of Mauritius, Reduit, Mauritius
e-mail: mbhuruth@uom.ac.mu

N. Thakoor
e-mail: n.thakoor@uom.ac.mu

D. Y. Tangman
e-mail: y.tangman@uom.ac.mu

D. Kumar Behera
Mechanical Engineering Department, Indira Gandhi Institute of Technology,
Sarang, Dhenkanal 759146, India
e-mail: dkb_igit@rediffmail.com

© Springer Nature Singapore Pte Ltd. 2019
H. S. Behera et al. (eds.), *Computational Intelligence in Data Mining*,
Advances in Intelligent Systems and Computing 711,
https://doi.org/10.1007/978-981-10-8055-5_13

The unavailability of a closed-form valuation formula for the American option makes the pricing of this financial derivative as one of the most challenging and most researched problems in computational finance. Many of the existing algorithms employ standard second-order approximations for the spatial derivatives in the pricing equation. A different approach proposed by McCartin and Labadie [1] employs a Crandall–Douglas [2, 3] scheme for the discretisation and the Elliott–Ockendon algorithm [4] for solving the discretised linear complementarity problem. This work proposes a superior approach to the McCartin–Labadie method by combining Howard's algorithm [5], also known as policy iteration [6] with the Crandall–Douglas discretisation. Numerical examples indicate that policy iteration is superior to the method based on the Elliott–Ockendon procedure.

The rest of the paper is organised as follows. In Sect. 2, the linear complementarity and free-boundary formulations of the American option pricing problem are given, and in Sect. 3, a free-boundary method is reviewed. The new computational procedure is then described in Sect. 4. Numerical examples to compare the performances of the different algorithms are described in Sect. 5, and Sect. 6 concludes this work.

2 The American Option Pricing Problem

We consider a financial market with a risky asset with price S_t at time t over the trading period $0 \leq t \leq T$ following the Black-Scholes diffusion process [7]

$$dS_t = (r - q)S_t + \sigma S_t \, dW_t, \tag{1}$$

under the risk-neutral measure $\mathbb{Q}$. In (1), r is the risk-free interest rate, q is the continuous dividend yield, σ is the instantaneous volatility, and W_t is a $\mathbb{Q}$-Brownian motion.

The time t price $V(S, t)$ of an American put is such that $V(S, t) \geq (K - S)^+ = \max(0, K - S)$. The critical stock price $S_f(t)$ at which early exercise is optimal is such that $V(S, t) > (K - S)^+$ for $S > S_f(t)$ and $V(S, t) = K - S$ for $0 < S < S_f(t)$.

Using no-arbitrage arguments, it can be shown that the free-boundary value formulation for computing the American put price requires the solution of [8]

$$\frac{\partial V}{\partial t} + \mathcal{L}_S V = 0, \quad S_f(t) \leq S < \infty, \quad 0 \leq t \leq T, \tag{2}$$

$$V(S, T) = (K - S)^+,$$

$$V(S_f(t), t) = (K - S_f(t)), \quad 0 \leq t \leq T,$$

$$\frac{\partial V}{\partial S}(S_f(t), t) = -1, \quad 0 \leq t \leq T,$$

$$V(S, t) \to 0, \quad \text{as} \quad S \to \infty, \quad 0 \leq t \leq T,$$

where $\mathcal{L}_S$ is the Black–Scholes operator given by

$$\mathcal{L}_S V = \frac{1}{2}\sigma^2 S^2 \frac{\partial^2 V}{\partial S^2} + (r-q)S\frac{\partial V}{\partial S} - rV. \tag{3}$$

The linear complementarity formulation for the American put option is given by

$$\frac{\partial V}{\partial t} + \mathcal{L}_S V \le 0, \quad S \ge 0,\, 0 \le t \le T, \tag{4}$$

$$V(S,\,t) \ge (K-S)^+, \quad S \ge 0,\, 0 \le t \le T,$$

$$\left[\frac{\partial V}{\partial t} + \mathcal{L}_S V\right] \cdot \left[V(S,\,t) - (K-S)^+\right] = 0, \quad S \ge 0,\, 0 \le t \le T,$$

$$\lim_{S \to +\infty} V(S,\,t) = 0.$$

with terminal condition $V(S, T) = (K - S)^+$.

3 Algorithms for Free-Boundary Problems

The idea behind the procedure of Han and Wu [9] is to obtain a computational domain which is small in order to avoid redundant computations. To this effect, an exact boundary condition is derived on an artificial boundary. Implementation details of the algorithm can be found in [10].

3.1 The Moving Boundary Method

The moving boundary method [11] computes the optimal exercise boundary using an iterative process. From Eqs. (2) and (3), it follows that the option price $V(S, t)$ satisfies

$$\min\left(\frac{\partial V}{\partial t} + \mathcal{L}_S V,\; V - (K-S)^+\right) = 0.$$

Starting with an initial guess $S_f^0(T)$ which lies below the optimal exercise boundary $S_f(T)$, the algorithm generates a sequence of exercise boundaries $S_f^1(T)$, $S_f^2(T)$, …. For an exercise boundary $S_f^p(T)$, let $V^p(S, t)$ be the option price in the exercise region R_e^p and first consider the computation of $V^0(S, t)$ by solving

$$\frac{\partial V^0}{\partial t} + \mathcal{L}_S V^0 = 0 \text{ in } R_e^0,$$

$$V^0(S_f^0(t),\, t) = (K - S_f^0(t))^+,\, 0 \le t \le T,$$

$$V^0(S, T) = (K - S)^+, \tag{5}$$

$$V^0(S, t) = 0, \quad \text{as } S \to +\infty, \quad 0 \le t \le T.$$

Problem (5) is solved using the Brennan–Schwartz algorithm [12] on a truncated computational domain. The moving boundary method is then implemented as follows:

1. Choose $\hat{T}$ such that $0 \leq T \leq \hat{T}$, tolerance parameters ϵ_v and ϵ_p and initial exercise boundary guess $S_f^0(T)$, and choose asset price computational domain $S_f^0(T) \leq S \leq S_{max}$.
2. Compute $V^0(S, 0)$.
3. Obtain $S_f^1(T)$ using the update condition

$$S_f^{p+1}(T) = \sup_S \left\{ S \geq S_f^p(T) \text{ and } \frac{\partial V(S_0, T)}{\partial S} < -1 \text{ for all } S_0 \in [S_f^p(T), S) \right\}.$$

4. Iterate the sequence and update $S_f^{p+1}(T)$ until either

$$\max_T(S_f^p(T) - S_f^{p-1}(T)) < \epsilon_s, \text{ or}$$

$$\max_{S,T}(V^p(S, T) - V^{p-1}(S, T)) < \epsilon_v.$$

4 Proposed Work

Let $\alpha = \left(\frac{1}{2} - \frac{(r-q)}{\sigma^2}\right)$, $\beta = -\left(\frac{1}{4}\left(\frac{2(r-q)}{\sigma^2} - 1\right)^2 + \frac{2r}{\sigma^2}\right)$. Using the substitutions $S = Ke^x$, $t = T - 2r/\sigma^2$ and $V(S, t) = Ke^{\alpha x + \beta \tau}u(x, \tau)$, the linear complementarity problem (4) becomes

$$\frac{\partial u}{\partial \tau} \geq \frac{\partial^2 u}{\partial x^2}, \quad -\infty < x < +\infty, \ 0 \leq \tau \leq \tau^*,$$

$$u(x, \tau) \geq g(x, \tau), \quad -\infty < x < +\infty, \ 0 \leq \tau \leq \tau^*,$$

$$u(x, 0) = g(x, 0), \quad -\infty < x < +\infty,$$

$$\lim_{x \to \pm\infty} u(x, \tau) = \lim_{x \to \pm\infty} g(x, \tau),$$

where $\tau^* = \sigma^2 T/2$ and $g(x, \tau) = e^{-\alpha x - \beta \tau}(1 - e^x)^+$.

4.1 The Obstacle Problem

The above is equivalent to solving the obstacle problem given by

$$\min\left(\frac{\partial u}{\partial \tau} - \frac{\partial^2 u}{\partial x^2}, u - g\right) = 0. \tag{6}$$

The solution to (6) is obtained by using the Crandall–Douglas discretisation and then solving the finite-dimensional problem using Howard's algorithm [13].

We start by localising the problem to a finite computational domain $[a, b]$ and shifting the far-field boundary conditions to the ends of the computational domain. Let $\Delta x = (b - a)/M$ be the grid spacing in the space direction, $\Delta \tau = \tau^*/N$ be the temporal time step, and denote $u(x_m, \tau_n)$ by u_m^n where $x_m = a + m\Delta x$ for $0 \leq m \leq M$. The resulting discrete form is given by

$$\min \left(\frac{1}{12} \left(f_{m+1}^{n+\frac{1}{2}} + 10 f_m^{n+\frac{1}{2}} + f_{m-1}^{n+\frac{1}{2}} \right) - \frac{1}{(\Delta x)^2} \delta_x^2 u_m^{n+\frac{1}{2}}, \, u_m^{n+1} - g_m^{n+1} \right) = 0, \quad (7)$$

where

$$\delta_x^2 u_m^{n+\frac{1}{2}} \approx \left(u_{m-1}^{n+\frac{1}{2}} - 2u_m^{n+\frac{1}{2}} + u_{m+1}^{n+\frac{1}{2}} \right), \quad u_m^{n+\frac{1}{2}} \approx \frac{1}{2} \left(u_m^{n+1} + u_m^n \right),$$

and

$$f_m^{n+\frac{1}{2}} = \frac{u_m^{n+1} - u_m^n}{\Delta \tau}.$$

In the following, let $g_m^n = g(x_m, \tau_n) = e^{-\alpha x_m - \beta \tau_n} (1 + e^{x_m})^+$ for $0 \leq m \leq M$, define $g^n = [g_1^n, g_2^n, \ldots, g_{M-1}^n]^T$, let $u^n = [u_1^n, u_2^n, \ldots, u_{M-1}^n]^T$ denote the vector of option prices at the time level n, and let $\gamma = 1/(12\Delta\tau)$ and $\eta = 1/(2(\Delta x)^2)$.

For vectors c and d, we denote by $\min(c, d)$ the vector whose components are given by $\min \left(c_i, d_i \right)$. The discretized problem (7) can be written in the form

$$\min \left(A u^{n+1} - b^n, \, u^{n+1} - g^{n+1} \right), \quad (8)$$

where the matrices $A \in \mathbb{R}^{(M-1)\times(M-1)}$ and $b \in \mathbb{R}^{M-1}$ are such that

$$A = \text{tridiagonal}[\gamma - \eta, \, 10\gamma + 2\eta, \, \gamma - \eta],$$

and $b^n = \left[b_1^n, b_2^n, \ldots, b_{M-1}^n \right]^T$ with

$$b_1^n = (\gamma + \eta) g_0^n + (10\gamma - 2\eta) u_1^n + (\gamma + \eta) u_2^n - (\gamma - \eta) g_0^{n+1},$$
$$b_m^n = (\gamma + \eta) u_{m-1}^n + (10\gamma - 2\eta) u_m^n + (\gamma + \eta) u_{m+1}^n, \quad 2 \leq m \leq M - 2,$$
$$b_{M-1}^n = (\gamma + \eta) u_{M-2}^n + (10\gamma - 2\eta) u_{M-1}^n + (\gamma + \eta) g_M^n - (\gamma - \eta) g_M^{n+1}.$$

4.2 Howard's Algorithm

Let $\mathcal{A} = \{0, 1\}$ and consider the solution to the problem

$$\text{Find } v \in \mathbb{R}^{M'}, \quad \min_{\hat{\alpha} \in \mathcal{A}^{M'}} (A(\hat{\alpha})v - c(\hat{\alpha})), \quad (9)$$

where, for $\hat{\alpha} \in \mathcal{A}^{M'}$, $A(\hat{\alpha})$ is a monotone matrix of dimension M' and $c(\hat{\alpha})$ is a vector of length M'. For the case when $\hat{\alpha} = (\hat{\alpha}_1, \hat{\alpha}_2, \dots, \hat{\alpha}_{M'}) \in \mathcal{A}^{M'}$ and the elements $A_{ij}(\hat{\alpha})$ depend only on $\hat{\alpha}_i$, by setting for every $\hat{a} \in \mathcal{A}$, $\hat{\alpha}^{\hat{a}} = (\hat{a}, \hat{a}, \dots, \hat{a}) \in \mathcal{A}^{M'}$, $A^{\hat{a}} = A(\hat{\alpha}^{\hat{a}})$ and $c^{\hat{a}} = c(\alpha^{\hat{a}})$, we find that (9) can be equivalently formulated as

$$\min_{\hat{a} \in \mathcal{A}} \left(A^{\hat{a}} v - c^{\hat{a}} \right). \tag{10}$$

We thus find that the obstacle problem in (8) can be written in the form (10) with $A^0 = A$, $A^1 = I_{M-1}$, $c^0 = b^n$ and $c^1 = g^{n+1}$. We then apply the following Howard's algorithm described in [13] for the solution of (10).

1. Initialise $\hat{\alpha}^0 \in \mathcal{A}^{M-1} = \{0, 1\}^{M-1}$.
2. Iterate for $k \geq 0$:

   (a) Find $v^k \in \mathbb{R}^{M'}$ such that $A(\hat{\alpha}^k) v^k = c(\hat{\alpha}^k)$.
       If $k \geq 1$ and $v^k = v^{k-1}$, then stop. Otherwise go to (b).
   (b) For every $i = 1, 2, \dots, M-1$,

   $$\text{Take } \hat{\alpha}_i^{k+1} = \begin{cases} 0, & \text{if} \left(\left(A^0 v^k - c^0 \right) \leq \left(A^1 v^k - c^1 \right) \right); \\ 1, & \text{otherwise.} \end{cases}$$

   (c) Set $k := k + 1$ and go to (a).

In the specific case of the American option pricing problem given by (7), we set $c^0 = b^n$, $c^1 = g^{n+1}$ and $v = u^{n+1}$. Since the discretisation matrix A is monotone, the above algorithm converges in at most $M - 1$ iterations [13].

5 Numerical Results

This section presents some numerical examples to illustrate the performance of the Howard–Crandall–Douglas method. All numerical experiments have been performed using MATLAB R2015a on a Core i5 laptop with 4GB RAM and speed 4.60 GHz.

Table 1 shows computed prices and the total run times for the computation of American puts with maturities $T = 0.5$ year and strike prices $K = 100$ for different set of parameters.

For this short maturity problem, we choose $a = -1$ and $b = 1$. For the Han–Wu, moving boundary and policy iteration methods, we use $M = 2^{10}$ spatial and $N = 2^{10}$ temporal time steps while for Elliot–Ockendon–Crandall–Douglas (EO-CD) [1] and the Howard–Crandall–Douglas schemes we use $M = 2^{10}$ and we observe that $N = 2^5$ temporal steps are sufficient to yield accurate solutions. The results indicate that the Howard–Crandall–Douglas method vastly improves over the computational times of the Han and Wu, moving boundary and Elliot–Ockendon methods.

Table 1 American put option prices for a small maturity problem ($T = 0.5$ year)

Parameters (r, q, σ)	Asset price	Benchmark	Existing algorithms				Howard–Crandall–Douglas
			Han–Wu	EO-CD	Moving boundary	Policy iteration	
(0.05, 0, 0.2)	80	20.00000	20.00000	20.00000	20.00000	20.00000	20.00000
	90	10.66631	10.66603	10.66606	10.66609	10.66609	10.66644
	100	4.65568	4.65552	4.65556	4.65551	4.65551	4.65556
	110	1.66799	1.66797	1.66798	1.66796	1.66796	1.66797
	120	0.49756	0.49758	0.49757	0.49757	0.49757	0.49756
	cpu (s)		3.458	2.009	0.859	0.215	0.184
(0.07, 0.03, 0.4)	80	21.87100	21.87096	21.87097	21.87095	21.87095	21.87100
	90	15.22975	15.22962	15.22964	15.22970	15.22970	15.22980
	100	10.23868	10.23853	10.23856	10.23834	10.23856	10.23854
	110	6.67800	6.67795	6.67793	6.67792	6.67792	6.67800
	120	4.24757	4.24753	4.24754	4.24747	4.24747	4.24752
	cpu (s)		3.839	2.245	0.992	0.233	0.207

Table 2 American put option prices for a long maturity problem ($T = 3$)

Parameters (r, q, σ)	Asset price	Benchmark	Existing algorithms				Howard–Crandall–Douglas
			Han–Wu	EO-CD	Moving boundary	Policy iteration	
(0.05, 0, 0.2)	80	20.27970	20.27985	20.27985	20.27990	20.27986	20.27988
	90	13.30852	13.30758	13.30758	13.30759	13.30809	13.30784
	100	8.71048	8.71055	8.71055	8.71039	8.71039	8.71044
	110	5.68249	5.68255	5.68254	5.68198	5.68253	5.68257
	120	3.69636	3.69641	3.69640	3.69502	3.69635	3.69637
	cpu (s)		3.372	2.099	0.954	0.214	0.177
(0.07, 0.03, 0.4)	80	28.90465	28.90453	28.90452	28.90451	28.90452	28.90452
	90	24.44837	24.44820	24.44818	24.44815	24.44851	24.44851
	100	20.79344	20.79317	20.79315	20.79282	20.79344	20.79344
	110	17.77146	17.77131	17.77129	17.77062	17.77142	17.77147
	120	15.25610	15.25597	15.25594	15.25474	15.25604	15.25609
	cpu (s)		3.831	2.456	0.972	0.302	0.189

In Table 2, a long maturity problem of $T = 3$ years is considered with both low and high volatilities. Here, we set $a = -1.2$ and $b = 1.2$ and use the same number of spatial and temporal steps as in the first example. The results lead to similar conclusions as for the first test example.

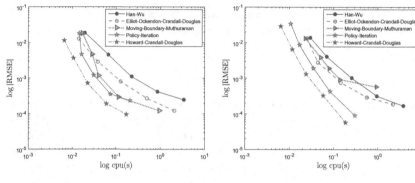

(a) Short maturity $T = 0.5$ year and low volatility $\sigma = 0.2$.

(b) Long maturity $T = 3$ years and high volatility $\sigma = 0.4$.

Fig. 1 Accuracy and speed comparisons

Figure 1 compares the accuracy of the different methods against their running times for $M = 2^6, 2^7, 2^8, 2^9, 2^{10}$. For both test problems, the Howard–Crandall–Douglas method outperforms the other numerical schemes.

6 Conclusion

Although the American option pricing problem has been extensively studied, the search for algorithms with fast run times continues. This work considered the Howard's algorithm for the solution of the linear complementarity problem where the pricing operator was approximated with a high-order finite difference scheme. Numerical results indicate that this new computational technique improves the run times of some of the currently available PDE methods for solving this challenging financial derivative valuation problem.

References

1. McCartin, B.J., Labadie, S.M.: Accurate and efficient pricing of vanilla stock options via the Crandall-Douglas scheme. Appl. Math. Comput. **143**, 39–60 (2003)
2. Crandall, S.H.: An optimum implicit recurrence formula for the heat conduction equation. Q. Appl. Math. **13**, 318–320 (1955)
3. Douglas Jr., J.: The solution of the diffusion equation by a high order correct difference equation. J. Math. Phys. **35**, 145–151 (1956)
4. Elliott, C.M., Ockendon, J.R.: Weak and variational methods for moving boundary problems. Research notes in Mathematics, Pitman, Boston, Mass. **59** (1982)
5. Howard, R.A.: Dynamic Programming and Markov Processes. The MIT Press, Cambridge, MA (1960)

6. Reisinger, C., Witte, J.H.: On the use of policy iteration as an easy way of pricing American options. SIAM J. Financial Math. **3**, 459–478 (2012)
7. Black, F., Scholes, M.: The pricing of options and corporate liabilities. J. Polit. Econ. **81**, 637–654 (1973)
8. Seydel, U.R.: Tools for Computational Finance. Springer-Verlag, Heidelberg (2006)
9. Han, H., Wu, X.: A fast numerical method for the Black-Scholes equation of American options. SIAM J. Numer. Anal. **41**, 2081–2095 (2003)
10. Saib, A.A.E.F., Tangman, D.Y., Thakoor, N., Bhuruth, M.: On some finite difference algorithms for pricing American options and their implementation in Mathematica. In: Proceedings of the 11th International Conference on Computational and Mathematical Methods in Science and Engineering, pp. 1029–1040. Alicante, Spain (2011)
11. Muthuraman, K.: A moving boundary approach to American option pricing. J. Econ. Dyn. Control **32**, 3520–3537 (2008)
12. Brennan, M.J., Schwartz, E.S.: The valuation of American put options. J. Finance **32**, 449–462 (1977)
13. Bokanowski, O., Maroso, S., Zidani, H.: Some convergence results for Howard's algorithm. SIAM J. Numer. Anal. **47**, 3001–3026 (2009)

Graph Anonymization Using Hierarchical Clustering

Debasis Mohapatra and Manas Ranjan Patra

Abstract Privacy preserving data publication of social network is an emerging trend that focuses on the dual concerns of information privacy and utility. Privacy preservation is essential in social networks as social networks are abundant source of information for studying the behavior of the social entities. Social network disseminates its information through social graph. Anonymization of social graph is essential in data publication to preserve the privacy of participating social entities. In this paper, we propose a hierarchical clustering-based approach for k-degree anonymity. The attack model focuses on identity disclosure problem. Our approach unlike other approach discussed in Liu and Terzi (Proceedings of ACM SIGMOD, 2008, [1]) generates k-degree anonymous sequence with the k value. Havel–Hakimi algorithm is used to check the sequence is graphic or not. Subsequently, the construction phase takes place with the help of edge addition operation.

Keywords Social graph · Hierarchical clustering · k-degree anonymity

1 Introduction

Social graph is a powerful nonlinear representation of social network. Social graph mining is an interesting knowledge discovery technique that discovers the relationships among the social entities, interaction patterns between the social entities,

D. Mohapatra (✉)
Department of Computer Science & Engineering, PMEC, Berhampur 761003,
Odisha, India
e-mail: devdisha@gmail.com

M. R. Patra
Department of Computer Science, Berhampur University, Berhampur 760007,
Odisha, India
e-mail: mrpatra12@gmail.com

© Springer Nature Singapore Pte Ltd. 2019   145
H. S. Behera et al. (eds.), *Computational Intelligence in Data Mining*,
Advances in Intelligent Systems and Computing 711,
https://doi.org/10.1007/978-981-10-8055-5_14

their societal connections and involvements, etc. This type of mining is sometimes directed toward disclosure of private information. It has been studied that the disclosure of identity of an entity in a social network may lead to deterministic or probabilistic inference of private information of concern individual. To obstruct the privacy disclosure, privacy preserving publication of the social graph is essential. Several anonymization techniques were proposed in last decade, each focuses on the solution of a specific type of predefined attack model. Our proposed method provides a solution to degree attack adopted in [1]. We mainly focus on identity disclosure [1, 2] of a node in a social graph. Privacy preserving data publication has been explored extensively in relational databases. Methods like secure multiparty computation [3] and data obscuration [4] are used in privacy preserving data publication. Various anonymization models like k-anonymity [5, 6], l-diversity [7], (α, k) anonymity [8], and anatomy [9] are proposed by the data scientists to facilitate anonymization in relational data. The direct application of these models in social graph is not possible because the dependencies between the social entities play an important role in social network that create problems in anonymization [10]. The societal extension of this anonymization problem makes it even more challenging and dynamic. Naïve anonymity [11] is a basic model of graph anonymization. This model is very much susceptible to attacks. The identity disclosure is possible even with simple background knowledge. Wentao et al. [11] proposed k-symmetry model, which modifies a naïve anonymized network such that for any vertex in the network, there are at least k-1 structurally equivalent counterparts. K-degree anonymization model [1] is a stronger model than naïve anonymity model. The operations like edge addition, edge deletion, swapping, vertex addition are used to convert a social graph to anonymous graph. In the edge addition problem, the k-label sequence anonymity of arbitrary labeled graph is hard [12]. To achieve k-anonymity by addition of minimum vertices, on vertex-labeled graph is NP-complete [13]. k-degree anonymization cannot ensure identity protection under structural knowledge domain. Deciding adversary background knowledge is an important step that must be carefully considered before proposing any solution. In [14], we discussed k-degree closeness anonymity model by assuming a k value in k-degree anonymous sequence. In this paper, we propose a level-cut heuristic-based hierarchical clustering to generate k value and k-degree anonymous sequence. Our approach uses Havel–Hakimi algorithm [15, 16] to realize degree sequence. If the degree sequence is not realizable, new degree sequence is generated using random noise addition. Our method supports only edge addition operation to do so.

Fig. 1 Sample graph

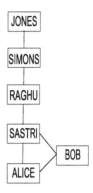

2   Background

2.1   Naïve Anonymization

Definition 1 (*Naïve Anonymization* [14]) A *graph* G(*V, E*) is mapped to an iso-morphic *graph* G(*V, E*), such that *Label*(*V*) is mapped to *Label*(*V*), where *Label* (*V*) ∈ *RS*(*Random Set*).

There is no change in any structural property of the original graph. Figure 1 shows a sample graph. Figure 2 is a naïve anonymized version of this graph, constructed by one-to-one mapping {Alice → 1 Bob → 2 Sastri → 3 Raghu → 4 Simons → 5 Jones → 6}.

2.2   k-degree Anonymity

Definition 2 (*K-degree Anonymous sequence* [14]) A graphic degree sequence *ds* (*G*) is k-degree anonymous if |*ds*(*G*)| / *ud*(*G*)| ≥ *k*, *k* = *min*(*F*), where *ud*(*G*) is the sequence of unique degrees and *F* = $f_1, f_2, ..., f_n$ is the frequency sequence of *ud*(*G*) in *ds*(*G*).

3   Problem Statement

In this section, we discuss the identity disclosure problem with degree attack prospective. Our focus in this paper is totally on identity disclosure problem, where there is a high probability of disclosure of identity of an individual. In Sect. 2, we discussed naïve anonymization which fit well only in a zero knowledge adversary model. A simple knowledge about the graph like degree of node can disclose the identity. Suppose the adversary knows that degree of *Sastri* is 3, then identification

Fig. 2 Naïve anonymous
graph

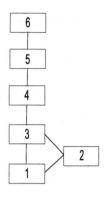

Fig. 3 two-degree
anonymous

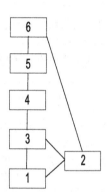

of *Sastri* is easy in Fig. 2. Because in naïve anonymized version given in Fig. 2, only one node with degree 3 is present that is node number 2. A preferable stronger model is k-degree anonymity. Figure 3 is a two-degree anonymous graph of Fig. 1. The probability of finding ***Sastri*** is reduced to 1/2 in Fig. 3 from 1 in Fig. 2. Hence, the structural uniqueness needs to be hidden during publication. In this context, we need to hide degree uniqueness in the published graph, so that the probability of successful mapping of original entity to the graph node is $1/k$, where k is the anonymity parameter.

We have explained the detailed approach in Sect. 4. Our solution approach is executed according to the four steps explained in Fig. 4. The original degree sequence in Fig. 1 is {3, 2, 2, 2, 2, 1}; when it passes through the hierarchical clustering, it generates a two-anonymous degree sequence {3, 3, 2, 2, 2, 2}. By investigating the graphic property of this degree sequence by using Havel–Hakimi algorithm, we found it to be graphic and we constructed the two-anonymous graph from the anonymous degree sequence. During the construction, we first construct the original graph. The residual sequence is generated by subtracting original degree sequence from the k-degree anonymous sequence, i.e., {3, 3, 2, 2, 2, 2}−{3, 2, 2, 2, 2, 1} = {0, 1, 0, 0, 0, 1}. So an edge can be connected to two nodes those nodes were having degree 2 and degree 1 originally, and after doing this, the

Fig. 4 Overall structure

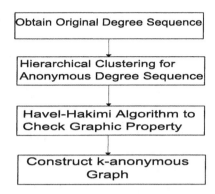

residual is a zero sequence and we constructed the resultant graph shown in Fig. 3 that is a two-anonymous graph.

3.1 Attack Model and Privacy Model

Our attack model is based on the degree of the node. It is named as degree attack. Based on the degree attack model, we have considered a privacy model to ensure k-degree anonymity.

Definition 3 (*Degree Attack* [11]) The attacker knows the degree of the target victim before mapping this information to published graph.

Definition 4 (*k-degree Anonymity* [14]) A graph G is called k-degree anonymous if it satisfies k-degree anonymous sequence.

Though we can use different operations like edge deletion, edge addition, vertex deletion, vertex addition, swapping, we have considered only edge addition for generating k-anonymous graph and anonymous sequence is generated to support this operation.

4 Proposed Approach of Anonymization

We propose an approach to convert a social graph G with degree sequence ds to a k-degree anonymous graph $G1$ with k-degree anonymous sequence $ds1$. The steps 1–4 of Fig. 4 explain our overall approach of anonymization.

4.1   Obtain Original Degree Sequence

The first and foremost job is to obtain the degree sequence of the original graph. The degree sequence of the graph G is ds, such that $ds = \{d_1, d_2, \ldots, d_n\}$ where n is the number of nodes and $d_i \geq d_{i+1}$, where $1 \leq i \leq n-1$. We can easily find the degree sequence ds from the adjacency matrix of the graph G. The degree of a node x is the addition of all 1 present in that particular row in the adjacency matrix. Then, we can sort the resultant degrees to obtain a degree sequence.

4.2   Hierarchical Clustering for Anonymous Degree Sequence

Hierarchical clustering adopts a hierarchical decomposition of the given set of data objects. It is either agglomerative or divisive. The dendrogram tree structure is used to represent the process of hierarchical clustering. In dendrogram, among different distance metrics, we have used *Cityblock* distance with *average* linkage. First, we partition the set of degree sequence into groups and then assign cluster labels to the groups. Here, we are using clustering to divide the given degree sequence into various groups. During cluster formation, in each ith cluster, the diversity of degree distribution is τ. It is formulated as Eq. (1).

$$|maxdegree(Cluster(\text{i})) - mindegree(Cluster(\text{i}))| = \tau \qquad (1)$$

4.3   Defining Level-Cut

We propose two algorithms: Algorithm 1, *KAnonSequence*() and Algorithm 2, *FormCluster*(). In Algorithm 2, the initial *level-cut* is set at the level 1 (step 2), that is the bottom level of dendrogram tree with nl number of levels as shown in Algorithm 2. We update the *level-cut* (*levelcut in Algorithm*) by moving to the upper level, up to the level where no cluster contains a single object (steps 3–9). The algorithm returns all the clusters present in the level *level-cut*.

Algorithm 1 generates k-degree anonymous degree sequence by the help of Algorithm 2. In Algorithm 1, step 2 calls a procedure *HeirarchicalCluster*(); this procedure returns a dendrogram tree. The step 3 stores all clusters at the *level-cut* in m by using Algorithm 2. The steps 4–7 compute the cluster label of all the objects. The function *Clusterof*(i) returns label of ith vertex. The matrix *cluster* contains all nodes' degree in column 1 and their corresponding clusters' label in column 2. The steps 8–10 find anonymous degree sequence, *Anonseq*. The function *maxdegree* (i) computes the maximum vertex degree in cluster i. All degree of *cluster*(i) is mapped to the maximum degree in *Cluster*(i). Each ith cluster is anonymized with its anonymous degree $ad(i)$, which is calculated by Eq. (2). The step 12 assigns the size of minimum size cluster to k. The function $minsizecluster(m)$ finds the minimum size cluster. The algorithm returns k value with $Anonseq$. The anonymous degree sequence only supports the edge addition operation to original graph to obtain anonymous graph, as the degrees are mapped to the maximum degree of the cluster.

$$ad(i) = maxdegree(cluster(i)) \qquad (2)$$

4.4 Havel–Hakimi Algorithm to Check Graphic Property

Before applying Havel–Hakimi algorithm [15], we use the necessary and sufficient condition proposed by Erdos and Gallai [1, 16]. In this, anonymous sequence $AS = \{d_1, d_2, d_3,..., d_n\}$ generated in Sect. 4.2 is sorted so that $d_i \geq d_{i+1}$. The anonymous sequence $\{d_i, d_{i+1}, d_{i+2},..., d_n\}$ is converted to a new sequence $\{d_x-1, d_{x+1}-1, d_{x+2}-1,..., d_y-1,..., d_n\}$ where $x = i + 1$, $y = i + d_i$. Then, the next vertex with nonzero degree is considered as v_i, after sorting. This process repeats until all elements of AS are 0. If it contains some negative elements at the last, then it is non-graphic. The non-graphic degree sequence is modified to new sequences by random noise addition. During the random addition, we select different part of the degree sequence randomly and add less random noise to generate multiple graphic anonymous degree sequences. Here, we measure the MD [4] of the new sequences with existing anonymous degree sequence and select the one with less MD. This step generates an anonymous closer degree sequence, which minimizes modifications. MD between the two degree sequences say A and B can be measured by using Eq. (3).

$$MD(A, B) = \sum_{i=1}^{n} (A_i - B_i) \qquad (3)$$

4.5 Construction of Anonymous Graph

In this work, we have considered only edge addition as a modification operation. Hence, there is no reduction in any degree, and only there is an increment in the degree to satisfy k-anonymity. The detailed of construction algorithm is omitted due to the limited space, but the steps are explained briefly below.

Step-1: Construct the original graph and subtract the original degree sequence from the k-anonymous degree sequence called as residual sequence. Label the graph for better understanding.

Step-2: From the residual sequence randomly choose any two vertices with positive degree and construct an edge between them. Reduce the degree by one of the end nodes of the constructed edge.

Step-3: Continue step 2 until all the degrees are converted to 0. At the last, we get the k-degree anonymous graph.

Algorithm 1.	Algorithm 2.
KAnonSequence (ds [ ], n)	FormCluster($Dendro$, nl)
Input:- ds [ ]: Degree Sequence of a social	Input:- $Dendro$:Dendrogram Tree
graph G	nl: Number of levels in $Dendro$
n: Number of nodes in G	Output:- Clusters at updated
Output:-$Kdegreeseq$ [ ]:	level-cut
k-anonymous degree sequence	1:**Begin**
	2: *levelcut*= 1
1:Begin	3: **if**(no single object in a cluster
2: $Dendro = HeirarchicalCluster$ (ds[ ], n)	in *levelcut*)
3: m= $FormCluster$ ($Dendro$, nl)	4: **return**(clusters in *levelcut*)
4: **for** i= 1 to n	5: **else**
5: $Cluster[i][1] = ds[i]$	6: **for** $i = levelcut$ to nl
6: $Cluster[i][2]$ = Clusterof(i)	7: **if**(no single object in a
7: **end**	cluster in level i)
8: $Anonseq[n]$=0;	8: **return**(clusters in
9: **for** j=1 to n	level i)
10: Anonseq[j]=maxdegree(Cluster[j][2])	9: **end**
11: **end**	10: **end**
12: k= sizeof(minsizecluster(m))	11: **end**
13: return(k,Anonseq)	12:**End**
14:**End**	

5   Result and Analysis

The algorithms are implemented in MATLAB on different synthetic data sets. We have created five different synthetic graphs in SocNetV and supplied corresponding adjacency matrix to MATLAB for analysis. The analysis result is shown in Table 1.

Table 1 Analysis on synthetic graphs

Graphs	N	E	NC	k	E′ − E	ILoss
SGraph1	19	31	5	2	2	6.45
SGraph2	30	88	5	3	8	9.09
SGraph3	15	36	3	3	9	25
SGraph4	20	53	7	2	0	0
SGraph5	25	93	3	3	34	36

In this table, N represents number of nodes, E is number of edges, NC is number of cluster formed, k is the anonymity value, $E' - E$ is the number of edge added to achieve anonymity, and $ILoss$ is the information loss measured according to Eq. (4).

$$ILoss = \frac{E' - E}{E} \times 100 \qquad (4)$$

where E' is the number of edges in $G1$ (*Anonymized Graph*) and E is the number of edges in G(*Original Graph*). Our method follows a different way to generate k-degree anonymous sequence than proposed in [1]. Method proposed in [1] does not work if the value of k is not predefined, but our method automatically generates a k value with the anonymous degree sequence. If the k value is given, then it is irrelevant to use our approach. It can be used only when we are not supplied with any k value and want to find it by running this heuristic. From *ILoss* column of Table 1, we observe that in case of **SGraph4**, no edge addition is required as the original graph is anonymous one. In case of **SGraph5**, we observe that the number of edge addition is more; hence, information loss is also more. This curse of the level-cut heuristic happens due to application of only one operation that is edge addition. For better result, we can also implement edge deletion and simultaneous edge addition and edge deletion.

6 Conclusion

In this paper, we propose a method for automatic generation of k value with k-degree anonymous sequence by using a level-cut on the dendrogram tree generated by hierarchical clustering. MD is used to select the anonymous sequence from set of sequences when the sequence generated from the heuristic is non-graphic. The future work may include the operations like simultaneous edge addition and deletion to reduce the information loss.

References

1. Liu K and Terzi E. Towards identity anonymization on graphs. In: Proceedings of ACM SIGMOD, 2008, 93–106.
2. Sweeney L. Achieving k-anonymity privacy protec-tion using generalization and suppression. International Journal on Uncertainty, Fuzziness, and Knowledge-Base Systems 2002, 10(5): 571–588.
3. Lindell Y, Pinkas B. Secure Multiparty Computation for Privacy-Preserving Data Mining. The Journal of Privacy and Confidentiality 2009, 1(1), 59–98.
4. Hann J, Kamber N. Data mining: Concepts and techniques. San Francisco: Morgan Kaufmann Publishers, 2001.

5. Aggarwal G, Feder T, Kenthapadi K, Motwani R, Panigrahy R, Thomas D, Zhu A. Anonymizing tables. In: ICDT, 2005, 246–258.
6. Sweeney L. k-ANONYMITY: A Model for Protecting Privacy. International Journal on Uncertainty, Fuzziness and Knowledge-based Systems 2002, 10 (5): 557–570.
7. Machanavajjhala A, Gehrke J, Kifer D, Venkitasubramaniam M. l-diversity: Privacy beyond k-anonymity. In: Proceeding of 22nd ICDE, 2006, 24.
8. Chi-Wing Wong R, Li J, Fu A W, Wang K. (α, k) Anonymity: An Enhanced k-Anonymity Model for Privacy Preserving Data Publishing, KDD'06, 754–759.
9. Xiao X, Tao Y. Anatomy: Simple and Effective Privacy Preservation. VLDB, 2006, 139–150.
10. Liu K, Das K, Grandison T, Kargupta H. Privacy-Preserving Data Analysis on Graphs and Social Networks. In: Kargupta H, Han J, Yu P, Motwani R, Kumar V, editors. Next Generation of Data Mining, Chapman & Hall/CRC, 2008, 419–437.
11. Wentao W, Yanghua X, Wei W, Zhenying H and Zhihui W. K-Symmetry Model for Identity Anonymization in Social Networks. In: Proceedings of the 13th International Conference on Extending Database Technology (EDBT'10), 2010, 111–122.
12. Kapron B.M, Srivastava G, Venkatesh S. Social Network Anonymization via Edge Addition. ASONAM, 2011, 155–162.
13. Chester S, Kapron B.M., Ramesh G, Srivastava G, Thomo A, Venkatesh S. k-Anonymization of Social Networks by Vertex Addition. ADBIS (2), 2011, 107–116.
14. Mohapatra D, Patra M.R. k-degree Closeness Anonymity: A Centrality Measure Based Approach for Network Anonymization. In: Proceedings of ICDCIT, 2015, 299–310.
15. West D.B. Introduction to Graph Theory. Second Edition: PHI, 2009.
16. Erdos P., Gallai T., Graphs with Prescribed Degrees of Vertices, Mat. Lapok (1960).

Reevaluation of Ball-Race Conformity Effect on Rolling Element Bearing Life Using PSO

S. N. Panda, S. Panda, D. S. Khamari, P. Mishra and A. K. Pattanaik

Abstract Longest fatigue life is one of the most decisive criteria for design of rolling element bearing. However, the lifetime of bearing will depend on more than one numbers of explanations like fatigue, lubrication, and thermal traits. Within the present work goals, specifically the dynamic load capability, life factors, and life of bearing have been optimized utilizing a optimization algorithm centered upon particle swarm optimization (PSO). Here, life factors are being represented based on reliability, materials, and processing and operating conditions. Also from the reliability concepts, strict series system is considered which depicts the total bearing system. A convergence study has been performed to make certain the most desirable factor in the design. The most suitable design outcome shows the effectiveness and efficiency of algorithm.

Keywords Particle swarm optimization · Life factors · Ball-race conformity

S. N. Panda (✉) · D. S. Khamari
Department of Production Engineering, V.S.S. University of Technology,
Burla 768018, Odisha, India
e-mail: suryanarayan.uce@gmail.com

D. S. Khamari
e-mail: debanshushekhar@gmail.com

S. Panda · P. Mishra
Department of Mechanical Engineering, V.S.S. University of Technology,
Burla 768018, Odisha, India
e-mail: sumanta.panda@gmail.com

P. Mishra
e-mail: priya.punya@gmail.com

A. K. Pattanaik
Department of Mechanical Engineering, Govt. College of Engineering,
Kalahandi 766002, Odisha, India
e-mail: ajitpuce@gmail.com

© Springer Nature Singapore Pte Ltd. 2019
H. S. Behera et al. (eds.), *Computational Intelligence in Data Mining*,
Advances in Intelligent Systems and Computing 711,
https://doi.org/10.1007/978-981-10-8055-5_15

1 Introduction

The design disorders for rolling element bearing have pleasant affect on the performance, fatigue life; wear life, and soundness of bearings. Thus, the effective design of bearing can affect the high quality operation further affecting financial system of machines. The accountability of choosing a most beneficial design from all feasible replacement designs is an extraordinarily tedious job for a bearing designer. This leads to the requirement of optimal design of rolling element bearings for effective bearing life.

Palmgren [1] depicted about bearing life measure and an analytical formula, i.e., L_{10} life. Weibull [2] described failure theory using contact stress concept. Lundberg and Palmgren [3] modeled the principal relation between bearing life and geometry of bearing, relating to variables, viz. ball diameter, pitch diameter, conformities of raceways of inner and outer, ball numbers, also contact angle. The Lundberg–Palmgren [4] equations include ball life to their analysis for bearing life prediction relating to the inner and outer races lives. The Lundberg–Palmgren theory depicts relation between the Hertz stresses with fatigue life as inverse 9th power for the ball bearings. Zaretsky [5] calculated the ball set life related to the races for a ball bearing which depends upon the relative contact (Hertz) stresses at the respected races. Analysis reported by Zaretsky et al. [6] determined bearing life as affected by race conformity and incorporated the life factors using the Lundberg–Palmgren theory. An algebraic relation proposed by Zaretsky et al. [6] calculates life factors (LF_c) and determines L_{10} life for ball bearing, considering series summation of inner and outer races conformity of ball bearing.

A GA-based constraint optimization technique is implemented by Chakraverty et al. [7] with five design variables for optimum design of bearing life. Tiwari et al. [8] also formulated methodology for optimization of fatigue life of tapered roller bearing using evolutionary-based algorithm. Waghole and Tiwari [9] described the use of metaheuristics for optimization of the dynamic load of needle roller bearing. Gupta et al. [10] proposed a multiobjective optimization approach using NSGA-II for roller bearing with load capacity in dynamic and static case and elasto-hydrodynamic minimum film thickness as multiple objectives. These researchers have considered the optimum designed approach as Changsen [11] predicted design model. Based on the stated review, it is noticeable that soft computing approach is the recent trends of the present researches but they have not emphasized their work basing on Lundberg and Palmgren theory.

The aim of this proposed work is an extension of the work proposed by Zaretsky et al. [6] where he includes relation between bearing fatigue life and maximum Hertz stress as 9th power, hence to find agreement of the results as using soft computing method as particle swarm optimization considering the total bearing system as a series system where inner race and outer race considered to be in series.

2 Bearing Geometry Formulation

The finish geometry (Fig. 1) of a deep score ball bearing is characterized by the different geometric variables, viz. bore diameter (d), outer diameter (D), bearing width (w), ball diameter (D_b), pitch diameter (D_m), inner and outer raceway curvature coefficients (f_i and f_o), and number of rolling component (Z).

Here design for entire inner geometry (i.e., D_b, D_m, fi, fo, and Z) of a bearing, at the same time optimizing its efficiency attributes and global fatigue life, is addressed.

2.1 Design Variables

The design variables are fundamentally geometric specifications and different factors, called primary specifications. Stated specifications are to be resolved in the bearing outline. The information specifications are as:

$$X = [D_b, Z, D_m, f_o, f_i, K_{Dmin}, K_{Dmax}, \varepsilon, e, \zeta] \tag{1}$$

where

$$f_0 = r_0/D_b \text{ and } f_i = r_i/D_b.$$

The parameters for bearing interior geometries are D_m, D_b, Z, f_i, and f_o, whereas K_{Dmin}, K_{Dmax}, ε, ζ, e are part of imperatives [9] and are typically kept constant during design of bearings [9]. For reward work, theses are also handled as variables.

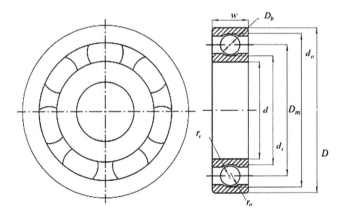

Fig. 1 Radial deep groove ball bearing internal geometries

Assembly attitude angle (ϕ_0) of a bearing additionally varieties an principal constraint on the performance of rolling elements. Established as analytical induction exhibited in [9], the following formulation is the assembly attitude

$$\phi_0 = 2\Pi - 2\cos^{-1}\frac{\left[\{(D-d)/2 - 3(T/4)\}^2 + \{D/2 - (T/4) - D_b\}^2 - \{d/2 + (T/4)\}^2\right]}{2\{(D-d)/2 - 3(T/4)\}\{D/2 - (T/4) - D_b\}}$$

(2)

where

$$T = D - d - 2D_b$$

(3)

2.2 Objective Function

The performance measure of rolling element bearing (deep groove ball bearing), namely dynamic load capacity (C_d) is optimized for best performance of bearing. The dynamic load capacity directly influences the exhaustion life of bearing. It is expressed as,

$$\text{Fatigue life in millions of revolutions, } L = \left(\frac{C_d}{F}\right)^a$$

(4)

where F is applied load and a = 3 for ball bearings

$$C_d = \begin{cases} max\left[-f_c Z^{2/3} D_b^{1.8}\right] & D_b \leq 25.4\,mm \\ max\left[-3.647 f_c Z^{2/3} D_b^{1.4}\right] & D_b > 25.4\,mm \end{cases}$$

(5)

$$f_c = 37.91\left\{1 + \left[1.04\left(\frac{1-\gamma}{1+\gamma}\right)^{1.72}\left(\frac{f_i(2f_o-1)}{f_o(2f_i-1)}\right)^{0.41}\right]^{10/3}\right\}^{-0.3}\left[\frac{\gamma^{0.3}(1-\gamma)^{1.39}}{(1+\gamma)^{1/3}}\right]\left[\frac{2f_i}{2f_i-1}\right]^{0.41}$$

(6)

where $\gamma = D_b\cos\alpha/D_m$ is not an impartial parameter, thus it does not show up in aim of plan parameters. Observed α is the free-contact attitude angle (in present case zero) that relies on bearing geometry. The dynamic load capacity is being determined on premise of most extreme octahedral stress developed between rolling element and races, where i represents number of rows, and it equals one for unit row deep score rolling bearing.

As scope of design of rolling aspect bearing, various practical design requirements are given by scientists so to cut back the parameter space for ease of design optimization. Thus, many constraints conditions [10] are being implemented on the objective function.

2.3 Bearing Life Prediction

Based on the work of researchers [2–4], the survival S is written as a relation of orthogonal shear stress τ_o, life η, depth to the maximum orthogonal shear stress Z_o, and stressed volume V as Eqs. (7), (8).

$$\ln\frac{1}{S} \sim \tau_o \frac{\eta^e}{Z_o^h} V \tag{7}$$

$$V = aZ_o l \tag{8}$$

where a, Z_o are functions of the maximum Hertz stress S, l is the length of rolling path of rolling elements, generally considered same as circumference.

Formulation for the bearing life [3] as Eq. (9),

$$L_{10} = \left(\frac{C_d}{p_{eq}}\right)^p \tag{9}$$

For deep groove, ball bearings exponent p = 3.

Reliability, materials, processing and operating conditions may affect the life of bearing, thus the aforesaid formula can be rewritten as Eq. (10) considering a_1, a_2, and a_3 coefficients.

$$L = a_1 a_2 a_3 Z_o L_{10} \tag{10}$$

The relation between ball bearing fatigue life and maximum Hertz stress S along with bearing load equivalent P_{eq} is presented as in Eq. (11).

$$L \sim \left(\frac{1}{p_{eq}}\right)^p \sim \left(\frac{1}{S_{max}}\right)^n \tag{11}$$

where as per Hertz theory, p = $n/3$ and accordance with Lundberg–Palmgren, Hertz stress life exponent $n = 9$.

2.4 Life Ratio of Outer to Inner Race

The fatigue life L of a respected race as inner and outer races can be determined as follows in Eq. (12),

$$L \sim \left(\frac{1}{S_{max}}\right)^n \left(\frac{1}{l}\right)^{\frac{1}{e}} \left(\frac{1}{N}\right) \tag{12}$$

where N is number of stress cycles per inner race revolution, and e is Weibull slope generally taken as 1.11.

The ratio X is outer race life to inner race life can be written as Eq. (13),

$$\frac{L_o}{L_i} \approx \left(\frac{S_{maxi}}{S_{maxo}}\right)^n \left(\frac{1}{k}\right)^{\frac{1}{e}} \tag{13}$$

2.5 Bearing Life Factor

The appropriate life factor using Eq. (14) can be determined based on life factors LF_i and LF_o of respective races conformity after normalizing the value of Hertz stress for respected races to a standard conformity of 0.52 [6].

$$LF = \left(\frac{S_{max} 0.52}{S_{max}}\right)^n \tag{14}$$

Value of $(\mu\nu_{0.52})$ is different for inner and outer races, called as transcendental functions varies with race conformity.

2.6 Series System Reliability

Lundberg and Palmgren [4] expressed the relation among system life and individual component life. A number of multiple components compose the bearing system, where each may have different life. Thus, system L_{10} life can be formulated using series system of each components fatigue lives Eq. (15).

$$\frac{1}{L_{10}^e} = \frac{1}{L_{10_i}^e} + \frac{1}{L_{10_o}^e} \tag{15}$$

2.7 Bearing Life Factors and Life of Bearing

The life factors of inner race life and outer race life as formulated [6] in Eqs. (16), (17)

$$LF_i = \left[\frac{\left(\frac{2}{D_m - D_b} + \frac{4}{D_b} - \frac{1}{0.52 D_b} \right)^{2/3} (\mu v)_i}{\left(\frac{2}{D_m - D_b} + \frac{4}{D_b} - \frac{1}{f_i D_b} \right)^{2/3} (\mu v)_{0.52}} \right]^n \tag{16}$$

$$LF_o = \left[\frac{\left(-\frac{2}{D_m - D_b} + \frac{4}{D_b} - \frac{1}{0.52 D_b} \right)^{2/3} (\mu v)_o}{\left(\frac{2}{D_m - D_b} + \frac{4}{D_b} - \frac{1}{f_o D_b} \right)^{2/3} (\mu v)_{0.52}} \right]^n \tag{17}$$

Life factor and life of bearing of the bearing accordance with Lundberg–Palmgren approach as Eqs. (18), (19) as follows:

$$LF_c = \left[\frac{(LF_i)^e (LF_o)^e (X^e + 1)}{(LF_o)^e X^e + (LF_i)^e} \right]^{1/e} \tag{18}$$

$$L_{10_m} = \frac{(LF_i)(LF_o) X L_i}{[(LF_o)^e X^e + (LF_i)^e]^{1/e}} \tag{19}$$

Life factor and life of bearing of the bearing accordance with Zeratsky approach as Eqs. (20), (21) as follows:

$$LF_c = \left[\frac{(LF_i)^e (LF_o)^e (X + 2)}{(LF_o)^e X^e + 2(LF_i)^e} \right]^{1/e} \tag{20}$$

$$L_{10_m} = \frac{(LF_i)(LF_o)(X L_i)}{[(LF_o)^e X^e + 2(LF_i)^e]} \tag{21}$$

3 Proposed Method

In this present work the different objectives like dynamic load capacity as Eq. (5), life of outer and inner races as Eq. (12), ratio of life of outer and inner races as Eq. (13), life factors as Eqs. (18), (20) and mean life Eqs. (19), (21) of bearing has been optimized under constraint conditions [10] using population-based algorithm as particle swarm optimization and optimum parameters as in Eq. (1) values were observed considering each objective under single objective problem.

3.1 Optimization Algorithm (PSO)

A population-centered evolutionary algorithm is proposed, and particle swarm optimization (PSO) has been developed by Kennedy and Eberhart [12]. Bound to the search space, each and every particle maintains track of its positions, which is associated with the most effective solution (fitness), it has observed up to now, pBest. Another best esteem followed by the global best version of the particle swarm optimizer is the global best value, gBest, and its position is obtained thus far by any particle in the population. The strategy for imposing the PSO is to initialize a population of particles, evaluate the fitness worth of each and every particle, evaluate each and every particle's evaluated fitness and the fitness evaluation with the population's total prior pleasant, and update the velocity and role of the particle as in Eq. (22) and Eq. (23), once more evaluation of fitness value of each and every particle until the stopping criterion is met commonly as the highest number of iterations. c1 and c2 are rated governing the cognitive and social components. The values are taken 0.5 each, respectively.

$$v[] = v[] + c1*rand()*(pBest[] - present[]) + \\ c2*rand()*(gBest[] - present[]) \tag{22}$$

$$present[] = present[] + v[] \tag{23}$$

4 Outcomes and Analysis

Table 1 gives an imperative perception of dynamic capacity of bearing utilizing PSO under constraint conditions [10]. The dynamic capacity value using PSO was found to be 5822.70 which is higher than the catalog value, and hence, it indicates better design. The value of life ratio is found to be 5.42 which is less than the value in Lundberg–Palmgren. Life ratio is less means life of inner race is more which indicates better design. Life factor for inner race is found to be 1.1998 which indicates that inner race is under more stress than outer race. The life factor of bearing system and mean life as resulted are minorly deviating which shows a close

Table 1 Optimization outcomes

D	d	w	D_b	D_m	Z	f_i	f_o	Φ_0	K_{Dmin}	K_{Dmax}
30	10	9	6.2	20.05	7	0.515	0.515	3.77	0.4296	0.6482
ε°	e	β	C_d	LF_i	LF_o	X_{life}	LF_c	LF_{cz}	L_{10m}	L_{10mz}
0.3	0.0659	0.743	5822.7	1.1998	1.2333	5.42	1.2116	1.2116	1.464×10^3	1.4643×10^3

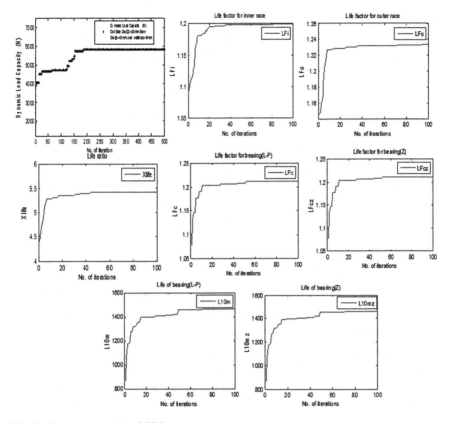

Fig. 2 Convergence traits of PSO

agreement of soft computing approach for bearing design. Further, Fig. 2 depicts the convergence characteristics of all performance parameter with no of iterations. It shows that the life prediction approach is quite same a Lundberg–Palmgren approach.

5 Conclusions

The PSO algorithm is applied to the constraint optimization problem involving dynamic load capacity along with life factors and life of bearing for deep groove ball bearing. The algorithm successfully handles mixed integer variables. The reported result indicates the preeminence of proposed PSO algorithm over design-based problems in prospects to rolling element bearing. The convergence characteristics show that race conformity is an important parameter that effects not only the contact stress but also life. Tiwari et al. also reported that the conformity ratios are the important geometric consideration for optimum design of rolling element bearings. So this study is a good agreement with work of Zaretsky et al. for calculating life of bearing. The content of this paper enables bearing design engineers to calculate changes in bearing life that include the life of race conformity.

References

1. Palmgren, A.: "Die Lebensdauer von Kugellagern (The Service Life of Ball Bearings)," *Z. Ver. Devt. Ingr.*, 68, 14, pp 339–341, (1924).
2. Weibull, W.: "The Phenomenon of Rupture in Solids," *Inge-niorsvetenskapsakademiens Handlingar*, 153. Stockholm, Sweden, (1939).
3. Lundberg, G., and Palmgren, A.: "Dynamic Capacity of Rolling Bearings," *Acta Polytechnica Mechanical Engineering Series*, **1**, 3, (1947).
4. Lundberg, G. and Palmgren, A.: "Dynamic Capacity of Roller Bearings," *Ingeniorsvetenskapsakademiens Handlingar*, 210. Stockholm, Sweden, (1952).
5. Zaretsky, E.V.: STLE Life Factors for Rolling Bearings, STLE SP–34, Society of Tribologists and Lubrication Engineers, Park Ridge, IL, (1992).
6. Zaretsky, E.V., Poplawski, J.V., and Root, L.E.: "Reexamination of Ball-Race Conformity Effects on Ball Bearing Life," *Tribology Trans.*, 50, 3, pp 336–349, (2007).
7. Chakraborthy I, Vinay K, Nair SB, Tiwari R.: Rolling element bearing design through genetic algorithms. Eng Optim; 35 (6):649–59, (2003).
8. Tiwari R, Kumar SK, Prasad PVVN.: An optimum design methodology of tapered roller bearing using genetic algorithm. Int J Comput Methods Eng Sci Mech;13(2):108–27, (2012).
9. Waghole V, Tiwari R.: Optimization of needle roller bearing design using novel hybrid method. Mech Mach Theory, 72(2):71–85, (2014).
10. Gupta S, Tiwari R, Nair SB.: Multi-objective design optimisation of rolling, (2007).
11. Changsen W.: Analysis of rolling element bearings. London: Mechanical Engineering Publications Ltd.; (1991).
12. Kennedy J, and Eberhart R.: Particle swarm optimization. In: Proceedings of the IEEE international conference on neural networks (ICNN). (1995)1942–1948, (1995).

Static Cost-Effective Analysis of a Shifted Completely Connected Network

Mohammed N. M. Ali, M. M. Hafizur Rahman, Dhiren K. Behera
and Yasushi Inoguchi

Abstract The computational power challenges have been increased in the contemporary era, and it motivated the scientist community to find alternative choices to replace the current ones. Systems using the conventional computing power became infeasible to cope with the grand computing problems. Therefore, building a system with special characteristics became the main concern of the research work in this area. As a result, multiprocessor systems have been revealed to manipulate the computing tasks in parallel and concurrently, leading to massively parallel computers (MPCs) which have been spread widely as an adopted solution to be used in solving the complex computing challenges. The structure of underlying interconnection network of these systems plays the main role in improving the overall performance, and in controlling the cost of the system. Thus, many topologies of these networks have been presented in order to find the optimal one. In this paper, we present the architecture of a new hierarchical interconnection network (HIN) called shifted completely connected network (SCCN). This network has been described previously, and the static network performance of this network has been evaluated in previous studies. The main focus of this paper is to analyze the static cost-effective parameter of SCCN which can be calculated from the relation between the static parameters.

M. N. M. Ali
Department of Computer Science, KICT, IIUM, 53100 Kuala Lumpur, Malaysia
e-mail: moh.ali.exe@gmail.com

M. M. Hafizur Rahman (✉)
CCSIT, King Faisal University, Hofuf, Kingdom of Saudi Arabia
e-mail: rahmanjaist@gmail.com

D. K. Behera
Indira Gandhi Institute of Technology, Sarang 759146, Odisha, India
e-mail: dkb_igit@rediffmail.com

Y. Inoguchi
School of IS, JAIST, Asahidai 1-1, Nomi-Shi, Ishikawa 923-1292, Japan
e-mail: inoguchi@jaist.ac.jp

© Springer Nature Singapore Pte Ltd. 2019
H. S. Behera et al. (eds.), *Computational Intelligence in Data Mining*,
Advances in Intelligent Systems and Computing 711,
https://doi.org/10.1007/978-981-10-8055-5_16

Keywords Network-on-chip · Interconnection network · Hierarchical
interconnection network · Static network performance · Shifted completely
connected network (SCCN)

1 Introduction

At the current time, multicomputer systems became the prominent solution for the
grand computational challenges. This technology is using numerous numbers of
processors to be connected and processing the data simultaneously. In addition, it
helps in increasing the execution speed of the computer system. To solve a computational problem in parallel, a group of processors will be connected to share the
data; the communication between these processors must be proficient in order to
avoid the wasting of the computational power. The interconnection networks are
being used to interconnect the processing elements (PE) within the system. These
networks are a time-effective way to keep the system efficiency. Therefore, it is
practically essential to design a dexterous interconnection network. Thereby, many
pieces of research have been conducted to design such network, and many new
topologies have been proposed. However, there is still no clear winner [1]. On the
other hand, the studies which have been done on the conventional networks
revealed poor network performance in case of increasing the network size.
Therefore, using hierarchical interconnection networks (HINs) is the alternative
way to maintain the efficiency of these networks. In HIN, several topologies are
being connected to produce a cost-effective network; in addition, these topologies
are integrated hierarchically to build the higher levels of the proposed system [2, 3].
Furthermore, connecting thousands or millions of nodes by using HIN will maintain
the system with good performance. Therefore, HIN is a promising solution to
replace the conventional networks in building the future generations of the massively parallel computer (MPC) systems. In addition, it is a credible way to decrease
the cost and to improve the system performance [3].

The dexterous cooperation between the intellectual property (IP) cores is the
assurance of delivering good and rich services through utilizing the available
resources [4]. The appropriate architecture to provide accommodation for a high
number of cores is the network-on-chip (NoC); it also guarantees the communication and the data transfer between these cores. The motivations of inventing the
NoC were to replace the conventional bus architecture and to solve the limitation of
the system performance in the case of long interconnect. In addition, NoC has a
crucial role in solving the global wire delay problem by integrating a high number
of IP cores in a single system-on-chip (SoC). Due to these reasons, NoC became the
trusted choice by the industrial sector to be used in designing the on-chip inter-
connects for the multiprocessor systems-on-chip (MPSoC) [4, 5]. Sharing the
wiring resources between many communication flows will make the usage of these
wires more efficient, and it will be clear when one client becomes idle then the
network resources will be used by other clients [6, 7]. Massively parallel computer

(MPC) system composed of either thousands or millions of processors. In addition, the advancements in very-large-scale integration (VLSI) and the network-on-chip (NoC) technology participated in building multiprocessor systems in three dimensions [8].

The shifted completely connected network (SCCN) is a hierarchical interconnection network. It is composed of multiple basic modules (BMs); the BMs of SCCN are connected in a hierarchical fashion to construct the higher levels of this network [9]. In previous studies, we have described in detail the architecture of SCCN; in addition, we have evaluated the static network performance parameters. In order to examine the superiority of SCCN, we have compared it to the most popular conventional interconnection networks, 2D mesh and 2D torus. SCCN revealed good results in all aspects. The main objective of this paper is to analyze the static cost-effective parameter of SCCN by exploiting the relationship between the static network performance parameters in a contemporary manner; these parameters include node degree, diameter, wiring complexity, and the total number of nodes. In Sect. 2 of this paper, we will describe the structure of SCCN; in Sect. 3, we will evaluate the static cost-effective parameter of SCCN; and Sect. 4 will be the conclusion of this work.

2 Proposed Architecture of Shifted Completely Connected Network

Shifted completely connected network (SCCN) is a hierarchical interconnection network. SCCN is composed of multiple basic modules (BMs) connected hierarchically to provide the higher levels of this network. In addition, each basic module (BM) contains six nodes connected completely.

2.1 Basic Module of SCCN

The nodes in the BM of SCCN are connected completely through electrical links. The number of nodes in each BM is six nodes; these nodes are connected in a certain shape to be placed two inside the chip. Using direct electrical wires to connect the nodes within the BM is important in degrading the network cost and improving the performance. In addition, the direct link between each two nodes is significant in decreasing the network congestion and increasing the communication speed. In the BM also the data takes only one hop to reach from the source node to the destination. The characteristics of SCCN are leading to short network diameter which is important in determining the network performance. Figure 1 portrayed the BM of SCCN, and it shows the structure of this network; each node is connected to the other nodes through direct links which will strengthen the performance of this

Fig. 1 BM of SCCN

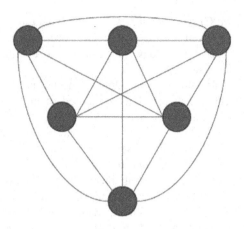

network. The diameter of SCCN is one which is significant to degrade the possibility of the network congestion, and it will be useful in avoiding the latency. As a result, these features are vital to obtaining a system with good overall performance. The relation between the nodes contained by the BM can be represented in pairs. For example, node (N_1) can be connected to these nodes as: (N_1, N_2), (N_1, N_3), (N_1, N_4), (N_1, N_5), (N_1, N_6). Similarly, we can represent the relation between the other nodes. In the BM of SCCN, the packet moves from Ni to reach $Ni + 1$ by traversing only one link such feature, making the interconnection between the nodes inside the BM easier and faster. Figure 1 represents the BM of SCCN which is referring to the Level 0 network.

2.2 The Higher Level of SCCN

Figure 2 illustrates the structure of Level 1 network of SCCN. This level is composing of 2^m groups, where m is an integer number. The reason of making the number of the groups equal to 2^m is to maintain the regularity of the network architecture. Each single group represents a basic module, and these groups are connected hierarchically to provide the higher level network of SCCN. In addition, each group composed of completely connected nodes; these nodes are six in each group. Equation (1) is used to calculate the number of nodes in each level of SCCN, and it has been derived from Fig. 2.

$$Total\ Number\ of\ Nodes = 6 \times 2^m \qquad (1)$$

where m is an integer number, and 2^m is the number of groups in each level. The groups in Level 1 are connected to each other by using either electrical or fiber optic wire. However, for long-distance connection, it is preferable to use the fiber optic wires which maintain the quality of the signal. In contrast, to degrade the cost of

Fig. 2 Level 1 of SCCN

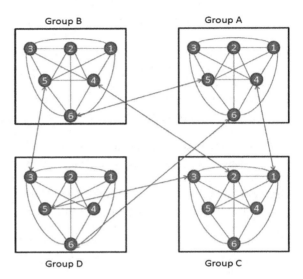

using the fiber optic links, we prefer to replace it by economical fiber wires which have been invented recently. In SCCN network, Level 0 is the basic module of Level 1, and Level 1 is a basic module of Level 2 network, and so on.

The nodes in Level 1 of SCCN are connected based on circular shifting. The shifting is on the binary digits of each node either to the left or to the right. Using the circular shifting is essential to keep a fixed number of nodes in each group. Therefore, the architecture of SCCN is suitable to place the network-on-chip. In contrast, if we use more than six nodes to construct the BM of SCCN, the shifting mechanism will be not applicable. For example, consider the number of nodes is seven and the binary of this number is (111). If we shift the digits of seven to the left or to the right, it will provide the same results. On the other hand, the connection type is determining the direction of the circular shifting. The connection could be either unidirectional or bidirectional connection. In addition, the connection between the groups is affected by the distance between these groups.

Figure 2 depicts the architecture of the Level 1 network, and it is clear that there are four groups A, B, C, and D which are connected to each other. In addition, it is clear that each group contains six completely connected nodes. These nodes can be represented as: $(A_1, A_2, A_3, A_4, A_5, A_6)$ nodes in group A, $(B_1, B_2, B_3, B_4, B_5, B_6)$ nodes in group B, $(C_1, C_2, C_3, C_4, C_5, C_6)$ nodes in group C, and $(D_1, D_2, D_3, D_4, D_5, D_6)$ nodes in group D. In order to interconnect two groups, we need to transform the decimal number of each node either from the source or destination group to their equivalent in binary, and then, we apply the circular shifting on the binary digits of each node in the destination group steps to the left. In this paper, we consider the shifting to the left as the default direction of the unidirectional connection. However, using the right circular shifting will be applied in the case of using the bidirectional connection in order to connect the nodes in two opposite directions; one to the left and the other one will be to the right.

The shifting steps are determined by the distance between the groups. Therefore, in this paper, we considered the sequence of the representative letter of each group to be used in measuring the number of the steps either to the left or to the right. For instance, Level 1 network of SCCN is composed of four groups A, B, C, and D. By applying the sequence of the representative letters between these groups, we will have one-hop distance between A and B, two-hop distance between A and C, and three-hop distance between A and D. In similar way, we can calculate the distance between any two different groups. Therefore, to establish the connection between two groups based on using the circular shifting, we need to consider the distance between these groups. For example, to connect node from group A as a source node to another node from group B as a destination node, we will transform the decimal number of the source node to it its equivalent in binary, and then, we will transform each node from the destination group to their bidirectional digits in binary. Consequently, we will apply the circular shifting mechanism on the binary digits from the bidirectional group B. Based on the distance between the groups A and B; the circular shifting will be one step to the left. In the next step, we will compare the binary numbers of each node from group B after shifting to the binary number of the source node in group A. If the binary digit of any node from the destination group is equal to that of the source node, then the two nodes will be connected. Similarly, we do to connect the nodes from group A to other nodes from other groups. However, we need to consider the distance between these groups according to Fig. 3 which illustrates the relationship between the groups in the higher levels of SCCN network.

From Fig. 2, we can notice that the groups are connected through bidirectional links. That means the packet moves from A_i where ($1 \leq i \leq 6$) to B_i by using the same link in two directions. In order to attain a bidirectional connection between two groups, we need to consider the circular shifting mechanism which we have discussed earlier in this paper. In addition, we need to take into account that the shifting direction could be to the left or to the right. Moreover, we need to assume the distance between the groups based on what we have illustrated in Fig. 3. Therefore, to achieve the bidirectional connection between two nodes from two

Fig. 3 Distance between groups in SCCN

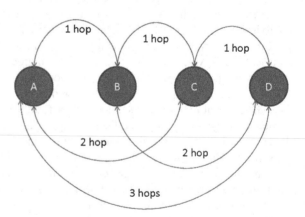

different groups, we will apply the circular shifting on the binary digits of each node from the destination group, also on the source node. The binary digits of each node from the destination group will be shifted steps to the left as a default direction, or steps to the right in order to achieve the connection in the opposite direction. Figure 4 depicts the bidirectional connection between two nodes; for example, A_5 from group A, and B_6 from group B. The relation between the nodes from group A and the other groups can be represented in pairs. From Fig. 2, we can derive the following nodes: (A_5, B_6), (A_4, C_1), and (A_6, D_6). For example, the interconnection between the first pair (A_5, B_6) happened when we transformed the decimal number of node 5 in group A to the equivalent in binary (101), and then, we transform each node from the destination group B to their binary digits. The third step was applying the circular shifting on the binary digits for each node from group B; due to the distance between A and B, the shifting will be one hop to the left. After that we will compare the binary digit of each node after shifting with that of node 5 in group A. Therefore, if any node from group B has the same binary digit of node 5 from group A, the two nodes will be connected. Similarly, we do to connect the second pair (A_4, C_1). However, we need to consider the distance between group A and group C during the shifting process. Thus, the shifting will be two hops to the left, leading to interconnect node number 4 from group A that will be connected to node number 1 in group C. Likewise we connect the last pair (A_6, D_6), and by considering the distance between group A and D, the circular shifting will be three hops to the left. Hence, node number 6 in group A will be connected to node number 6 in group D.

It is possible to connect each node in Level 1 of SCCN completely, and each node from the different groups has a counterpart node to be connected with. However, to decrease the wiring complexity and the cost of applying SCCN, we chose only one global link to be used in connecting the different groups as shown in Fig. 2. The global links between the groups could be either electrical or optical optic wires; however, the usage of the optical wires will be considered for the long-distance connections. In contrast, connecting the groups through more than

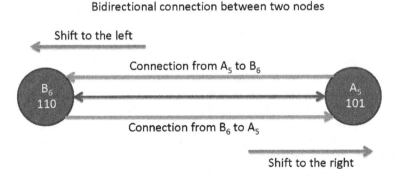

Fig. 4 Bidirectional connection between two nodes

one global link will be effective in controlling the high bandwidth, avoiding the network congestion, increasing the connection speed, and it will be useful for the communication diversity. The decision of increasing the number of global links between the groups will be left to the nature of each system, the available bandwidth, the environmental factors, the number of users, and the conditions surrounding the system area.

3 Static Cost-Effective Analysis of SCCN

The performance of the massively parallel computer (MPC) systems is affected widely by the design of the interconnection network which is the backbone of these systems. Moreover, it has a significant influence on either increasing or decreasing the cost of these systems [10]. The static network performance parameters have been used to measure the effectivity of the interconnection network topology. Based on the results of the evaluation process, the research community determines whether the network is eligible to be used or it needs to be a subject for more improvements. In previous studies, we have evaluated the static network performance of SCCN in terms of node degree, diameter, average distance, wiring complexity, and many of these parameters. SCCN revealed the superiority in almost all aspects over the most popular topologies 2D mesh and 2D torus networks.

In this paper, we will focus on analyzing the static cost-effective factor of SCCN by calculating the relation between some of the static network parameters which we have evaluated in previous work. Some of these parameters have a crucial role in either increasing or decreasing the network cost, and the other is related to the network performance and it has a serious effect in either enhancing or degrading the performance. For example, the network diameter has a critical effect on the network performance and is defined as the maximum number of nodes the packet needs to traverse to reach from the source node to the destination node by choosing the short way. Network with short diameter indicates good performance, while a network with long diameter will provide weak performance [11]. On the other hand, the network cost can be affected by parameters such as the node degree and the network wiring complexity which has the main role in increasing or decreasing the network cost. The node degree defined as the maximum number of links emanating from this node, and it is a measure of the node's I/O complexity [12]. In contrast, the network wiring complexity has an important effect on the network cost by either increasing or decreasing the number of interconnection links inside the network. In addition, the wiring complexity defined as the total number of wires needed to connect each node to the other nodes. Furthermore, the wiring complexity is affected by the node degree of the interconnection network [13]. The network cost is also affected by the number of nodes in each network which participates in increasing the number of the computational devices inside the network. In this paper, we will consider SCCN with 24 nodes and it will be compared to (5×5) 2D mesh and 2D torus which contains 25 nodes. It is quite difficult to compare networks with different shape by considering the same number of nodes; therefore, we have used (5×5) 2D mesh and 2D torus to make the comparison as fair as possible. The static cost-effective analysis for each network can be calculated by using particular criteria to determine the relation between some of the static network parameters; these parameters are the total number of nodes, the node degree, the network diameter, and the wiring complexity. The actual cost of each network will be the result of determining the relationship between these parameters. The higher result of analyzing the static cost-effective for the network is better than the lower one. In this paper, we will use the relation between the previously mentioned parameters to analyze the static cost-effectivity of SCCN; in addition, we will compare the results to those results from analyzing the static cost of 2D mesh and 2D torus network. The cost-effective analysis for each network can be determined from Eq. (1).

$$Static\ Cost\ effective = \frac{Node\ Degree \times Wiring\ complexity}{Diameter \times Total \neq of\ Nodes} \qquad (2)$$

The statical comparison between SCCN, 2D mesh and 2D torus is shown in Table 1. As we mentioned earlier, the number of nodes in SCCN is equal to 24 nodes, and to make the comparison as fair as possible, we used (5×5) nodes in 2D mesh and 2D torus networks. From Table 1, it is clear that the node degree of 2D mesh and 2D torus networks is equal to 4. However, the node degree of SCCN is 6. Higher node degree is useful in improving the network performance by achieving connection diversity which will help in avoiding the network congestion. Table 1 also is showing that the network diameter of SCCN is lower than that of both 2D mesh and 2D torus networks. The short diameter indicates network with good performance and less congestion. The wiring complexity of SCCN is higher than that of the other networks. Higher wiring complexity will increase the network cost; however, we can mitigate the high cost by applying the electrical wires which are less price compared to the fiber optic wires. From Table 1, it is interesting to notice that the result of analyzing the static cost-effectiveness of SCCN is extremely higher than that of the other networks 2D mesh and 2D torus. This indicates that SCCN is a cost-effective network compared to the other networks. Furthermore, SCCN achieved good performance in many aspects compared to these networks. On the other hand, Table 1 shows the static cost analysis result of 2D torus is better than that of the 2D mesh network. The static cost analysis of each network has been calculated from Eq. (1).

Table 1 Comparison between the networks

Network	Total number of nodes	Node degree	Network diameter	Wiring complexity	Static cost-effective analysis
2D mesh	25	4	8	40	0.8
2D torus	25	4	5	50	1.6
SCCN	24	6	3	66	5.5

Based on these results, we can say that the static cost-effective analysis of SCCN is extremely higher than that of both networks, 2D mesh and 2D torus, as shown in Table 1. In addition, SCCN provides good performance over these networks with low network diameter and completely connected nodes. These features of SCCN are great in providing a network with high-speed communication and less congestion. These features will help in building a parallel computer system with good performance.

4 Conclusion

The architecture of the shifted completely connected network (SCCN) has been discussed in this paper. We have described the basic module (BM) of this network, and the higher level network which indicates to Level 1 of the system. Level 1 is composed of multiple BM connected to each other in hierarchical fashion. In addition, we have tabulated the static network performance parameters of SCCN which have been evaluated in previous work, in order to analyze the static cost-effective parameter of this network. The obtained results from evaluating SCCN have been compared to those results of 2D mesh and 2D torus networks. We have used the node degree, diameter, wiring complexity, and the total number of nodes to calculate the cost-effective parameter of the different networks which have been shown in Table 1. The static cost-effective parameter of SCCN was higher than that of the other networks. That means the static cost analysis of SCCN is better than that of the other networks. In addition, SCCN has the best network diameter which indicates network with better performance. Therefore, SCCN yields good features in many aspects which will lead to a parallel computer system with high performance. This paper focused on analyzing the static cost-effective of the shifted completely connected network (SCCN), which proposed to improve the performance of network-on-chip (NoC) system. For the future work, we will have further exploration which includes (1) 3D NoC implementation and (2) dynamic communication performance evaluation by using dimension-order routing algorithm.

References

1. Cheng, S. Y., and Chuang, J. H. 1994. Varietal hypercube-a new interconnection network topology for large scale multicomputer. Proceeding of the International Conference on Parallel and Distributed Systems 1994. pp. 703–708, IEEE.
2. Al Faisal, F., Rahman, M. M. H., and Inoguchi, Y. 2016. Topological analysis of low-powered 3D-TESH network. *IEICE Tech. Report* 2016. Vol. *115, no.* 399, pp. 143–148.
3. Awal, M. R., Rahman, M. M. H., and Akhand, M. A. H. 2013. A new hierarchical interconnection network for future generation parallel computer. Proc. of the 16th

International Conference on Computer and Information Technology (ICCIT). pp. 314–319, IEEE.

4. Anagnostopoulos, I., Bartzas, A., Vourkas, I., and Soudris, D. 2009. Node resource management for DSP applications on 3D network-on-chip architecture. Proceeding of the 16th International Conference on Digital Signal Processing. pp. 1–6, IEEE.

5. Goossens, K., Dielissen, J., and Radulescu, A. 2005. Æthereal network on chip: concepts, architectures, and implementations. IEEE Design & Test of Computers. Vol. 22, no. 5, pp. 414–421.

6. Dally, W. J., and Towles, B. 2001. Route packets, not wires: On-chip interconnection networks. Proceedings of Design Automation Conference 2001. pp. 684–689, IEEE.

7. Bjerregaard, T., and Mahadevan, S. A survey of research and practices of network-on-chip. ACM Computing Surveys (CSUR). Vol. 38, no. 1, pp. 1.

8. Miura, Y., Kaneko, M., Rahman, M. M. H., and Watanabe, S. 2013. Adaptive routing algorithms and implementation for TESH network. Communications and Network. 5(1), 34.

9. Ali, M. N. M., Rahman, M. M. H., & Tengku Sembok, T. M. 2016. SCCN: a cost effective hierarchical interconnection network for network-on-chip. International Journal of Advancements in Computing Technology (IJACT). 8(5), 70–79.

10. Kim, J., Dally, W. J., Scott, S., and Abts, D. 2008. Technology-driven, highly-scalable dragonfly topology. In ACM SIGARCH Computer Architecture News. Vol. 36, no. 3, pp. 77–88. IEEE Computer Society.

11. Rahman, M. M. H., Inoguchi, Y., Yukinori, S. A. T. O., and Horiguchi, S. 2009. TTN: a high performance hierarchical interconnection network for massively parallel computers. IEICE transactions on information and systems. Vol. 92, no. 5, pp. 1062–1078.

12. Rahman, M. M. H., and Horiguchi, S. 2003. HTN: a new hierarchical interconnection network for massively parallel computers. IEICE Trans on Inf. and Syst. Vol. 86, no. 9, pp. 1479–1486.

13. Ali, M. N. M., Rahman, M. M. H., Nor, R. M., and Sembok, T. M. B. T.2016. A high radix hierarchical interconnection network for network-on-chip. In the Recent Advances in Information and Communication Technology, 2016. pp. 245–254. Springer.

Dynamic Notifications in Smart Cities for Disaster Management

Sampada Chaudhari, Amol Bhagat, Nitesh Tarbani
and Mahendra Pund

Abstract The smart city is how citizens are shaping the city by using technology, and how citizens are enabled to do so by getting the support of city's government. Diverse data are collected on a regular basis by satellites, wireless and remote sensors, national meteorological and geological departments, NGOs, and various other international, government, and private bodies, before, during, and after the disaster. Data analytics can leverage such data deposit and produce insights which can then be transformed into enhanced services. Disasters are sudden and calamitous events that can cause severe and pervasive negative impacts on society and huge human losses. It causes enormous evil impact on society. The proposed system is based on disaster management scenario for avoiding negative impacts on society and huge human losses. The system is providing an alert to people leaving in particular area as well as in nearby area. The system is based on the activities on social media during disaster. This system tries to help society by using the information revealed by them only by collecting the data or messages spread by the people suffering from the disaster or the people who have an idea about its occurrence. It will help people to save themselves as well as possibly other living and non-living things that come in society. The system will send the alert message to a particular area or the people who come under that particular area so that people can save their lives as well as their time and other things depending on the types of disaster occur.

S. Chaudhari (✉) · A. Bhagat
Innovation and Entrepreneurship Development Centre, Prof Ram Meghe College
of Engineering and Management, Badnera, Amravati 444701, Maharashtra, India
e-mail: sampada.chaudhari7@gmail.com

A. Bhagat
e-mail: amol.bhagat84@gmail.com

N. Tarbani · M. Pund
PG Department of Computer Science and Engineering, Prof Ram Meghe Institute
of Technology and Research, Badnera, Amravati 444701, Maharashtra, India
e-mail: nmtarbani@mitra.ac.in

M. Pund
e-mail: mapund@mitra.ac.in

© Springer Nature Singapore Pte Ltd. 2019
H. S. Behera et al. (eds.), *Computational Intelligence in Data Mining*,
Advances in Intelligent Systems and Computing 711,
https://doi.org/10.1007/978-981-10-8055-5_17

Keywords Data analytics · Digital technologies · Disaster management Smart cities · Social networks

1 Introduction

Cities are seeking to become a smart city by using networked and digital technologies faces the number of issues such as improving service delivery, growing the local economy, becoming more sustainable, producing better mobility, enhancing the quality of life, and increasing safety and security. A city can be defined as smart when it provides advanced applications and services to the community through cutting-edge information and communication technologies (ICTs) [1]. To make city secure and safe, we can use data mining and analytical techniques designed so far for prediction, detection, and development of an appropriate disaster management strategy based on the collected data from the disasters. To avoid this, there is a requirement to get alert before such disaster destroys all. To provide safety and security to the smart city, it is important to secure city from such natural calamities. For this purpose, we can use data mining. There are data mining models which have been used to predict the time, place, and severity of the disasters.

City data emerge from a wide variety of governmental departments, private and public stakeholders, and individual citizens and visitors and are collected, analyzed, and stored without any kind of central coordination or collaboration. There have been many efforts to predict the disaster based on various sources of data [2]. Hence, sources may not be used in prediction of the disasters, but they have contributed significantly to early detection and adoption for an appropriate disaster response [1]. The smart city scenario is a fertile application domain for different sciences and technologies, in particular for those related to the information and communication areas [3]. Social media Web sites are an integral part of many people's lives in delivering news and other emergency information. This is especially true during natural disasters. The process of collecting and analyzing social media data from Twitter and other social media gives a variety of visualizations that can be generated by the tool in order to understand the public sentiment [4].

The complete process can get divided into collecting information about residences and businesses in the affected area, as well as the people who live there, is collected from public records or ingested from online sources such as Google Places, Foursquare. Data of recent activities are collected from social media platforms such as Twitter, Foursquare, and LinkedIn that provide additional information about the current activity. To avoid the consequence that occurs due to disaster, it is important to manage it. Rapid disaster relief is important to save human lives and reduce property loss. Disaster management is the discipline responsible to mitigate, prepare for, response to, and recover from the disasters with the ultimate goal to save lives, property, and the environment, which ultimately help society. This motivates me to perform research in the sector.

The objectives are to save people and reduce property loss by managing the disaster, as well as to collect data of recent activities from social media platforms such as Twitter, Foursquare, and LinkedIn that provide additional information about the current activity. This will be useful for alerting people to make them safe from the disaster consequences. The objective is to save human lives as well as other living and non-living things. This can be done by analyzing information and taking proper action to help people who may get suffered from the disaster. The remainder of this paper discusses existing approaches for disaster management in smart cities, the proposed approach for the alert generation in smart cities, and experimental results.

2 Existing Disaster Management Systems

U-city is a smart city with the intelligent convergence system that integrates IT and urban engineering technology such as environment, road traffic, safety, and GIS. The network used in U-city should provide services without transmission failure and support functionality which enlarges the role of sensors used to collect information. It proposes a ubiquitous network management system (uNMS) that mobile devices in it are suitable in a mobile network which goes in and out U-city frequently and can manage lots of agents. uNMS minimizes a usage of bandwidth [5].

As shown in Fig. 1, u-City is a system which manages a specific area automatically based on the information such as water pollution, air pollution, and road traffic collected by sensors, and Fig. 2 shows that when a mobile node loses a connection to AP and cannot find any available AP, mobile node turns on ad hoc mode to communicate with other mobile nodes.

Network congestion during disasters and big events is a major issue, especially in metropolitan areas. SPArTaCuS is a framework used to prioritize the network traffic adaptively for such situations in the smart cities using a software-defined network (SDN) approach, where services that require priority are placed in virtualized networks and the mechanism is accomplished through a priority management layer in SDN architecture [6].

Figure 3 gives an overview of SDN. The control is centralized in software-based SDN controllers, thereby giving it the global view of the network due to which the network is viewed as a single, logical switch to the applications and policy engine. Figure 4 shows the network after the network priority is done for the public VN3. Nodes and links marked in white in VN2 denote resources that have been de-allocated, while the red nodes and links in VN3 denote the resources that are being added to VN3. The priority can be stated as VN3_VN2_VN1. SPArTaCuS is used to prioritize traffic based on SDNs for different service classes, divide the traffic based on different organizations, and prioritize them using the priority management layer in the middle box [6, 7].

Figure 5 shows that citizens around a town have installed application interfaces (APIs) on their smart mobiles phones. These will be now capable of scanning the

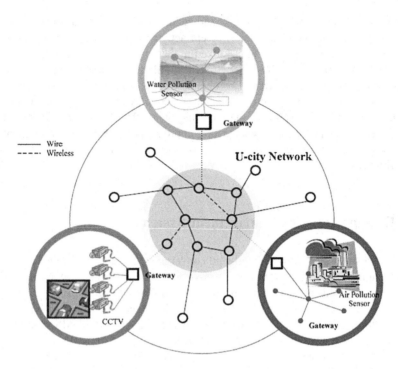

Fig. 1 uNMS structure in u-City [5]

environment, and thus, it starts collecting various information about it. Information could be of personal nature or about their environment. These could include but not limited to positioning, health status, weather data, traffic data, radioactivity levels, noise, and other environmental conditions. Smart cities, smart buildings and various participant users including critical infrastructures, cars, buildings, and humans could be connected via sensors and mobile APIs in order to capture data about their surrounding environment [8–10].

3 Proposed Mechanism for Disaster Management

Social network data are given as an input, which will be shared by users, and that data get stored, filtered, and examined using geo-location. As an output, these alert messages are sent to nearby people for their safety, in which, for good network, ubiquitous network management system is used to alert people in a disaster situation. A disaster management system helps to alert people by sending alert message through network in a disaster situation. Ubiquitous network management system (uNMS) is used in the smart city, which is an existing system, where it is designed for managing disaster network. uNMS [11–13] is a network management system

Fig. 2 Message transmission in disaster situation [5]

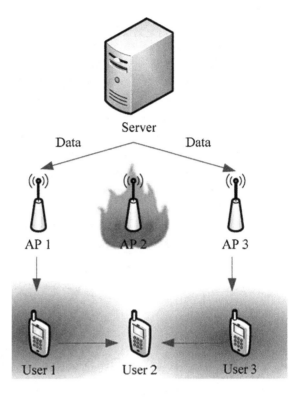

Fig. 3 Software-defined network (SDN) architecture [6]

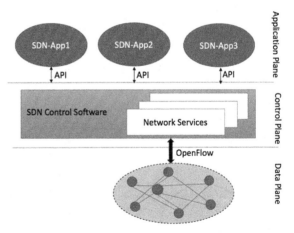

that serves mobile agents with proper quality information by monitoring information and status of mobile nodes in u-city. uNMS is used here in proper network to provide alert message in a disaster situation. After extracting disaster information from social media and by examining the data according to geo-location, the alert

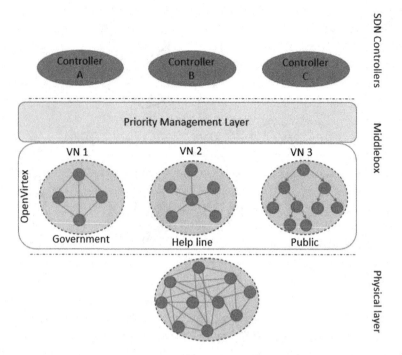

Fig. 4 Modeling smart city networks in SPArTaCuS and network prioritization during big events (VN3_VN2_VN1) [6]

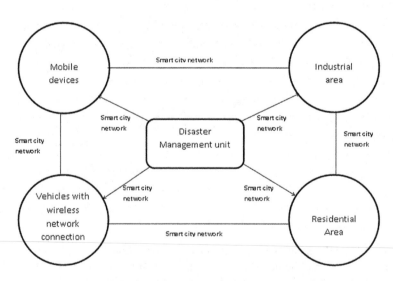

Fig. 5 Block diagram for smart cities and crowdsourcing [7]

Fig. 6 Proposed mechanism
for disaster management

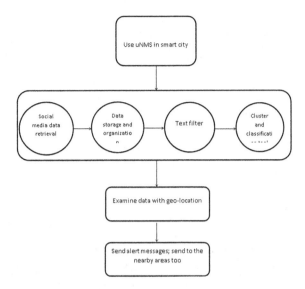

messages get conveyed to nearby people who may get affected due to disaster. The
following steps are used as per the proposed mechanism Fig. 6:

Step 1: Apply ubiquitous network management system in the smart city.
Step 2: Retrieve disaster data from social media.
Step 3: Filter data related to disaster.
Step 4: Examine disaster information according to geo-location.
Step 5: Send alert message to nearby people.

Here, as per the first step, ubiquitous network management system is applied to
the smart city which serves mobile agents with proper quality information by
monitoring information and status of mobile nodes in u-city. u-City is a system
which manages a specific area automatically based on the sensors. Then, as per the
third step, information related to disaster is filtered. After that, as per the fourth step,
filtered disaster information is examined according to geo-location and finally alert
messages are sent to the people in nearby areas about disaster to save their life and
for providing help after disaster. The proposed system gives the facility to add
different organizations which belong to different areas under which the number of
branches could be there according to the parts of the area, or geo-location of the
branch is shown on map. Organizations are denoted by its name and will have an
acronym by which it gets recognized. The organization could be governmental; it
could be an NGO [14]. The organization could be of supplier type which supplies
needful thing to disaster-affected people [15–17]. The country of the organization
and other information related to organization has to be set correctly; for example,
contact number and the Web site of the organization have to be submitted while
adding any organization. The organizations have the office according to branches.
Figure 7 shows the organization module of disaster management system.

Module1

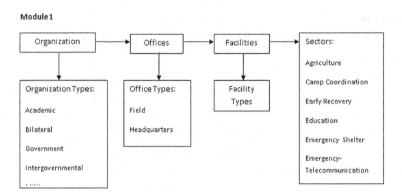

Fig. 7 Organization module of proposed disaster management system

The proposed system covers the staff and volunteers of the organization who work separate or work as a team of volunteers for organization. A team of staff or volunteers are formed according to their area or field of specialization and on the basis of organization. Staff and volunteers are getting decided on the basis of their permanent location. The country they belong to, state of the staff from which they are, and their district and town are considered for the staff and volunteers. Proposed disaster module combines the hazards, weather details, events, disaster victims, and facility provided to disaster-affected people. Disaster would be of any type in this system the hazards like earthquakes: recent event show on the map as well as the sea level is shown on map for the precaution. Weather details are also get displayed as current weather, weather station, cloud forecast and precipitation forecast comes in weather. Event is categorized into incident and type of incident. Incident could be of any types such as aircraft crash, bomb, overflow of dam, and chemical hazards. Incident can be categorized into flood, landslide, building collapsed, people trapped, power failure, bridge closed, and road closed.

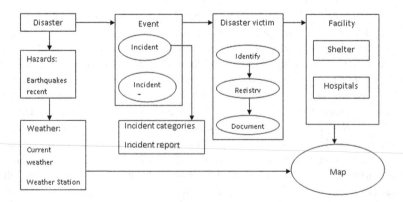

Fig. 8 Proposed disaster management module

This gets managed by disaster module. Disaster victim identification, registry, and documentation come under the disaster module. And facility would provide under the disaster module to the disaster-affected people such as shelter and hospitality. All disaster-related information is shown on map in the system. Figure 8 shows the complete structure of disaster module.

4 Experimental Evaluations and Results

Large-scale disasters bring together a massive amount of heterogeneous data that must be managed by the system and could help the society. For this, firstly, the information related to the area must get collected. For managing disaster, it is important to collect data about people living in the particular area. Firstly, the disaster data are collected and analyzed. In this system, different organizations are found who may help people which are suffered or suffering from the disaster. Organizations will have different branches in nearby area of the organization. There will be offices available for the branches which will be located on map to help volunteers of organization. Each organization will have its own type that provides which type of help it will provide to the people affected due to disaster. System provides the map of whole organization under which its branches, offices, warehouses, staff, and volunteers location can be seen on map to help people in disaster situation.

The proposed system is developed using Sahana Open Source Disaster Management Software (https://sahanafoundation.org/), PHP, and MySQL and is tested on Intel Core i3 with 4 GB RAM under Windows 10 operating system. Figure 9 shows the screen after adding an organization. Figure 10 shows the skill catalog which covers the skills of the staff and volunteers of the organization. Figure 11

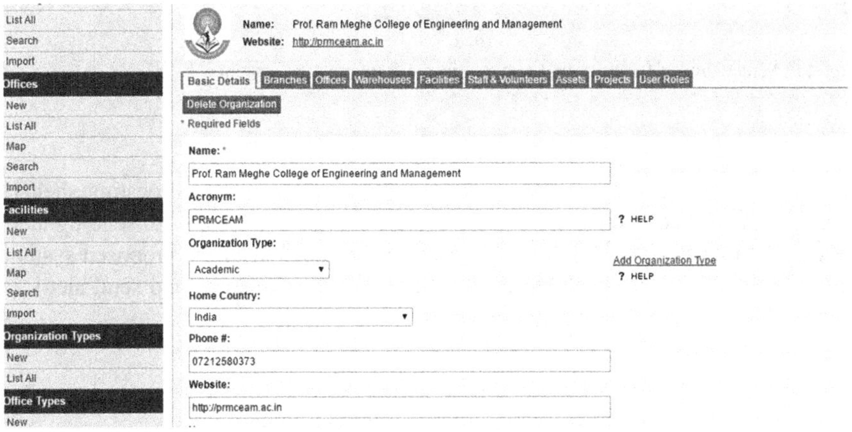

Fig. 9 Screenshot of added organization in the system

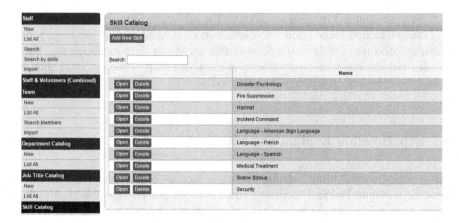

Fig. 10 Screenshot of skill catalog covering skills of staff and volunteers which will be helpful during disasters

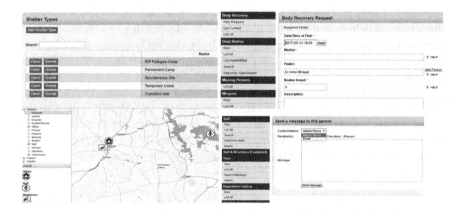

Fig. 11 Screenshot of facilities and features in the implemented system

shows the facilities and features provided during disaster such as locating shelter, finding unidentified bodies, displaying disaster locations on map, and sending alert messages to nearby persons. Table 1 shows the evaluation of the proposed system on the basis of disaster generated and time required to generate and send alerts.

Table 1 Different types of disaster with generated notification and time required to deliver notifications

Disaster	Causes	Notification generated	Notifications sent to	Time required to generate notification	Time required to send/ receive notification	Benefits provided by the notification
Air pollution	Dangerous industrial impurities in air	Due to dangerous impurities in air, area is suffering from air pollution	People in industrial area and in nearby area which could get affected, and volunteers belong to area	3–5 min after it happens	15–20 s	People will be safe and will take proper precaution, and volunteers will provide needful
Bridge closed	Work in progress	Due to the work in progress of road/ bridge, it will be closed	People using that bridge or people use to travel by the road, and volunteers belong to area	10–15 min after knowing work is in progress	15–20 s	People get aware, and their time will be saved
Bridge collapsed	Flood, and use of low grade or faulty material	Bridge is collapsed, and use another way to travel	People who usually travel by that road and who used the bridge, and volunteers belong to area	5–10 min after it happens	15–20 s	Accidents will get avoided due to awareness and time will get saved, and volunteers will provide needful
Child abduction emergency	Money	Children are getting abducted, and please take care of your children	People living in that area where it is happening and in nearby area, and volunteers belong to area.	12–24 h after it happens	20–25 s	People will take decisions more carefully about their children, and volunteers will provide needful
Crime	Money, robbery, serial killing	Continuous robbery incidents are noticed, and please be safe	People living in the area and in nearby area, and volunteers belong to area	10–12 h/days after it happens	15–25 s	People take required precaution after having the notification, and volunteers will provide needful

(continued)

Table 1 (continued)

Disaster	Causes	Notification generated	Notifications sent to	Time required to generate notification	Time required to send/ receive notification	Benefits provided by the notification
Dam overflow	Heavy rainfall, dam is overflowed	Due to the heavy rainfall, dam is overflowed and please be safe	People living in nearby dam area who may get affected due to the dam overflow, and volunteers belong to area	10–15 min after prediction	15–20 min	People get aware of the condition of dam, will do required., and will get saved, and volunteers will provide needful
Earthquake	Prediction, earth shell shakings	Earthquake is predicted in the area, and please be careful	People living in the predicted area and nearby area, and volunteers belong to area	10–15 min after prediction	15–20 min	People will be aware of penalties and will take required action immediately, and volunteers will provide needful
Flood	Heavy rainfall, dam collapsed	Due to flood, there is a risk to people living nearby river	People living nearby flooded river and volunteers belong to area	10–15 min after prediction	15–20 s	People will take proper precaution and take required action, and volunteers will provide needful
Forest fire	Natural fires, human-made fires	There is a wildfire in forest, and please avoid going in that area	People living nearby forest and people went there for some reasons, and volunteers belong to area	10–15 min after it happens	15–25 s	People will make proper decisions and take care of them self, and volunteers will provide needful
Road closed	Accidents, work in progress	Road is closed due to reason and please avoid using that road	People who regularly use that road and people going from that road, and volunteers belong to area	5–10 min after announcement or accident	15–25 s	People will make decision to go by other routes and will save their time as well, and volunteers will provide needful

5 Conclusion and Future Scope

In this paper, a disaster management system is designed, is capable of providing disaster-related information in an emergency condition, made people secure, and can save their time as well. Disaster could be of any types such as small scales and large scales; because of this, there are many problems have to be faced by people. It could be aircraft crash, road accident, civil emergency, and crime according to the area. The system could work at any location with particular data needed about the place and people living in the area. And all the related information is shown on the map. The system provides the location of offices, shelter, and hospitals on map, and this information can get share with the needy people according to their area of existence.

Staff location, warehouses, and offices of disaster management system are also showed on the map so that the related information can get shared when needed. The system tries to help people for avoiding disaster and to save their time. Disaster management system will be helpful to people to save themselves and to save their family. However, the system helps people after disaster by alerting them for saving their time as well as their life. There must be provision that this work should be done more faster. The major problem with disaster management is system is to predict disaster which could never happen. The known large-scale disasters could get predicted, but the disasters such as road accident, bomb spots, and civil accidents could never be predicted. There should be a solution to this, and it should provide faster help to the disaster-affected people. The robustness and estimation of performance boundary are currently under investigation.

References

1. Saptarsi Goswamia, Sanjay Chakrabortya, Sanhita Ghosha, Amlan Chakrabartib, Basabi Chakraborty, "A review on application of data mining techniques to combat natural disasters", proceeding in Ain Shams Engineering Journal, 16 January 2016.
2. Katarzyna Nowicka, "Smart City logistics on cloud computing model", 1st International Conference Green Cities 2014—Green Logistics for Greener Cities, PP. 266–281, 2014.
3. Giovanni Merlino, Dario Bruneo, Salvatore Distefano, Francesco Longo, Antonio Puliafito and Adnan Al-Anbuky, "A Smart City Lighting Case Study on an Open Stack-Powered Infrastructure", proceeding in open access sensors, ISSN 1424-8220 PP. 16314–16335, 2015.
4. Han Dong, Milton Halem, and Shujia Zhou, "Social Media Data Analytics Applied to Hurricane Sandy", Proceeding Social Com/PASSAT/Big Data/Econ Com/Bio Med Com IEEE, PP. 963–966, 2013.
5. Yong-Hong Ku, Janny M.Y. Leung, Helen M. Meng, Kelvin K.F. Tsoi, "A Real-Time Decision Support Tool for Disaster Response: A Mathematical Programming Approach", *Proceeding in International Congress on Big Data, IEEE*, PP. 639–648, 2015.
6. Jin Goo Kang*, Ju Wook Jang*, Chang Ho Yun**, Yong Woo LEE, "A Network Management System for u-City", PP. 279–283, Feb. 13–16, 2011.
7. Eleana Asimakopoulou, Nik Bessis, "Buildings and Crowds: Forming Smart Cities For More Effective Disaster Management", IEEE, PP. 229–234, 2011.

8. Lara Zomer, Winnie Daamen, Sebastiaan Meijer and Serge Hoogendoorn, "Managing Crowds: The Possibilities and Limitations of Crowd Information During Urban Mass Events," Planning Support Systems and Smart Cities Springer Lecture Notes in Geoinformation and Cartography, pp. 77–97, 2015.
9. Maggie X. Cheng and Wei Biao Wu, "Data Analytics for Fault Localization in Complex Networks", IEEE Internet of Things Journal, 2015.
10. Zhong Fan, Qipeng Chen, Georgios Kalogridis, Siok Tan, and Dritan Kaleshi, "The Power of Data: Data Analytics for M2 M and Smart Grid", 3rd IEEE PES Innovative Smart Grid Technologies Europe, PP. 1–8, 2012.
11. A. Ganz et al., "Real-time Scalable Resource Tracking Framework (DIORAMA) for Mass Casualty Incidents", International Journal of EHealth and Medical Communications, 4(2), April-June 2013, PP. 34–49.
12. A. Ganz, J. Schafer, Z. Yang, J. Yi, G. Lord, and G. Ciottone, "Mobile DIORAMA-II: Infrastructure less Information Collection System for Mass Casualty Incidents", IEEE Engineering in Medicine and Biology Society, Chicago, IL, August 2014.
13. Aura Ganz, James Schafer, Jingyan Tang, Zhuorui Yang, Jun Yi, Gregory Ciottone, "Interactive Visual Analytic Tools for Forensic Analysis of Mass Casualty Incidents using DIORAMA System", IEEE, 2015.
14. Karen España, Rhia Trogo, James Faeldon, Juanito Santiago, Delfin Jay Sabido, "Analytics-Enabled Disaster Preparedness and Emergency Management", IEEE International Conference on Data Mining, PP. 1068–1075, 2014.
15. Yu Hua, Wenbo He, Xue Liu, Dan Feng, "SmartEye: Real-time and Efficient Cloud Image Sharing for Disaster Environments", IEEE Conference on Computer Communications (INFOCOM), PP. 1616–1624, 2015.
16. Jonathan Silvertown, "A new dawn for citizen science," Trends in ecology & evolution, 24 (9), pp. 467–71, October 2009.
17. Winter Mason and DJ. Watts, "Financial incentives and the performance of crowds," ACM SIGKDD Explorations Newsletter, 11(2), pp. 77–85, 2010.

Application of Classification Techniques for Prediction and Analysis of Crime in India

Priyanka Das and Asit Kumar Das

Abstract Due to dramatic increase of crime rate, human skills for accessing the massive volume of data is about to diminish. So application of several data mining techniques can be beneficial for achieving insights on the crime patterns which will help the law enforcement prevent the crime with proper crime prevention strategies. This present work collects crime records for kidnapping, murder, rape and dowry death and analyses the crime trend in Indian states and union territories by applying various classification techniques. Analysing the crime would be much easier by the prediction rates shown in this work, and the effectiveness of these techniques is evaluated by accuracy, precision, recall and F-measure. This work also describes a comparative study for different classification algorithms used.

Keywords Crime prediction · Classification · Naïve Bayes · Random Forest Precision · Recall

1 Introduction

Crime is a social nuisance which has been on rise in almost all parts of the world including India. Criminologists analyse the data with varying degrees of success. But with the increasing crime rate, human skills tend to fail when they are provided with huge volume of data sets. Application of data mining techniques can be used to facilitate the task that can extract the hidden knowledge from the massive data sets and provide the crime investigation department a new edge for crime analysis. Collecting crime information from government portals employs data mining

P. Das (✉) · A. K. Das
Department of Computer Science and Technology,
Indian Institute of Engineering Science and Technology,
Shibpur, Howrah 711103, West Bengal, India
e-mail: priyankadas700@gmail.com

A. K. Das
e-mail: akdas@cs.iiests.ac.in

© Springer Nature Singapore Pte Ltd. 2019   191
H. S. Behera et al. (eds.), *Computational Intelligence in Data Mining*,
Advances in Intelligent Systems and Computing 711,
https://doi.org/10.1007/978-981-10-8055-5_18

techniques that predicts the future crime trends. Past crime records accumulated from government portals constitute the crime type, time, location, information about the victims, their genders, ages, social status and many more. Thus, crime prediction, a subtask of crime analysis, considers all the past crime records, classifies the crime categories and predicts the future crime. Numerous research works exist in the literature that employs different data mining techniques for crime analysis of different countries and cities. Crime prediction using pattern and association rule mining determines the chances of performing crime by the same criminal [1]. Given a time and place, type of crime occurring in San Fransisco City is predicted in [2]. Again, association rule mining task was incorporated for detecting crime locations in Denver and Los Angeles in [3]. Analysis of crime has been done by mapping, and similarities have been found with the past crime trend when compared to present scenario [4]. This task was an approach to determine places where maximum numbers of crime incidents take place. Application of data mining techniques has proven to be crucial for crime detection and prevention task. A comparative study was conducted for different crime patterns which exhibited better results for linear regression than other classification methods [5]. An architecture was implemented that collects the raw data and categorises the data into crime types, locations and places. Then, existing classification algorithms were used, and the most effective technique was chosen resulting in crime prediction [6]. Among all the available classification techniques, particularly two methods such as Naïve Bayes and backpropagation algorithm were compared for predicting the crime type in distinct states of America. This experiment shows that Bayes classification technique provides better accuracy than the backpropagation method [7]. Apart from the classification techniques, a clustering-based model [8] is used for anticipating the crimes that may help the law enforcement agencies preventing the crimes at a faster pace. An interdisciplinary approach was introduced in [9] that incorporated the knowledge of computer science and criminology to prepare a crime prediction model that focuses on crime factor for each day. Though most of the above-mentioned works have been done on crime data set from USA, none have done an extensive crime analysis for Indian states and in its union territories. The present work demonstrates crime prediction for 28 states (Andhra Pradesh inclusive of Telengana) and 7 union territories of India. It has considered the collection of crime records from 2001–2014 containing information about four different types of crime like kidnapping or abduction, murder, rape and dowry death. The data set for kidnapping comprises information about the number of victims, their gender, age, whereas the data for dowry death contains records of the number of cases pending investigation, cases discharged, cases claimed false, etc. Data set for rape holds information about the victims and types of rape occurred, and for murder cases, a different data set provides list of male and female victims with their ages. The present work has considered all the data sets for 2001–2012 as training data as input to several learning techniques like KNN, Random Forest, Naïve Bayes, AdaBoost and Classification Tree, and the predictive model has been learnt. Then, this predictive model has been used to predict the future crime trends. For kidnapping cases, the purpose of kidnapping has been chosen as the class label, whereas for rape cases, the types of rape (incest or others), the gender (male/female) for murdered victim and for dowry death, three different aspects have been chosen for prediction.

The rest of the paper is organised as follows: Sect. 2 describes the proposed work in detail. Section 3 shows the results of the proposed method followed by conclusion and future work in Sect. 4.

2 Proposed Framework

This section describes the present work elaborately in the following subsections:

2.1 Crime Data Collection and Preparation

Input data play a crucial role in the field of crime data mining. The crime data for the present work has been collected from National Crime Records Bureau (NCRB) and Open Government Data Platform India which provide documents and applications for research purpose as well as for public use. Collected data contain information about 28 states and 7 union territories of India. The collected data did not contain any missing values so as necessary preprocessing, the raw data have been converted to computer-readable format for further analysis. Most frequent crimes like kidnapping, murder, rape and dowry death have been chosen to deal with. For all the states, each data set for each crime type contains several attributes like name of the state, year, number of victims, number of cases. Though there exist significant methods for attribute selection, the present work has acknowledged the most probable attributes for crime types depending on human perception. Table 1 shows all the details of the integrated attributes from the data sets for the present task.

2.2 Classification Techniques

Classification techniques involve assignment of any object to one of the multiple predefined classes. Here, the predictive modelling is separately used for each crime type for all the states. Five different classification techniques have been used in this work, and they are briefly introduced as follows:

Decision Tree: It is simple yet widely used classifier with three types of nodes. The root and other nodes hold the test conditions for the features, and each leaf node is assigned with a class label.

K-Nearest Neighbour: It thus is an instance based learning where K defines the number of nearest neighbours and a proximity measure is needed for determining the similarity between the instances.

Table 1 Details of the collected data for kidnapping, murder, rape and dowry

Attribute	Description
State/UT	It refers to the names of the states or union territories
Year	Year of crime
Gender	Refers to male or female
Age	It illustrates age-wise male/female victims of crime
Purpose	Demonstrates nine different reasons of kidnapping like begging, adoption, prostitution
Type	It describes types of rapes like incest or other cases occurring in the states
CPIP	Cases pending investigation from previous year
CRY	Cases reported during the year
CWG	Cases withdrawn by the Govt. during investigation
CNI	Cases not investigated or in which investigation was refused
CDF	Cases declared false on account of mistake of fact or of law
CC	Cases in which chargesheets were laid
CCN	Cases in which chargesheets were not laid but final report submitted during the year
CPIY	Cases pending investigation at the end of the year
CPTP	Cases pending trial from the previous year
CST	Cases sent for trial during the year
TCT	Total number of cases for trial
CW	Cases withdrawn
CTC	Cases in which trials were completed
CCon	Cases convicted
CD	Cases acquitted or discharged
CPTY	Cases pending trial at the end of the year

Naïve Bayes: Provided a class label, this classifier assumes the features to be independent and determines their class conditional probability. Let the class label be y, then the conditional independence can be defined as (1).

$$P(x|y) = \sum_{i=1}^{k} P(x_i|y) \tag{1}$$

where feature set $x = (x_1, x_2, \ldots, x_k)$ and k is the number of features.

Random Forest: Random Forest comprises several classification trees, where each tree classifies an object and a voting is done for that particular class. Now, the classification with highest number of votes is selected by the forest. Random Forest is very efficient when dealing with large data sets, and depending on the applications, it often provides better accuracy than other classification techniques.

AdaBoost: Here, adaptive boosting algorithm is used with other classification techniques for achieving optimal performance on the crime data sets.

2.3 Crime Prediction Task

Once the data are collected, they have been divided into two parts, namely training and test data. Data for 2001–2012 have been chosen as training data with all attributes having known class labels. They have been given as input to the learning techniques, and the result of the trained predicting model is applied on the test data with unknown class labels. Data sets for 2013 and 2014 have been used as test data in this task, and trained model classifies the instances present in the test set. Figure 1 shows the basic layout of the crime prediction task, and Table 2 shows that though reason for abduction is the sole task in case of kidnapping, the present work predicts several extreme cases for dowry deaths. It emphasises on the issues regarding the cases convicted, cases that were claimed false and cases pending investigation. Thus, various issues regarding the crime get discovered and crime prediction is done for Indian states. This crime prediction task helps in analysing the future crime trends for crimes like rape, kidnap, murder and dowry deaths.

Table 2 Details of the predicted classes

Crime type	Prediction
Kidnapping	Purpose of kidnapping (adoption, begging, camel racing, etc.)
Rape	Types of rape (incest or others)
Murder	Gender of the victim
Dowry death	Cases convicted, cases pending investigation from previous years, cases declared false on account of mistake of fact or of law

Fig. 1 Layout for crime prediction task

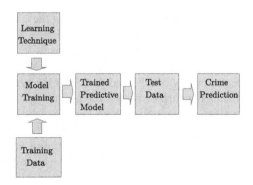

3 Experimental Results

Once the predictive model has been prepared based on the crime data for 2001–2012, the testing and scoring are done based on tenfold cross-validation on test data of 2013–2014. The performance of the classifiers has been measured by evaluation techniques, namely precision, recall and F-measure using (2–4). Confusion matrix denotes the number of classified and misclassified instances in the crime data. True positive (TP) and True negative (TN) represent correctly classified instances, whereas false positive (FP) and false negative (FN) denote the incorrectly classified instances.

$$Precision\,(P) = \frac{TP}{TP + FP} \tag{2}$$

$$Recall\,(R) = \frac{TP}{TP + FN} \tag{3}$$

$$F - measure = \frac{2PR}{P + R} \tag{4}$$

Tables 3 and 4 show the performance for each classifier with correctly classified and misclassified instances for the year 2013–2014. The present method has predicted the purpose of kidnapping; i.e., it classifies the instances as 'adoption', 'begging', 'camel racing', 'prostitution', 'marriage', 'illicit intercourse', 'slavery'. The training data for crime related to murder contain all the instances, and it has been learnt to the predictive model which classifies the gender of the murdered victims. Table 5 shows the performance measures for murder cases. Likewise, Table 6 shows the classification accuracy for predicting the types of rapes such as incest rape cases or others from the separate data set of rape victims. Now for dowry death cases, the present work has focused on predicting three different issues regarding the cases, investigation, etc. It predicts from the instances if the case is pending from the previous year, case convicted and case that is declared false on account of mistake of fact or law. The data set for dowry death cases contains many more attributes but as the motive is to predicting and analysing the crime trend, the present work has focused on the most important attributes. Tables 7, 8 and 9 show the results for different prediction accuracies in dowry death cases.

It has been observed from the results that Random Forest classifier provides the best accuracy of 90% and 95.6% for the year 2013 and 2014, respectively, for predicting the purpose of kidnapping and 95.2% for classifying the gender of murdered victims, whereas Naïve Bayes provides the highest F-measure of 85.9% for predicting the types of rape that occurs mostly in the Indian states. Decision tree based classifier provides the best F-measure of 86.1 for predicting the instances related to cases that have been declared false by law, and Random Forest, Naïve Bayes and AdaBoost all provide same F-measure of 98% for classifying the cases that have been convicted. Here, we have also shown the classification accuracy (CA) and area under the curve (AUC). Figure 2 shows receiver operating characteristic (ROC) curves for

Table 3 Performance measure in (%) for prediction in kidnapping 2013

Classifier	AUC	CA	P	R	F	Correctly classified	Misclassified
KNN	92.8	91.0	91.1	91.0	86.6	90.7	9.3
Tree	88.5	84.9	85.0	84.9	76.1	84.4	15.6
Random Forest	94.3	90.0	90.0	90.0	90.0	91.6	8.4
Naive Bayes	95.0	89.0	88.9	89.0	88.9	90.2	9.8
AdaBoost	90.0	90.4	90.4	90.4	86.1	91.4	8.6

Table 4 Performance measure in (%) for prediction in kidnapping 2014

Classifier	AUC	CA	P	R	F	Correctly classified	Misclassified
KNN	95.1	91.0	91.1	91.0	91.0	95.6	4.4
Tree	93.3	94.7	94.6	94.7	94.5	96.6	3.4
Random Forest	97.6	95.6	94.9	95.0	94.8	96.6	3.4
Naive Bayes	96.5	88.5	91.4	88.5	90.0	97.0	3.0
AdaBoost	82.4	94.7	94.6	94.7	94.5	96.6	3.4

Table 5 Performance measure in (%) for prediction in murder cases

Classifier	AUC	CA	P	R	F	Correctly classified	Misclassified
KNN	95.3	94.3	94.3	94.3	94.3	94.2	5.8
Tree	94.3	92.4	92.4	92.4	92.5	92.3	7.7
Random Forest	98.9	95.2	95.3	95.2	95.3	94.3	5.7
Naive Bayes	97.7	94.3	94.3	94.3	94.3	94.2	5.8
AdaBoost	94.7	93.3	93.2	93.5	93.3	90.9	9.1

Table 6 Performance measure in (%) for prediction in rape cases

Classifier	AUC	CA	P	R	F	Correctly classified	Misclassified
KNN	91.5	77.5	78.3	77.5	75.8	73.2	26.8
Tree	81.2	84.5	85.2	84.5	83.6	80.0	20.0
Random Forest	94.7	83.1	83.6	83.1	82.4	78.4	21.6
Naive Bayes	94.0	85.9	86.1	85.9	85.9	85.7	14.3
AdaBoost	78.3	84.5	84.1	84.8	84.1	81.6	18.4

Table 7 Performance measure in (%) for prediction of the dowry death cases that have been declared false

Classifier	AUC	CA	P	R	F	Correctly classified	Misclassified
KNN	85.0	83.3	84.2	83.3	84.1	89.5	10.5
Tree	82.5	86.1	87.5	86.1	86.4	94.4	5.6
Random Forest	82.5	83.3	84.1	83.3	81.2	89.5	10.5
Naive Bayes	90.0	83.3	84.1	83.3	81.2	89.5	10.5
AdaBoost	80.0	86.1	83.9	86.4	86.1	90.0	10.0

Table 8 Performance measure in (%) for prediction of the dowry death cases that have been convicted

Classifier	AUC	CA	P	R	F	Correctly classified	Misclassified
KNN	98.0	97.0	97.0	97.0	98.0	98.0	2.0
Tree	97.0	96.0	97.0	97.0	97.0	97.9	2.1
Random Forest	98.0	96.0	98.0	98.0	98.0	98.0	2.0
Naive Bayes	98.0	97.0	98.0	98.0	98.0	97.0	3.0
AdaBoost	97.0	98.0	98.0	98.0	98.0	98.0	2.0

Table 9 Performance measure in (%) for prediction of the dowry death cases that are pending investigation from previous years

Classifier	AUC	CA	P	R	F	Correctly classified	Misclassified
KNN	91.2	88.9	88.9	88.9	88.9	88.9	11.1
Tree	92.5	91.7	91.4	91.8	91.7	89.5	10.5
Random Forest	97.5	91.7	91.4	91.8	91.7	89.5	10.5
Naive Bayes	97.5	88.9	88.9	88.9	88.9	88.9	11.1
AdaBoost	95.0	91.7	91.4	91.8	91.7	89.5	10.5

few crime instances. The ROC curve shows the trade-off between sensitivity and specificity, and most of the curves are to the top of the ROC space which denotes better classification accuracy for learning techniques like Random Forest and Naïve Bayes.

As an outcome of the research work, this study provides knowledge on how many people are getting victimised each year, and it reflects the motivation behind the crimes, as it predicts that several males are being kidnapped for slavery or any other unlawful activity, whereas most of the females are being kidnapped and sent to the middle east part of the globe for prostitution or most often their body parts are being sold. Most of the children are kidnapped for adoption and begging. For murder cases, most of the females of age 18–30 years are raped and murdered every year.

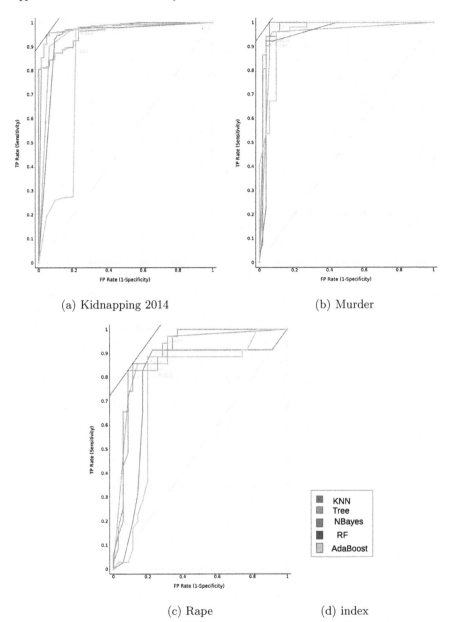

(a) Kidnapping 2014

(b) Murder

(c) Rape

(d) index

Fig. 2 ROC curves for predicting different instances of crime

Regarding prediction in rape cases, it is observed that though number of rape cases has gone up till date, but more number of girls are being raped by their relatives (incest) and friends. The prediction in dowry death cases emphasises on three extreme issues that reflect how government takes action against the victims and their relatives. Though cases are being filed everyday, timely actions are not taken resulting into huge number of cases that are pending both investigation and trial. Very few cases have been convicted through all these years 2001–2014. Due to lack of facts and evidences, some cases have been declared false by the law. Thus, the present crime analysis is an effective approach so that law enforcement section can consider it and take preventive measures to reduce the crime rate.

4   Conclusion and Future Work

Application of several data mining techniques can be beneficial for achieving insights on the crime patterns which will help the law enforcement prevent the crime with proper crime prevention strategies. The present work has used a Python-based software called 'Orange' [10] for the learning techniques and demonstrates a simple yet effective approach for crime prediction from Indian crime data. It utilises few existing learning algorithms that provides an insight of the attributes present in the data set and later helps in predicting the crime trend in India for the year 2001–2014. The results show high prediction accuracy for most of the cases. The present work also demonstrates a comparative study of the classification algorithms used for crime analysis. Not only it analyses the crime trend, it also reflects on the victims, their ages, number of people getting victimised every year, and most importantly it reflects on the actions taken by the government dealing with dowry death cases. Though the present work reflects the crime scenario of Indian states, this method can also be applied for analysing the global crime scenario. As a future work, other crime types can be considered and other classification methods can also be employed for analysing the crime in an extensive manner.

References

1. Anisha Agarwal, Dhanashree Chougule, A.A.D.C.: Application for analysis and prediction of crime data using data mining. In: Proceedings of IRF-ieeeforum International Conference. (2016) 35–38
2. Chandrasekar, A., Raj, A.S., Kumar, P.: Crime prediction and classification in San Francisco city
3. Subhash Tatale, N.B.: Crime prediction based on crime types and using spatial and temporal criminal hotspots. International Journal of Data Mining & Knowledge Management Process 5(4) (2015) 1–19
4. Subhash Tatale, N.B.: Criminal data analysis in a crime investigation system using data mining. Journal of Data Mining and Management 1(1) (2016) 1–13

5. Lawrence McClendon, N.M.: Using machine learning algorithms to analyze crime data. Machine Learning and Applications: An International Journal (MLAIJ) **2**(1) (2015) 1–12

6. Yu, C.H., Ding, W., Chen, P., Morabito, M.: Crime forecasting using spatio-temporal pattern with ensemble learning. In: Pacific-Asia Conference on Knowledge Discovery and Data Mining. (2014) 174–185

7. Abba Babakura, Md Nasir Sulaiman, M.A.Y.: Improved method of classification algorithms for crime prediction. In: Proceedings of International Symposium on Biometrics and Security Technologies (ISBAST). (2014) 250–255

8. S. Yamuna, N.B. Chang: Datamining techniques to analyze and predict crimes. The International Journal of Engineering And Science (IJES) **1**(2) (2012) 243–247

9. Sathyadevan, S., Devan, M.S., Surya Gangadharan, S.: Crime analysis and prediction using data mining. In: 2014 First International Conference on Networks Soft Computing (ICNSC2014). (Aug 2014) 406–412

10. Demšar, J., Curk, T., Erjavec, A., Črt Gorup, Hočevar, T., Milutinovič, M., Možina, M., Polajnar, M., Toplak, M., Starič, A., Štajdohar, M., Umek, L., Žagar, L., Žbontar, J., Žitnik, M., Zupan, B.: Orange: Data mining toolbox in python. Journal of Machine Learning Research **14** (2013) 2349–2353

Improving Accuracy of Classification Based on C4.5 Decision Tree Algorithm Using Big Data Analytics

Bhavna Rawal and Ruchi Agarwal

Abstract C4.5 is an algorithm of decision tree that broadly used classification technique. There are many challenges in the era of big data like size, time, and cost for building a decision tree. Aim of the decision tree construction is to boost up the accuracy on the training data. In predictive modeling, it requires to split the training datasets for this MATLAB is a good choice. Also analysis of data is done easily by decision tree instead of heterogeneous data. In this paper, C4.5 is implemented with the help of MATLAB using four different datasets which provides a confusion matrix in terms of target and output classes. At the end, it compared the features of datasets. The main objective of this research is to boost up the classification accuracy and roll back timing to build a classification model. We have reduced input space using Bhattacharya distance. The proposed method shows better performance for the data file. With the help of BD, improved C4.5 is performing better than original C4.5 in every test case.

Keywords Bhattacharya distance · Big data analytics · C4.5 · Decision tree

1 Introduction

Data mining is recent research area that is used to solve various problems in different domains. It extracts the knowledge from hidden patterns [1]. Data mining consists of various approaches like machine learning and statics. In data mining, classification is a popular task for knowledge discovery and plans. Classification technique has various algorithms such as decision tree, genetic, nearest neighbor, and support vector machine (SVM) [2]. It monitors new patterns and information

B. Rawal (✉) · R. Agarwal
Department of Computer Science & Engineering, School of Engineering & Technology, Sharda University, Greater Noida 201306, India
e-mail: rawalbhavna2@gmail.com

R. Agarwal
e-mail: ruchi.agarwal@sharda.ac.in

© Springer Nature Singapore Pte Ltd. 2019
H. S. Behera et al. (eds.), *Computational Intelligence in Data Mining*,
Advances in Intelligent Systems and Computing 711,
https://doi.org/10.1007/978-981-10-8055-5_19

from the datasets by removing the hidden relationships between features. The main focus of this paper is on C4.5 decision tree algorithm for classification. Huge amounts of data being collected and stored in a database which come from social sites and information industry. It requires the decision tree as knowledge representation using the dataset. Decision tree is most widely used process for the decision making (Quinlan).

1.1 Experimental Criteria

"The non trivial extraction of inherent, potentially functional data from data" is described as the data excavation. An iterative procedure to coil data is just the commencing into data, data into vision, and vision into action that is removing functional outlines from data. With the help of data mining process, we can obtain the useful information for extracting the knowledge. The data mining process is also known as knowledge discovery in databases [3].

1. Datasets are used for knowledge discovery and for mining the data; it is also used for selecting the data and understanding of concern variety of variables. Goal statistics is selected based on consumers.
2. This step is used to remove the noise and inconsistent data. The target information has to be preprocessed to remove the bad data now and manage the time sequencing.
3. After the multiple data sources are combined, the data are retrieved from the database. Information transformation permits the powerful range of variable that is useful for creating the mining functions.
4. Then, it helps to apply a method to extract the data patterns and needs to check the kind of information discovery.
5. Used to interpret again sample and evaluate the patterns and expertise received by way of records mining technique.
6. It is far feasible to behave in addition iteration primarily based on translation of mined styles.

2 Decision Tree

Decision tree is developed by J. R. Quinlan in 1980. Here, the decision tree is used to classify the dataset and predict the target classes. For the classification purpose, C4.5 decision tree-based algorithm is used here because of C4.5 feature and its computational efficiency. Features of the C4.5 algorithm are categorization of continuous attributes and handle missing values [4]. Decision tree provides human-readable classification rules, fast construction of the tree, and the better accuracy. C4.5 is a decision tree-based algorithm that is an extension of ID3 algorithm. Decision tree is used to calculate the numeric weight of connections between the nodes. While information point falls in a partitioned vicinity, a section tree classifies it as belonging to most frequent class. So the decision tree is very effective for the classification purpose. Evaluating the data, a decision tree is a predictive model that both classifiers and regression model can be used. There are some types of nodes in a decision tree like root node, branch node, and leaf node. "Root" is a node where exactly no incoming edge is there [5]. There are three nodes in decision tree named as root node, internal node and decision node. End nodes of the tree are the leaves nodes or the decision nodes that are used to predict something. There are some conditions in this research for the decision tree that can be focused for the classification.

2.1 Tree Size

Decision tree that is not complicated is apt to be more comprehensible favor. Its accuracy is an important result according to the complexity and size of tree [6]. It is used to eliminate the use of a number of properties:

Normally, the complexity of tree is measured by these metrics like pruning of the tree and handling continuous data.

2.2 Hierarchical Nature of Decision Tree

The hierarchical nature of decision tree is one of the characteristics. According to patients after treatment it is required to identify the disease to develop a health system with the help of decision tree [2]. So that with the help of this, the health examinations are so sought after just performed the price is reduced.

3 Big Data Analytics

Big data analytics is huge amount of data being collected from different sources. There are three types of big data like structured data, unstructured data, and semi-structured data. Both have their own importance [3]. Its importance can be explained with help of V's like volume, velocity, and variety, where volume means amount of data, velocity means speed of data, and variety means different types of data [7]. In this paper, wholesale customers' dataset is used for boosting up the sales. This dataset refers to client of wholesale distributor. There is huge amount of data so that it is impossible to analyze. For implementing large amount of data, this paper is using the MapReduce function in MATLAB. Big data can be used for handing the online transaction data.

4 Related Work

There are many researchers that have made use of statistical classifier to improve the classification criterion. Some of the researchers like Harvinder Chauhan find the accuracy based on ID3 and C4.5 classifiers. In 2016 by the Boonchuay, kesinee and krung experimented with well-known classifiers that can predict balanced dataset but misclassify the imbalanced dataset [8]. In 2010, Zighed and Djamel experimented on various algorithms of machine learning that use entropy measure as optimization criterion. Then in 2011, Chumphol experimented on an application which operates on an imbalanced datasets that loses classification performance on a minority class which is rare and effective [7].

5 Methodology

5.1 Bhattacharya Distance

Bhattacharya distance is used to check the similarity of the data. It is a splitting criterion for the decision tree [9]. Bhattacharya distance measures input space with help of Bhattacharya coefficient. In this paper, Bhattacharya distance set as default in MATLAB which works in Bins.

5.2 C4.5 Decision Tree Algorithm

C4.5 is one of most popular and effective classifier which is also known as the statistics classifier. This classifier is used to build data mining systems commonly that are used in devices [10, 11]. A classifier produces a new class that performs as predictor belongs to garb. C4.5 is the advance version of ID3 algorithm.

5.3 Information Theory

Information theory is used to check the uncertainty of data. For selecting the best split, choose the information gain, and for changing the state, choose the entropy. Information gain helps to improve the accuracy of C4.5 decision tree algorithm using big data analytics. For importing the big data analytics, the MapReduce function is required [12]. Map function is used to mapping whole data, and Reduce function is used to aggregate that data [5].

Fig. 1 Steps for
classification

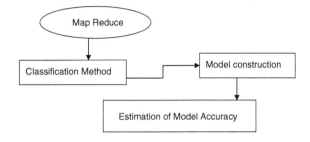

6 Proposed System

The main goal of this paper is to classify the dataset and predict the classes of the
dataset. So here, the predictive model is developed for prediction as shown in
Fig. 1. There are two main functionalities:

- Predict the target class.
- Generate the classification rules.

  There are some phases in the purposed system such as:

- Preprocess the data for loading the dataset.
- Apply the information gain and Bhattacharya distance for constructing the C4.5
  algorithm.
- Then, construct the model and use it at last.

  Big data analytics is used for re-evaluation of system and huge amount of data
management. For implementing the big data, we perform computation in MATLAB
using MapReduce function. To do this, first we need to produce a data store, and
then, with the help of Map and Reduce functions find the total error of the dataset.
First Map function is mapping all the data, and then, Reduce function aggregates
that data.

7 Results and Analysis

The output of the training dataset accuracy is shown in Fig. 4 that shows the
accuracy of improved C4.5 algorithm. For this accuracy, first we need to find total
error of dataset with the help of MapReduce function shown in Fig. 2. C4.5 is better
in pruning rather than ID3 [13].

Fig. 2 MapReduce progress
in MATLAB on wholesale
customers' dataset

```
*******************************
*        MAPREDUCE PROGRESS        *
*******************************
Map    0%  Reduce    0%
Map   50%  Reduce    0%
Map  100%  Reduce    0%
```

7.1   Dataset

Here, we are using the wholesale customers' dataset which contains eight attributes, namely fresh, milk, grocery, frozen, detergents paper, delicatessen, channel, and region as given in Table 1. Due to huge amount of data authors are using MapReduce function for implementing this data in MATLAB. MapReduce reads chucks of data and Map function is worked on chucks [14]. After that MapReduce is used to group all data in intermediate phase, and at last, Reduce function is used to aggregate that data.

Table 1 shows the training dataset in which we are predicting the classes. So that it is easy to enhance the sale of the wholesaler. There are two classes such as (hotels/restaurants/cafes) and retail.

Category process: To classify instance, we have a tendency to begin with idea of building tree and observe direction of attribute in the tree. That technique is sustained till a leaf is encountered. Eventually, we have tendency to use connected label to rear expected category cost of instance to handy.

A very common data mining task is to build for predicting the classes of dataset. Here is an object named wholesale customer, and with the help of classification tree, we predict two classes of wholesale customers' data as hotels/restaurants/cafes and second class is retail (Fig. 3).

Based on information theory and Bhattacharya distance, we find the accuracy of C4.5 using big data analytics as shown in (Fig. 4). A MapReduce function is used for mapping and aggregating the big data where Map function helps to read data in

Table 1 Wholesale customers' dataset with the annual spending (m.u)

Channel	Region	Fresh	Milk	Grocery	Frozen	Detergent paper	Delicatessen
2	3	12669	9656	7561	214	2674	1338
2	3	7057	9810	9568	1762	3293	1776
2	3	6353	8808	7684	2405	3516	7844
1	3	13265	1196	4221	6404	507	1788
2	3	22615	5410	7198	3915	1777	5185
2	3	9413	8259	5126	666	1795	1451
2	3	12126	3199	6975	480	3140	545

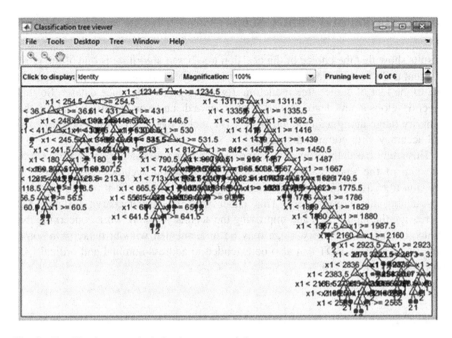

Fig. 3 Classification tree of wholesale customers' dataset

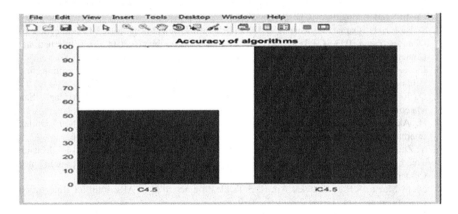

Fig. 4 Accuracy of C4.5 and improved C4.5 decision tree algorithms on wholesale customers' data

chucks and combine the key and values as keystore; then, it passes to Reduce function that is used to aggregate that data. Now, apply the classification method on aggregate data and construct the model; then finally, estimate the model accuracy.

8   Conclusion and Future Scope

Results show that the choice of the decision tree-based algorithm is one of the most frequent and effective classifiers. Decision tree is not working on imbalance dataset due to the fact of choice tree induction it works well on a balance dataset. So this paper develops a new degree called Bhattacharya distance on the instances from the minority range along a single attribute. BDT yields overall the better performance and accuracy with computation time.

This study basically proposed to improve a rule-based category algorithm C4.5. The aim of the study is to boost up the classification accuracy based on C4.5 decision tree algorithm using big data analytics and reduce the time consumption for running the big datasets. This paper used the Bhattacharya distance for decreasing the input space and improving the accuracy. Future work concerning the choice of tree, the minority range may be implemented without delay with some impurity measures. BDT can also be extended to address nominal and ordinal fact types consisting of Gini.

References

1. S. Desai, S. Roy, B. Patel, S. Purandare and M. Kucheria, "Very Fast Decision Tree (VFDT) algorithm on Hadoop", 2016 International Conference on Computing Communication Control and automation (ICCUBEA), 2016.
2. S. Bashir, U. Qamar, F. Khan and M. Javed, "An Efficient Rule-Based Classification of Diabetes using ID3, C4.5 & amp; amp; CART Ensembles", 2014 12th International Conference on frontiers of Information Technology, 2014.
3. Jiawei Han and MichelineKamber-Data Mining: Concepts and Techniques, 3rd edition, first volume, 2011.
4. Q. Ross, Morgan Kaufmann Publishers, "C4.5: Programs for Machine Learning", San MateoInc (1993).
5. H. Akash, Kiran Bhowmick "A MapReduce based approach for classification" Online International Conference on Green Engineering and Technology (IC-GET) 2016.
6. Y. Zhen, Q. Yong and L. Jing, "The application of short classification based on C4.5 decision Tree in video retrieval", 2011 6th IEEE Joint Information Technology Artificial Intelligence Conference, 2011.
7. M. M Mazid, A.B.M Shawkat Ali, K. S Tickle, "Improved C4.5 Algorithm for Rule Based Classification", vol. 13, pp 296–301, 2010.
8. Yuan Z. "An improved network traffic classification algorithm based on Hadoop Decision tree", Vol. 3, No. 1, March 2016.
9. X. Bao and X. Guan, "A Method of Predicting Crude Oil Output Based on RS-C4.5 Algorithm", 3rd International Conference on Information Science and Control Engineering (ICISCE), 2016.
10. X. Zhao and J. Yang, "An improved TANC classification algorithm based on C4.5", The 26th Chinese Control and Decision Conference (2014 CCDC), 2014.
11. S. Soliman, S. Abbas and A. Salem, "Classification of thromobosis collagen diseases based on C4.5 algorithm", 2015 IEEE Seventh International Conference on Intelligent computing and Information System (ICICIS), 2015.

12. Z. Yuan and C. Wang, "An improved network traffic classification algorithm based on Hadoop decision tree", 2016 IEEE Interntional Conference of Online Analysis and Computing Science (ICOACS), 2016.
13. B. Hssina, A. Merbouha, H. Ezzikouri, M. Erritali, "A comparative study of decision tree ID3 and C4.5", vol. 1, No. 1, 2010.
14. Gongging Wu-haiguang Li-Xuegang Hu-yuanjun Bi-jing Zhang-XindongWu-"MReC4.5 Ensemble Classification with MapReduce" 4rt ChinaGrid Annual Conference-2009.

Comparative Study of MPPT Control of Grid-Tied PV Generation by Intelligent Techniques

S. Behera, D. Meher and S. Poddar

Abstract A grid-tied photovoltaic (PV) system with boost converter is considered for study here. The maximum power point tracking (MPPT) control on the duty cycle of the boost converter is achieved by intelligent techniques such as grey wolf optimization (GWO), Moth-Flame optimization (MFO) and compared with perturb and observe (P&O) method. The proposed approach of MFO reduces the ripples in power, voltage and current and imparts better efficiency under different configurations as compared to latest literature for a similar approach.

Keywords P&O · MFO · GWO · PV · Boost · Grid-tied

1 Introduction

The technologically advancing world can sustain and grow if the required power is met by the generation. But, time has come to rethink on harvesting power from alternative sources than thermal power generation as the fuel stock and pollution both are alarming. Solar energy is profusely available that can be converted to electricity by PV cell. But, the energy conversion efficiency of PV panel is 35–45%. The generation also varies with irradiance level. As the PV cell is basically a photodiode, the output is DC. The PV output is inverted to AC and utilized by the end user or supplied to grid. As the sun is available in the daytime only, isolated systems charge a battery and utilize during the night. For any irradiance, maximum power point tracking (MPPT) control is achieved to maximize generation. Various MPPT control algorithms have been brought out [1] such as hill climbing,

S. Behera (✉) · D. Meher · S. Poddar
Veer Surendra Sai University of Technology, Burla 768018, India
e-mail: sasmitabehera2000m@gmail.com

D. Meher
e-mail: dibya.meher967@gmail.com

S. Poddar
e-mail: poddarsidq1w2e3@gmail.com

© Springer Nature Singapore Pte Ltd. 2019
H. S. Behera et al. (eds.), *Computational Intelligence in Data Mining*,
Advances in Intelligent Systems and Computing 711,
https://doi.org/10.1007/978-981-10-8055-5_20

fractional open circuit voltage control, perturb and observe (P&O) [2, 3], incremental conductance (INC) [4], neural network control, fuzzy control based [5] and direct MPPT algorithm [6]. Though P&O is popular [7] as it is simple and fast, but on approaching the MPP, it still perturbs leading to oscillations [8]. Also for fast variation of irradiation level (due to which MPP shifts), it takes it as a change in MPP due to perturbation and finishes in incorrect MPP. So, further improvement in MPPT is a current trend of research.

Taking maximizing power as an optimization problem, all the soft computing approaches have been well reviewed and indicated application of variants of PSO for improvement of MPPT by various researchers. Adding on this area, Sundareswaran et al. [9] discovered firefly algorithm (FA) performing better than P&O and PSO under partial shading condition. In a similar study [10], re-tracking speed to new MPP was better by cuckoo search (CS) followed by PSO, whereas P&O failed. Also, the CS provided less oscillation in the transient and steady state. PSO has three parameters to be adjusted whereas in CS only two. Another such derivative-free optimization technique, grey wolf optimization (GWO) [11] has been successfully applied to an isolated system that is faster and more accurate than conventional ones [12] and previous similar researches. It is a very simple global search, and one has to adjust only one of its control parameters, which makes it easy to implement. Grid-tied PV system generates in KW to MW range. So, when grid connected, improvement in MPPT imparts better utilization. A recent algorithm Moth-Flame optimization (MFO) [13] also has one parameter to be adjusted and gives promising results as compared to PSO and CS algorithms and many others. Hence, MFO has been implemented for MPPT control and compared to GWO and conventional method of P& O in this problem for the grid-tied system.

Rest part of the paper constitutes of five sections. Section 2 is on modelling of the grid-tied PV system. Section 3 is on maximum power point tracking implementation by various algorithms. In Sect. 4, simulation results on different configurations are discussed, and finally, the paper concludes with significant outcomes for future research.

2 Model of Grid-Tied Photovoltaic System

In a photovoltaic cell, energy of photon from light when exceeds band gap energy, then the electrons are emitted, and their flow creates current output I of the module expressed as [14]:

$$I = N_p I_{sc} - N_p I_0 \left\{ e^{[q_e(V + R_s I)/kN_s T_k)]} - 1 \right\} - (V + R_s I)/R_p \qquad (1)$$

where,

V is output voltage of a PV module (V)
T_k is the standard temperature = 298 K

I_0 is the PV module saturation current (A)
k is Boltzmann's constant = 1.3805×10^{-23} J/K
q_e is the charge of an electron = 1.6×10^{-19} C
R_s is the series resistance of a PV module
R_p is the resistance in parallel with the diode in the cell model
I_{sc} is the short-circuit current at 25 °C and 1 kW/m^2
N_s is the number of cells in series to increase voltage
N_p is the number of cells to increase current

Here in these work, mono-crystalline PV module is used. It has the advantage over polycrystalline cell for higher efficiency (15–20%), less space, longer life and better performance than a polycrystalline cell at low light condition [4]. But, these cells are expensive and partial shading leads to the breakdown of the entire circuit if improperly controlled.

The output voltage is stepped up from its input by a DC–DC boost converter. For grid connectivity at high voltage, and high-frequency switching converter, insulated gate bi-polar transistor (IGBT) is used as a switch as it has the smallest turn-on and turn-off time. The voltage boost is achieved by an inductor that opposes a change in current by building and wiping out magnetism. A boost converter is shown in Fig. 1a. When the switch is gated, current flows from left to right and the inductor magnetizes. When the switch is turned off, current decreases due to high impedance. The magnetic field is destroyed, and the current flows to the load. Thus, the polarity reverses, and left side of inductor becomes negative. Consequently, the two sources cause more voltage across the capacitor charged through the diode D than the input.

For high-frequency switching, the magnetism of inductor retains partly in between on-off states, and the load for all time experiences a voltage higher than input source alone in off state of the switch. To suppress voltage ripple, filters are added on both load-side and supply-side.

A schematic of a grid-tied PV generation is given in Fig. 1b. It has solar panels, one or several inverters, a power conditioning unit and grid connection equipment. Unlike stand-alone power systems, a grid-connected system excludes battery and supplies the power to the utility grid [1]. The PV system is connected to the utility company executing interconnection agreement. It deals with the various safety

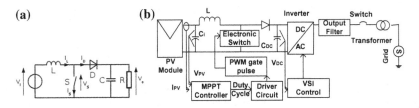

Fig. 1 a Circuit of the boost converter. **b** Schematic of grid-tied PV generation

standards to be maintained. Thus, grid interconnection of PV generation provides efficient utilization of generated power and less pollution hazard because there is no storage involved.

3 Maximum Power Point Tracking

According to maximum power transfer theorem, the power output of a circuit is maximum when the Thevenin impedance of the circuit (source impedance) matches with the load impedance. In the source side, a boost converter is used. By changing the duty cycle of the boost converter, impedance matching can be done [14].

The duty cycle,

$$D = (V_0 - V_i)/V_i \qquad (2)$$

where, V_i is the input voltage from PV module and V_0 is the output from the boost converter. Thus, adjusting the duty cycle drives the PV generation at MPP. Thus, the MPPT control is formulated as an optimization problem. The duty cycle is assigned to the population within the minimum and maximum bounds, and the individual in the population is evaluated for maximum power. The maximum power P is the best fitness, and the corresponding duty cycle is the best individual or best solution. So, the problem is to find iteratively maximum fitness J.

$$\text{Maximize } J = mean(P) \, \forall \, D_{min} \leq D \leq D_{max} \qquad (3)$$

3.1 Grey Wolf Optimization (GWO)

In GWO algorithm proposed by Mirjalili et al. [11], the hunting mechanism of grey wolves is imitated. In descending order of leadership, grey wolves are termed alpha, beta, delta and omega. The hunting comprises of searching, encircling and attacking prey that are mathematically implemented to perform optimization. It is stochastic and hence performs better. The flowchart can be referred from [12].

3.2 Moth-Flame Optimization (MFO)

Moth-Flame optimization algorithm proposed by Mirjalili [13] is inspired by the inclined movement of moths towards flame or source of light. When far off the light, they move in a straight line but when nearby move spirally around the flame and converge. The flow chart is depicted in Fig. 2.

Fig. 2 Flow chart of
proposed MFO-based MPPT
control

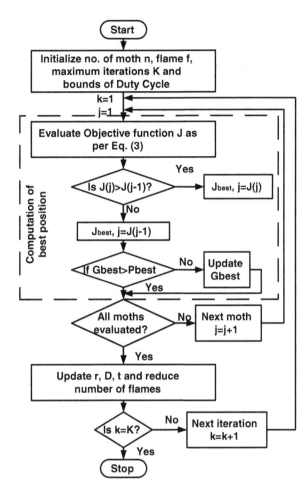

4 Simulation Results

The studied module has nominal parameters of the single diode model contributed
by 5 cells (Sun Power SPR-305) in series and 66 such modules in parallel to form
array contributing to $P_{max} = 100.72$ KW, $V_{oc} = 327.4$ V, $I_{sc} = 393.35$ A, $V_{mp} = 273.5$ V and $I_{mp} = 368.28$ A. The model built in MATLAB Simulink is shown in
Fig. 3. In a similar manner, for 4S each module has 66 cells in series with four such
ones in series; for 2S2P, each module has 32 cells with two such ones in series and
two such combinations of modules in parallel. This modification is done to maintain
the power output to grid within limit. The PV curve of one module and the array are
given in Fig. 4a and b, respectively. The parameters of GWO and MFO algorithm
are upper and lower bounds ub = 1, lb = 0, respectively; a number of moths/
wolves N = 5, iterations K = 7 and simulation time is 3 s.

The MFO-based MPPT algorithm has been compared with P&O and
GWO MPPT algorithms. As seen from Fig. 5, MFO sets the MPP precisely from

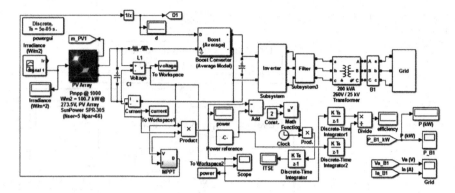

Fig. 3 Model of grid-tied PV system in MATLAB Simulink

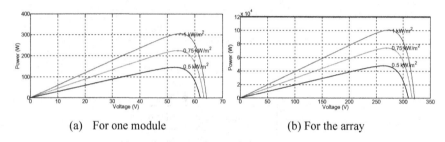

(a) For one module   (b) For the array

Fig. 4 PV graph of 1S configuration **a** and **b**

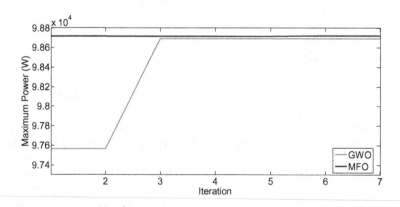

Fig. 5 Convergence performance of algorithms

Pattern 1: Single PV Module (1S)

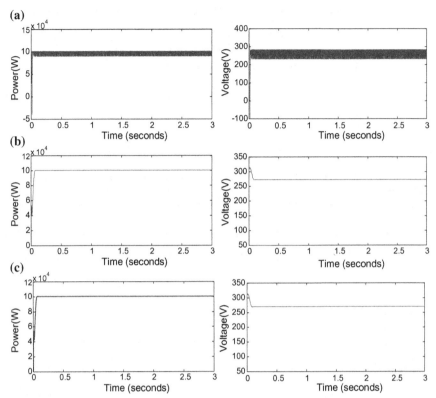

Fig. 6 Tracking curves for MPPT controllers for single module **a** P&O, **b** GWO, **c** MFO

1st iteration and improves minimally in 6th iteration as compared to the large variation of GWO. With more iterations and moths, it has also been observed that there is less variation of power in successive iterations and runs of MFO. So, as compared to previous results [12], this algorithm with reduced population size and iteration maintain good accuracy. The MPPT is implemented under normal irradiance for 1S, 4S and 2S2P configurations. The power and voltage for the 1S configuration deploying MFO, GWO and P&O are plotted in Fig. 6. At a constant irradiance of 1 kW/m^2, from Table 1, it is observed that MFO-based MPPT tracks a mean of 100.04 kW, GWO tracks the GP of 100.01 kW and the P&O algorithm gives 97.61 kW. It is so due to the inefficiency of P&O to distinguish local and global peak (GP). Thus, it reduces the efficiency of the PV system. Also, the MFO-based MPPT has higher tracking speed seen from Fig. 6 and reduced oscillations than the other two, namely GWO and P&O.

Table 1 Comparative evaluation of studied MPPT methods for 1S configuration

Rated power of array (kW)	Control algorithm	Mean power (kW)	Mean current (A)	Mean voltage (V)	%Tracking efficiency [15]	ITSE
100.72	P&O	97.61	367.92	266.11	96.92	8.17×10^7
	GWO [12]	100.01	368.52	271.65	99.3	1.32×10^7
	MFO	100.04	367.02	272.84	99.33	1.40×10^7

Table 2 Comparative evaluation of studied MPPT methods for 4S configuration

Rated power of array (kW)	Control algorithm	Mean power (kW)	Mean current (A)	Mean voltage (V)	%Tracking efficiency [15]	ITSE
80.57	P&O	77.37	367.28	211.50	96.03	8.98×10^7
	GWO [12]	79.85	362.72	220.43	99.11	1.52×10^7
	MFO	79.90	366.30	218.42	99.17	1.32×10^7

Table 3 Comparative evaluation of studied MPPT methods for 2S2P configuration

Rated power of array (kW)	Control algorithm	Mean power (kW)	Mean current (A)	Mean voltage (V)	%Tracking efficiency [15]	ITSE
80.57	P&O	77.37	367.28	211.50	97.01	8.23×10^7
	GWO [12]	79.79	370.23	215.80	104.7	6.76×10^7
	MFO	87.18	387.84	227.62	109.4	7.00×10^7

In 4S configuration as per results in Table 2, the MFO-based MPPT reaches 79.9 kW, GWO gives 79.85 kW, but P&O algorithm reaches 77.37 kW anonymously as it tracks the 1st peak. In 2S2P results tabulated in Table 3, the MFO-based MPPT reaches GP of 87.18 kW, GWO tracks GP of 79.79 kW and P&O reaches 77.37 kW. So from the plots in Figs. 7 and 8 for 4S and 2S2P, it is

Pattern 2: Four Series PV modules (4S)

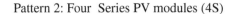

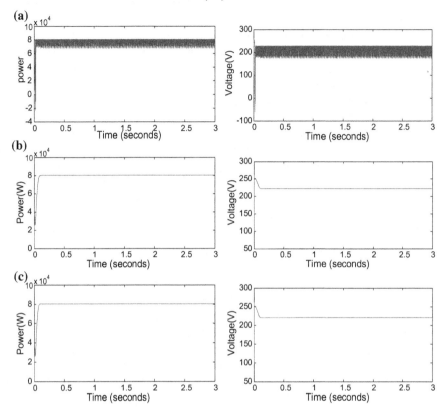

Fig. 7 Tracking curves for MPPT controllers in 4S **a** P&O, **b** GWO, **c** MFO

clear that MFO tracks efficiently than the rest two. The MPPT tracking efficiency is calculated as the ratio of integral of output power obtained to the maximum available power of the PV array under certain irradiance [15]. The quality of the algorithm is studied and depicted in Table 4 for 1S configuration. As compared to qualitative comparison [12], it is quantified in terms of integral time square error (ITSE) index that quantifies the MPP tracking time and oscillations which figures out to be six times higher for P&O than both MFO and GWO. The optimization run-time of MFO is also less compared to GWO.

Pattern 3: Two Series Two Parallel PV modules (2S2P)

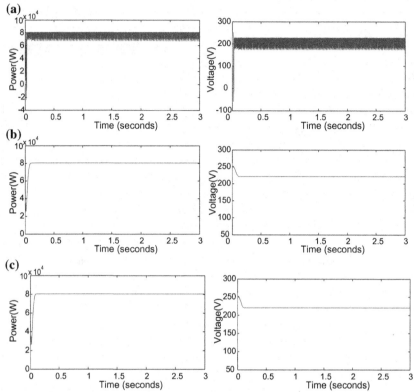

Fig. 8 Tracking curves for MPPT controllers in 2S2P **a** P&O, **b** GWO, **c** MFO

Table 4 Comparisons of different MPPT techniques

Type	P&O	GWO	MFO
Tracking speed	Slow	Fast	Very fast
Transient power fluctuation	Low	Low	Very low
Number of tuning parameters	1	1	1
Steady state oscillations	Large	Zero	Zero
ITSE of power	8.17×10^6	1.32×10^6	1.4×10^6
Power efficiency	Low	High	High
Implementation complexity	Low	Medium	Medium
Dynamic response	Poor	Good	Good
Average optimization algorithm run-time (S)	–	124	116

5 Conclusions

From simulation result, it is found that the efficiency of MFO- and GWO-based MPPT is more as compared to conventional P&O based MPPT. These simple optimization algorithms' accuracy is achieved with less optimization time. In case of MFO, there is more reduction of ripples in output voltage and current, which improves quality of DC output and efficiency of MPPT than GWO. Thus, MFO helps to find the optimized values very fast to generate constant output voltage improving the power output. This leads to better performance of the system, and the panel becomes more reliable in terms of providing a stable output which is required by the utility. Improved and faster optimization algorithms can be implemented for the MPPT in future.

References

1. Salam, Z., Ahmed, J., Merugu, B.S.: The application of soft computing methods for MPPT of PV system: A technological and status review. Applied Energy, Vol. 107 (2013) 135–148.
2. Elgendy, M. A., Zahawi, B., Atkinson, D. J.: Assessment of perturb and observe MPPT algorithm implementation techniques for PV pumping applications. IEEE Trans. Sustain. Energy, Vol. 3(1) (Jan. 2012) 21–31.
3. Elgendy, M. A., Zahawi, B., Atkinson, D. J.: Operating characteristics of the P&O algorithm at high perturbation frequencies for standalone PV systems. IEEE Trans. Energy Convers., Vol. 30(1) (Jun. 2015) 189–198.
4. Hsieh, G.C., Hsieh, H.I., Tsai, C.Y., Wang, C.H.: Photovoltaic power-increment-aided incremental-conductance MPPT with two-phased tracking. IEEE Transactions on Power Electronics, 28(6) (2013) 2895–2911.
5. Gheibi, A., Mohammadi, S.M.A, Farsangi, M.M.: A proposed maximum power point tracking by using adaptive fuzzy logic controller for photovoltaic systems. Scientia Iranica. Transaction D, Computer Science & Engineering, Electrical, 23(3) (2016) 1272–1281.
6. Dallago, E., Liberale, A., Miotti, D., Venchi, G.: Direct MPPT algorithm for PV sources with only voltage measurements. IEEE Transactions on Power Electronics, Vol. 30(12) (2015) 6742–6750.
7. Hohm, D.P., Ropp, M.E.: Comparative study of maximum power point tracking algorithms. Progress in photovoltaics: Research and Applications, Vol. 11(1) (2003) pp. 47–62.
8. Ishaque, K., Salam, Z., Amjad, M., Mekhilef, S.: An improved particle swarm optimization (PSO)–based MPPT for PV with reduced steady-state oscillation. IEEE transactions on Power Electronics, Vol. 27(8) (2012) 3627–3638.
9. Sundareswaran, K., Peddapati, S., Palani, S.: MPPT of PV systems under partial shaded conditions through a colony of flashing fireflies. IEEE transactions on energy conversion, 29 (2) (2014) 463–472.
10. Ahmed, J., Salam, Z.: A Maximum Power Point Tracking (MPPT) for PV system using Cuckoo Search with partial shading capability. Applied Energy, Vol. 119(2014) 118–130.
11. Mirjalili, S., Mirjalili, S. M., Lewis, A.: Grey wolf optimizer. Adv. Eng. Software, Vol. 69 (2014) 46–61.

12. Mohanty, S., Subudhi, B., Ray, P.K.: A new MPPT design using grey wolf optimization technique for photovoltaic system under partial shading conditions. IEEE Transactions on Sustainable Energy, Vol. 7(1) (2016) 181–188.
13. Mirjalili, S.: Moth-flame optimization algorithm: A novel nature-inspired heuristic paradigm. Knowledge-Based Systems, Vol. 89 (2015) 228–249.
14. Villalva, M.G., Gazoli, J.R., Ruppert Filho, E.: Comprehensive approach to modeling and simulation of photovoltaic arrays. IEEE Transactions on Power Electronics, Vol. 24(5) (2009) 1198–1208.
15. Jantsch, M., Real, M., Häberlin, H., Whitaker, C., Kurokawa, K., Blässer. G., Kremer, P., Verhoeve, C.W.: Measurement of PV maximum power point tracking performance. Netherlands Energy Research Foundation ECN, 30 Jun 1997.

Stability Analysis in RECS-Integrated Multi-area AGC System with SOS Algorithm Based Fuzzy Controller

Prakash Chandra Sahu, Ramesh Chandra Prusty and Sidhartha Panda

Abstract This paper aims toward coordination between generation and demand of electric power, which is termed as automatic generation control (AGC). A wind energy conversion system (WECS)-based doubly fed induction generator (DFIG) integrated with two equal areas conventional thermal generation was proposed. A fuzzy-Proportional Integral Derivative (fuzzy-PID) controller was used for stabilizing deviation in frequency (Δf) and tie-line power (ΔP_{tie}). The gains of fuzzy-PID and DFIG controller are tuned optimally using a multi-objective optimization technique called symbiotic organism search (SOS) algorithm. In addition, the dynamic response and accuracy of system under study was investigated using integral of time multiplied absolute error (ITAE). The performance of fuzzy-PID controller was compared with conventional PID, PI, and fuzzy-PI controller in terms of settling time and peak overshoot. Finally, it was observed experimentally that the proposed SOS optimized fuzzy-PID controller gives superior dynamic and robust performance as compared to other controllers under various operating conditions.

Keywords AGC · Renewable Energy Conversion System (RECS) Symbiotic Organism Search (SOS) · Fuzzy-PID · Doubly Fed Induction Generator (DFIG) · Wind Energy Conversion System (WECS)

P. C. Sahu (✉) · R. C. Prusty · S. Panda
Department of Electrical Engineering, VSSUT, Burla 768018, Odisha, India
e-mail: prakashsahu.iter@gmail.com

R. C. Prusty
e-mail: ramesh.prusty82@gmail.com

S. Panda
e-mail: panda.sidhartha@rediffmail.com

© Springer Nature Singapore Pte Ltd. 2019
H. S. Behera et al. (eds.), *Computational Intelligence in Data Mining*,
Advances in Intelligent Systems and Computing 711,
https://doi.org/10.1007/978-981-10-8055-5_21

1 Introduction

In today's scenario, there is a large demand for electricity. Due to this, various generating stations capacity must be expanded. This is done by integrating some external source with existing system. Here basically, the external source is wind energy conversion system (WECS)) [1–5]. Though by using wind energy, generation level increased, but it is necessary to control frequency and voltage deviation, when WECS is integrated with existing generating system. So to do operation in secure manner in an integrated system, research has been carried out in the area of AGC of power system with penetration of WECS. Among different types WECS, the most developed type is based on variable speed wind turbine (VSWT) [6]. It is equipped with power electronics converters, which enable to decouple the rotational speed with grid frequency by controlling the pitch angle. Normally, wind farms respond differently to the variation of network frequencies compared with thermal system. Wind farms are capable of doing ramp increase/decrease real power generation for maintaining reverse margin [1]. Therefore, it has been implemented in system frequency support. VSWTs driving DFIG [6] are developed to control the rotational speed in wide variation by utilizing the stored kinetic energy, which enhances to control the torque and active power and also to frequencies [3].

The natural inertial responding capability of a system reduces, when there will be large penetration of wind farm-based DFIG. For reliability, the inertial capability of system is restored by implementing an additional control mechanism in DFIG [4–7]. A proposed current controller mechanism gives inertial response from a power system integrated with DFIG. There is a chance for greater leverage in the rotor speed variation, which enhances more utilization of kinetic energy of turbine-DFIG system [5]. The proposed method modifies inertial control strategy by utilizing stored kinetic energy of WECS and also helps to communicate WECS response among all generating units, so that load imbalance may be rectified. The effect of different level of wind penetration from DFIG on its frequency response is discussed in two thermal area AGC system [8]. In this paper, the DFIG and proposed fuzzy-PID [9] controller gains are turned in such that, it gives better frequency regulation without interfering AGC control performance. Under each and every condition, the controller exhibits consistency and robustness. Here the gains are tuned by using optimization techniques. In power system, different optimization techniques have been utilized [10–14]. The techniques like GA [10], PSO [11], cuckoo search [12], and LP have been utilized for this purpose. By considering the different constraints in AGC and difference in the nature of WECS and thermal system, a multi-objective optimization technique, i.e., SOS algorithm is used [13]. It has been seen that it is most robust, consistent, and gives better performance over all above optimization techniques.

2 Two Area Thermal System with DIFG-Based Wind Power Generation

2.1 Thermal Model

The model based on transfer function of two equal area thermal systems with DFIG is shown in Fig. 1. Here in each area, the thermal system comprises governor, prime mover (reheat-type turbine), and generator [15]. Governor is modeled with single time constant transfer function, by using Eq. (1).

$$\Delta P_g(s) = \Delta P_{ref}(s) - \frac{\Delta f(s)}{R} \tag{1}$$

Hydraulic actuator (governor) transfer function can be modeled using Eq. (2).

$$G_g(s) = \frac{\Delta P_v(s)}{\Delta P_g(s)} = \frac{1}{1 + sT_g} \tag{2}$$

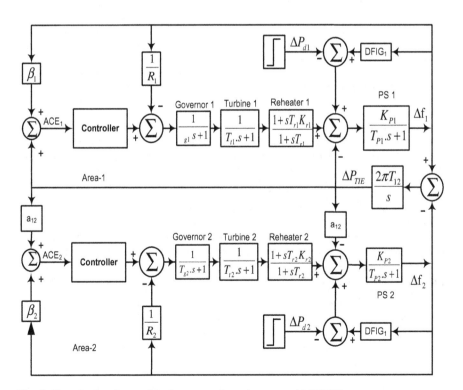

Fig. 1 Transfer function model of two area thermal system with WECS

Reheat-type turbine dynamics transfer function can be represented mathematically in Eq. (3).

$$G_T(s) = \frac{\Delta P_T(S)}{\Delta P_V(S)} = \frac{1}{1+sT_T} \cdot \frac{1+s.T_R K_R}{1+s.T_R} \tag{3}$$

Generator output to power system can be modeled using Eq. (4).

$$G_p(s) = \frac{K_p}{1+sT_p} \tag{4}$$

2.2 Wind Energy Conversion System (WECS)

For this article, a WECS is composed of VSWT [6]-based DFIG. The DFIG delivers power to common grid through stator and rotor part of DFIG. For this, a back-to-back power electronics converter is connected between rotor of DFIG and grid. To control frequency deviation and tie-line power deviation in system, a proposed DFIG controller is implemented for this article.

3 Proposed Controller

3.1 Fuzzy-PID Controller

For providing zero deviation in frequencies and tie-line powers of different areas, controllers are required. In this article, comparative performance analysis of fuzzy-PID controller (Fig. 2) is compared with that of fuzzy-PI and conventional PID and PI controllers. Structure of the proposed fuzzy-PID is shown in Fig. 2. It has two inputs, one is area control error (ACE), i.e., '$e(t)$' and other is its derivative, i.e., '$e'(t)$'. The controlled outputs u_1, u_2, and u_3 are given as input to power setting points (i.e., ΔP_r).

The parameters K_P, K_I, and K_D are the gains of controller and K_1 and K_2 are the scaling factors of controller. Triangular membership functions along with five linguistic variables are implemented for this controller. These variables far negative (FN), near negative (NN), zero (Z), near positive (NP), and far positive (FP) are considered for both input and output variables of this proposed controller.

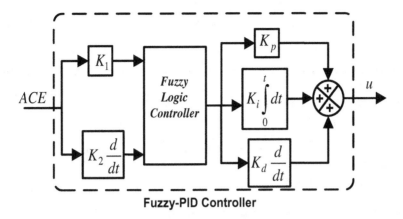

Fuzzy-PID Controller

Fig. 2 Fuzzy-PID controller

4 SOS Algorithm

SOS algorithm is developed by Cheng and Prayog [13] and is specially based on symbiotic nature between two organisms in ecosystem. Symbiotic co-relations are two categories. These are obligate and facultative. In obligate category the organisms are dependent on each other for their survival. In nature, there are mostly three types of symbiotic relationship. These are mutualism, commensalism, and parasitism.

Define objective function $f(x)$; $x = (x_1\ x_2\ x_3 \ldots x_d) \ldots$ (d is the dimension of the problem).

Initialize an ecosystem having n number of organisms with random solution.

    While (t < Maxgeneration)
        For i= 1: n (n= number of organisms)
In ecosystem get best organism'X_{best}'
    % Mutualism Phase

Select organism X_j as randomly such that $X_J \neq X_i$ (i^{th} organism in ecosystem)

Calculate $mutual\_vector = (X_j + X_i)/2$, Benefit Factors ($BF_1$ & BF_2) are either 1 or

2
X_j and X_i are updated by equations
$X_{inew} = Xi + rand\ (0,1)*(Xbest-Mutual-Vector*BF1)$
$X_{jnew} = Xj + rand\ (0,1)*(Xbest-Mutual-Vector*BF2)$

If present organism gives better fitness value than previous, then update in ecosystem.

If present organism gives better fitness value than previous then update in ecosystem.
   % Commensalism Phase
Select organism X_j as randomly such that $X_J \neq X_i$

Modify X_i using equation
Xi new =Xi +rand (-1,1)*(Xbest-Xj)
If present organism gives better fitness value than previous then update in ecosystem.
   % Parasitism Phase

Select organism X_j as randomly such that $X_J \neq X_i$

From organism X_i produce parasitic vector
If parasitic vector gives better fitness value than X_j then update parasitic vector.
   end
Global best solution is saved as optimal solution.
   end

5   Results and Analysis

For this research work, the simulation of proposed model along with related programming is done in MATLAB/SIMULINK environment. For simulation, a model of two interconnected areas along with different controllers like PI, PID, fuzzy-PI, and fuzzy-PID were taken with different optimization techniques (like GA, PSO, and SOS). While simulating, a SLP of 2% is considered only in area 1. The different responses like Δf_1, Δf_2, ΔP_{tie}, and convergence curve were presented for best solution as shown in Figs. 3, 4, 5, 6, and 7. From results in Table 1, it was seen that SOS optimized fuzzy-PID controller-based response has less overshoot and settling time compared with other implemented controllers. The fuzzy-PID controller exhibits robust performance, and its gains need not to change again for any system parameter variation.

For PI controller, the gains are

$KP_1 = 0.2770$, $KP_2 = 0.8453$, $KI_1 = 0.4829$, $KI_2 = 1.0809$; $KDf_1 = 1.9501$; $KDf_2 = 0.6941$; $KPf_1 = 1.1553$; $KPf_2 = 0.4171$;

For PID controller, gains are

$KP_1 = 0.6487$, $KP_2 = 1.3821$, $KI_1 = 0.9702$, $KI_2 = 0.0965$, $KD_1 = 0.5131$, $KD_2 = 1.0420$, $KDf_1 = 1.1699$; $KDf_2 = 0.7743$; $KPf_1 = 1.5680$; $KPf_2 = 0.3569$;

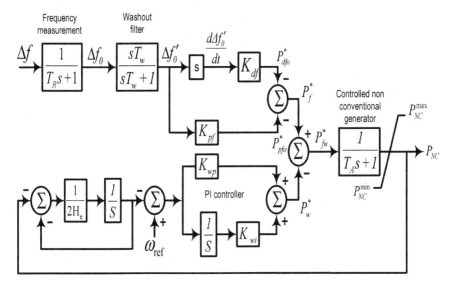

Fig. 3 Proposed controller in DFIG system

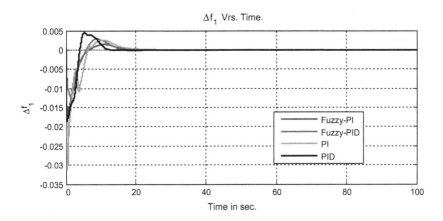

Fig. 4 Deviation of frequency in area 1 versus time

For fuzzy-PI controller

$KP_1 = 0.3617$, $KP_2 = 0.2597$, $KI_1 = 2.0000$, $KI_2 = 2.0000$, $KDf_1 = -0.3247$; $KDf_2 = -0.2431$; $KPf_1 = 1.7419$; $KPf_2 = -0.3907$; with scaling factor $K_1 = 0.7548$, $K_2 = 0.4320$.

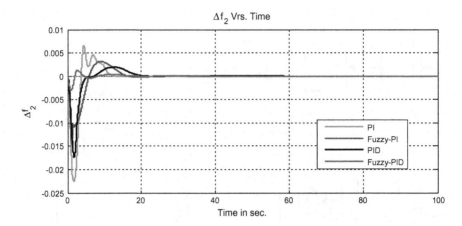

Fig. 5 Deviation of frequency in area 2 versus time

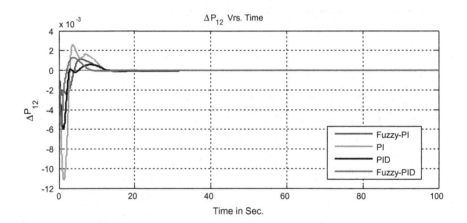

Fig. 6 Deviation of tie-line power between area 1 and area 2 versus time

Fitness value = 0.7770;

For fuzzy-PID controller

$KP_1 = 0.2452$, $KP_2 = 0.0382$, $KI_1 = 1.0790$, $KI_2 = 0.7988$, $KD_1 = 0.2096$, $KD_2 = 1.1221$, $KDf_1 = 0.2678$; $KDf_2 = -0.6404$; $KPf_1 = -0.4876$; $KPf_2 = 0.0139$

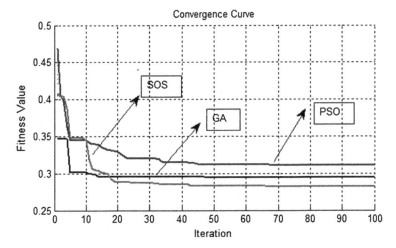

Fig. 7 Convergence curve

Table 1 Settling times of Δf_1, Δf_2, and ΔP_{tie} in different controllers

Error	PI	PID	Fuzzy-PI	Fuzzy-PID
Δf_1	17.5434	16.8262	16.5258	15.7877
Δf_2	16.5461	15.2600	14.6430	12.4040
ΔP_{12}	15.2488	14.2468	11.6464	9.8450

6 Conclusion

In this research work, a fuzzy-PID controller is applied as main controller for stabilizing the frequency and tie-line power dynamics of two area thermal-integrated wind energy conversion system. To make it easy, a nature-inspired optimization algorithm called SOS algorithm is implemented to optimize the parameters of different controllers along with fuzzy-PID controller. Comparison of different controller shows that fuzzy-PID controller gives superior dynamic response than other controllers. SOS algorithm is compared with GA and PSO. The parameters and scaling factors of fuzzy-PID controller are optimized using SOS technique, and its superiority is compared with conventional GA and PSO. It has been seen that SOS takes advantage of global exploration and local exploitation capabilities of PSO. To show superiority of SOS, its results are compared with the results of PSO and GA through convergence curve.

Appendix

A. Nominal parameter of system

i = subscript referred to area (1, 2)
Hi = Inertia constant = $H_1 = H_2 = 5$
f_i = Incremental change in frequency (Hz)
P_{di} = Incremental load change
$D_i = 8.33 \times 10^{-3}$ p.u MW/Hz
$R_1 = R_2 = 2.4$ Hz/p.u MW
$Tg_1 = Tg_2 = 0.08$ s (steam governor time constant)
$Kr_1 = Kr_2 = 0.5$ (steam turbine reheat coefficient)
$Tr_1 = Tr_2 = 10$ s (steam turbine reheat time constant)
$Tt_1 = Tt_2 = 0.3$ s (steam turbine time constant)
$\beta 1 = \beta 1 = 0.425$ (frequency bias)
$f_1 = f_2 = 60$ Hz
$Tp_1 = Tp_2 = 20$ s (power system time constant)
$Kp_1 = Kp_2 = 120$ Hz/p.u MW (power system gain constant)
T_{12} = synchronizing coefficient
$Pr_1 = Pr_2 = 2000$ MW (area capacity)

B. Nominal parameter of system with DFIG

$\beta 1 = \beta 2 = 0.314$ (frequency bias)
$H_e = 3.5$ s (equivalent WECS time constant)
$T_w = 6$ s (washout filter time constant)
T_R = frequency transducer time constant
T_A = controlled WECS time constant
K_{pw} = speed regulator proportional time constant
K_{iw} = speed regulator integral constant
$P_{NC}^{min}/P_{nc}^{max}$ = WECS output power limit = 0/1.2 p.u.

References

1. Eltra, E. "Wind turbines connected to grids with voltages above 100 kV." Regulation document TF 3.5 (2004).
2. Van Hulle, F. J. L. "Large Scale Integration of Wind Energy in the European Power Supply: Analysis, Issues and Recommendations": Executive Summary. European Wind Energy Association, 2005.
3. Mullane, Alan, and Mark O'Malley. "The inertial response of induction-machine-based wind turbines." IEEE Transactions on power systems 20.3 (2005): 1496–1503.
4. Lalor, Gillian, Alan Mullane, and Mark O'Malley. "Frequency control and wind turbine technologies." IEEE Transactions on Power Systems 20.4 (2005): 1905–1913.

5. Anaya-Lara, O., et al. "Contribution of DFIG-based wind farms to power system short-term frequency regulation." IEE Proceedings-Generation, Transmission and Distribution 153.2 (2006): 164–170.

6. Mauricio, Juan Manuel, et al. "Frequency regulation contribution through variable-speed wind energy conversion systems." IEEE Transactions on Power Systems 24.1 (2009): 173–180.

7. Bhatt, Praghnesh, Ranjit Roy, and S. P. Ghoshal. "Dynamic participation of doubly fed induction generator in automatic generation control." Renewable Energy 36.4 (2011): 1203–1213.

8. Elgerd, Olle I., and Charles E. Fosha. "Optimum megawatt-frequency control of multiarea electric energy systems." IEEE Transactions on Power Apparatus and Systems 4 (1970): 556–563.

9. Ghoshal, Sakti Prasad. "Optimizations of PID gains by particle swarm optimizations in fuzzy based automatic generation control." Electric Power Systems Research 72.3 (2004): 203–212.

10. Ghoshal, S. P., and S. K. Goswami. "Application of GA based optimal integral gains in fuzzy based active power-frequency control of non-reheat and reheat thermal generating systems." Electric Power Systems Research 67.2 (2003): 79–88.

11. R. K. Sahu, S. Panda and G.T. Chandra Sekhar. "A novel hybrid PSO-PS optimized fuzzy PI controller for AGC in multi area interconnected power systems", Int J Electr Power Energy Syst, Vol. 64, pp. 880–93, 2015.

12. P. Dash, L. C. Saikia and N. Sinha, "Comparison of performances of several Cuckoo search algorithm based 2DOF controllers in AGC of multi-area thermal system" Int. J Electric Power Energy Syst, Vol. 55, pp. 429–36, 2014.

13. M.Y. Cheng, D. Prayogo, Symbiotic organisms search: a new metaheuristic optimization algorithm, Comput. Struct". 139 (2014) 98–112.

14. P. Dash, L. C. Saikia and N. Sinha, "Automatic generation control of multi area thermal system using Bat algorithm optimized PD–PID cascade controller" Int J Electr Power Energy Syst, 68, pp. 364–78, 2015.

15. Nanda, A. Mangla and S. Suri, "Some new findings on automatic generation control of an interconnected hydrothermal system with conventional controllers," IEEE Trans Energy Convers, Vol. 21 (1), pp. 187–94, 2006.

Log-Based Reward Field Function for Deep-Q-Learning for Online Mobile Robot Navigation

Arun Kumar Sah, Prases K. Mohanty, Vikas Kumar and Animesh Chhotray

Abstract Path planning is one of the major challenges while designing a mobile robot. In this paper, we implemented Deep-Q-Learning algorithms for autonomous navigation task in wheel mobile robot. We proposed a log-based reward field function to incorporate with Deep-Q-Learning algorithms. The performance of the proposed algorithm is verified in simulated environment and physical environment. Finally, the accuracy of the performance of the obstacle avoidance ability of the robot is measured based on hit rate metrics.

Keywords Path planning · Obstacle avoidance · Mobile robot
Wheel robot · Reward function · Q-learning

1 Introduction

Navigation is one of the characteristic features of a mobile robot, and the path planning is one of the most challenging tasks. Path planning problem can be defined as the problem of finding a path from a starting location (S) to a final location (G) in presence of obstacles and goal while avoiding collision with obstacles. Path planning problem by two approaches is the classical approaches and heuristic

A. K. Sah · P. K. Mohanty (✉) · V. Kumar
Department of Mechanical Engineering, National Institute of Technology,
Yupia 791112, Arunachal Pradesh, India
e-mail: pkmohanty30@gmail.com

A. K. Sah
e-mail: arunkumar626@gmail.com

V. Kumar
e-mail: vikasmenitap2013@gmail.com

A. Chhotray
Department of Mechanical Engineering, National Institute of Technology,
Rourkela 769008, Odisha, India
e-mail: chhotrayanimesh@gmail.com

© Springer Nature Singapore Pte Ltd. 2019
H. S. Behera et al. (eds.), *Computational Intelligence in Data Mining*,
Advances in Intelligent Systems and Computing 711,
https://doi.org/10.1007/978-981-10-8055-5_22

approaches [1]. The classical approach encompasses the methods purely based on mathematical techniques such as road map, cell decomposition, potential fields, and mathematical programming. Heuristic-based approaches are based on intuition or inspired by nature [2]. There are three primary types of heuristic-based algorithms, namely nature-inspired techniques, fuzzy logic technique, and neural network-based technique.

Genetic algorithms, Ant colony optimization, Particle swarm optimization, Honey bee algorithm, Firefly algorithm, Cuckoo Search algorithm, Artificial bee algorithm, etc. [2] are some of popular nature-inspired optimization algorithms that find application in path planning problem.

The fuzzy mathematics was presented by Zadeh [3] and the theory was applied for motion planning by Vachtsevanos and Hexmoor [4]. In this approach, the navigation problem is broken into simpler task and the decision making of robot navigation is phrased by a set of IF–THEN rules. Valdez et al. [5] surveyed on nature-inspired optimization algorithms with fuzzy logic for dynamic parameter adaptation.

Zacksenhouse and Johnson [6] presented the use of neural networks for motion planning. Neural networks are computational approach inspired from biological neurons. Apart from robot motion planning, neural network has been used for interpreting the sensory data, obstacle avoidance, path planning, and many other tasks. A neural network solves the path and time optimization problem of mobile robots that deals with the cognitive tasks such as learning, adaptation, generalization, and optimization. A back-propagation algorithm is used to train the network.

Reinforcement learning has been a successful machine learning paradigm that learns by interacting in an environment [7, 8]. Lin [9] applied reinforcement learning for robot-learning problem. Motlagh et al. [10] presented a technique based on utilization of neural networks and reinforcement learning to enable a mobile robot to learn constructed environments on its own. The robot learns to generate efficient navigation rules automatically without initial settings of rules by experts. Jaradat et al. [11] used Q-learning approach for mobile robot path planning in an unknown dynamic environment.

Fuzzy logic is used for solving various kind of problem but they lack generalizability when it comes to autonomous navigation. Neural network in the form of multi-layer perception is been widely used for solving various kinds of problem in this decade [12–15].

In this paper, we presented the used of Deep-Q-Learning for path planning problem, incorporating log-based reward field function. For this purpose, we have developed a simulation and a hardware robot. The simulation provides a platform to train the robot in virtual environment and the train model is uploaded to the hardware robot. The hardware robot used the training information from the simulator and further gets trained in real physical environment. We compare the performance of the implemented algorithms with other heuristic methods and finally validate the hardware performance with the simulation agent performance.

2 Methodology

2.1 Theory

Environment is a region of physical or virtual space with characteristic state space where entities can exist. An agent is an entity that perceives and acts in an environment. There are basically two kinds of agents: (1) purposive agent and (2) non-purposive agent. Non-purposive agents do not favor any world's state while purposive agents prefer some world's more than the other world's states. Agent's preferences over possible outcomes are defined by utility function that maps outcomes to a real-value number such that higher the number the more the agent prefer that outcomes. For example, according to economists, human are utility-maximizing agents, i.e., humans try to maximize their internal utility function. Agent chooses actions based on their outcomes. Suppose, o_1 and o_2 are outcomes such that o_1 is greater than or equal to o_2 means o_1 is at least as desirable as outcome o_2, i.e., o_1 is weakly preferred over o_2. A relational agent selects actions that maximize its expected utility. Most commonly the reinforcement learning task is modeled as Markov Decision Process (MDP) [16]. MDP is used for problems in which outcomes are partly random and partly depends on the action of agent [17].

MDP is defined by (S, A, P, R, γ) where

- S is a set of spaces.
- A is a set of actions.
- P is defined as $P_a(s, s') = P(s_{t+1} = s' | s_t = s, a_t = a)$ which is the probability that action a in state s at time t will leads to state s' at time $t + 1$.
- R is $R_a(s, s')$ is the immediate rewards or expected immediate rewards received after transitioning from state s to state s' due to action a.

To decide which action to take, the agent needs to take account the immediate and the future rewards since the future rewards depend on the current action.

For an episode of MDP, total reward of one episode is given by Eq. (1).

$$R = \sum_{i=1}^{i=n} r_i \qquad (1)$$

Total future reward from time t is given by Eq. (2).

$$R_t = \sum_{i=0}^{i=n} r_{t+i} \qquad (2)$$

In stochastic environment, the sequence of reward varies for same actions and more we see into the future from time t, more it may diverge. Thus, we use discounted future reward which is given by $R_t = \sum_{i=0}^{i=n} \gamma^i r_{t+i}$ or $R_t = r_t + \gamma R_{t+1}$ where γ is discount factor whose values lie between 1.0 and 0.0. To have balance between immediate and future rewards, we use discount factor to $\gamma = 0.9$.

Q-learning is a reinforcement learning approach that acts optimally in constrained markovian domains Q-learning works by improving evaluations of the quality of an action at particular states iteratively [8]. Watkins and Dayan [9] proved a theorem for Q-learning convergence based on Watkins [8] outline of the approach. The idea behind Q-learning is to successively approximate the Q-function in accordance with Bellman equation which says that the expected long-term rewards for a given action are equal to the immediate reward from the current action combined with expected reward from the best future action taken at the following state.

A function $Q(s, a)$ (given by Eqs. 3 and 5) represents the maximum discounted future reward if we perform action a in state s and continue optimally. Equation (4) is the policy of selecting action in state s.

$$Q(s_t, a_t) = \max R_{t+1} \tag{3}$$

$$\pi(s) = \mathbf{argmax}_a Q(s, a) \tag{4}$$

$$Q(s, a) = r + \gamma \, \mathbf{max}_a Q(s', a') \tag{5}$$

where

r	*rewards received in current state*
γ	*discount factor, value between 0.0 and 1.0*
$Q\ (s,\ a)$	*Q-value for state (s) and action (a)*
$Q\ (s',\ a')$	*Q-value for next state (s') and next action (a')*

Many MDP problem may have infinite numbers of state space. In such cases, Q-function is approximated by neural network called as Q-network. Neural network performs well as features extraction for highly structured data [14]. Q-network takes state as input and returns the Q-value for each possible action as output for that state. Q-values are real values, which make them a regression task. The network can be optimized with square error loss function given by (6) or (7).

$$J = \frac{1}{2} [target\ value - predicted\ value]^2 \tag{6}$$

$$J = \frac{1}{2} \left[(r + \gamma \, max_{a'} Q(s', a')) - Q(s, a) \right]^2 \tag{7}$$

2.2 Architecture

In our implementation (shown in Fig. 1), the agent performs an action a from set of available action A. We had $A = \{move\ forward, turn\ right, turn\ left\}$. The action performed by agent changes the environment state and the agent observes new state

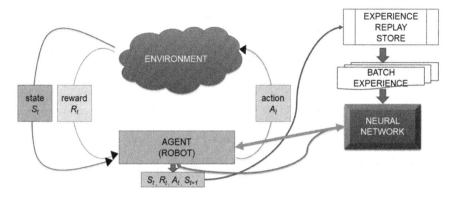

Fig. 1 Deep-Q-Network architecture

s' as observation. The agent also receives a reward r for previous action a. The training of Q-network is not stable using nonlinear functions. To improve the stability of prediction and convergence, at every iteration the agent stores the set of values $[s, a, r, s']$ in Experience Replay Memory (ERM). ERM is parametrized by maximum number of experience pool size and threshold experience pool size. During training period, random mini-batch of experiences is fetched from the experience memory and feed into Q-network. During test period, observed state s is passed through Q-network and it returns optimal action that could maximize the expected future reward. Q-learning solves the credit assignment problem by propagating the rewards back through time until it reaches the actual deciding moment that leads to the reward. We initialize the network with random weights, and thus, the prediction made by the Q-network initially is not correct. As the Q-function converges, the Q-network becomes better at predicting actions but the amount of exploration decreases. Thus, the Q-learning performs exploration but it settles when it discovered first effective strategy it finds. In practice, ε-greedy exploration approaches are used according to which random action is chosen by the agent instead the action predicted by Q-network. The value of ε is kept between 1.0 (initially) to 0.1 (after convergence).

2.3   Reward Value Field Function

We implemented a reward value field function based on Euclidian distance and log function.

In two-dimensional plain, if $p = (p_x, p_y)$ and $q = (q_x, q_y)$, then Euclidian distance is given by Eq. (8).

$$d(p,\ q)\ or\ d(q,\ p) = \sqrt[2]{(q_x - p_x)^2 + (q_y - p_y)^2} \tag{8}$$

We generated the reward value field function by using *numpy* module of Python.

$X = np.arange(0,\ width,\ gridsize)$
$Y = np.arange(0,\ height,\ gridsize)$
$X,\ Y = np.meshgrid(X,\ Y)$

$Z_{positive,\ i} = positivelogfield\left(X, Y, x_{goal,\ i}, y_{goal,\ i}\right)$

$Z_{negative,\ j} = negativelogfield\left(X, Y, x_{obstacle,\ j}, y_{obstacle,\ j}\right)$

$Z_{positive,\ total} = \Sigma Z_{positive,\ i}$
$Z_{negative,\ total} = \Sigma Z_{negative,\ j}$
$Z_{positive,\ normalized} = normalized\left(Z_{positive,\ total}\right)$
$Z_{negative,\ normalized} = normalized\left(Z_{negative,\ total}\right)$
$Z_{field} = Z_{positive,\ normalized} + Z_{negative,\ normalized}$
def logfield $\left(X_{field}, Y_{field}, x, y\right)$:

$Z = -numpy.\log\left(numpy.sqrt\left((X_{field} - x)^2 + (Y_{field} - y)^2 + \varepsilon\right)\right);$

return $(Z/(-\alpha * min(Z.flatten\ (\ ))) + (-\alpha * min(Z/(-\alpha * min(Z.flatten\ (\ ))).flatten\ (\ ))));$

where: $\varepsilon = 1.0e - 5$ for numerical stability; and

$$\alpha = \left\{ \begin{array}{ll} -1.0, & \text{for positivelogfield }(\ ) \\ 1.0, & \text{for negativelogfield }(\ ) \end{array} \right\}$$

3 Simulation and Experimental Results

In our experiment, we developed a simulated agent and robot equipped with range sensors such as IR-sensors, ultrasonic sensors, and a camera. The simulation has similar arrangement to simulate range sensors and is able to distinguish targets and goal. The robot detects obstacles with the help of distance sensors and distinguishes target from obstacle with a simple convolutional neural network [14] architecture. We perform five set of experimental and simulation experiment. The details of representational figures used in simulation and experimental setup are presented in Table 1. The real time experiments we conducted on a platform with four obstacles and a goal point. For different experiments, we change the positions of the obstacles and goal points.

Table 1 Various representational image used

Agent	Obstacle	Target	Path traced

We recreate the experimental scenario in the simulation software to compare the performance. The Deep-Q-Network ran on a laptop with 4 GB RAM and *i5* processors.

Three sets of experimental and simulation result are presented in figures [2–9]. Each experiment is presented with three sets of images. Experimental setup is shown in Figs. 2, 3, and 4. Simulation setup is shown in Figs. 5, 6, and 7. Reward Field function is shown in Figs. 8, 9, and 10. One important thing to note that the graph for reward field function represents the reward value due to obstacles and the

Fig. 2 Experimental setup 1

Fig. 3 Experimental setup 2

Fig. 4 Experimental setup 3

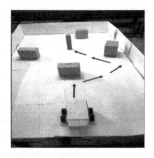

Fig. 5 Simulation setup 1

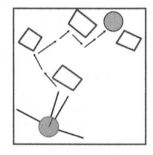

Fig. 6 Simulation setup 2

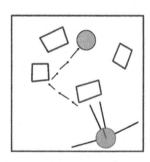

Fig. 7 Simulation setup 3

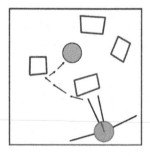

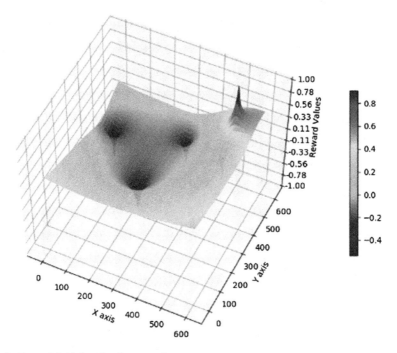

Fig. 8 Reward field function for setup 1

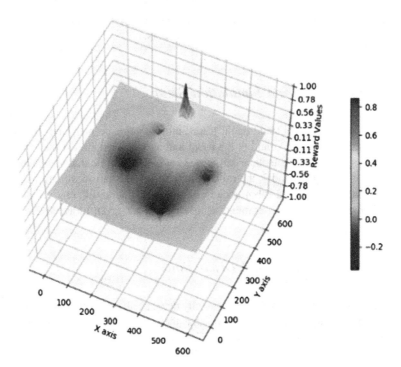

Fig. 9 Reward field function for setup 2

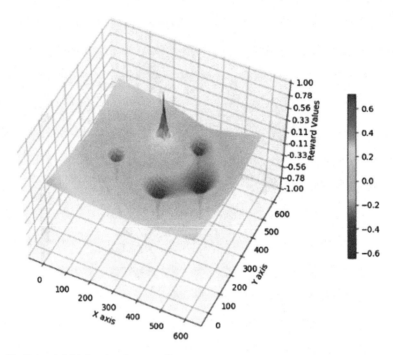

Fig. 10 Reward field function for setup 3

target only while the reward for walls is not shown in the image as the presence of walls directly impacts the path traced by the robots.

3.1 Experimental and Simulation Setup

In Figs. 2 and 5, the robot and agent traced almost opposite path due to different initial conditions. In Figs. 3 and 6, the robot and agent traced almost the same path while the robot hits an obstacle at initial stage which might be due to oversteps of robot motion. In Figs. 4 and 7, the robot escaped from the narrow situation to reach the target but the agent and robot traced different path due to different initial states. We performed two more experiments whose results are given in Table 2.

3.2 Result Validation and Discussion

We developed a mobile robot that successfully detects and avoids the obstacles. After the experiments on the robot, we found that the robot is able to detect and avoid obstacles with an average accuracy of 90%. The accuracy is calculated by Eq. (9).

Table 2 Average accuracy of the robot

Scenario	Number of obstacles	Number of obstacles detected	Number of obstacles avoided	Accuracy (%)
Hardware Test 1	4	4	4	100
Hardware Test 2	4	4	3	75
Hardware Test 3	4	4	4	100
Hardware Test 4	4	4	3	75
Hardware Test 5	4	4	4	100
			Average accuracy	90

$$Accuracy = \frac{\sum \textbf{Number of Successful Avoidances}}{\sum \textbf{Number of Test Cases}} \qquad (9)$$

In the above-given formula, the numerator denotes the total number of times the robot was able to avoid the obstacles. The denominator denotes the total number if test cases.

4 Conclusion

In this paper, we introduced log-based reward field function for the obstacle and target. Using the log-based reward function, the agent gets a reward value between −1.0 and +1.0 in two-dimensional plane. According to our experiment, after the simulation agent gets trained to the environment for hours, the agent performance for avoiding the obstacles increases with respect to training time. The same result is validated with physical robot which becomes better at avoiding the obstacles in the vicinity of goal points.

The limitation of the presented reward field function is the training time. As log function are highly nonlinear in nature, its takes huge amount of time to train the robot as well as simulation agent. The other point is that we only considered obstacle avoidance problem and the method is not optimized for the shortest path so this reward field function can be supplemented with some heuristic-based method for shortest path traversal.

References

1. Mohanty, P. K., & Parhi, D. R.: Controlling the motion of an autonomous mobile robot using various techniques: a review. Journal of Advance Mechanical Engineering, 1(1), 24–39 (2013)
2. Mac, T. T., Copot, C., Tran, D. T., & De Keyser, R.: Heuristic approaches in robot path planning: A survey. Robotics and Autonomous Systems, 86, 13–28 (2016)
3. Zadeh, L. A.: Fuzzy sets. Information and control, 8(3), 338–353 (1965)
4. Vachtsevanos, G., & Hexmoor, H.: A fuzzy logic approach to robotic path planning with obstacle avoidance. In Decision and Control, 1986 25th IEEE Conference on (Vol. 25, pp. 1262–1264) (1986)
5. Valdez, F., Melin, P., & Castillo, O.: A survey on nature-inspired optimization algorithms with fuzzy logic for dynamic parameter adaptation. Expert systems with applications, 41(14), 6459–6466 (2014)
6. Zacksenhouse, M., & Johnson, D. H.: A neural network architecture for cue-based motion planning. In Decision and Control, 1988. Proceedings of the 27th IEEE Conference on (pp. 324–327) (1988)
7. Watkins, C. J. C. H.: Learning from delayed rewards (Doctoral dissertation, University of Cambridge) (1989)
8. Watkins, C. J., & Dayan, P.: Q-learning. Machine learning, 8(3–4), 279–292 (1992)
9. Lin, L. J.: Reinforcement learning for robots using neural networks (Doctoral dissertation, Fujitsu Laboratories Ltd) (1993)
10. Motlagh, O., Nakhaeinia, D., Tang, S. H., Karasfi, B., & Khaksar, W.: Automatic navigation of mobile robots in unknown environments. Neural Computing and Applications, 24(7–8), 1569–1581 (2014)
11. Jaradat, M. A. K., Al-Rousan, M., & Quadan, L.: Reinforcement based mobile robot navigation in dynamic environment. Robotics and Computer-Integrated Manufacturing, 27 (1), 135–149 (2011)
12. Dahl, G. E., Yu, D., Deng, L., & Acero, A.: Context-dependent pre-trained deep neural networks for large-vocabulary speech recognition. IEEE Transactions on Audio, Speech, and Language Processing, 20(1), 30–42 (2012)
13. Krizhevsky, A., Sutskever, I., & Hinton, G. E.: Imagenet classification with deep convolutional neural networks. In Advances in neural information processing systems (pp. 1097–1105) (2012)
14. Mnih, V., Kavukcuoglu, K., Silver, D., Graves, A., Antonoglou, I., Wierstra, D., & Riedmiller, M.: Playing atari with deep reinforcement learning. arXiv preprint arXiv:1312.5602 (2013)
15. Vasquez, D., Okal, B., & Arras, K. O.: Inverse reinforcement learning algorithms and features for robot navigation in crowds: an experimental comparison. In Intelligent Robots and Systems (IROS 2014), 2014 IEEE/RSJ International Conference on (pp. 1341–1346) (2014)
16. Poole, D. L., & Mackworth, A. K.: Artificial Intelligence: foundations of computational agents. Cambridge University Press (2010)
17. Puterman, M. L.: Markov decision processes. Handbooks in operations research and management science, 2, 331–434 (1990)

Integrated Design for Assembly Approach Using Ant Colony Optimization Algorithm for Optimal Assembly Sequence Planning

G. Bala Murali, B. B. V. L. Deepak, B. B. Biswal
and Bijaya Kumar Khamari

Abstract To reduce the assembly efforts and cost of the assembly, researchers are motivated to reduce the part number by applying design for assembly (DFA) concept. The so far existed literature review has no generalized method to obtain optimum assembly sequence by incorporating the DFA concept. Even though the DFA concept is applied separately, still it demands high-skilled user intervention to obtain optimum assembly sequence. As the assembly sequence planning (ASP) is NP-hard and multi-objective optimization problem, it requires more computational time and huge search space. In this paper, an attempt is made to combine DFA concept along with ASP problem to obtain optimum assembly sequence. Ant colony optimization algorithm (ACO) is used for combining DFA and ASP problem by considering directional changes as fitness function to obtain optimum feasible assembly sequences. Generally, the product with 'N' parts consists of N − 1 levels during assembly, which are reduced by applying DFA concept. Later on, optimum assembly sequence can be obtained for the reduced levels of assembly using different assembly predicates.

Keywords Design for assembly · Assembly sequence planning
Ant colony optimization algorithm · Multi-objective optimization

G. Bala Murali (✉) · B. B. V. L. Deepak · B. B. Biswal · B. K. Khamari
Department of Industrial Design, National Institute of Technology, Rourkela
769008, Odisha, India
e-mail: bmgunji@gmail.com

B. B. V. L. Deepak
e-mail: bbv@nitrkl.ac.in

B. B. Biswal
e-mail: bbbiswal@nitrkl.ac.in

B. K. Khamari
e-mail: 514id1002@nitrkl.ac.in

© Springer Nature Singapore Pte Ltd. 2019

249

H. S. Behera et al. (eds.), *Computational Intelligence in Data Mining*,
Advances in Intelligent Systems and Computing 711,
https://doi.org/10.1007/978-981-10-8055-5_23

1 Introduction

As the assembly is the major cost-contributing process in manufacturing [1–3], DFA is applied to reduce the number of levels of the assembly by reducing the number of parts. Even though the modified topology assembly obtained from the DFA concept has the problem of obtaining the optimum assembly sequence, this is due to extraction of assembly predicates and optimization constraints, which affect the quality of the solution. At the initial stages, researchers implemented knowledge-based methods to solve ASP problems [4–6], which require lot of computational time and huge search space, especially for complex products. Later, researchers turned the attention towards artificial intelligence (AI) techniques to avoid above-said problems. However, these techniques are quiet impressive up to certain extent; the major limitation with these techniques is local optimal solution.

In most of the AI techniques, to test the feasibility of the generated sequence, manually extracted assembly predicates are used, which are time consuming. Hence, various computer-aided design (CAD) exchanging data formats are used to extract such data. Bahubalendruni [7–9] introduces an algorithm to test the assembly predicates (stability data, liaison data, mechanical feasibility data) using CAD interface.

The integration of CAD-extracted assembly predicates with AI techniques is still in developing stage. Many AI techniques are introduced to solve ASP problem, and out of that genetic algorithm (GA) uses most [10, 11] because it suits best for multi-objective problem like ASP. Moreover, this algorithm has some limitations like complexity in achieving global optimal solution; quality of the solution is less because of considering one or two objective functions only. Later on, [12–14] introduces artificial immune system (AIS), which is based on immune system of the body. As this algorithm uses immune concept, high-skilled users are required to implement the algorithm. Moreover, it uses only one assembly predicate as objective function, which affects the solution. It has the limitation in achieving the global optima for complex parts. Swarm intelligence technique like ACO is introduced by [15, 16] to obtain optimal assembly sequence by considering all assembly predicates. ACO is the next best algorithm used after GA because it is best suited for discrete optimization problems like ASP. ACO is also used for disassembly planning [17] as it is best suited for multi-objective optimization problem. In order to achieve the quality assembly sequence, an improved ACO algorithm is introduced by [18, 19], which reduces the local optimal solution.

Even though many algorithms are present to obtain the optimal assembly sequence, most fail to achieve global solution and some may have the problem of search, etc. By keeping this thing in mind, researchers are motivated towards hybrid algorithms [20] to improve the quality of the solution. Meanwhile, automated CAD-based assembly sequence planning is introduced by [21, 22] to obtain quality solution.

In this paper, an attempt is made by integrating DFA concept along with CAD assistance for extracting assembly information using ACO algorithm to achieve optimum assembly sequence used. By this, number of levels of the assembly can be reduced, which saves lot of computational time as well as cost of the assembly.

2 Assembly Information Extraction from CAD Environment

Manual extraction of assembly information is a time-consuming process, which requires obtaining the optimal assembly sequence. This section deals with automatic extraction of assembly information from CAD environment. CATIA V5R17 is used for programming the macros in Visual Basic (VB) scripting to extract the information. A machine frame assembly consisting of eight parts (four are physical connectors) is considered in Fig. 1.

2.1 Liaison Data Extraction

Liaison data give contact information of the parts in assembly. Weights 1 and 0 are assigned for the mating and no mating parts, respectively. In the present methodology, a four parts assembly is considered; matrix of size 4×4 is obtained from CATIA (V5R17). The liaison matrix for the assembly structure is as follows:

Fig. 1 A mechanical frame assembly 1-Base, 2-L shape frame, 3-L shape frame, and 4-Rod

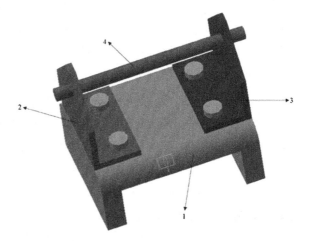

$$\text{Liaison data matrix} = \begin{array}{c} \\ 1 \\ 2 \\ 3 \\ 4 \end{array} \begin{array}{cccc} 1 & 2 & 3 & 4 \\ \begin{bmatrix} 0 & 1 & 1 & 0 \\ 1 & 0 & 0 & 1 \\ 1 & 0 & 0 & 1 \\ 0 & 1 & 1 & 0 \end{bmatrix} \end{array}$$

2.2 Material Data Extraction

Material data matrix gives the material information of contact parts. In the matrix weights, 1 and 0 are assigned for same material and not same material. The material data matrix is obtained from CATIA (V5R17) for the machine frame assembly. The matrix is as follows:

$$\text{Material data matrix} = \begin{array}{c} \\ 1 \\ 2 \\ 3 \\ 4 \end{array} \begin{array}{cccc} 1 & 2 & 3 & 4 \\ \begin{bmatrix} 0 & 1 & 1 & 0 \\ 1 & 0 & 0 & 0 \\ 1 & 0 & 0 & 0 \\ 0 & 0 & 0 & 0 \end{bmatrix} \end{array}$$

2.3 Functionality Matrix

Functionality matrix gives functionality disturbance information for merging the parts. The functionality of the assembly is to insert the rod between two L shape-frames. Weight functions 1 and 0 are considered to specify the functionality satisfies and failure conditions for merging the parts. The functionality matrix is obtained from CATIA V5R17 for the machine frame assembly. The matrix is as follows:

$$\text{Functionality data matrix} = \begin{array}{c} \\ 1 \\ 2 \\ 3 \\ 4 \end{array} \begin{array}{cccc} 1 & 2 & 3 & 4 \\ \begin{bmatrix} 0 & 1 & 1 & 0 \\ 1 & 0 & 0 & 0 \\ 1 & 0 & 0 & 0 \\ 0 & 0 & 0 & 0 \end{bmatrix} \end{array}$$

2.4 Relative Motion Matrix

Relative motion matrix provides the information about the relative motion between the parts of the assembly. Weights 1 and 0 have been considered for relative motion and no relative motion between the parts. The relative motion matrix is obtained from CATIA V5R17 for the machine frame assembly. The matrix is as follows:

$$
\text{Relative motion matrix} =
\begin{array}{c}
\\ 1 \\ 2 \\ 3 \\ 4
\end{array}
\begin{array}{cccc}
1 & 2 & 3 & 4 \\
\left[\begin{array}{cccc}
0 & 0 & 0 & 0 \\
0 & 0 & 0 & 1 \\
0 & 0 & 0 & 1 \\
0 & 1 & 1 & 0
\end{array}\right]
\end{array}
$$

The above four data matrices (liaison, functionality, material and relative motion) are used in DFA concept to reduce the levels of the assembly.

2.5 Interference Data Matrices

These matrices give the information about all possible six directions interference of the components during the assembly operations. Weights 1 and 0 are assigned for the no interference and interference, respectively. Six matrices of 4 × 4 size are obtained from CATIA (V5R17). The obtained interference matrices for the considered assembly structure are as follows:

$$
\begin{array}{ccc}
+X & -X & +Y \\
\begin{array}{cccc}
1 & 2 & 3 & 4
\end{array} &
\begin{array}{cccc}
1 & 2 & 3 & 4
\end{array} &
\begin{array}{cccc}
1 & 2 & 3 & 4
\end{array} \\
\begin{array}{c}1\\2\\3\\4\end{array}
\left[\begin{array}{cccc}
0 & 1 & 1 & 1 \\
1 & 0 & 1 & 1 \\
1 & 0 & 0 & 1 \\
1 & 1 & 1 & 0
\end{array}\right] &
\begin{array}{c}1\\2\\3\\4\end{array}
\left[\begin{array}{cccc}
0 & 1 & 1 & 1 \\
1 & 0 & 0 & 1 \\
1 & 1 & 0 & 1 \\
1 & 1 & 1 & 0
\end{array}\right] &
\begin{array}{c}1\\2\\3\\4\end{array}
\left[\begin{array}{cccc}
0 & 0 & 0 & 0 \\
1 & 0 & 1 & 0 \\
1 & 1 & 0 & 0 \\
1 & 0 & 0 & 0
\end{array}\right] \\
-Y & +Z & -Z \\
\begin{array}{cccc}
1 & 2 & 3 & 4
\end{array} &
\begin{array}{cccc}
1 & 2 & 3 & 4
\end{array} &
\begin{array}{cccc}
1 & 2 & 3 & 4
\end{array} \\
\begin{array}{c}1\\2\\3\\4\end{array}
\left[\begin{array}{cccc}
0 & 1 & 1 & 1 \\
0 & 0 & 1 & 0 \\
0 & 1 & 0 & 0 \\
0 & 0 & 0 & 0
\end{array}\right] &
\begin{array}{c}1\\2\\3\\4\end{array}
\left[\begin{array}{cccc}
0 & 1 & 1 & 1 \\
1 & 0 & 1 & 0 \\
1 & 1 & 0 & 0 \\
1 & 0 & 0 & 0
\end{array}\right] &
\begin{array}{c}1\\2\\3\\4\end{array}
\left[\begin{array}{cccc}
0 & 1 & 1 & 1 \\
1 & 0 & 1 & 0 \\
1 & 1 & 0 & 0 \\
1 & 0 & 0 & 0
\end{array}\right]
\end{array}
$$

2.6 Stability Matrix

Stability matrix gives the information about the parts which are incomplete stable, partial stable and complete stable. Three weights, 2, 1 and 0 have been allotted for complete stable, partial stable and incomplete stable assemblies, respectively. A matrix of size 4 × 4 is obtained from CATIA (V5R17).

$$
\text{stability data matrix} =
\begin{array}{c}
\ \\
1 \\
2 \\
3 \\
4
\end{array}
\begin{array}{cccc}
1 & 2 & 3 & 4 \\
\left[\begin{array}{cccc}
0 & 2 & 2 & 0 \\
1 & 0 & 0 & 2 \\
1 & 0 & 0 & 2 \\
0 & 2 & 2 & 0
\end{array}\right]
\end{array}
$$

2.7 Mechanical Feasibility Matrix

Mechanical feasibility matrix gives the information about the joining of two parts with the physical connectors in the presence of other part. Weights 0 and 1 are considered for representing collision and no collision in the presence of other part. For example, in the below matrix '0' represents there is no collision with other parts in the presence of part 1 during joining with physical connectors, similarly for the remaining parts also no collision during joining with physical connectors.

Mechanical feasibility matrix=

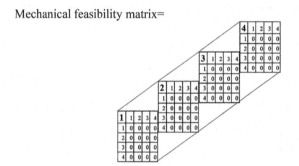

The above three data matrices (interference, stability and mechanical feasibility) are used to generate optimum assembly sequence for the assembly obtained from the DFA concept.

In this research to reduce the cost of the assembly, directional changes have been considered as the fitness function.

Directional changes are represented as (D_C)

$$\text{Fitness function } F = D_C * W_C. \tag{1}$$

3 Integrated DFA Approach for Optimal Assembly Sequence Planning Using Ant Colony Algorithm

An integrated DFA approach is applied using ACO to obtain the optimal assembly sequence by reducing the levels of the assembly sequences. In this approach, four criteria liaison, material, functionality and motion have been considered to reduce

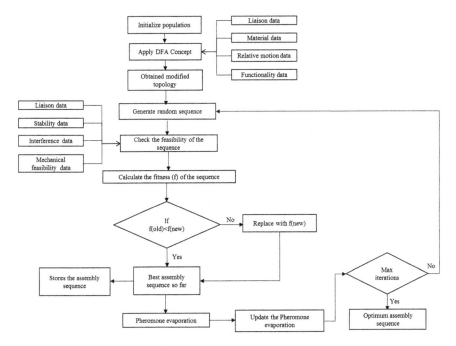

Fig. 2 Detailed flow chart of the integrated DFA concept using ACO to obtain optimum assembly sequence

the levels of the assembly sequence. CAD interface is used to extract the assembly constraints automatically. Later on, ACO is used to obtain the optimum assembly for the modified topology obtained from the DFA concept. The parameters considered for evaluating using ACO algorithm is as follows: population size $= 100$; $\alpha = 0.7$; $\beta = 0.5$; $T_{ij} = 0.9$; $n_{ij} = 0.8$. The detailed flow chart of integrated DFA concept using ant colony algorithm is shown in Fig. 2.

In the ant colony optimization algorithm [15], the equations used for updating the pheromones are as follows:

$$p(c_{ij})^k = \frac{(T_{ij})^\alpha * (n_{ij})^\beta}{\sum c_{ij} \in N(s^p)[(T_{ij})^\alpha * (n_{ij})^\beta]}, c_{ij} \in N(s^p) \qquad (2)$$

where T_{ij} and n_{ij} are, respectively, the pheromone value and the heuristic value associated with the component c_{ij}.

α and β are the positive real parameters whose values determine the relative importance of pheromone versus heuristic information.

$N(s^p)$ is the construction step done probabilistically at each step.

Equation (1) represents the transition probability for ant k to go from place i to place j while building its route.

$$\Delta(T_{ij})^k = \{1/L_k \text{ if ant k used edge } (i,j) \text{ in its tour, otherwise zero} \tag{3}$$

Equation (3) represents the pheromone concentration calculation.
Where L_k is the tour length of kth ant.

4 Results and Discussions

The proposed algorithm has been implemented on a machine frame assembly to obtain optimum assembly sequence by reducing the number of levels of the assembly. The considered assembly consists of four parts as shown in Fig. 1, and the possible levels in the assembly are $4 - 1 = 3$ levels. By the application of the DFA concept, the possible assembly levels are reduced to 1, which is easy to generate the assembly sequence. Considering the assembly sequences generated from the proposed methodology shown in Table 1 by considering directional changes as fitness equation.

In this article, analysis has been done in two stages according to the industry requirement.

Stage-1: Grouping the parts of the assembly which are in contact and having the same direction during assembly.

Stage-2: Check the functionality, material and relative motion of the grouped parts in the stage-1 and eliminate the grouped parts which are not satisfying the conditions. After elimination process, the set of assembly sequences which satisfies the DFA assembly constraints are listed in Table 2.

Table 1 A set of twelve assembly sequences

S. no	Assembly sequence	Directional sequence	Directional changes	No. of levels
1	1-2-3-4	2-2-2-2	0	3
2	1-3-2-4	1-1-1-1	0	3
3	1-3-4-2	1-1-1-1	0	3
4	2-1-3-4	2-2-2-2	0	3
5	2-1-4-3	2-2-2-2	0	3
6	2-4-1-3	1-1-1-1	0	3
7	2-4-3-1	1-1-1-1	0	3
8	3-1-2-4	2-2-2-2	0	3
9	3-1-4-2	2-2-2-2	0	3
10	3-1-2-1	1-1-1-1	0	3
11	4-2-1-3	2-2-2-2	0	3
12	4-2-3-1	2-2-2-2	0	3

Table 2 Set of assembly sequences obtained after application of DFA concept

S. no	Assembly sequence	Directional sequence	No. of levels
1	(1-2-3)-4	(2-2-2)-2	1
2	(1-3-2)-4	(1-1-1)-1	1
3	(2-1-3)-4	(2-2-2)-2	1
4	(3-1-2)-4	(2-2-2)-2	1
5	(3-2-1)-4	(1-1-1)-1	1

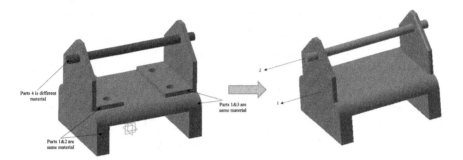

Fig. 3 Modified machine frame after application of DFA concept

In the above Table 2, (1-2-3), (1-3-2), (2-1-3), (3-1-2) and (3-2-1) represent one set of parts, satisfying DFA assembly constraints. The modified assembly with level-1, obtained from the DFA concept is shown in Fig. 3. Obtaining the optimum assembly sequence is very easy for the modified topology assembly as the assembly is reduced to level 1, by which cost of the assembly and time to assemble is drastically reduced.

Now, the optimum assembly sequence is of either 1-2 or 2-1.

5   Conclusion

An integrated DFA concept using ACO algorithm has been implemented in this paper to obtain optimum assembly sequence for a given assembly. The proposed methodology has been explicated by considering a machine frame assembly. In this, two stages are being introduced for reducing the levels of the assembly and to obtain the optimal assembly sequence. In stage-1, contact data and directional changes data have been considered for grouping the parts as single set, reduce the levels of the assembly. In stage-2, the grouped parts have been checked for the functionality, material and motion data, by which the unsatisfied assembly sequences are eliminated to obtain the final modified assembly. The algorithm is found to be effective in attaining optimum assembly sequence to the modified topology of machine frame assembly. As a future work, the developed algorithm can be effectively implemented for more part number assemblies.

References

1. De Fazio, T., Whitney, D. Simplified generation of all mechanical assembly sequence: Journal on Robotics and Automation, vol. 3(6), pp. 640–658, 1987.
2. Boothroyd, G., Design for assembly the key to design for manufacture. The International Journal of Advanced Manufacturing Technology, vol. 2(3), pp. 3–11, 1987.
3. Bala Murali, G., et al. "Optimal Assembly Sequence Planning Towards Design for Assembly Using Simulated Annealing Technique." (2017).
4. Bahubalendruni, M. V. A., Biswal, B. B., and Khanolkar, G. R. A Review on Graphical Assembly Sequence Representation Methods and Their Advancements, Journal of Mechatronics and Automation, vol. 1(2), pp. 16–26, 2015.
5. Bahubalendruni, M. R., & Biswal, B. B. A review on assembly sequence generation and its automation. Proceedings of the Institution of Mechanical Engineers, Part C: Journal of Mechanical Engineering Science, vol. 230(5), pp. 824–838, 2016.
6. Pan, C., Smith, S.S.F. and Smith, G.C. Determining interference between parts in CAD STEP files for automatic assembly planning. Journal of Computing and Information Science in Engineering, vol. 5(1), pp. 56–62, 2005.
7. Bahubalendruni, M.V.A. and Biswal, B.B. An algorithm to test feasibility predicate for robotic assemblies. Trends in Mechanical Engineering & Technology, vol.4 (2), pp. 11–16, 2014.
8. Raju Bahubalendruni M. V. A, Biswal B B. Computer Aid for stability testing between parts towards automatic assembly sequence generation. Journal of computer technology & applications., vol. 7(1), pp. 11–16, 2016.
9. Bahubalendruni, M.R., Biswal, B.B., Kumar, M. and Deepak, B.B.V.L. A Note on Mechanical Feasibility Predicate for Robotic Assembly Sequence Generation. In CAD/CAM, Robotics and Factories of the Future (pp. 397–404). Springer India, 2016.
10. Smith, S.S.F. Using multiple genetic operators to reduce premature convergence in genetic assembly planning. Computers in Industry, vol. 54(1), pp. 35–49, 2004.
11. Chen, S.F. and Liu, Y.J. An adaptive genetic assembly-sequence planner. International Journal of Computer Integrated Manufacturing, vol. 14(5), pp. 489–500, 2001.
12. Chang, C.C., Tseng, H.E. and Meng, L.P. Artificial immune systems for assembly sequence planning exploration. Engineering Applications of Artificial Intelligence, vol. 22(8), pp. 1218–1232, 2009.
13. Deepak, B.B, Bahubalendruni, M. R., & Biswal, B. B. An Advanced Immune Based Strategy to Obtain an Optimal Feasible Assembly Sequence. Assembly Automation, vol. 36 (2), pp. 127–137, 2016.
14. Biswal, B. B., Deepak, B. B., & Rao, Y. Optimization of robotic assembly sequences using immune based technique. Journal of Manufacturing Technology Management, vol. 24(3), pp. 384–396, 2013.
15. Failli, F. and Dini, G. Ant colony systems in assembly planning: a new approach to sequence detection and optimization. In Proceedings of the 2nd CIRP international seminar on intelligent computation in manufacturing engineering (pp. 227–232), June 2000.
16. Wang, J.F., Liu, J.H. and Zhong, Y.F. A novel ant colony algorithm for assembly sequence planning. The international journal of advanced manufacturing technology, vol. 25(11–12), pp. 1137–1143, 2005.

17. McGovern, S.M. and Gupta, S.M. Ant colony optimization for disassembly sequencing with multiple objectives. The International Journal of Advanced Manufacturing Technology, vol. 30(5–6), pp. 481–496, 2006.
18. Sharma, S., Biswal, B.B., Dash, P. and Choudhury, B.B. Generation of optimized robotic assembly sequence using ant colony optimization. In Automation Science and Engineering, 2008. CASE 2008. IEEE International Conference on (pp. 894–899). IEEE, August 2008.
19. Shi, S.C., Li, R., Fu, Y.L. and Ma, Y.L. Assembly sequence planning based on improved ant colony algorithm. Computer Integrated Manufacturing Systems, vol. 16(6), pp. 1189–1194, 2010.
20. Gunjia, Bala Murali, et al. "Hybridized genetic-immune based strategy to obtain optimal feasible assembly sequences."
21. Bahubalendruni, M. R., Biswal, B. B., Kumar, M., & Nayak, R. Influence of assembly predicate consideration on optimal assembly sequence generation. Assembly Automation, vol. 35(4), pp. 309–316, 2015.
22. Bahubalendruni, M. R., & Biswal, B. B. A novel concatenation method for generating optimal robotic assembly sequences. Proceedings of the Institution of Mechanical Engineers, Part C: Journal of Mechanical Engineering Science, 0954406215623813. 2015, https://doi.org/10.1177/0954406215623813.

Design and Performance Evaluation of Fractional Order PID Controller for Heat Flow System Using Particle Swarm Optimization

Rosy Pradhan, Susmita Pradhan and Bibhuti Bhusan Pati

Abstract The purpose of this paper is to apply a natured inspired algorithm called as Particle Swarm Optimization (PSO) for the design of fractional order proportional-integrator-derivative (FOPID) controller for a heat flow system. For the design of FOPID controller, the PSO algorithm is considered as a designing tool for obtaining the optimal values of the controller parameter. To obtain the optimal computation, different performance indices such as IAE (Integral Absolute Error), ISE (Integral Squared Error), ITAE (Integral Time Absolute Error), ITSE (Integral Time Squared Error) are considered for the optimization. All the simulations are carried out in Simulink/Matlab environment. The proposed method has shown better result in both in transient and frequency domain as compared to other published works.

Keywords FOPID · Performance indices · Time domain specification
PSO

1 Introduction

The advances in control engineering have kept momentum since past few decades [1–5]. Despite advances in control technique, the PID controller is still widely used in the industry due to its demonstrated advantages such as easy to understand due to its only three tunable parameters, simple structure and easy to implement [6–10]. In recent days, modern control theories have made much advancement in the fields of

R. Pradhan (✉) · S. Pradhan · B. B. Pati
Department of Electrical Engineering, Veer Surendra Sai University of Technology,
Burla 768018, Odisha, India
e-mail: rosypradhan_ee@vssut.ac.in

S. Pradhan
e-mail: susmita14electrical@gmail.com

B. B. Pati
e-mail: pati_bibhuti@rediffmail.com

© Springer Nature Singapore Pte Ltd. 2019   261
H. S. Behera et al. (eds.), *Computational Intelligence in Data Mining*,
Advances in Intelligent Systems and Computing 711,
https://doi.org/10.1007/978-981-10-8055-5_24

designing of PID controller by using the idea of fractional calculus to improve its performance in various industrial systems. The fractional-order-proportional-integral-derivative (FO-PID) controller is a generalized controller of the classical PID controller. FO-PID is the highly improving the robustness of the system along with model uncertainties [11]. FO-PID provides a better response than the integer order PID both for integer order system and fractional order systems in many industries [12]. Tuning of FO-PID is more difficult because it has five parameters to select instead of three parameters in a standard classical PID controller. Therefore many advanced controller parameter tuning methods have been developed in recent years to solve the difficulties arises in FO-PID controller design using fractional calculus [11–14]. Some researchers have made contribution to tune the controller parameters using evolutionary optimization algorithms [15–17] and frequency domain methods [18, 19]. In this paper, we formulate an optimization problem to control the FOPID controller parameters for heat flow system. The most popular nature inspired algorithm named PSO is used for designing FOPID controller for heat flow system. The PSO has been considered in this paper due to its proven advantages indesigning various PID and FOPID controllers reported in the literature [20–25]. This work is organized is as follows. Section 2 presents the modeling of the heat flow system and preliminaries of fractional order PID. In Sect. 3, the problem formulation and proposed algorithm are discussed. Section 4 contains the result and analysis of the proposed method. Finally, Sect. 5 concludes the work followed by references.

2 Preliminaries

2.1 Modeling the Heat Flow System

The Basic objective of the heat flow system (HFS) which is provided by Quanser [26, 27] is to maintain the air temperature at a certain point inside the fiber optic chamber. The HFS is consists of a fiber optic chamber with three sensor present equidistance from each other, a heater to increase the chamber temperature and a blower to maintain it into a specified point at one end. Figure 1 shows the Quanser HFS. The temperature at a particular sensor depends on the magnitude of input voltage to the heater and blower, the distance between the sensors and blower and the ambient temperature. Thus, the thermodynamics model of the system will be complex one which need be simplified to design the desired controller. The open loop model of the heat flow system is described by Eq. (1).

$$T_n = f(V_h, V_b, T_a, X_s) \tag{1}$$

where T_n shows the temperature of sensor n, V_h and V_b are represent the voltage applied at heater and blower, T_a gives the ambient temperature and X_s represents the distance of sensor n and heater.

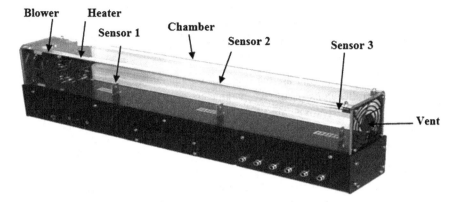

Fig. 1 Heat flow system [26]

As, the dynamics of a temperature controlled system can be well captured by a first order delay transfer function, here, the complex dynamics of the heat flow system is approximated to a first order delay system [28]. The first-order transfer function of the HFS based on sensor 1 data is given in Eq. (2).

$$P(s) = \frac{9.5}{1 + 30.5s} e^{-0.3s} \tag{2}$$

The plant transfer function consists of first order with time delay, to make the system simplified there are three different ways to represent the delay i.e. neglect the delay by making $\theta = 0$, Taylor series expansion and first order Pade approximation. In this paper, the first order delay with Taylor series expansion is considered. The transfer function with Taylor series expansion is presented in Eq. (3).

$$P(s) = \frac{9.5}{1 + 30.5s} (1 + 0.3s) \tag{3}$$

2.2 The Fractional Order PID Controller

Fractional order PID controller comes from the area of fractional calculus. There are three most common approximate definitions for fractional order integro-differential operation such as Grunwald-Letnikov, Riemann-Liouville and Caputo [29]. The generalized form of fractional calculus is represented by an operator aD_t^α and the details are described in Eq. (4).

$$aD_t^\alpha = \begin{cases} \frac{d^\alpha}{dt^\alpha} & \alpha > 0 \\ 1 & \alpha = 0 \\ \int_a^t (dt)^\alpha & \alpha < 0 \end{cases} \tag{4}$$

where α is the order, a and t are the limits of the operator.

FOPID controller consists of five parameters such as proportional gain (K_p), integral gain (K_i), derivative gain (K_d), integer order (λ) and derivative order (μ). Equation 5 represents the mathematical representation of FOPID where an order of integrator and derivative value lies between 0 and 1.

$$P_c(s) = \frac{U(s)}{E(s)} = K_p + \frac{K_i}{S^\lambda} + K_d S^\mu \tag{5}$$

As fractional order PID controller is generalized representation of integer order controller. Therefore FOPID with $\lambda = 1$ and $\mu = 1$ is represent the standard PID controller. The time domain expression for FOPID controller is given in Eq. (6).

$$u(t) = K_p e(t) + K_i D^{-\lambda} e(t) + K_d D^\mu e(t) \tag{6}$$

3 The Proposed Algorithm

3.1 Performance Criteria and Problem Formulation

The objective functions for this algorithm are the various time domain integral performance indices which are represented by the Eqs. (7–10). Optimum values of the controller can be calculated by minimizing the indices functions. The objective function is chosen for minimizing the time response characteristics due to the dependency of error on time. Figure 3 show block diagram of proposed algorithm based FOPID controller. The proposed optimization problem for heat flow system is presented in Eq. (11).

$$J_1 = \text{IAE (Integral Absolute Error)} = \int_0^\infty |e(t)| dt \tag{7}$$

$$J_2 = \text{ISE (Integral Square Error)} = \int_0^\infty e^2(t) dt \tag{8}$$

$$J_3 = \text{ITAE (Integral with Time Absolute Error)} = \int_0^\infty t|e(t)| dt \tag{9}$$

$$J_4 = \text{ITSE (Integral with Time Square Error)} = \int_0^\infty te^2(t)dt \qquad (10)$$

The problem can be represented as

$$\text{Minimize } J \qquad (11)$$

Subjected to

$$K_{pmin} < K_p < K_{pmax}$$
$$K_{imin} < K_i < K_{imax}$$
$$K_{dmin} < K_d < K_{dmax}$$
$$\lambda_{min} < \lambda < \lambda_{max}$$
$$\mu_{min} < \mu < \mu_{max}$$

Here, J is the objective function $(J_1, J_2, J_3,$ and $J_4)$
and e(t) is the error $e = max|r(t) - y(t)|$
where $r(t)$ and $y(t)$ are defined as system output and the desired input of the system respectively.

3.2 Particle Swarm Optimization (PSO) Based FOPID Controller Design

Input: Heat flow system (HFS) with adjustable FOPID controller parameter including swarm population size and their respective velocity.

Output: The optimum value of FOPID controller parameter with the help of global best solution.

Step1. Specify the lower and upper bounds of the five controller parameters such as K_P, K_I, K_D, λ, μ, randomly initialize position (x) and velocity (v) of the five parameters each having a population of n and also specify number of iterations N. The position and velocity are presented in Eq. (12).

$$x = \begin{bmatrix} K_{p1} & K_{i1} & K_{d1} & \lambda_1 & \mu_1 \\ K_{p2} & K_{i2} & K_{d2} & \lambda_2 & \mu_2 \\ \vdots & \vdots & \vdots & \vdots & \vdots \\ K_{pn} & K_{i3} & K_{d3} & \lambda_n & \mu_n \end{bmatrix} \quad v = \begin{bmatrix} v_{p1} & v_{i1} & v_{d1} & v_{\lambda1} & v_{\mu1} \\ v_{p2} & v_{i2} & v_{d2} & v_{\lambda2} & v_{\mu2} \\ \vdots & \vdots & \vdots & \vdots & \vdots \\ v_{pn} & v_{in} & v_{dn} & v_{\lambda n} & v_{\mu n} \end{bmatrix}$$

$$(12)$$

Step2. Set the error index function as a fitness function $f(x)$ (presented in Eq. (13)), calculate the fitness function for every population and find the global best solution (g_{bst}).

$$f(x) = \begin{bmatrix} f(K_{p1}, K_{i1}, K_{d1}, \lambda_1, \mu_1) \\ f(K_{p2}, K_{i2}, K_{d2}, \lambda_2, \mu_2) \\ \vdots \\ f(K_{pn}, K_{in}, K_{dn}, \lambda_n, \mu_n) \end{bmatrix} \tag{13}$$

Step3. For every iteration, update velocity of particle according to the local best and global best position of the particles as shown in Eq. (14). Consequently, using Eq. (15) the updated particle position is calculated.

$$v_{updated(i)} = v_{old(i)} + c_1 r_1 (x_{local} - x_i) + c_2 r_2 (x_{global} - x_i) \tag{14}$$

$$x_{updated(i)} = x_{old(i)} + v_{updated(i)} \tag{15}$$

where c_1, c_2 are the constant values and r_1, r_2 are the random numbers.

Step4. The fitness functions are evaluated from the updated corresponding position value.

Step5. The local best position is updated based on the fitness function.

Step6. Step 3 and 4 are repeated until the number of iterations reaches the maximum.

Step7. The latest best fitness value saved as an optimum controller parameter and terminates the algorithm.

4 Result and Analysis

4.1 Parameters of Proposed Algorithm

The lower bound and upper bound are set zero and 2 respectively for all five control parameters (proportional gain (K_p), integral gain (K_i), derivative gain (K_d), integer order (λ) and derivative order (μ)). During the design of PSO based FOPID controller for HES the following parameter values has been considered.

Swarm size = 20, Maximum number of iterations = 100, Inertia weight = 0.9, swarm best weight = 2, particle best weight = 0.5, $c_1 = 2$, $c_2 = 2$, Number of evaluations = 20.

Table 1 PSO-FOPID controller for HFS for various objective functions

Parameters/objective functions	IAE	ISE	ITAE	ITSE
K_P	4.5384	**4.3872**	4.9474	4.3755
K_I	0.4286	**0.4178**	0.9687	1.8103
K_D	1.8441	**1.7414**	1.9197	1.4032
λ	0.7059	**0.9198**	0.7794	0.3654
μ	0.2780	**0.3748**	0.6647	0.7231

4.2 Transient Response Analysis

Simulation studies are carried out to make a comparison between different stated objective functions. The best parameters of FOPID controller with different objective functions are shown in Table 1. The step response with optimized FOPID controller is depicted in Fig. 2. A comparative analysis of proposed PSO-FOPID algorithm with other recent published method fractional order filter with IMC-PID technique [28] are shown in Table 2. The comparative analysis are based on the transient performance, e.g., rise time (T_r), settling time (T_s), peak time (T_p), maximum overshoot $(\%M_p)$ and steady state performance i.e. steady state error (E_{ss}). It is clear from Table 2 that the proposed PSO optimized FOPID controller shows better result as compared fractional order filter with IMC-PID method in terms of transient and steady state performance. The step response is given in Fig. 3 for FO-PI, FO-PID and PSO tuned FOPID controller.

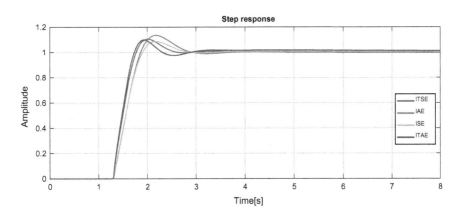

Fig. 2 Comparison of different objective functions tuned by PSO

Table 2 Performance of various methods

Method	T_r	T_s	Overshoot	Peak	Peak time
FO-filter + PID [10]	–	19.6	–	–	–
FO-filter + PI [10]	–	19.6	–	–	–
PSO-FO-filter + FOPID-IAE	0.4228	2.782	13.9023	1.135	2.15
PSO-FO-filter + FOPID-ISE	**0.4494**	**2.614**	**7.935**	**1.086**	**2.20**
PSO-FO-filter + FOPID-ITAE	0.335	2.762	8.81713	1.098	1.91
PSO-FO-filter + FOPID-ITSE	0.3618	2.502	10.8004	1.105	2.01

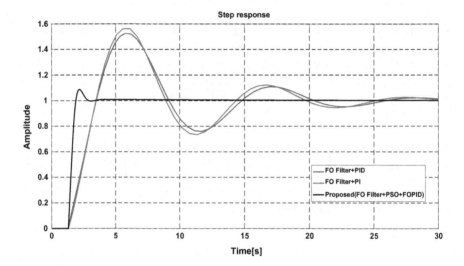

Fig. 3 Comparison of FO-filter + PI, FO-filter + PID and PSO-FOPID

4.3   Bode Analysis

The frequency response analysis by using bode plot for PSO tuned FOPID for the HFS system is shown in Fig. 4. In Table 3, delay margin, phase margin and peak gain are presented for the PSO algorithm. From bode plot, the minimum peak gain, maximum phase margin, delay margin and bandwidth are obtained. Therefore we conclude that, the PSO algorithm results the best frequency response.

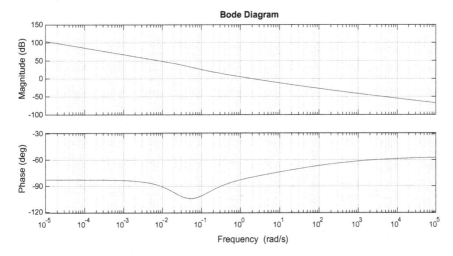

Fig. 4 Bode Plot of the HFS tuned by PSO

Table 3 Bode analysis of various methods

Algorithms	Peak gain (dB)	Phase margin (deg)	Delay margin (s)
FO- filter + PID	120	35.0428	Inf
FO-filter + PI	120	34.1886	Inf
PSO-FO-filter + FOPID	**101**	**100.2454**	**Inf**

5   Conclusion

This paper introduces an application of an optimization algorithm named Particle Swarm Optimization to regulate the performance indices for heat flow system. The PSO is used to evaluate the optimal tuning parameters for FOPID controlled heat flow system. The important contribution of the work includes

(i)   Comparison of application of four objective functions such as IAE, ISE, ITAE and ITSE in the process to obtain the control parameters. From the results, it is evident that the ISE is a better choice for FOPID to optimize the control parameters.

(ii) In terms of convergence characteristics, PSO exhibits promisingly better results for PID in terms of rise time, settling time, overshoot and steady state error.

(iii) The transient analysis and bode analysis has been carried out for the proposed system to show the supremacy in performance of the proposed algorithm over other recently reported methods.

References

1. Lin, H., Su, H., Shu, Z., Wu, Z. G., & Xu, Y. (2016). Optimal estimation in UDP-like networked control systems with intermittent inputs: stability analysis and suboptimal filter design. IEEE Transactions on Automatic Control, 61(7), 1794–1809.
2. Pradhan, R., Patra, P., & Pati, B. B. (2016, September). Comparative studies on design of fractional order proportional integral differential controller. In Advances in Computing, Communications and Informatics (ICACCI), 2016 International Conference on (pp. 424–429). IEEE.
3. Pradhan, J. K., Ghosh, A., & Bhende, C. N. (2017). Small-signal modeling and multivariable PI control design of VSC-HVDC transmission link. Electric Power Systems Research, 144, 115–126.
4. Pradhan, J. K., & Ghosh, A. (2015). Multi-input and multi-output proportional-integral-derivative controller design via linear quadratic regulator-linear matrix inequality approach. IET Control Theory & Applications, 9(14), 2140–2145.
5. Sain, D., Swain, S. K., & Mishra, S. K. (2016). TID and I-TD controller design for magnetic levitation system using genetic algorithm. Perspectives in Science, 8, 370–373.
6. Åström, K. J., & Hägglund, T. (2001). The future of PID control. Control engineering practice, 9(11), 1163–1175.
7. Ogata, Katsuhiko. "Modern control engineering." (2002): 1.
8. Visioli, A. (2012). Research trends for PID controllers. Acta Polytechnica, 52(5).
9. Tan, W., Liu, J., Chen, T. and Marquez, H. (2006). Comparison of some well-known PID tuning formulas. Computers & Chemical Engineering, 30(9), pp. 1416–1423.
10. Hägglund, T., & Åström, K. J. (2002). Revisiting The Ziegler-Nichols Tuning Rules For Pi Control. Asian Journal of Control, 4(4), 364–380.
11. Oustaloup, A., 1991. La commande CRONE, Commande robuste d'ordre non entier, Hermes (Traité des Nouvelles Technologies-Série Automatique), Paris. ISBN 2-86601-289-5.
12. Samko, S. G., Kilbas, A. A. and Marichev, O. I., 1993. Fractional integrals and derivatives. Theory and Applications, Gordon and Breach, Yverdon, 1993.
13. Zamani, M., Karimi-Ghartemani, M., Sadati, N. and Parniani, M., 2009. Design of a fractional order PID controller for an AVR using particle swarm optimization. Control Engineering Practice, 17(12), pp. 1380–1387.
14. Hamamci, S. E., 2007. An algorithm for stabilization of fractional-order time delay systems using fractional-order PID controllers. IEEE Transactions on Automatic Control, 52(10), pp. 1964–1969.
15. Tang, Y., Cui, M., Hua, C., Li, L., & Yang, Y. (2012). Optimum design of fractional order PI λ D μ controller for AVR system using chaotic ant swarm. Expert Systems with Applications, 39(8), 6887–6896.
16. Biswas, A., Das, S., Abraham, A., & Dasgupta, S. (2009). Design of fractional-order PI λ D μ controllers with an improved differential evolution. Engineering applications of artificial intelligence, 22(2), 343–350.
17. Lee, C. H., & Chang, F. K. (2010). Fractional-order PID controller optimization via improved electromagnetism-like algorithm. Expert Systems with Applications, 37(12), 8871–8878.
18. Luo, Y., & Chen, Y. (2009). Fractional order [proportional derivative] controller for a class of fractional order systems. Automatica, 45(10), 2446–2450.
19. Wang, D. J., & Gao, X. L. (2012). H∞ design with fractional-order PDμ controllers. Automatica, 48(5), 974–977.
20. Zamani, M., Karimi-Ghartemani, M., Sadati, N., & Parniani, M. (2009). Design of a fractional order PID controller for an AVR using particle swarm optimization. Control Engineering Practice, 17(12), 1380–1387.
21. Gaing, Z. L. (2004). A particle swarm optimization approach for optimum design of PID controller in AVR system. IEEE transactions on energy conversion, 19(2), 384–391.

22. Panda, S., & Padhy, N. P. (2008). Optimal location and controller design of STATCOM for power system stability improvement using PSO. Journal of the Franklin Institute, 345(2), 166–181.
23. Ramezanian, H., Balochian, S., & Zare, A. (2013). Design of optimal fractional-order PID controllers using particle swarm optimization algorithm for automatic voltage regulator (AVR) system. Journal of Control, Automation and Electrical Systems, 24(5), 601–611.
24. Bingul, Z., & Karahan, O. (2011, April). Tuning of fractional PID controllers using PSO algorithm for robot trajectory control. In Mechatronics (ICM), 2011 IEEE International Conference on (pp. 955–960). IEEE.
25. Bouarroudj, N. (2015). A Hybrid Fuzzy Fractional Order PID Sliding-Mode Controller design using PSO algorithm for interconnected Nonlinear Systems. Journal of Control Engineering and Applied Informatics, 17(1), 41–51.
26. Quanser Innovative edutech. Heat flow laboratory. Ontario: Quancer Inc., 2012.
27. Quanser Innovative edutech. Heat flow experiment: User Manual. Ontario: Quanser Inc., 2009.
28. Al-Saggaf, U., Mehedi, I., Bettayeb, M., & Mansouri, R. (2016). Fractional-order controller design for a heat flow process. Proceedings of the Institution of Mechanical Engineers, Part I: Journal of Systems and Control Engineering, 230(7), 680–691.
29. Gutiérrez, R. E., Rosário, J. M., & Tenreiro Machado, J. (2010). Fractional order calculus: basic concepts and engineering applications. Mathematical Problems in Engineering, 2010.

Land Records Data Mining: A Developmental Tool for Government of Odisha

Pabitrananda Patnaik, Subhashree Pattnaik and Prashant Kumar Pramanik

Abstract 'Data mining' is the method of extracting valuable information from the large data sets. It may be called as knowledge mining from data. Nowadays, Data Analytics and Business Intelligence are focused on exploring useful information from the databases created for different purposes. One such database created for Land Records System of Odisha is 'Bhulekh.' The data of land properties are safeguarded by the government in Revenue and Disaster Management department. These data are very sensitive, voluminous, and quite unstructured in nature. Regional language 'Odia' is used for preparation of Record of Rights (RoR). Thus, Bhulekh Database of Odisha contains data in Odia language. Government of India at national level takes steps to provide better service in Land Records area to the public through its Digital India Land Records Management Programme (DILRMP). Earlier this programme was known as National Land Records Modernisation Programme (NLRMP). With the support from Government of India, Government of Odisha started computerizing its Land Records. The Bhulekh database created for the purpose contains 1.47 crore Khatiyans, 3.23 crore Tenants, and 5.47 crore Plots for 51681 villages of Odisha. Besides, the textual data, it also contains cadastral maps in another database known as 'BhuNaksha.' There is linkage between Bhulekh and BhuNaksha for spatial and non-spatial data integration for better service of the citizens. This helps to get the data easily from any corner in the globe. This paper discusses how data mining approach is used on Bhulekh for socioeconomic development of the society. Further, this helps the Government to take decisions, better manage government lands and resolving issues in time.

P. Patnaik (✉) · P. K. Pramanik
National Informatics Centre, Unit – 4, Bhubaneswar, Odisha, India
e-mail: p.patnaik@nic.in

P. K. Pramanik
e-mail: pk.pramanik@nic.in

S. Pattnaik
Capital Institute of Management and Science, Sundarapada Jatani Road,
Mundala 751002, Odisha, India
e-mail: spattnaik12@gmail.com

© Springer Nature Singapore Pte Ltd. 2019   273
H. S. Behera et al. (eds.), *Computational Intelligence in Data Mining*,
Advances in Intelligent Systems and Computing 711,
https://doi.org/10.1007/978-981-10-8055-5_25

Keywords Land Records · Bhulekh · BhuNaksha · Data mining
Record of Rights (RoR)

1 Introduction

The history of data mining dates back to 1763 on Bayes Theorem for relating current probability to prior probability. In 1970s, with the development of relational database management system, the storing of huge amount of data in database made it easy. Gradually, the data mining concept of exploring valuable information from the stored data was started. Now, the concepts of Big Data analysis, Data Analytics, Business Intelligence have come up. The development of Data warehousing concepts made the Data Mining job more successful. Presently, data mining is a branch in Computer Science field. It is used for Market Analysis, Business growth, Education Research as well as in Government activities. In this paper, the e-Governance applications which are emphasizing on government action for society development will be considered [1–3].

Now, popularly data mining technique is used to generate hidden but useful information from the data stored in a database. It is a process of extracting non-trivial or implicit or previously unknown information from data in large databases, data warehouse, or in flat files [4–6].

In Odisha Land Records system, distributed database architecture is followed. In this case, the database is divided into 31 databases, one for each district of the 30 districts available in Odisha and one for the State level. All the 317 Tehsils are covered in 30 databases maintained by the districts. This makes the query processing reliable, optimum, and faster [7, 8].

1.1 Data Mining in e-Governance Applications

Day to day, the number of e-Governance applications are increasing in Odisha. Land Records, which belongs to Revenue and Disaster Management Department, is one of the key e-Governance applications in the State and it is maintained as a Mission Mode Project (MMP) in view of its sensitivity and accuracy [9, 10]. The objective of each e-Governance application is to provisioning of service to the public. To provide such benefits, government needs to analyze the data in different ways before taking the decisions. Thus, data mining plays a very important role in analyzing the data and finding conclusions. The statistical analysis helps the government for strategy formulation and execution. The online analytical processing assists to do such data analysis in various perspectives and concluding the action to be taken. All stakeholders explore benefits from this analysis.

The major stakeholders of Bhulekh are Public, Government, Financial Institutions, Builders, Private Organisations and NGOs and Researchers and Social Scientists.

Thus, data mining is an useful tool for e-Governance projects and service provider of good governance. In Government of Odisha, different projects are implemented in different core sectors, in which data mining makes these systems more efficient and meaningful [7, 8, 11].

1.2 Bhulekh: Land Records System of Odisha

Government of Odisha has created two databases, one for the textual data and another for spatial data. The spatial database contains all the individual and mosaic maps of villages containing individual plots. The textual database for Land Records System is called as the 'Bhulekh' and spatial database is called as 'BhuNaksha.' Accordingly, the applications are also named Bhulekh and BhuNaksha, respectively. The objective of these two applications is to provide facilities to the public for viewing their own RoRs with cadastral maps linkage to the textual data of Bhulekh. There are 30 numbers of Districts, 317 numbers of Tehsils, 2,413 numbers of Revenue Inspector (RI) Circles, and 51,681 numbers of villages in Odisha. These villages have 1,47,77,930 numbers of Khatiyans and 51,643 numbers of Cadastral Maps.

The khatiyans and maps are available in the Website. The cadastral mosaic maps of revenue villages integrated with the textual data meet the requirement of public and other stakeholders [7, 8].

2 Spatial and Non-spatial Data Mining in Land Records

2.1 Evolution of Data Warehousing and Data Mining in Odisha

Government of Odisha desired to develop Bhulekh database for storing the details of RoRs available in Tehsil. During nineties, Internet connectivity was not being used in full fledge. Very minimum numbers of networking systems were used mainly for e-mail exchange. Pictures, Audios, and Video data were not possible to store as these data take more space in the storage devices. Thus, in those earlier stages, the process conversion of manual updation of RoR was taken up to create an electronic database for data storage and effective usage of those data. In 2004, with the use of Internet technologies, a Website named as 'Bhulekh' was created and hosted for public use. This helped a lot to the people who were not able to see their RoRs. Then gradually the hits to this site were increased by different stakeholders such as Citizens, Builders, Financial institutions, Researchers, Government Departments. Now, the visits to this Website are around 30 lakhs in a year which is approximately 8000 per day. Presently, the spatial data of all cadastral maps available in government are digitized and hosted in Website for public access. A link between textual data of Bhulekh was established with the map data for providing better transparency to the public. Now, many Management Information System (MIS) reports are added to make the s/w complete and fulfilling requirement of the stakeholders. Other value added information is also included in the system. The database became more robust and capable of handling data mining in Bhulekh application [4, 7].

The maps of the landed properties of Odisha are preserved with care, for future references. These paper maps are very old. Even some maps are aging more than 60 or 70 years. Many of them are already in powder form and may get destroyed soon. Thus, the Government of Odisha planned for digitizing those maps for saving in soft format and using for the public usage. This action made capable for creating a database activity with facilities of data mining. The maps were integrated with the textual data to provide access to public.

Thus, the data mining of both textual and spatial data of Odisha Land Record System is built up [12–14].

2.2 Literature Review

Land is the basic property of human beings. For safeguarding the landed property, government takes responsibilities to the maximum extent. So, at different levels of State and Central government, a lot of efforts are done to give justice to the actual owner of the land. Still, most of the times, the land disputes arise and the land of weaker sections is forcibly occupied by stronger parties by different means. The Land Record system is continuing since British time, and the owners have paper documents for claiming their rights on the lands. But, in that earlier system, the owners had very little scope to know about the actual possession of their properties. They were dependent on some others to know about the validity and authenticity of their documents. As a result, the land was occupied or resold to different parties without the knowledge of real land owners. Further, the successors were finding lot of difficulties to get their properties on inheritance. These situations were bringing lot of disputes and court cases. On the other hand, Government responsibility is again increasing to sort out those court cases. Thus, government decided to display all the Records of Rights to the public using ICT, so that anyone can view it from any place on the globe. This would protect the property and maintain transparency on transactions on those lands [7, 11, 15].

Hence, the Land Records system is implemented in all the States of India. The Records of Right are displayed in the Website, so that, one can view it using the Internet service. It would not be required to run to the Government offices for getting the information.

To meet the requirements of public, Government of Odisha launched Bhulekh project during late nineties and hosted the Website in 2004. The textual data helped a lot to the common man to view their documents. Gradually, the use of data mining was implemented on that database and it was used in many different ways for the better service of the citizens [13, 16].

In this paper, decision table and statistical analysis are used to find the socioeconomic impact of the data mining of Bhulekh application.

2.3 Benefits of Non-spatial Data Mining

The first and basic service of Bhulekh is to provide the RoR view to the public. A person, at any place in the world, is able to view his RoR and map. In addition to RoR view, many other services as described below are also provided by the system through MIS [7].

i. View of Record of Rights from anywhere at anytime.

ii. Intelligent searching facility either by supplying Khatano or Plot No., or Tenant Name is possible. It is not necessary to remember the Khatiyan number, Plot number, or Tenant name to fetch a record. With very minimum knowledge on RoR data, one may access the records from the Website.

iii. During property registration, it provides the status of land to the buyers before making any transaction on that land within the State of Odisha.

iv. One may get certified copy of RoR from the Tehsil office. It helps the public as the Tehsildars may do quicker references for issuing Caste certificate, Residential/Nativity certificates, etc. It saves lot of times.

v. The weaker sections and low-literate people can easily access the Website as it is in Odia language.

vi. Builders, Scholars, and Researchers get the benefit from Bhulekh database for doing different statistical analysis and hypothesis testing. The Online Analytical Processing (OLAP) of Bhulekh is a good tool for the above-mentioned stakeholders.

vii. It helps financial institutions to ascertain the ownership, land type, and area of land for sanctioning loan to the tenants for different purposes.

viii. It helps government to locate/identify land for industrialization/social projects. This service makes the Land Acquisition process simpler.

ix. The analytical facility of Bhulekh database helps the government for taking citizen-oriented strategic decisions. This Land Governance protects the weaker sections such as single woman, tribals, and other socially and economically backward citizens of Odisha.

x. It helps the government to protect the Government lands like 'Patita,' 'Rakshit,' 'Sarbasadharana' from the land grabbers. These lands can be better utilized by the government for benefit of the society.

xi. The base data of Bhulekh help the Government for Hi-Tech, Town Planning, Irrigation facilities, and other surveys.

xii. In a single day, more than 8000 hits are done to this Website by the different stakeholders in a single day.

xiii. Approximately, more than Rs. 7.29 crores revenue are generated from Bhulekh s/w in one year.

2.4 Benefits of Spatial Data Mining

In Fig. 1, the village map and individual plot map are displayed. In a village map, if someone clicks on a plot then it automatically links to the textual database of Bhulekh and fetches the RoR. At the same time, it also finds other plots available in the same mouza for that tenant. These plots are marked by other color shading. Some of the OLAP activities provided by BhuNaksha are listed below [17, 18].

i. The digitized Cadastral maps of 51,681 villages are presently made available on the Web for public view. The paper maps which get damaged due to aging factor are now stored in soft digitized format and the retrieval of those maps is easy.

ii. The citizen can take printout of the village maps and plot maps for their references.

iii. While viewing RoR in Bhulekh, one can view or print plot maps available in the concerned RoR.

iv. If a single user has multiple plots in a village, then Bhunaksha can easily track those plots with color shadings.

v. It helps the town planning authorities for better planning by easily referring to the mouza maps.

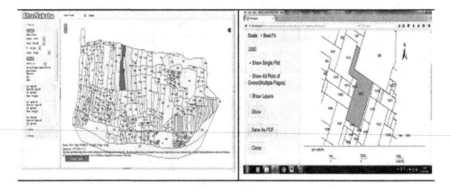

Fig. 1 Village map and individual plot map

vi. The Buyers and Sellers can refer the BhuNaksha Website for land transactions.

vii. It helps the government for various decision making purposes.

3 Citizen Centric Benefits from Land Records Data Mining

The spatial and Textual databases of BhuNaksha and Bhulekh now provide many tangible and intangible socioeconomic benefits to the public. This database was initially designed for providing computerized Record of Rights to tenants. But, with application of data mining, immense benefits are derived for present and in future. One such analysis is explained in this paper below [19, 20].

Figure 2 explains the Knowledge Discoveries/Target Groups and their Impacts in Bhulekh data mining. These are given in Table 1 [20, 21].

From the data in Table 1, the socioeconomic benefit in present and future days is shown in the chart below.

In Fig. 3, it is depicted that the social and economic benefits from the basic data of Land Records both spatial and textual are many. Lot of qualitative and quantitative benefits are derived from the databases used in Bhulekh and BhuNaksha.

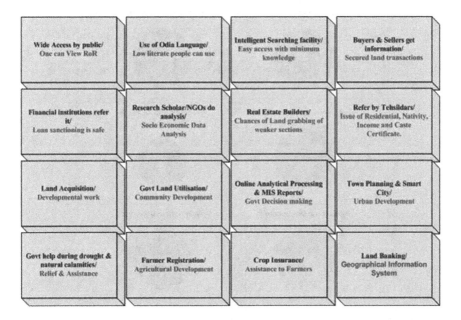

Fig. 2 Knowledge discoveries and target groups and impacts of Bhulekh data mining

Table 1 Decision table for drawing the socioeconomic benefits from Bhulekh and BhuNaksha

Sl no	Knowledge discoveries/target groups	Impacts
1	Wide access by public	One can view RoR
2	Use of Odia language	Low-literate people can use
3	Intelligent searching facility	Easy access with minimum knowledge
4	Buyers and sellers get information	Secured land transactions
5	Financial institutions refer it	Loan sanctioning is safe
6	Research scholar/NGOs do analysis	Socioeconomic data analysis
7	Builders	Chances of land grabbing of weaker sections
8	Refer by Tehsildars	Issue of residential, nativity, income, and caste certificate
9	Land acquisition	Developmental work
10	Government land utilization	Community development
11	Online analytical processing/MIS reports	Government decision making
12	Town planning and smart city	Urban development
13	Government help during drought and natural calamities	Relief and assistance
14	Farmer registration	Agricultural development
15	Crop insurance	Assistance to farmers
16	Land banking	Geographical information system

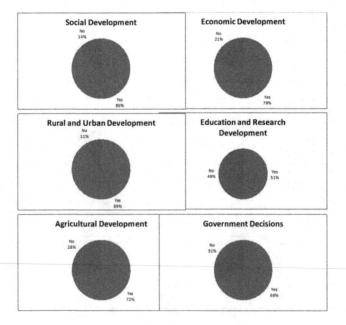

Fig. 3 Graphs showing development for citizens from Bhulekh data mining

The figures show that the socioeconomic impact of Bhulekh is high. It has both qualitative and quantitative impact on the society.

3.1 Hypothesis Testing

Hypothesis-1: (Chi-square Test)

Null Hypothesis: There is no development by Land Records Data Mining.
Alternate hypothesis: There exists development by Land Records Data Mining.

The level of significance is taken as 5% ($\alpha = 0.05$).
Table 2 of the observed frequency has 87 respondents. The expected values are given in Table 3.

Calculation: Chi square $= \chi^2 = \sum \left(\frac{(O-E)^2}{E} \right)$

where $\sum$—Summation
O—Observed frequency
E—Expected frequency

The calculated value of Chi-square $= (75 - 64.7)^2/64.7 + (69 - 64.7)^2/64.7 + (77 - 64.7)^2/64.7 + (44 - 64.7)^2/64.7 + (63 - 64.7)^2/64.7 + (60 - 64.7)^2/64.7 + (12 - 22.3)^2/22.3 + (18 - 22.3)^2/22.3 + (10 - 22.3)^2/22.3 + (43 - 22.3)^2/22.3 + (24 - 22.3)^2/22.3 + (27 - 22.3)^2/22.3$
$= 43.97847$

The critical value or the tabulated value of chi-square (obtained from the tables) at level of significance 5% and degree of freedom $((m - 1) * (n - 1) = (6 - 1) * (3 - 1)) = 10$ has been obtained $\chi2$ tabulated $= 18.307$.

Conclusion: Since χ^2 calculated value (i.e., 43.97847) is greater than the χ^2 tabulated value (i.e., 18.307), the null hypothesis is rejected. *Therefore, there exists development by Land Records Data Mining.*

Table 2 Citizen centric developmental activities of Land Records

Sl no	Area of development	Yes	No	Total
1	Social development	75	12	87
2	Economic development	69	18	87
3	Rural and urban development	77	10	87
4	Education and research development	44	43	87
5	Agricultural development	63	24	87
6	Government decisions	60	27	87
	Total	388	134	522

Source Bhulekh Project [7, 8]. Data collected through 87 numbers of questionnaires filled up by Stakeholders of Bhulekh Project

Table 3 Expected values

Sl no	Area of development	Yes	No	Total
1	Social development	64.7	22.3	87
2	Economic development	64.7	22.3	87
3	Rural and urban development	64.7	22.3	87
4	Education and research development	64.7	22.3	87
5	Agricultural development	64.7	22.3	87
6	Government decisions	64.7	22.3	87
	Total	388	134	522

3.2 Online Analytical Processing (OLAP)

The Online Analytical Processing (OLAP) is an approach for performing analytical queries online and statistical analysis of multidimensional data generating required information. The OLAP focuses on good query evaluation and query optimization algorithms for better and efficient results [22, 23]. Different OLAP queries are available in the Bhulekh database for the assistance of public and Government.

4 Key Findings of the Paper

This paper focuses on the data mining aspects of Land Records project in Odisha and ensures how data mining is helpful to all the stakeholders of Bhulekh [7, 8, 11, 15].

4.1 Social Development

Odisha is rich in natural resources, heritage, culture, and traditions. But, the developments in Odisha are still lacking. Presently, 47.04% people are still under below poverty line. The literacy rate of Odisha is lesser than the national literacy percentage. Most of the tribal population are living in remote pockets of the State. In such situations, the RoRs available in Odia language help a lot to the public. Further, the weaker sections such as socially and economically backward classes, single woman tenants, low-educated citizens and tribals are getting benefits to protect their lands. Issue of Caste certificate, income certificate, residential/nativity certificate and others are done basing on this database.

4.2 Economic Development

Government takes decisions to implement different economic development schemes basing on the data available in Bhulekh database. The reliefs during flood, cyclone, and other natural devastations are distributed according to the records available in Land Record database. Particularly during 'Failin' and 'Hoodhood,' it was significantly noticed about the usage of Bhulekh database to give compensation to the citizens. Besides this, the data mining of this project helps to get loans from the banks also.

4.3 Rural and Urban Development

The village maps and RoR details available in the public Website help different institutions and agencies to take developmental actions for rural as well as for urban areas. It helps the builders and town planning officers to plan accordingly.

4.4 Education and Research Development

It is observed that the Website is extensively used by the Scholars and Researchers for different analysis. The OLAP and MIS reports help for the research too. The NGOs make different analysis on these data and conducts conferences, workshops, and trainings for the citizens' awareness and participation.

4.5 Agricultural Development

The farmers and agricultural projects would get the base data from Bhulekh database. The actual tenants with agriculture profession can be identified.

4.6 Government Decisions

This data mining activity has become the base for many other projects such as Agriculture, Land Acquisition, Disaster Management, Hi-Tech survey, and Panchayatiraj. The government land management is effectively done using the data warehouse created for Bhulekh and BhuNaksha project.

Thus, the findings of this paper, is that, multiple benefits are drawn from Bhulekh and BhuNaksha implemented by Government of Odisha. It is one of the helpful tools in good governance.

5 Conclusion

Land Governance is one of the important activities of Government of Odisha. The land governance is done in the following manner.

  i. Government Land Governance: The main resources of government are government lands. It takes lot of social activities like school, hospital, religious monuments, or land distribution for homestead lands. During these actions, government needs to know about the government lands in the category of 'Patit,' 'Rakshit,' 'Sarbasadharana,' etc., and takes the actions to execute the developmental activities of society.
 ii. Gender Equitable Land Governance: The rights of women are to be given. It is seen in the society that the sons are getting their shares from their parental properties, but daughters are deprived of that property. Now, government wishes to give equal share to both sons and daughters. The Bhulekh database is helpful to provide such information to the decision makers [24].
iii. Weaker Section Governance: The weaker section of the society is always exploited by the stronger ones. In such cases, the government protects the rights of the weaker sections during land grabbing, unauthorized possessions, and land transactions. The tribals, backward, and low-literate people are the main priorities of government. For this purpose, the documents are in Odia as they would be able to read it easily.
 iv. Farmer Support: The farmers are able to get different assistance and subsidies from the government. For this, government has issued land passbooks to identify the actual tenants. Bhulekh database is the base for providing data for land pass book. It is the base database for agricultural projects.
  v. Town Planning and Smart cities: The land identification for Town planning and Smart city project, Bhulekh would be able to provide the correct data. The Hi-Tech Survey is also based on this Bhulekh data.
 vi. Tool for Assistance during Natural calamities and Disasters: During Super cyclone, Failin and Hoodhood, government distributed compensations basing on Land Records data. Even during floods, cyclones, and other natural calamities, this database is very useful.

Thus, the Bhulekh database is the base for almost all other development sector projects. The Data Warehousing and Data Mining which are emerging fields in Computer Science help a lot to the society now, through e-Governance applications. The impact of data mining is better felt in e-Governance applications as the benefits are directly related to the common man. Day to day, new technologies are coming up.

But, the data mining concept developed by Bill Inmonin nineties is truly useful tool for governance. With development and penetration of mobile and Internet technologies, the data mining usage is also increasing [17, 23].

Presently, the public are accessing Bhulekh and BhuNaksha even from the mobiles in remote areas. Thus, in this paper, how data mining activities are created in Revenue Department, its usage and benefits are discussed thoroughly. The 'BhuNaksha' provides Spatial output and 'Bhulekh' provides Textual output and in combination it serves the public a lot.

Acknowledgements We are very much thankful to National Informatics Centre which is a pioneer organization in developing and implementing e-Governance applications 'Bhulekh' in Odisha. Further, we are thankful to Government of Odisha and Government of India for providing valuable information in the Websites.

References

1. K. Cios, W. Pedrycz, R. Swiniarski, L. Kurgan, Data Mining: A Knowledge Discovery Approach, Springer, ISBN: 978-0-387-33333-5, 2007
2. Mehmed Kantardzic, Data Mining: Concepts, Models, Methods, and Algorithms, ISBN: 0471228524, Wiley-IEEE Press, 2002
3. Olivia Parr Rud, Data Mining Cookbook, modeling data for marketing, risk, and CRM. Wiley, 2001
4. David J. Hand, HeikkiMannila and Padhraic Smyth, Principles of Data Mining, MIT Press, Fall 2000
5. Ian Witten, Eibe Frank, Mark Hall, Data Mining: Practical Machine Learning Tools and Techniques, 3rd Edition, Morgan Kaufmann, ISBN 978-0-12-374856-0,2011
6. Silberschartz Abraham, Korth Henry F, S. Sudarshan Database System Concepts, Fourth Edition
7. http://bhulekh.ori.nic.in
8. http://dolr.nic.in/
9. http://egovstandards.gov.in/
10. http://seminarprojects.com/tag/grid-computing-ppt
11. http://www.nic.in
12. Glenn J. Myatt, Making Sense of Data: A Practical Guide to Exploratory Data Analysis and Data Mining, John Wiley, ISBN: 0-470-07471-X, November 2006
13. Margaret Dunham, Data Mining Introductory and Advanced Topics, ISBN: 0130888923, Prentice Hall, 2003
14. Ojha A., E-Governance in Practice, GIFT Publishing, Pages 33–41
15. http://www.mit.gov.in
16. Michael Berry & Gordon Linoff, Mastering Data Mining, John Wiley & Sons, 2000
17. Robert Nisbet, John Elder, IV and Gary Miner, Handbook of Statistical Analysis and Data Mining Applications, Elsevier, 2009. ISBN: 978-0-12-374765-5
18. S. Džeroski, A. Kobler, V. Gjorgijoski, P. Panov Using Decision Trees to Predict Forest Stand Heght and Canopy Cover from LANDSAT and LIDAR data. 20th Int. Conf. on Informatics for Environmental Protection – Managing Environmental Knowledge – ENVIROINFO 2006
19. Hughes B., Cotterell M. (2001), Software Project Management, Tata McGraw-Hill, Second Edition, Pages 235–259
20. http://www.sciencedirect.com/science/journal/01678191
21. http://www.springer.com

22. Pang-Ning Tan, Michael Steinbach, Vipin Kumar, Introduction to Data Mining, Pearson Addison Wesley (May, 2005). Hardcover: 769 pages. ISBN: 0321321367
23. Patricia Cerrito, Introduction to Data Mining Using SAS Enterprise Miner, ISBN: 978-1-59047-829-5, SAS Press, 2006
24. SoumenChakrabarti, Earl Cox, Eibe Frank, Ralf G ting, Jiawei Han, Xia Jiang, Micheline Kamber, Sam Lightstone, Thomas Nadeau, Richard E. Neapolitan, Dorian Pyle, Mam-douhRefaat, Markus Schneider, Toby Teorey, and Ian Witten, Data Mining: Know It All, Morgan Kaufmann, 2008
25. Pressman Roger S. (1992), Software Engineering, A Practitioner's Approach, McGraw Hill International Editions
26. Sholom M. Weiss and Nitin Indurkhya, Predictive Data Mining: A Practical Guide, Morgan Kaufmann, 1997

Piecewise Modeling of ECG Signals Using Chebyshev Polynomials

Om Prakash Yadav and Shashwati Ray

Abstract An electrocardiogram (ECG) signal measures electrical activity of the heart which is used for cardiac-related issues. The morphology of these signals is affected by artifacts during acquisition and transmission which prevents accurate diagnosis. Also a typical ECG monitoring device generates massive volume of digital data which require huge memory and large bandwidth. So there is a need to effectively compress these signals. In this paper, a piecewise efficient model to compress ECG signals is proposed. The model is designed to perform three successive steps: denoising, segmentation, and approximation. Preprocessing is done through total variation denoising technique to reduce noise, while bottom-up time-series approach is implemented to divide the signals into various segments. The individual segments are then approximated using Chebyshev polynomials. The proposed model is compared with other compression models in terms of maximum error, root mean square error, percentage root mean difference, and normalized percentage root mean difference showing significant improvements in performance parameters.

Keywords ECG · Total variation denoising · Bottom-Up approach · Chebyshev approximation

1 Introduction

An electrocardiogram (ECG) records electrical activity and rhythm of heartbeat. It is obtained as voltage variations (mV) by placing sensors on the skin. Each ECG cycle consists of 5 waves: P, Q, R, S, T corresponding to different phases of the heart activities. Shape and interval of each wave and interaction between individual waves are diagnostically important. ECG signals are low-magnitude (1 mV or less) signals in the presence of high offsets and noises. Other factors like high-frequency

O. P. Yadav (✉) · S. Ray
Bhilai Institute of Technology Durg, Chhattisgarh 491001, India
e-mail: opyadav@csitdurg.in

S. Ray
e-mail: shashwatiray@yahoo.com

© Springer Nature Singapore Pte Ltd. 2019
H. S. Behera et al. (eds.), *Computational Intelligence in Data Mining*,
Advances in Intelligent Systems and Computing 711,
https://doi.org/10.1007/978-981-10-8055-5_26

power supply interference, RF interference, electrode movement, and physiological monitoring systems also affect the morphology of ECG signals [1]. So these factors need to be reduced to facilitate proper diagnosis.

Also, ECG signals are often recorded for long interval for the purpose of identifying reasons for abnormalities in the heart rhythm. As a result, the produced ECG recording amounts to huge data sizes. Transmission of these signals through networks is also a big challenge. Therefore, they need to be compressed for efficient storage and transmission.

In this paper, a model is designed to piecewise approximate ECG signals using Chebyshev polynomials. To retain the diagnostics features and to increase compression ratio, three consecutive steps, i.e., noise reduction, segmentation, and approximation, are performed.

2   Noise Reduction Techniques

The timing pattern of non-stationary ECG signals has always been used for analysis. They get distorted due to the presence of noises; therefore, it is essential to reduce noise intelligently [2].

Several algorithms to reduce noise from ECG signals have been found in the literature. To reduce high-frequency noise from ECG signals, low-pass filters [3] were used. These filters introduced artifacts on the QRS which make them less popular. Adaptive filters have been used for the noise cancelation of ECG signals containing baseline wander, electromyography (EMG) noise, and motion artifacts [4]. The least mean square adaptive algorithm is also designed to improve signal quality. Lu et al. in [5] successfully implemented recursive-least-square adaptive filter to track EMG noises from ECG signals. Alfaouri and Daqrouq in [6] reduced noise from ECG signals using wavelet transform by reducing minimum error between detailed coefficients of the original and the noisy signal. A soft thresholding method to reduce noise components through wavelet for high-resolution ECG denoising was proposed in [7]. In this paper, we have reduced artificially generated noise from ECG signals using total variation approach and found it suitable for denoising.

2.1   Total Variation Denoising

The total variation (TV) of a signal measures variation between signal values. According to TV principle, signals with excessive and possibly spurious details have high total variation. Total variation denoising (TVD) is an optimization problem [8] which minimizes the cost function stated in Eq. (1) for reduction of noise and preservation of sharp edges.

$$arg \min_{x} = \left\{ F(x) = \frac{1}{2} \sum_{n=0}^{N-1} |y(n) - x(n)|^2 + \lambda \sum_{n=1}^{N-1} |x(n) - x(n-1)| \right\} \quad (1)$$

where $x(n)$ is a denoised signal and $y(n)$ noisy signal. The first term in Eq. (1) represents error in original and reconstructed signals, and second term refers to variation within the signal. The regularization parameter λ is used to control the degree of smoothening. Increasing λ gives more weight to the TV of the signal [9].

Majorization-minimization (MM) algorithm [10] minimizes optimization problem $F(x)$ indirectly by solving a sequence of optimization problems, $G_k(x)$, $k = 0, 1, 2, \ldots$, where $G_k(x)$ should be convex. The algorithm produces a sequence x_k, which is obtained by minimizing $G_{k-1}(x)$.

3 ECG Segmentation

The aim of segmentation is to divide a given signal into consecutive sections that can further be used independently for analysis purpose. In case of ECG, segmentation is performed to determine the onset and offset points of P, QRS, T waves and the peak locations of R waves. Automatic segmentation of the ECG is challenging due to its non-stationary nature [11]. Wavelet transforms have been used to detect discontinuities in ECG signals and have proven to provide good results for ECG segmentation [12]. Hidden Markov Model (HMM) with the coefficients of an ECG wavelet transform was found to be suitable for segmentation of ECG signals in [13]. Akhbari et al. in [14] used Multi-Hidden Markov Model (MultiHMM) to locate fudicial points on ECG signals. Other techniques for segmentation of ECG signals like dynamic time warping [15], artificial neural networks [16], and Bayesian models [17] are also found in the literature.

In this paper, we have performed time-series segmentation by dividing a denoised time-series signal into a sequence of discrete segments in such a way that the individual segments retain the underlying properties of the source signal.

For this, we have used the bottom-up algorithm [18] which initially divides a signal of length n into a large number of equal segments. Each pair of consecutive segments is then compared; the pair that has lowest increase in the error is identified and consequently merged to form new bigger segment. The algorithm terminates when the required number of segments is obtained or approximation error is less than the specified threshold.

4 ECG Compression

There are basically two reasons for ECG compression, viz. requirement for efficient storage and efficient real-time transmission. The main aim in ECG signal compression is to achieve highest compression ratio while preserving clinically important features. Polynomial approximation is an efficient way where polynomial coefficients only are used to approximate signal. It is concerned with the approximation with the lower order function. Raid et al. [19] modeled ECG signal by alignment of heartbeats in a matrix and then fitting polynomial curves to column elements. Polynomials of degree 3 have been proposed for ECG interpolation in [20]. In [21], ECG signals were reconstructed using cubic interpolation. Nygaard et al. in [22] represented ECG signals with second-degree quadratic polynomials. Hermite basis functions of order 5 were used to model segmented ECG beats in [23]. In [24], ECG compression was done using discrete Chebyshev transform by segmenting the signal into blocks consisting of multiple cardiac cycles.

Here, we have used Chebyshev approximation method where each segment of the ECG signal is represented by a Chebyshev polynomial.

4.1 Polynomial Approximation Using Chebyshev Polynomials

The Chebyshev polynomials [25] of first type and degree n are defined by Eq. (2) as:

$$T_n(x) = cos(ncos^{-1}x) \tag{2}$$

ECG segment $p(\mathbf{x})$ of length N consisting of $p(x_i)$ samples can be expressed by Eq. (3):

$$p(\mathbf{x}) = \{p(x_0), p(x_1), \ldots, p(x_N)\}, \mathbf{x} \in [a, b] \tag{3}$$

The interpolation polynomial f of degree N is given by Eq. (4):

$$f(x) = a_N x^N + a_{N-1} x^{N-1} + \cdots + a_2 x^2 + a_1 x + a_0, x \in [a, b] \tag{4}$$

The derived polynomial will then satisfy Eq. (5) for a given set of $N + 1$ data points $(x_i, p(x_i))$:

$$f(x_i) \approx p(x_i), i = 0, 1, \ldots, N \tag{5}$$

The required polynomial can be derived by Eq. (6) using Lagrange's formula [26] as

$$f(x) = \sum_{i=0}^{N} f(x_i) \prod_{\substack{j=0 \\ j \neq i}}^{N} \frac{x - x_j}{x_i - x_j}, x \in [a, b] \tag{6}$$

Any arbitrary function $p(x) \in [-1, 1]$ can be approximated by Eq. (7):

$$p(x) = \sum_{k=0}^{n} c_k T_k(x), x \in [-1, 1] \tag{7}$$

where the coefficients c_j are defined in Eq. (8).

$$c_0 = \frac{1}{n+1} \sum_{j=1}^{n+1} p(x_j)$$

$$c_k = \frac{2}{n+1} \sum_{j=1}^{n+1} p(x_j)T_k(x_j), k = 1, \ldots, n \qquad (8)$$

Since individual segments are approximated independently, their interval domain $[a, b]$ can be transformed in the interpolation interval $y \in [-1, 1]$ by Eq. (9).

$$x = \frac{(b-a)y + (a+b)}{2} \qquad (9)$$

The roots [27] of the Chebyshev polynomial $T_n(x)$ in the interval $x \in [-1, 1]$ are calculated using Eq. (10).

$$x_j = \cos\left(\frac{2j+1}{2n+1}\pi\right), 0 \le j \le n \qquad (10)$$

In this paper, the function $f(x)$ is constructed using Eq. (6) with all the N ECG samples of one segment. Then, we calculate the Chebyshev nodes and subsequently the interpolating polynomial using these nodes. Since an ECG signal sampled value may not be available at all the Chebyshev nodes, we derive these values by linear interpolation using adjacent ECG sampled values. All the segments are then independently modeled using the same technique.

5 Implementation and Results

Ten ECG signals with 1000 samples each, sampled at 360 Hz from MIT-BIH arrhythmia database [28] with 11 bits per sample of resolution, were so chosen so as to test our proposed model and compare the obtained results with those reported in the literature. We have divided the modeling process into three consecutive stages: denoising, segmentation, and approximation. The complete algorithm is implemented in MATLAB.

The signals are initially added with artificial noise $w(n)$ modeled by Eq. (11):

$$w(n) = x_1(n) + x_2(n) + x_3(n) \quad \text{for} \quad n = 0, \ldots, N-1 \qquad (11)$$

where $x_1(n)$ represents powerline noise of 50 Hz; $x_2(n)$ is baseline wander noise; and $x_3(n)$ is white Gaussian noise. The noisy signal $y(n)$ is obtained as $y(n) = x(n) + w(n)$ where $x(n)$ is original signal and $w(n)$ is noisy signal. These noisy signals $y(n)$ are

Table 1 Performance metrics for denoised model

Record number	MSE	PRD	SNRimp (dB)	PSNR (dB)	CC
100	1.4831	7.7734	61.3644	60.0084	0.9997
101	1.4089	7.4494	61.7741	60.7836	0.9996
102	0.8233	5.5272	65.1166	64.2024	1.0000
103	1.0407	6.2971	63.4671	62.7432	0.9999
104	1.1245	7.0306	62.9895	61.9280	0.9998
105	1.2352	7.0865	62.4388	61.2809	0.9998
106	1.7043	8.6075	61.5563	60.8459	0.9999
107	0.9655	5.2827	63.8710	62.4781	0.9999
108	1.7339	9.1023	60.2040	57.8326	0.9992
109	1.1987	7.0560	62.3283	60.9401	0.9999

then passed through the TVD-MM filter to get filtered signal $x(n)$. In this paper, smoothening parameter λ for reducing noise is set to two. The performance of the denoised model is computed in terms of mean square error (MSE), percentage root mean square difference (PRD), improvement in signal-to-noise ratio (SNR), peak signal-to-noise ratio (PSNR), and cross-correlation (CC).

Performance parameters of TVD-MM filter applied to the signals are shown in Table 1.

Original ECG (Record No. 109 from MIT database), artificial noise modeled by Eq. (11), noisy ECG, and TVD-MM filtered ECG are shown in Fig. 1.

In [29], the best CC values, i.e., 0.9466 and 0.9779, for soft and hard thresholding using wavelets, respectively, were reported which are lower than that reported in this paper. Even the lowest value CC obtained, i.e. 0.9992, by the proposed model is considered to be good for similarity measures. PRD obtained by the proposed method are less than those reported in [30]. In [31], the highest SNR obtained as 12.67 dB is much lesser than SNR obtained in this paper. Also, the PSNR value obtained in this paper is quite high as compared to other methods existing in the literature.

These samples are then segmented into 25 segments. Here, our intention is to find out the fudicial points of ECG signals not P, QRS, T, and U waves. The individual segments are approximated using Chebyshev polynomials of suitable order. In this paper, we have used fourth-order Chebyshev polynomial to approximate all the segments of all the signals and evaluated in terms of MaxError, MSE, PRD, and Normalized PRD (PRDN). Table 2 shows the results of approximation.

As shown from the entries of Table 2, the values of the performance parameters are relatively low. Even though ECG signals approximation errors less than 10% are considered to be medically accepted [32]. MaxError obtained by the proposed method is 0.0077. PRD greater than 10% indicates that the reconstructed signal is too distorted to be useful for diagnostic purposes [33]. Highest PRD obtained here is of the order 10^{-3}. Jokic et al. [34] developed polynomial models of ECG sig-

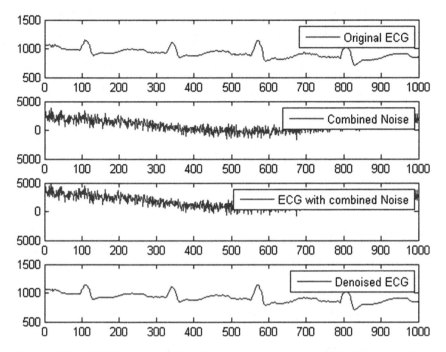

Fig. 1 **a** MIT-BIH ECG record no. 109, artifical noise, noisy signal, and TVD-MM results obtained at $\lambda = 2$

Table 2 Performance metrics for approximated model

Record number	MaxError	RMSE	PRD $\times 10^{-4}$	PRDN
100	0.0015	0.0005	0.5867	0.0074
101	0.0021	0.0015	1.5661	0.0528
102	0.0077	0.0030	4.5178	0.0016
103	0.0013	0.0006	0.5702	0.0020
104	0.0031	0.0011	1.1194	0.0058
105	0.0006	0.0003	0.3265	0.0101
106	0.0022	0.0013	1.3567	0.0201
107	0.0002	0.0002	0.2229	0.0003
108	0.0008	0.0006	0.6151	0.0075
109	0.0011	0.0004	0.4959	0.0029

nals and obtained PRD in the range of 3.5 and 10.8. Ktata et al. in [35] developed ECG codec based on the set partitioning in hierarchical trees (SPIHT) on same set of data and reported least PRD as 1.17 for record 107. For the same record, we obtained PRD as 0.0002. Istepanian et al. in [36] developed optimal zonal wavelet-based ECG data compression method and reported PRD as 0.70. In [37, 38], same set of ECG

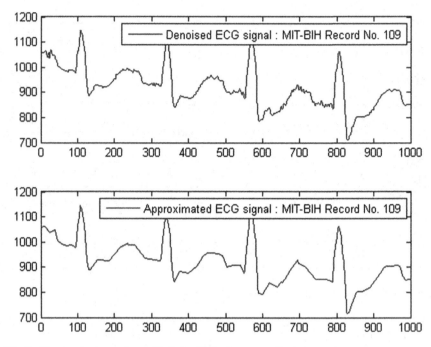

Fig. 2 The denoised record and 4th Order Chebyshev approximated record

signals were approximated using cubic interpolation and piecewise cubic spline, respectively, and the results were inferior than the results obtained by the proposed method. Compression ratio (CR) is the ratio of number of bytes required by original signal to number of bytes required by approximated signal. The first 1000 samples of our test signal required 8000 bytes which resulted in CR as 20, which can be conveniently increased by considering lower order polynomials at the cost of error.

Figure 2 shows the original denoised signal and approximated ECG signal.

6 Conclusion

ECG produces electrical activity of heart which is useful for diagnosis of heart-related diseases. These signals are affected by various types of noises present in and around the measuring devices. These signals also consume huge memory and large bandwidth. In this paper, an efficient piecewise modeling of ECG signals is proposed which is executed in three consecutive stages: denoising, segmentation, and approximation. These steps are achieved through TVD, bottom-up, and Chebyshev approximations successively. Results obtained are validated and compared with existing compression models and are found to be superior and within medically accepted region.

References

1. Walraven, Gail: Basic arrhythmias, Pearson Higher Ed (2014).
2. Mainardi, L. T., Bianchi, A. M., Baselli, G., Cerutti, S.: Pole-tracking algorithms for the extraction of time-variant heart rate variability spectral parameters. J. Mol. Biol. IEEE Transactions on Biomedical Engineering 42(3), 250–259 (1995).
3. Poungponsri, S., Yu, X. H.: Electrocardiogram (ECG) signal modeling and noise reduction using wavelet neural networks. In: ICAL'09. IEEE International Conference on Automation and Logistics, pp. 394–398, (2009).
4. Levkov, C., Mihov, G., Ivanov, R., Daskalov, I., Christov, I., Dotsinsky, I.: Removal of power-line interference from the ECG: a review of the subtraction procedure. BioMedical Engineering OnLine, 4(1), 50 (2005).
5. Lu, G., Brittain, J. S., Holland, P., Yianni, J., Green, A. L., Stein, J. F., Aziz, T. Z. and Wang, S.: Removing ECG noise from surface EMG signals using adaptive filtering. Neuroscience letters 462(1), 14–19 (2009).
6. Alfaouri, M., Daqrouq, K.: ECG signal denoising by wavelet transform thresholding. American Journal of Applied Sciences, 5(3), 276–281 (2008).
7. Donoho, D. L., Johnstone, I. M.: Threshold selection for wavelet shrinkage of noisy data. In: In Engineering in Medicine and Biology Society, 1994. Engineering Advances: New Opportunities for Biomedical Engineers. Proceedings of the 16th Annual International Conference of the IEEE, pp. A24–A25. IEEE Press, (1994).
8. Rudin, L. I., Osher, S., Fatemi, E.: Nonlinear total variation based noise removal algorithms. Physica D: Nonlinear Phenomena, 60(1–4), 259–268 (1992).
9. Solo, V.: Selection of regularisation parameters for total variation denoising. In: In Acoustics, Speech, and Signal Processing, IEEE International Conference pp. 1653–1655. IEEE Press, New (1999).
10. Bioucas-Dias, J. M., Figueiredo, M. A., Oliveira, J. P.: Total variation-based image deconvolution: a majorization-minimization approach In: IEEE International Conference on Acoustics, Speech and Signal Processing Conference (2), (2006).
11. Beraza, I., Romero, I.: Comparative study of algorithms for ECG segmentation. Biomedical Signal Processing and Control 34, 166–173 (2017).
12. Khawaja, A., Sanyal, S., Dossel, O.: A wavelet-based multi-channel ECG delineator In: The 3rd European Medical and Biological Engineering Conference, (2005).
13. Clavier, L., Boucher, J. M., Polard, E.: idden Markov models compared to the wavelet transform for P-wave segmentation in EGC signals. In: In Signal Processing Conference (EUSIPCO'1998), 9th European, pp. 1–4. IEEE (1998).
14. Akhbari, M., Shamsollahi, M. B., Sayadi, O., Armoundas, A. A., Jutten, C.: ECG segmentation and fiducial point extraction using multi hidden Markov model. Computers in Biology and Medicine 79, 21–29 (2016).
15. Zifan, A., Saberi, S., Moradi, M. H., Towhidkhah, F.: Automated ECG segmentation using piecewise derivative dynamic time warping. International Journal of Biological and Medical Sciences 1(3) (2006).
16. Bystricky, W., Safer, A.: Modelling T-end in Holter ECGs by 2-layer perceptrons. Computers in Cardiology, 105–108 (2002).
17. Sayadi, O., Shamsollahi, M. B.: A model-based Bayesian framework for ECG beat segmentation. Physiological measurement 30(3), 335 (2009).
18. Keogh, E., Chu, S., Hart, D., Pazzani, M.: Segmenting time series: A survey and novel approach. Data Mining in Time Series Databases 57, 1–22 (2004).
19. Borsali, R., Nait-Ali, A., Lemoine, J.: ECG compression using an ensemble polynomial modeling: Comparison with the DCT based technique. Cardiovascular Engineering 4(3), 237–244 (2004).
20. Karczewicz, M., Gabbouj, M.: ECG data compression by spline approximation. Signal Processing, 59(1), 43–59 (1997).

21. Fira, C. M., Goras, L.: An ECG signals compression method and its validation using NNs. IEEE Transactions on Biomedical Engineering 55(4), 1319–1326 (2008).
22. Nygaard, R., Haugland, D.: Compressing ECG signals by piecewise polynomial approximation. In: Proceedings of the 1998 IEEE International Conference in Acoustics, Speech and Signal Processing, pp. 1809–1812 (1998).
23. Abdoli, M., Ahmadian, A., Karimifard, S., Sadoughi, H., Yousefi Rizi, F.: An Efficient Piecewise Modeling of ECG Signals Based on Critical Samples Using Hermitian Basis Functions. In: In 4th European Conference of the International Federation for Medical and Biological Engineering, pp. 1188–1191. Springer Berlin Heidelberg (2009).
24. Tchiotsop, D., Ionita, S.: ECG Data Communication Using Chebyshev Polynomial compression methods. Telecommunicatii Numere Publicate, AN XVI. 2010(2):22–32.
25. Mason, J. C., Handscomb, D. C.: Chebyshev polynomials. CRC Press (2002).
26. Yang, W. Y., Cao, W., Chung, T. S., Morris, J.: Applied numerical methods using MATLAB. John Wiley and Sons (2005).
27. Cheney, E. W.: Approximation theory III (Vol. 12). Academic Press, New York (1980).
28. Moody, G. B., Mark, R. G.: The MIT-BIH arrhythmia database on CD-ROM and software for use with it. In: 1 In Computers in Cardiology, Proceedings, pp. 185–188. IEEE (1990).
29. Ustundag, M., Sengur, A., Gokbulut, M., ATA, F.: Performance comparison of wavelet thresholding techniques on weak ECG signal denoising. Przeglad Elektrotechniczny 89(5), 63–66 (2013).
30. Kabir, M. A., Shahnaz, C.: Denoising of ECG signals based on noise reduction algorithms in EMD and wavelet domains. Biomedical Signal Processing and Control 7(5), 481–489 (2012).
31. Georgieva-Tsaneva, G., Tcheshmedjiev, K.: Denoising of electrocardiogram data with methods of wavelet transform. In: International Conference on Computer Systems and Technologies, pp. 9–16 (2013).
32. Sandryhaila, A., Saba, S., Puschel, M., Kovacevic, J.: Efficient compression of QRS complexes using Hermite expansion. IEEE Transactions on Signal Processing 60(2), 947–955 (2012).
33. Nygaard, R., Haugland, D., Husoy, J. H.: Signal Compression by Second Order Polynomials and Piecewise Non-Interpolating Approximation (1999).
34. Jokic, S., Delic, V., Peric, Z., Krco, S., Sakac, D.: Efficient ECG modeling using polynomial functions. Elektronika ir Elektrotechnika 110(4), 121–124 (2011).
35. Ktata, S., Ouni, K., Ellouze, N.: A novel compression algorithm for electrocardiogram signals based on wavelet transform and SPIHT. International Journal of Signal Processing 5(4), 32–37 (2009).
36. Istepanian, R. S., Petrosian, A. A.: Optimal zonal wavelet-based ECG data compression for a mobile telecardiology system. IEEE Transactions on Information Technology in Biomedicine, 4(3), 200–211 (2000).
37. Negoita, M., Goras, L.: On a compression algorithm for ECG signals. IEEE Signal Processing Conference, 1–4, 2005.
38. Luong, D. T., Duc, N. M., Linh, N. T., Ha, N. T., Thuan, N. D.: Advanced Two-State Compressing Algorithm: A Versatile, Reliable and Low-Cost Computational Method for ECG Wireless Applications. American Journal of Biomedical Sciences, 8(1), 2016.

Analysis of Supplementary Excitation Controller for Hydel Power System GT Dynamic Using Metaheuristic Techniques

Mahesh Singh, R. N. Patel and D. D. Neema

Abstract There are different types of disturbances in power system such as switching, transient, load variations, which affect stability and efficiency of the power system. These disturbances cause fluctuation at low frequency that are unacceptable, which decreases the power transfer capability in the transmission line and unstable mechanical shaft load. In order to compress low-frequency oscillations, a common solution is to use the Power System Stabilizer (PSS). The Proportional, Derivative, and Integral (PID) controller has the ability to minimize both settling time and the maximum overshoot. In this paper, design of a Proportional, Derivative, and Integral (PID)-based Power System Stabilizer (PSS) and different techniques for tuning of PID-PSS controller are proposed. The parameter of the PID-PSS has been tuned by the Genetic Algorithm (GA), Ant Colony Optimization (ACO), and Firefly Algorithm (FFA) based optimization techniques. Solution results indicate that the performance of Firefly Algorithm (FFA) based PID-PSS controller is much better than the GA and Ant Colony Optimization based PID-PSS controller.

Keywords Ant colony optimization · Firefly algorithm · Power system optimization · Power system stability · Low-frequency oscillations

M. Singh (✉) · R. N. Patel
Shri Shankarcharya Technical Campus, Bhiali 491001, Chhattisgarh, India
e-mail: singhs004@gmail.com

R. N. Patel
e-mail: ramnpatel@gmail.com

D. D. Neema
Chhattisgarh Institute of Technology, Rajnandgaon 491441, Chhattisgarh, India
e-mail: neemadd@gmail.com

© Springer Nature Singapore Pte Ltd. 2019
H. S. Behera et al. (eds.), *Computational Intelligence in Data Mining*,
Advances in Intelligent Systems and Computing 711,
https://doi.org/10.1007/978-981-10-8055-5_27

1   Introduction

In a modern power system, stability is one of the most important criteria for performance analysis of the network. Power system stability is an ability to regain synchronism on an occurrence of small and large disturbances. Stability can be classified [1] into mainly two parts: transient stability and small signal stability. Automatic Voltage Regulator (AVR) is used for improving transient stability and controls the oscillation of frequency ranging from 0.1 to 2 Hz. AVR and generator field dynamics introduce a phase lag so that resulting torque is out of phase with both rotor angle and speed deviation. Positive synchronizing torque and negative damping torque often result, which can cancel the small inherent positive damping torque available, leading to instability. This type of regulator is designed for specific operating conditions, hence limits the output results. Different type of controllers like Power System Stabilizer (PSS), Fuzzy Logic Controller, a Proportional, Derivative, and Integral (PID) are used for the solution to these problems. Here, we design a hybrid PID-PSS controller and investigate the performance of this controller using different optimization tuning techniques such as Bacterial Foraging optimization (BG), Harmony Search algorithm (HS), and hybrid BG-PSO optimization techniques using different performance index [2–4]. The performance of Genetic Algorithm-based PID-PSS, Ant Colony Optimization (ACO)-based PID-PSS and Firefly Algorithm (FFA)-based PID-PSS controllers used for a single-machine connected infinite-bus (SIMB) hydel power system is analyzed and compared. The approach investigated in this paper clears the downside of Genetic Algorithm and Ant Colony Optimization techniques. Thus, the Firefly Algorithm (FFA) has become an evolving platform to increase the performance of the conventional tuning algorithm.

The main objective is to recommend the best PID-PSS designed by a simple procedure such that it increases the damping ratio and minimizes the overshoots and settling time of the parameter deviations, thus enhancing the power system stability. So we apply Firefly Algorithm in order to achieve above-mentioned goals by finding a set of selected PID-PSS parameters.

The organization of the paper is as follows. In Sect. 2, a single-machine infinite-bus system with hydel GT dynamics is elaborated by stability analysis. The principles of PID-PSS and the objective function are described in Sect. 3, whereas in Sect. 4, evolutionary ant colony and Firefly optimization algorithms are explained. In Sect. 5, results of proposed techniques on test systems are discussed and lastly, in Sect. 6 the best optimum technique is given.

2 System Modeling

The main purpose of this paper is to utilize the different intelligence techniques for tuning of PID-PSS controller parameter. To complete the tuning process, objective function should be defined to give satisfactory results.

2.1 SMIB Power System

The system under concern is single-machine connected infinite-bus (SMIB) power system network. In operation of power system network, a synchronous machine is essential. The system under study is one machine connected to infinite-bus system through a transmission line having resistance R_T and inductance X_T [5] through a transformer. The connection of Automatic Voltage Regulator (AVR), the transmission line to generator and power system stabilizer, excitation system [2] is shown in this Fig. 1. The liberalized model called Heffron–Phillips model is presented in Fig. 2.

2.2 Hydel GT Dynamics

In stability analysis of single-machine systems, generally the mechanical input power is assumed as constant. But in this work, mechanical power input, in terms of hydel governor and turbine model, is included along with the hydel generator model for modeling, simulating, and analyzing of a system which is variable. Here, hydel generator is equipped with hydel governor and turbine. The transient droop compensation block is not included in the hydel governor model. The IEEE Type 1 excitation (rotating exciter type) is taken in this model [6, 7].

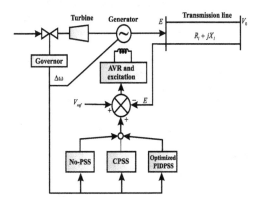

Fig. 1 Single-line diagram of SMIB power system

Fig. 2 Heffron–Phillips model with PID-PSS

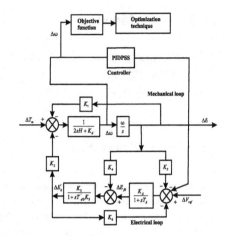

Fig. 3 Hydel governor turbine with SMIB

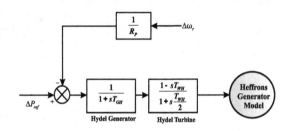

The hydel governor and turbine model are shown in Fig. 3, where T_{GH} and T_{WH} are the time constants of a hydel governor and turbine, respectively. The output of hydel governor turbine model is given as input to the single-machine infinite-bus (Heffron–Phillips generator) model. The state-space equation of hydel models for open-loop analysis is given by (1).

$$[\dot{x}_{M_1}] = [A_{M_1}]_n \times [x_{M_1}] + [B_{M_1}]_n \times [u_{M_1}] \tag{1}$$

where $n =$ number of hydel generators, here $n = 1$. A_{M_1}, $B_{M_1} =$ state matrix and input matrix of hydel model, and $x_{M_1} =$ state variable vector of hydel model.

$$[x_{M_1}] = \left[\Delta\omega, \Delta\delta, \Delta E_q', \Delta E_{FD}, \Delta V_R, \Delta V_E, \Delta X_e, \Delta T_M \right]^T \tag{2}$$

Similarly, (2) indicates the state variables for closed-loop analysis.

$$[x_{M_1}] = \left[\Delta\omega, \Delta\delta, \Delta E_q', \Delta E_{FD}, \Delta V_R, \Delta V_E, \Delta X_e, \Delta T_M, \Delta P_1, \Delta P_2, \Delta U_E \right]^T \tag{3}$$

In (3), ΔP_1, ΔP_2 and ΔU_E represent the state variables involved in the closed-loop PID-PSS model.

3 Excitation Controllers

The basic function of an excitation damping controller is to provide the required damping torque to damp the power oscillations [8]. The excitation controllers are discussed in next section.

3.1 Conventional PSS Controller

Conventional power system stabilizer consists of the three blocks, namely gain, washout, and the cascaded identical phase compensation block. The gain (K_s) and the time constants $(T_1, T_2, T_3$ and $T_4)$ of the phase compensation block are tuned effectively, so that the damping controller provides the required damping torque to damp the power oscillations.

3.2 PID-PSS Controller

In this work, optimal design of PID controller connected to power system stabilizer (PSS) [9] is designed for damping low-frequency power oscillations of a Single-Machine Infinite-Bus (SMIB) hydel power system as shown in Fig. 4.

The input to the controller is the deviation of rotor speed $\Delta\omega_r$ and output is the damping control signal (Δv_s). This output is given to generator-excitation system feedback loop. The transfer function of the PID-PSS controller is given by (4).

$$\frac{\Delta v_s}{\Delta \omega_r} = \left[\frac{sT_w}{1 + sT_w}\right] \times \left[K_P + \frac{K_I}{s} + sK_D\right] \tag{4}$$

3.3 Objective Function

The primary goal of formulated objective function is to minimize the error signal for improving the system stability. Input is applied to the control system, and the error performance criteria are being applied for reducing the error. For a specified minimum damping factor and objective value, J_1 is described as given in (5).

Fig. 4 Proposed PID-PSS controller

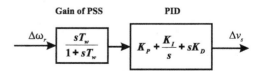

$$J_1 = \sum_{j=1}^{N} \sum_{\xi_1 \ge \xi_0} (\xi_0 - \xi_1)^2 \tag{5}$$

We are using Integral Time Absolute Error (ITAE) criteria for reducing the error [10] for system considered. The washout time constant of PSS (T_w) is taken as 20 s in this work. The expression of ITAE is described as given by (6); subjected to the constraints given by (7).

$$ITAE = \int_{t=0}^{t=t_{sim}} t|e(t)|dt \tag{6}$$

$$K_P^{min} \le K_P \le K_P^{max}, \ K_I^{min} \le K_I \le K_I^{max}, K_D^{min} \le K_D \le K_D^{max} \tag{7}$$

4 Optimization Techniques

By using biological-inspired optimization algorithms, it is found that the closed-loop system has very fast rise time, settling time, and zero maximum overshoot to sustain the system stability under different conditions [11]. Here, we have implemented the Firefly Algorithm and compared it with GA and ACO.

4.1 Firefly Algorithm

The Firefly Algorithm (FFA) has been proved to be efficient at solving optimization problems and can be more efficient than other metaheuristic algorithms when applied to nonlinear optimization problems [12, 13]. Firefly Algorithm is as follows:

Step 1. Generate initial population of fireflies x_i where i = 1, 2, ..., n. $n =$ number of fireflies light absorption coefficient γ, initially $\gamma = 1$. Randomization parameter α, initial $\alpha = 0.2$. Initial attractiveness $\beta_0 = 1.0$.
Step 2. Define the objective function (x).
Step 3. Select the range for the objective function parameters.
Step 4. Generate the initial value x_i where ($i = 1$: number of model parameters) of the model parameters randomly.
Step 5. Determine function value I_i where ($j = 1$: number of model parameters) by the value of $f(x_i)$.
Step 6. On the basis of function values, model parameters are set.
Step 7. Assume iteration = 1.

Step 8. If $I_i < I_j$ where $i, j = 1: n$, the parameter values will change according to the Equation (8).

$$x_i = x_i + \beta_0 e^{-\gamma r_{ij}^2}(x_j - x_i) + \alpha(rand - 1/2) \tag{8}$$

Step 9. At new values of x_i, determine function value.
Step 10. If condition, iteration > maximum iteration satisfies, then go to step 13.
Step 11. Create new randomness parameter using expression given as (9).

$$\alpha = \delta \times \alpha \tag{9}$$

Step 12. Set iteration count as iteration = iteration + 1 and go to step 9.
Step 13. Determine the optimum value of the function and its associated parameters.

5   Results and Discussion

In this analysis, performance of hydel power system stability is analyzed, whenever the model parameters and operating conditions are varied with two different conditions. For the modern power systems, this analysis is highly effective in analyzing the stability of the system.

5.1   Stability Analysis

Stability analysis of the system model under consideration with controllers involved in our work is performed based on two main categories.

1. Analysis of eigenvalue and damping ratio for weakly damped closed-loop system.
2. Analysis of time-domain response for deviation in rotor speed and power angle.

The different operating conditions [14] considered are given in Table 1.

Table 1 Different operating conditions

S. N.	Operating conditions	
1	Case 1	$P = 0.5$, $Q = -0.015$ p.u, $\Delta P_d = 0.02$ p.u
2	Case 2	$P = 0.89$ p.u, $Q = 0.14$ p.u., $\Delta P_d = 0.04$ p.u, 15% decrease in M and T'_{do}

Table 2 Tuned controller parameter

S. N.	Operating condition	Tuned PID-PSS controller parameters $(K_P, K_I$ and $K_D)$		
		GA-PIDPSS	ACO-PIDPSS	FFA-PIDPSS
1	Case 1	10.254, 7.434, 2.228	23.254, 1.027, 12.824	43.625, 8.826 6.665
2	Case 2	13.351, 0.823 9.120	12.144, 5.234, 11.727	55.366, 4.353 10.586

The different damping controllers optimized by Genetic Algorithm, Ant Colony, and Firefly Algorithm PID-PSS have been implemented in the hydel system models considered [15]. For better stability enhancement, optimally tuned controller parameters, namely K_P, K_I and K_D are to be computed using different optimization techniques shown in Table 2.

5.2 Time Response Stability Analysis

In this section, analysis is carried out to minimize the error using ITAE [16, 17] criteria involved in speed and power angle deviations. This is analyzed in terms of minimizing the settling time and overshoots of the speed and power deviation responses.

Fig. 5 Responses of speed deviation in case 1

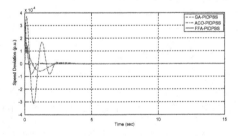

Fig. 6 Responses of power angle deviation in case 1

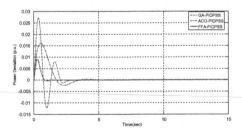

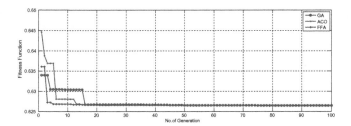

Fig. 7 Performance index versus iteration curve for case 1

Table 3 Maximum overshoot and settling time with different operating conditions

S. N.	Operating conditions	Tuning method	Overshoot (p.u.)		Settling time (s)	
			For $\Delta\omega_r$	For power angle	For $\Delta\omega_r$	For power angle
1.	Case 1	GA-PIDPSS	3.72×10^{-4}	0.028	2.98	3.15
		ACO-PIDPSS	3.25×10^{-4}	0.016	3.01	3.28
		FFA-PIDPSS	1.56×10^{-4}	0.009	1.37	1.95
2.	Case 2	GA-PIDPSS	3.25×10^{-4}	0.078	6.38	5.21
		ACO-PIDPSS	3.10×10^{-4}	0.074	3.25	4.45
		FFA-PIDPSS	2.80×10^{-4}	0.066	2.21	2.38

The response of deviations in speed and power angle is obtained by implementing the three damping controller parameters $(K_P, K_I$ and $K_D)$ in the system for case 1 which is shown in Figs. 5 and 6. These responses show that the Firefly Algorithm-based controller (FFA-PIDPSS) provides improved damping to minimize oscillations when compared with the performance of GA-PIDPSS and ACO-PIDPSS. Figure 7 also represents that the objective function value corresponding to FFA converges faster than the other methods.

In Table 3, the maximum deviation overshoot obtained for GA-PIDPSS is 3.72×10^{-4} p.u, for ACO-PIDPSS 3.25×10^{-4} and for FFA-PIDPSS; overshoot obtained is only 1.56×10^{-4} p.u. Also, in power angle, maximum overshoots obtained for GA-PIDPSS is 0.028 p.u, for ACO-PIDPSS 0.016 and it is only 0.009 p.u for ACO-PIDPSS, so the oscillations settle (at 5% of the steady state value) in 2.98 s for GA-PIDPSS, 3.01 s for ACO-PIDPSS, and it is only around 1.37 s for ACO-PIDPSS, as evident in Fig. 6 in case 1.

Similarly, responses of speed deviation and power angle deviation in second operating conditions are presented in Figs. 8 and 9. The convergence curve is shown in Fig. 10, for fitness function versus number of iterations. It shows that the fitness of FF Algorithm converges faster and to a lower value in comparison to

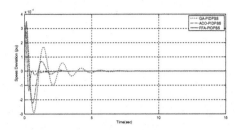

Fig. 8 Responses of speed deviation in case 2

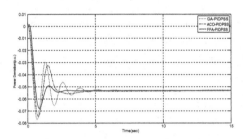

Fig. 9 Responses of power angle deviation in case 2

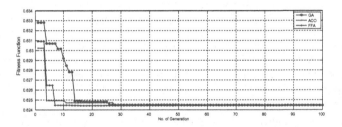

Fig. 10 Performance index versus iteration curve in case 2

other techniques. Time-domain specifications of different operating conditions are obtained and listed in Table 3.

It is observed that FFA-PIDPSS has reduced settling times and minimum overshoots when compared to GA-PIDPSS and ACO-PIDPSS at most of the operating conditions for all generators.

5.3   Eigenvalue and Damping Ratio Stability Analysis

The computed optimal controller parameters of the system are used in the closed-loop state matrices of the hydel models to calculate the stable eigenvalues and damping ratios of the system [17]. Table 4 provides the eigenvalues and

Table 4 Eigenvalues and damping ratio for different operating conditions

S. N.	Operating condition	Weakly damped values		
		Eigenvalues and damping ratio for closed-loop system		
		GA-PIDPSS	ACO-PIDPSS	FFA-PIDPSS
1	Case 1	$-0.2932\pm j6.925$ 0.0423	$-0.7532\pm j8.991$ 0.0834	$-0.9238\pm j8.752$ 0.1049
2	Case 2	$-0.4633\pm j7.338$ 0.0710	$-0.6983\pm j6.688$ 0.1038	$-0.8986\pm j8.263$ 0.1081

damping ratio for weakly damped closed-loop system of the system calculated for two different operating conditions using the computed optimal damping controller parameters.

It can be seen that the electromechanical mode eigenvalues generated by the system having bio-inspired optimized PID-PSS have the highest negative real parts and thus improve system stability.

6 Conclusion

The proposed method for stability improvement has been implemented on two test cases of the hydel power system which consists of single machine connected to an infinite bus. All the results discussed in this paper establish that Firefly Algorithm in hydel power system optimizes the parameters of PID-PSS, and stability is greatly enhanced by the implementation of damping controllers. The magnitude of the damping ratio obtained is minimum for Firefly Algorithm; also the eigenvalue lies in the left-hand side of 's plane' with negative maximum value. The other optimized controllers enhance the performance of the hydel power system. But, they do not give optimal solution required for stability as Firefly does. For conventional PI controllers, as the gain and time constants are calculated using mathematical formulae, the computed solution is not the best solution, for wide ranges of different operating conditions which bound to happen in a dynamic power system. Hence, the proposed bio-inspired controllers provide better damping to the system and make the system more stable. It is found that the Firefly Algorithm based optimization controllers provide the best possible positive damping to the system than the other optimized controllers (GA and ACO) for different operating conditions considered in our work.

References

1. Kundur, P.: Power System Stability and Control. McGraw-Hill, Vol-7, New York (1994).
2. Sambariya, D.K., Prasad, R.: Design of optimal proportional integral derivative based power system stabilizer using bat algorithm. Applied Computational Intelligence and Soft Computing.1 (2016) 5.
3. Sebaa, K., Boudour, M.: Optimal locations and tuning of robust power system stabilizer using genetic algorithms. Electric Power Systems Research 79 (2009) 406–416.
4. Sambariya, D. K., Prasad, R.: Optimal tuning of fuzzy logic power system stabilizer using harmony search algorithm. International Journal of Fuzzy Systems 17 (2015) 457–470.
5. Sauer, Peter W., Pai, M. A.: Power system dynamics and stability. Urbana (1998).
6. Shivakumar, R., Lakshmipathi, R.: An Innovative Bio Inspired PSO Algorithm to Enhance Power System Oscillations Damping. Advances in Recent Technologies in Communication and Computing. ARTCom'09, International Conference. IEEE (2009) 260–262.
7. Shivakumar, R., Lakshmipathi, R.: Implementation of an innovative bio inspired GA and PSO algorithm for controller design considering steam GT dynamics. arXiv preprint arXiv (2010) 1002–1184.
8. Pal, B., Chaudhuri, B.: Robust Control in Power Systems. Springer Series (2005).
9. Hameed, K. A., Palani, S.: Robust design of power system stabilizer using harmony search algorithm. automatika 55 (2014) 162–169.
10. Singh, M., Patel, R. N., Jhapte, R.: Performance comparison of optimized controller tuning techniques for voltage stability. Control, Measurement and Instrumentation (CMI). IEEE First International Conference. IEEE (2016) 11–15.
11. Shayeghi, H., Safari, A. Shayanfar, H. A.: Multimachine power system stabilizers design using PSO algorithm. International Journal of Electrical Power and Energy Systems Engineering 1.4 (2008) 226–233.
12. Colorni, A., Dorigo, M., Maniezzo.: Distributed Opt Colonies. Proc. imizat of the ion 1st by V Ant European Conf. on Artificial Life. Paris, France. Elsevier Publishing. (1991) 134–142.
13. Bendjeghaba, O., Boushaki, S. I., Zemmour, N.: Firefly algorithm for optimal tuning of PID controller parameters. Power Engineering, Energy and Electrical Drives (POWERENG), Fourth International Conference. IEEE (2013) 1293–1296.
14. Alkhatib, H., Duveau, J.: Dynamic genetic algorithms for robust design of multi-machine power system stabilizers. International Journal of Electrical Power & Energy Systems, 45 (2013) 242–251.
15. Hsu, Y.Y., Chen, C.L.: Identification of optimum location for stabiliser applications using participation factors. IEE Proceedings for C-Generation, Transmission and Distribution. 134 (1987) 238–244.
16. Singh, M., Brahmin, K., Neema, D. D.: Performance analysis of improving stability using tuned PID and Hybrid controllers. Electrical Power and Energy Systems (ICEPES), International Conference on. IEEE (2016) 83–87.
17. Abido, Ali, M., Lotfy Y., Magid, A.: Eigenvalue assignments in multi-machine power systems using tabu search algorithm. Computers & Electrical Engineering 28.6 (2002) 527–545.

Farthest SMOTE: A Modified SMOTE Approach

Anjana Gosain and Saanchi Sardana

Abstract Class imbalance problem comprises of uneven distribution of data/instances in classes which poses a challenge in the performance of classification models. Traditional classification algorithms produce high accuracy rate for majority classes and less accuracy rate for minority classes. Study of such problem is called class imbalance learning. Various methods are used in imbalance learning applications, which modify the distribution of the original dataset by some mechanisms in order to obtain a relatively balanced dataset. Most of the techniques like SMOTE and ADASYN proposed in the literature use oversampling approach to handle class imbalance learning. This paper presents a modified SMOTE approach, i.e., Farthest SMOTE to solve the imbalance problem. FSMOTE approach generates synthetic samples along the line joining the minority samples and its 'k' minority class farthest neighbors. Further, in this paper, FSMOTE approach is evaluated on seven real-world datasets.

Keywords SMOTE · ADASYN · FSMOTE · Borderline SMOTE
Safe-level SMOTE · CIP

1 Introduction

Several machine learning techniques have been implemented to develop predictive models to deal with real-world datasets. However, imbalanced datasets pose a serious challenge and have a negative effect on the performance of machine

A. Gosain (✉) · S. Sardana
University School of Information, Communication and Technology,
GGS Indraprastha University, Dwarka, New Delhi 110078, India
e-mail: anjana_gosain@hotmail.com

S. Sardana
e-mail: saanchi.sardana@gmail.com

© Springer Nature Singapore Pte Ltd. 2019
H. S. Behera et al. (eds.), *Computational Intelligence in Data Mining*,
Advances in Intelligent Systems and Computing 711,
https://doi.org/10.1007/978-981-10-8055-5_28

309

learning algorithms [1]. The imbalanced learning problem comes into sight when the distribution of one class, i.e., the majority class, has a higher ratio than the other class, i.e., the minority class [2]. In simpler terms, majority class (negative) has large number of instances as compared to minority class (positive) having less number of instances. The unbalanced dataset problem appears in many real-world applications like software defect prediction, target detection, video mining, text mining, medical and fault diagnosis, anomaly detection, telecommunication, the Web and email classification, ecology, biology and financial services, credit card frauds, shuttle system failure, oil spill detection, Web spam detection, risk management and nuclear explosion, helicopter gearbox fault monitoring, intrusion detection, information retrieval. In these domains, minority class is of more concern and significant than the predominant class, i.e., the majority class [3–5]. Conventional classification algorithms do not work properly in such cases. This problem is called class imbalance learning. Performance of traditional classification algorithms is mostly analyzed using overall accuracy, whereas accuracy is highly influenced by majority class in case of binary classification or two-way classification. But real-life applications are mostly concerned in detecting rare cases/minority class which is the region of interest. Thus, accuracy is not considered as the basis of measuring classifier performance in case of class imbalance learning as it provides insufficient information about the minority class, thus leading to inaccurate and delusive information regarding classifiers' performance.

Sampling of datasets is a common solution to handle the imbalance problem in which either existing samples are removed or new samples are added; i.e., data distribution of each class is changed [6]. These two forms of sampling are under-sampling of majority/prevalent class and oversampling of minority class, respectively. Some of the oversampling techniques are described as below.

Synthetic minority oversampling technique (SMOTE) proposed by Chawla et al. [7] is the most popular algorithm used to overcome the problem of overfitting due to replication of the same samples [8]. In SMOTE approach, new synthetic samples are generated from minority instances and their k-nearest neighbors which are randomly selected. Overgeneralization is one of the drawbacks of SMOTE as it generates new instances without considering the nearest instances from the majority class [9].

ADASYN algorithm overcomes the overgeneralization problem of SMOTE. It decides the number of artificial data samples that need to be generated for each minority instance based on the majority instances in the neighborhood; hence, the probability of oversampling increases in proportion with majority nearest neighbor [10].

Borderline SMOTE [11] is an advance oversampling approach which identifies the boundary between two classes and generates minority class samples close to or on the decision boundary. The chance of misclassifying borderline samples belonging to a minority class is much higher than misclassifying a sample far from borderline.

Safe-level SMOTE assigns a safe-level value to each minority sample so synthetic samples can be generated closer to the largest safe-level region [12]. Safe-level value is defined by the number of minority/positive instances in k-nearest neighbor.

In this paper, we have discussed four oversampling algorithms (SMOTE, ADASYN, borderline SMOTE, safe-level SMOTE) with a proposed algorithm Farthest SMOTE. We have experimentally showed the performance comparison of all these algorithms using different prediction models which are Naïve Bayes and SVM classifier. Different performance metrics are evaluated over seven datasets.

Remaining paper is organized as follows: Sect. 2 describes the performance metrics used for evaluating the performance of the classification algorithms. Section 3 describes our proposed algorithm Farthest SMOTE (FSMOTE). Section 4 provides the details of datasets used. Section 5 presents the experimental results by comparing FSMOTE to SMOTE, ADASYN, borderline SMOTE, and safe-level SMOTE. Section 6 describes the conclusion of this work.

2 Performance Metrics

As predictive/overall accuracy is not suitable for evaluating the performance of classifiers in case of imbalanced data distribution, so instead of accuracy other performance metrics are used [13]. In case of binary classification problem, confusion matrix is used as a basis to evaluate the performance of classifiers.

Confusion matrix comprises of correctly and incorrectly classified instances of the dataset as shown in Table 1. Minority class instances are referred to as positive, whereas majority class instances as negative. To assess the performance of classifiers, confusion matrix keeps a count of TN (True Negative), FP (False Positive), FN (False Negative), and TP (True Positive) instances. FP refers to the number of negative instances incorrectly classified as positive, whereas TN refers to the number of negative instances correctly classified as negative. Therefore, FP and TN both contribute to the total negative instances represented in Eq. (2). FN refers to the number of positive instances incorrectly classified as positive instances, whereas TP refers to the number of positive instances correctly classified as positive. Therefore, FN and TP both contribute to the total positive instances represented in Eq. (1).

$$P = FN + TP, \tag{1}$$

Table 1 Datasets description

		Predicted	
		Negative	Positive
Actual	Negative	True Negative (TN)	False Positive (FP)
	Positive	False Negative (FN)	True Positive (TP)

And

$$N = TN + FP, \tag{2}$$

Accuracy is defined as the ratio of correctly classified instances to the total number of instances. It can be expressed as given in Eq. (3)

$$Accuracy = \frac{TP + TN}{P + N}, \tag{3}$$

Sensitivity refers to the ratio of positive instances classified as positive. Sensitivity is also referred as recall and calculated as in Eq. (4),

$$Sensitivity = \frac{TP}{P}, \tag{4}$$

Specificity refers to the ratio of negative instances classified as negative and calculated as in Eq. (5),

$$Specificity = \frac{TN}{N}, \tag{5}$$

Precision [14] determines that how many instances are actually positive out of all the instances that are classified as positive and is calculated as Eq. (6)

$$Precision = \frac{TP}{TP + FP}, \tag{6}$$

F-measure, otherwise called as F-score, is weighted average of precision and recall; i.e., harmonic mean of precision and recall [15] can be expressed as given in Eq. (7)

$$F - measure = \frac{(1 + \sigma^2) * Precision * Recall}{\sigma * (Recall + Precision)}, \tag{7}$$

where value of σ is set to be one; i.e., precision and recall are given equal weightage in the balanced case.

G-mean represents the geometric mean of specificity and sensitivity; i.e., assessing the degree of biasness in terms of ratio of positive class accuracy and negative class accuracy [14] can be formulated as given in Eqs. (8) and (9)

$$G - Mean = \sqrt{TP*TN}, \tag{8}$$

OR

$$G - Mean = \sqrt{Sensitivity*Specificity}, \tag{9}$$

Receiver operating characteristic (ROC) Curve is a curve achieved by plotting True Positive (TP) rate and False Positive (FP) rate on y-axis and x-axis, respectively. FP rate can be calculated as (1-specificity), and TP rate is also known as sensitivity. Each prediction result/instance from a confusion matrix is represented in the ROC space. Area under curve (AUC) is the area that lies under the ROC curve. AUC is the arithmetic mean of TP rate and TN rate which is formulated as given in Eq. (10)

$$AUC = \frac{TP\,rate + TN\,rate}{2}, \tag{10}$$

3 Proposed Algorithm

Farthest SMOTE (FSMOTE) is the modified SMOTE technique in which synthetic samples are generated along the line joining the minority samples and its 'k' minority class **farthest neighbors**. This approach differs from the original SMOTE approach by considering the farthest neighbors. In this report, the proposed method FSMOTE increases the decision area due to which minority samples closer to the boundary are considered.

The algorithm randomly selects a data point from the k-farthest neighbor for minority samples and then generates synthetic samples using the farthest neighbors.

Table 2 Datasets description

Datasets	Total instance	Attributes	Minority class #	Majority class #	Minority class	Majority class	IR
Pima Indian diabetes	768	9	268	500	Class '1'	Class '0'	0.53
Breast cancer	699	11	241	458	Malignant	Benign	0.52
Heart	270	14	120	150	Class '2'	Class '1'	0.8
Ionosphere	351	35	126	225	Bad	Good	0.56
Spam base	4601	58	1813	2788	Class '1'	Class '0'	0.65
German	1000	25	300	700	Class '2'	Class '1'	0.42
Mammography	11183	7	260	10923	Class '1'	Class '0'	0.02

Algorithm is given below:

```
Algorithm FSMOTE(T, N, k)

Input: T - Number of minority class samples;
       N% - Amount of FSMOTE;
       k - Number of Farthest neighbors;

Output: (N/100)* T; Number of synthetic minority samples

/* If Amount of oversampling (FSMOTE) i.e. N is less than
100%, then randomize the minority class samples */
  if N <100
     then Randomize the T minority class samples
            T = (N/100) * T
            N = 100
  endif
  N = (int)(N/100) /* Amount of oversampling */
  k = Number of Farthest neighbors
  Nattrs = Number of attributes
  Sample[ ][ ]: array for original minority class samples
  Nindex: keeps a count of synthetic samples generated,
  Synthetic[ ][ ]: array for synthetic samples
/* For each sample of minority class Generate k farthest
neighbors. */
  for i = 1 to T
      Populate(N, i, nnarray) /* Evaluate nnarray and save
indices for k Farthest neighbors of i */
  endfor
  Populate( N, i, nnarray) // Used to compute the
synthetic samples. //
    while N != 0
// Let nn be an arbitrary number between 1 and k. This
procedure selects one of the k farthest neighbors of i.//
     for attrs = 1 to Nattrs
     Compute:  diff  =  Sample[nnarray[nn]][attrs]  -
Sample[i][attrs]
       Compute: gap = random number between 0 and 1
       Synthetic[Nindex][attrs] = Sample[i][attrs] + gap *
diff
     endfor
     Nindex++
     N = N - 1
  endwhile
  return (* End of Populate function. *)
```

4 Datasets Description

In this section, we have discussed about the datasets used in our experimental analysis, in a tabular form (Table 2).

Table 3 Naïve Bayes classifier on different datasets

Naïve Bayes							
Dataset	Methods	Overall accuracy	TP rate	Precision	F-measure	G-mean	ROC area
Pima Indian diabetes	NONE	0.763	0.616	0.676	0.645	0.720	0.825
	SMOTE	0.723	0.612	0.806	0.696	0.718	0.823
	BSMOTE	0.727	0.573	0.729	0.642	0.695	0.817
	ADASYN	0.682	0.513	0.755	0.611	0.657	0.789
	SLSMOTE	0.726	0.617	0.805	0.699	0.721	0.831
	FSMOTE	**0.779**	**0.720**	**0.830**	**0.711**	**0.779**	**0.869**
Breast cancer	NONE	0.960	0.971	0.918	0.944	0.962	0.987
	SMOTE	0.969	0.983	0.958	0.970	0.968	0.988
	BSMOTE	0.963	0.974	0.948	0.961	0.964	0.987
	ADASYN	0.967	0.981	0.956	0.968	0.967	0.987
	SLSMOTE	0.968	0.981	0.957	0.969	0.967	0.987
	FSMOTE	**0.969**	**0.983**	**0.958**	**0.970**	**0.968**	**0.988**
Heart	NONE	0.852	0.808	0.851	0.829	0.847	0.916
	SMOTE	0.831	0.796	0.918	0.853	0.840	0.908
	BSMOTE	0.854	0.824	0.885	0.853	0.855	0.916
	ADASYN	0.843	0.794	0.860	0.825	0.839	0.907
	SLSMOTE	0.853	0.832	0.921	0.874	0.859	0.921
	FSMOTE	**0.854**	**0.833**	**0.922**	**0.875**	**0.860**	**0.925**
Ionosphere	NONE	0.829	0.865	0.717	0.784	0.836	0.940
	SMOTE	0.820	0.829	0.829	0.829	0.819	0.938
	BSMOTE	0.818	0.831	0.756	0.792	0.820	0.924
	ADASYN	0.763	0.719	0.794	0.755	0.763	0.870
	SLSMOTE	0.835	0.859	0.832	0.845	0.833	0.931
	FSMOTE	**0.843**	**0.873**	**0.837**	**0.854**	**0.840**	**0.944**
Spam base dataset	NONE	0.795	0.956	0.668	0.786	0.813	0.941
	SMOTE	0.844	0.962	0.802	0.875	0.815	0.945
	BSMOTE	0.807	0.956	0.707	0.813	0.813	0.941
	ADASYN	0.820	0.955	0.747	0.838	0.812	0.938
	SLSMOTE	0.843	0.961	0.800	0.873	0.815	0.946
	FSMOTE	**0.850**	**0.974**	**0.804**	**0.880**	**0.820**	**0.955**
German	NONE	0.766	0.533	0.630	0.578	0.679	0.801
	SMOTE	0.714	0.537	0.774	0.634	0.682	0.748
	BSMOTE	0.727	0.518	0.719	0.603	0.670	0.800
	ADASYN	0.678	0.483	0.777	0.596	0.646	0.788
	SLSMOTE	0.720	0.545	0.773	0.639	0.687	0.819
	FSMOTE	**0.781**	**0.683**	**0.813**	**0.743**	**0.769**	**0.865**

(continued)

Table 3 (continued)

Naïve Bayes

Dataset	Methods	Overall accuracy	TP rate	Precision	F-measure	G-mean	ROC area
Mammography	NONE	0.956	0.715	0.308	0.431	0.830	0.920
	SMOTE	0.951	0.721	0.473	0.572	0.833	0.926
	BSMOTE	0.953	0.751	0.456	0.568	0.850	0.940
	ADASYN	0.947	0.604	0.403	0.484	0.763	0.882
	SLSMOTE	0.954	0.766	0.473	0.585	0.858	0.942
	FSMOTE	**0.957**	**0.844**	**0.513**	**0.638**	**0.901**	**0.957**
Winning times	NONE	0	0	0	0	0	0
	SMOTE	0	0	0	0	0	0
	BSMOTE	0	0	0	0	0	0
	ADASYN	0	0	0	0	0	0
	SLSMOTE	0	0	0	0	0	0
	FSMOTE	7	7	7	7	7	7

Table 4 SVM classifier on different datasets

SVM

Dataset	Methods	Overall accuracy	TP rate	Precision	F-measure	G-mean	ROC area
Pima Indian Diabetes	NONE	**0.775**	0.534	0.749	0.623	0.695	0.719
	SMOTE	0.692	0.541	0.858	0.664	0.699	0.723
	BSMOTE	0.735	0.508	0.797	0.621	0.678	0.706
	ADASYN	0.679	0.441	0.813	0.572	0.631	0.672
	SLSMOTE	0.726	0.553	0.859	0.673	0.707	0.729
	FSMOTE	0.743	**0.593**	**0.869**	**0.705**	**0.732**	**0.749**
Breast cancer	NONE	0.970	0.963	0.951	0.957	0.968	0.968
	SMOTE	0.975	0.977	0.975	0.976	0.975	0.975
	BSMOTE	0.973	0.972	0.969	0.971	0.973	0.973
	ADASYN	0.973	0.972	0.974	0.973	0.973	0.973
	SLSMOTE	0.974	0.975	0.975	0.975	0.975	0.974
	FSMOTE	**0.977**	**0.979**	**0.975**	**0.977**	**0.977**	**0.977**
Heart	NONE	0.852	0.817	0.845	0.831	0.848	0.848
	SMOTE	0.846	0.804	0.915	0.856	0.841	0.842
	BSMOTE	0.848	0.818	0.878	0.847	0.848	0.849
	ADASYN	0.843	0.802	0.854	0.827	0.840	0.841
	SLSMOTE	0.850	0.832	0.917	0.872	0.856	0.856
	FSMOTE	**0.864**	**0.854**	**0.919**	**0.886**	**0.867**	**0.867**

(continued)

Table 4 (continued)

SVM

Dataset	Methods	Overall accuracy	TP rate	Precision	F-measure	G-mean	ROC area
Ionosphere	NONE	0.915	0.802	0.953	0.871	0.886	0.890
	SMOTE	0.885	0.802	0.976	0.880	0.886	0.890
	BSMOTE	0.901	0.794	0.962	0.870	0.881	0.886
	ADASYN	0.837	0.701	0.970	0.814	0.828	0.840
	SLSMOTE	0.898	0.827	0.976	0.895	0.899	0.902
	FSMOTE	**0.922**	**0.873**	**0.978**	**0.922**	**0.924**	**0.925**
Spam base dataset	NONE	0.908	0.838	0.920	0.877	0.894	0.896
	SMOTE	0.896	0.853	0.959	0.903	0.902	0.903
	BSMOTE	0.900	0.833	0.932	0.880	0.891	0.893
	ADASYN	0.885	0.815	0.943	0.874	0.881	0.884
	SLSMOTE	0.904	0.866	0.959	0.911	0.908	0.910
	FSMOTE	**0.923**	**0.900**	**0.961**	**0.929**	**0.926**	**0.926**
German	NONE	0.788	0.513	0.700	0.592	0.682	0.710
	SMOTE	0.718	0.498	0.819	0.620	0.672	0.702
	BSMOTE	0.743	0.499	0.779	0.608	0.672	0.702
	ADASYN	0.704	0.495	0.835	0.622	0.670	0.700
	SLSMOTE	0.735	0.532	0.825	0.647	0.694	0.719
	FSMOTE	**0.779**	**0.632**	**0.852**	**0.725**	**0.757**	**0.769**
Mammography	NONE	**0.980**	0.142	0.949	0.247	0.377	0.571
	SMOTE	0.963	0.148	0.973	0.256	0.385	0.574
	BSMOTE	0.964	0.129	0.968	0.227	0.359	0.564
	ADASYN	0.962	0.081	0.950	0.150	0.285	0.541
	SLSMOTE	0.963	0.148	0.973	0.256	0.385	0.574
	FSMOTE	0.973	**0.410**	**0.991**	**0.580**	**0.640**	**0.705**
Winning times	NONE	2	0	0	0	0	0
	SMOTE	0	0	0	0	0	0
	BSMOTE	0	0	0	0	0	0
	ADASYN	0	0	0	0	0	0
	SLSMOTE	0	0	0	0	0	0
	FSMOTE	5	7	7	7	7	7

5 Experimental Results

Result of different oversampling algorithms using Naïve Bayes and SVM classifier is shown in Table 3 and Table 4, respectively. Performance of classifiers is represented through various performance metrics. The best results are highlighted in bold. Results show that our proposed technique Farthest SMOTE outperforms other oversampling techniques according to the winning times. Performance metrics used are overall accuracy, TP rate, precision, F-measure, G-mean, and ROC area.

In this section, we also present ROC curves for Pima Indian Diabetes Dataset and Spam Base Dataset using SVM classifier. The curves depict two oversampling algorithms Farthest SMOTE and ADASYN having highest and lowest area under the curve (AUC) value, respectively.

Figures 1 and 2 depict ROC curves for Pima Indian Diabetes and Spam Base Datset.

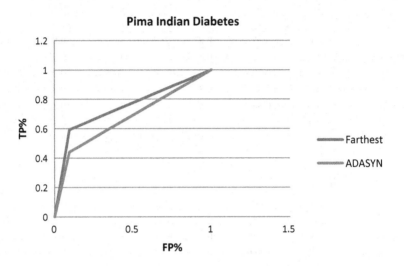

Fig. 1 ROC curve for Pima Indian diabetes dataset using SVM classifier

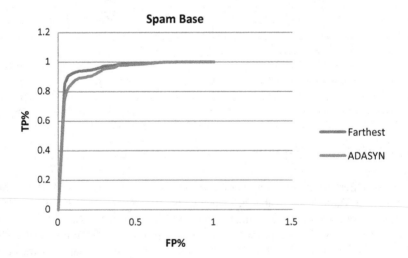

Fig. 2 ROC curve for Spam Base dataset using SVM classifier

6   Conclusion

This paper has discussed the class imbalance learning and the approaches used by researchers to alleviate the problem. In this paper, we have proposed Farthest SMOTE (FSMOTE) approach which is basically the modification of SMOTE approach. We have compared the performance of our proposed approach FSMOTE with oversampling methods, namely SMOTE, ADASYN, borderline SMOTE, and safe-level SMOTE on seven publically available datasets using two traditional classification algorithms, namely Naïve Bayes and SVM. The experimental results demonstrate that our technique outperforms the existing techniques. This can be observed by various performance metrics, such as overall accuracy, sensitivity, precision, F-measure, geometric mean (G-mean), and receiver operating curve (ROC) area, i.e., area under the curve (AUC) value.

References

1. S. Maheshwari, J. Agrawal and S. Sharma, "New approach for classification of highly imbalanced datasets using evolutionary algorithms," Int. J. Sci. Eng. Res., vol. 2, no. 7, pp. 1–5, 2011.
2. A. Amin, S. Anwar, "Comparing Oversampling Techniques to Handle the CIP: A Customer Churn Prediction Case Study", IEEE Translations and content mining, Vol. 4, 2016.
3. G. Weiss, "Mining with Rarity: A Unified Framework", SIGKDD Explorations, Vol. 6, No. 1, pp. 7–19, 2004.
4. X. Guo, Y. Yin, C. Dong, "On the class imbalance problem", Natural Computation, 2008. ICNC'08. Fourth International Conference on. 2008.
5. K. P. N. V. Satyashree, and J. V. R. Murthy, "An Exhaustive Literature Review on Class Imbalance Problem", Int. Journal of Emerging Trends and Technology in Computer Science Vol. 2, No. 3, pp. 109–118, 2013.
6. N. Chawla, N. Japkowicz and A. Kolcz, "Editorial: Special Issue on Learning from Imbalanced Data Sets", SIGKDD Explorations, Vol. 6, No. 1, pp. 1–6, 2004.
7. N. Chawla et al., "SMOTE: Synthetic Minority Over-Sampling Technique", Journal of Artificial Intelligence Research, Vol. 16, pp. 321–357, 2002.
8. N. Chawla et al., "Data mining for imbalanced datasets: An overview", in Data Mining and Knowledge Discovery Handbook, Springer, pp. 853–867, 2005.
9. B. X. Wang and N. Japkowicz, "Imbalanced Data Set Learning with Synthetic Samples", Proc. IRIS Machine Learning Workshop, 2004.
10. H. He, Y. Bai, E. A. Garcia, and S. Li, "ADASYN: Adaptive synthetic sampling approach for imbalanced learning", Proc. IEEE Int. Joint Conf. Neural Netw., IEEE World Congr. Comput. Intell., pp. 1322–1328, 2008.
11. H. Han, W. Wang, and B. Mao, "Borderline-SMOTE: A new oversampling method in Imbalanced Data-sets Learning", In. ICIC 2005. LNCS, Vol. 3644, pp. 878–887, Springer, Heidelberg, 2005.
12. C. Bunkhumpornpat, K. Sinapiromsaran, and C. Lursinsap, "Safe-Level-SMOTE: Safe Level-Synthetic MI Over-Sampling Technique for handling the Class Imbalance Problem", PADD2009, LNAI, Vol. 5476, pp. 475–482, Springer, 2009.
13. J. Huang and C. X. Ling, "Using AUC and Accuracy in Evaluating Learning Algorithms", IEEE Transactions on Knowledge and Data Engineering, Vol. 17, No. 3, March 2005.

14. A. Gosain and S. Sardana, "Handling Class Imbalance Problem Using Oversampling Techniques: A Review", communicated in International Conference on Advances in Computing, Communications and Informatics (ICACCI) 2017, Manipal, Karnataka, India, September 2017.
15. Buckland, M., Gey, F., "The Relationship between Recall and Precision", Journal of the American Society for Information Science 45(1), pp. 12–19, 1994.

DKFCM: Kernelized Approach to Density-Oriented Clustering

Anjana Gosain and Tusharika Singh

Abstract In this chapter, we have proposed a new clustering algorithm: density-oriented kernel-based FCM (DKFCM). It uses kernelized approach for clustering after identifying outliers using density-oriented approach. We have used two types of kernel functions for the implementation of DKFCM—Gaussian function and RBF function—and compared its result with other fuzzy clustering algorithms such as fuzzy C-means (FCM), kernel fuzzy C-means (KFCM), and density-oriented fuzzy C-means (DOFCM) to show the effectiveness of the proposed algorithm. We have demonstrated the experimental performance of these algorithms on two standard datasets: DUNN and D15.

Keywords Fuzzy clustering · FCM · KFCM · Density-oriented approach
Distance metric

1 Introduction

Clustering is an unsupervised classification process, which groups a set of data points into clusters so that the elements in the same cluster have high similarity and elements in other clusters are dissimilar [1]. Clustering can be classified into two types: hard clustering and fuzzy clustering. When data points of a dataset are divided into different clusters and each data point belongs to only one cluster, then this type of clustering is called hard clustering [2]. Fuzzy clustering allows each data point to belong to multiple clusters. Hard clustering finds natural boundaries in

A. Gosain (✉) · T. Singh
University School of Information and Communication Technology,
Guru Gobind Singh Indraprastha University, Dwarka, New Delhi 110078, India
e-mail: anjana_gosain@hotmail.com

T. Singh
e-mail: tusharikasingh170@gmail.com

© Springer Nature Singapore Pte Ltd. 2019   321
H. S. Behera et al. (eds.), *Computational Intelligence in Data Mining*,
Advances in Intelligent Systems and Computing 711,
https://doi.org/10.1007/978-981-10-8055-5_29

the data, but for real-world application, it is nearly impossible to find a reasonable criterion that includes some data objects into a cluster but exclude others. Fuzzy clustering handles vague boundaries of clusters and hence solves 'crisp boundary' problem of hard clustering [3]. An element in fuzzy clustering shares some fraction of membership in a number of clusters [4]. Strength of association between a cluster and an element is indicated by these membership levels. Fuzzy clustering uses the concept of fuzzy logic where a membership grade is associated with each data point, between the range of 0 and 1 [5].

Most used algorithm for fuzzy clustering was proposed by Bezdek [5], called fuzzy C-means (FCM). FCM uses Euclidean distance metric, which calculates the distance between cluster centroid and data points of the dataset. A constraint is followed by FCM, which states that the sum of memberships of a data point across all clusters must be equivalent to one [1]. This algorithm works well on most noise-free data, but it is completely unsuitable for datasets that include noise and outliers [6]. It has another limitation of equal partition trend for datasets and can only detect hyperspherical clusters [7].

Kernel fuzzy C-means (KFCM) clustering algorithm overcomes the 'non-spherical data' problem of FCM, which arises due to distance metric it uses, i.e., Euclidean distance. KFCM, introduced by Zhang and Chen [8], replaced the original Euclidean norm metric in FCM with a new kernel-induced metric. By swapping the conventional distance measure with a suitable 'kernel' function, without increasing the number of parameters, a nonlinear mapping can be performed to a high-dimensional feature space [8], which is effective in clustering 'non-spherical' clusters. It gives better result than FCM in the absence of noisy data and exhibits the attraction of centroid toward outliers.

Another variant of FCM is density-oriented fuzzy C-means (DOFCM). It uses density of dataset to identify outliers before applying clustering [9]. The idea is to assign zero membership to outliers, which are not required during clustering process. It modifies the membership of FCM and results in 'n' good clusters and one invalid cluster of outliers.

Detection of noise and outliers in datasets is the main motive of all algorithms discussed till now as its presence results in deterioration of performance in centroid computation. Except DOFCM approach, all the above techniques do not produce efficient clusters because they take outliers into consideration for the result of conclusive clusters and are not able to obtain noiseless clusters. In this chapter, we try to ameliorate the performance of DOFCM and present density-oriented kernel-based fuzzy c-means (DKFCM) algorithm by including kernel functions (Gaussian function and radial basis function) as the distance measure. Later, we compare the results of new proposed technique with other clustering algorithms to show the effectiveness of DKFCM.

This chapter is organized as follows: Sect. 2 briefly discusses various algorithms used in our work. Section 3 gives a detail description of the proposed algorithm. Section 4 presents the experimental result of datasets with respect to algorithms used, in form of figures and tables. Section 5 concludes with a short summary.

2 Related Work

This segment explores the FCM and its variants: FCM, KFCM, DOFCM, and DKFCM. Let $\{x_1, x_2, x_3, \ldots\ldots x_n\}$ define the dataset X specifying 'n' points in two-dimensional space. v_k represents cluster centroid where k denotes the cluster. d_{ik} is the distance between data point x_i and cluster center v_k. c denotes the number of clusters present in the dataset. 'm' represents the weighting exponent.

2.1 Fuzzy C-Means (FCM)

Among all fuzzy clustering algorithms, FCM is the most promising one. Bezdek [5] extended the work of fuzzy ISODATA algorithm, proposed by Dunn [10] and introduced first ever fuzzy clustering algorithm, named FCM. FCM minimizes the objective function as represented in Eq. (1) [11]:

$$J_{FCM} = \sum_{k=1}^{c} \sum_{i=1}^{n} u_{ik}^m d_{ik}^2 \tag{1}$$

With respect to membership function u_{ik} of a data point x_i in cluster k.

Euclidean distance d_{ik}, shown in Eq. (1), between x_i and v_k, is computed in the A-norm as expressed in Eq. (2):

$$d_{ik}^2 = \|x_i - v_k\|_A^2 \tag{2}$$

$$\sum_{k=1}^{c} u_{ik} = 1 \ i = 1, 2, 3 \ldots n \tag{3}$$

Equation (3) depicts the constraint followed by FCM, which means the total of all the membership grades of a data point belonging to all clusters must be equal to one.

An alternate optimization technique is used for the minimization of J_{FCM} with the continuous update of cluster centers v_k and memberships u_{ik} given in Eqs. (4) and (5):

$$v_k = \frac{\sum_{i=1}^{n} \left(u_{ik}^m x_i \right)}{\sum_{i=1}^{n} \left(u_{ik}^m \right)} \ 1 \leq k \leq c \tag{4}$$

$$u_{ik} = \frac{1}{\sum_{j=1}^{c} \left(\frac{d_{ki}}{d_{ji}} \right)^{\frac{2}{m-1}}} \tag{5}$$

2.2 Kernel Fuzzy C-Means (KFCM)

Zhang and Chen [8, 11] introduced a new kernel-induced metric in the data space, instead of conventional Euclidean norm metric in FCM, in order to deal with high-dimensional dataset. For estimating the distance between the data points and centroids, algorithm uses kernel function [12].

KFCM modifies the objective function of FCM with the mapping ϕ as expressed in Eqs. (6) and (7):

$$J = \sum_{k=1}^{c} \sum_{i=1}^{n} u_{ik}^{m} \| \Phi(x_i) - \Phi(v_k) \|^2 \tag{6}$$

$$\| \Phi(x_i) - \Phi(v_k) \|^2 = K(x_i, x_i) + K(v_k, v_k) - 2K(x_i, v_k) \tag{7}$$

where $K(x, y)$ is an inner product kernel function. If kernel width is a positive number, then $K(x, x) = 1$. Equations (6) and (7) can be combined and revised to form Eq. (8):

$$J_{KFCM} = 2 \sum_{k=1}^{c} \sum_{i=1}^{n} u_{ik}^{m} (1 - K(x_i, v_k)) \tag{8}$$

2.3 Density-Oriented Fuzzy C-Means (DOFCM)

DOFCM, proposed by Prabhjot Kaur and Anjana Gosain [9], separates noise into different clusters. The density of dataset is used to identify outliers. Neighborhood membership is described in Eq. (9) [13]:

$$M_{neighborhood}^{i}(X) = \frac{\eta_{neighborhood}^{i}}{\eta_{\max}} \tag{9}$$

with respect to point 'i' in dataset 'X', $\eta_{neighborhood}^{i}$ defines the total number of points in neighborhood of point 'i'. Maximum point of neighborhood is formulated in Eq. (10):

$$\eta_{\max} = \max_{i=1...n} (\eta_{neighborhood}^{i}) \tag{10}$$

Equation (11) expresses the condition to be satisfied in order to call point 'p' as in neighborhood of points 'i':

$$p \in X, i \in X | dist(p, i) \le r_{neighborhood} \tag{11}$$

where $dist(p, i)$ is Euclidean distance between points 'p' and 'i' and $r_{neighborhood}$ is the neighborhood's radius.

After detecting outliers, algorithm uses FCM technique to create clusters. DOFCM modified FCM objective function and minimizes as defined in Eq. (12):

$$J_{DOFCM} = \sum_{k=1}^{c+1} \sum_{i=1}^{n} u_{ik}^m d_{ik}^2 \tag{12}$$

3 The Proposed Technique, DKFCM

Density-oriented kernel version of FCM (DKFCM) is developed with a purpose to eliminate the effect of outliers on cluster centroid locations, instead of minimizing the influence of outliers on clustering process. The idea is to give zero membership value to the data points (outliers), which should not be involved during clustering. DKFCM finds 'n' noiseless clusters and one cluster which is not valid consisting of all the outliers present in a dataset, resulting in total of '$n + 1$' clusters. Like DOFCM, it identifies outliers before clustering and then applies clustering algorithm to find noiseless clusters.

DKFCM is a hybrid of KFCM and DOFCM. It combines the features of KFCM and DOFCM. On the basis of the density of data points in a dataset, algorithm identifies outliers. Neighborhood membership is a density factor used in this algorithm to measure density of an object in relation to its neighborhood. Neighborhood membership is defined as in Eq. (13):

$$M_{neighborhood}^{i}(X) = \frac{\eta_{neighborhood}^{i}}{\eta_{max}} \tag{13}$$

where $\eta_{neighborhood}^{i}$ is the total number of points in neighborhood of point 'i'. Maximum point of neighborhood is formulated in Eq. (14):

$$\eta_{max} = \underset{i=1...n}{max}(\eta_{neighborhood}^{i}) \tag{14}$$

A threshold value 'α' is selected from the complete range of neighborhood membership values defined in Eq. (13). If neighborhood membership of a point is less than 'α', then it is considered as an outlier and will not be included in the clustering process, as expressed in Eq. (15).

$$M^i_{neighborhood} = \begin{cases} < \alpha & outlier \\ \geq \alpha & non-outlier \end{cases} \qquad (15)$$

Once the outliers are identified, clustering technique can be applied. Unlike DOFCM, which uses FCM technique, the proposed method uses kernel-based approach. Kernel function replaces the Euclidean distance metric of FCM in order to enhance the performance of clustering [14].

Kernel function is of two types:

(i) Gaussian function: $K(x,y) = \exp(-\|x-y\|^2/2\sigma^2)$
(ii) Radial basis function (RBF): $K(x,y) = \exp(-\|x-y\|^2/\sigma^2)$

Objective function of DOFCM is modified and rewritten in Eq. (16):

$$J_{DKFCM} = \sum_{k=1}^{c+1} \sum_{i=1}^{n} u_{ik}^m (1 - K(x_i, v_k)) \qquad (16)$$

Its cluster center v_k and membership u_{ik} are represented in Eq. (17) and Eq. (18):

$$v_k = \frac{\sum_{i=1}^{n} u_{ik}^m K(x_i, v_k) x_i}{\sum_{i=1}^{n} u_{ik}^m K(x_i, v_k)} \qquad (17)$$

$$u_{ik} = \begin{cases} \dfrac{(1/(1 - K(x_i, v_k)))^{1/(m-1)}}{\sum_{j=1}^{c}(1/(1 - K(x_i, v_j)))^{1/(m-1)}} & M^i_{neighborhood} \geq \alpha \\ 0 & M^i_{neighborhood} < \alpha \end{cases} \qquad (18)$$

4   Result and Simulation

We have compared proposed technique with other algorithms like FCM, KFCM, and DOFCM using three datasets. For all datasets, we have considered m = 2, $\varepsilon = 0.0001$, and maximum number of iterations as 100.

Example 1 Dataset: D15 [13] is a diamond dataset consisting of actual data points and outliers.

Algorithms used: FCM, KFCM, DOFCM, DKFCM

Number of clusters: 2

Number of data points in respective clusters: 6, 5

Number of outliers: 4

Figure 1 depicts the clustering result of discussed algorithms. D15 is implemented using FCM, KFCM with B = 2.4, and DOFCM and DKFCM using

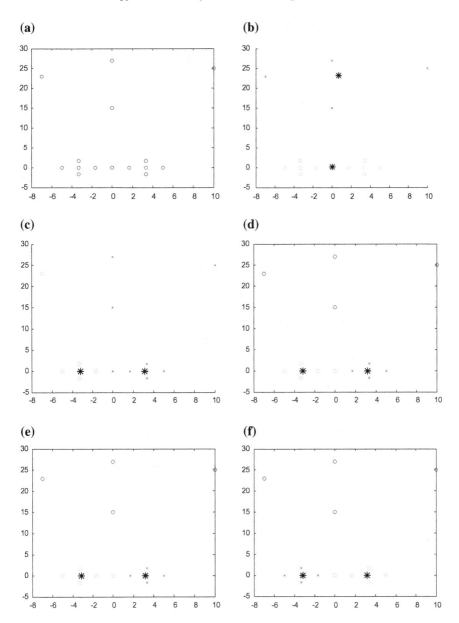

Fig. 1 **a** Original dataset with outliers **b** FCM **c** KFCM **d** DOFCM **e** DKFCM using Gaussian function **f** DKFCM using RBF

Gaussian function and radial basis function with B = 2.5. Symbols 'x' and 'o' represent two clusters. Centroids of two clusters are plotted by '*', and outliers are plotted using symbol 'o' in blue. We examined and observed that FCM could not

Table 1 Centroid coordinates produced on standard datasets

Dataset	Algorithm	Cx	Cy	No. of clusters
D15	FCM	0.6757	23.1738	2
		0.0047	0.1227	
	KFCM	−3.2205	0.0033	
		3.1256	0.0026	
	DOFCM	−3.1672	0.0000	
		3.1675	0.0000	
	DKFCM using Gaussian function	−3.1838	−0.0003	
		3.1669	0.0003	
	DKFCM using RBF	−3.1868	0.0001	
		3.1663	0.0002	
DUNN	FCM	15.3149	0.3322	2
		5.7652	0.1165	
	KFCM	15.2015	0.1724	
		5.4841	0.0784	
	DOFCM	5.4870	0.1719	
		15.3848	0.0086	
	DKFCM using Gaussian function	15.3041	0.0102	
		5.3598	0.1374	
	DKFCM using RBF	15.264	0.0001	
		5.3279	0.106	

detect original clusters and produces centroids, which are more attracted toward the outliers. KFCM gives centroids, which exhibit no attraction toward outliers but allocate noise points to both clusters, and hence, performance is affected. DOFCM and DKFCM both identified outliers and produce final cluster without considering them, but the performance of DKFCM is better than DOFCM. When Gaussian function and RBF are used as kernel functions in DKFCM, the performance of DKFCM using RBF is slightly better than Gaussian function. Table 1 shows the centroid coordinates produced by each algorithm on D15 dataset.

The ideal centroids of dataset D15 are:

$$V_{ideal} = \begin{bmatrix} 3.34 & 0 \\ -3.34 & 0 \end{bmatrix}$$

We have calculated the error to show the effectiveness of the proposed technique.

$$E = \| V_{ideal} - V_{observed} \|^2$$

Table 2 Error percentage

Algorithm	Cluster 1	Cluster 2	Average error
*D*15			
FCM	553.1508529	11.13928138	282.1450672
KFCM	0.01429114	0.04597412	0.030132630
DOFCM	0.02985984	0.02975625	0.029808045
DKFCM using Gaussian function	0.02439853	0.0299637	0.027181115
DKFCM using RBF	**0.02347025**	**0.03017173**	**0.026820990**
DUNN			
FCM	0.12355885	0.14694329	0.135251070
KFCM	0.02972401	0.01321937	0.021471690
DOFCM	0.03422500	0.03711861	0.035671805
DKFCM using Gaussian function	0.01094085	0.0204948	0.015717825
DKFCM using RBF	**0.00409601**	**0.01643441**	**0.010265210**

where V_{ideal} is the ideal centroid of a cluster and $V_{observed}$ is the observed centroid of a cluster with respect to each algorithm applied. Table 2 reveals the error percentage.

Example 2 Dataset: DUNN [10] is a two-dimensional dataset representing two square-shaped clusters.

Algorithms used: FCM, KFCM, DOFCM, DKFCM

Number of clusters: 2

Number of data points in respective clusters: 81, 53

Number of outliers: 21

Figure 2 depicts the clustering result of discussed algorithms. DUNN is implemented using FCM, KFCM with B = 2.4, and DOFCM and DKFCM with B = 2.5. We examined that the result of FCM is inaccurate. KFCM could not detect outliers but performs better than FCM. DOFCM and DKFCM identified outliers. DKFCM using RBF found almost original centroid location and outperforms all other algorithms used including DKFCM implemented using Gaussian function. Table 1 shows the centroid coordinates produced by each algorithm on DUNN dataset.

The ideal centroids of dataset DUNN are:

$$V_{ideal} = \begin{bmatrix} 15.2 & 0 \\ 5.4 & 0 \end{bmatrix}$$

Table 2 reveals the error percentage, which tells the effectiveness of proposed algorithm.

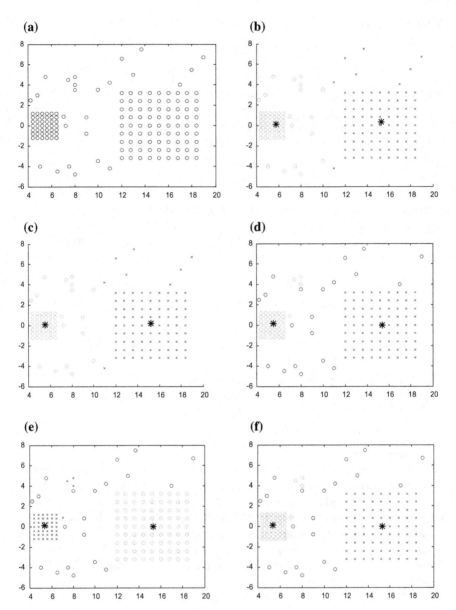

Fig. 2 **a** Original dataset with outliers **b** FCM **c** KFCM **d** DOFCM **e** DKFCM using Gaussian function **f** DKFCM using RBF

5 Conclusion

In this chapter, we have proposed a new fuzzy clustering algorithm DKFCM. It uses kernelized approach for clustering after identifying outliers using density-oriented approach. We have compared proposed algorithm with other conventional algorithms and applied them on two datasets considering noise and outliers. We observed that FCM does not perform well in the presence of noise and outliers, whereas KFCM exhibits attraction of centroid toward outlier but do not produces desirable results. DOFCM and DKFCM identified outliers, but cluster produced by DKFCM produces more accurate centroid location compared to all other algorithms. Among Gaussian function and RBF function used in the implementation of DKFCM, the performance of RBF is slightly better than Gaussian function. In future, we will try to come up with an algorithm, which offers more appropriate results.

References

1. C.C. Hung, S. Kulkarni, and B. Kuo, "A new weighted fuzzy c-means clustering algorithm for remotely sensed image classification", IEEE Journal of Selected Topics in Signal Processing, Vol. 5, No. 3, pp. 543–553, 2011.
2. N. Grover, "A study of various Fuzzy Clustering Algorithms", International Journal of Engineering Research (IJER), Vol. 3, No. 3, pp. 177–181, 2014.
3. P. Kaur, et al. "Novel intuitionistic fuzzy C-means clustering for linearly and nonlinearly separable data", WSEAS Transaction on Computers, Vol. 11, No. 3, pp. 65–76, 2012.
4. A. Gosain and S. Dahiya, "Performance Analysis of Various Fuzzy Clustering Algorithms: A Review", Procedia Computer Science, pp. 100–111, 2016.
5. J. Bezdek, "Pattern Recognition with Fuzzy Objective Function Algorithms", New York: Plenum, 1981.
6. M. Gong, et al. "Fuzzy c-means clustering with local information and kernel metric for image segmentation", IEEE Transactions on Image Processing, Vol. 22, No. 2, pp. 573–584, 2013.
7. D.M. Tsai and C.C. Lin, "Fuzzy C-means based clustering for linearly and nonlinearly separable data", Pattern recognition, Vol. 44, No. 8, pp. 1750–1760, 2011.
8. D. Zhang and S.C. Chen, "Kernel-based fuzzy and possibilistic c-means clustering", Proceedings of the International Conference Artificial Neural Network, 2003.
9. P. Kaur, I.M.S. Lamba and A. Gosain, "DOFCM: A Robust Clustering Technique Based upon Density", IACSIT International Journal of Engineering and Technology, Vol. 3, No. 3, pp. 297–303, June 2011.
10. J. Dunn, "A fuzzy relative of the ISODATA process and its use in detecting compact well-separated clusters", J. Cybern., Vol. 3, No. 3, pp. 32–57, 1974.
11. D.Q. Zhang and S.C. Chen, "Clustering incomplete data using kernel-based fuzzy c-means algorithm", Neural Processing Letters, Vol. 18, No. 3, pp. 155–162, 2003.
12. D. Singh and A. Gosain, "Robust new distance kernelized approach to distributed clustering", International Journal of Engineering Science & Advanced Technology (IJESAT), Vol. 3, Issue 4, pp. 165–169, 2013.
13. P. Kaur and A. Gosain, "Density oriented approach to identify outliers and get noiseless clusters in fuzzy c-means", IEEE International Conference on Fuzzy Systems, pp. 1–8, 2010.
14. D. Graves and W. Pedrycz, "Kernel-based fuzzy clustering and fuzzy clustering: A comparative experimental study", Fuzzy sets and systems, Vol. 161, No. 4, pp. 522–543, 2010.

Comparative Study of Optimal Overcurrent Relay Coordination Using Metaheuristic Techniques

Shimpy Ralhan, Richa Goswami and Shashwati Ray

Abstract For the protection of power system, Directional Overcurrent Relays (DOCRS) are generally used as an economical means for protection in sub-transmission and distribution systems. Thus, it is required that relay has minimum operating time on the occurrence of fault keeping the operation coordinated with other relays. In our paper, a comparative analysis for optimal overcurrent relay coordination based on different metaheuristic techniques such as Genetic Algorithm (GA), Particle Swarm Optimization (PSO), Cuckoo Search Algorithm (CSA), and Firefly Algorithm (FFA) is analyzed. The various heuristic techniques are executed on different number of relays for constrained systems and the method which gives the operating time with minimum iterations and time is found.

Keywords Metaheuristic techniques · Genetic algorithm · Particle swarm optimization · Cuckoo search algorithm · Firefly algorithm

1 Introduction

In a protection scheme, one of the basic requirements of any power system is reliability. Power system consists of primary and backup protection using Directional Overcurrent Relays (DOCRS). Any power system is a consistent process, where primary relays of the protection scheme must work during a fault as swiftly

S. Ralhan (✉) · R. Goswami
Shri Shankarcharya Technical Campus, Bhiali 490023, Chhattisgarh, India
e-mail: shimpys@gmail.com

R. Goswami
e-mail: richagos23@gmail.com

S. Ray
Bhilai Institute of Technology, Durg 491001, India
e-mail: shashwatiray@yahoo.com

© Springer Nature Singapore Pte Ltd. 2019
H. S. Behera et al. (eds.), *Computational Intelligence in Data Mining*,
Advances in Intelligent Systems and Computing 711,
https://doi.org/10.1007/978-981-10-8055-5_30

as possible to isolate the unhealthy section and if the primary relay fails to operate; the backup relay should work automatically within the certain time to dodge the outage [1]. This situation is the most desirable function of any protection scheme designed as it leads to isolation of only unhealthy section whereas the backup or secondary protection isolates even the healthy section if former fails. Hence, the coordination between the relays is essential to prevent the complete outage. DOCRs are generally used as primary relays for distribution systems and backup relays for sub-transmission systems [1, 2]. Directional overcurrent relays are broadly used to protect the system fed from two ends, ring/mesh networks, and parallel/cascaded parallel feeders. A Directional Overcurrent relay is characterized by two settings, namely time setting and current setting. The time setting is carried out using Time Multiplier Setting (TMS), and current setting is carried out using Plug Setting Multiplier (PSM). For the various protection schemes, numerous approaches have been established. The overcurrent relays are best suited for the protective scheme and are optimally coordinated using many algorithms. Thus, to ensure their operating characteristics, the adaptive setting of overcurrent relays and their optimum coordination by using algorithms have been tested for load and structural changes [1, 3].

Metaheuristic optimization techniques have shown the great inherent capability to solve protection coordination problems. A new objective function by using the constraints on TMS because of bounds on relay operating time as the boundaries of variables are being tested for a different number of relay systems using GA [4]. The results obtained from the Genetic Algorithm used to coordinate overcurrent relay system show that GA technique contributes the lowermost objective value as compared to the quasi-Newton process [5]. The algorithms are applied on several medium and small-sized systems in all these samples. Even though the tactics are slightly time-taking, they offer highly excellent solutions for the protection of radial feeders. Particle Swarm Optimization (PSO) is a heuristic search technique, which is found on the clue of concerted behavior and swarming in biological populations. PSO is analogous to GA in intellect that they are together population-based search tactics and both are subject to information distribution among their population members to improve their search methods using a mixture of deterministic and probabilistic rules [6]. The new metaheuristic techniques, which mimic the biological processes such as Cuckoo Search Algorithm and Firefly Algorithm, have been studied and implemented for relay coordination in various test systems. The code is developed in MATLAB to solve the systems having several relays and number of primary and backup pairs.

Our paper is organized as follows: in Sect. 2, the relay coordination problem with coordination constraints and operating bounds is described. In Sect. 3, a detail of various algorithms of metaheuristic optimization methods investigated in our work is described. Lastly, in Sect. 4, comparison of the results obtained on different test systems is discussed, and in Sect. 5, the most effective method is recommended.

2   Problem Identification

2.1   Objective Function

The coordination among the Directional Overcurrent Relays in protection scheme of the power system is considered as a main optimization problem which is highly constrained [3, 8]. The sum of the operating time of the primary relays is minimized in order to coordinate the backup relays for which the objective function is given by (1).

$$z = \sum_{p=1}^{m} W_p \times T_{p,k} \tag{1}$$

where m signifies the number of relays, Wp is the weight assigned for operating time. We know that in distribution system, the length of lines is short and is nearly equal, so we can say that existence of the fault is equivalent for all the lines; therefore, equivalent weights are assumed and allocated to all the relays, i.e., $W = 1$, for $p = 1,...,m$. Also, $T_{p,k}$ is the operating time of the primary relay.

2.2   Coordination Constraints

The equations governing the coordinated operation of the relays in the system are (2) and (3).

$$T_{q,k} - T_{p,k} \geq CTI \tag{2}$$

where, $T_{p,k}$ is the operating time of the primary relay at k, near fault and $T_{q,k}$ is the operating time of the backup relay for the same fault. The minimum time of operation discriminating the operation of primary/backup pair (P/B pair) is termed as discrimination time or coordination time interval (CTI). It is considered on the basis of speed with which the circuit breakers operate. CTI is generally taken between 0.1 and 0.5 s. In our case, CTI is taken as 0.3s [2].

2.3   Bounds on Relay Operating Time

It is mandatory that relay must operate within the minimum time during the fault and it should be ensured that it does not take the very long duration of time to operate [2, 7]. Thus, the time constraint is given by (3).

$$T_{p\max} \geq T_p \geq T_{p\min} \tag{3}$$

where $T_{p,min}$ is the minimum operation performing time and $T_{p,max}$ is the maximum operation performing time of relay near the location of the fault.

2.4 Relay Characteristics

Here, we have considered the relays as identical with normal inverse definite minimum time (IDMT) characteristics given by (4).

$$T_{op} = \frac{0.14*(TSM)}{(PSM)^{0.2-1}} \tag{4}$$

The values of 0.14 and 0.02 are standard values in IDMT characteristics. T_{op} is the relay operating time, TMS is time multiplier setting is and PSM is plug setting multiplier described by (5).

$$PSM = \frac{I_{OP}}{TSM} \tag{5}$$

where Iop is the operating current. The problem of relay coordination is stated as linear as plug setting multiplier is considered having a fixed value [9]. Thus, (4) reduces to the equations given by (6) and (7).

$$T_{OP} = a_p*TSM \tag{6}$$

where,

$$a_p = PSM^{0.4} - 1 \tag{7}$$

Hence, the objective function gets modified as a function to be minimized as given by (8).

$$z = \sum_{p=i}^{n} a_p (TMS)_P \tag{8}$$

3 Optimization Techniques

The metaheuristic comes from two words, i.e. meta: in an upper level, heuristic: to find, thus, an optimization technique to determine a set of precise quality approximations leading to an exact solution. A heuristic technique is used to provide better computational performance as compared to conventional optimization techniques. A metaheuristic is defined as an iterative search method, which guides a subsequently heuristic by appending processes intelligently for the various concepts in exploring and exploiting the closest search space, in order to obtain an efficiently approximated optimal solution. There are two types of techniques for optimization techniques [8]:

1. Evolutionary algorithms (EA): They have evolved from the various biological processes and their characteristics. Genetic algorithm is the EA tactics used in our work.
2. Swarm intelligence (SI): They have evolved from the various organisms and their behaviors. Particle Swarm Optimization (PSO), Cuckoo Search (CSA), and Firefly Algorithm (FFA) are used in our work.

We had done the coding of the metaheuristic techniques, viz GA, PSO, CSA, and FFA in MATLAB 2012b by using optimization toolbox for GA, and rest of the algorithms are implemented in m-file coding. The following SI techniques are implemented on different relay systems in our work.

3.1 Genetic Algorithm

Genetic algorithm (GA) was first introduced by John Holland. This is the optimization technique which uses the natural selection procedure for calculating its parameters [9]. The basic algorithm for GA to solve optimization problems involves the following steps.

Step1—At first, the initial population of the function is generated.
Step2—Iteration process for different sub-step is being carried out until we achieve the termination.

(i) Here, each program in the population is being executed, and the fitness value is being applied.
(ii) After that a new population of programs is created by applying the following operations: (a) reproduction (b) crossover (c) mutation.

Step3—Result for each individual is being designated. This result may be a solution or an approximate solution to the problem.

3.2 Particle Swarm Optimization

Particle Swarm Optimization (PSO) is an optimization technique developed by Eberhard and Kennedy in 1995. PSO has been derived from the concept of fish schooling and bird flocking [10]. The steps involved in PSO are as follows:

Step1—At first, the minimum and maximum values of the parameters are being specified. This is done by selecting the population of the individual which includes the searching point, P_{best} which is its individual best value and g_{best} which is known as the global best value.

Step2—After that the fitness value is being calculated for each individual using the evaluation function.

Step3—Comparison of each individual is being done which is known as g_{best}. The best value from P_{best} is denoted as g_{best}.

Step4—After that the member velocity is being modified for each individual k.

$$v_{j,g}^{(t+1)} = w * v_j^{(t)} + c_1^* \, rand() * \left(p_{best_{j,g}} - k_{j,g}^{(t)} \right) + c_2^* \, rand() * \left(g_{best_g} - k_{j,g}^{(t)} \right)$$

where, $j = 1, 2, 3....n$ and $g = 1, 2, 3....n$ and w is a known value.

Step5—If $v_{j,g}^{(t+1)} > V_g^{max}$, then $v_{j,g}^{(t+1)} = V_g^{max}$, If $v_{j,g}^{(t+1)} > V_g^{min}$, then $v_{j,g}^{(t+1)} = V_g^{min}$

Step6—Modified the member of each individual k.

$$k_{j,g}^{(t+1)} = k_{j,g}^{(t)} + v_{j,g}^{(t+1)}, k_g^{min} \leq k_{j,g}^{(t+1)} \leq k_g^{max}$$

where k_g^{min} and k_g^{max} represent the minimum and maximum, respectively, of the individual member g.

Step 7—If the maximum value is reached through a number of iteration then proceed to Step 8 or else proceed to Step 2.

Step 8—The latest individual which is now generated becomes the optimal value.

3.3 Cuckoo Search Algorithm

The Cuckoo Search Algorithm (CSA) is inspired by brood obligate parasite cuckoo species bird which lay their egg in other bird's nest (host/other species) [7]. They use levy flights random walk rather than simple isotropic random walk. Its criteria are based on the fact that cuckoo lays one egg and dumps it in the random nest, the egg laid by bird is investigated by host birds with probability P $\in$ [0, 1]. The number of host nests is fixed, and the host will either get rid of the egg or will abandon it and rebuild new nest for next generation. Nest with best egg will carry over for a future generation [11].

The pseudo code for CSA is as follows:

Step1—Objective function $f(u)$, $u = (u_1, \ldots, u_d)^T$.
Step2—Produce primary population of n host nests u $(i = 1, 2, \ldots, n)$ **While** $(t < \text{Max Generations})$ or (stop criterion).
Step3—Contract a cuckoo casually by Lévy flights.
Step4—Estimate its quality/fitness F.
Step5—Pick a nest between n (say j) casually **If** $(Fi > Fj)$, Swap j by the new solution.
Step6—**End if** a segment nest is uninhibited, and new ones are constructed.
Step7—Retain the best solutions (Or cases with excellence solutions)
Step8—Rank the results and find the present best.
Step9—**End while** post method grades and conception.
Step10—End.

3.4 Firefly Algorithm

The Firefly Algorithm is similar to particle swarm optimization inspired by the flashing pattern and the behavior of fireflies. It has three main considerations [12], i.e. for a minimization problem, the brightness is proportional to the objective function.

Step1—Objective function $f(u)$, $u = (u_1, \ldots, u_d)^T$.
Step2—Create primary populace of fireflies u_i $(i = 1, 2, \ldots n)$.
Step3—Light intensity I_i at u_i is resolute by $f(u)$.
Step4—Specify light absorption coefficient γ.
Step5—While $(t < \text{MaxGeneration})$, for $i = 1: n$ all n fireflies,
for $j = 1: i$ all n fireflies, if $(I_j > I_i)$,
Step6—Transfer firefly i towards j in d-dimension;
Step7—**End if** attractiveness fluctuates by distance "r" via exp $[-\gamma r]$.
Step8—Estimate new resolutions and bring up-to-date light concentration.
Step9—End for j; end for I; Rank the fireflies and find the present best value.
Step10—End while rest of the method results and visualization.
Step11—End

Table 1 Initial values of parameters

GA parameters	PSO parameters	CS parameters	FF parameters
Population size: 100	Population size: 100	No. of nests: 100	No. of fireflies: 100
Mutation rate: 0.25	$W_{max} = 0.9$ $W_{min} = 0.4$	Discovery rate of alien eggs = 0.25	Alpha = 0.5
Arithmetic crossover	Best generation	Best nest	Brightest fly the best one
Iteration: 100	Iteration: 100	Iteration: 100	Iteration: 100

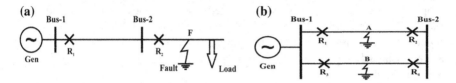

Fig. 1 **a** A two-bus radial system. **b** A two bus parallel feeder system

4 Results

The parameters for GA, PSO, CSA, and FFA, which are used for initialization, are described in Table 1. In proposed work, a radial system, a parallel feeder, and ring main system have been considered as test systems consisting of 2, 4, 6, and 10 relays, respectively. A radial system as shown in Fig. 1a is fed from one end and consists of relays R_1 and R_2 at bus1 and bus2, respectively. We have coordinated the relay operation for the fault as shown where relay R_1 acts as primary and R_2 as secondary [2, 12, 13]. The single-line diagram of the two-bus radial system is consists of the generator at the source side with two buses and two DOCRs to ensure the safety from the faulty condition in line.

The parallel feeder system [12] considered consists of 4 relays and a generator as shown in Fig. 1b. Relays 2 and 3 are directional overcurrent relays whereas 1 and 4

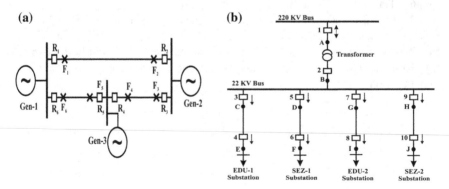

Fig. 2 **a** A three bus test system. **b** System for Serum Substation, Pune

Table 2 Optimal objective function in seconds

Relay system	GA	PSO	CS	FF
2 relay system	0.553	0.530	0.400	0.399
4 relay system	0.819	0.4874	0.502	0.487
6 relay system	1.864	1.707	0.594	0.594
10 relay system	1.544	2.198	1.265	1.154

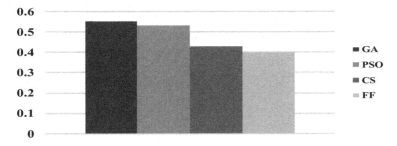

Fig. 3 Best objective function

are non-directional. For fault at A, relay 4 acts as backup for relay 2 and for fault at B, relay 1 will act as backup for relay 3. A 3-bus ring main system [14] is considerably supplied by generators G_1, G_2, and G_3. There are six relays R_1, R_2, R_3, R_4, R_5, and R_6 as shown in Fig. 2a. A 220 kV, 30 MVA real system of 10 relays network fed from Serum Substation with network topology is shown in Fig. 2b [15].

In Table 2, the value of objective function obtained is shown, and it is found that the Firefly Algorithm gives the best optimal value in comparison with other techniques for the optimal coordination of directional overcurrent relays [16].

The bar graph in Fig. 3 shows the best value of an objective function of different algorithms for two relay systems. The results for the time multiplier settings of the relays for the various test systems described as shown in Table 3.

The convergence is shown in Fig. 4 for two relay systems. Similar results are obtained for 4, 6, and 10 relay test systems also.

The statistical methods to assess agreement of four techniques such as GA, PSO, CSA, and FFA in the prediction of optimum operating time have been presented. The results obtained by FFA have been compared with other results. The Bland-Altman is a most popular method followed by correlation coefficient and mean comparison. It has been proposed in [17], a method for analysis of agreement by Bland-Altman plot and limit of agreement (LOA). In this approach, the result of difference against average of two methods is plotted, and the magnitude of disagreement (bias, error) and spot outliers is evaluated to assess the trend. Figure 5 shows comparison result of FFA versus GA, PSO, and CSA for operating time. If

Table 3 TMS of test relay systems

Relay system	Number of relays	TMS in seconds			
	P/B pair	GA	PSO	CS	FF
2 relay system	1	0.104	0.104	0.062	0.052
	2	0.516	0.056	0.056	0.050
4 relay system	1	0.067	0.033	0.033	0.023
	2	0.035	0.025	0.025	0.020
	3	0.035	0.025	0.025	0.020
	4	0.067	0.033	0.033	0.027
6 relay system	1	0.107	0.101	0.041	0.021
	2	0.186	0.185	0.057	0.050
	3	0.119	0.119	0.049	0.039
	4	0.122	0.122	0.038	0.032
	5	0.152	0.153	0.055	0.050
	6	0.112	0.120	0.047	0.040
10 relay system	1	0.043	0.003	0.003	0.002
	2	0.020	0.018	0.018	0.002
	3	0.020	0.018	0.018	0.018
	4	0.020	0.018	0.018	0.018
	5	0.044	0.032	0.024	0.020
	6	0.030	0.024	0.024	0.020
	7	0.020	0.018	0.018	0.018
	8	0.020	0.018	0.018	0.018
	9	0.030	0.024	0.024	0.020
	10	0.093	0.086	0.024	0.020

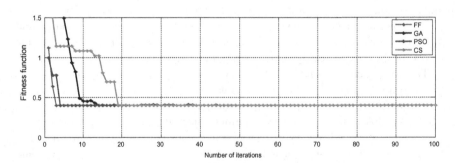

Fig. 4 Convergence for two relay system using GA, PSO, FFA, and CSA

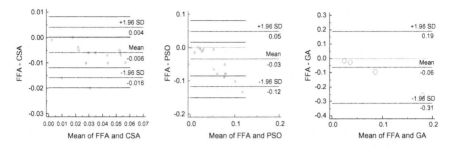

Fig. 5 Comparative analysis using Bland-Altman statistical method

Table 4 Comparison of multivariate analysis for operating times

Parameters	Genetic algorithm	Particle swarm optimization	Cuckoo search
Mean	0.06282	0.03432	0.005909
SD	0.09787	0.04250	0.005042
Median	1.64070	0.46740	0.020000
95% of CI	0.1793 to 0.3299	0.08492 to 0.1503	0.01191 to 0.01967

the measurements are comparable, they will be tightly scattered about the line as in Fig. 5. Here, the mean of FFA versus CSA is 0.0059 with 95% confidence interval 0.01191 to 0.01967 as shown in Table 4.

5 Conclusions

In this paper, the four different metaheuristic algorithms, namely GA, PSO, CSA, and FFA have been implemented to check their performances on protection coordination problem of DOCRs as an investigation problem. The execution of each algorithm has been set to 100 times keeping the initial conditions as same for all the four test systems as considered above. FF algorithm has been found to be the best one among the four algorithms based on the results obtained in our work. The best value obtained by FFA is always the lowest among the best values obtained by the other algorithms considered here. Hence, FFA can be considered to be the most suitable algorithm for the coordination of DOCRs among the other algorithms studied in this work.

References

1. Hassan A. Kubba., Samir Sami Mahmood: Genetic algorithm based load flow solution problem in electrical power systems. Journal of Engineering. IEEE Volume 15 (2009).
2. Madhumitha R., Parul Sharma., Deepika Mewara: Optimum Coordination of Overcurrent Relays using Dual Simplex and Genetic Algorithms. IEEE International Conference on Computational Intelligence and Communication Networks (2015).
3. Dusit Uthitsunthom., Thanatchai Kulworawanichpong: Optimal Overcurrent Relay Coordination using Genetic Algorithms. IEEE International Conference on Advances in Energy Engineering (2010).
4. Dusit Uthitsunthom., Thanatchai Kulworawanichpong: Optimal Overcurrent Relay Coordination using Genetic Algorithms. IEEE International Conference on Advances in Energy Engineering (2010).
5. S. S. Gokhale., V. S. Kale: Time overcurrent relay coordination using the Levy flight Cuckoo search algorithm. TENCON 2015 IEEE Region 10 Conference (2015).
6. S. S. Gokhale., Dr. V. S. Kale: Application of the Firefly Algorithm to Optimal Over-Current Relay Coordination. IEEE International Conference on Optimization of Electrical and Electronic Equipment (OPTIM) (2014).
7. Ritesh Rawat., Vijay S. Kale: Application of Nature Inspired Metaheuristic Techniques to Overcurrent Relay Coordination. IEEE FTC - Future Technologies Conference (2016).
8. Xin-She Yang: Nature-Inspired Optimization Algorithms. Book Aid international publications.
9. Rania Hassan., Babak Cohanim., Olivier de Weck: A comparison of particle swarm optimization and the genetic algorithm. Conference on Proceedings of the 46th AIAA/ASME/ASCE/AHS/ASC Structures, Structural Dynamics & Material Conference (2005).
10. Dharmendra Kumar Sing., Shubhrata Gupta: Use of Genetic Algorithms (GA) for Optimal Coordination of Directional Over Current Relays. IEEE Students Conference on Engineering and Systems (SCES) (2012).
11. G.U. Darji., M.J. Patel., V.N. Rajput., K.S. Pandya: A tuned cuckoo search algorithm for optimal coordination of directional overcurrent relays. IEEE International Conference on Power and Advanced Control Engineering (ICPACE) (2015).
12. Divya S Nair., Reshma S: Optimal coordination of protective relays. IEEE International Conference on Power, Energy and Control (ICPEC) (2013).
13. Seyed Hadi., Mousavi Motlagh., Kazem Mazlumi: Optimal Overcurrent Relay Coordination Using Optimized Objective Function. ISRN Power Engineering vol. 2014, Article ID 869617 (2014).
14. Shimpy Ralhan., Shashwati Ray: Directional Overcurrent Relays Coordination using Linear Programming Intervals: A Comparative Analysis. IEEE Annual India Conference (INDICON) (2013).
15. Jagdish Madhukar Ghogare., V. N. Bapat: Field Based Case Studies On Overcurrent Relay Coordination Optimization Using GA-NLP Approach. IEEE CONNECT (2015).
16. Yaser Damchi., Javad Sadeh., Habib Rajabi Mashhadi: Optimal coordination of distance and overcurrent relays considering a non-standard tripping characteristic for distance relays. IET gen. trans. Distri. vol. 6, pp. 1448–1457 (2016).
17. T. V. Dixit, Anamika Yadav., S Gupta: Optimization of PV array inclination in India using ANN estimator: Method comparison study. Sadhana Vol. 40, Part 5, pp. 1457–1472, Indian Academy of Sciences (2015).

Prediction of Gold Price Movement Using Discretization Procedure

Debanjan Banerjee, Arijit Ghosal and Imon Mukherjee

Abstract Accurate prediction of commodity prices by using machine learning techniques is considered as a significant challenge by the researchers and investors alike. The main objective of the proposed work is to highlight that discretized features provide more accuracy compared to the continuous features for predicting the gold price movement in either positive or negative direction. This work utilizes three unique techniques for measuring performance of the discretization procedure. These techniques are "percentage of accuracy", "receiver operating characteristics or ROC" and "the area under the ROC curve or AUC."

Keywords Feature discretization · Logistic regression · Random forest
Machine learning

1 Introduction

In the contemporary world, the aftermath of many spectacular events such as the burst of the sub-prime bubble in the USA, the 2008 financial crisis, etc., has witnessed significant volatility in the financial markets. This particular phenomenon has received a lot of attention from both the academic as well as the trading

D. Banerjee
Department for the Management Information Systems, Sarva Siksha Mission Kolkata,
Kolkata 700042, India
e-mail: debanjanbanerjee2009@gmail.com

A. Ghosal (✉)
Department of Information Technology, St. Thomas' College of Engineering
& Technology, Kolkata, Kolkata 700023, India
e-mail: ghosal.arijit@yahoo.com

I. Mukherjee
Department of Computer Science and Engineering, Indian Institute of Information
Technology, Kalyani, Kalyani 741235, India
e-mail: mukherjee.imon@gmail.com

© Springer Nature Singapore Pte Ltd. 2019
H. S. Behera et al. (eds.), *Computational Intelligence in Data Mining*,
Advances in Intelligent Systems and Computing 711,
https://doi.org/10.1007/978-981-10-8055-5_31

communities across the world. The commodity gold is used as an investment hedge against any supposed upward or downward tendency in the market prices. This work applies machine learning techniques such as the random forest and logistic regression for predicting whether the prices of gold would go upward or downward based on given historical data. The current work uses historical data from the Indian markets from January 2010 to June 2017 utilizing the web resource in.investing.com in order to make predictions whether the price of gold would increase or decrease. The features that this work employs are market indices, government bonds, commodity prices and currency prices.

2 Related Work

The first major challenge of predicting the movement of gold prices is finding a set of features that will be the optimum set for predicting the gold price movement. In [1], various financial market commodity prices including index prices were used as a set of features from different commodity exchanges across the world. In [2], it has been observed by the author that there are strong relations between major exchange currency blocks and the gold price. According to that work, any abrupt disruption in any major exchange currency blocks, for example, the US Dollar, could lead to instability in the gold price direction. In [3], it has been observed by the authors that the historical relationship between gold and silver prices with the help of the Tokyo stock exchange historical data has become somewhat ambiguous from the late twentieth century. The insights from these significant past works have aided the effort to find an optimum number of features for the present work.

The second major challenge is to find out the appropriate machine learning technique for obtaining the accurate prediction. In [1], logistic regression and SVM have been utilized by the authors with mostly continuous numerical features. The proposed work is unique in the sense that instead of using continuous value features, the work utilizes the technique of discretization of features to improve the accuracy. In [4], it has been observed by the authors that applying neural networks increase the potentiality of features such as currency exchange rates and interest rates for accurate price prediction. In [5], artificial bee colony algorithm was utilized for gold price prediction. In [6, 7], various machine learning techniques have been compared by the authors to predict the gold price, accurately. The proposed work is different from these works as the proposed work considers the problem as a classification problem. In [8], various state-of-the-art techniques were discussed by the authors that do impact the price of the stocks. In [9], the logistic regression technique has been used to help users select from a set of exchange traded funds (ETFs) based upon a probability score of possibility of the growth of the stocks. In [10], stock market data from Saudi Arabia has been utilized by the authors to predict the potential movement for a particular stock in a certain direction. These experiments have shown that with an optimum amount of features the logistic regression technique produces almost similar accuracy to that of a neural network model.

In [11], logistical regression has been used by the authors in the context of Australian stock markets to create a stock selection strategy for stock users. In [12], the concept of random forest has been utilized for predicting the direction of the market prices. In this work, technical indicators such as relative strength index (RSI), stochastic oscillator, on balance volume were introduced by the authors as features. In [13], predictive machine learning techniques have been utilized by the authors for the equity data from stock markets situated in different parts of the world. The primary difference between the proposed work and other works is that this work utilizes discretization technique which provides better accuracy values than most of the above-mentioned works. Another key difference is that while some of the works predict exact prices of the commodity in question, the proposed work predicts whether the gold price will increase or decrease from the price on the previous day.

3   Procedure for Predicting Gold Price Direction

The present approach for predicting the direction of gold price can be described in the following steps.

(a) **Feature definition**: This is the very first step, whereby the necessary features for the work are identified and collected. Potoski [1] uses gold price-only features as well as inter-market features.
(b) **Feature selection**: Pearson's correlation coefficient is a major tool used for selecting features.
(c) **Discretization**: Once feature selection is completed, then the procedure of discretization is applied upon selected features. We describe this procedure in detail later.
(d) **Application of machine learning techniques**: The current work uses machine learning techniques such as logistic regression, SVM, random forest, and neural network for the purpose.
(e) **Performance measurement**: The performance of the classification is measured using accuracy measured as a percentage, ROC curves, and AUC.

3.1   Feature Definition

Potoski [1] represents the daily gold price change as a classification issue in terms of whether the price will increase or decrease for the next day. She uses feature definition procedures by using technical indicators such as trend lines, rate of change, ratios between rates of change, and stochastic oscillator. Inter-market variables such as stock market indices as well as various currency exchange rates are taken into consideration. Then a feature selection method is applied for selecting the most appropriate features. After this, the machine learning techniques are applied. Potoski mentions that logistic regression performs better than SVM as SVM has definite bias toward those outcomes that are majority in number in the training set.

3.2 Feature Selection

The Pearson's correlation coefficient for measuring the correlation of the continuous features with respect to the gold future price is being used for the purpose of feature selection. An R function "*cor.test*" has been utilized to help in this regard. The Pearson's coefficient rule states that the range of correlation always stays between the range [−1, 1]. Thus, Pearson's coefficient assumes that the higher the correlation coefficient value gets toward 1, the two variables on which the Pearson's coefficient was generated become more strongly positive correlated. On the other side, if that Pearson's coefficient value turns toward −1, the variables will become strong negatively correlated. We refer correlation coefficient value as CCV. The general assumption behind selecting the 0.5 value threshold is that the highly correlated values will allow greater accuracy. So we follow this rule as if CCV 0.5, then only we select the feature else we reject it. Similarly for negative correlation, we apply the similar principle, when CCV is less than −0.5. The correlation coefficient values as depicted in Table 1 have been constructed with the help of "*cor.test*" function from the R-programming language. Based on this ranking, we consider the variables silver future price and copper future price as these variables possess the highest correlation coefficient in the matter. Silver is given priority in terms of feature selection since silver has got the highest CCV amongst all these variables that we have considered. We also come to observe that other features like US government 10-year bond price are having strong negative correlation with the gold price. Crude oil future price, copper future price, US Dollar–Indian rupee exchange values also have positive correlation with the gold price, whereas some of the primary stock exchanges such as DAX future price, Hang Seng future price, and Nikkei are negatively correlated with the gold price.

Table 1 Correlation coefficient values for various features

Feature	Correlation value
Silver future price	0.84
Copper future price	0.65
Crude oil future price	0.54
Nasdaq composite price	0.14
FTSE 100 price	−0.02
Hang Seng price	−0.24
Nikkei 225 price	−0.47
KOSPI 200 future price	0.41
DAX future price	−0.04

3.3 Discretization Algorithm

Discretization is the technique that takes as input continuous values from a feature and converts these into discrete values. In this work, we utilize this technique and in our knowledge we have not come across any other work in this domain using the same technique for improvement of accuracy. The discretization algorithm transforms continuous values of any feature into discrete values by performing the following steps.

Step 1. Validate whether $P(X) > P(X - 1)$. If so then the value of discretized variable will be 1.

Step 2. Else the value of discretized variable will be 0. Here $P(X)$ represents the price of a commodity on a given day X, whereas $P(X - 1)$ represents the price of the same commodity on the previous day. So if the price of the commodity on a given day is lesser than the previous day, then the discretized variable will hold 0 and on the other days where the price of the commodity on the previous day is lesser than the price of the commodity on the current day it will hold the value 1 (Table 2).

3.4 Applying Machine Learning Techniques

In this work, four different types of machine learning algorithms were utilized. These are (a) logistic regression, (b) support vector machine, (c) random forest, and (d) neural network.

3.4.1 The Logistic Regression Algorithm

The logistic regression function equation is implemented in the following way:

$$p(g(p)) = (exponential(a + bx))/(1 + exponential(a + bx)) \quad (1)$$

where $p(g(p)) = 1$ indicates the probability of increase in the gold price from that of the previous day and x represents independent variables such as silver price, copper

Table 2 Impact of discretization of a value on continuous feature

Pre-discretization continuous value	Post-discretization discrete value
233.434	1
123.43	0
324.123	1
543.234	1

price and *a, b* represent intercept and slope coefficients for these features, respectively.

The above formula for logistic regression has been constructed in such a way that the value $p(g(p))$ will always result in a value between 0 and 1. Once we derive the values, a threshold (if the value $\geq$ 0.5) is applied. If we observe that the value is ≥ 0.5 then, we replace that value with 1, else we replace that value with 0.

The logistic regression has been implemented by using the R-package *"glm."*

3.4.2 The Support Vector Machine Algorithm

The support vector machine algorithm creates a linear discriminant function with the maximum margin to differentiate classes of data elements. The margin is defined as the width that the boundary could be increased by, before hitting a data point.

(a) In our work, we use *Python Canopy* software and *"scikit"* package for implementing support vector machine.
(b) We use the *"rbf"* kernel and the variables *c* and *gamma* whose values we use as a range from 0.1 to 1000.
(c) The technique provides the best accuracy when we use the c and gamma value as 10. From 10 to 1000, the accuracy remains static.

3.4.3 The Random Forest Algorithm

We explore the R-programming language library *"RandomForest"* for our present work. The algorithm for the random forest works in the following way—(a) *Bootstrapping*: A subset of the whole feature space is taken in a random manner. This subset of features is called bootstrapping. (b) *Tree formation*: For each bootstrap sample, a regression tree is formed. (c) *Aggregation*: The trees that were obtained in the previous tree formation process are now selected and the best among them are aggregated together to form the random forest.

3.4.4 The Neural Network

The current work utilizes *"avNNet"* function from the *"caret"* package in the R framework for using neural network. This function treats the independent features and applies some weight values with it. Then it estimates these neural networks and selects the one with the best average. The neural networks are very useful when nonlinear relationship is present between the dependent and independent variables. The independent variables are used as inputs by the neural networks algorithms. Beneath the input levels, there reside certain hidden layers that employ weights on these variables. The output layer is the one which generates output in the form of the dependent variable. The output is derived after integrating all previous hidden and input layers outputs. In the case of the "*avNNet*" algorithm, a number of neural networks are generated by using the above process and then the average of these networks are derived for the optimum performance. The parameters of the "*avN-Net*" are controllable by the users so that they can manipulate the number of hidden layers for operating with the input values. The input layers here are $n + 1$. Here n represents the total number of features. The total number of hidden layers is variable depending upon how the users want to operate the same. The output layer here is 1 as this is the case of a regression.

3.5 Performance Measurement

In this phase, the performance of the models is validated using techniques such as accuracy measurement, ROC, and AUC.

3.5.1 Accuracy Measurement

The accuracy measurement is calculated by taking a percentage of the grand total of all the true predicted high values as well as all the true predicted low values from the set of all high values and all low values. The equation for the accuracy measurement has been described here. This work describes a high value as $P(x) > P(x_1)$ whereby $P(x)$ is the value of gold price on a given day and $P(x_1)$ denotes the value of gold price for the previous day. The set of all the high values can be obtained by using the discretization algorithm on the set of gold prices and is denoted by 1. The set of all the low values can be obtained in the similar manner and is denoted by 0. The low value is described as $P(x) < P(x_1)$. Accuracy can be described by the following formula.

If we consider $h = t_h$, $l = t_l$, $fh = f_h$, and $fl = f_l$ we can deduce the following equations:

$$d = ((h+l)*100) \tag{2}$$

$$b = (h+l+f_h+f_l) \tag{3}$$

$$A = d = b \tag{4}$$

The variable A does represent accuracy.

We can denote the notations used in the above formula as:

1. Here we can represent t_h as all correctly identified high values
2. Here we can represent t_l as all correctly identified low values. Here we can represent f_h as all incorrectly identified high values

3. Here we can represent f_l as all incorrectly identified low values

The above equation can be further simplified as

$$Accuracy = ((t_{hl}) * 100)/(t_{hl} + f_{hl}) \tag{5}$$

We can denote the notations used in the above formula as:

1. Here we can represent t_{hl} as all correctly predicted values
2. Here we can represent $(t_{hl} + f_{hl})$ as the set of all values.

3.5.2 Receiving Operating Characteristic

Receiving operating characteristic or ROC is a very well-known technique for validating the performance potential of any binary predictor. The procedure was first adopted by the British military department during the Second World War when they were working with the radars for identifying correctly a plane or a bird. ROC curve is a plotted graph between the true positives and false positives as identified by the binary classifier.

In the case of the current work, the true positives are the days correctly identified by the predictor when the gold price increased from the close on the previous day and these are indicated by 1. The false positives are those days which the predictor incorrectly predicts as 0 where gold prices ending lower than the previous day and these are incorrect since these are the days when the gold price increased from the close on the previous day.

3.5.3 Area Under the Curve

Area under the curve is measured by the integration of the differences between all distinct points on the ROC. The AUC is a value which usually exists between 0 and 1. The more the AUC value, the better is the performance of the mathematical model on whose performance the ROC has been plotted. Higher AUC values indicate better model performance.

4 Experimental Results

The work applies logistic regression, SVM, random forest, and neural network techniques, respectively, over both non-discretized as well as discretized features. The accuracy measurements are also noted down. The experimental reuslts for accuracy measurement are tabulated in Table 3.

Table 3 Impact of discretization of a value on continuous feature

Algorithm	Feature engineering	Accuracy (in %)
Logistic regression	W/o discretization	51.32
SVM	W/o discretization	50.15
Random forest	W/o discretization	50.57
Neural network	W/o discretization	53.37
Logistic regression	With discretization	77.32
SVM	With discretization	80.15
Random forest	With discretization	80.57
Neural network	With discretization	80.37

The discretization procedure is applied over the same features and then previously utilized machine learning techniques such as logistic regression, SVM, random forest, and neural network are again applied over the same features. So from Table 3, we can observe that irrespective of machine learning technique, it is possible to achieve accuracy improvement while using discretization.

4.1 Performance Comparison Using AUC

In an approach similar to the process described in the above, AUC is utilized as a performance measurement criterion instead of accuracy percentage. It can be deduced from Table 4 that discretization leads to important improvement while using the logistic regression, random forest, neural network as well as the SVM algorithm.

The usage of the receiver operating characteristics procedure as a performance measurement criterion clearly depicts the amount of performance improvement while using discretization. Figure 1 shows that without discretization the ROC plotted line moves alongside the middle of the graph, thereby the performance of

Table 4 Comparison between various approaches using AUC

Algorithm	Feature engineering	AUC
Logistic regression	W/o discretization	0.5132
SVM	W/o discretization	0.5232
Random forest	W/o discretization	0.5112
Neural network	W/o discretization	0.5362
Logistic regression	With discretization	0.8132
SVM	With discretization	0.7832
Random forest	With discretization	0.8032
Neural network	With discretization	0.8133

Fig. 1 ROC curve for
logistic regression without
implementing discretization

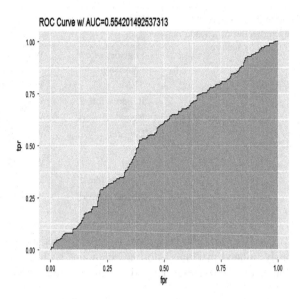

Fig. 2 ROC curve for
logistic regression after
implementing discretization

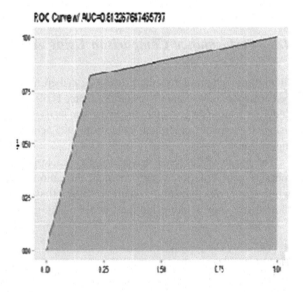

the model can be termed as average. The work utilizes "*ggplot2*" package in R
language for this purpose. From Figs. 1 and 2, we can deduce that after discretization the receiver operating characteristics curve plotted line covers a much
larger area; therefore, it is indicative of the improvement of performance using ROC
as a measurement criterion.

Table 5 Comparison between existing and proposed methods

Machine learning technique	Accuracy for existing (%)	Accuracy for proposed (%)
Logistic regression	69.3	81.44
SVM	69.1	80.24

4.2 Performance Comparison with Previous Works

Table 5 illustrates that in comparison with the work of Potoski [1] how the adoption of the discretization procedure has improved the accuracy of the work. From the table, we can clearly observe that the usage of discretization does improve the accuracy of the prediction compared to what the existing work achieved.

5 Conclusion

We conclude from experimental results that feature discretization is something that produces more accuracy if used alongside machine learning algorithms. After using discretization, we can observe a significant improvement compared to what was achieved in the previous work while applying various performance parameters such as accuracy percentage measurement, ROC, and AUC. From this work, it can be deduced that silver future price is the most significant feature with which we can work to improve accuracy of our mathematical models, whereas adding other features to silver does not necessarily improve the accuracy performance of the model.

References

1. Potoski, M.: Predicting gold prices. Stanford University 2013 publication. (2013) 1(1), 2–4.
2. Sjaastad, L.A.: The price of gold and the exchange rates: Once again. Resources Policy. (2008) 33(2), 118–124.
3. Ciner, C.: A note on the long run relationship between gold and silver prices. Global Finance Journal. (2000) 12(2), 299–303.
4. Grudnitski, G., Osburn, L.: Forecasting S and P and Gold Futures Prices: An Application of Neural Networks. Journal of Futures Markets. (1993) 13(6), 631–643.
5. Li, B.: Research on WNN modeling for gold price forecasting based on improved artificial bee colony algorithm. Computational Intelligence and Neuroscience. 2014 (1), 2–10.
6. Shen, S., Jiang, H., Zhang, T.: Stock market forecasting using machine-learning algorithms. Stanford: Stanford University. (2012) 1(1), 26–32.
7. Shah, V.: Machine learning techniques on stock prediction. Foundations of Machine Learning| Spring. (2007) 1(1), 6–12.
8. Agarwal, J.G., Chourasia, V.S., Mitra, A.K.: State-of-the-art in stock prediction techniques. Electronics and Instrumentation Engineering. (2013) 2(4), 3–5.

9. Bilberry, J.K., Riley, N.F., Sams, C.L.: Short-term prediction of exchange traded funds (ETFs) using logistic regression generated client risk profiles. Journal of Finance and Accountancy. (2015) 3(5), 5–6.
10. Zaidi, M., Amanat, A.: Forecasting stock market trends by logistic regression and neural networks. International Journal of Economics, Commerce and Management. (2016) 4(6), 4–7.
11. Hargreaves, C., Hao, Y.: Prediction of stock performance using analytical techniques. Journal of Emerging Technologies In Web Intelligence. (2000) 5(2), 136–143.
12. Khaidem, L., Saha, S. Dey, S.R.: Predicting the direction of stock market prices using random forest. Applied Mathematical Finance. (2016) 1(5), 1–20.
13. Imandoust, S.B., Bolandraftar, M.: Forecasting the direction of stock market index movement using three data mining techniques: the case of Tehran stock exchange. International. Journal of Engineering Research and Applications. (2014) 4(6), 106–117.

Language Discrimination from Speech Signal Using Perceptual and Physical Features

Ghazaala Yasmin, Ishani DasGupta and Asit K. Das

Abstract Humans are the most authoritative language identifiers in the province of speech recognition. They can determine within a glimpse of second whether the language of the hearing speech is known to them or not. This eventuality has come true because of the basic discrimination in the sound pattern characteristics based on frequency, time and perceptual domain. This certitude has provoked the motivation to introduce a potent scheme for the proposed plan. The proposed work includes the identification of three well-spoken languages in India that is English, Hindi and Bengali. The scheme has been encountered using some well known perceptual feature such as pitch along with some physical features like zero-crossing rate (ZCR) of the audio signal. To generate the feature set more efficient, the proposed effort has adopted mel-frequency cepstral coefficients (MFCCs) and the statistical textural features by calculating co-occurrence matrix from MFCC.

Keywords Language identification system (LID) · Zero-crossing rate (ZCR) Mel-frequency cepstral coefficient (MFCC)

G. Yasmin (✉)
Department of Computer Science and Engineering, St Thomas' College of Engineering and Technology, Kolkata 700023, India
e-mail: me.ghazaalayasmin@gmail.com

I. DasGupta
Department of Electronics and Communication Engineering, St Thomas' College of Engineering and Technology, Kolkata 700023, India
e-mail: ishanidasgupta.96@gmail.com

A. K. Das
Department of Computer Science and Technology, Indian Institute of Engineering Science and Technology, Shibpur, Howrah 711103, India
e-mail: akdas@cs.iiests.ac.in

© Springer Nature Singapore Pte Ltd. 2019   357
H. S. Behera et al. (eds.), *Computational Intelligence in Data Mining*,
Advances in Intelligent Systems and Computing 711,
https://doi.org/10.1007/978-981-10-8055-5_32

1 Introduction

Speech is determined as the most natural communication mode for human being where language acts as an interface between two speakers. Automatic language identification is the process of determining the language from a sample speech being uttered by an unknown speaker [1]. Numerous application domains are demanding for language identification services such as telephonic companies for handling foreign-language calls, radio production house including forensic department. This increasing need has made this problem more challenging. The main objective of this work is to propose new feature set for multilingual speech data and discriminate it with respect to the variability of the speakers and their languages. English and Hindi are the largest spoken language in India, and it also bears a third rank and fourth rank in world's largest spoken languages. Rather than any other languages in India, Bengali occupies a seventh rank in this list [2]. This is why Bengali has got prime attention to be introduced in the proposed work compared to other languages spoken in India.

1.1 Related Work

Researchers have gone through the survey related to language discrimination techniques and its applications [1, 3, 4]. Behravan et al. [5] have introduced an effective scheme of out-of-set i-vector selection for open-set language identification. Vatanen et al. [6] have identified different languages for short text segments with n-gram models. Torres-Carrasquillo et al. [7] have proposed an approach for language identification using Gaussian mixture model tokenization. Lakhani and Mahadev [8] have initiated language classification using deep neural network from the standard feature called mel-frequency cepstral coefficients (MFCCs). Dehak et al. [9] and Yu et al. [10] both have proposed the language recognition via i-vectors. Gwon et al. [11] have introduced a scheme for language recognition based on sparse coding. Chandrasekhar et al. [12] have introduced Gaussian mixture models and shifted delta cepstral features for the recognition of language. Itrat et al. [13] have discriminated Pakistani languages using MFCC and vector quantization.

2 Proposed Methodology

The three main components involved in speech-making are vocal cords, voice box (larynx) and the vocal tract. The sound that is produced is governed by the shape of the vocal tract and the vibration of the vocal cords. These characteristics can be properly reflected in physical features like zero-crossing rate and spectral shape

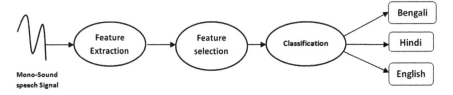

Fig. 1 Schematic block diagram for language identification (LID) system

features like MFCCs. The certitude has been inspired through the proposed work. The methodology involves a combination of these two features including some other well-known feature. Figure 1 reflects the overall architecture of the proposed system.

2.1 Feature Extraction

2.1.1 Zero-Crossing Rate (ZCR)

By the reference of discrete time signal, a zero-crossing rate is defined as the number of zero crossing or the rate of sign changes along a signal. Greater number of zero crossing shows the signal is changing rapidly resulting in high-frequency information. Thus, it can be concluded that ZCR shows indirect information about frequency content of the given signal. In the suggested methodology, the mean and standard deviation have been found out. Each speech is split into P frames $\{x_r(m): 1 \leq i \leq P\}$. After that, for rth frame, zero-crossing rates has been calculated as ZCR, say Z_r. This may be defined by Eq. (1).

$$z_r = \sum_{m=1}^{n-1} sign[x_r(m-1) * x_r(m)] \tag{1}$$

Here, n indicates the number of samples present in the rth frame and

$$sign[\mathbf{x}] = \begin{cases} \mathbf{1}, & \text{if } x > 0 \\ \mathbf{0}, & \text{otherwise} \end{cases} \tag{1.1}$$

The two parameters mean and standard deviation have been calculated. Those are summarized in Eqs. (1.a) and (1.b), respectively,

$$\text{Mean:} \quad \frac{Sum\ of\ ZCRs\ of\ all\ signals}{Total\ number\ of\ signals\ considered} \tag{1.a}$$

Standard Deviation:

$$\sigma = \sqrt{\frac{1}{M} \sum_{k=1}^{M} (z_k - m)} \tag{1.b}$$

where

M   no. of signals
m   Mean
z_k ZCR value of the kth signal.

2.1.2   Pitch

Pitch is termed as a measure of its degree of highness or lowness. Formally, the quality of any tone can be dictated by the rate of vibrations through which is it generated. The fundamental frequency of the glottal pulse is called pitch. A shrill speech with a frequency acquires higher pitch whereas vice versa for bass speech. This is one of the factors which changes with respect to different language. This is why this feature has been incorporated in this methodology. Here, the speech signal has been broken up into 88 frequency band. Now each band has got divided into short-time frame. From these frames, short-time mean-square power (STMSP) has been calculated followed by the averaging value of STMSP for each band (Fig. 2).

2.1.3   Mel-Frequency Cepstral Coefficients

MFCCs are derived from cepstral representation of audio clip. It is a nonlinear "spectrum-of-a-spectrum". Mel-frequency cepstral (MFC) differs from cepstrum because the frequency bands are equally spaced on the mel scale [12]; this approximates the human auditory system's response more closely compared to the linearly spaced frequency bands used in the normal cepstrum. This notion has been found to be efficient for the language classification because of two properties: one is to be proportional to the audio energy and other is that there is no correlation between different coefficients. In the suggested LID system, we have generated 13 coefficients of MFCC based feature where each coefficient has given a precise value for each of the three languages (Fig. 3).

The mel-frequency can be calculated as the following equation.

$f_n = 2595 * \log_{10}(1 + f)$; f is the frequency of the given speech signal.

Fig. 2 **a**, **b**, **c** ZCR plot for
English, Hindi and Bengali,
respectively

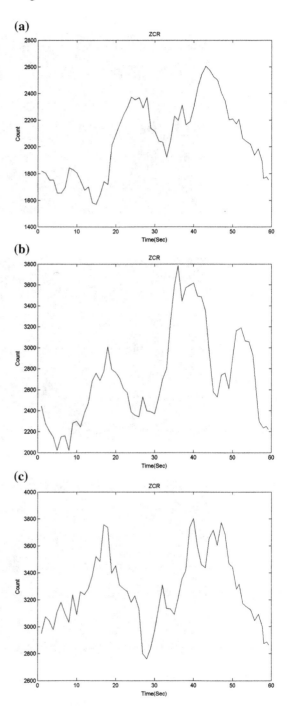

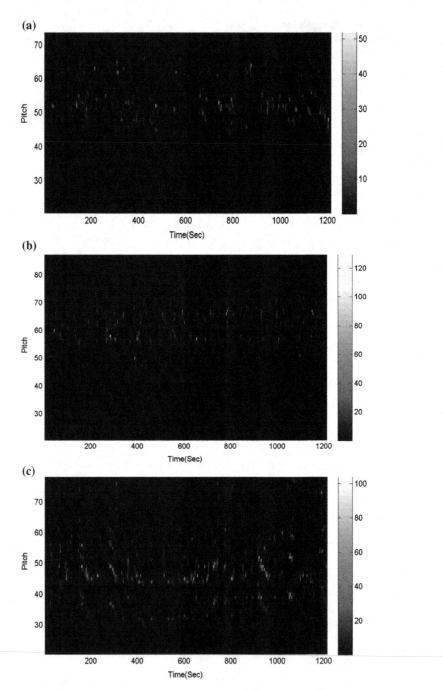

Fig. 3 **a**, **b**, **c** STMSP distribution plot for English, Hindi and Bengali, respectively

2.1.4 Co-occurrence Matrix of Mel-Frequency Cepstral Coefficients

This notion is generally used for finding the features related to image. The co-occurrence pattern shows the fine changes in the features related to audio as well. Henceforth, this variant is found to be suitable to incorporate in the proposed LID scheme.

Figure 4a–c shows the surface plot for co-occurrence matrix generated from 13 coefficients of the MFCC of each of the English, Hindi and Bengali. Co-occurrence matrix is defined as the co-occurrence distribution of a value with respect to its neighbouring value. This checks how frequently a pair of value occurs in an array or a matrix. The checking is to be done with respect to four angles: $0°$, $90°$, $45°$, and $135°$. Since, MFCC is having 13 coefficients. It is a one-dimensional array. Thus, the comparison is supposed to be one only with respect to $0°$ angle.

2.1.5 Textural Features

To make the system more efficient, some standard statistical features have been calculated from the co-occurrence matrix generated from 13 MFCC's coefficients. The five textural features are mathematically represented by Eqs. (2–5)

(i)
$$Entropy = - \sum_i \sum_j M_{CO}[i][j] log_2 M_{CO}[i][j] \tag{2}$$

(ii)
$$Energy = \sum_i \sum_j [M_{CO}[i][j]]^2 \tag{3}$$

(iii)
$$Inverse\, Difference = \sum_i \sum_j M_{CO}[i][j]/|i-j| \quad where\, i \neq j \tag{4}$$

(iv)
$$Correlation = (1/\sigma_x \sigma_y) \sum_i \sum_j (i-\mu_x)(j-\mu_y) M_{CO}[i][j] \tag{5}$$

where

$M_{CO}[i][j]$ Value of co-occurrence matrix at position [i][j]

μ_x $\sum_i i \sum_j M_{CO}[i][j]$

μ_y $\sum_i j \sum_j M_{CO}[i][j]$

σ_x^2 $\sum_i (1-\mu_x)^2 \sum_j M_{CO}[i][j]$

σ_y $\sum (1-\mu_y)^2 \sum M_{CO}[i][j].$

Fig. 4 **a**, **b**, **c** Surface plot for
of MFCC co-occurrence
English, Hindi and Bengali,
respectively

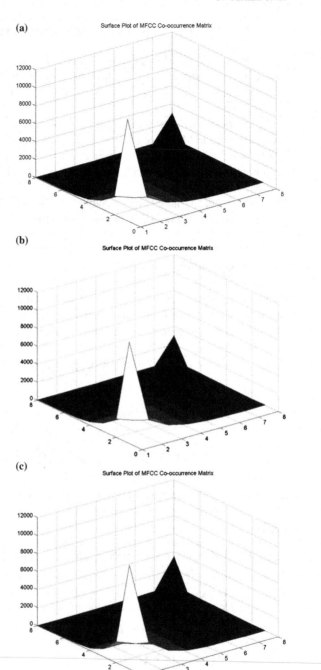

2.2 Feature Selection

The feature selection has been done by reducing the dimension by incorporating a statistical tool called principal component analysis (PCA). PCA has been performed by calculating covariance matrix obtained by column vector holding features of the speech file. Here, 108 features got reduced to the final set of feature vector of dimension 50 and 35, respectively.

2.3 Classification

Classification is termed as the procedure of the identification of the given observation into a set of categories. This terminology falls into two types of learning process; one is supervised learning where the set of class for identification is given. Another is unsupervised learning also termed as clustering which groups the data based on some inherent similarity. The classification algorithm is known as classifier. Classifiers like Naïve Bayes, multilayer perceptron have been taken into account. Furthermore, random forest has been chosen as it has been noticed that this classifier gives the better result among all the decision tree algorithms for the proposed system. For each classifier, ten-fold cross-validation has been done.

50% of the whole data has been chosen for training set; rest of the 50% has been chosen for test set. The classification has been taken into account with these sets. Furthermore, another set classification has been done by reversing these sets. Henceforth, the average result of the set of classification has been being acquired.

3   Experimental Result

The verification for the proposed experiment has been conducted for wide range of heterogeneity in aural data set. These data sets have been made up with 600 speech signals acquiring 200 each of English, Hindi and Bengali, each file having duration of 80 s. These data sets have been gathered from live recording of speeches spoken by political leaders and parliament speakers. Moreover, recorded voice of radio jockey's in radio production house has been taken into consideration. Some files have been downloaded from the Internet. To determine the feature more precisely, each speech signal has been broken into multiple frames consisting of 8 s each. To avoid miss of information, 50% of these frames are overlapped with the previous frame.

The proposed system has been preceded with the two sets of feature vector having dimension 50 and 35. Among these two sets, the feature vector with dimension 50 has been come up with better accuracy compared to the other. Table 1 summarizes the result of classification.

Table 1 Classification accuracy (in %) for proposed work

Classification scheme	English (%)	Hindi (%)	Bengali (%)
Naïve Bayes	98	96.3	97.1
Neural network	91.2	88.1	87.8
Random forest	89.3	84.1	83.4

Table 2 Comparative study (in %) for the substantial work

Precedent approach	English (%)	Hindi (%)	Bengali (%)
Chandrasekhar, Sargin and Ross	89	87	85.6
Itrat, Ali, Asif, Khanzada and Rathi	87.7	88.2	86.7

3.1 Comparative Analysis

The methodologies of the past work have been chosen for the comparison with the proposed LID system. Chandrasekhar et al. [12] and Itrat et al. [13] suggested a methodology that has been implemented and applied on the collected data set of the proposed methodology. As they have imposed the concept of MFCC, the proposed work has been came up with better accuracy by using the common concept of MFCC including some additional effective features. Table 2 reflects the average accuracy gained after applying the mentioned classifiers. The classification is performed by using all the classifiers as mentioned above, and the result is tabulated by doing the average of all those classifiers.

4 Conclusion

The presented work has been taken a firm attempt to rejuvenate the approach for language identification. The system has been come up with a feature after breaking pitch. However, it is informative to note that the proposed system can be explored with the advent of multilingual corpus for the seven major classes based on seven continents such as Asian languages, African languages, etc. These languages can be contrasted with the time, frequency and perceptual domain. Moreover, some new feature apart from MFCC will be aiming to introduce, As of the fact that MFCC is now become one of the common feature in this area of research. We are expecting that the revived LID system would come up with a promising result.

References

1. Garg, Archana, Vishal Gupta, and Manish Jindal. "A survey of language identification techniques and applications." *Journal of Emerging Technologies in Web Intelligence* 6.4 (2014): 388–400.
2. https://www.babbel.com/en/magazine/the-10-most-spoken-languages-in-the-world.
3. Karpagavalli, S., and E. Chandra. "A Review on Automatic Speech Recognition Architecture and Approaches." *International Journal of Signal Processing, Image Processing and Pattern Recognition* 9.4 (2016): 393–404.
4. Grothe, Lena, Ernesto William De Luca, and Andreas Nürnberger. "A Comparative Study on Language Identification Methods." *LREC*. 2008.
5. Behravan, Hamid, Tomi Kinnunen, and Ville Hautamäki. "Out-of-Set i-Vector Selection for Open-set Language Identification." *Odyssey 2016* (2016): 303–310.
6. Vatanen, Tommi, Jaakko J. Väyrynen, and Sami Virpioja. "Language Identification of Short Text Segments with N-gram Models." *LREC*. 2010.
7. Torres-Carrasquillo, Pedro A., Douglas A. Reynolds, and John R. Deller. "Language identification using Gaussian mixture model tokenization." *Acoustics, Speech, and Signal Processing (ICASSP), 2002 IEEE International Conference on.* Vol. 1. IEEE, 2002.
8. Lakhani, Vrishabh Ajay, and Rohan Mahadev. "Multi-Language Identification Using Convolutional Recurrent Neural Network." *arXiv preprint* arXiv:1611.04010(2016).
9. Dehak, Najim, et al. "Language Recognition via i-vectors and Dimensionality Reduction." *Interspeech*. 2011.
10. Yu, Chengzhu, et al. "UTD-CRSS system for the NIST 2015 language recognition i-vector machine learning challenge." *Acoustics, Speech and Signal Processing (ICASSP), 2016 IEEE International Conference on.* IEEE, 2016.
11. Gwon, Youngjune L., et al. "Language recognition via sparse coding." *INTERSPEECH*. 2016.
12. Chandrasekhar, Vijay, Mehmet Emre Sargin, and David A. Ross. "Automatic language identification in music videos with low level audio and visual features." *Acoustics, Speech and Signal Processing (ICASSP), 2011 IEEE International Conference on.* IEEE, 2011.
13. Itrat, Madiha, et al. "Automatic Language Identification for Languages of Pakistan." *International Journal of Computer Science and Network Security (IJCSNS)* 17.2.

Constructing Fuzzy Type-I Decision Tree Using Fuzzy Type-II Ambiguity Measure from Fuzzy Type-II Datasets

Mohamed A. Elashiri, Ahmed T. Shawky and Abdulah
S. Almahayreh

Abstract One of the most tools of data mining techniques is decision trees for both learning and reasoning from the crisp dataset. In a case of fuzzy dataset, the fuzzy decision tree must be established to extracted fuzzy rules. The paper illustrates an approach to establish fuzzy type-I decision tree from fuzzy type-II dataset using the ambiguity measure in fuzzy type-II form.

Keywords Data mining · Fuzzy decision tree · Fuzzy type-II
Ambiguity measure

1 Introduction

One of the main fields which are used to extract implicit and patterns from the large database is the data mining. It is used in a wide variety of application and for varied purposes; decision tree is one of the data mining classification methods. Fuzzy set theory is popularization of traditional or crisp set theory to model both of vagueness and uncertainty through a continuous generalization of set characteristic functions, for example, the attribute values can be linguistic terms (i.e., fuzzy sets), such as an attribute "Temperature" may be given the values of "Hot," "Mild," and "Cloudy" [1–6].

M. A. Elashiri (✉)
Faculty of Computers and Information System, Computer Science Department,
Beni-Suef University, Beni Suef, Egypt
e-mail: melashiry@fcis.bsu.edu.eg

A. T. Shawky
Management Information System Department, El Madina High Institute of Administration
and Technology, Giza, Egypt
e-mail: ah_taisser@hotmail.com

A. S. Almahayreh
Computer Science Department, Community College, Hail University, Hail, Saudi Arabia
e-mail: a.almahaira@uoh.edu.sa

© Springer Nature Singapore Pte Ltd. 2019   369
H. S. Behera et al. (eds.), *Computational Intelligence in Data Mining*,
Advances in Intelligent Systems and Computing 711,
https://doi.org/10.1007/978-981-10-8055-5_33

The rest of the paper is structured as follows: the structure of fuzzy type-II dataset is presented in Sect. 2. Section 3 introduces comparison between fuzzy type-I and type-II, in Sect. 4, creating fuzzy type-I decision tree using proposed algorithm is discussed. Section 5 presents the experiments, and conclusions are displayed in Sect. 6.

2 Structure of Fuzzy Type-II Dataset

2.1 Fuzzification Process

a. Determination number of intervals on each dimension or attributes and interval locations, such as center and width of each interval.
b. Assign membership function for each interval such as left most intervals, the right-most interval, the internal intervals, and general case of triangular membership.
c. Class label assignment for each decision region [7, 8].

2.2 Fuzzy Type-II Dataset Forms

Consider a dataset S consisting of n attributes $S = \{A_1, ..., A_n\}$ to be chosen $\forall k$, $1 \leq k \leq n$. Each attribute A_k has m_k values of fuzzy subsets $A_k = \{A_{k,1}, ..., A_{k,m}\}$, and the sub-attribute $A_{k,m}$ takes w values of fuzzy subsets t, where $1 \leq t \leq w$, $A_{k,m} = \{A_{k,m,1}, A_{k,m,2}, ..., A_{k,m,t}, ..., A_{k,m,w}\}$, and fuzzy classification set has m attributes as $A_C = \{A_{C,1}, ... A_{C,b}, ..., A_{C,j}\}$. Forms of both fuzzy type-I and fuzzy type-II training datasets are shown in Tables 1 and 2.

3 Comparison Between Type-I and Type-II Fuzzy Set

3.1 Fuzzy Type-I

1. Extend crisp sets, where $x \in A$ or $x \notin A$.
2. Represent membership (degree of belong) by real numbers in range [0, 1].

Table 1 Fuzzy type-I training datasets

Cases	A_1			A_k		A_c		
	$A_{1,1}$	...	$A_{1,i}$	$A_{k,1}$	$A_{k,i}$	$A_{c,1}$	$A_{c,2}$	$A_{c,j}$
x_1	$\mu_{A_{1,1}}(x_1)$					$\mu_{A_{C,1}}(x_1)$		
								$\mu_{A_{C,j}}(x_2)$
x_N					$\mu_{A_{k,i}}(x_N)$			

Table 2 Fuzzy type-II training dataset

Cases	A_1				A_k						A_c		
	A_{1,1}	…	A_{1,i}		A_{k,1}		…	A_{k,i}			A_{c,1}	A_{c,2}	A_{c,j}
	$A_{1,1,1}$	…	$A_{1,i,1}$	…	$A_{k,1,1}$	$A_{k,1,t}$	…	$A_{k,i,1}$	…	$A_{k,i,t}$			
x_1	$\mu_{A_{1,1,1}}(x_1)$										$\mu_{A_{c,1}}(x_1)$		
x_N								$\mu_{A_{k,i,1}}(x_N)$					$\mu_{A_{c,j}}(x_2)$

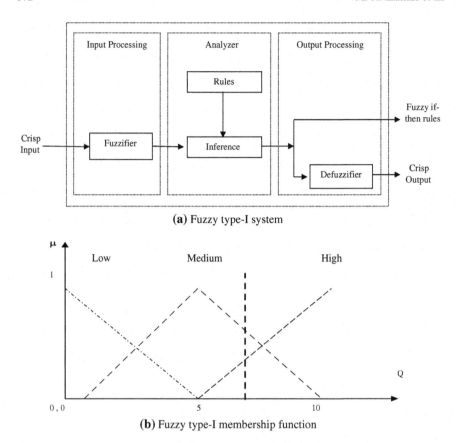

(a) Fuzzy type-I system

(b) Fuzzy type-I membership function

Fig. 1 **a** Fuzzy type-I system. **b** Fuzzy type-I membership function

3. Fuzzy type-I logic system contains three stages are: Input processing which fuzzifier the crisp input. Analyzer to inferences rules and last one is output processing which have two outputs fuzzy output or crisp output after defuzzifier, as shown in Fig. 1a [5, 9, 10].

Memberships function of fuzzy type-I
Fuzzy set type-I elements have membership grade as a crisp number in [0, 1]. It means fuzzy set type-I has two dimensions, as shown in Fig. 1b.

3.2 Fuzzy Type-II

1. It provides the capability of handling in cases, such as higher level of uncertainty, cannot determine the exact membership function for a fuzzy set and with the number of missing components.

2. It is an extension of type-I fuzzy sets in which uncertainty is represented by three dimensions; the third dimension gives more degrees of freedom for better representation of uncertainty compared to type-I fuzzy sets.

3. Fuzzy logic system of type-II contains the three stages are: input processing which fuzzifier two times the crisp input or fuzzifier one time the fuzzy input, analyzer to inferences rules and last one is output processing which have three outputs such as fuzzy if–then rule in type-II or fuzzy if–then rule in type-I after type reducer or crisp output after defuzzifier of type reducer outputs, as shown in Fig. 2a.

Memberships function of fuzzy type-II:
Fuzzy set type-II elements have membership grade that is a fuzzy set in [0, 1] but in three dimensions, as shown in Fig. 2b.

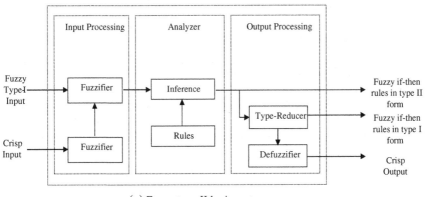

(**a**) Fuzzy type-II logic system

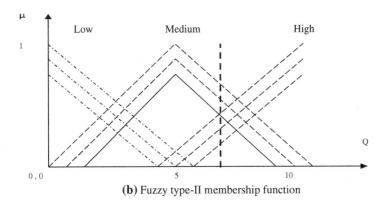

(**b**) Fuzzy type-II membership function

Fig. 2 **a** Fuzzy type-II logic system. **b** Fuzzy type-II membership function

4 Creating Fuzzy Type-I Decision Tree Using Proposed Algorithm

Next subsection presents steps for creating fuzzy type-I decision tree using proposed algorithm [6, 9, 11–13] and some issues such as selecting root node based on fuzzy type-II ambiguity measure, controlling criteria, and classification ambiguity of fuzzy partition.

4.1 Steps for Creating Fuzzy Type-I Decision Tree

Input: Dataset in Fuzzy Type-II form

Step 1: Measuring the classification ambiguity for each main attributes

Determine decision node root by selecting an attribute which has the smallest value of ambiguity.

Step 2.A: Drop all empty branches of the selecting attribute and measure the fuzzy classification ambiguity of its fuzzy sub attributes

Step 2.B: Calculate the tree threshold of classifying all objects within each sub attributes of the root attribute into each classes

Step 2.C: Check for each sub attributes of the root attribute

If the tree threshold of classifying into one class is greater than given threshold $j_{threshold}$ then terminate this sub attribute as a leaf.

Else If another attribute with the smallest value of classification ambiguity among recent attributes will further partition the branch and decrease the value of classification ambiguity measure, it will be a new decision node for each sub attributes of the root attribute.

Else close this branch as a leaf and Label it to a class with the highest Truth rule degree.

End If

End If

Step 3: Repeating the steps 1 and 2 until generating decision nodes and extracting fuzzy classification rules from the fuzzy decision tree.

Output: Group of fuzzy classification rules extracted from the generated fuzzy decision tree T.

4.2 Selecting Root Node Based on Fuzzy Type-II Ambiguity Measure

Truth rule degree
Truth rule degree $S(A_{k,i,t}, A_{C,j})$ measures the degree to which $A_{k,i,t}$ is a subset of $A_{C,j}$ in rule "IF $A_{k,i}$, THEN $A_{C,j}$," is defined as:

$$S(A_{k,i,t}, A_{C,j}) = \frac{M(\mu_{A_{k,i,t}}(u) \cap \mu_{A_{C,j}}(u))}{M(\mu_{A_{k,i,t}}(u))} \qquad (1)$$

Sub-possibility distribution of an attribute $A_{k,i,t}$
Normalized possibility distribution of sub-attribute $A_{k,i,t}$ on subclasses $A_C = \{A_{C,1}, A_{C,2}, ..., A_{C,j}\}$ is given as $\Pi(A_C|A_{k,i,t}) = \{\Pi(A_{C,1}|A_{k,i,t}), \Pi(A_{C,2}|A_{k,i,t}), ..., \Pi(A_{C,j}|A_{k,i,t})\}$ and defined as:

$$\prod(A_{C,j}|A_{k,i,t}) = \frac{S(A_{k,i,t}, A_{C,j})}{MAX_j(S(A_{k,i,t}, A_{C,j}))} \qquad (2)$$

Normalized of possibility distribution of an attribute $A_{k,i,t}$
Normalized of possibility distribution Π_i^* is sorted from $\Pi(A_{j,e}^C|A_{i,t}^k)$ so that $\Pi_i^* \geq \Pi_{i+1}^*$ and $\Pi_{n_c+1}^* = 0 \ \forall 1 \leq i \leq n_c, n_c = m_c$.

Possibility of ambiguity measure normalized
The possibility measure of ambiguity of sub-attribute $A_{k,i,t}$ on subclass $A_{C,j}$ is defined as:

$$G(A_{k,i,t}, A_{ce}) = g\left(\prod(A_{ce,j}|A_{k,i,t})\right) = \sum_{s=1}^{nce}\left(\Pi_s^* - \Pi_{s+1}^*\right)\ln s \qquad (3)$$

Measuring the fuzzy classification ambiguity of fuzzy sub-attribute $A_{k,i,t}$
$G(A_{k,i,t})$ is fuzzy classification ambiguity of fuzzy sub-attribute $A_{k,i,t}$ which is computed as follows:

$$G(A_{k,i,t}) = g\left(\prod(G(A_{k,i,t}, A_{ce}))\right) = \sum_{s=1}^{nc}\left(\Pi_s^* - \Pi_{s+1}^*\right)\ln s \quad \forall 1 \leq e \leq j \qquad (4)$$

Measuring the fuzzy classification ambiguity of fuzzy sub-attribute $A_{k,i}$
$G(A_{k,i})$ is fuzzy classification ambiguity of fuzzy sub-attribute $A_{k,i}$ which is computed as follows:

$$G(A_{k,i}) = \sum_{t=1}^{h_t}\left(\frac{M(A_{k,i,1} \cap \cdots \cap A_{k,i,t} \cdots \cap A_{k,i,h_t})}{\sum_{t=1}^{h_t} M(A_{k,i,1} \cap \cdots \cap A_{k,i,t} \cdots \cap A_{k,i,h_t})}\right) G(A_{k,i,t}) \qquad (5)$$

Measure of fuzzy classification ambiguity G(A$_k$)

The fuzzy classification ambiguity with fuzzy attribute A$_k$ is:

$$G(A_k) = \sum_{i=1}^{m_k} \left(\frac{M(A_{k,1} \cap \cdots \cap A_{k,i} \cdots \cap A_{k,m_k})}{\sum_{t=1}^{m_k} M(A_{k,1} \cap \cdots \cap A_{k,t} \cdots \cap A_{k,m_k})} \right) G(A_{k,i}) \qquad (6)$$

For example, if an attribute $A_1 = \{A_{1,1}, A_{1,2}, \ldots, A_{1,i}\}$, and one of its sub-attributes is $A_{1,1} = \{A_{1,1,1}, \ldots, A_{1,1,t}\}$, then class attribute A$_c$ with three sub-classes $A_c = \{A_{c,1}, A_{c,2}, A_{c,3}\}$.

4.3 Controlling Criteria

Tree threshold $\beta_{threshold}$ controls of tree size, increasing the value of $\beta_{threshold}$ leads to a larger tree, from Eq. 7, computing degree of truth of the rule of branch any branch A$_{k,i}$. If truth level of any subclass $\beta(A_{k,i}, A_{c,j}) \geq \beta_{threshold,}$ then the branch A$_{k,i}$ terminates to be a leaf and class A$_{c,j}$ is selected as the label for it. Otherwise, if truth level of any subclass $\beta\ (A_{k,i}, A_{c,j}) \leq \beta_{threshold,}$ then of then needed to partition the branch A$_{k,i}$

$$\beta(A_{k,i}, A_{C,j}) = \frac{M(\mu_{A_{k,i}}(u) \cap \mu_{A_{C,j}}(u))}{M(\mu_{A_{k,i}}(u))} \qquad (7)$$

4.4 Tree Extension and Generating New Decision Node

This section introduces ambiguity measure of fuzzy partition G(Ak|F) for generating new node in fuzzy decision tree

G(A$_k$|F) is classification ambiguity of fuzzy partition A$_k$ on fuzzy evidence F by computing the weight average of classification ambiguity with each subset of partition where attributes and sub-attributes of F do not belong to A$_k$ is defined as:

$$G(A_k|F) = \sum_{i=1}^{m_k} \left(w(A_{k,i}|F) \cdot G(A_{k,i} \cap F) \right) \qquad (8)$$

F denotes to set of A$_{k,i,t}$, where G(A$_{k,i}$ ∩ F) is the classification ambiguity with fuzzy evidence A$_{k,i}$ ∩ F, and w(A$_{k,i}$|F) is the weights which represents the relative size of subset $A_i^k \cap F$ in F.

Ambiguity with fuzzy evidence G(A$_{k,i}$ ∩ F)

G(A$_{k,i}$ ∩ F) is the classification ambiguity with fuzzy evidence A$_{k,i}$ ∩ F is defined as:

Table 3 Dataset of iris

Method	Fuzzy type-I (fuzzy ID3)	Fuzzy type-I (Ambiguity Yuan's method)	Fuzzy type-II (proposed)
No. of rules	9	8	9
Training accuracy	91.2	91.9	91
Testing accuracy	96	95.2	95

$$G(A_{k,i} \cap F) = \sum_{t=1}^{n_i^k} \left(w(A_{k,i,t}|F) \cdot G(A_{k,i,t} \cap F) \right) \tag{9}$$

Weight of fuzzy evidence $w(A_{k,i}|F)$

$w(A_{k,i}|F)$ is the weights of fuzzy evidence which represents the relative size of subset $A_i^k \cap F$ in F is defined as:

$$w(A_{k,i}|F) = \sum_{t=1}^{n_i^k} M(A_{k,i,t} \cap F) / \sum_{t=1}^{n_i^k} \left(M(A_{k,i,t} \cap F) \right) \tag{10}$$

5 Experiments

In the process of experiment, the data is divided into two parts for experimental convenience. Our experiment applying on data of the iris flower dataset from UCL "Center for Machine Learning and Intelligent Systems" after applying fuzzification process two times to generate fuzzy type-II dataset [14, 15]. Iris data has four features, and each feature is divided into two parts. It has 4-dimensional data of each flower is expressed as the fuzzy type-II dataset to be 12-dimensional. Table 3 summarizes the number of rules, training, and testing accuracy for iris dataset in form of fuzzy type-I dataset using fuzzy Id3, fuzzy ambiguity. Also in our proposed applying over fuzzy type-II dataset using fuzzy ambiguity in type-II form.

6 Conclusion

This technique has been proposed to build fuzzy type-I decision tree from fuzzy type-II dataset, and the main target is to deal with huge complexity dataset with many dimensions to obtain same result using another fuzzy decision tree algorithm such as Fuzzy Id3 and Yuan's Ambiguity. The future work is how to deal with dataset from type-III or greater.

References

1. Yao, Y. Y.: A comparative study of fuzzy sets and rough sets. Information sciences 109. 227–242. (1998).
2. Buell, D. A.: A general model of query processing in information retrieval systems. Information Processing and Management. 17(5), 249–262 (1981).
3. Lee, M. C., Chang, T.: Rule extraction based on rough fuzzy sets in fuzzy information systems. Transactions on computational collective intelligence III. Springer Berlin Heidelberg. 115–127 (2011).
4. Cai Z., Shao, Y., Cao, Y., Dun, Y.: A New Method of Information System Processing Based on Combination of Rough Set Theory and Pansystems Methodology. In Emerging Research in Artificial Intelligence and Computational Intelligence. Springer Berlin Heidelberg. 225–233 (2012).
5. Zarandi, M. F., Gamasaee, R., Castillo, O.: Type-1 to Type-n Fuzzy Logic and Systems. In Fuzzy Logic in Its 50th Year. Springer International Publishing. 129–157 (2016).
6. Elashiri, M. A., Hefny, H. A., Elwhab, A. H.: Reduction Fuzzy Data Set based on Rough Accuracy Measure. In International Conference on Advances in Computer Science, AETACS. Elsevier (2013).
7. Sinha, Divyendu, and Edward R. Dougherty. "Fuzzification of set inclusion: theory and applications." Fuzzy sets and systems 55.1 (1993): 15–42.
8. Sinha, D., Dougherty, E. R.: Fuzzification of set inclusion: theory and applications. Fuzzy sets and systems. 55(1), 15–42 (1993).
9. Elashiri, M. A., Hefny H. A., Elwhab A. H.: Induction of fuzzy decision trees based on fuzzy rough set techniques. In Computer Engineering and Systems International Conference (ICCES) on IEEE (2011).
10. Agüero, J. R., Vargas, A.: Using type-2 fuzzy logic systems to infer the operative configuration of distribution networks. In Proceedings IEEE Power Engineering Society General Meeting. 2379–2386 (2005).
11. Rondeau, L., et al.: A defuzzification method respecting the fuzzification. Fuzzy sets and systems 86.3, 311–320 (1997).
12. Zhai, Jun-hai.: Fuzzy decision tree based on fuzzy-rough technique. Soft Computing 15.6 1087–1096 (2011).
13. Elashiri, M. A., Hefny H. A., Elwhab A. H.: Construct fuzzy decision trees based on roughness measures. In International Conference on Advances in Communication, Network, and Computing. Springer Berlin Heidelberg (2012).
14. Yuan, Y., Shaw, M.J.: Induction of fuzzy decision trees. Fuzzy Sets and Systems, 69(2), 125–139 (1995).
15. Center for Machine Learning and Intelligent Systems at the University of California, Irvine, https://archive.ics.uci.edu/ml/datasets/Iris.

Aspect-Level Sentiment Analysis on Hotel Reviews

Nibedita Panigrahi and T. Asha

Abstract Sentimental analysis is a part of natural language processing which extracts and analyzes the opinions, sentiments, and emotions from written language. In today's world, every organization always wants to know public and customer's feedback about their products and also about their services that gives very important for business or organization about their product in the market and their services to perform better. Aspect-level sentiment analysis is one of the techniques which find and aggregate sentiment on entities mentioned within documents or aspects of them. This paper converts unstructured data into structural data by using scrappy and selection tool in Python, then Natural Language Tool Kit (NLTK) is used to tokenize and part-of-speech tagging. Next the reviews are broken into single-line sentence and identify the lists of aspects of each sentence. Finally, we have analyzed different aspects along with its scores calculated from a sentiment score algorithm, which we have collected from the hotel Web sites.

Keywords Opinion analysis · Aspects mining · Machine learning
Natural language processing (NLP) · POS tagging

1 Introduction

Opinions are very important to all human activities. Sentiment analysis and opinion mining give the information about sentiments of opinions, emotions, and reactions. Since 2000, opinion analysis had become the most important research area in NLP. Sentiment analysis is mainly used in data mining. Due to importance in computer science field, sentiment analysis is widely used in management services,

N. Panigrahi (✉) · T. Asha
Department of Computer Science & Engineering, Bangalore Institute of Technology,
Bangaluru 560004, India
e-mail: nibedita.kuni@gmail.com

T. Asha
e-mail: asha.masthi@gmail.com

© Springer Nature Singapore Pte Ltd. 2019   379
H. S. Behera et al. (eds.), *Computational Intelligence in Data Mining*,
Advances in Intelligent Systems and Computing 711,
https://doi.org/10.1007/978-981-10-8055-5_34

social science also. NLP plays a vital role of user actions, and for this reason every users' decision is based on others opinions. The basic task of sentimental analysis is to find the difference of a given user data and text from a data set and give output as positive, negative, or neutral. The sentimental analysis types are document level, aspect level, and sentence level. The final output in document level is identifying whether a whole document gives positive, negative, or neutral opinion. Here each document gives opinions on a single entity, so this is not good for those types of document that contains more than one entity like hotel reviews. In sentence-level analysis, every sentence expresses a positive opinion, negative opinion, or neutral opinion. Positive opinion means the sentence will have some positive sense or similar feelings, negative opinion means the sentence will have some negative sense or similar feelings, and neutral opinion means the sentence does not have any sentiment. Here, the sentence that expresses factual information is found first, and it is known as subjective sentence and examines sentiment value for each sentence. This kind of analysis is better than the document level of opinion analysis. Both document and sentence levels could not give proper understanding, what the user is trying to tell. So far that another analysis that is aspect-level analysis where aspects inside the sentence will be identified just and then finding out the polarity is whether positive, negative, or neutral. The analysis of this kind gives clear result of sentiment score. For example, in the sentence "Hotel rooms are not good; wifi internet facility is good", here the opinion looks like positive but that is combination of positive and negative opinions. Here the analysis is positive for "hotel facilities" but negative for the "hotel rooms" which gives two different aspects, where the aspect gives negative polarity and the second aspect gives positive polarity. So, the main aim of aspects level is to discover sentiment on various aspects.

2 Related Work

The most recent two decades have seen change in the field of opinion mining or sentiment analysis. A couple of experimentation papers have also been published and issued showing new methods and original plans to perform sentiment analysis. Still there required many ideas for the field of corpus creation and data extraction. According to Kim et al. [1], the opinion on new movies can be analyzed in three phases: The first phase is to building the sentiment word list for analyzing opinions of the user, then organizing certain contractions and phrases for performing the process of opinion mining, and finally managing a new movie features, for example, the actors. According to D'Aranzo and Pilato [2], user opinion analysis is done from specific sort of business sectors. The analysis practices Vygotsky's zone of proximal development model and the model introduces Bayesian Learner and TF-IDF grounded chooser. The procedure has been useful on pages of Facebook mobile device and style marketplaces. Another author Monti et al. [3] analyzes about disaffection from political process. So here the creator accumulates a great number of Twitters from the Italian Twitter database and utilizes an adaptable machine learning method to deal with deliver a time series in regard to Italian political disillusionment. Denecke [4] presented sentiment analysis and multilingual sentiment analysis methodologies on the basis of SentiWordNet. The previous one demonstrates that opinion mining presents diverse difficulties, once connected to a multilingual setting. By and large lexical methodologies require language particular lexical and linguistic assets. Producing these assets is exceptionally tedious, and it regularly requires labor-intensive work. The later one depends on SentiWordNet.

Baccianella et al. [5] investigation depends on a lexical asset that partners three scores showing objectivity obj(s), positivity pos(s), and negativity Neg(s) to a gathering of subjective equivalent words called synset. Every synset coordinate set is comprised of things, verbs, and descriptive words, and each of these gatherings communicates an unmistakable idea. The approach speaks to an advancement of the lexical database WordNet. The scores that are ascribed to single synset are the aftereffect of a blend of the outcomes delivered by eight ternary classifiers, altogether portrayed via genuinely comparative accuracy stages, yet unique in relation to conduct arrangement. Each score related to every synset fluctuates in the vicinity of 0.0 and 1.0, and the whole of the three markers is constantly equivalent to union value. Artale et al. [6] analyze various disambiguates regarding the SentiWordNet which is an issue for the computational utilization of WORDNET. According to Pang and Lee [7], online review sites and personal blogs, new opportunities, and challenges can be classified using unsupervised lexicon approaches and other unsupervised approaches to search out and comprehend the sentiments of others. Generally, the aspect entity recognition techniques use machine learning and linguistic approach. In machine learning approach, a set of collection of data is used to perform automatic rule-based approach on new input data, and this approach does not require any predefined rules. This approach requires large collection of annotated corpus. The supervised learning technique and semi-supervised learning technique are the two techniques that are mainly used for machine learning process. In linguistic approach, predefined rules are used by the user, and input defines a pattern which contains scientific features and some rules that contain dictionary features. This approach is also known as knowledge- or rule-based approach.

3 Issues in Sentimental Analysis

The words which expresses positive or negative sense are called sentiment words or also known as opinion words. For example, good, awesome, amazing are the positive opinion words and bad, worst, poor are the negative opinion words. Apart from individual sentiment words, the phrases and idioms that also give positive opinion or negative sense are known as sentiment lexicon or opinion lexicon. Opinion lexicons play very important role for opinion analysis but is it not sufficient for opinion analysis because of the following issues.

1. A positive or negative sentiment word may have different meaning in sentences in dissimilar domains. For example, "This vacuum cleaner sucks", thus sentence indicates a positive opinion about vacuum cleaner.
2. Sentences that are sarcastic which does not contain any sentiment words these types of sentences are hard to deal. For example, "what a nice food! I stopped eating".
   These types of sentences are common in political discussion. When customer gives review about any product and services, they use very less sarcastic word.
3. There are many sentences that contain factual information with no sentiment words, and these types of sentences contain some useful information. Those sentences are objective sentences that are used to give certain useful evidence, and there are numerous of such kinds of sentences. For example, "This hotel charges lot of money for food".

Above sentence implies a negative sentiment about "food" that is provided by "hotel", and this sentence does not contain sentiment word but overall this is negative sentiment.

4 Problem Definition

The paper's fundamental aim is to identify aspects of entities and sentiment expressed for each aspect, and finally the goal is to summarize all the aspects and their sentiment values. The final outcome will be average opinion for each aspect of an entity. Here input is taken as real hotel review from a hotel located at New Delhi.

5 Methodology

Aspect-level sentimental analysis task:

(1) Extraction and categorization of entity: In this task, extracting all the entities from dataset, i.e., hotel reviews by customers, and then categorizing into similar groups with a group name, where each group gives a similar entity.
(2) Extraction and categorization: for each entity in above task, extracting aspect for each entity, into similar group with group name, where one group or on cluster represents one type of aspects.
(3) Extraction and categorization of opinion holder: This task is parallel to above two tasks and extraction opinion holder of those opinions and also save the time.
(4) Classification of aspect sentiment: In this task, performing main calculation for sentiment value of each opinion that is found in the user review sentence by using a sentiment score algorithm. That may be positive value, negative value,

or neutral, i.e., zero value, based on this numeric value sentences that have positive opinion, negative opinion, or neutral opinion.

Sentiment score algorithmic steps:

for each single Sentence s
Assign P = 0 and N = 0

Step 1: Check for the presence of idiom in s
Set s = 1 if exist
and s = 1 without idiom
Based on idiom update P and N

Step 2: If not exist check for the presence of token
tokenize = 1

    (a) For each token t, check for the negative word

    (b) If the token exists, then check for the emotion word
If the emotion word exist,
extract and invert the scores and also based on magnitude of scores update values of P, N

    (c) If the token exists, then check for the presence of next emotion word

    (d) Then extract score and verify whether the score is positive or negative
If positive, add one to emotion word score otherwise subtract one from emotion word score.
Again update P and N values based on scores' magnitudes,

    (e) Check whether token is booster word or negative word or an emotion word if matches, then
extract scores and assess the values of P, N on the basis of scores' magnitudes

Step 3: Check for the positive and negative words' values if anyone is nonnegative
Then enter the line into the output file in a table format and end up the while-loop decision tree and also perform well with all the datasets. The accuracy of classifiers decreases when using Bank data due to the presence of categorical attributes in the dataset.

The accuracy of classifiers could be enhanced by developing a fraud detection model on some selected attributes of the dataset and by using the datasets which have less categorical attributes. It would decrease the computational time or time taken to build a model. In future, more analysis could be done using other combination of classifiers. Other ensemble classifiers for different datasets and methods for handling diverged variety of attributes.

6 Classification of Aspect Sentiment

Supervised learning approach and lexicon-based approach are two main approaches to find out the opinion value for each aspect in a given customer review (Fig. 1).

a. Machine learning approach: It depends on certain famed algorithms for solving sentimental analysis as a systematic text classification problem that utilizes syntactic and/or linguistic features. The supervised learning procedures hinge on presence of labeled training documents for finding the aspect and opinion value in given sentence. So supervised learning approach is relying on the small set of training data. These trained data may not give correct result for large applications which give poor result as compared to lexicon-based approach.

b. Lexicon-based approach: Lexicon-based methods are unsupervised. The lexicon-based approach gives better result in large number of domains. The list of sentiment words and phrases are recycled for finding the sentiment orientation on every aspect in the given input sentence. Opinion shifters are also used which may affect opinions. Lexicon-based approach has mainly four steps:

- Identify opinion words
- Apply opinion changer
- Handle but clauses
- Aggregate sentiments

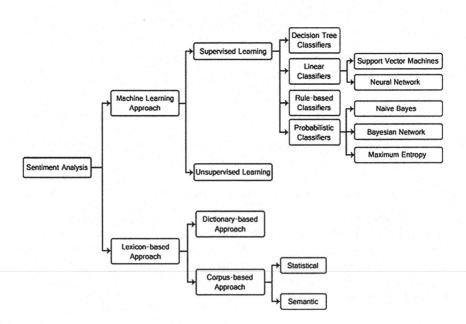

Fig. 1 Sentiment classification technique

Identify Opinion Words

Here first customer review is taken as input, then the review is broken into single-line sentence, then each sentence that has one or more aspects is identified. Now the total numbers of aspects that are available in a given sentence are listed where all positive word is set with a sentiment score +1 and −1 is allocated for all the negative word.

Apply Opinion Changer

Opinion changer or sentiment changer are the words and phrases that can swing user opinion from positive to negative or negative to positive. Most common opinion shifters are not, none, neither, nobody, none, nowhere, and cannot.

Handle But Clauses

"But" is mostly used in English sentences to changes the opinion of given sentence. The words and phrases which contain "but" changes the meaning and orientations of sentences and gives different output. The rules to handle "but" are before but and after "but", if the sentiment word cannot be found then both sides have opposite sentiment.

Aggregate Sentiments

Here sentiment score of all the opinion words is aggregated, and total number of aspects along with their sentiment scores will be displayed.

7 Natural Language Tool Kit (NLTK)

Natural Language Tool Kit makes us to write simple program in Python that works with large quantities of text. NLTK extracts keywords and phrases from the structured test, gives useful meaning, and saves that meaningful data into database for further use. NLTK treats text as raw data and performs operation in an interesting way. NLTK is free and open source, and it is used as a good tool and stunning library to work with natural language. It provides functionality that can convert input text into tokenized form and also classifies the words, and labeling can be done by part-of-speech tagging (POS tagging). POS tagger takes input as tokenized form of sentence and gives output as tag for each word (Table 1 and Fig. 2).

Part-of-speech-based features

- Classify total of adjectives in the sentences.
- Find out total of adverbs.
- Total number of interjections in the sentence (e.g., "hey", "hello", "wow").
- All verbs in the sentence.
- All nouns in the sentence.
- All proper nouns in the sentence.

Table 1 Universal POS

No	Tag	Description
1	CC	Coordinating_Conjunction
2	CD	Cardinal_Number
3	DD	Determiner
4	EX	Existential_There
5	FW	Foreign_Word
6	IN	Preposition
7	JJ	Adjective
8	JJR	Adjective, Comparative
9	JJS	Adjective, Superlative
10	LS	List_Item_Marker
11	MD	Model
12	NN	Noun, singular
13	NNS	Noun, plural
14	NNP	Proper_Noun, singular
15	NNPS	Proper_Noun, plural
16	PDT	Pre_Determiner
17	POS	Possessive_Ending
18	PRP	Personal pronoun
19	PRP$	Possessive pronoun
20	RB	Adverb
21	RBR	Adverb_Comparative
22	RBS	Adverb_Superlative
23	RP	Participle
24	SYN	Symbol
25	TO	To
26	UH	Interjection
27	VB	Verb
28	VBD	Base-Verb
29	VBG	Verb-Present-Participle
30	VBN	Verb-Past-Participle
31	VBP	Verb-Non-3rd Person-Singular-Present
32	VBZ	Verb-3rd-Person-Singular-Present
33	WDT	Wh-Determiner
34	WP	Wh-Pronoun
35	WP$	Possessive-wh-Pronoun
36	WRB	Wh-Adverb

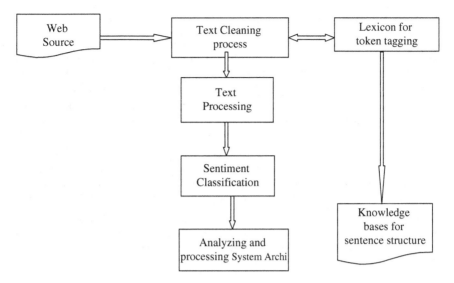

Fig. 2 System architecture

8 System Architecture

NLTK provides different libraries to find out the subjective and objective in sentences. The necessary steps of the aspect-based sentiment analysis are given below (Fig. 3).

- Break the customer review into sentences and make in tokenized form.
- Remove unwanted symbols from the sentences and use part-of-speech for individual word of the above tokenized form of sentence.
- Identify important aspect inside sentence with part-of-speech tagging help.
- Arrange the sentences into subjective and objective with the help of lexicon approach.
- With the help of lexical directory, identify the sentiment score for each positive, negative, or neutral sentence.
- Analyze the final output of different aspect versus sentiment score.

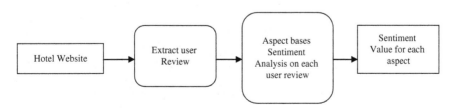

Fig. 3 Steps of aspect-level analysis

9 Results and Analysis

This chapter is showing results from different modules. Final result of this project contains four different parts. First part is scrappy module that converts unstructured data into structured data and saves that data into text file. The structured data are used as input for the next module, i.e., break long user reviews into sentences and these sentences are saved into separate files.

Structured data: All unstructured data are converted form. First, data will be crawled from the Web site. Data crawling is done by improving scrappy spider. In this paper, spider is Python code that crawls all the unstructured data and saves into the text file as structured form, and later this data is taken as next module, i.e., sentiment analysis module. Sentiment analysis module takes that structured data and extracts aspects from that structured data. Next, one sentiment score algorithm is used to find the score values of each aspect. Finally, the result is analyzed by using bar chart and pie chart by taking the aspect count and sentiment score (Fig. 4).

Fig. 4 Bar chart and pie chart of sentiment scores

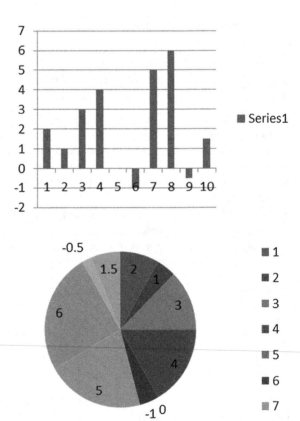

10 Conclusion and Future Work

Aspect-based sentiment analysis is new topic to the academics, as the customer's reviews play a central role of user's actions. Online users, different discussion group, online forums, and user blogs are growing very fast; all users share their information through these means of Internet on daily basis. So that is very necessary to design an efficient and effective. In aspect-based sentiment analysis system for online user data, there are many challenges in the field of sentiment analysis which will give better understanding of user's data. Hence, sentiment analysis gives very important impact on natural language processing and also gives great understanding on political science, management science, and social science because these all are affected by the user's opinions.

References

1. D. Kim et al., 'A user opinion and metadata mining scheme for predicting box office performance of movies in the social network environment', New Review of Hypermedia and Multimedia, 2013.
2. E. D'Avanzo, G. Pilato, 'Mining Social Network users Opinions to Aid Buyers shopping Decisions', Procedia Computer Science 118, 2014.
3. C. Monti, A Rozza, G. Zappela, A. Arvidsson, E. Colleoni, 'Modelling Political Disaffection from Twitter data', WISDOM'13 proceedings of the second international Workshop on Issues of Sentiment Discovery and Opinion Mining, 2013.
4. K. Denecke, 'Using SentiWordNet for Multilingual Sentiment Analysis' ICDEW, 2008.
5. S. Baccianella, A. Esuli and F. Sebastiani, 'SENTIWORDNET 3.0; An enhanced lexical Resources for Sentiment Analysis and Opinion Mining'. ELRA 2010.
6. A. Artale, A. Goy, B. Magnini, E. Pianta, C. Strapparava, 'Coping with WORDNET Sense Proliferation', ELRA, 1998.
7. B. Pang, and L. Lee, "Opinion Mining and Sentiment Analysis," Foundations and Trends in Information Retrieval, vol. 2, pp 1–135, 2008.

Node Grouping and Link Segregation in Circular Layout with Edge Bundling

Surbhi Dongaonkar and Vahida Attar

Abstract Every industry is producing a huge amount of data today, which is analyzed and used for future predictions and making business decisions. Networked data can be analyzed node-link diagrams, which give different trends with different layouts to analyze network data. Many of these layouts have complex algorithms. Thus, construction of alternative layouts used is the topic of research for many organizations and industries. Many real-time examples require grouping of nodes, separation of links, simple layout, and abstract visuals of data. This paper proposes a technique which will tend to meet the above requirements of real data. The essence of this technique is the use of simple circular layout with node grouping and link segregation. View level abstraction is achieved with the concepts of edge bundling and node abstraction. Edge-bundling algorithm also reduces the clutter in the graph. Thus, above techniques will lead to viewing the networked data with new trends coming out by grouping nodes, link segregation, and compare data by focusing on different attributes of data at different levels of view (i.e. abstract and detailed).

Keywords Network graph · Node-link diagram · Edge bundling
Node abstraction · Circular layout

1 Introduction

Analysis of data helps in predictions, getting the patterns out, emphasizing on the sensitive part, prioritizing the modules, and concentrating on the most valued part of the given data. More pattern formation and less cognitive efforts lead to superior

S. Dongaonkar (✉) · V. Attar
Department of Computer Engineering, College of Engineering, Pune,
Shivaji Nagar, Pune 411005, Maharashtra, India
e-mail: surbhidongaonkar@gmail.com

V. Attar
e-mail: vahida.comp@coep.ac.in

© Springer Nature Singapore Pte Ltd. 2019
H. S. Behera et al. (eds.), *Computational Intelligence in Data Mining*,
Advances in Intelligent Systems and Computing 711,
https://doi.org/10.1007/978-981-10-8055-5_35

predictions. The pictorial view of the data is beneficial for analysis as more patterns can be observed with less cognitive efforts.

For a pictorial representation of any data, various visualization techniques are in use. Visualization of information is abstraction of that information in some schematic form. Graph drawing and information visualization communities have developed many sophisticated techniques for visualizing network data, often involving complicated algorithms that are difficult for the uninitiated to learn and require a research work for simplification and performance upgradation [1]. Graphs are used to visualize networks, which are increasingly encountered in numerous fields of study.

The network graph visualization must clarify the actual data points and how they are related to each other. This can be shown with various types of network graphs as matrix representation, node-link diagrams, hive plot, etc. Node-link diagram can cover all types of networks to visualize with different kind of layouts for different applications. Existing layouts are complex and sometime do not meet many aesthetic criteria of graph drawing. Circular layout is a simple layout which can prove better visual results with improvement. Integration of layout and node aggregation [2] as well as bundling algorithms will change the insight quality of the graph.

Thus to improve the performance, we chose the circular layout to plot node-link diagrams. Grouping and link segregation helped us to come out with new trends. Further, the cost-based edge-bundling algorithms applied to it helped to reduce clutter and make the view more elegant. All above concepts are applied at different levels of view so that user can first overview the network and then go into details of required part of network or complete network [3].

The rest of the paper is organized as below. Section 2 surveys the network graphs and various layouts for node-link diagrams. It also studies various edge-bundling techniques. Section 3 deals with the proposed methodology. Section 4 introduces experiments carried out and results obtained. Section 5 concludes the paper.

2 Literature Review

Network graphs have many types of matrix-based diagrams [4], hive plots, and most commonly used node-link diagrams which to meet different aesthetic criteria [5].

Node-link diagrams plot data points as nodes and show the connectivity in the network with links (edges). Placement of the nodes and links on the display is managed with various layout algorithms such as force-directed layout, radial layout, dominance drawing, and circular layout [6]. Force-directed algorithm provides a mental map and constructed using physical analogies but have variable running time. Many improvements are done in this algorithm by Kamada and Kawai, Davidson and Harel, Furchterman and Reingold algorithm [7]. Orthogonal layout methods, which allow the edges of the graph to run horizontally or vertically, are parallel to the coordinate axes of the layout. Dominance drawing places vertices in such a way that one vertex is upwards, rightwards, or both of another if and only if it is reachable from the other vertex [8]. Circular layout is another popular layout in which as name suggests, nodes are plotted on a circle and links are drawn as chords of the circle [9]. This layout is O(n) time algorithm. But high-scale network results in cluttered view and large edge crossing. Multiple improvements to reduce this crossing involve placing highly connected nodes together, drawing of shorter links, etc. [9]. Another improvement as suggested by Fabian Beck, Martin Puppe, Patrick Braun, Michael Burch, and Stephan Diehl is the implementation of edge-bundling algorithm over the layout [10].

Also, DOSA technique on force-directed and scatterplot algorithm to allow the user to go from detailed to overview via selection and aggregation [3]. It was also observed that the data in the friendship network can be grouped based on certain property like sex and grade [11]. Visualization of the different groups and their connectivity is better if drawn as a graph using circular layout. Further study suggested that almost all the networks such as communication networks, transportation networks, and social networks can be internally classified into multiple groups. Edge bundling is considered as a good approach to reduce hair ball caused by dense edges. One approach is cost-based edge bundling, where the ink or energy cost is used to determine the shapes of curved edges. In this approach, total four points control the shape of the edge. Using an iterative process, some layout amounts to a sequence of edge clustering and image processing operations [12]. Other edge-bundling techniques as hierarchical will bundle the edge by using subtree and foresting concept [13]. Force-directed edge-bundling works on laws in physics [14].

Above literature suggests the development of new faster layouts with lesser complex algorithms is required which will simultaneously provide more or equal benefits as that of existing once. Circular layouts have good performance as the single iteration is sufficient to plot the graph [9]. Minimization of edge crossing and clutter reduction will make the circular layout more powerful. Also, to get more insights into network, grouping of nodes, abstract view, filtering of data plays an important role.

3 Proposed Methodology

Main research area in network graphs is to increase the readability of node-link diagram which analysts use to analyze network [16]. The depth to which user wants to see and analyze data varies with user and data. Thus, our layout consists of two views (abstract and detailed) producing different trends. For both views, the circular layout is used, data is classified into different groups, and links are segregated as intragroup and intergroup links to help in analysis.

The first one is an abstract view where each group is shown with single node [17]. The property on which grouping is done can be changed based on user's requirement. Further, the links are segregated, and the single link is shown between every two groups. The width of the link is proportional to an actual number of links between those two groups. All the links which have source and target node belonging to same group (intragroup) are shown as single self-loop on that group. The thickness of loop is also proportional to the count of actual links. Thus, this layout will give an overview of data. Though it is an abstract view, it analyzes the comparative count of nodes in each group, i.e. total values of property, the number of data points in each group, and bonding strength between the groups. Then, the user can select the groups he is interested in or all groups if required which will be then displayed in details.

Selected groups are expanded into detailed view. Every data point is mapped as node on the circle of specified radius. The different groups are identified by keeping angular distance between two groups. Links are again segregated as intragroup links and intergroup links. Intragroup links are plotted outside the circle, and intergroup links are plotted inside. Further, to reduce the edge crossing and clutter inside the circle, similar inside the circle are rendered together as bundles whose width varies with the actual number of links. Links are considered as similar one if they belong to same source and target group (Fig. 1).

The improvements in detailed view are shown in Fig. 2.

Bundling is done using cost-based edge-bundling techniques for each group. Using this detailed view specifications of every groups, pattern connectivity of every node and in between groups and bond of internal connectivity is analyzed.

Node Placement: We looked for the placement of the n nodes of a graph G ($V = \{1, ..., n\}$, E) which are classified into m groups. By convention, in the basic circular layout, we assume nodes are equally spaced on the circle which is carried in an abstract view. So as every node represents one group, the separation of nodes in abstract view is determined by

$$\text{Angular separation } = 360/m \tag{1}$$

But in detailed view, to identify groups, spacing is modified as

$$\text{Angular separation } = 360/(n + 2 * m) \tag{2}$$

Fig. 1 Abstract and detailed view of improved circular layout

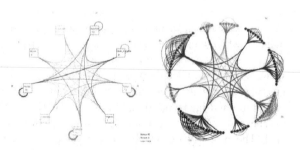

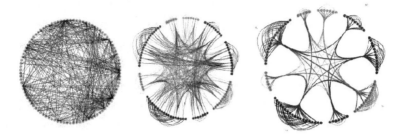

Fig. 2 Improvements in detailed view

where extra 2 * m space is utilized to show separation between groups. It introduces extra space between two groups which is equal to twice the diameter of a node added with trice the space between two nodes to the space between the nodes in one group. Using above formulae, the angle of every node is determined as ang$i$ for every node i $\in$ n.

We denote the coordinates of a node $i \in n$ by $(Xi, Yi) \in$ R. The nodes are arranged on the unit circle centered at the origin. The coordinates are calculated as

$$Xi = radius * \sin(angi) \tag{3}$$

$$Yi = radius * \cos(angi) \tag{4}$$

Link placement: To render the links first, the segregation process is carried out by checking whether the grouping property of source and target matches or not. Intergroup links are passed to edge bundling to see in which bundle it fits.

Edge bundling: The idea of bundling edges is related to the work on confluent drawing [3], where edge crossings are eliminated by grouping edges in tracks. For bundling, two control points reassigned to each edge, the meeting point of the sources and the meeting point of the targets. Hence, the shape of an edge is controlled by four points: the source node, convergence point of the sources, the meeting point of the targets, and lastly, the target node. The positions of the source node and the target node are fixed, whereas other two control points can be moved freely. The edge-bundling algorithm bundles edges by placing the control points of similar edges close together. Ink minimization technique computes the optimized positions of the control points, which is solved by using a numerical method. Let the angle of the first node is Sj, and the last node is Ej for every group j where j belongs to m. Thus, the angle of control point is calculated as

$$\text{Angle of control point } cj = (Sj + Ej)/2 \tag{5}$$

To pass the edges through these control points, the edge is divided into three parts, from source to control point of source group, from control point of source group to control point of target group, from control point of target group to target,

Fig. 3 Work flow

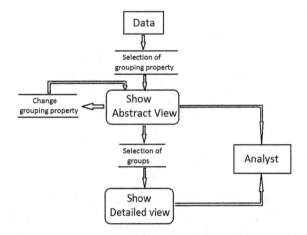

which reduces the clutter in the graph. Curve from source control point to target control point is again controlled by single control point which is center of the circle.

Further, the width of the bundle is kept proportional to a number of actual links present in a particular bundle.

The detailed view specifically shows an isolated node. It also provides insight to a node which is not connected to other nodes of same group but connected in other groups or vice versa. So the overall system flow will be as shown in flow diagram (Fig. 3).

After entering networked data, the user has to select the property for grouping which is rendered as the abstract view. Observing the trends coming out in overview, the user can change grouping property or select the groups which he wants to analyze in details. Change in the property will again provide an overview. This process is continued until the user has observed all trends with different grouping property.

Thus, the observation of different properties and filtering of data before analyzing detailed view helps analysts.

4 Experiments and Results

Dataset: Social networking site profiles.

Software required—Gephi [18].

The dataset is analyzed in force-directed layout with Gephi tool and in a circular layout with our system.

Results: The dataset in the improved circular layout is viewed at abstract and detailed level with different grouping properties. As discussed, the data points in the dataset are classified into groups based on the certain property.

Let the location of the individual is taken as a grouping attribute. Thus, every individual is placed in a single group based on the city to which the person belongs. All the individuals belong to one of the six cities as shown in the graph.

The abstract view shows that there are total six cities across which this network is distributed. It is observed that the highest connectivity is between the individuals belonging to Pune and Mumbai. Also, individuals in Pune have good connections with individuals in every other city. Whereas the individuals in Bangalore are less connected, hence campaigning in Bangalore will be less transitive to spread the business or product (Fig. 4).

Also, by observing abstract view, it is clear that the individuals in Mumbai have good connectivity with people in Pune only. Also, the people in Pune can be contacted through individuals in any other city (by observing connectivity with a width of the bundle in abstract view). Thus, the data is filtered, and the Mumbai group is dropped out.

Detailed view: Every city has a good number of individuals as a part of this network except Kolkata. There is one individual each in Bangalore and Chennai which are not connected with a single individual in the same city. So the product/business will not be advertised in the city through such individuals. Campaigning with these individuals does not improve the sell, and hence the selection of other people will be advantageous though it can be costly.

Similarly, the different trends can be observed by changing the property. If the organization in which the individual is considered as a grouping property, then it will guide the user so that he/she can campaign in proper organization with low cost and greater benefits. Thus, the developed layout helps in deciding the firms and places to sell any product or to make other business predictions.

The data is also visualized in Furchterman and Reingold algorithm which is improved force-directed layout using Gephi. The comparative analysis is listed in Table 1.

Fig. 4 Visualization of social media profiles with improved circular layout

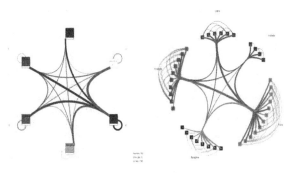

Table 1 Comparative analysis

Sr. no.	Observation	Force-directed layout	Improved circular layout
1.	Nodes	Observed	Observed
2.	Links	Observed	Observed
3.	No. of groups	Not observed	Observed (7)
4.	Intergroup links	Not observed	Observed
5.	Intragroup links	Not observed	Observed
6.	Cluttered view	Observed	Not observed
7.	Highly connected nodes	Observed	Observed
8.	Running time	Variable	Fixed

5 Conclusion

This research aims at effective graph drawing with simpler layout and effective visualization without performance degradation. The circular layout with O(n) complexity achieved the performance requirements. Further improvements in a graph such as grouping of nodes, link separation, abstract view, and clutter reduction were targeted in this work. Drawing of two separate views gives better insights into the graph. Experimental results show that the grouping of nodes so as to observe one extra property of network is achieved. Segregation of links helps users to observe the intragroup and intergroup connectivity and its strengths. This segregation of links reduces the inside edge crossing. Further, clutter in the graph is reduced with the application of edge-bundling algorithm. Also, the user can filter out the groups in abstract view which allow them to focus on the particular part of network dataset and analyze it more specifically.

In future, sorting of nodes and use of multiple circles will improve visualization.

Acknowledgement We would like to express my sincere gratitude to **Mr. Vijay Chougule** and **Mr. Vineet Raina** for their constant support and inspiration.

References

1. Shixia Liu, Weiwei Cui, Yingcai Wu, Mengchen Liu.: A survey on information visualization: recent advances and challenges, Springer-Verlag Berlin Heidelberg (2014) 7–16
2. Aleks Aris and Ben Shneiderman.: A Node Aggregation Strategy to Reduce Complexity of Network Visualization using Semantic Substrates, Information Visualization 6(4):281–300, (2010) 2–7
3. Stef van den Elzen, Jarke J. van Wijk.: Multivariate Network Exploration and Presentation: From Detail to Overview via Selections and Aggregations, InfoVis, (2014) 1–9
4. Joris Sansen, Romain Bourqui, Bruno Pinaud, Helen Purchase.: Edge Visual Encodings in Matrix-Based Diagrams, International Conference on Information Visualisation (iV), (2015) 1–6
5. http://guides.library.duke.edu/datavis/vis_types

6. Michael J. McGuffin.: Simple Algorithms for Network Visualization: A Tutorial, ISSN Volume 17, Number 4, (2012) 1–16
7. *Boštjan Pajntar.:* OVERVIEW OF ALGORITHMS FOR GRAPH DRAWING, 1–6
8. https://en.wikipedia.org/wiki/Graph_drawing
9. Janet M. Six, Ioannis G. Tollis.: Circular Drawing Algorithms, Journal of Discrete algorithms, vol. 4, issue 1 (2006) 1–5
10. Fabian Beck, Martin Puppe, Patrick Braun, Michael Burch, Stephan Diehl.: Edge Bundling without Reducing the Source to Target Traceability, InfoVis, (2011) 1–2
11. Tarik Crnovrsanin, Chris W. Muelder, Kwan-Liu Ma, Bob Faris, Diane Felmlee,: VISUALIZATION OF FRIENDSHIP AND AGGRESSION NETWORK, People Research Publication Gallary About VIDI, (2014) 3–11
12. O. Ersoy, C. Hurter, F. Paulovich, G. Cantareira, A. Telea.: Skeleton-based edge bundling for graph visualization, IEEE Trans Vis Comput Graph, (2011) 10
13. Yuntao Jia, Michael Garland and John C. Hart.: Hierarchical Edge Bundles for General Graphs, (2009) 3–8
14. David Selassie, Brandon Heller and Jeffrey Heer.: Divided Edge Bundling for Directional Network Data, *IEEE Trans. Visualization & Comp. Graphics (Proc. InfoVis)*, (2011) 3–6
15. Danny Holten, Jarke J.van Wijk.: Force-Directed Edge Bundling for Graph Visualization, IEEE-VGTC Symposium on Visualization, Vol 28, (2009) 1–6
16. Wei Peng, Matthew O. Ward and Elke A. Rundensteiner.: Clutter Reduction in Multi-Dimensional Data Visualization Using Dimension, INFOVIS '04 Proceedings of the IEEE Symposium on Information Visualization (2004) 2–6
17. Sun GD, Wu YC, Liang RH *et al.:* A survey of visual analytics techniques and applications: State-of-the-art research and future challenges, JOURNAL OF COMPUTER SCIENCE AND TECHNOLOGY 28(5), (2013) 1–8
18. Guo-Dao Sun, Rong-Hua Liang, Shi-Xia Liu.: A Survey of Visual Analytics Techniques and Applications: State-of-the-Art Research and Future Challenges, Journal of Computer Science and Technology 28(5):852–867, (2013) 1–5
19. *Mathieu Bastian, Sebastien Heymann, Mathieu Jacomy.:* Gephi: An Open Source Software for Exploring and Manipulating Networks, ICWSM, (2009) 3–4

Fuzzy-Based Mobile Base Station Clustering Technique to Improve the Wireless Sensor Network Lifetime

R. Sunitha and J. Chandrika

Abstract A wireless sensor network is an emerging paradigm in the present era of computer communication technology. Sensor nodes are minute, lightweight, and autonomously distributed over the network; these nodes are not rechargeable. So energy consumption of the sensor node is a crucial constraint in the wireless sensor network. Sensor nodes are clustered to reduce the communication overhead. This paper proposes a new fuzzy-based mobile base station clustering technique. This technique uses fuzzy approach for the base station movement to decrease energy consumption of the sensor nodes and increases the lifetime of the network. Proposed work is implemented in the MATLAB software. Comparatively, it reduces the energy consumption of the sensor nodes.

Keywords Fuzzy system · Wireless sensor network · Clustering
Energy efficiency · Mobile base station

1 Introduction

Wireless sensor network (WSN) [1–3] is an important part of Internet of Things. It senses different environmental parameters like humidity, pressure, temperature, and gas. There exists a wide range of applications of WSN including military application, health monitoring, fraud detection, surveillance, transport monitoring, fire detection, and flood detection. Sensor nodes are autonomously distributed over a particular area to gather data.

Several clustering techniques [4–6] are used in the WSN to decrease the energy consumption of the sensor nodes. Nodes in the network are grouped into several

R. Sunitha (✉) · J. Chandrika
Department of CSE, Malnad College of Engineering, Hassan 573202,
Karnataka, India
e-mail: sunisunitha594@gmail.com

J. Chandrika
e-mail: chandrikaramesh@gmail.com

© Springer Nature Singapore Pte Ltd. 2019   401
H. S. Behera et al. (eds.), *Computational Intelligence in Data Mining*,
Advances in Intelligent Systems and Computing 711,
https://doi.org/10.1007/978-981-10-8055-5_36

clusters based on the distance between them. Each cluster selects one node as cluster head based on the different prototypes. Each sensor node senses the data from its surroundings and sends it to cluster head. Cluster head (CH) collects all the data from other nodes and en route for the base station (BS).

Fuzzy approach [7, 8] creates a multidisciplinary platform of discussion on recent trends in WSN, data mining, as well as advanced applications to biological studies, economics, ecological study, engineering trends, finance field, management, and medicine field. Fuzzy set theory intended for modeling uncertainty is associated with ambiguity and imprecision. Linguistic variables, rather than quantitative variables, are used in this theory to represent vague concepts.

Proposed fuzzy-based mobile base station enhances the features of LEACH [5, 9] protocol, and it overcomes the limitations of leach. Leach is a hierarchical routing protocol based on clustering technique. In this protocol, cluster head is selected based on the probabilistic model and all sensor nodes will get an equal chance of electing as cluster head. Fuzzy-based mobile base station is a dynamic protocol, in which the base station is dynamic in nature. The base station moves toward the sensor nodes to accumulate data.

2   Related Work

In this section, several protocols and techniques are discussed for energy efficient data transmission but not efficiently reduced energy consumption. Power efficient gathering in sensor information systems (PEGASIS) [10] protocol transmits the data to the neighboring nodes which in turn transmits to the base station. It mainly concentrates on the node failure constraint. In this data transmission phase, all the sensor nodes will participate and consume more energy.

Multilayer LEACH protocol [5, 6] divides the network into different levels and transmits the data. In each level, there exist several clusters with the cluster head. Lower level cluster head collects the data from the sensor nodes of the cluster it belongs to and forwards it to the next higher level cluster head, which in turn sends it to base station followed by several layer cluster heads. Each cluster head participates more in the data transmission so that the energy of the cluster head will be more.

In existing distributed energy efficient clustering protocol [2] cluster heads are selected based on the residual energy of the node and the average energy of all the nodes in the cluster. In partition-based LEACH protocol [5, 9], the network is partitioned into several sectors. Node with the maximum residual energy is selected as the cluster head in each sector.

Clustering schema [11, 12] introduced to design energy efficient wireless sensor network for the data transmission. In this method, the network is divided into two parts on the basis of distance. First part is core part, and second part is border part.

The area from d to the border is known as border part, and the area from d to the center is known as core part.

CHEF [8] protocol is a cluster head energy efficient method [13, 14] of data communication. In this method, the cluster head is selected based on the residual energy of the node and distance. Fuzzy approach is used to elect the cluster with highest residual energy of the node, and it should be locally optimal. The simulation result shows that it is more efficient than existing protocols. Section 3 explains the proposed work, Sect. 4 explains the results and implementation, and Sect. 5 concludes the paper.

3 Proposed System

In the fuzzy-based mobile base station (FBMBS) clustering technique, the wireless sensor network with autonomously distributed remote nodes to monitor the surrounding environment is considered. Following are some network assumptions

- Sensor nodes are fixed in the network except BS
- Mobile base station
- Identical sensor nodes with equally distributed energy over all the sensor nodes
- Based on the signal potency of the sensor node, the distance between the base station and mobile station is estimated.

A fuzzy-based mobile base station clustering technique uses basic concepts of LEACH [9] protocol for clustering the sensor nodes in the network. In the LEACH protocol, cluster heads are selected based on the probability value selected by the sensor nodes. In each round, all the sensor nodes will select the random number between 0 and 1. If the node is with value greater than the threshold value (Ts), then that node will be elected as cluster head. All of the nodes will get an equal chance of getting elected as cluster head. The proposed FBMBS approach increases the functionality of LEACH by adding new stages in both the setup and steady phase.

This protocol has two stages:

- Setup stage: In this stage, the sensor nodes form a cluster.
- Steady stage: In this stage, data transmission takes place from sensor node to cluster head.

In each round of protocol forms C is number of clusters from the P is number of nodes in the selected network area. This C value can be calculated analytically by using the energy communication model of the clustering protocol. If P is the number of sensor nodes in the network, $Q \times Q$ is the region in which the sensor nodes are distributed. C is the number of clusters formed with P/C nodes per cluster in each round. Each cluster with one Cluster Head and (P/C)-1 non-cluster head nodes.

3.1 Fuzzy-Based Mobile Base Station Clustering Algorithm

Fuzzy-based mobile base station clustering algorithm steps are shown below.

1. Select the network area and deploy the sensor nodes.
2. Base station divides the network area into grids.
3. Sensor nodes are distributed randomly over the grids and form a cluster in each grid based on LEACH protocol.
4. Round = 1;
5. Select the cluster head based on the threshold value.
6. Run the fuzzy rules at the base station in order to move BS toward the cluster head to collect the data (fuzzy-based mobile base station).
7. Base station moving toward the cluster head satisfies the following three criteria compared to the other cluster heads in the network.

   - Distance = far
   - Energy = low
   - Proximity = high

8. Cluster head collects the data from sensors and transmit to the BS.
9. Round = Round + 1 & go to step 5.
10. If simulation time = Tsim, then stop the excecution

FBMBS clustering algorithm implements the three fuzzy parameters at base station. Here, we are considering distance, energy, and proximity as fuzzy parameters because distance metric is very important to reduce the communication overhead and energy dissipation in communication is more with node deployed far away from the base station. Residual energy of all sensor nodes is also very important for the communication, initially all the nodes are having equal amount of energy. In each round, sensor node sense the data and transfers it to the cluster head. Cluster head aggregate this data and transmit it to the base station. This process consumes energy of sensor nodes in the network.

Centrality of the sensor node is calculated by using the distance between the nodes as shown in Formula (2), and the distance is calculated by Formula (1). So the residual energy of the sensor nodes and it is also very important metric for the network's lifetime. Third parameter is proximity of the cluster head; proximity depends on the number of sensor nodes which are present nearby that cluster head or the cluster head with the more number of sensor nodes in its cluster. If maximum sensor nodes send the data to the cluster head, then that cluster head and sensor nodes energy get dissipated, and the proximity is calculated by Formula (2). In Formulas (1) and (2), xi, xj, yi, and yj are the distance parameters, and dist is the distance between the nodes. In Formula (2), CHx and CHy are cluster heads instances, and BSx and BSy are base station instances. The base station moves toward the node with maximum distance, minimum residual energy, and maximum proximity.

Following is formula (1) to calculate the centrality.

$$centrality(xi) = sum(dist) \qquad (1)$$

Proximity of the cluster head is calculated based on the number of nodes which are present in that cluster. Following is formula (2) to calculate the proximity

$$proximity = Sqrt(CH\_x - BSx)^2 + (CH\_y - BSy) \qquad (2)$$

Fuzzy-based mobile base station (FBMBS) clustering algorithm divides the selected network into several grids as shown in Fig. 1. Sensor nodes are distributed randomly all over the grids. Each grid nodes form a cluster by using the LEACH protocol, each cluster elect one node as cluster head. Sensor nodes start sense the data and transmit to the cluster heads. In each round, a new sensor node will elect as cluster head so all the nodes will get an equal opportunity of electing as cluster head.

The base station is dynamic in nature, and based on the fuzzy parameters, it moves near cluster head to collect the data. There are three fuzzy parameters: distance between the cluster head and the base station must be far, proximity of the cluster head must be high, and energy of the cluster head must be low. Base station moves toward the cluster head which satisfies all the three criteria to collect the data. The energy of the sensor node mainly depends on the distance and the data transmission; it reduces communication overhead.

Fig. 1 FBMBS clustering and data transmission

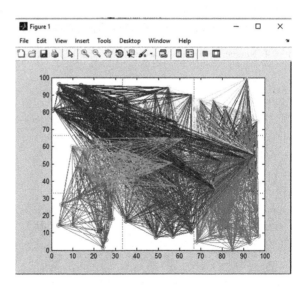

4 Result Discussion

The fuzzy-based mobile base station (FBMBS) clustering approach is implemented in MATLAB software. Following is the result of the proposed work compared to the existing LEACH protocol. In the following three graphs, x-axis represents the number of rounds and y-axis represents live nodes, residual energy, and energy variance.

Figure 2 shows the number of live nodes in the network after certain period of data transmission. Red colored line shows the FBMBS clustering, and blue line shows the LEACH protocol. In the proposed system, node's death ratio is less over long period of data transmission.

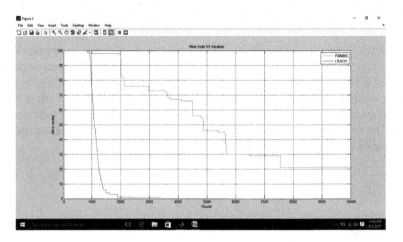

Fig. 2 Number of live nodes when the number of rounds is 10000

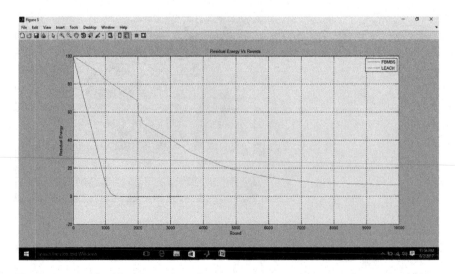

Fig. 3 Amount of residual energy in each sensor node when the number of rounds is 10000

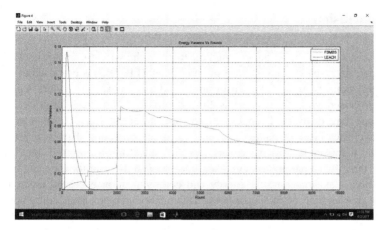

Fig. 4 Amount of energy variance in the cluster nodes when the number of rounds is 10000

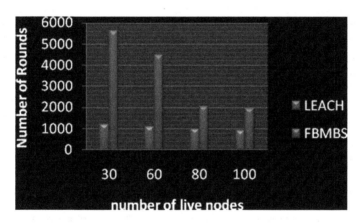

Fig. 5 Number of live nodes per rounds

Figure 3 shows the amount of remaining energy in each sensor node after a certain period of data transmission in the network. Red colored line shows FBMBS clustering, and blue line shows the LEACH protocol. In the proposed system, the node energy decreases gradually over a long period of data transmission.

Figure 4 shows the amount of energy variance in each sensor node after a certain period of data transmission in the network. Red colored line shows FBMBS clustering, and blue line shows the LEACH protocol. In the proposed system, the node energy decreases gradually over a long period of data transmission.

Figure 5 shows the number of live nodes per number of rounds. Here, we are taken 30, 60, 80, and 100 nodes. As graph shows, in FBMBS clustering approach after the 5600 rounds of data transmission, 30 nodes are still alive but in LEACH protocol 30 nodes are alive until 1200 rounds of data transmission. This shows the gradual improvement in the sensor node's lifetime and also network's lifetime.

5 Conclusion

Energy consumption is a challenging constraint in the wireless sensor network. In this paper, fuzzy-based mobile base station system (FBMBS) is proposed to reduce the energy consumption of the sensor nodes and to increase the network's lifetime. In this proposed work, the network is divided into M × N rectangular grids. In each grid, nodes are distributed randomly and form a cluster. Each cluster elects one node as cluster head, It gathers the data from sensor nodes. The base station is dynamic; it moves toward the cluster head to collect the aggregated data. In this system, three fuzzy parameters are taken into consideration for the base station. The sensor node data transmission energy and communication overhead are reduced, and it increases lifetime of sensor node. This method can be enhanced by applying any intelligent technique to detect the outliers. It reduces the energy consumption of outlier node communication in the network.

References

1. D. Bhattacharjee, S. Kumar, A. Kumar, S. Choudhury, "Design and Development of Wireless Sensor Node", (IJCSE) International Journal on Computer Science and Engineering 02, No. 07, 2431–2438, 2010.
2. M. C. M. Thein, T. Thein., "An Energy Efficient Cluster-Head Selection for Wireless Sensor Network", International Conference on Intelligent System, Modelling and Simulation, 287–291, 2010.
3. J. Yick, B. Mukherjee, D. Ghosal, "Wireless sensor network survey", in Computer Networks-52 2292–2330, 2008.
4. Dan Liu, Qian Zhouy, Zhi Zhangz, Baoling Liux, "Cluster-Based Energy-Efficient Transmission Using a New Hybrid Compressed Sensing in WSN", 2016 IEEE Conference on Computer Communications Workshops (INFOCOM WKSHPS): 2016 IEEE Infocom MiseNet Workshop- 978-1-4673-9955-5/16/$31.00 ©2016 IEEE.
5. R.U. Anitha, P. Kamalakkannan, "Energy Efficient Cluster Head Selection Algorithm in Mobile Wireless Sensor Network", in ICCCI-2013, 1–5, 2013.
6. Sunitha R, Chandrika J, "Distance based Data Mining by Multi-Level Clustering in Wireless Sensor Network", ITSI Transactions on Electrical and Electronics Engineering (ITSI-TEEE), ISSN (PRINT): 2320–8945, Volume 3, Issue 2, 2015.
7. Padmalaya Nayak, D. Anurag, "A Fuzzy Logic based Clustering Algorithm for WSN to extend the Network Lifetime", https://doi.org/10.1109/jsen.2015.2472970, IEEE Sensors Journal Sensors-12824-2015.R1.
8. Jong-Myoung Kim, Seon-Ho Park, Young-Ju Han, TaiMyoung Chung, "CHEF: Cluster Head Election mechanism using Fuzzy logic in Wireless Sensor Networks" ICACT, PP. 654–659, Feb. 2008.
9. Omar Banimelhem, Moad Mowafi, Eyad Taqieddin, Fahed Awad, Manar Al Rawabdeh "An Efficient Clustering Approach using Genetic Algorithm and Node Mobility in Wireless Sensor Networks" 978-1-4799-5863-4/14/$31.00 ©2014 IEEE.
10. A. S. Lindsey and C. S. Raghavendra, "PEGASIS: Power-efficient gathering in sensor information systems," in Proceedings of the IEEE Aerospace Conference, March 2002.

11. Suparna Biswas1, Jayita Saha1, Tanumoy Nag1, Chandreyee Chowdhury2, Sarmistha Neogy2, "A Novel Cluster Head Selection Algorithm for Energy-Efficient Routing in Wireless Sensor Network" 2016 IEEE 6th International Conference on Advanced Computing-978-1-4673-8286-1/16 $31.00 © 2016 IEEE, https://doi.org/10.1109/iacc.2016.114.
12. G.Y. Park, H. Kim, H.W. Jeong, H.Y. Youn, "A Novel Cluster Head Selection Method based on K-Means Algorithm for Energy Efficient Wireless Sensor Network", 27th International Conference on Advanced Information Networking and Application Workshops-2013.
13. Dr. L.M. Varalakshmi R. Srividhya, "Enhanced Energy-Efficient and Reliable Routing for Mobile Wireless Sensor Networks", International Conference at MVCE-2015.
14. V. Devasvaran, N. M. Abdul Latiff, and N. N. Nik Abdul Malik, "Energy Efficient Protocol in Wireless sensor Networks using Mobile Base station", 2nd International Symposium on Telecommunication Technologies (ISTT), Langkawi, Malaysia (24–26 Nov 2014) - 978-1-4799-5982-2/14/$31.00 ©2014 IEEE.

Hydropower Generation Optimization and Forecasting Using PSO

D. Kiruthiga and T. Amudha

Abstract Deriving optimal operation rules for maximizing the hydropower generation in a multi-purpose reservoir is relatively challenging among the various other purposes such as irrigation and flood control. This paper addresses the optimal functioning of a multi-purpose reservoir for improving hydropower generation. Efficient bio-inspired optimization techniques were proposed for hydropower optimization and hydrological variables forecasting. A particle swarm optimization (PSO)-based methodology is proposed for maximal hydropower generation through optimal reservoir release policies of Aliyar reservoir, located in Coimbatore district of TamilNadu state in India. The reservoir release is also optimized by Global Solver LINGO and compared with PSO, and it is explored that PSO-based model is powerful in hydropower maximization. To handle the uncertain behavior of hydrologic variables, artificial neural networks model is also applied for forecasting reservoir inflow and hydropower generation. The results obtained through the optimal reservoir release patterns suggested in this work have shown that the Aliyar Mini Hydel Power Station has a huge potential in generating considerably more hydropower than the actual generation observed from the power plant over the past years.

Keywords Hydropower optimization · Bio-inspired methods · Optimal release policies · Hydropower forecasting · Artificial neural network

1 Introduction

Hydropower is the most eco-friendly way of power generation, and many of the Indian reservoirs operate hydropower plants. The Asian Development Bank (ADB) report states that due to lack of planning, almost 78% of the hydropower

D. Kiruthiga · T. Amudha (✉)
Department of Computer Applications, Bharathiar University, Coimbatore 641046,
TamilNadu, India
e-mail: amudhaswamynathan@buc.edu.in

D. Kiruthiga
e-mail: kiruthigadevaraj04@gmail.com

© Springer Nature Singapore Pte Ltd. 2019
H. S. Behera et al. (eds.), *Computational Intelligence in Data Mining*,
Advances in Intelligent Systems and Computing 711,
https://doi.org/10.1007/978-981-10-8055-5_37

capability is untapped and relatively lesser potential is being consumed [1]. Hydropower is heavily dependent upon the irrigation release in many reservoirs, and a poorly planned irrigation release leads to limited hydropower production. Certain reservoirs release water separately for hydropower and irrigation, thereby reducing the water supply for the plant, which in turn reduces the power generation. [8]. It is the need of the hour to identify and employ optimal operational techniques to utilize the unexplored potential of the existing hydropower plants in India.

Maximization of hydropower generation through optimal operation of multi-purpose reservoir is quite challenging because of the conflicting objectives of various purposes, multiple decision variables, and physical, technical, and operational constraints. In past, many traditional optimization techniques like linear programming, nonlinear programming, and chance-constrained linear programming have been used to derive optimal operation rules to maximize the hydropower generation in multi-purpose reservoir, for example, M.G. Devemane et al. [6], R. Arunkumar and V. Jothiprakash [4, 5], K.R. Sreenivasan and S. Vedula [15]. Review on various techniques used for multi-purpose reservoir operation can be found in Yakowitz [17], Simonovic [14], Wurbs [16], and Yen [18].

In the past, to manage complex water resources systems, many traditional optimization algorithms have been used. But while originating the traditional optimization techniques model to solve real-time problems, it led to nonlinearities and nonconvexities, often got wrapped up with local optimal solutions [10] and also have massive computational requirements, and hence, it has become quite complex to work with real-world problems. To overcome these issues, bio-inspired techniques have been proposed and employed. In recent times, the bio-inspired algorithms are popularly applied to solve many water recourses problems especially in hydropower maximization, for example, D. Nagesh Kumar and M. Janga Reddy used bio-inspired techniques like ACO [10], MODE [11], and EMPSO [12] for optimal operation of reservoirs for hydropower maximization in Hirakud reservoir.

Among the available types of power generation schemes, hydropower is one of the clean and green schemes. It has many advantages like quick response to meet demands, absence of pollution, and free of fuel cost. To satisfy the power demand of a country, hydropower generation maximization is a vital task [5]. This research deals with maximization of hydropower generation through optimal reservoir release in a multi-purpose reservoir. This work was carried out with the idea of diversion of the entire reservoir release to power plants for power production and then to other purposes. Different release policies were also proposed to assess and improve the hydropower generation. In this work, reservoir operation model for hydropower generation is solved by using nonlinear programming (NLP) technique and particle swarm optimization (PSO). The optimal release attained by PSO and NLP is compared and was found that PSO gave best optimal release, whose results were in turn applied to optimize the hydropower generation. From the optimal release, hydropower generation for the future years was also forecasted using artificial neural network (ANN) technique.

2 Study Area Description

The study area considered in this research is Aliyar reservoir located in TamilNadu, India. The Aliyar reservoir is one of the reservoirs in Parambikulam-Aliyar project (PAP), which is an interstate project [13] of TamilNadu and Kerala states of India. Aliyar reservoir is a multi-purpose scheme, and the water available in the dam is used for the following purposes: irrigation, drinking water, power generation, and flood control. Also as per interstate agreement of TamilNadu and Kerala, Government of Tamil Nadu should release 7250 Mcft water annually from Aliyar reservoir to Kerala for irrigation. The installed capacity of power generation is 1250 kW with two units (2 × 1250 kW). The powerhouse generates the energy by utilizing the water discharged directly from the reservoir. Then, the water passes for other purposes. The reservoir inflow, utilization pattern, and reservoir details were collected from Public Works Department and Water Resource Organization, Pollachi, TamilNadu.

3 Reservoir Release Optimization for Hydropower Generation

Hydropower generation maximization is a highly essential operation in a multi-purpose reservoir among other purposes. Since, electricity demand from sectors like industry, domestic, and agriculture is increasing. Hydropower production is expressed in terms of kWh as given as Eq. (1).

$$PH_t = Z * R_t * NH_t * \eta \tag{1}$$

where Rt is the release to the powerhouse, NHt is the net head, η is the overall plant efficiency during the time period t, and Z is the constant for converting the product hydropower into kilowatt hours (kWh) [9]. It takes the value 2725. NHt is calculated by subtracting the TWL from average head H_t during time period t.

In this research, the objective function is framed to maximize the hydropower production over a year. The objective function is expressed as given in Eq. (2).

$$\text{Max} Z = \sum_{t=1}^{n} PH_t \tag{2}$$

The constraints given below of net head level, power production limit, release, storage limits, evaporation, water balancing, and overflow are in Eq. (3) to Eq. (10), respectively, subjected to the above objective function.

$$NH_t \geq MDDL \tag{3}$$

$$Pmin \leq PH_t \leq Pmax \tag{4}$$

$$Rmin \leq Rt \leq Rmax \tag{5}$$

$$Sdead \leq St \leq Smax \tag{6}$$

$$EV_t = a_t + b_t \left(\frac{S_t + S_{t+1}}{2} \right) \tag{7}$$

$$S_{t+1} = S_t + I_t - R_t - O_t - EV_t \tag{8}$$

$$O_t = S_{t+1} - S_{max} \tag{9}$$

$$O_t \geq 0 \tag{10}$$

where NH_t—net head; P_{min}, P_{max}—minimum and maximum power production limits; R_{min}, R_{max}—minimum and maximum reservoir release; S_{dead}, S_{max}—minimum, maximum storage level; I_t—inflow; I_{min}, I_{max}—minimum, maximum inflow; TWL—tail water level; EV_t—evaporation; O_t—overflow; $MDDL$—minimum drawn down level;

Alcigeimes B. Celesteis [2] recommended an ISO optimization model which is used to obtain optimal allocation of water. The equation model is given in Eq. (11).

$$R(t) = D(t) \left[\frac{\sqrt{S_t^2 + I_t^2} - \sqrt{S_{dead}^2 + I_{min}^2}}{\sqrt{S_{max}^2 + I_{max}^2} - \sqrt{S_{dead}^2 + I_{min}^2}} \right]^m \tag{11}$$

4 Particle Swarm Optimization (PSO)

PSO algorithm developed based on birds and fish simulation. The core idea of PSO is interaction behavior among each particle (birds) in a group. This behavior helps the group members in the search space to search or to reach the most promising area. Also, each particle changes its best position within the search space based on the swarm best position [7]. PSO has been successfully applied to various water resources problems like optimal release, water distribution system, unit commitment in power station, and hydrothermal scheduling. Considering the advantages and the previous works, PSO was taken to maximize the hydropower generation at Aliyar reservoir.

```
Begin (Initialization, i=1,...,N)
   Generate Initial position X_i (0)
   Generate Initial velocity V_i (0)
End                            .
Set n=0 (n is the iteration number)
While (termination criteria not met) do
   For (i=1,...,N)
      Compute the fitness function value
      Compute (pbest)
   End For
Compute (gbest)
   For (i=1,...,N)
      Calculate V_i^{n+1} using equation (13)
      Calculate X_i^{n+1} using equation (14)
End For
   Set next iteration (n=n+1)
End While
```

Every decision variable involved in the problem is represented by a single dimension. Initially, a set of solution is randomly generated for each particle in the swarm. In D-dimensional search space, the position of *ith* particle can be denoted as $Xi = (xi1, xi2, \ldots, xiD)$. Based upon the velocity, the position of the particle changes, which can be denoted as $Vi = (vi1, vi2, \ldots, viD)$. The best position visited by the particle is calculated and stored while the model is executed. $Pi = (pi1, pi2, \ldots, piD)$ is the best position of the *ith* particle. Particles' velocity and position can be updated by using the following Eqs. (12) and (13).

$$V_{id}^{n+1} = V_{id}^{n} + C1 * Rnd(0,1) * \left(pb_{id}^{n} - X_{id}^{n}\right) + C2 * Rnd(0,1) * gb_{gd}^{n} + X_{id}^{n} \quad (12)$$

$$X_{id}^{n+1} = X_{id}^{n} + V_{id}^{n} \quad (13)$$

where $C1$, $C2$—cognitive and social parameters of PSO; D—index for decision variables; I—index of particles, gd—index of best particle; $Rnd(0,1)$—random function.

5 Result and Discussions

5.1 Reservoir Releases Optimization for Hydropower Generation Maximization

This research work was focused on maximization of hydropower generation in Aliyar reservoir using the model given in the Sect. 3. The formulated model was solved by using particle swarm optimization algorithm coded in Java and also by nonlinear programming technique solved by using LINGO/Global solver 14.0.

Hydropower production needs more water to maximize the hydropower production. But large quantity of water could not be utilized for power production in multi-purpose reservoir, because of disputes among various stakeholders. Since the quality and quantity of water will be retained as such after power generation, the present research was carried out with the idea of diversion of the entire optimal reservoir release to power plants for power production and then to other purposes. Different release policies based on the idea were framed to maximize the hydropower generation. Earlier, D. Kiruthiga and T. Amudha [8] have framed two release policies based on this idea to maximize the hydropower production in Aliyar reservoir. In this paper, two more release policies were also included. In addition, the performances of different release patterns were analyzed. These different release policies gave promising results in exploiting the full potential of hydropower generation in Aliyar reservoir.

The first step in the research was to find the optimal release based on the reservoir demands. Aliyar reservoir has a standard annual demand which includes Kerala irrigational demand, irrigation demand of new and old ayacuts of Tamil-Nadu, and drinking water [13]. The demands are shown in Table 1.

The optimal release of Aliyar reservoir is obtained by PSO technique and also by NLP technique using LINGO/Global solver. To solve the hydropower maximization problem by using PSO algorithm, proper selection of parameters like number of particles and acceleration constants $C1$, $C2$ helps in getting quicker optimal solution. The value of the parameter usually differs depending upon the problem and the number of decision variables. After the conduct of experimental analysis with various population sizes and iterations, it is found that the population size of 5, $C1$ and $C2$ values of 1.0 and 0.5, and 100 iterations gave good performance for yearly timescale. The parameter selection of PSO is given in Table 2. The optimal release of Aliyar reservoir for the years, 2009–2013 obtained through PSO and NLP is compared in Fig. 1.

It was observed from the results that PSO yields better optimal reservoir release for the given demand than the NLP. Hence, the optimal release given by PSO was considered for maximization of hydropower generation of Aliyar reservoir. In order

Table 1 Demands of Aliyar reservoir (Annually)

Nature of demand	Demand in (mm^3)
Kerala demand (As per interstate agreement [5])	205
Irrigation—Old Ayacut	70
Irrigation—New Ayacut	35
Drinking water and industrial purpose	45

Table 2 Parameter Values of PSO

Variables	No. of particles	No. of runs	Constants [$C1$, $C2$]
Values	5	100	[1.0, 0.5]

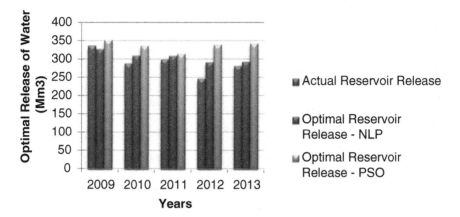

Fig. 1 Annual optimal release of Aliyar obtained through NLP and PSO

to improve the hydropower production, four different release policies were proposed. These different policies will be helpful to explore the full potential of power generation in Aliyar reservoir. The proposed policies considered in this present research are as follows:

Reservoir Policy 1: Diversion of Kerala irrigational water through powerhouse, presently Aliyar reservoir follows this policy;

Reservoir Policy 2: Diversion of both Kerala irrigational water and old ayacut irrigation water through powerhouse;

Reservoir Policy 3: Diversion of Kerala irrigational water, old and new ayacut irrigation water of TamilNadu through powerhouse;

Reservoir Policy 4: Diversion of entire optimal water release got through PSO.

In order to explore the full potential of the Aliyar reservoir in power production, the above proposed four policies are discussed in the following sections. In case of release policy, 1, 58% of optimal release was diverted through powerhouse for power production, rest of the water is directly sent to the respective demands without utilization for power production. In this pattern, both the power production capacity constraint (Eq. (5)) and release constraint (Eq. (6)) of reservoir were considered. This is the actual policy being followed in Aliyar reservoir. From the result of proposed release policy 1, it was observed that the power production could be considerably increased through optimal reservoir release with the confined plant capacity and release capacity. The other three proposed patterns are framed to explore the potential of Aliyar reservoir with the relaxation of constraints.

The actual capacity of Aliyar reservoir is 14 Giga Watt hour (GWh). In release policy 2, the power production exceeded the plant capacity when diversion of water is done to satisfy Kerala irrigation demand and old ayacut demand of TamilNadu irrigation. Hence, in this policy 2, the plant capacity constraint (Eq. (5)) was not considered. Nearly, 68% of optimally released water was utilized for this policy. The release policy 3 and policy 4 did not consider both the release and power plant

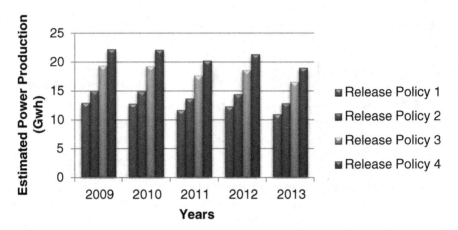

Fig. 2 Annual power production obtained through proposed released policies

capacity constraints. These two proposed policies are helpful to explore the full potential of Aliyar reservoir in power generation. In release policy 3, the demands of Kerala irrigation, TamilNadu irrigation of new ayacut and old ayacut were sent through the powerhouse for power production. In policy 4, the entire optimally released water was sent through the powerhouse. In policy 4, the drinking and industrial demands were included compared with policy 3. The annual power production for the years 2009–2013 from proposed policies are shown in Fig. 2.

This research has estimated a maximal power production within the confined plant capacity in release policy 1, whereas the estimated maxima in other policies have exceeded the plant capacity and release capacity. The release policy 4 has performed well when compared with other policies. Since, it has no constraint in water release and plant capacity.

This research has identified that power production could be considerably increased in Aliyar reservoir power plant, than the actual power production. This was achieved through optimal release and diversion of entire water to powerhouse and then to the respective demand areas. If this is implemented in all the reservoirs and corresponding power plants in TamilNadu, power production could be increased than the current scenario, which in turn increases the contribution of hydropower in the overall power contribution. The important reason to maximize the hydropower generation is that it is highly environment-friendly method of power generation.

5.2   Reservoir Inflow and Hydropower Forecast Using Artificial Neural Network

Hydrological variables forecasting plays an important role in water resource management. Since, these are required to make decisions for numerous purposes in reservoir operations. Especially, inflow forecast is a key component in planning, development, design, operation, and maintenance of water resources [3]. Inflow forecasting is the basis to forecast other water resources application like hydropower generation, and flood control. In this present research, the inflow of Aliyar reservoir is forecasted by using back-propagation artificial neural network. The historical inflow data from January 2009 to June 2014 were used. To train the ANN, 56 months of data set were selected, and then to test the performance of the trained model, the remaining 10 months of data were used. Using the trained data set, the inflow of Aliyar reservoir forecasted for the years 2014–2019. Using the predicted values of inflow, the corresponding storage and release of the reservoir for the years 2014–2019 were estimated. By using the values of predicted inflow, storage, and release, the hydropower production for the years 2014–2019 was estimated.

In this research, multilayered feed-forward ANN procedure is adopted. The back-propagation training algorithm is used to train the ANN. The hyperbolic tangent activation function is used to transfer the input values to next layers. In this research work, the number of hidden layers and neurons are selected based on the trial and error method. In this research, the architecture of the network is one input layer, number of hidden layer is one, number of neuron in hidden layer is eight, and the number of output layer is one (1–8–1). The performance of the network is measured by using mean square error (MSE) and correlation coefficient. In this research, the values of η and α were taken as 0.01 and 0.01, respectively. The maximum number of epochs is 1000. The MSE of the trained network is 0.0189, and it forecasts the testing data set with the accuracy of 0.0283 MSE. The correlation coefficient of training and testing is 0.7371 and 0.7345, respectively.

The ANN models are developed using neural network tool in MATLAB. The ANN estimated the values of inflow for the time series January 2009 to June 2014. Using the predicted values of inflow, the corresponding storage and the release of the reservoir for 2014–2019 were calculated. By using the values of predicted inflow, storage, and release, the hydropower production for the years 2014–2019 was estimated.

Hydropower generation for the years 2014–2019 was calculated with the estimated release. Since demand was not predicted, water from the reservoir was released without the demand constraint. But the reservoir release constraint was considered. That is, the minimum drawn down Level of Aliyar reservoir is 8.90 mm^3, in additional to that 10 mm^3 was also stored for future use. So, 18.90 mm^3 was not released from the total estimated storage. The hydropower estimation using ANN is shown in Fig. 3.

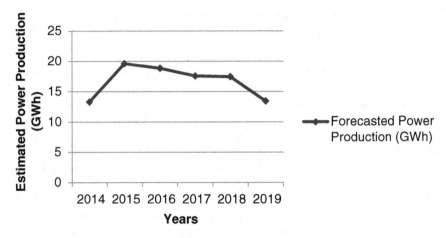

Fig. 3 Forecasted power production of Aliyar reservoir

6 Conclusion

In this research, Aliyar Mini Hydel Power Station is considered as the study area to maximize the hydropower generation using particle swarm optimization. In order to maximize the hydropower in Aliyar power station, release from the Aliyar reservoir was optimized using particle swarm optimization. Then various reservoir release policies (release policies I, II, III, IV) were proposed to optimally release the water to power station. In the release policies, the optimally released water was diverted to power station first then to the respective demand areas. Based on these release policies proposed in this research work, it was proven that the hydropower generation in Aliyar Hydel power station could be maximized. Aliyar reservoir power plant has the capacity of 14 Giga Watt hour (GWh). This research work could be suggested that there is a possibility of maximizing the power production not only through the optimal release of available water, but also through increasing the plant capacity. To estimate the hydropower production for the years 2014–2019, inflow for the Aliyar reservoir was predicted using back-propagation artificial neural network (ANN). To train the ANN, historical data of 2009–2014 were used. From the trained network, next time series data was predicted. Using the predicted inflow, the corresponding hydropower was also calculated. Through this research, it was identified that in Aliyar reservoir power plant, power production could be increased than the actual power production. If the findings of this research are implemented in all the reservoirs and corresponding power plants in TamilNadu, hydropower production could be increased to a greater extent than the current scenario, which will lead to enormous improvement in the contribution of hydropower in TamilNadu's overall power production requirement.

References

1. ADB.: Hydropower Development in India: A Sector Assessment. Asian Development Bank, Publication Stock No. 031607, Philippines (2007).
2. Alcigeimes, B., Celeste., Wilson, F., Curi., Rosires, C., Curi.: Implicit Stochastic Optimization For Deriving Reservoir Operating Rules In Semiarid Brazil. Operations Research. 29, 223–234 (2009).
3. Arunkumar, R., Jothiprakash, V.: Artificial Neural Network Models for Shivajisagar Lake Evaporation Prediction. National Journal on Chembiosys. 2(1), 35–42 (2011).
4. Arunkumar, R., Jothiprakash, V.: Multi-reservoir Optimization for Hydropower Generation Using NLP Technique. KSCE Journal of Civil Engineering. 18(1), 344–354 (2014).
5. Arunkumar, R., Jothiprakash, V.: Optimal Reservoir Operation for Hydropower Generation Using Non-Linear Programming Model. Journal of The Institution of Engineers, India Ser A. 93 (2), 111–120 (2012).
6. Devamane, M.G., Jothiprakash, V., Mohan, S.: Non-linear Programming Model for Multipurpose Multi-reservoir Operation. Hydrology Journal. 29(3–4) (2006).
7. Ghimire, B.N.S., Reddy, M.J.: Optimal reservoir operation for hydropower production using particle swarm optimization and sustainability analysis of hydropower. ISH. J. Hydraul. Eng. **19**(3), 196–210 (2013).
8. Kiruthiga, D., Amudha, T.: Optimal Reservoir Release for Hydropower Generation Maximization Using Particle Swarm Optimization. Innovations in Bio-Inspired Computing and Applications, Advances in Intelligent Systems and Computing, Springer, 424, 577–585 (2015).
9. Loucks, D.P., Stedinger, J.R., Haith, D.A.: Water Resources Systems Planning and Analysis. Prentice Hall Inc, Englewood Cliffs, New Jersey (1981).
10. Nagesh, Kumar, D., Janga, Reddy, M.: Ant Colony Optimization for Multi-Purpose Reservoir Operation. Water Resources Management. 20, 879–898 (2006).
11. Nagesh, Kumar, D., Janga, Reddy, M.: Multiobjective Differential Evolution with Application to Reservoir System Optimization. Journal of Computing in Civil Engineering. (2007).
12. Nagesh, Kumar, D., Janga, Reddy, M.: Performance Evaluation of Elitist-Mutated Multi-Objective Particle Swarm Optimization for Integrated Water Resources Management. Journal of Hydroinfomatics. 11.1, 79–88 (2009).
13. Public Works Department: Aliyar Reservoir. Water Resource Organization, Pollachi Region, Coimbatore (2013).
14. Simonovic, S.P.: Reservoir System Analysis: Closing Gap Between Theory and Practice. Journal of Water Resources Planning and Management. 118(3), 262–280 (1992).
15. Sreenivasan, K.R., Vedula, S.: Reservoir Operation for Hydropower Optimization: A Chance-Constrained Approach. Sadhana, 211(4), 503–510 (1995).
16. Wurbs, R.A.: Reservoir-system Simulation and Operation Models. Journal of Water Resource Planning and Management. 119(4), 455–472 (1993).
17. Yakowitz, S.: Dynamic Programming Applications in Water Resources. Water Resources Research. 18(4), 673–696 (1982).
18. Yeh, W.W.G.: Reservoir Management and Operation Models: A State-of-the-Art Review. Water Resources Research. 21(12), 1797–1818 (1985).

Automatic Identification and Classification of Microaneurysms, Exudates and Blood Vessel for Early Diabetic Retinopathy Recognition

Vaibhav V. Kamble and Rajendra D. Kokate

Abstract Diabetic retinopathy (DR) is vital concern that leads to blindness in adults around the world. In this paper, we proposed a system for early identification and classification of retinal fundus images as DR or non-DR. The ophthalmic features like blood vessels, microaneurysms and exudates are extracted and calculated by applying morphological of 2D median filter, multilevel histogram analysis and intensity transformation, respectively. The proposed system is executed on DIARETDB0 130 and DIARETDB1 89 fundus images dataset using artificial neural networks (ANNs). Result analysis is completed by calculating mean, variance, standard deviation, and correlation. We trained the proposed system model by multilayer perceptron with back-propagation, and system achieved sensitivity 0.83 and specificity 0.045 for DIARETDB0 and sensitivity 0.95 and specificity 0.2 for DIARETDB1.

Keywords Diabetic retinopathy · Blood vessels · Microaneurysms
Exudates · Tortuosity · ANN · Fundus images

1 Introduction

Diabetic retinopathy associated with the study of retinal alterations seen in cases with diabetic mellitus (sugar diabetes). With escalations in the anticipation of diabetics, the frequency of DR has increased. It is the prime reason of blindness. Hazards of diabetic retinopathy are as follows: (1) Duration of diabetes is the most

V. V. Kamble (✉)
Department of Electronics and Tele-communication, Dr. Babasaheb Ambedkar
Marathwada University, Aurangabad 431004, Maharashtra, India
e-mail: kamblevv@gmail.com

R. D. Kokate
Department of Instrumentation Engineering, Government College of Engineering,
Jalgaon 425001, Maharashtra, India
e-mail: rdkokate3394@gmail.com

© Springer Nature Singapore Pte Ltd. 2019   423
H. S. Behera et al. (eds.), *Computational Intelligence in Data Mining*,
Advances in Intelligent Systems and Computing 711,
https://doi.org/10.1007/978-981-10-8055-5_38

significant determining factor. Generally in the cases (patients), DR grows 50% after 10 Yr, 70% after 20 Yr and 90% after 30 Yr. (2) In females, DR ratio (4:3) is more than males. (3) Deprived metabolic control is less significant, but pertinent to the growth of DR. (4) The proliferative retinopathy is more due to heredity. (5) Pregnancy may hasten the alteration of DR. (6) DR changes happen to be hypertension. (7) Smoking, anaemia, obesity and hyperlipidemia are some hazardous aspects [1].

DR has been categorized as (a) non-proliferative diabetic retinopathy (NPDR) (b) Proliferative diabetic retinopathy (PDR). NPDR is diagnosis of early diabetic retinopathy by analysing blood vessels in the retina, little swells in the blood vessels and swelling of the macula. PDR occurs as advanced form of the DR disease [1, 2].

Moreover on the basis of severity, NPDR has been classified as mild, moderate and severe under ophthalmic features shown in Table 1.

Table 1 Severity NPDR has classified under ophthalmic features

No DR			Features: Microaneurysms are seen near tiny blood vessels in the macular area. It looks as red spot. It is larger than 125 μm in size
Mild NPDR	At least one microaneurysm or intraretinal haemorrhage Hard/soft exudate may or may not be present		Hard exudates were yellowish white waxy-appearing spots that are arranged in clumps or in circinate pattern Cotton wool spots or soft exudates were small whitish fluffy superficial lesions
Moderate NPDR	Microaneurysm/ intraretinal haemorrhages in 2 or 3 quadrants Exudates may or may not be present		Intraretinal microvascular abnormalities (IRMA) seen as fine irregular red lines connecting arterioles with venules represent arteriolar-venular shunts
Severe NPDR	Four quadrants of severe microaneurysms/ intraretinal haemorrhages Two quadrants of venous beading		Oedema characterized by retinal thickening is caused by capillary leakage

2 Related Work

Bo Wu et al. [3] studied the fundus image; total 27 features are extracted which enclose local and profile features for KNN classifier to diagnose DR. These techniques were implemented on two dataset, i.e. e-ophtha and ROC.

Rosas-Romero et al. [4] worked on image processing applied to detection of microaneurysms in fundus images; principal component analysis (PCA) and the Radon transform were used to recognize real microaneurysms and false positives (FP). Implemented system reaches to sensitivity, specificity of 92.32, 93.87% for DIARETDB1 dataset and 88.06, 97.47% for ROC dataset.

Sharath Kumar P N et al. [5] studied optical disk, blood vessels, white lesion and red lesions extracted from fundus images of the Regional Institute of Ophthalmology (RIO). The proposed system recognized retina as DR or non-DR. Their system yielded sensitivity and specificity were 80% and 50%, respectively.

G. Mahendran et al. [6] worked on to extract exudate from fundus images of MESSIDOR database. This system observed 97.89% and 94.76% recognition rate for SVM and PNN classifiers, respectively.

Jose Tomas Arenas Cavalli et al. [7] have developed; NN classifier is used to categorize fundus image regions Retinopathy Screening', in this research blood vessel and optic disk (OD) localization along with bright into exudates and nonexudates. Proposed system analysis is done at Penalolen Health Reference Center 550 images database and got in 91.89% specificity and 65.24% sensitivity for DR.

Harry Pratt et al. [8] developed CNN methodology on Kaggle dataset. Clinical features, i.e. microaneurysms, exudate and haemorrhage, were extracted from fundus images to do classification. Final results are estimated in (95%) specificity and (30%) sensitivity for DR recognition.

Pavle Prentasic et al. [9] tested experimental process on our method on the DRiDB database and extracted ophthalmic retinal features like blood vessels, optic disk, exudates and hemorrhages. Experimental test result has been calculated as true positives (TPs), false positives (FPs) and false negatives (FNs) using NN.

3 System Overview

Our proposed system consists of database, feature extraction techniques, ANN classifier and results. Figure 1 shows complete flow diagram of diabetic retinopathy recognition system.

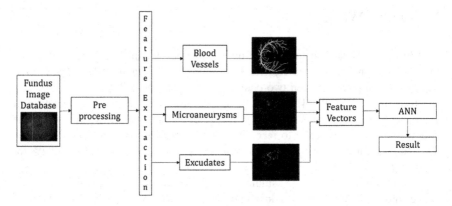

Fig. 1 Architecture of proposed diabetic retinopathy recognition system

3.1 Database

For the experimental work to detect severity NPDR, we use the publically available database, i.e. DIARETDB0 and DIARETDB1. This database is available from the Web site http://www.it.lut.fi/project/imageret. DIARETDB0 and DIARETDB1 databases consist of 130 and 89 colour fundus images, respectively [10].

3.2 Feature Extraction

There are various ophthalmic/clinical DR features likewise optic disk, blood vessels, exudate, microaneurysms, haemorrhage, IRMA and macular oedema. To do experimental work, we extracted blood vessels, microaneurysms and exudate.

Pre-processing: First step of feature extraction is pre-processing, and it consists of three steps: (i) Convert RGB image to gray level (ii) filtering (low-pass filtering (Wiener filtering using a 3×3 filter)) (iii) contrast enhancement [6, 11]. The optical disk identification in fundus image is very significant assignment as its resemblance to exudates (intensity, color and contrast). It always seems in exudate exposure results and hence therefore necessity to mask it out [12].

Blood vessel Extraction: Next step of pre-processing is 2D Gabor matched filter to do segmentation of blood vessel. The retinal blood vessel shape is similar to tube structure; its cross section can be mathematically related to a Gabor response. So, the Gabor response filter is stated as in Eqs. 1 and 2 [13].

$$G(x,y) = \exp\left\{ -\left(\frac{x_1^2}{\sigma_x^2} + \frac{y_1^2}{\sigma_y^2}\right) \right\} \cos\left(\frac{2\pi x_1}{\lambda}\right) \tag{1}$$

where $x_1 = y_1 = x\cos\theta + y\sin\theta$ and $\sigma_x = \sigma_y = \sigma$

$$G(x,y) = \exp\left(\frac{2x_1^2}{\sigma^2}\right) \cos\left(\frac{2\pi x_1}{\lambda}\right) \tag{2}$$

A 15 × 15 window size gives an accurate result. As the images have blood vessels of varying width, σ has two values, i.e. for small vessels and the large vessels. Let r(x) mean the peak of the truncated Gabor function. The threshold value "T" is stated at the points x = (x + d) and (x − d) as in the following inequalities Eqs. 3 and 4 [13].

$$r(x) - r(r - d) > T \tag{3}$$

$$r(x) - r(r + d) > T \tag{4}$$

where "d" is found by multiplying "σ" by c_d as shown in Eq. 5 [13].

$$d = c_d \times \sigma \tag{5}$$

where c_d is a constant. However, if c_d is large, there will be interference between neighbouring vessels. If c_d is small, it means the width of the line will be underestimated. c_d is locating around two struck a well balance among spotting the entire line and minimizing the interference from neighboring vessels as stated in Eqs. 6, 7 and 8 [13].

$$t = r(x) - r(x + d) \tag{6}$$

$$= r(x) - r(x - d) \tag{7}$$

$$t = 1 - \left(\exp^{\frac{d^2}{2\sigma}}\right) \cos\left(\frac{2\pi d}{\lambda}\right) \tag{8}$$

Since, every convolution outcome produces more intensity pixels indicating vessels in that direction. Hence, amount of total number of pixels is the part of blood vessels [13].

Microaneurysms Extraction: Microaneurysms (MAs) is also known as red lesion. We want to extract red lesion so we need to eliminate other lesions from the image. Its circular shape has diameter less than 125 mm and is extracted in local window size (n). The n value can adjust to detect red lesion. Intensity, size and shape are the characteristics to detect MAs. The maximum diameter of the biggest MAs in pixel equal to size of n was used. Thresholding is a mode to calculate the intensity. Vascular image is eliminated from the processed image to get red lesion image. We have obtained binary image with red lesion by thresholding the match-filtered image [14].

Exudates Extraction: Exudates is also known as white lesion; it appears like yellowish region in fundus image. To detect exudates, image is pre-processed by doing green channel conversion, greyscale conversion and binary conversion. After that morphological closing and finally removing blood vessels to get exudates. Exudates look more contrast in the green channel. Therefore, the green channel component is used to detect exudates. Subsequently, deciding threshold level image was converted into binary with pixel values 1 and 0. Next stage is morphological closing, i.e. optic disk has alike features associated with exudates, and thus, it is necessary to remove it. Finally, all other components are eliminated and the rest of the binary images having exudates are obtained [13].

3.3   ANN Classifier

The most frequently used feed-forward artificial neural network is the multilayer perceptron (MLP). MLP forms a pattern that maps the input neurons to the computational output, and it is shown in Fig. 2. MLP appears as the multiple layers of nodes in a graph, with each layer completely linked to the next one. MLP utilizes back-propagation method to train the network which is classified as a supervised learning technique. From linear perceptron, the back-propagation is reformed and it employs three or more hidden layers with nonlinear activation function [15, 16]. The main activation function used is sigmoid and is described as in Eq. 9 [16].

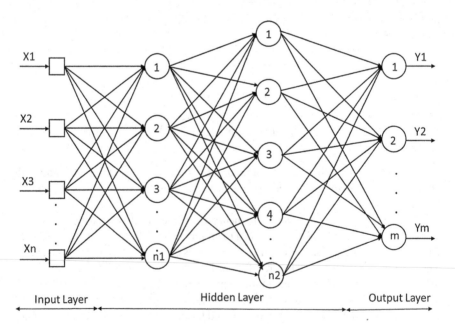

Fig. 2 Architecture of feed-forward artificial neural network with MLP

$$y(v_i) = (1 + e^{-v_i})^{-1} \qquad (9)$$

This logistic function ranges in between 0 to 1. Here, y_i means output of the i_{th} node (neuron) and v_i means weighted sum of the input synapses.

The error at the output is compared with expected result; a learning process occurs in the perceptron by changing the joining weights after each node. The error in output node j in the n_{th} data point shown in Eq. 10 [16].

$$e_j(n) = d_j(n) - y_j(n) \qquad (10)$$

where d indicates the target value and y indicates the value originated by the perceptron. Improving the weights of the nodes is made to minimize the error occurred in the output and given by Eq. 11 [16].

$$\varepsilon(n) = \frac{1}{2} \sum_j e_j^2(n) \qquad (11)$$

Using gradient descent, the change is observed in each weight and described in Eq. 12 [16].

$$\Delta w_{ji}(n) = -\eta \frac{\partial \varepsilon(n)}{\partial v_j(n)} y_i(n) \qquad (12)$$

where η indicates the learning rate and y_i indicates the outcome of the previous neuron.

4   Performance Analysis

Performance analysis of proposed system is executed on MATLAB platform. A graphical user interface (GUI) is designed. Database is classified into training and testing samples. Testing images are browsed from the GUI. Blood vessels, microaneurysms and exudate are the features extracted from test image.

According to training sample, classifier ANN is trained to classify the test image into DR and NPDR. Figure 3 shows the proposed system GUI.

We calculate the mean, variance, standard deviation and correlation for all test datasets using following illustrated formulas in Eqs. 13, 14, 15 and 16 [17, 18]. The coefficient of correlation (r) can range from +1 to −1. A value of 0 denotes that there is no relationship between the two variables. A value greater than 0 represents a positive relationship; that is, as the value of one variable increases, other variable value also increases. A value less than 0 represents a negative relationship; i.e. as the value of one variable increases, other variable value decreases [17].

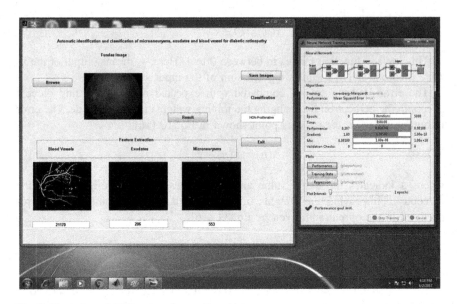

Fig. 3 Illustration of GUI system for result analysis

$$\text{Mean} = \frac{\text{sum of all elements}}{\text{total no of elements}} \qquad (13)$$

$$\text{Variance} = \frac{\sum(x - \bar{x})}{N} \qquad (14)$$

$$\text{Standard Deviation} = \sqrt{\text{Variance}(x)} \qquad (15)$$

$$\text{Correlation } (r) = \frac{\sum(x - \overline{X}) \sum(y - \overline{Y})}{\sqrt{\sum(x - \overline{X})^2 \sum(y - \overline{Y})^2}} \qquad (16)$$

In Tables 2, 3 and 4, we represent statistical parameter for blood vessels, microaneurysms and exudates, respectively, for 5 test samples. In this way, we can calculate more test dataset.

The performance of the proposed system is calculated using sensitivity and specificity using following formulas by Eqs. 17 and 18 [17, 18]. Sensitivity means detection is positive, i.e. for DR cases. Specificity means detection is negative, i.e. for Non DR cases.

$$\text{Sensitivity} = \text{TP}/(\text{TP} + \text{FN}) \qquad (17)$$

$$\text{Specificity} = \text{TN}/(\text{TN} + \text{FP}) \qquad (18)$$

Table 2 Statistical parameter for blood vessels

Sr. no	Image name	Diameter (x)	Tortuosity (y)	$(x - \overline{X})$	$(y - \overline{Y})$	$(x - \overline{X}) * (y - \overline{Y})$	(xy)
1	05_test	15	2	14.53	1.86	27	30
2	06_test	13	3	12.53	2.86	35.82	39
3	07_test	15	3	14.53	2.86	41.54	45
4	08_test	10	2	9.53	1.86	17.71	20
5	09_test	13	2	12.53	1.86	23.29	26

Table 3 Statistical parameter for microaneurysms

Sr. no	Image name	Manual counting of microaneurysm (x)	Microaneurysm by algorithm (y)	$(x - \overline{X})$	$(y - \overline{Y})$	$(x - \overline{X}) * (y - \overline{Y})$	(xy)
1	19_test	136	136	119.61	119.24	14262.3	18496
2	20_test	143	143	126.61	126.24	15983.25	20449
3	21_test	149	149	132.61	132.24	17536.35	22201
4	22_test	31	31	14.61	14.24	208.05	961
5	23_test	88	89	71.61	72.24	5173.11	7832

Table 4 Statistical parameter for exudates

Sr. no	Image name	Manual counting of exudates (x)	Exudates by algorithm (y)	$(x - \overline{X})$	$(y - \overline{Y})$	$(x - \overline{X}) * (y - \overline{Y})$	(xy)
1	49_test	302	309	300.24	307.19	92231.04	93318
2	50_test	37	37	35.24	35.19	1240.13	1369
3	51_test	66	66	64.24	64.19	4123.64	4356
4	52_test	65	65	63.24	63.19	3996.21	4225
5	53_test	39	39	37.24	37.19	1385	1521

where

TP   True positive = retina correctly identified
FN   False negative = retina wrongly rejected
TN   True negative = retina correctly rejected
FP   False positive = retina wrongly identified

Contingency Table 5 shows the performance analysis for DIARETDB1 and DIARETDB0 database, respectively.

From Table 5, we observed the sensitivity 0.955 and specificity 0.2 for DIARETDB1, and for DIARETDB0, we observed sensitivity 0.838 and specificity 0.045.

Table 5 Performance analysis for DIARETDB1 and DIARETDB0 database

Test	Disease present	n		Disease absent	n			Total		
		DIARETDB1	DIARETDB0		DIARETDB1	DIARETDB0			DIARETDB1	DIARETDB0
Positive	True positive	a = 85	a = 109	False positive	c = 4	c = 21			a + c = 89	a + c = 130
Negative	False negative	b = 4	b = 21	True negative	d = 1	d = 1			b + d = 5	b + d = 22
Total		a + b = 89	a + b = 130		c + d = 5	c + d = 22				

5 Conclusion

In this paper, we have studied an automatic identification and classification of microaneurysms, exudates and blood vessel for diabetic retinopathy. The proposed system is implemented on DIARETDB0 and DIARETDB1 databases to early detection of retina as DR or NPDR through ANN classifier. In this paper, we concentrate on ophthalmic features like microaneurysms, blood vessel and exudates. These extracted feature vectors employed to multilayer perceptron with back-propagation neural network classifier to train the network. The performance of the system was evaluated on the merits of sensitivity and specificity. The proposed system achieved sensitivity 0.83 and specificity 0.045 for DIARETDB0 and for DIARETDB1 sensitivity 0.95 and specificity 0.2.

References

1. Book: A K Khurana;Comprehensive Opthalmology.
2. M. R. Krishnan Mookiah, U. R. Acharya, C. K. Chua, C. M. Lim, E.Y.K. Ng, A. Laude; Computer-aided diagnosis of diabetic retinopathy: A review; Computers in Biology and Medicine 43 (2013)2136–2155.
3. Bo Wu, W. Zhu, Fei Shi, S. Zhu, X. Chen; Automatic detection of microaneurysms in retinal fundus images; Computerized Medical Imaging and Graphics 55 (2017) 106–112.
4. Roberto R. R., Jorge M-C., Jonathan H-C., Laura J. Uribe V.; A method to assist in the diagnosis of early diabetic retinopathy: Image processing applied to detection of microaneurysms in fundus images; Computerized Medical Imaging and Graphics 44 (2015) 41–53.
5. P N Sharath K., Deepak R U, A. Sathar, Sahasranamam V, Rajesh K. R; Automated Detection System for Diabetic Retinopathy Using Two Field Fundus Photography; Procedia Computer Science 93 (2016) 486–494.
6. R. Dhanasekaran, G. Mahendran; Investigation of the severity level of diabetic retinopathy using supervised classifier algorithms; Computers and Electrical Engineering 45 (2015) 312–323.
7. Jose T. A-C., S. A. Rıos, M. Pola, R. Donoso; A Web-Based Platform for Automated Diabetic Retinopathy Screening; Procedia Computer Science 60 (2015) 557–563.
8. H. Pratt, Frans C., D. M Broadbent, S. P Harding, Y. Zheng; Convolutional Neural Networks for Diabetic Retinopathy; Procedia Computer Science 90 (2016) 200–205.
9. P. Prentasic, S. Loncaric; Detection of exudates in fundus photographs using deep neural networks and anatomical landmark detection fusion; computer methods and programs in biomedicine 137 (2016) 281–292.
10. http://www.it.lut.fi/project/imageret/.
11. Priyakshi B., Dr. S.R. Nirmala and Jyoti P. M.; Detection of Hemorrhages in Diabetic Retinopathy analysis using Color Fundus Images; 2015 IEEE 2nd International Conference on Recent Trends in Information Systems (ReTIS); 978-1-4799-8349-0/15/$31.00 ©2015 IEEE.
12. Saiprasad R., Arpit J., A. Mittal; Automated Feature Extraction for Early Detection of Diabetic Retinopathy in Fundus Images; 978-1-4244-3991-1/09/$25.00 ©2009 IEEE.
13. M.R.K. Mookiah, C. K.Chua, U. R. Acharya, R. J. Martis, A. Laude, C.M. Lim, E.Y.K. Ng; Evolutionary algorithm based classifier parameter tuning for automatic diabetic retinopathy grading: A hybrid feature extraction approach; Knowledge-Based Systems 39 (2013) 9–22.

14. M. Tavakoli, R. P. Shahri, H. Pourreza, Alireza M., Touka B., Mohammad H. B. T.; A complementary method for automated detection of microaneurysms in fluorescein angiography fundus images to assess diabetic retinopathy; Pattern Recognition 46(2013)2740–2753.
15. P.V.Rao, Gayathri.R, Sunitha.R; A Novel Approach for Design and Analysis of Diabetic Retinopathy Glaucoma Detection using Cup to Disk Ration and ANN; Procedia Materials Science 10 (2015) 446–454.
16. H.C.Vijayalakshmi and H. M. Saifuddin; PREDICTION OF DIABETIC RETINOPATHY USING MULTI LAYER PERCEPTRON; International Journal of Advanced Research (2016), Volume 4, Issue 6, 658–664.
17. B. P. Gaikwad, S. N. Kayte, R R. Manza, S. B. Dabhade; Design and Development for Detection of Blood Vessels, Microneurysms and Exudates from the Retina; AEMDS - 2nd National Conference on Advancements in the Era of Multi-Disciplinary Systems 2013.
18. Manjiri B. Patwari, Ramesh R. Manza, Yogesh M. Rajput et al., "Classification and Calculation of Retinal Blood vessels Parameters", IEEE's International Conferences for Convergence of Technology 2016.

Performance Analysis of Tree-Based Approaches for Pattern Mining

Anindita Borah and Bhabesh Nath

Abstract Extracting meaningful patterns from databases has become a significant field of research for the data mining community. Researchers have skillfully taken up this task, contributing a range of frequent and rare pattern mining techniques. Literature subdivides the pattern mining techniques into two broad categories of level-wise and tree-based approaches. Studies illustrate that tree-based approaches outshine in terms of performance over the former ones at many instances. This paper aims to provide an empirical analysis of two well-known tree-based approaches in the field of frequent and rare pattern mining. Through this paper, an attempt has been made to let the researchers analyze the factors affecting the performance of the most widely accepted category of pattern mining techniques: the tree-based approaches.

Keywords Frequent patterns · Rare patterns · Pattern mining
Data structure

1 Introduction

Mining compelling and meaningful patterns from databases have widespread applications and have become one of the significant fields of research in recent years. Frequent and rare patterns are distinctive in their nature and provide disparate information depending upon the application domain in which they are applied. Frequent patterns on one hand can be applied in planning marketing strategies and drug design, and rare patterns on the other hand can be used for fraud and rare disease detection. Researchers have contributed abundant techniques to retrieve

A. Borah (✉) · B. Nath
Department of Computer Science & Engineering, Tezpur University,
Napaam, Sonitpur 784028, Assam, India
e-mail: anindita01.borah@gmail.com

B. Nath
e-mail: bnath@tezu.ernet.in

© Springer Nature Singapore Pte Ltd. 2019   435
H. S. Behera et al. (eds.), *Computational Intelligence in Data Mining*,
Advances in Intelligent Systems and Computing 711,
https://doi.org/10.1007/978-981-10-8055-5_39

both kinds of patterns that can be helpful in solving many data mining tasks. Attempts for discovering patterns from databases are initiated with the extraction of frequent patterns, moving on to the extraction of rare patterns considering their significance.

The primer endeavor [2] for the generation of frequent patterns was based on a level-wise strategy that involves tedious steps and enormous number of candidate generation. The intricacy and complications involved in level-wise approaches compelled the researchers to employ efficient data structures to reduce the overhead of mining process. The data structure-based approaches popularly known as the pattern growth approaches justified effectiveness over the former ones with their outstanding performance. The most popular among the pattern growth approaches is the tree-based mining technique developed by Han et al. [7]. Their eminent technique called FP-Growth has shown excellent performance over the level-wise approaches, thus becoming the most accepted pattern mining approach for extracting the frequent and rare patterns. FP-Growth stores the entire information of the database into a compact tree-like representation, prominently known as the FP-Tree. This not only minimizes the overhead caused by multiple database scans but also makes the mining process much faster.

Further, the paper is systematized as: Some existing tree-based pattern mining techniques are illustrated in Sect. 2. Section 3 presents the experimental analysis of two well-known tree-based approaches for frequent and rare pattern mining followed by the factors affecting their performance in Sect. 4. The paper finally ends with a conclusion in Sect. 5.

2 Literature Review

Due to the superiority of tree-based approaches of pattern mining over the level-wise ones, literature has been bestowed with numerous pattern mining techniques that employ tree data structures during the mining process. This section discusses some of the well-known approaches that deal with frequent and rare pattern mining.

To cater the memory limitation of FP-Growth, a disk-based approach was developed by Adnan and Alhajj [1]. Keeping in view the need of handling incremental datasets, Koh and Shieh [8] extended the FP-Growth algorithm. To find a special class of patterns called frequent closed patterns, several FP-Tree-based techniques like CLOSET [10] and FPClose [6] were proposed. Lin et al. [9] generated the high utility frequent patterns. Giannella et al. [5] extracted frequent patterns from data streams using tree-based approach. Significance of tree-based approach can be observed in mining big data as well. A parallel approach for the same was proposed by Chen et al. [4]. Being an emerging area, rare pattern mining has attempted only few endeavors based on tree data structure. The most popular among them is the RP-Tree algorithm developed by Tsang et al. [11] that was further enhanced by Bhatt and Patel [3].

3 Performance Analysis of Tree-Based Approaches

The tree-based approaches show striking performance over the level-wise approaches on various grounds. However, there are still some performance gaps in tree-based approaches that need to be resolved. This section provides the performance analysis of two eminent tree-based approaches for extracting frequent and rare patterns. The effectiveness of tree-based approaches is established through experiments on several real-life and synthetic datasets. Synthetic datasets were generated using the data generation technique described in [2], while real-life datasets were obtained from UCI repository. The implementation has been done in Java on a machine of 2.90 GHz Intel i5 processor having 4 GB main memory and 64-bit Windows 8 Pro operating system.

3.1 Frequent Pattern Extraction

To gauge the efficiency of tree-based frequent pattern mining techniques, the best method would be the FP-Growth algorithm discussed in the previous sections. The criteria considered to judge the effectiveness of tree-based approaches are the compactness of tree structure with respect to the original database and the respective execution time. It is to be noted that the execution time considered is for the entire mining process and not only for tree construction. Table 1 and Table 2 illustrate the reduction rate of FP-Tree with respect to the original database for dense and sparse datasets, respectively. Density of database plays an important role on compactness of the tree structure and the execution time. Hence, both dense and sparse datasets have been considered for evaluation. It is to be noted that the execution time for entire mining operation has been considered and not only for itemset generation. The tables show the compactness of tree data structure for a particular dataset. Results given in Tables 1 and 2 illustrate the fact that the reduction rate of FP-Tree tends to decrease and execution time tends to increase when the sparseness of dataset increases.

For better understanding, a graphical analysis of the results obtained is given in Fig. 1. The red bar in the graph represents the nodes generated in the tree in comparison to the total number of items present in the database, which is represented by the blue bar.

3.2 Rare Pattern Extraction

From the tree-based approaches for rare pattern mining, the most eminent is the RP-Tree algorithm. Therefore, to measure the efficiency of tree-based rare pattern mining techniques, RP-Tree algorithm has been taken into account. Similar to

Table 1 Compactness evaluation on dense datasets

Dataset	Category	Transactions in the dataset	Average item per transaction	Number of items in dataset	Nodes generated	Reduction ratio	Execution time (s)
Zoo	Real	101	17	1717	512	3.35	32
Tumor	Real	339	17	5736	1568	3.68	216
Soya bean	Real	631	17	10,727	2435	4.4	1832
Anneal	Real	812	43	34,916	6815	5.12	2321
Chess	Real	3196	17	115,056	38609	2.9	5152
Hypothyroid	Real	3247	44	142,868	11507	4.3	8637
Mushroom	Real	8124	23	186,852	27,349	6.83	15632
T10I20D10K	Synthetic	10,000	10	100,000	19,686	5.08	17712

Table 2 Compactness evaluation on sparse datasets

Dataset	Category	Transactions in the dataset	Average item per transaction	Number of items in dataset	Nodes generated	Reduction ratio	Execution time (s)
Hepatitis	Real	137	35	4795	4144	1.157	348
Lymph	Real	148	28	4144	2015	2.05	236
T10I00D10K	Synthetic	10,000	10	98,764	69,038	1.43	9142
T25I10D10K	Synthetic	10,000	25	409061	216111	1.9	11321
T40I10D15K	Synthetic	15,000	40	598,634	452,836	1.32	22364
T30I10D20K	Synthetic	20,000	10	401,762	278,426	1.45	36547

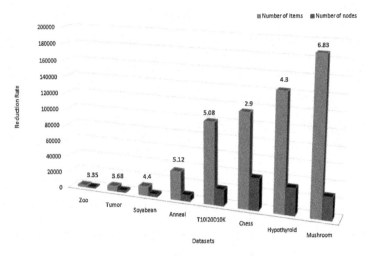

(a) Compactness evaluation on dense datasets

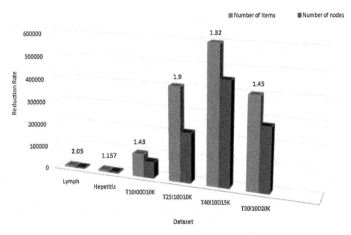

(b) Compactness evaluation on sparse datasets

Fig. 1 Compactness evaluation of FP-Growth on dense and sparse datasets

FP-Growth, effectiveness of RP-Tree algorithm has been evaluated using dense as well as sparse datasets. Table 3 illustrates the experimental results obtained for RP-Tree algorithm in case of dense datasets, and Table 4 illustrates the same for sparse datasets. A graphical analysis of the results obtained is given in Fig. 2.

Table 3 Compactness evaluation on dense datasets

Dataset	Category	Transactions in the dataset	Average item per transaction	Number of items in dataset	Nodes generated	Reduction ratio	Execution time (s)
Zoo	Real	101	17	1717	383	4.48	18
Tumor	Real	339	17	5736	1020	5.62	143
Soya bean	Real	631	17	10,727	1815	5.91	1076
Anneal	Real	812	43	34,916	5283	6.6	3143
Chess	Real	3196	17	115,056	32,471	3.54	7643
Hypothyroid	Real	3247	44	142,868	25,642	5.57	11453
Mushroom	Real	8124	23	186,852	23,323	8.01	38564
T10I20D10K	Synthetic	10,000	10	100,000	15,462	6.46	21436

Table 4 Compactness evaluation on sparse datasets

Dataset	Category	Transactions in the dataset	Average item per transaction	Number of items in dataset	Nodes generated	Reduction ratio	Execution time (s)
Hepatitis	Real	137	35	4795	4023	1.91	256
Lymph	Real	148	28	4144	1842	2.24	116
T10I00D10K	Synthetic	10,000	10	98,764	65,427	1.5	9182
T25I10D10K	Synthetic	10,000	25	409061	196,235	2.08	38645
T40I10D15K	Synthetic	15,000	40	598,634	411,245	1.46	51918
T30I10D20K	Synthetic	20,000	10	401,762	243,628	1.65	78645

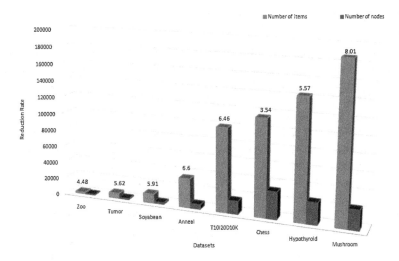

(**a**) Compactness evaluation on dense datasets

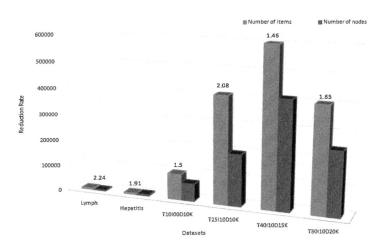

(**b**) Compactness evaluation on sparse datasets

Fig. 2 Compactness evaluation of RP-Tree on dense and sparse datasets

3.3 Observation and Explanation

From experimental evaluation, it has been observed that FP-Growth achieves good compression in case of dense datasets. In case of dense datasets, the transactions are very much similar to each other. A tree data structure-based algorithm is suitable for dense datasets since many transactions will share common prefixes, and henceforth, the database could be compactly represented.

However, in case of sparse data, FP-Growth could not achieve good compression. Since sparse data contains lesser number of frequent items, tree data structures will be large and thicker as there will be fewer shared common prefixes. Let us illustrate this fact with the help of a suitable example given in Fig. 3.

Figure 3a represents a dense dataset that contains many common items appearing frequently. From the resultant tree given in Fig. 3b, it can be observed that many common elements represented by nodes are being shared. This kind of sharing minimizes the nodes generated in the tree for dense dataset making it more

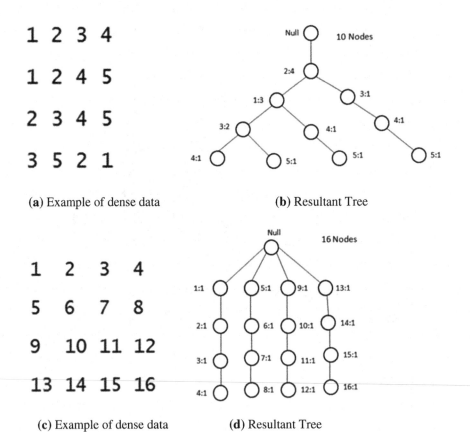

(a) Example of dense data   (b) Resultant Tree

(c) Example of dense data   (d) Resultant Tree

Fig. 3 Example of dense and sparse data

compact. Thus for 16 items in the dense dataset, only 10 corresponding nodes have been generated. On the contrary, all the 16 items in the example sparse dataset given in Fig. 3c are different. Due to this, the resultant tree in Fig. 3d will have 16 corresponding nodes, one node for each item in the dataset.

4 Factors Affecting Performance of Tree-Based Approaches

The performance range of tree-based approaches is demonstrated through experiments on a range of datasets by varying their dimensions. Experiments were performed for FP-Growth algorithm. Synthetic datasets are used for experiments that were generated using the data generation technique developed in [2]. The dataset name shows its characteristics: T → average items present per transaction, I → distinct items present in the database, D → number of transactions in the database. For instance, T10I10D1K indicates that there are 10 items per transaction, 10 distinct items, and 1000 transactions in the dataset.

4.1 Distinct Items Present in the Database

The analysis of the effect of distinct items present in the database on the compactness of FP-Tree is done by performing experiments on several synthetic datasets. Distinct items present in the synthetic datasets were changed keeping the other two factors, average items present per transaction and number of transactions in the dataset, constant.

From experimental evaluation given in Fig. 4, it has been observed that for the same number of transactions in dataset and average items present per transaction, compactness of FP-Tree decreases with the increase in distinct items present in the

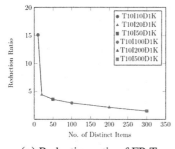

Dataset	Number of Transactions	Number of Items per Transaction	Number of Distinct Items	Reduction Ratio
T10I10D1K	1000	10	10	15.1
T10I20D1K	1000	10	20	4.4
T10I50D1K	1000	10	50	3.56
T10I100D1K	1000	10	100	2.91
T10I200D1K	1000	10	200	2.15
T10I500D1K	1000	10	300	1.5

(a) Reduction ratio of FP-Tree   (b) Characteristics of dataset used

Fig. 4 Compactness affected by distinct items in the database

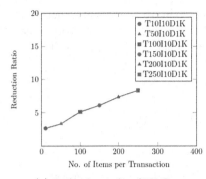

(a) Reduction ratio of FP-Tree

Dataset	Number of Transactions	Number of Distinct Items	Number of Items per Transaction	Reduction Ratio
T10I100D1K	1000	10	10	2.67
T50I100D1K	1000	10	50	3.35
T100I100D1K	1000	10	100	5.13
T150I100D1K	1000	10	150	6.11
T200I10D1K	1000	10	200	7.36
T250I10D1K	1000	10	250	8.35

(b) Characteristics of dataset used

Fig. 5 Compactness affected by average items present per transaction

database. For instance, it can be seen in Fig. 4a that upon varying the distinct items in the database from 10 to 300, reduction ratio of the FP-Tree reduces from 15.1 to 1.5.

4.2 Average Items Present Per Transaction

The effect of average items present per transaction on the compactness of FP-Tree is evaluated by varying that factor and keeping rest of the two factors constant as shown in Fig. 5.

Figure 5a shows that compactness of FP-Tree increases with increase in average items present per transaction. For example, upon increasing the average items present per transaction from 10 to 250, reduction ratio of FP-Tree increases from 2.67 to 8.35.

4.3 Number of Transactions Present in the Database

The numbers of transactions were changed keeping the other two factors constant given in Fig. 6, in order to verify its effect on the compactness of tree structure.

From Fig. 6a it can be observed that, when the number of transaction is increased from 1000 to 3000 with other two factors constant, reduction ratio of FP-Tree increases from 15.1 to 50.6.

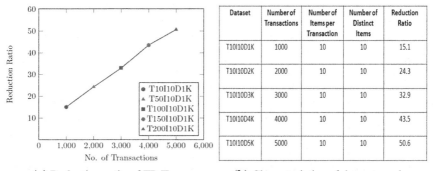

Dataset	Number of Transactions	Number of Items per Transaction	Number of Distinct Items	Reduction Ratio
T10I10D1K	1000	10	10	15.1
T10I10D2K	2000	10	10	24.3
T10I10D3K	3000	10	10	32.9
T10I10D4K	4000	10	10	43.5
T10I10D5K	5000	10	10	50.6

(a) Reduction ratio of FP-Tree   (b) Characteristics of dataset used

Fig. 6 Compactness affected by number of transactions in the database

5 Conclusion

Preeminence of tree-based approaches over level-wise pattern mining approaches can be established from the fact that numerous tree-based techniques were developed for solving the pattern mining challenges. However, the inconsistency of tree-based approaches at certain instances cannot be overlooked and needs utmost attention. Through this paper, an attempt has been made to exemplify the performance of tree-based pattern mining approaches under various conditions.

From the performance evaluation of a frequent and a rare pattern mining technique given in Sect. 3, it can be stated that the efficiency of tree-based approach depends on the characteristics of data to a great extent. The approaches perform better in case of dense datasets while for the sparse ones, the results are not appreciable. Furthermore, the experimental analysis given in Sect. 4 illustrates the factors affecting the performance of tree-based approaches. This study is therefore an attempt to let the researchers analyze the performance bottlenecks of tree-based approaches for developing efficient pattern mining techniques employing data structures.

References

1. Adnan, M., Alhajj, R.: Drfp-tree: disk-resident frequent pattern tree. Applied Intelligence 30 (2), 84–97 (2009)
2. Agrawal, R., Srikant, R., et al.: Fast algorithms for mining association rules. In: Proc. 20th int. conf. very large data bases, VLDB. vol. 1215, pp. 487–499 (1994)
3. Bhatt, U., Patel, P.: A novel approach for finding rare items based on multiple minimum support framework. Procedia Computer Science 57, 1088–1095 (2015)
4. Chen, M., Gao, X., Li, H.: An efficient parallel fp-growth algorithm. In: Cyber Enabled Distributed Computing and Knowledge Discovery, 2009. CyberC'09. International Conference on., pp. 283–286. IEEE (2009)

5. Giannella, C., Han, J., Pei, J., Yan, X., Yu, P.S.: Mining frequent patterns in data streams at multiple time granularities. Next generation data mining 212, 191–212 (2003)

6. Grahne, G., Zhu, J.: Fast algorithms for frequent itemset mining using fp-trees. IEEE transactions on knowledge and data engineering 17(10), 1347–1362 (2005)

7. Han, J., Pei, J., Yin, Y.: Mining frequent patterns without candidate generation. In: ACM Sigmod Record. vol. 29, pp. 1–12. ACM (2000)

8. Koh, J.L., Shieh, S.F.: An efficient approach for maintaining association rules based on adjusting fp-tree structures. In: International Conference on Database Systems for Advanced Applications. pp. 417–424. Springer (2004)

9. Lin, C.W., Hong, T.P., Lu, W.H.: An effective tree structure for mining high utility itemsets. Expert Systems with Applications 38(6), 7419–7424 (2011)

10. Pei, J., Han, J., Mao, R., et al.: Closet: An efficient algorithm for mining frequent closed itemsets. In: ACM SIGMOD workshop on research issues in data mining and knowledge discovery. vol. 4, pp. 21–30 (2000)

11. Tsang, S., Koh, Y.S., Dobbie, G.: Rp-tree: rare pattern tree mining. In: Data Warehousing and Knowledge Discovery, pp. 277–288. Springer (2011)

Discovery of Variables Affecting Performance of Athlete Students Using Data Mining

Rahul Sarode, Aniket Muley, Parag Bhalchandra, Sinku Kumar Singh and Mahesh Joshi

Abstract Contemporary researches in stress performance analysis have given a lot of emphasis on the issues of college-going students. These studies discovered social, emotional, and financial conditions at large affect the academic performance of students. Similarly, the academic stress and sports performance have been associated with various factors belonging to personality attributes, cognitive competencies, concentration level, socioeconomic background, locality, etc. However, these were hidden and no attempts were made to discover them. In the underlined research work, these aspects were discovered using data mining techniques. We have devised out our own dataset for the work, and experimentations were carried out in SPSS platform.

Keywords Data mining · Stress patterns · Behavior patterns

R. Sarode · S. K. Singh · M. Joshi
School of Educational Sciences, S.R.T.M. University, Nanded 431606, Maharashtra, India
e-mail: rahulsarode243@gmail.com

S. K. Singh
e-mail: drsinkukumarsingh@gmail.com

M. Joshi
e-mail: maheshmj25@gmail.com

A. Muley (✉)
School of Mathematical Sciences, S.R.T.M. University, Nanded 431606, Maharashtra, India
e-mail: aniket.muley@gmail.com

P. Bhalchandra (✉)
School of Computational Sciences, S.R.T.M. University, Nanded 431606, Maharashtra, India
e-mail: srtmun.parag@gmail.com

© Springer Nature Singapore Pte Ltd. 2019
H. S. Behera et al. (eds.), *Computational Intelligence in Data Mining*,
Advances in Intelligent Systems and Computing 711,
https://doi.org/10.1007/978-981-10-8055-5_40

449

1 Introduction

Data mining is the art of insights into large databases to find valuable information, which otherwise is invisible. This valued information can be used for various future processes of the organization. This is a process used to turn raw data into useful information. Data mining helps us to look for patterns in large data which enables us to learn more about customer behavior, more effective marketing strategies, product life cycle, etc. Thus, it gives us knowledge insights, which otherwise are invisible in large data [1, 2]. Data mining needs effective data collection and computer processing abilities. Educational industry can be a sunshine sector for data mining applications in order to explore hidden knowledge in crucial aspects of educational scenario. Since all educational institutes have been digitalized today, their accumulated databases can be data mined to learn about academic reformations. Such applications are very interesting and challenging. This leads to a new arena, called Educational Data Mining [2, 3]. The output knowledge gained is in language of patterns. The discovered patterns can be used for characterizations of models for prediction, association, grouping, classification, and better resource management of academia [4].

When our work started, we realized that there are five components in fitness [5], muscular endurance, flexibility, aerobic capacity, muscle strength, and body composition. Hence, we measure the serviceable status of these variables in order to understand status of physical fitness. Today, college students have declined physical activities due to number of factors. College sports students have to perform academic assignments as well as physical activities demanding large part of their mental and physical energy. This creates imbalance between academics and sports performance which resulted in academic stress among student players. Academic has been studied widely as an important predictor of health, physical fitness, and performance of both player and non-player students. Stress is manifested as an alarm for physical fitness. It is in plural forms like unease, dejection, unwillingness, and different forms of negative thoughts. It is common phenomenon in student life which may affect his occupational life ahead. According to Bernstein [6], stress is viewed as a negative psychophysical effect of situation that affects an individual's life and daily activities. Thus, enhancing the ability of coping with stress is very important during college life.

On this background, many researchers have identified stress as an important predictor for performance of college students in academics and sports. Academic stress and sports performance have been associated with various factors belonging to personality attributes, cognitive competencies, concentration level, socioeconomic background, locality, etc [7]. The negative impact of stress is seen as anxiety, physical, and psychological weakness. The remedies for stress management includes several techniques like time management, good study habits, social support, positive thinking, enhancing hobbies, exercise. Students perceive that to manage the time for big volume of content, difficulty level of content, and pressure of achieving higher grade creates stress among them [8]. It has further revealed that stressors alone are not only responsible for stress but also the mechanism of person to react the stress and his perception of stress is important factor in forming stress. The surveyed literature has found that stress can lead to academic decline, poor relationships with peers and family members, and overall dissatisfaction with life [9, 10]. Students frequently get fatigue, uncomfortable mind, restlessness due to decreased physical fitness. Fitness in physical form is very necessary to live a quality life. There are many contemporary studies related to physical fitness [11] across the globe. All these highlight compulsory contribution of the physical fitness and health education in the growth of total fitness. Some studies have claimed failure of establishing a relation among stress, physical fitness, and academic performance. Petrie and Stoever [12] revealed that life events stress would not be identified as an associate of academic performance for student-athletes. According to Sandler [13], perceived stress is not significantly associated with attendance of adult students in the colleges and their performance.

On this backdrop above discussions, it is very necessary to examine associations among the research discovered variables with the stress and performance. Since these associations are invisible, we need to implement data mining algorithms for their discovery. The School of Computational Sciences, School of Educational Sciences, and School of Mathematical Sciences of our university have collaboratively taken this as an interdisciplinary opportunity to introduce data mining techniques to personally devise out dataset. A student's primary dataset was collected with actual physical tests on grounds and using standard structured questionnaire method. The factor analysis methods and their algorithms were implemented using SPSS 22.0v software [2]. The primary objective of this study is to introduce educational mining applications to collected datasets. The secondary research objective is to investigate association of variables related to physical fitness of students. A student's dataset was created with actual physical tests on grounds and using standard questionnaire method. The factor analysis, cluster analysis and their algorithms were implemented using SPSS 22.0v software [2]. The secondary research objective, as stated above, comes in consideration mainly because the physical fitness and general health of student-athletes have been more and more alarming in recent days. The students are under significant training stress which can cause subjective stress and influence health outcomes. Due to participation in sports, academic commitments, financial pressures and lack of time management skills; athletes experiences high stress level. Further, it is observed that, all-around performances are negatively associated with health and academics [5]. The importance of physical fitness program is linked to a refined quality of life as well as academic progression.

In the light of the above discussions, there is compelling interest in all investigators in determining the health-related physical fitness and general health of student-athletes. The School of Educational Sciences has Department of Physical Education which has helped us to get concise data about fitness of students. A brief overview of collected information is given in next section.

2 Methodology

As the purpose of this study is to determine health-related physical fitness on general mental health of students, we did some prior work before actual experiments could start. In order to devise out our own dataset, we did assessment of students using fitness tests. These tests were actually performed on the ground. All the data were personally collected from the fieldwork. The proper enactment of dataset took six months time. Some distinguished activities in data collection are elaborated as below: Setting up target population was the first step to start with. In all, 200 urban and 200 rural student-athletes participated in the study and their age ranged between 18 and 28 years.

In order to have our own dataset for data mining purpose, the study insisted on primary source of data. The data was collected through respondents in the form of physical fitness test and questionnaires during the intercollegiate games/sports event under the jurisdiction of Swami Ramanand Teerth Marathwada University, Nanded, MS, India. This data was primarily collected for Ph.D. research work of the principal author and was approved by the Research Recognition Committee of the University. There were considerations related to the sampling method and sample size. The purposive sampling method was chosen for students with a specific purpose as it is a non-random method of sampling design. The sample size of the study was to 200 urban student-athletes those who hail from urban area and 200 rural students those who are from rural area. The sample also categorized in biological maturity of student-athletes into two age groups including 18–22 years and 23–28 years. The collected data also included demographic information about age, height, weight, and rural or urban residential information.

For proper scientific analysis, we have decided to use inclusion and exclusion criteria. The inclusion criteria were

a. The participant agreed to participate in the study via an informed consent.
b. The participants must be resident in urban area are pursuing graduate and postgraduate degree within the age group of 18 to 28 years.
c. The participants were not rotating through other health facility at the time of study.
d. The participants must be free from smoking, drug abuse and alcohol consumption during the data collection period.

Similarly, the exclusion criteria were decided to be

a. The participants were advised not to participate if they are under active physical illness or any injuries and management within 2 weeks of study.
b. Inability to obtain the consent of the respondent.
c. The presence of chronic medical conditions such as asthma, heart disease, or any other severe conditions.

The assessment of physical fitness tests was done through following means. Flexibility was assessed using the sit and reach test to measure lower back and hamstring flexibility. A simple strategy was used which included the participants to sit on the floor, with their shoes off, their legs straight, and feet against the flexometer foot stop. The nine (9)-minute run test was used for measuring cardiovascular fitness. For measuring handgrip strength, the hand grip dynamometer was used. The General Health Questionnaire (GHQ-12) [6, 12, 13] was used to ascertain mental health. It consisted of six positive and six negative factors. The questionnaire prepared for this had question 1, 3, 4, 7, 8, and 12 as positively worded and the remainder is negatively worded. Responses were coded using a non-weighted four-point Likert scale (0, 1, 2, 3). The Pearson's product moment coefficients [1, 2] were used to determine correlations between inventory categories and initial overall perceived stress rating.

3　Results and Discussions

Taking into consideration the facts and terminology in Sect. 2 above, this section describes discovery of stress-affecting variables. After all the data collection, it was processed through the cycle of data processing where answers to questions were coded into numerical figures. The collected data was investigated as a whole and trash. The accuracy and completeness were checked through the coded data with SPSS 22.0v software's descriptive statistical tool for all studied variables. Further, SPSS software is used to perform cross-tabulation, comparative and inferential study of the students. This is witnessed in Table 1 through 5. Initially, Chi-square test was performed to test the significance of physical fitness of students with their habits. Further, to test the significance of students age with their performance. Table 1 represents morphological characteristics of urban and rural students. Table 2 represents student's age-wise distribution. Table 3 represents location-wise distribution of number of smokers and non-smokers. According to Table 4, it is observed that there is significant difference among rural and urban student's smoking behavior. It was also observed that student's smoking is independent of their regional differences/locality. Further, we were interested to investigate, whether smoking is independent of their exercise type? From Tables 4 and 5, the data reveals that there is significant difference among rural and urban students smoking number of cigarettes.

Table 1 Morphological characteristics of urban and rural student-athletes

Sr. no.	Morphological characteristics	Urban		Rural	
		Mean	Standard dev.	Mean	Standard dev.
1	Age (Year)	24.21	3.87	23.31	3.45
2	Weight (Kg)	69.67	14.89	71.67	15.34
3	Height (Cm)	173.30	25.50	169.30	24.91

Table 2 Students age-wise distribution

Class interval	Frequency	Percentage	Valid (%)	Cumulative (%)
21.00–23.00	8	2.0	2.0	2.0
24.00–26.00	178	44.5	44.5	46.5
27.00–29.00	198	49.5	49.5	96.0
30.00+	16	4.0	4.0	100.0
Total	400	100.0	100.0	

Table 3 Locality versus smoker students

Locality	Smoking		Total
	No	Yes	
Rural	136	64	200
Urban	66	134	200
Total	202	198	400

Table 4 Chi-square tests

	Value	df	Asymp. sig. (2-sided)	Exact sig. (2-sided)	Exact sig. (1-sided)
Pearson chi-square	49.005[a]	1	0.000		
Continuity correction[b]	47.615	1	0.000		
Likelihood ratio	50.059	1	0.000		
Fisher's exact test				0.000	0.000
N of valid cases	400				

NB [a]0 cells (0.0%) have expected count less than 5. The minimum expected count is 99.00
[b]Computed only for a 2 × 2 table

Table 5 Chi-square tests

	Value	df	Asymp. sig. (2-sided)
Pearson chi-square	59.611[a]	4	0.000
Likelihood ratio	62.194	4	0.000
No. of valid cases	400		

NB [a]0 cells (0.0%) have expected count less than 5. The minimum expected count is 18.50

The study further provides a scope for a rotated factor analysis, which can be performed on the data as we can devise out six components which are more significant in the study. It is observed that, total variation among the six component is found to be 58.47%. The scree plot shown in Fig. 1 is the way of identifying a number of useful factors. This component analysis is based on *Kaizen Mayer* method [1, 4, 14]. Table 6 reveals that the first component's physiological and

Fig. 1 Component analysis

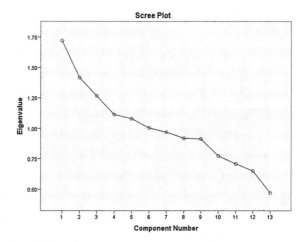

Table 6 Rotated component matrix

Parameters	Component					
	1	2	3	4	5	6
Age (Binned)	−0.086	0.064	−0.087	−0.028	−0.036	0.816
Living arrangement	0.130	0.712	−0.257	0.130	−0.180	−0.002
Exercise time	0.058	−0.052	−0.050	0.740	0.020	0.157
No_of_cigarette_day	−0.070	0.209	0.464	0.106	−0.229	−0.054
Frustrations	−0.003	−0.197	0.665	−0.144	−0.058	−0.184
Conflicts	−0.031	0.819	0.159	−0.055	0.121	0.068
Pressures	−0.019	−0.108	−0.024	−0.709	−0.021	0.200
Changes	0.089	−0.029	−0.012	0.019	0.865	0.050
Self-imposed	0.068	−0.008	0.586	0.029	0.234	0.408
Physiological	0.755	0.153	−0.171	−0.145	0.109	−0.143
Emotional	−0.431	0.054	−0.215	0.099	0.489	−0.281
Behavioral	0.686	−0.167	−0.061	0.139	−0.005	0.149
Cognitive_appraisals	−0.624	−0.106	−0.124	−0.092	0.030	0.120

behavioral parameters affect the students stress level. Second component shows conflicts; in third component, the frustrations and self-imposed parameters like exercise time show significance on student's performance. The components four to sixth, it is observed that, the changes in the students and their age were found to be significant parameters. In Figs. 1 and 2, a hierarchical cluster analysis is performed with wards linkage and Euclidean distance method for the collected data. Fig. 2 represents the number of clusters. The dendrogram in Fig. 3 represents the

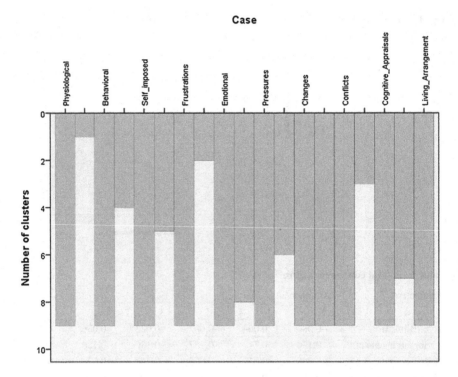

Fig. 2 Cluster analysis

diagrammatic representation of associated variables. It clearly states that the conflicts, changes, pressures, emotional and living arrangement parameters show most association among each other. There is also a joint association of these variables with the second cluster having parameters like frustrations, self-imposed, behavioral parameters. These two clusters are jointly associated with physiological parameters.

4 Conclusion

This study has hypothesized to extract the parameters which are likely influencing the performance of the athlete students, using data mining techniques. The hypothesized parameters were invisible and their association with other variables was unknown. To make them visible, we have devised out our own dataset and data mining algorithms were implemented on SPSS platform. It is seen that the smoking habit differences are more among the urban students. The study was extended

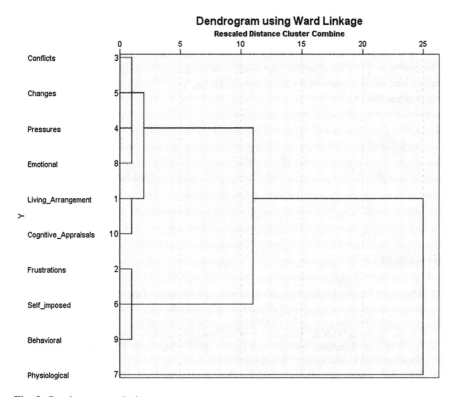

Fig. 3 Dendrogram analysis

further to apply rotated factor analysis and hierarchical cluster analysis. These clearly showed that the physiological and behavioral variables significantly affect the performance of students.

Statement on Ethical Approval

This is to certify that the study conducted for the research paper entitled "Discovery of variables affecting performance of Athlete students using Data Mining" has the ethical approval for performing the study. Moreover the nature and purpose of the study was well communicated to the participants.

Name of the Ethics Committee: Dr.(Mrs) V.N.Patil , Mr. Ramthirthe (Deputy Registrar)

Rahul Sarode   Aniket Muley   Parag Bhalchandra

Statement on Informed Consent of the Participants

This is to certify that all the participants in the study conducted for the research paper entitled "Discovery of variables affecting performance of Athlete students using Data Mining" are fully aware of the purpose and nature of the study and have consented to be the part of this study as participants.

Rahul Sarode   Aniket Muley   Parag Bhalchandra

References

1. Margaret Dunham, Data Mining: Introductory and Advanced Topics, Pearson publ, 2002.
2. Han, J. and Kamber, M., (2006) "Data Mining: Concepts and Techniques", 2nd edition.
3. Jim Gray. Behrouz. et al., (2003) Predicting Student Performance: An Application of Data Mining Methods With the Educational Web-Based System Lon-CAPA © 2003 IEEE.
4. Data Mining: Concepts and Techniques by Jiawei, Han and Micheline Kamber, San Francisco, California, Morgan Kauffmann, 2001.
5. Patil, R.B., Doddamani, B.R., Milind, Bhutkar., S.M., Awanti. A. (2012) Comparative Study of Physical Fitness among Rural Farmers and Urban Sedentary Group of Gulbarga District. AJMS Al Ameen J Med Sci 5(1).
6. Bernstein, D.A; Penner, L.A; Stewart, A.C and Roy, E.J (2008) Psychology (8th edition). Houghton Mifflin Company Boston New York.
7. Fairbrother, K. & Warn, J. (2003). Workplace Dimensions, Stress and Job Satisfaction. Journal of Managerial Psychology, 18(1): 8–21.
8. Abouserie, R. (1994). Sources and levels of stress in relation to locus of control and selfesteem in university students. Educational Psychology, 14(3), 323–330.
9. Bruinings, A.L., Van Den Berg-Emons, H.J., Buffart, L.M., Van Den Heijden-Maessen, H.C., Roebroeck, M.E., Stam, H.J. (2007). Energy cost and physical strain of daily activities in adolescents and young adults with myelomeningocele. Dev. Med. Child. Neurol., 49: 672–677.
10. Salmon, J., Owen, N., Crawford, D., Bauman, A., Sallis, J.F. (2003) Physical activity and sedentary behavior A population-based study of barriers, enjoyment and performance. Health Psychology. 22: 178–188.
11. Campbell R. L. & Jarvis G. K. (1992). Perceived level of stress among university undergraduate students in Edmonton, Canada. Perceptual and Motor Skills, 75, 552–554.
12. Petrie, Trent A.; Stoever, Shawn. (1997) Journal of College Student Development, v38.
13. Sandler, M. E. (2000). Career decision-making self-efficacy, perceived stress, and an integrated model of student performance: A structural model of finances, attitudes, behavior, and career development. Research in Higher Education 41(5): 537–578.
14. Yitzhak, W. (2000). Physical activity and health. 6th Sport Sciences Congress, Ankara.

Implementation of Non-restoring Reversible Divider Using a Quantum-Dot Cellular Automata

Ritesh Singh, Neeraj Kumar Misra and Bandan Bhoi

Abstract The CMOS-based integrated circuit may scale down to nanometer range. The primary challenge is to further downscale the device and high-energy dissipation. Reversible logic does not dissipate energy and no information loss. In this way, the state-of-the-art technology such as QCA was forced toward high-speed computing with negligible energy dissipation in the physical foreground. This work targets the design of non-restoring reversible divider circuit and its implementation in QCA. We have utilized few 2×2 FG and 4×4 HNG gates as the block construction and also show the QCA implementation having cost-efficient approach. Further, the divider circuit has synthesized with FG and HNG gates and QCA implementation. This divider circuit inherits many benefits such as fewer garbage outputs, reduce quantum cost are achieved, and also reduced QCA primitives can be improved by using efficient QCA layout scheme. Simulation investigations have been verified by QCA Designer. The proposed non-restoring divider also compares the reversible metrics results with some of other existing works.

Keywords Quantum-dot cellular automata · Nanoelectronics · Clocking Reversible computing · High-speed nanoelectronics

R. Singh · N. K. Misra
Department of Electronics Engineering, Institute of Engineering and Technology, Lucknow 226021, India
e-mail: ritesh.singh089@gmail.com

N. K. Misra (✉)
Department of Electronics and Communication Engineering, Bharat Institute of Engineering and Technology, Hyderabad 501510, India
e-mail: neeraj.mishra@ietlucknow.ac.in

B. Bhoi
Departement of Electronics and Telecommunication, Veer Surendra Sai University of Technology, Burla 768018, India
e-mail: bkbhoi_etc@vssut.ac.in

© Springer Nature Singapore Pte Ltd. 2019
H. S. Behera et al. (eds.), *Computational Intelligence in Data Mining*, Advances in Intelligent Systems and Computing 711, https://doi.org/10.1007/978-981-10-8055-5_41

1 Introduction

Due to the increased density of the chip with high-energy dissipation, the chip design has become complex and faces lots of challenges [1]. The reversible logic circuit seems to tackle these difficulties such as no information loss, energy-free computation, and application of quantum computing [2]. This synthesis approach is an emerging area of the domain because it recovered input information from the output information. Reversible gates have a bijective mapping from the input to the output vectors and essential of balanced inputs and outputs. Reversible circuits are necessary of feedback free, fan-out free, and no loops [3]. Reversible circuit has same inputs and outputs, and there are restricted criteria of balanced. If non-balanced criteria exist in the reversible circuit, then garbage output is required to be balanced. The standard design of the reversible logic circuit is to explicit its figure of merits such as quantum cost, delay, garbage outputs, and ancilla inputs [4]. The construction of a gate or a circuit is essential for low figure of merits belonging to the efficient quantum computing regime. In focus on reversible logic, two criteria exist such as logical reversibility and physical reversibility [5]. In the logic reversibility, a one-to-one mapping is established. Physical reversibility computes the information without the energy dissipation as the second level of criteria. These two rules are necessary for reversible circuit design.

QCA is a prominent technology that employs nanometric scale level to perform logic computing [6]. QCA technology is focused on a concept that position of electron is possible by Coulomb repulsion, whereas CMOS device required voltage for computation [7]. The computing advantages are high density, high speed, and low power. QCA design has always preferred the lower value of primitives such as latency, area usages, and complexity [8]. The information such as binary '0' and '1' is stored by polarization value of the cell, and clock zones achieve flow of information. QCA design interacts the cells with the nearest neighbor cells, and polarization changes to stored information. QCA was first originated by Lent and implemented in the physical foreground in the year 1997 and proves the physical existence possible in nanoelectronics area [9].

The divider is the popular operation that is used mostly in arithmetic and logic unit and central processing units [10]. The proposed non-restoring divider circuits are better than the prior circuits in the sense of the architecture complexity. In terms of reversible figure of metrics, this optimization is considered regarding a number of inputs and outputs, the number of garbage output (GO), constant input (CI), quantum cost (QC), and the count of one and two-qubit gates [11]. This work targets to design toward non-restoring divider based on reversible technique. The design of non-restoring divider consists of two blocks such as FG and HNG gates. The individual blocks are tested in QCADesigner tool. Further, the cascaded-based approach is to design the higher order non-restoring divider. The design of non-restoring divider is a target on a single layer of QCA technology. The proposed single-layer design of the non-restoring divider in QCA has better primitives' results based on a simulation study. The proposed design claims the lower area and high computation speed as compared to prior work. QCADesigner tool was adopted to simulation and verification of result in the physical foreground. In addition, some improvement is made in the primitives' results based on QCA design.

The following are the attempts of the proposed work:

- We presented a compact non-restoring divider which is adopted a reversible approach. The design has been achieved by utilizing existing FG and HNG gates.
- We synthesize non-restoring block in QCADesigner which works on the principle of Columbia interaction. The achievement of this implementation such as complexity, area, and latency is pretty good and suitable for nanoelectronics application.

This work is organized in the following sections: Sect. 2 deals with the necessary background related to QCA technology to understand. Section 3 provides a necessary existing work related to the non-restoring divider. This section also deals with an existing work with its pros and cons. Section 5 details the architecture, including the QCA design and explication of results. Section 7 provides the comparative results based on prior work, and Sect. 8 discusses their conclusion.

2 QCA Concept

QCA is established on the four quantum dots in which maximum distance settles the two electrons [12]. A polarization outline in the cell is presented in Fig. 1a. The two polarizations are shown as $P = -1$, i.e., binary '0,' and $P = +1$, i.e., binary '1.' The activity of cell is changed due to the Coulombic interaction between cells. Majority gate and inverter are the essential gates that are devoted to building any circuits, which are shown in Fig. 1b, 1c respectively. The two same logic information is transferred into two output nodes that are shown in Fig. 1d. Normally, QCA cells are provided by four-phase clock. The four-phase clock is known as a switch, hold, release, and relax as presented in Fig. 1e. The clock zones provide the information flow in cells. The simple architecture of majority gate in experimental view is drawn in Fig. 1f. The simple QCA cell architecture is shown in Fig. 1g.

3 Preliminary Work

Several approaches have been reviewed in the literature to achieve restoring and non-restoring divider designs[13–18]. In the existing divider [16] based on non-restoring and restoring, there is more computing QCA metrics. However, the designs have attempted in non-reversible approach. Ali Bolhassani et al. presented the reversible divider-based different modules [10]. The modules used in the design

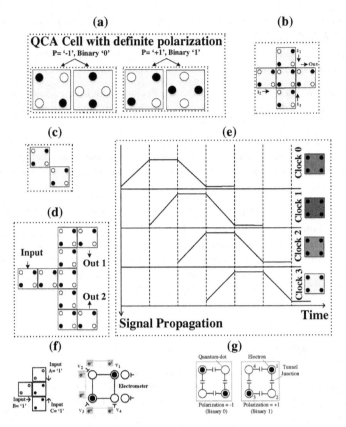

Fig. 1 QCA concept: **a** Polarization, **b** majority gate architecture, **c** inverter architecture, **d** fan-out, **e** clocking, **f** experiment of cell, **g** cell

are a multiplexer, left-shift register, and parallel adder/subtraction. All the modules are cascaded to meet the design requirement of the divider. These designs are more complex and high reversible figure of metric regarding gate count, garbage output, and quantum cost. These designs of 1-bit divider utilize CI of 19, GO of 26, and QC of 90, whereas n-bit divider uses $(10n + 9)$ CI, $(11n + 15)$ GO, and $(52n + 38)$ QC. These designs required more functional units. It is essential that the design required a fewer functional unit with a large size of divider unit. We compared the proposed design against the existing state-of-the-art design in [10]. Some other works have been presented by T. N. Sasmal et al. [17]. These 2×2 and 3×3 type non-restoring design was employed divider functionality to nanocomputing QCA based. The 4×4 non-restoring divider in QCA design implementation can be realized with 3180 number of cell count, the latency of 12, and area of 6.5 μm^2. These designs are implemented in a single layer with coplanar technique. Reversible synthesis of the divider, which is a popular technique in the recent area, is studied in limited state-of-the-art designs. The noted work in this area reduced the reversible figure of merits.

4 Popular Reversible Gates

Feynman gate (FG) and Haghparast (HNG) are the widely used gates in the design of Ex-OR, multiplier, full adder, and full subtraction because of its low quantum cost [3]. The following section shows these gate schematics, quantum equivalent, and nanoscale implementation in QCA to prove the efficiency in the physical foreground. We also presented these gates' quantum circuit along with Toffoli gate and elemental gate.

Feynman gate (FG): Inputs and outputs for a 2 × 2 FG are shown as follows: $I_v = A, O_v = A \oplus B$ The schematic of FG and quantum circuit based on Toffoli gate is presented in Fig. 2a. Feynman gate is used as "Ex-OR" and "repeating one bit" with no junk outputs. Figure 2b, c presents the QCA implementation and the corresponding simulation results of FG. The cell layout of FG has zero latency and consists of 14 cell count and area of $0.02\mu m^2$.

Haghparast gate: Inputs and outputs of HNG gate are presented in the schematic diagram Fig. 3a. Figure 3b, c represents Toffoli gate block as well as elemental quantum gate-based representation of the HNG gate. To realize the QCA implementation of HNG gate, cell complexity of 14, area of $0.02\ \mu m^2$, and zero latency are needed as depicted in Fig. 4a. The simulation result of HNG is shown in Fig. 4b.

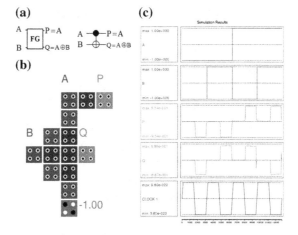

Fig. 2 Feynman gate: **a** Block diagram and its quantum circuit, **b** QCA implementation, **c** simulation result

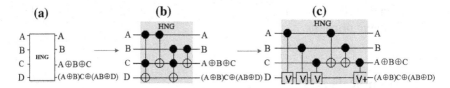

Fig. 3 HNG gate: **a** Block diagram, **b** reversible circuit, **c** quantum circuit

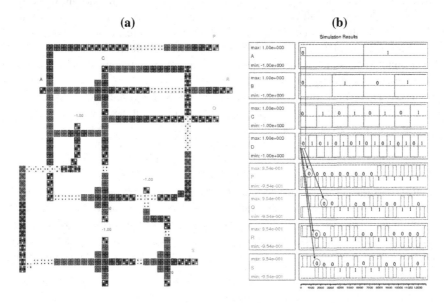

Fig. 4 QCA implementation of HNG: **a** Layout, **b** simulation result

5 The Proposed Architecture of Non-restoring Divider

By cascading the FG and HNG gates, the reversible non-restoring divider can simply be constructed as shown in Fig. 5a. Reversible FG and HNG gates are chosen to optimize the quantum cost of the design. The non-restoring cell is shown in Fig. 5a. The Toffoli gate and elemental quantum gate-based representation of non-restoring divider are presented in Fig. 5b, c, respectively. The design is constructed two outputs S and C. Recently, lots of existing work [13–18] have been available showing the non-reversible synthesis approach.

Figure 5a shows the schematic representation of the one-by-one non-restoring reversible divider. Note that, in order to obtain an output, individual block provides the intermediate outputs. These intermediate outputs are applied to other blocks and synthesize the required outputs of the divider. However, the individual block architecture is a general solution for achieving the one-by-one non-restoring

(a)

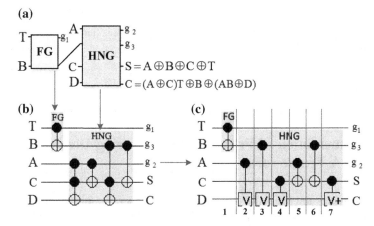

(b) **(c)**

Fig. 5 Proposed one-by-one non-restoring reversible divider **a** basic block, **b** Toffoli implementation, **c** elemental quantum gate diagram

divider. Therefore, the topology selection and the synthesis of the non-restoring divider are the key points in the design construction.

The FG and HNG gates are used to design the two-by-two non-restoring reversible divider. By using the 3 × FG and 4 × HNG gate, a new architecture of non-restoring divider is designed which is drawn in Fig. 6. In the architecture, T is used as a control input, A_0, A_1, A_2 are used as inputs, R_0, and R_1 denote the remainder, Q_0 and Q_1 represent the quotient. The synthesis of four inputs based on non-restoring divider for the solution of divider construction is shown in Fig. 6. The four inputs based on non-restoring divider are designed using 8 gate count (4 × FG, and 4 × HNG), 28 quantum cost (4 × 1 + 4 × 6 = 28), and 13 garbage output.

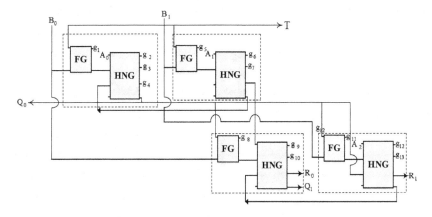

Fig. 6 Proposed architecture of a two-by-two non-restoring reversible divider

Table 1 QCA primitives' result in statistics as compared to existing works

Non-restoring divider	Attempt	Complexity	Area (μm^2)	Technique
Proposed	R	268	0.54	Coplanar
[16]	NR	235	0.35	Multilayer
[17]	NR	60	0.08	Coplanar
[18]	NR	147	0.27	Coplanar

6 Implementation of Proposed Non-restoring Divider in QCA Framework

Table 1 provides the results obtained in QCADesigner. In spite of simplicity, this synthesis approach for non-restoring reversible divider succeeded its achieving cell complexity of 269 and area of 0.54 μm^2 by the layout design as shown in Fig. 7. Most of the existing work [13–18] in non-restoring divider has emphasized on non-reversible approach. The non-reversible approach often lacks recovered outputs from inputs and loss of inputs information as the outputs obtained. This work introduces a one-by-one non-restoring reversible divider which utilizes a 7 quantum cost and 3 garbage output. The important statistical features of QCA design used were complexity, area, and latency. Latency was employed to perform clock delay.

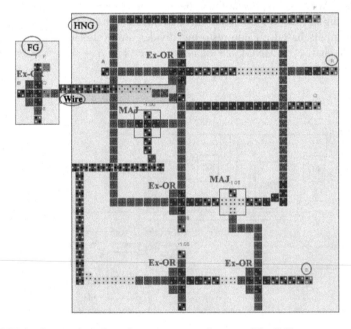

Fig. 7 QCA implementation of two-by-two non-restoring reversible divider

Table 2 Comparison of statistics results with existing works

Designs	R	GC	CI	GO	QC	HC	QE
New	Yes	2	0	3	7	$6\alpha + 2\beta$	Yes
[10]	Yes	–	19	26	90	$65\alpha + 7\beta + 52\delta$	No
[13]	Yes	–	21	29	111	$59\alpha + 67\beta + 33\delta$	No
[14]	Yes	–	26	32	149	$84\alpha + 82\beta + 42\delta$	No
[15]	Yes	–	28	41	104	$91\alpha + 98\beta + 50\delta$	No

Note GC gate count, *CI* ancilla input, *GO* unwanted output, *QC* quantum cost, *HC* hardware complexity, *R* reversible, *NR* non-reversible, *QE* quantum equivalent

The robust design of non-restoring divider is that it used less statistical features for fast computation and efficiency.

7 Comparison Result and Discussion

Most of the existing work has designed the non-reversible approach, whereas existing work deals with the reversible technique. The present work deals with the robust design approach of the divider. Such an efficient, simple, and unique design model will successfully evaluate the behavior of a design which is established to be most suitable for nanoelectronics application. The garbage outputs, quantum cost, gate count, and ancilla inputs of design based on the quantum equivalent circuit of the proposed design are compared through Table 2. This comparison result shows the proposed reversible implementation of the non-restoring divider that is very effective in quantum computing paradigm with very low quantum cost, fewer garbage outputs, and gate count.

8 Conclusions

In this paper, a non-restoring reversible divider is realized on QCA. Design gates for the non-restoring divider are presented in QCA in addition to the complete design adopted in coplanar technique. Unique design as compared to existing makes these cost-efficient approaches. Therefore, the less critical delay of the non-restoring divider is very beneficial to reversible computing application. Thus, the non-restoring adder is usually implemented as two-by-two that is used to construct large size order of non-restoring adder. The study presented here shows that this synthesis of the non-restoring divider is a less reversible figure of merit in the sense of quantum computing framework. A simulation carried out on QCA-Designer tool reveals that the proposed structure of non-restoring binary divider outperformed the state-of-the-art designs and demonstrated promising ability against nanoelectronics technology. Here, in this work, both the better primitive's results and correct simulation pattern of individual blocks have been achieved simultaneously. The non-restoring divider can be used in many computing application inside the ALU, microcontroller, and low power devices.

References

1. Misra, N.K., Sen, B. and Wairya, S.: Towards designing efficient reversible binary code converters and a dual-rail checker for emerging nanocircuits. Journal of Computational Electronics, 16(2), 442–458 (2017).
2. Misra, N.K., Wairya, S. and Sen, B.: Novel Conservative Reversible Error Control Circuits Based On Molecular-QCA, International Journal of Computer Applications in Technology, Inderscience Publishers, vol. 56, no. 1, 2017.
3. Sen, B., Dutta, M., Goswami, M. and Sikdar, B.K.: Modular Design of testable reversible ALU by QCA multiplexer with increase in programmability. Microelectronics Journal, 45(11), 1522–1532 (2014).
4. Misra, N.K., Sen, B., Wairya, S. and Bhoi, B.: Testable Novel Parity-Preserving Reversible Gate and Low-Cost Quantum Decoder Design in 1D Molecular-QCA. Journal of Circuits, Systems and Computers, 26(09), p. 1750145 (2017).
5. Misra, N.K., Sen, B. and Wairya, S.: Designing conservative reversible n-bit binary comparator for emerging quantum-dot cellular automata nano circuits. Journal of Nanoengineering and Nanomanufacturing, 6(3), 201–216 (2016).
6. Chabi, A.M., Roohi, A., Khademolhosseini, H., Sheikhfaal, S., Angizi, S., Navi, K. and DeMara, R.F.: Towards ultra-efficient QCA reversible circuits. Microprocessors and Microsystems, 49, 127–138 (2017).
7. Misra, N.K., Wairya, S. and Singh, V.K.: Approach to design a high performance fault-tolerant reversible ALU. International Journal of Circuits and Architecture Design, 2(1), 83–103 (2016).
8. Sen, B., Dutta, M., Mukherjee, R., Nath, R.K., Sinha, A.P. and Sikdar, B.K.: Towards the design of hybrid QCA tiles targeting high fault tolerance. Journal of Computational Electronics, 15(2), 429–445 (2016).
9. Tougaw, P.D. and Lent, C.S.: Logical devices implemented using quantum cellular automata. Journal of Applied physics, 75(3), 1818–1825 (1994).
10. Bolhassani, A. and Haghparast, M.: Optimised reversible divider circuit. International Journal of Innovative Computing and Applications, 7(1), 13–33 (2016).
11. Thapliyal, H., Ranganathan, N. and Kotiyal, S.: Reversible logic based design and test of field coupled nanocomputing circuits. In Field-Coupled Nanocomputing Springer Berlin Heidelberg, 133–172 (2014).
12. Walus, K., Dysart, T.J., Jullien, G.A. and Budiman, R.A.: QCADesigner: A rapid design and simulation tool for quantum-dot cellular automata. IEEE transactions on nanotechnology, 3(1), 26–31 (2004).
13. Nayeem, N.M., Hossain, A., Haque, M., Jamal, L. and Babu, H.M.H.: Novel reversible division hardware. 52nd IEEE International Midwest Symposium on Circuits and Systems MWSCAS'09. 1134–1138 (2009).
14. Dastan, F. and Haghparast, M.: A novel nanometric fault tolerant reversible divider. International Journal of Physical Sciences, 6(24), 5671–5681 (2011).
15. Dastan, F. and Haghparast, M.: A novel nanometric reversible signed divider with overflow checking capability. Research Journal of Applied Sciences, Engineering and Technology, 4(6), 535–543 (2012).

16. Sayedsalehi, S., Azghadi, M.R., Angizi, S. and Navi, K.: Restoring and non-restoring array divider designs in quantum-dot cellular automata. Information sciences, 311, 86–101 (2015).
17. T.N., Singh, A.K. and Ghanekar, U.: Design of non-restoring binary array divider in majority logic-based QCA. Electronics Letters, 52(24), 2001–2003 (2016).
18. Cui, H., Cai, L., Yang, X., Feng, C. and Qin, T.: Design of non-restoring binary array divider in quantum-dot cellular automata. Micro & Nano Letters, 9(7), 464–467 (2014).

Depth Estimation of Non-rigid Shapes Based on Fibonacci Population Degeneration Particle Swarm Optimization

Kothapelli Punnam Chandar and Tirumala Satya Savithri

Abstract In this paper, we address the problem of recovering the shape and motion parameters of non-rigid shape from the 2D observations considering orthographic projection camera model. This problem is nonlinear in nature and the gradient-based optimization algorithms may easily stick in local minima on the other hand and the generic model fitting may result inexact shape. We propose Fibonacci population degeneration particle swarm optimization (fpdPSO) algorithm and used to estimate the shape and motion. We report the shape estimation results on face and shark data set. Pearson Correlation Coefficient is used to measure the accuracy of depth estimation.

Keywords Non-rigid structure from motion · Depth estimation
Orthographic projection · Particle swarm optimization

1 Introduction

Recovering shape parameters from the finite period 2D observations is a classical problem of structure from motion. The recovered shape, i.e., reconstructed 3D model, can be used in virtual reality simulations, 3D games, face recognition [1], face reconstruction [2], face morphing, human–computer interactions and animations. Estimating the shape and motion of rigid body is well studied in structure from motion (SfM) and the field is matured in both theory and practice. Extending SfM to recover the shape and motion of non-rigid scenario has turned

K. Punnam Chandar (✉)
Department of E.C.E, University College of Engineering, Kakatiya University, Kothagudem 507118, Telangana, India
e-mail: k_punnam@kakatiya.ac.in

T. Satya Savithri
Department of E.C.E, Jawaharlal Nehru Technological University, Hyderabad 500085, Telangana, India
e-mail: tirumalasatya@gmail.com

© Springer Nature Singapore Pte Ltd. 2019
H. S. Behera et al. (eds.), *Computational Intelligence in Data Mining*,
Advances in Intelligent Systems and Computing 711,
https://doi.org/10.1007/978-981-10-8055-5_42

out to be difficulty. This is due to difficulty in modeling the deformations and missing data.

Ullman [3] proposed the concept of structure from motion to recover the 3D shape given 2D features of the object that are tracked in available multiple frames of an image. Given the measurement noise-free orthographic projections, a robust factorization algorithm was proposed to decompose the measurement matrix into shape and motion parameters [4]. In [5], general non-rigid 3D shape is recovered by representing the observations as a linear combination of a few basis shapes. In [6] a different formulation, trajectory space model instead of linear subspace model is proposed to recover 3D shape assuming Gaussian prior and solved using Expectation Maximization algorithm. In [7], 2D features from multiple 2D images are used as input to the similarity transform and estimation of camera rotation parameters is modeled as optimization problem and solved using genetic algorithm. The recovered shape conforms to the structure of the face shape. In addition, the model was used to perform face recognition to alleviate pose variation. The use of genetic algorithm to recover the camera rotation parameters encounters heavy computational burden and requires feasible chromosome and fine tuning of genetic operators—selection, crossover, and mutation. In [8], the rotation and translation process from a frontal view face image to a non-frontal view face image is formulated as a constrained Independent Components analysis problem (cICA). In the cICA algorithm, the negentropy is used as a contrast function, and the cICA is formulated as a constrained optimization problem. Newton-like learning rule was used to recover the 3D model. Further, in [9] one or more 2D images with different poses are used to reconstruct the 3D structure of human face based on nonlinear least squares model by means of similarity transform. However, nonlinear least squares algorithm suffers from local minima problem as that of the other global optimization algorithms.

In the above-mentioned techniques [8, 9], the shape estimation is performed based on optimization utilizing gradients. As the shape estimation problem is nonlinear in nature, the gradient-based methods may not work well. In [10], a reference face shape, CANDIDE model [11] is used to initialize structure and the optimization algorithm is carried out in two phases using differential evolution. In the first step, motion parameters are estimated, and in second step, utilizing the estimated motion parameters the shape parameters are estimated minimizing the reprojection error in successive iterations.

In this work, we propose Fibonacci population degeneration particle swarm optimization (fpdPSO). The shape and motion parameters are considered as the variables of the optimization problem formulated based on the orthographic projection model and the optimal solution is estimated using the fpdPSO. The 3D shape is assumed to be lying near a low-dimensional subspace, having a Gaussian prior on each shape [6] as a result prior shape template is not required. Empirical results are reported on 3D Bosphorus Face Database and synthetic data set consisting of 3D animation of Shark. To qualitatively measure the estimation accuracy, we have used Pearson Correlation Coefficient which is commonly used to measure the similarity of two signals [12].

The remainder of the paper is organized as follows. In Sect. 2, we present orthographic projection camera model. In Sect. 3, the details of the proposed Fibonacci population degeneration particle swarm optimization are presented. Experimental results are given in Sect. 4 and conclusion in Sect. 5.

2 Orthographic Projection Camera Model

The Point P in $R^{3 \times 1}$ space is projected on to the point p in $R^{2 \times 1}$ space. This process is modeled either as orthographic projection or using weak perspective projection. Compared to weak perspective projection model, the orthographic projection yields easily to solve equations. The process of orthographic projection is given in Eq. 1.

$$
\underbrace{p_{j,t}}_{2 \times 22} = \underbrace{c_t}_{1 \times 1} \underbrace{R_t}_{2 \times 3} \left(\underbrace{s_{j,t}}_{3 \times 22} + \underbrace{d_t}_{3 \times 22} \right) + \underbrace{n_t}_{2 \times 22}
\tag{1}
$$

where $p_{j,t} = [u_{j,t}, v_{j,t}]^T$ is the 2D projection of feature point j at time t, d_t is a 3×1 translation vector, R_i is a 2×3 orthographic projection matrix, c_t is the weak perspective scaling factor, and n_t is a vector of zero-mean Gaussian noise with variance σ^2 in each dimension. The rotation matrix R_t between the frontal view image and the non-frontal view image at time t can be specified as three successive rotations around the x-, y-, and z-axes, by angles ϕ, Ψ, θ, respectively, and can be written as the product of these three rotations and is given in Eq. 2.

$$
R = \begin{bmatrix} \cos\phi & \sin\phi & 0 \\ -\sin\phi & \cos\phi & 0 \\ 0 & 0 & 1 \end{bmatrix} \begin{bmatrix} \cos\Psi & 0 & -\sin\Psi \\ 0 & 1 & 0 \\ \sin\Psi & 0 & \cos\Psi \end{bmatrix} \begin{bmatrix} 1 & 0 & 0 \\ 0 & \cos\theta & \sin\theta \\ 0 & -\sin\theta & \cos\theta \end{bmatrix} = \begin{bmatrix} r_{11} & r_{12} & r_{13} \\ r_{21} & r_{22} & r_{23} \\ r_{31} & r_{32} & r_{33} \end{bmatrix}
\tag{2}
$$

$s_{j,t} = (X_i, Y_i, Z_i)$, where (X_i, Y_i) is measured from the frontal view image and Z_i is the shape parameter and is lost during the projection process. If the shape parameters and motion parameters fit the non-frontal view face image, the following equation will be a minimum:

$$
D_{min} = \min_{z_1, \dots z_{22}, \, \theta, \psi, \phi} \left\| p - cR_{2x3} \begin{bmatrix} x_1 & \cdots & x_{22} \\ y_1 & \cdots & y_{22} \\ z_1 & \cdots & z_{22} \end{bmatrix} - T_i \right\|_F^2
\tag{3}
$$

Eq. 3 is nonlinear in nature and classical gradient algorithm easily stuck in local minima and the estimated structure is inexact. To search the large search space for optimal solution, we propose to use Fibonacci population degeneration particle swarm optimization.

3   Fibonacci Population Degeneration Particle Swarm Optimization

We propose fibonacci population degeneration Particle Swarm Optimization a modification to the PSO proposed by Kennedy and Eberhart [13] in 1995 based on swarm behavior in nature, such as fish and bird schooling. Similar to PSO, the fpdPSO algorithm is initiated with initial swarm agents, and here, the initial population is chosen a Fibonacci number F_n. The agents search the space of an objective function by adjusting the trajectories of individual agents. Let x_i and v_i be the position vector and velocity for particle I, respectively. The new velocity is determined by Eq. 4.

$$v_i^{t+1} = C * \left(v_i^t + \alpha \varepsilon_1 \left[g^* - x_i^t \right] + \beta \varepsilon_2 \left[x_i^{*(t)} - x_i^t \right] \right) \tag{4}$$

The parameter 'α' is the social parameter, 'β' is the cognitive parameter, and C is the constriction factor. The initial population is updated with new computed velocities. Next, we compute the cost of each swarm agent and sort the cost in ascending order. The swarm agents $F_{n-1} - F_{n-2}$ with higher cost are degenerated retaining F_{n-1} best swarm agents. The advantage of using Fibonacci population degeneration is that the search space approximates the golden spiral and is shown in Fig. 1, i.e., the proposed algorithm exhibits diversification in the initial stages and gradually exhibits intensification. The pseudo code of the proposed fpdPSO algorithm is mentioned below:

Fig. 1 Diversification and intensification of search space by fpdPSO algorithm

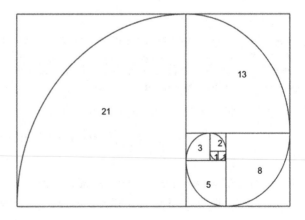

Fibonacci Population Degeneration Particle Swarm Optimization Algorithm

---

Objective function: $f(x)$

Initialize locations x_i and velocity v_i of Fibonacci number (n) particles.

Find x^* from $\min\{f(x_1),\quad ...,\quad f(x_n)\}$ (at t=0)

While (criterion)
For loop over all n particles and all d dimensions

Generate new velocity v_i^{t+1}

Calculate new locations $x_i^{t+1} = x_i^t + v_i^{t+1}$

Compute the cost of the objective function at new locations x_i^{t+1}

Find the current best for each particle x_i^*

end for

Find the current global best g^*

>> Sort the cost and drop the agents with the high cost and retain the best agents
(**Population Degeneration**).
Update t = t+1 (pseudo time or iteration counter)
End while
Output the final optimal estimated shape and motion parameter:

$$x_i^* = \{z_1,\quad ...\quad z_n,\quad \phi\quad \psi\quad \theta\}$$

---

4 Experimental Results

The parameter set used for fpdPSO in estimating the shape and motion parameters
is given in Table 1. The solution vector for the fpdPSO algorithm is represented as
shown in Fig. 2. The shape and parameter estimation results are reported after

Table 1 Experimental setup for the fpdPSO algorithm

Parameter	Value	Degenerated population	Generations
Swarm size	Fibonacci number (610)	610	G1
Cognitive parameter (β)	2	377	G2
Social parameter (α)	4	233	G3
Constriction factor (C)	1	144	G4
		:	:
		:	:
		3	G11 (STOP)

Z_1	Z_2	Z_3	...	Z_{n-1}	Z_n	θ	Ψ	Φ
Shape Parameters						**Rotation Parameters**		

Fig. 2 Solution vector representation for fpdPSO algorithm

performing the allowed number of degenerations that is dependent on the initial population. The proposed optimization algorithm is implemented in MATLAB R2013a running on a Laptop with 4 GB RAM.

4.1 Experiment-I

For the experiment-I, we have used the Bosphorus database, and this database consists of 3D face data along with corresponding 2D images of 105 subjects with rich expressions, systematic head poses, and varieties of occlusions. In this work, we have considered 2D images shown in Fig. 3 as training images where all the 22 features are visible and can represent the shape. The unique feature of this database is that a comparison can be made with the estimated depth values with the ground truth value of the depths of the features.

The 22 feature marked on the frontal face is shown in Fig. 4. The 22 features marked on the non-frontal view face image PR_U are shown in Fig. 5. The depths of the features are estimated using Eq. 3 considering five non-frontal images one at

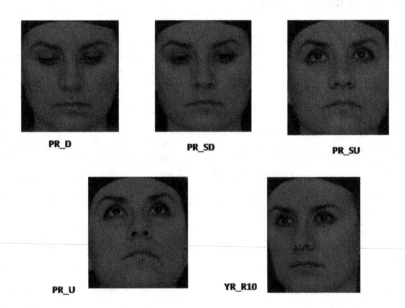

Fig. 3 Sample non-frontal images of Bosphorus database

Fig. 4 Marked 22 features
on PR_D

Fig. 5 Marked 22 features
on PR_U

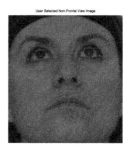

a time using fpdPSO algorithm. The features from PR_D and PR_U are used as input to the Eq. (3), and the optimal shape is estimated using fpdPSO algorithm. The estimated shape and true shape using PR_U are shown in Fig. 6. Further, the progress of the proposed algorithm is shown in Fig. 7 with computed minimum cost, mean cost (Population Average), and best cost. Pearson Correlation Coefficient is computed between the estimated depth and true depth and is given in Table 2 and compared with cICA [8].

4.2 Experiment-II

For the second experiment, we have used the synthetic data set [14], and this data consists of the 3D animation sequence of shark. The shape parameters of the shark are represented by 91 points. The shape deformation is available over 240 frames. In these frames, the shark undergoes different motion around x-, y-, and z-axes. In this work, we have considered the frames where the 91 features of the shark are visible. The sample frames are shown in Fig. 8.

For the estimation of the shape of the shark, the frames 1 and 2 are used and are shown in Fig. 9. fpdPSO is used to find the optimal shape of the shark represented using 91 points based on Eq. (3). The estimated shape and true shape are shown in Fig. 10. The Pearson Correlation Coefficient of the estimated depth and true depth considering frames 2, 3, 4, and 5 is computed and given in Table 3.

Fig. 6 fpdPSO estimated
depth and true depth using
non-frontal view PR_U

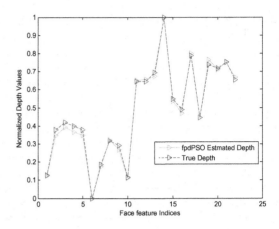

Fig. 7 Computed cost (min,
mean, and best) versus
generation

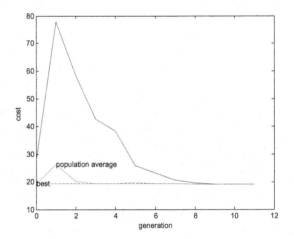

Table 2 Pearson Correlation
Coefficients using fpdPSO
using Bosphorus subject
bs000

Training sample	Pearson Correlation Coefficient	
	fpdPSO	cICA
PR_D	0.96	0.8822
PR_SD	0.98	0.8805
PR_SU	0.98	0.8775
PR_U	0.98	0.8758
YR_R10	0.99	0.8789

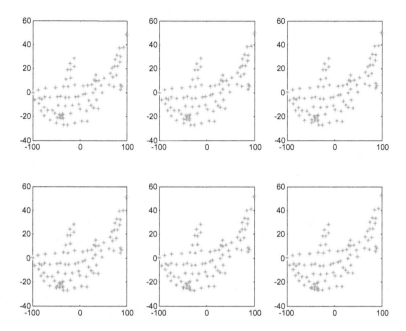

Fig. 8 Frames of shark at t = 1, 2, 3, 4, 5, 6

Fig. 9 Frames used for estimation of shape of shark

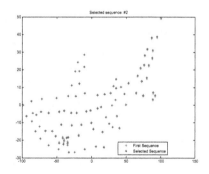

Fig. 10 fpdPSO estimated depth and true depth of the shark data

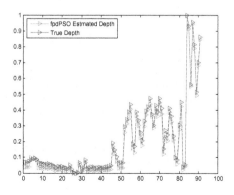

Table 3 Pearson Correlation Coefficients using fpdPSO for shark data

Training sample	Pearson Correlation Coefficient fpdPSO
Frame2	0.41
Frame3	0.38
Frame4	0.37
Frame5	0.36

5 Conclusion

The projection of the 3D shape on to the 2D is considered as an orthographic projection model. The observed shape over finite frames is used as input to the model and the reprojection error is minimized by estimating the shape and motion parameters using proposed Fibonacci population degeneration particle swarm optimization. The initial large population is a Fibonacci number and during progression of the generations the population is degenerated retaining the best-performing agents. This feature will be useful for high-dimensional parameter vector reducing the evaluation of agents in the subsequent generations. The results are reported on one subject of Bosphorus database and also on synthetic data of 3D motion of shark.

References

1. Bowyer, Kevin W., Kyong Chang, and Patrick Flynn. "A survey of approaches and challenges in 3D and multi-modal 3D + 2D face recognition." *Computer vision and image understanding* 101.1 (2006): 1–15.
2. Levine, Martin D., and Yingfeng Chris Yu. "State-of-the-art of 3D facial reconstruction methods for face recognition based on a single 2D training image per person." *Pattern Recognition Letters* 30.10 (2009): 908–913.
3. Ullman, Shimon. "The interpretation of visual motion." (1979), MIT Press.
4. Tomasi, Carlo, and Takeo Kanade. Shape and motion from image streams: a factorization method: full report on the orthographic case. Cornell University, 1992.
5. Bregler, Christoph, Aaron Hertzmann, and Henning Biermann. "Recovering non-rigid 3D shape from image streams." *Computer Vision and Pattern Recognition, 2000. Proceedings. IEEE Conference on*. Vol. 2. IEEE, 2000.
6. Torresani, Lorenzo, Aaron Hertzmann, and Christoph Bregler. "Nonrigid structure-from-motion: Estimating shape and motion with hierarchical priors." *Pattern Analysis and Machine Intelligence, IEEE Transactions on* 30.5 (2008): 878–892.
7. Koo, Hei-Sheung, and Kin-Man Lam. "Recovering the 3D shape and poses of face images based on the similarity transform." *Pattern Recognition Letters* 29.6 (2008): 712–723.
8. Sun, Zhanli, and Kin-Man Lam. "Depth estimation of face images based on the constrained ICA model." *Information Forensics and Security, IEEE Transactions on* 6.2 (2011): 360–370.
9. Sun, Zhan-Li, Kin-Man Lam, and Qing-Wei Gao. "Depth estimation of face images using the nonlinear least-squares model." *Image Processing, IEEE Transactions on* 22.1 (2013): 17–30.

10. Chandar, Kothapelli Punnam, and Tirumala Satya Savithri. "3D Structure Estimation Using Evolutionary Algorithms Based on Similarity Transform." *Modelling Symposium (AMS), 2014 8th Asia*. IEEE, 2014.

11. Ahlberg, Jörgen. "An active model for facial feature tracking." *EURASIP Journal on applied signal processing* 2002.1 (2002): 566–571.

12. M. Hollander and D. A. Wolfe, Nonparametric Statistical Methods. New York: Wiley, 1973. Kirkpatrick Jr, S. "CDG, and Vecchi." *MP Optimization by simulated annealing* 220 (1983): 671–680.

13. Kennedy, James, et al. *Swarm intelligence*. Morgan Kaufmann, 2001.

14. http://www.cs.dartmouth.edu/~lorenzo/nrsfm.html.

Connecting the Gap Between Formal and Informal Attributes Within Formal Learning with Data Mining Techniques

Shivanshi Goel, A. Sai Sabitha and Abhay Bansal

Abstract Formal and informal attributes are two distinct forms of learning which famed on the basis of the learning content, by where, when, and how learning happened. Formal attributes is a traditional learning which has official course work which should be completed in specified time. This study aimed at evaluating the challenges that students face while working for achieving good grades in exams. Data mining techniques are used to identify the challenges. The methods of collection working in this study were qualitative which involved testing and comparing.

Keywords Formal attributes · Informal attributes · Data mining
Rapid miner · PCA · K-Means clustering

1 Introduction

Learning analytics is a combination of different field like learning science, pedagogy, computer science, and web science [1]. This field is consistently used to find out new ways to analyze educational data. Learning analytics can be used for converting educational data into beneficial activities to foster learning [2, 3]. Learning analytics dimensions can be considered in four categories like collection

S. Goel (✉) · A. Sai Sabitha
Department of CSE, Amity University Noida, Noida, India
e-mail: shivanshigoel.27@gmail.com

A. Sai Sabitha
e-mail: saisabitha@gmail.com

A. Bansal
Amity University Noida, Noida, India
e-mail: abhaybansal@hotmail.com

© Springer Nature Singapore Pte Ltd. 2019   483
H. S. Behera et al. (eds.), *Computational Intelligence in Data Mining*,
Advances in Intelligent Systems and Computing 711,
https://doi.org/10.1007/978-981-10-8055-5_43

of data and its use for analysis to targeted group, need of analysis, and performance [4]. The focus of this research is to consider the first dimension (data, environments, and context). The learning analytics context can collect the data from traditional education system like learning management system (LMS) and less formal educational system like massive open online courses (MOOC). Though digital, technologies are increasing day by day and children's used internet to gain more knowledge about their choice of subject and through internet they get proper and deep knowledge of their subject. It leads children to higher qualification as it is systematic and time bounded, whereas informal attributes is interest obsessed. It is an arranged or precise approach that sets the learning targets initially, and plans and leads suitable strategies that empower representatives to ace foreordained results [5]. Informal attributes is characterized as discovering that is not exceedingly organized or planned, but the learning is controlled by the learners. The activities are learning exercises that people start in the working environment, include the consumption of physical, psychological, or enthusiastic exertion, and result in the improvement of expert or long-lasting information and abilities [6]. This research tries to bridge the gap between the context learning of informal and formal attributes and to recognize the issues that are connected with these behaviors and to find out that what the basic requirements for children are to gain good marks in school. Is informal attributes very much important even when formal attributes is going-on. Characteristics of openness learning, mooc are that there is no specified place for study, learning is spontaneous, no time bound, no assessment needed, attendance is depend on interest. Characteristics of traditional learning are that there is a school or institution for learning, specified course, time-bound learning, worksheets for assessments, attendance is compulsory. The current research need to understand and identify the parameter that connects the formal and informal attributes activities. In this research work, our goal is to focus on identifying formal and informal attributes in traditional learning context (TLC). Learning analytics and educational data mining are fairly related to each other as they both emphasize on educational domain and work with data inventing from educational environments, etc. with the objective of refining the learning practices. Siemens and Baker [7]. In this research work, data mining techniques are used to identify the above-said objective. The paper structured in six sections are literature review, data mining techniques used, methodology, experimental setup, results, and conclusion.

2 Literature Review

Formal attributes is related to a formal course and often leads to a qualification, though informal attributes is generally characterized as interest-driven and happening by chance suggested by Dabbagh and Kitsantas [8], Hall [9]. The hang of programming and all learning strategies should be tweaked by the basic characteristics/necessities of the learners. Learners have particular prerequisites and qualities, i.e., learning conduct, learning styles. Further, formal properties occur in an instructive foundation amid authority course time, (for instance: school, school), while casual traits are consistently done time permitting [8–10]. Formal characteristics are ordinarily guided by a teacher, while casual qualities are mainly self-facilitated and learner-focused [11, 12]. Seddon and Biasutti [13] make an endeavor in formal-casual properties, where a teacher masterminded a class (formal) yet was absent for when the understudies learned (easygoing). A couple of understudies delighted in working at a slower pace, while others expected to progress yet were not allowed. This could show to us that an instructor has a place in the casual actualized into the formal classroom. Regardless of the way that understudies will fill in as both suppliers and beneficiaries of companion evaluation, it cannot be acknowledged that understudies will be legitimately sorted out to comprehend the obligation of their own specific learning. This is especially legitimate if the territory customarily has an educator as the fundamental information source [14]. Thus, it will be the commitment of the teacher to be a guide and exhibit the right systems for partner composed learning and to demonstrate that the learners' can be compelling in drawing with the work as the pleasing while still an understudy themselves [15].

3   Data Mining Concepts

Data mining is utilized to mine the data from a dataset and revamp into coherent structure for further utilization which includes data pre-preparing, model and surmising considerations, representation, and so on. A portion of the procedures utilized as a part of data mining is classification, clustering, principal component analysis, and so on. In classification the commission of summing up known structure to apply to new data. While clustering is the task of realizing groups and structures in the data that are somehow or another "comparative," without utilizing known structures in data. PCA is for the most part utilized as a tool in exploratory data analysis and for making predictive models. Factors analysis is a strategy for clarifying the structure of data by explaining the relationships between factors and outlines data into few dimensions by gathering large variables into a smaller set of variables. Data mining techniques can change the learning processes and methods like classification [16, 17], prediction, association, and clustering [18, 19] are utilized as a part of e-learning environment.

4   Methodology

Collecting the dataset for preprocessing (like finding missing tuples, cleaning the data, reduction of data into important parameters) in which principal component analysis (PCA) is used. Clustering techniques are applied where (identifying different clusters, creating clusters of formal attributes and informal attributes) then clustering technique K-means applied for analysis. There are two different kinds of variables used which are listed in formal attributes and informal attributes. Variables used in formal attributes are school_support and activities, and variables used in informal attributes are family_sup, paid, internet, romantic, and health.

5   Experimental Setup

5.1   Data Collection

The dataset used for this study is taken from Portuguese schools dataset from (https://archive.ics.uci.edu/ml/datasets/Student+Performance) released on Nov 27, 2014. The dataset has 33 columns and 396 rows. Each row represents a learner who is engaged in somewhere in formal activities and informal activities. Screen-shot of the dataset (Refer Fig. 1).

6   Results and Analysis

6.1   PCA Test

PCA test is applied on formal attributes to gather the needful data for the research. These are the results of the PCA test used for further clustering process.

As the graph (Refer Fig. 2.) shows maximum cumulative proportion of variance between two variables (school_support, activities) that is why these two variables are chosen for further clustering process.

As the graph shows (Refer Fig. 3) maximum cumulative proportion of four variables (paid_classes, romantic, health, fam_support, and internet) in informal attributes. So, these variables are taken for clustering.

	A	B	C	D	E	F	G	H	I	J	K	L	M	N	O	P	Q	R	S	T	U
1	school	sex	age	address	famsize	Pstatus	Medu	Fedu	Mjob	Fjob	reason	guardian	traveltime	studytime	failures	schoolsup	famsup	paid	activities	nursery	higher
2	GP	F	18	U	GT3	A		4	4 at_home	teacher	course	mother	2	2	0	yes	no	no	no	yes	yes
3	GP	F	17	U	GT3	T		1	1 at_home	other	course	father	1	2	0	no	yes	no	no	no	yes
4	GP	F	15	U	LE3	T		1	1 at_home	other	other	mother	1	2	3	yes	no	yes	no	yes	yes
5	GP	F	15	U	GT3	T		4	2 health	services	home	mother	1	3	0	no	yes	yes	yes	yes	yes
6	GP	F	16	U	GT3	T		3	3 other	other	home	father	1	2	0	no	yes	no	yes	yes	yes

Fig. 1 Screenshot of dataset

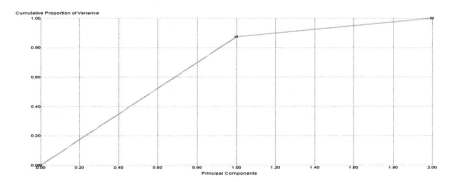

Fig. 2 Screenshot of PCA variance graph for formal attributes

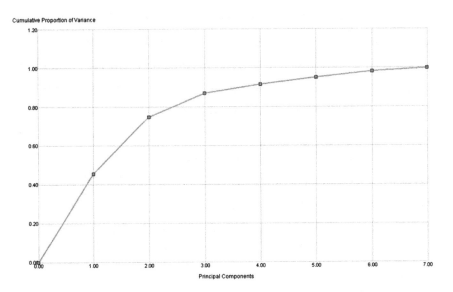

Fig. 3 Screenshot of PCA variable graph in informal attributes

6.2 K-Means Clustering Test

According to Zhang et al. [20], clustering is an essential method for comprehension and interpreting data. The objective of clustering is to group input objects or "clusters" to such an extent that items inside a group are like each other and questions in various groups are most certainly not. The result of clustering are shown in Table 1.

The above study concluded that activities from formal attributes and paid classes, fam_support, health from informal attributes connects the gap within traditional learning and students can achieve good grades (Refer in Table 2).

Table 1 Results of clustering

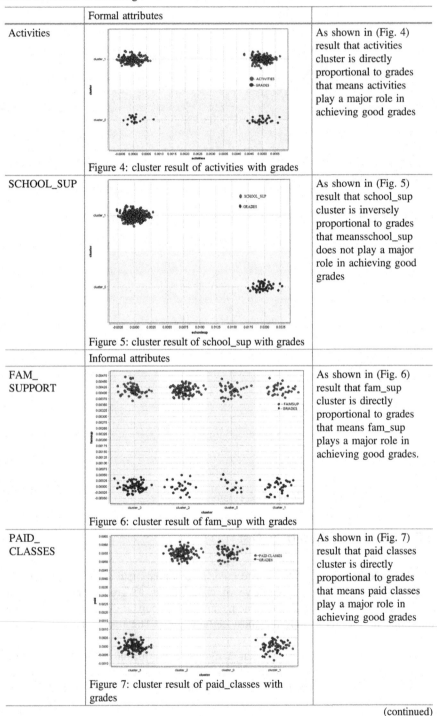

	Formal attributes	
Activities	Figure 4: cluster result of activities with grades	As shown in (Fig. 4) result that activities cluster is directly proportional to grades that means activities play a major role in achieving good grades
SCHOOL_SUP	Figure 5: cluster result of school_sup with grades	As shown in (Fig. 5) result that school_sup cluster is inversely proportional to grades that meansschool_sup does not play a major role in achieving good grades
	Informal attributes	
FAM_ SUPPORT	Figure 6: cluster result of fam_sup with grades	As shown in (Fig. 6) result that fam_sup cluster is directly proportional to grades that means fam_sup plays a major role in achieving good grades.
PAID_ CLASSES	Figure 7: cluster result of paid_classes with grades	As shown in (Fig. 7) result that paid classes cluster is directly proportional to grades that means paid classes play a major role in achieving good grades

(continued)

Table 1 (continued)

	Formal attributes	
Internet	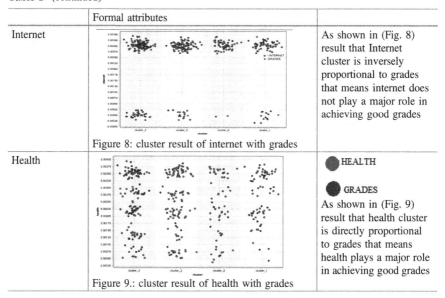 Figure 8: cluster result of internet with grades	As shown in (Fig. 8) result that Internet cluster is inversely proportional to grades that means internet does not play a major role in achieving good grades
Health	Figure 9.: cluster result of health with grades	● HEALTH ● GRADES As shown in (Fig. 9) result that health cluster is directly proportional to grades that means health plays a major role in achieving good grades

Table 2 Matrix distribution of results

	Grades	Activities	School_sup	Fam_sup	Health	Paid classes	Internet
Grades	–	Ok		Ok	Ok	Ok	
Activities	Ok	–					
School Sup			–				
Fam_sup	Ok			–			
Health	Ok			Ok	–	Ok	
Paid classes	Ok			Ok		–	
Internet							–

7   Conclusion and Future Scope

The results presented are just an example provide in order to connecting the gap between formal and informal attributes. Formal attributes is a traditional learning which has official course work which should be completed in specified time whereas informal attributes is characterized as discovering that is not exceedingly organized or planned, but the learning is controlled by the learners. Data collected from Portuguese schools could be used to analyze the gap between in-school and out-school learning and what factors depend on it.

According to the above research, the most important factors are activities, school_support, health, family_sup, paid classes, internet. After doing clustering, the result is shown in matrix table where good grades are dependent on school activities from formal attributes and paid classes, fam_support, health from informal attributes.

References

1. Dawson, S., Gašević, D., Siemens, G., & Joksimovic, S. (2014). Current state and future trends: A citation network analysis of the learning analytics field. In Proceedings of the Fourth International Conference on Learning Analytics and Knowledge (pp. 231–240). New York, NY, USA: ACM. https://doi.org/10.1145/2567574.2567585.
2. R. Ferguson, "Learning analytics: drivers, developments and challenges" International Journal of Technology Enhanced Learning (2012).
3. D. Clow, E. Makriyanni, "iSpot analysed: participatory learning and reputation" Proceedings of the 1st International Conference Learning Analytics and Knowledge (2011).
4. Chatti, M. A., Lukarov, V., Thüs, H., Muslim, A., Yousef, A. M. F., Wahid, U., & Schroeder, U. (2014). Learning analytics: Challenges and future research directions. E-learn Educ (Eleed) J, 10, 1–16.
5. Jacobs, R. L. (2003). Structured on-the-job training: Unleashing employee expertise in the workplace. San Francisco: Berrett-Koehler.
6. Lohman, M. C. (2000). Environmental inhibitors to informal attributes in the workplace: a case study of public school teachers. Adult Education Quarterly, 50(2), 83–101.
7. Siemens, G., & d Baker, R. S. (2012, April). Learning analytics and educational data mining: towards communication and collaboration. In Proceedings of the 2nd international conference on learning analytics and knowledge (pp. 252–254). ACM.
8. Dabbagh, N., & Kitsantas, A. (2012). Personal learning environments, social media, and self-regulated learning: A natural formula for connecting formal and informal attributes. *Internet and Higher Education, 15*(1), 3–8. https://doi.org/10.1016/j.iheduc.2011.06.002.
9. Hall, R. (2009). Towards a fusion of formal and informal attributes environments: The impact of the read/write web. *Electronic Journal of E-Learning, 7*(1), 29–40. Retrieved from http://www.ejel.org/volume7/issue1.
10. Sefton-Green, J. (2004). *Literature review in informal attributes with technology outside school.* Retrieved from https://www.nfer.ac.uk/publications/FUTL72/FUTL72.pdf.
11. Boustedt, J., Eckerdal, A., McCartney, R., Sanders, K., Thomas, L., & Zander, C. (2011). Students' perceptions of the differences between formal and informal attributes. In *Proceedings of the SeventhInternational Workshop on Computing Education Research* (pp. 61–68). New York, NY: ACM. https://doi.org/10.1145/2016911.2016926.
12. Lai, K. W., Khaddage, F., & Knezek, G. (2013). Blending student technology experiences in formal and informal attributes. *Journal of Computer Assisted Learning, 29*(5), 414–425. https://doi.org/10.1111/jcal.12030.
13. Seddon, F., & Biasutti, M. (2009). Participant approaches to and reflections on learning to play a 12-bar blues in an asynchronous e-learning environment. International Journal of Music Education, 27(3), 189–203.
14. Lebler, D. (2007). Student-as-master? Reflections on a learning innovation in popular music pedagogy. International Journal of Music Education, 25(3), 205–221.
15. Rust, C., O'Donovan, B., & Price, M. (2005). A social constructivist assessment process model: How the research literature shows us this could be best practice. Assessment & Evaluation in Higher Education, 30(3), 231–240.

16. Sabitha, A. S., Mehrotra, D., Bansal, A., & Sharma, B. K. (2016a). A naive bayes approach for converging learning objects with open educational resources. *Education and Information Technologies*, *21*(6), 1753–1767.
17. Sabitha, A. S., Mehrotra, D., & Bansal, A. (2016b). An ensemble approach in converging contents of LMS and KMS. *Education and Information Technologies*, 1–22.
18. Sabitha, A. S., Mehrotra, D., & Bansal, A. (2015). Delivery of learning knowledge objects using fuzzy clustering. Education and information technologies, 1–21.
19. Sabitha, S., Mehrotra, D., & Bansal, A. (2014). A data mining approach to improve re-accessibility and delivery of learning knowledge objects.Interdisciplinary Journal of E-Learning and Learning Objects, 10, 247–268.
20. Zhang, Y., Tangwongsan, K., & Tirthapura, S. (2017). Streaming Algorithms for k-Means Clustering with Fast Queries. *arXiv preprint* arXiv:1701.03826.

Multiple Linear Regression-Based Prediction Model to Detect Hexavalent Chromium in Drinking Water

K. Sri Dhivya Krishnan and P. T. V. Bhuvaneswari

Abstract This paper discusses the dependency between various water quality parameters (WQPs), namely pH, TDS, and conductivity that are determined to estimate the presence of hexavalent chromium compounds in drinking water. Multiple linear regression (MLR)-based prediction model is proposed to estimate the above parameters. The changes in WQPs are analyzed under both instant and stable conditions. The deviation between the measured and the estimated WQP is computed and added as the correction factor in order to improve the detection accuracy.

Keywords Water quality parameters · Hexavalent chromium
Multiple linear regression · Correction factor and detection accuracy

1 Introduction

The distribution of safe drinking water free from contamination is essential for a healthy living. World Health Organization (WHO) has reported that 780 million people are deprived of pure drinking water [1]. The major causes of contaminations identified are sewage disposal, sediments in reservoirs, and toxic chemical contaminants, namely chromium (Cr), sulfates (So_4^{2-}), and Lead (Pb) released from leather and tannery industries [2].

Water quality is analyzed by measuring the stochastic WQPs (Water Quality Parameters), namely pH, total dissolved solids (TDS), conductivity, total suspended solids, hardness, temperature [3]. pH is the measure of hydrogen ion concentration present in water [4]. It is used to represent acidity and alkalinity characteristics of

K. Sri Dhivya Krishnan · P. T. V. Bhuvaneswari (✉)
Department of Electronics Engineering, Madras Institute of Technology, Anna University,
Chennai 600044, Tamilnadu, India
e-mail: ptvbmit@annauniv.edu

K. Sri Dhivya Krishnan
e-mail: sridhivya.sridhivya@gmail.com

© Springer Nature Singapore Pte Ltd. 2019
H. S. Behera et al. (eds.), *Computational Intelligence in Data Mining*,
Advances in Intelligent Systems and Computing 711,
https://doi.org/10.1007/978-981-10-8055-5_44

the water. TDS represent the inorganic salts and organic matter that are dissolved in drinking water. The maximum allowable range of TDS as prescribed by WHO is 600 mg/l [5]. Conductivity indicates ionic activity of water which is measured in terms of electrical current. It is directly related to the concentration of ions present in water. Temperature determines the metabolic characteristics of water. Presence of highly dissolved minerals, especially calcium and magnesium, causes hardness in water [6].

Chromium is classified into two forms, namely trivalent (Cr^{+3}) and hexavalent chromium (Cr^{+6}). Cr^{+3} is the dietary nutrient and is essential to the human health which is in vegetables, fruits, and grains. Cr^{+6} is a highly toxic chemical compound produced by leather industrial effluents, textile manufacturing, wood preservation, electroplating, and steel production. It causes cancer even at the low concentration. The allowable limit of the total chromium in drinking water as prescribed by environmental protection agency is 0.1 mg/L [7]. National Institute of Environmental Health Science has analyzed the impact on hexavalent chromium compounds and has reported that the sodium dichromate dihydrate causes lung cancer and tumors and dermatitis [8]. Hence, in this paper, the dependency between WQPs such as pH, TDS, and conductivity is analyzed through multiple linear regression (MLR) model. Initially, to obtain the WQPs, an experimental analysis is performed in the laboratorial environment with synthetic sample of hexavalent chromium. The WQPs are measured by appropriate sensors. Further, the deviations between the estimated and the measured WQPs are determined to improve the accuracy of the proposed MLR model.

The rest of this paper is organized as follows: Sect. 2 presents the literature survey on the various methods used for detection of the hexavalent chromium in water. Section 3 explains the proposed MLR-based prediction model that detects the hexavalent chromium in drinking water. Further, it details the analysis performed to validate the proposed model under both instant and stability conditions. Section 4 discusses the results, and Sect. 5 concludes the paper with future work.

2 Literature Review

In [9], Alexopoulos has investigated the concentration of toluene levels and developed the prediction model using MLR. The author has validated the model by computing the deviation between actual and estimated toluene concentration. The accuracy of the model is improved by incorporating the deviation factor to the prediction model.

In [10], Hana vaskova and Karel Koloznik have used two spectroscopic methods to measure the concentration of Cr^{+3} and Cr^{+6}. The samples considered are 10 different types of leather and shaving cream. Leather samples are soaked in 0.1, 1, and 5% of $K_2Cr_2O_7$ solutions. Ultraviolet-Visible (UV-Vis) spectrometric method provides the concentration of the considered samples, while Raman spectroscopic method provides the peak changes occurring in both trivalent and hexavalent chromium due to concentration variation. For the considered samples, the spectral changes are acquired at the 550 cm^{-1} for trivalent chromium and 900 cm^{-1} for hexavalent chromium. From the two analyses, the authors have concluded that compared to UV-Vis spectrometric method, Raman spectroscopy results with better accuracy.

In [11], S. Xu et al. have developed a micro-electrochemical sensor to trace the level of hexavalent chromium ions in their samples. They have used bismuth film and mesoporous carbon electrode to develop the sensor. The authors have also analyzed the effect of hexavalent chromium on trivalent chromium in the solutions. It showed that reduction in Cr^{+3} concentration results in addition of Cr^{+6} to it. The developed sensor is tested for different concentrations of hexavalent chromium solutions. The analysis showed that the sensor had the ability to detect even 0.05 µg/L of hexavalent chromium concentration. The authors have concluded that the developed device detects the hexavalent chromium concentration between 1 and 20 µg/L.

In [12], the authors have assessed the water quality of Bailongjiang and Jialingjiang rivers which serve as the main water resources for Wuwei City. Water quality of the rivers has been degraded due to the discharge of waste water from industrial plants located near the rivers. They have analyzed the quality of the rivers using various WQPs, namely dissolved oxygen, chemical oxygen demand, permanganate index, NH$_3$-N, and total ammonium. The concentration values of these parameters are compared with environmental quality standard (EQS) [13]. It is inferred that the examined samples are drinkable as they fall under class II category of EQS. The authors have further studied the feasibility of comprehensive water quality identification index of the rivers using two different analytical tools, namely index of water quality (Iwq) and Nemerow index, that provide global vision of the overall water quality of the river. Iwq index ranges from 1.2 to 1.4 while Nemerow index from 1.47 to 1.59. Both these indexes are found to increase linearly. Further, the correlation between the two assessment tools is estimated using MLR. From the results, it is found that 84% of correlation exists between them.

In this research, an experimental work has been carried out with four different hexavalent chromium compounds to measure the WQPs, namely pH, TDS, and conductivity. The relationships between these parameters are modeled using MLR-based prediction model. The performance of the proposed MLR-based prediction model is analyzed for both instant and stable conditions of contaminants.

3 Proposed Work

This section details the procedure involved in the measurement of WQPs such as pH, TDS, and conductivity and the dependence existing between them. To obtain the statistical data of WQPs, four different hexavalent chromium compounds are considered. They are potassium dichromate ($K_2Cr_2O_7$), lead chromate ($PbCrO_4$), sodium chromate (Na_2CrO_4), and sodium dichromate dihydrate ($Na_2Cr_2O_4 \cdot 2H_2O$).

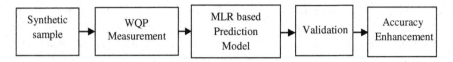

Fig. 1 Proposed MLR-based prediction model

The proposed MLR-based prediction model analyzes the instant change in WQPs when the contamination is added to the pure drinking water. It is also extended to stable condition. Then, the deviation between the estimated and the measured WQP is computed to enhance the accuracy of the estimation [14]. The procedure involved in the proposed research is illustrated in Fig. 1. It consists of the following stages.

3.1 Preparation of Synthetic Sample

Initially, the WQPs of one liter pure drinking water are measured as prescribed by WHO standards. Then, the synthesized sample is prepared by adding 0.25 mg of Na_2CrO_4 to one liter of drinking water. Color of the sample changes immediately to orange indicating the presence of contamination. The WQPs are measured using appropriate sensors. Then, the same process is repeated by varying the concentration of Na_2CrO_4 in the drinking water. The procedure is repeated for the remaining compounds such as $K_2Cr_2O_7$, $Na_2Cr_2O_4 \cdot 2H_2O$, and $PbCrO_4$ [15].

3.2 Measurement of WQPs

The WQPs, namely pH, TDS, and conductivity of the prepared synthetic samples are measured as per the standard experimental procedure in the laboratory environment. pH is measured using glass electrode. TDS is measured using HM digital TDS tester [16]. Conductivity is measured using Vernier conductivity probe [17]. It is connected to the data logger to store the measured conductivity.

3.3 Proposed MLR-Based Prediction Model

The dependency between WQPs is modeled using MLR model under instant and stable conditions. The correlation coefficients (r_1, r_2, r_3) between pH and TDS, TDS and conductivity, conductivity and pH are determined.

Let X_i be the measured pH, Y_i be the TDS, and Z_i be the conductivity. Then, the correlation coefficient (r_1) between pH (X_i) and TDS (Y_i) is computed using Eq. (1):

$$r_1 = \sum_{i=1}^{n} \left(\frac{(X_i - \overline{X})(Y_i - \overline{Y})}{\sqrt{(X_i - \overline{X})^2 (Y_i - \overline{Y})^2}} \right) \tag{1}$$

Similarly, the correlation coefficient (r_2) between TDS (Y_i) and conductivity (Z_i) is computed using Eq. (2):

$$r_2 = \sum_{i=1}^{n} \left(\frac{(Y_i - \overline{Y})(Z_i - \overline{Z})}{(Y_i - \overline{Y})^2 (Z_i - \overline{Z})^2} \right) \tag{2}$$

Then, the correlation coefficient (r_3) between conductivity (Z_i) and pH (X_i) is calculated using Eq. (3):

$$r_3 = \sum_{i=1}^{n} \left(\frac{(Z_i - \overline{Z})(X_i - \overline{X})}{(Z_i - \overline{Z})^2 (X_i - \overline{X})^2} \right) \tag{3}$$

where $\overline{X}$ = mean value of pH, $\overline{Y}$ = mean value of TDS, and $\overline{Z}$ = mean value of conductivity of the considered 'n' number of samples.

Initially, the proposed MLR-based prediction model is trained with the WQPs measured for the prepared synthetic samples. The relationships between these measured parameters are determined based on the least square method [18]. Then, the prediction model is formulated.

Let 'Y' be the measured TDS, then 'x' the pH is found using Eq. (4)

$$x = a_1 + a_0 Y \tag{4}$$

Let 'Z' be the conductivity, then 'y' the TDS is found from Eq. (5)

$$y = a_3 + a_2 Z \tag{5}$$

Similarly, the 'z' in terms of 'X' is found using Eq. (6)

$$z = a_4 + a_5 X \tag{6}$$

Here, a_0 to a_5 are called regression parameters. They are determined using (7)–(12):

$$a_0 = \overline{Y} - a_1 X \tag{7}$$

$$a_1 = \sum_{i=1}^{n} \left(\frac{(X_i - \overline{X})(Y_i - \overline{Y})}{(X_i - \overline{X})^2} \right) \tag{8}$$

$$a_2 = \overline{Z} - a_3 X \qquad (9)$$

$$a_3 = \sum_{i=1}^{n} \left(\frac{(Y_i - \overline{Y})(Z_i - \overline{Z})}{(Y_i - \overline{Y})^2} \right) \qquad (10)$$

$$a_4 = \overline{X} - a_5 X \qquad (11)$$

$$a_5 = \sum_{i=1}^{n} \left(\frac{(Z_i - \overline{Z})(X_i - \overline{X})}{(Z_i - \overline{Z})^2} \right) \qquad (12)$$

The above steps are repeated for instant and stable conditions of synthesized samples.

3.4   Validation of MLR Modeling

The accuracy of the proposed MLR-based prediction model is determined by computing the deviation between measured and estimated WQPs. The deviation between estimated pH (x_i) and measured pH (X_i) is calculated using Eq. (13)

$$E_1 = \sqrt{\sum \frac{(x_i - X)}{m - 2}} \qquad (13)$$

Similarly, the deviation between estimated TDS (y_i) and measured TDS (Y_i) is computed by Eq. (14) and the estimated conductivity (z_i) and the measured conductivity (Z_i) by Eq. (15):

$$E_2 = \sqrt{\sum \frac{(y_i - Y)}{m - 2}} \qquad (14)$$

$$E_3 = \sqrt{\sum \frac{(z_i - Z)}{m - 2}} \qquad (15)$$

Here, m is the number of considered parameters and m-2 is the degree of freedom of the considered WQPs.

4 Results and Discussion

This section analyzes the results obtained by applying the proposed MLR-based prediction model on WQPs. It investigates the instant changes in WQPs caused due to the injection of contaminant in drinking water. Further, it also examines the gradual changes in the WQPs until the stable condition is attained. Error analysis is performed to improve the accuracy of the proposed model. Error reduction is achieved by computing the deviation between the measured and the estimated parameters.

Figure 2 shows the dependency existing between TDS and conductivity of considered hexavalent chromium samples. It is measured using appropriate sensors. The correlation between them is computed for various concentrations of $K_2Cr_2O_7$, $Na_2Cr_2O_4 \cdot 2H_2O$, and Na_2CrO_4. The correlation coefficients (r_1, r_2, r_3) are computed. The dependence of TDS with conductivity for $K_2Cr_2O_7$ is found to be 98%. For $Na_2Cr_2O_4 \cdot 2H_2O$, it is 69%, and for Na_2CrO_4, it is 86%.

Figure 3 shows the correlation between TDS and conductivity of $K_2Cr_2O_7$, $Na_2Cr_2O_7 \cdot 2H_2O$, and Na_2CrO_4 under stable condition. From the computed correlation coefficients, it is inferred that TDS and conductivity of $K_2Cr_2O_7$ exhibit 99.46% correlation, $Na_2Cr_2O_4 \cdot 2H_2O$ exhibits 99.46% correlation, while Na_2CrO_4 exhibits 100% correlation which indicates that the estimated conductivity is equal to the measured conductivity. Hence, 100% accuracy is achieved by the proposed model. Then, deviations between the measured and the estimated conductivity of $K_2Cr_2O_7$, $Na_2Cr_2O_7 \cdot 2H_2O$, and Na_2CrO_4 are computed. They are found to be 29.46, 42.33, and 48.79, respectively, which is added to the proposed model in order to improve the accuracy.

Figure 4 shows the correlation estimated between TDS and conductivity of $PbCrO_4$ under both instant and stable conditions. It is found that 99.36% accuracy is achieved on adding 1.5865 deviation obtained by the proposed model. Figure 5 shows the correlation between TDS and pH for $K_2Cr_2O_7$, Na_2CrO_4, and $Na_2Cr_2O_7 \cdot 2H_2O$. The correlation coefficient of $K_2Cr_2O_7$ is found to be 48%. For Na_2CrO_4, it is 13%, and for $Na_2Cr_2O_7 \cdot 2H_2O$, it is 0.73%. Thus, it is inferred that it exhibits minimum dependence between the pH and TDS for $Na_2Cr_2O_7 \cdot 2H_2O$.

Fig. 2 Correlation estimation between TDS and conductivity at the instant of contaminant injection

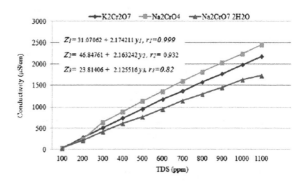

Fig. 3 Correlation estimation between TDS and conductivity under stable condition

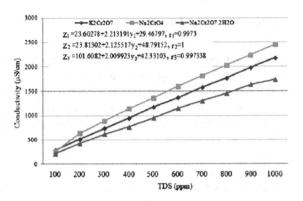

Fig. 4 Correlation estimation between TDS and conductivity of PbCrO$_4$ at the instant of contaminant injection and under stable condition

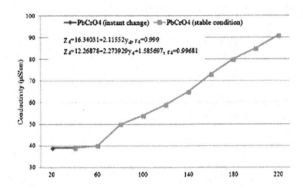

Fig. 5 Correlation estimation between TDS and pH at instant contaminant injection

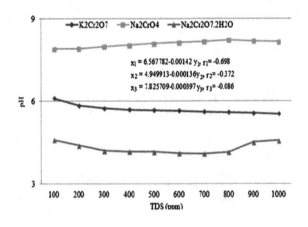

Figure 6 shows the dependency of TDS on pH for K$_2$Cr$_2$O$_7$, Na$_2$CrO$_4$, and Na$_2$Cr$_2$O$_7 \cdot$ 2H$_2$O under stable condition of contaminant. The dependency of TDS on conductivity for K$_2$Cr$_2$O$_7$ is found to be 80%. For Na$_2$CrO$_4$, it is 83.77% while for Na$_2$Cr$_2$O$_7 \cdot$ 2H$_2$O, it is 99.97%. The deviation between the measured and the estimated parameters is computed. They are 0.0778, 0.04912, and 0.55451,

Fig. 6 Correlation estimation between TDS and pH under stable condition

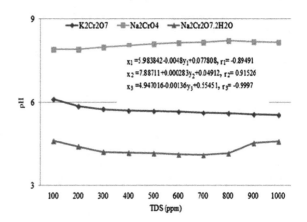

$x_1 = 5.983842 - 0.0048y_1 + 0.077808, r_1 = -0.89491$
$x_2 = 7.88711 + 0.000283y_2 + 0.04912, r_2 = 0.91526$
$x_3 = 4.947016 - 0.00136y_3 + 0.55451, r_3 = -0.9997$

Fig. 7 Correlation estimation between TDS and pH of PbCrO$_4$O at the instant contaminant injection and stable condition

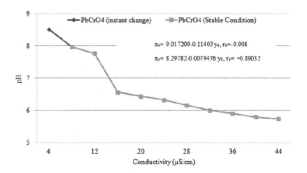

$x_4 = 9.017209 - 0.11403y_4, r_4 = -0.908$
$x_4 = 8.29782 - 0.0079476y_4, r_4 = -0.89032$

respectively, which is then added to the proposed MLR model to improve the accuracy.

Figure 7 shows the dependency of TDS on pH for PbCrO$_4$ under both instant and stable conditions. It is found that 81% correlation exists. Deviation of 0.1425 is added to the proposed MLR model to enhance the accuracy.

The dependency of conductivity on pH predicted by the proposed MLR model for K$_2$Cr$_2$O$_7$, Na$_2$CrO$_4$, and Na$_2$Cr$_2$O$_7$ · 2H$_2$O on instant contaminant injection is shown in Fig. 8. It is found that 49% dependency exists for K$_2$Cr$_2$O$_7$, while 17% for Na$_2$CrO$_4$ and 0.73% for Na$_2$Cr$_2$O$_7$ · 2H$_2$O. Figure 9 shows the correlation estimated between conductivity and pH for K$_2$Cr$_2$O$_7$, Na$_2$CrO$_4$, and Na$_2$Cr$_2$O$_4$ · 2H$_2$O under stable condition of contaminant. From the results, it is inferred that K$_2$Cr$_2$O$_7$ exhibits 91% correlation, while Na$_2$CrO$_4$ exhibits 83.77%, and Na$_2$Cr$_2$O$_7$ · 2H$_2$O as 99.74%.

The deviations between the estimated and the measured pH are found to be 0.07904, 0.046867, and 0.27506, respectively. Figure 10 shows the correlation relationship between conductivity and pH of PbCrO$_4$ at the time of contaminant injection and under stable condition. It is found that 75% correlation is observed for instant contamination while 78% correlation is exhibited under stable condition.

Fig. 8 Correlation estimation between conductivity and pH at instant contaminant injection

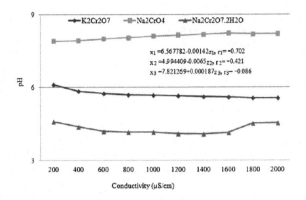

Fig. 9 Correlation estimation between conductivity and pH under stable condition

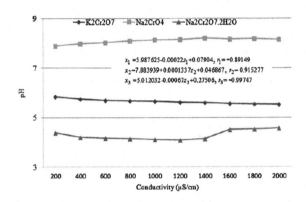

Fig. 10 Correlation estimation between conductivity and pH at instant contaminant injection and stable injection

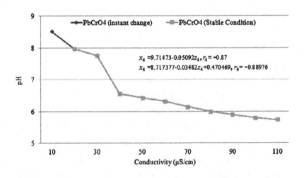

The deviation between the measured and the estimated pH is found to be 0.47 under stable contamination.

From the analysis, it is inferred that the correlation coefficients of $K_2Cr_2O_7$, Na_2CrO_4, $Na_2Cr_2O_7 \cdot 2H_2O$, and $PbCrO_4$ between TDS and conductivity are higher than 0.9 because the TDS and conductivity increase linearly. The dependency between pH and TDS of the considered samples lies between 0.086 and 0.999 at the time of contamination injection. But the correlation coefficients between the three considered WQPs are found to be closer to 0.99 under stable condition of contamination. From the comparison analysis, it is inferred that accuracy in WQPs estimation is enhanced under stable condition by the inclusion of deviation between the measured and estimated values.

5 Conclusion

In this research, the WQPs, namely pH, TDS, and conductivity are measured through standard experiments to detect the presence of hexavalent chromium compounds in synthesized sample of contaminated drinking water. The main objective of the work is to analyze the changes in WQPs under both instant and stable conditions of contamination. The obtained values of the correlation coefficients between WQPs are found to be higher than 0.89 for all the considered compounds. For Na_2CrO_4, the correlation coefficient is 1 which indicates that the proposed MLR-based prediction model is accurate. From the analysis, it is concluded that the dependency between WQPs is higher under the stable condition. The proposed work can be extended in the direction of developing the device to detect the hexavalent chromium.

References

1. https://www.cdc.gov/healthywater/global/wash_statistics.html.
2. http://www.conserve-energy-future.com/sources-and-causes-of-water-pollution.php.
3. https://www.epa.ie/pubs/advice/water/quality/Water_Quality.pdf.
4. http://www.who.int/water_sanitation_health/dwq/chemicals/ph_revised_2007_clean_version. pdf.
5. http://www.who.int/water_sanitation_health/dwq/chemicals/tds.pdf.
6. http://www.dwa.gov.za/iwqs/wq_guide/Pol_saWQguideFRESH_vol1_Domesticuse.PDF.
7. https://www.epa.gov/ground-water-and-drinking-water/national-primary-drinking-water-regulations.
8. https://www.nih.gov/news-events/news-releases/hexavalent-chromium-drinking-water-causes-cancer-lab-animals.
9. E C Alexopoulous, "Introduction to Multivariate Analysis", Int. J. on Hippokartia, 14 (Suppl 1), 2010, 23–28.
10. Hana Vaskova, and Karel Koloznik, "Spectroscopic Measurement of trivalent and hexavalent chromium", 17th Int. Carpathian Control Conference, 2016, 775–778.
11. S. Xu, X. Wang and C. Zhou, "A micro electrochemical sensor based on bismuth-modified mesoporous carbon for hexavalent chromium detection," IEEE sensors conference, 2015, 1–4.
12. Y. Jiang and Z. Ma, "An assessment of water quality from a reach of Bailongjiang River, Gansu province," 2011 Int. Conference on Electric Technology and Civil Engineering, 2011, 998–1000.
13. http://www.who.int/water_sanitation_health/resourcesquality/wpcchap2.pdf.

14. http://nptel.ac.in/courses/111104074/Module3/Lecture\df.
15. K. Sri Dhivya Krishnan and P.T.V. Bhuvaneswari, "Detection of Hexavalent Chromium contamination in drinking water using water quality sensor", 9th Int. Conference on Trends in Industrial Measurements and Automation, 2017, 177–181.
16. http://hmdigital.com/product/ap-1/.
17. https://www.vernier.com/products/sensors/conductivity-probes/con-bta.
18. http://www.itl.nist.gov/div898/handbook/toolaids/pff/pmd.pdf.

Data Engineered Content Extraction Studies for Indian Web Pages

Bhanu Prakash Kolla and Arun Raja Raman

Abstract The recent innovations in the Internet and cellular communications have opened many interesting and exciting areas of social and research activity, and one of the basic driving forces for this is the Web page containing data in different forms. Data can be in mobile or Internet based and can be online or off-line and normally of sizes ranging from kilo to terabytes. In the Indian context, these can relate to computer-generated, printed, or archived data in different languages and dialects. The present study is focused on applying engineering aspects to data so that a smart set is used to generate content in a short period, so that further developments can be easier. After a brief overview on the complexities of Indian Web pages and current approaches in data mining, a basic pixel-based approach is developed along with data reduction and abstraction to be used with classification approaches for content extraction. During data reduction, engineering approach based on organizing and adapting for suitable inputs for classification is highlighted, and a case study is given here for analysis.

Keywords Classification · Data engineering · Reduction · Pixel based
Knowledge · Mining

1 Introduction

Data has become increasingly the mainstay for all activities with developments in computers, Internet, and cellular communications picking up momentum in the last two decades [1, 2]. Now, a stage is reached where availability, access, and volume

B. P. Kolla (✉)
Department of Computer Science Engineering, Koneru Lakshmaiah Education Foundation, Green Fields, Vaddeswaram, Guntur 522502, Andhra Pradesh, India
e-mail: bhanu_prakash231@rediffmail.com

A. R. Raman
Department of Structural Engineering, IIT Madras, Chennai 600036, Tamil Nadu, India
e-mail: arraman41@yahoo.com

© Springer Nature Singapore Pte Ltd. 2019
H. S. Behera et al. (eds.), *Computational Intelligence in Data Mining*,
Advances in Intelligent Systems and Computing 711,
https://doi.org/10.1007/978-981-10-8055-5_45

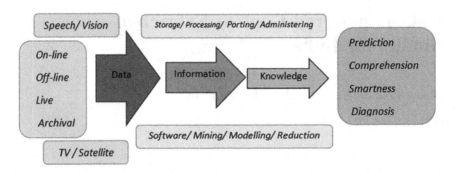

Fig. 1 Data evolution

of data are driving many aspects of day-to-day social, scientific, and living needs, sometimes making one wonder whether anything can be done without data! So, the last decade witnessed data evolving from accessing, storing, and processing to handling large volumes, manipulating them to get information and knowledge through data mining, data managing, and warehousing, the recent concepts of data science and engineering. Figure 1 gives an idea of the evolution of data with its various facets and its role in human comprehension, thinking, and intelligence. Here, one can see the different sources of acquiring data in different forms from live to archival, on the left and the ways it is handled through storage to the administration on top and the ways of dealing with it scientifically at the bottom and with aims on the right [3]. Human decision making in many instances depends on this evolution and how data supplements and sometimes even overrides human perception can be cited in many situations in recent times. So a separate discipline with data as the core is evolving, and this brings the need for science and engineering to be applied to data [4].

2 Need for Data Engineering

Data storage and processing have been going on ever since computers in different forms and sizes came on the scene; but in recent times, the volume, pattern, and sources from which data is accessed brought in the need for a higher level with science and engineering aspects to make a smarter and efficient way of handling and administering data. This brings in the role of data engineer to serve as a bridge between administrator, who organizes and scientist who analyzes making him to get the nuances of both. The various aspects of data, a data engineer needs to take care of, are the size—running in terabytes—forms ranging from binary to video form, nature of acquisition, form of porting and storing, and lastly, the background platform for processing in terms of hardware and software. All these aspects get more involved when one looks at the Indian context, and some features are presented next to highlight the complexity [5].

3   Indian Web Pages

India's entry into the age of data generation and accumulation began with the introduction of computers which were confined to mostly selected pockets of learning, but this got a tremendous boost with cellular phones and Internet access. Now, Indian Web pages can be in any languages from any region, and with any dialect, thanks to the proliferation of cell phones and this has led to a new era of business, communication, security, and social awareness with every Indian looking for or storing or even processing data.

Herein comes the complexity as Indian Web pages are bringing in data from live TV to displaying archival information in languages and dialects in different forms. Figures 2 and 3 give the two spectrums of Indian Web pages, with one, displaying news both online and off-line and, another an archival data, taken from a Web site

(a)   (b)

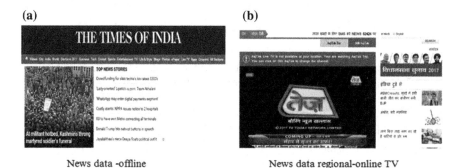

News data -offline   News data regional-online TV

Fig. 2 Typical current data types in India

(a)   (b)

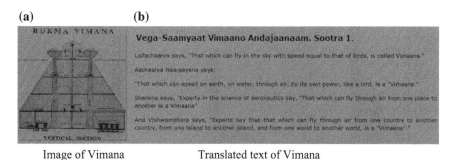

Image of Vimana   Translated text of Vimana

Fig. 3 Traditional data on aircraft. Ref: http://www.hinduwebsite.com/sacredscripts/hinduism/vimana/vchp01.asp

shown there. The complexities in terms of content are many; prominent among them is the mix-up of different forms like image text with different fonts and use of different languages. So, in extracting content from such Indian Web pages, data has to be properly organized both in terms of reducing the size and bringing out salient features for further mathematical processing. Here, a brief presentation of a method to handle different sizes of Indian data and extract the content and some case studies to highlight the application in the context of education is presented.

4 Pixel-Based Data Engineering

As the basic premise on which computing is dependent is either bytes or pixels, a method is developed based on pixels so that its usage will be easier. But the pixel maps are normally big in size, and it is necessary to reduce, extract salient features, and represent in scalar or vector form for further mathematical manipulations. Typically in the context of even basic text form, variations are many, and as an example, the letter in English 'I' can have equivalent in letter form or word form as shown below in three Indian languages, Hindi, Tamil, and Telugu and this means

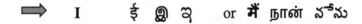

that there are letter–letter or letter–word equivalents both ways like 'x' unique in English versus ‘कौ’ a letter in Indian language needing three letters 'kau' in English. So to have a uniform approach, all can be translated into pixel maps and using different forms of reduction and some features, a comparison can be made. Here, the letter 'I' and its equivalent words ‘ मैं நான் న²ను in Indian languages are converted to pixel maps, and using nonzero, row, and column features, comparison is made with a full title **ELECTROMAGNETIC INDUCTION** in English. The pixel sizes vary from 456 to 7532 and three graphs showing variation in pixel map sizes, salient

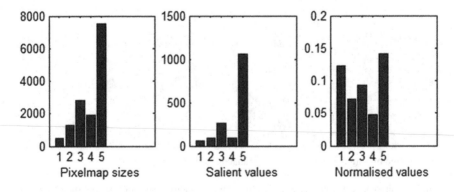

Fig. 4 Reduced data values for 5 cases

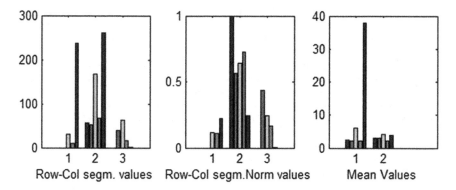

Fig. 5 Segmented values for 5 cases

values in pixel maps, and normalized ones irrespective pf sizes, in Fig. 4. It is seen that even though sizes increase depending on letter or word, normalized values are tractable, ranging from 0.1228 to 0.1407. Further improvements in terms of segmenting into three parts row can be made and these are shown in Fig. 5 while mean values can be evaluated and their values are shown in the last chart.

Having organized the data into tractable reduced values methodologies for classification can now be used, and this is shown next.

5 Classification for Content Extraction

To extract content from a Web page, salient words and images are chosen and grouped together and using reduction methods explained earlier, attributes can be generated either as mean values, salient values in vector form, or as matrices using segmented approach. An example of an image and word equivalents for 'circle' is given for illustration. Here with image, words in English, Hindi, Tamil, and Telugu are chosen along with 'circle' written as it is in Hindi as सिर्कल. The reduced values are shown in Fig. 6 where again, the normalized and mean values are

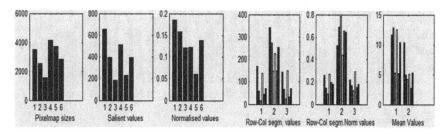

Fig. 6 Results for image–word case

(a)   **(b)**

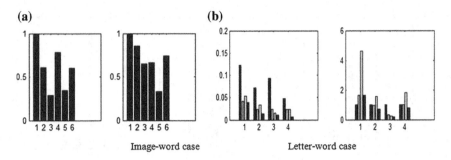

Image-word case   Letter-word case

Fig. 7 Patterns with base pixel map

tractable and very much reduced in size. Also one can see significant variation in segmented values.

Now, ratios with parent pixel map will give an idea of the pattern, and these are shown in Fig. 7 for both letter–word and image–word combinations.

6 Content Extraction

Using algorithms for classification like the scalar comparison, pattern recognition with statistical values and neural network, inputs in the form of reduced scalar vector and matrix form can now be used. Since for classification, any of the three algorithms as shown in flowchart can be used, and results are shown for typical ones here as details are available elsewhere [6, 7]. Different sets were used in the neural model and by varying the weighted matrices according to content relation; training of the inputs is done to achieve the target output as binary one with content

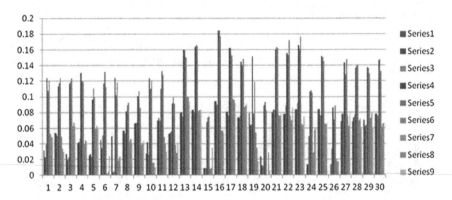

Fig. 8 Datasets chosen for classification

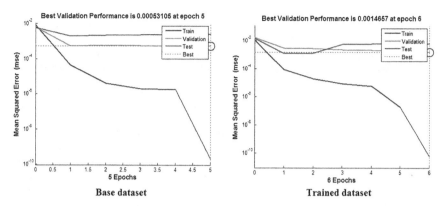

Fig. 9 Typical results with neural model

related being one and not related being zero. Three sets of pixel maps representing content words are used in the basic model with 'magnet,' 'iron,' 'filings,' etc. with corresponding regional ones. Results are shown in Figs. 8 and 9.

7 Conclusions

Features specific to Indian Web pages are presented to bring out the complexity and size of data one has to deal with. Using pixel-based modeling data is reduced through modeling, and feature extraction and comparison on reduction and size of scalar and vector values used for different cases are highlighted. A case study for classification is shown to bring out the generality and applicability of the method on any platform.

References

1. A. Busch, W. W. Boles and S. Sridharan, "Texture for Script Identification". IEEE Transactions on Pattern Analysis and Machine Intelligence, Vol. 27, No.11, IEEE Computer Society, 2005, pp. 1720–1732.
2. Deng Cai, Yu Shipeng and Wen Jirong, (2003) "VIPS: a vision-based page segmentation algorithm", Microsoft Technical Report, MSR-TR-2003-79, 406–417.
3. S. Kavitha, P. Shivakumara, G. Hemantha Kumar and C. L. Tan, "A Robust Script Identification System For Historical Indian Document Images", Malaysian Journal of Computer Science. Vol. 28(4), 2015, pp 283–300.
4. P. Krishnan, N. Sankaran, A. K. Singh and C. V. Jawahar, "Towards a robust OCR system for Indic scripts". Document Analysis Systems, IEEE, April 2014, pp. 141–145.
5. Maha Al-Yahya, Sawsan Al-Malak, Luluh Aldhubayi, "Ontological Lexicon Enrichment: The Badea System For Semi-Automated Extraction Of Antonymy Relations From Arabic Language Corpora", Malaysian Journal of Computer Science. Vol. 29(1), 2016, pp 56–73.

6. Kolla Bhanu Prakash, Dorai RangaSwamy, M, A, Raja Raman, Arun (2012), ANN for Multi-lingual Regional Web Communication, ICONIP 2012, Part V, LNCS 7667, pp. 473–478.
7. Kolla Bhanu Prakash, Dorai RangaSwamy, M, A, Raja Raman, Arun (2012), Statistical Interpretation for Mining Hybrid Regional Web Documents, ICIP 2012, CCIS 292, pp. 503–512.

Steganography Using FCS Points Clustering and Kekre's Transform

Terence Johnson, Susmita Golatkar, Imtiaz Khan,
Vaishakhi Pilankar and Nehash Bhobe

Abstract Steganography is the process of concealing one form of data within the same or another form of data. A cover medium is used to hide information within itself. Steganography is done using the Fuzzy C Strange Points Clustering Algorithm and the Kekre's Transform in this paper. Fuzzy C Strange Points Clustering Algorithm is used to provide security and robustness as this algorithm is found to give a better quality of clusters. Kekre's Transform is performed on the image, and the secret message is hidden in the LSBs of the transform coefficients. These two provide better hiding capacity and successful retrieval of the secret information.

Keywords Steganography · Clustering · Fuzzy C Strange Points Clustering Algorithm · Kekre Transform · LSB technique · Transform technique

1 Introduction

With the speedy expansion of communication technology, multimedia, and the sustained use of the Internet, dissemination of information needs to be secure as there could be many threats to the information. The communication can be strengthened by cryptography which scrambles the information. But one can easily

T. Johnson (✉) · S. Golatkar · I. Khan · V. Pilankar · N. Bhobe
Department of Computer Engineering, Agnel Institute of Technology & Design,
Assagao 403507, Goa, India
e-mail: ykterence@rediffmail.com

S. Golatkar
e-mail: 13co02@aitdgoa.edu.in

I. Khan
e-mail: 13co35@aitdgoa.edu.in

V. Pilankar
e-mail: 13co38@aitdgoa.edu.in

N. Bhobe
e-mail: 13co46@aitdgoa.edu.in

© Springer Nature Singapore Pte Ltd. 2019   513
H. S. Behera et al. (eds.), *Computational Intelligence in Data Mining*,
Advances in Intelligent Systems and Computing 711,
https://doi.org/10.1007/978-981-10-8055-5_46

comprehend that some kind of message has been encrypted and communicated which requires a key to be decoded. On the contrary, steganography techniques minimize the possibility of information being identified [1]. Steganography is used to insert the clandestine information into some concealed media like images, text, or audio visuals in a manner that intruders and hackers do not know about the original communication that the media may have and the methodologies used to implant or reclaim it [2]. More specifically, steganography deals with hiding the presence of data into some medium so that intruders do not know about its existence [3]. This paper proposes a system in which we use Fuzzy C Strange (FCS) Points Clustering [4] for clustering of the image and Kekre's Transform [5] for hiding the text in the image. In an improved steganography method [6] which used the Enhanced K Strange Points Clustering Algorithm [7], it was observed that the purity of Enhanced K Strange Points Clustering Algorithm was marginally lesser than the Fuzzy C Strange (FCS) Points Clustering Algorithm used in this paper as was earlier proved in the quantitative performance analysis [8] done on the family of enhanced strange points clustering algorithms. The FCS points clustering method also allows a value in the input collection to belong into several clusters in some measure thereby providing a soft grouping for a given input collection. Least significant bit (LSB) embedding is a well-known and trivial tactic of implanting information into cover images. In this technique, the least significant bit of a few or majority of the bytes in images is altered to a bit of the clandestine information [9]. LSB works in the spatial domain whereby it directly manipulates image intensities without any prior transformation, and therefore it is a simple method from which the extraction of secret data is easy. Spatial-to-frequency domain transformation techniques were found to provide the best immunity to attacks, and hence the Kekre's Transform was used as the transformation technique to improve the security. The coefficients of the Kekre's Transform are used to hide the clandestine text within the image. The reason for choosing the Kekre's Transform was that it was fairly unexplored in the field of steganography.

2 Review of Fuzzy C Strange (FCS) Points Clustering Algorithm

This technique starts by computing the first C strange value, C_{mn} from the origin, which is the smallest valued point from the dataset. The technique then finds the maximum value, i.e., C_{mx} which is extremely apart from C_{mn}. The technique then computes the third point C_s which is equidistant from C_{mn} and C_{mx}. If it is found that the distance from C_s and C_{mn} is smaller than the distance from C_s and C_{mx} or vice versa, respective Eqs. (1) and (2) are used to get the C_s point in the middle and allocate the pixels to their respective clusters using membership function in Eq. (3) so that the clusters are formed perfectly. A very high purity value is found when this algorithm is used on a particular dataset; it also has good clustering quality.

Steps:

1. Locate C_{mn}, the minimum of the input collection.
2. Then, trace a value C_{mx} which is at the most extreme distance from C_{mn} with the Euclidean distance measure.
3. Now look for a third value C_s which is farthest from C_{mn} and C_{mx}.
4. if $(d(C_{mn}, C_s) == d(C_{mx}, C_s))$

$$C_{strnge} = C_s$$

else if $(d(C_{mn}, C_s) < d(C_{mx}, C_s))$

$$Cstrnge = Cstrnge_{prv} + Tm\left(\frac{|C_{mx} - Cstrnge_{prv}|}{C-1}\right) \tag{1}$$

else if $(d(C_{mx}, C_s) < d(C_{mn}, C_s))$

$$Cstrnge = C_{mn} + Tm\left(\frac{|Cstrnge_{prv} - Cmn|}{C-1}\right) \tag{2}$$

5 The above task is redone until $C - 2$ strange points are traced from previous strange points.
6 Allocate the remainder of the points in the input collection into clusters using the membership function given in Eq. (3).
7. Output C clusters.
   The overall time complexity (TC) of the proposed algorithm is computed as

$$TC = O(m + m + m + ((C-2)f) + n) \text{ which gives}$$
$$Time = O(4m + (C-2)f)$$

where

  m   number of points in input collection and
  f   number of features for each of the points

$$\mu_{C_s}(x) = \frac{1}{\sum_{t=1}^{k}\left(\frac{\|x-u_s\|^2}{\|x-u_t\|^2}\right)^{\frac{1}{p-1}}} \quad 1 \leq s \leq k, x \in E \tag{3}$$

3 Kekre's Transform

In this paper, an unexplored image transform for the steganography process called the Kekre's Transform is used, which has earlier been employed successfully in various applications like compression, filtering, enhancement, feature extraction.

$$K_{NxN} = \begin{pmatrix} 1 & 1 & 1 & . & . & 1 & 1 \\ -n+1 & 1 & 1 & . & . & 1 & 1 \\ 0 & -n+2 & 1 & . & . & 1 & 1 \\ . & 0 & -n+3 & . & . & 1 & 1 \\ . & . & 0 & . & . & . & . \\ . & . & . & . & . & . & . \\ . & . & . & . & . & . & . \\ 0 & 0 & 0 & . & . & 1 & 1 \\ 0 & 0 & 0 & . & . & -n+(n+1) & 1 \end{pmatrix}$$

Fig. 1 Generalized N × N Kekre's Transform matrix

Transform domain techniques are mainly used in image steganography. It is possible to implant a clandestine text in different frequency bands of the cover image using transform domain techniques. Kekre's Transform is used on the full image. Clandestine data is implanted into the lower energy blocks of the image after applying Kekre's Transform. The stego-image is obtained by applying inverse Kekre's Transform to the image after inserting the clandestine data. If we verify the original image with the stego-image, its quality is almost the same. Kekre's Transform matrix can be of size N × N. As can be seen, the upper right diagonal and diagonal elements itself of Kekre's Transform matrix are 1, while the lower left diagonal part except the elements just below diagonal is 0 [10]. The line just below the diagonal elements is computed as follows (Fig. 1).

The formula for generating the element K × y of Kekre's Transform matrix is,

$$k \times y = \begin{cases} 1 & ; x \leq y \\ -n+(x-1) & ; x = y+1 \\ 0 & ; x > y+1 \end{cases}$$

4 Proposed Methodology—Embedding and Extracting Process

In image steganography, the data would be hidden in images. A stego-system is one where the clandestine information can be hidden so that no one knows about its existence as shown in Figs. 2 and 3. The output image from this system is called a stego-image, and this is similar to the cover image. Encoding is that part of the process which inserts the clandestine information into the cover medium. Decoding is that part of the process which extracts the clandestine information from the cover medium.

Clustering is a collection of similar objects. Each cluster contains useful information. They are categorized based on color, size, distance, etc. Image steganography is divided into two sections namely spatial domain where we directly insert

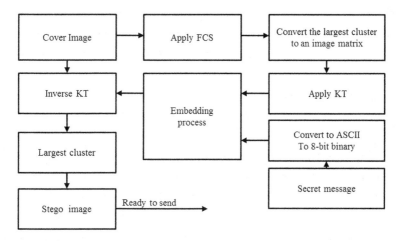

Fig. 2 Embedding workflow

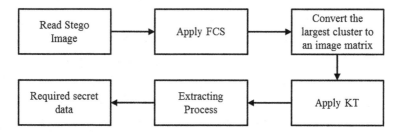

Fig. 3 Extracting workflow

the text in the intensity of the pixels and the frequency domain in which the images are initially transformed, and this is followed by the implanting of the clandestine text within the image. In this technique, the clandestine information is entrenched in the subset of the LSB plane of the cover image.

The embedding process is carried out in the following way:

1. Read the cover image, and apply FCS on the cover image, and then choose the largest cluster to entrench the clandestine text.
2. Read the clandestine text, and convert the characters of the secret message to ASCII values, and then convert these ASCII values to 8-bit binary form.
3. Elicit the first [1 + Length(message)] * 9 pixels from the largest cluster to form [1 + Length(message)] matrices of size 3 × 3, and then apply Kekre's Transform (KT) to the above matrices.
4. Entrench 8-bits (from step 2) into the last two bits of the lower energy blocks of the transformed matrices.
5. Apply inverse Kekre's Transform (IKT) on all the resulting matrices.

6. Reconstruct the resulting matrices (pixels are placed back to its initial position in the image) to form an image which is sent to the receiver.

The extracting process is carried out in the following steps:

1. Read the received stego-image.
2. Apply FCS on the stego-image.
3. Choose the largest cluster to extract the clandestine message.
4. Elicit the first nine pixels from the largest cluster and form an image matrix.
5. Apply Kekre's Transform (KT) on this image matrix.
6. Convert each value of the transformed image matrix into binary, and then extract the last two bits of the lower energy blocks. This gives the length 'n'.
7. Now elicit the next n * 9 pixels to form n matrices of size 3 × 3.
8. Repeat step 5 and 6 for all the matrices.
9. Convert each set of 8-bits into characters.
10. Combine all the characters, and this gives the secret information which is required.

Fig. 4 Cover image

Fig. 5 Sender side clustering output

```
SENDER
Enter image name:
img1
Cmin: (0, 0)
Cmax: (254, 255)
Cs: (0, 242)
Cstr: (127, 248)
Total pixels: 50246
Cluster 1: 0,38  0,39  0,40  0,41  0,42  0,43  0,45
Cluster 2: 16,111  16,112  16,115  17,108  17,110  :
Cluster 3: 0,0  0,1  0,2  0,3  0,4  0,5  0,6  0,7  (
Total no of pixels in Cluster1: 23166
 Total no of pixels in Cluster2: 5119
 Total no of pixels in Cluste3: 21961
```

5   Experimental Results

Steganography was implemented using Kekre's Transform on groupings formed by Fuzzy C Strange Points Clustering Algorithm. The image used to perform steganography is shown in Fig. 4.

The above cover image is used to perform clustering. Clustering the cover image into three clusters gives us the following results as shown in Figs. 5, 6, 7, and 8.

Now Kekre's Transform is applied to the largest cluster from the three clusters obtained, and embedding is successfully achieved. The message is then sent to the

Fig. 6 Cluster 1 (Sender side)

Fig. 7 Cluster 2 (Sender side)

Fig. 8 Cluster 3 (Sender side)

Fig. 9 Sender side process
snapshot of message insertion

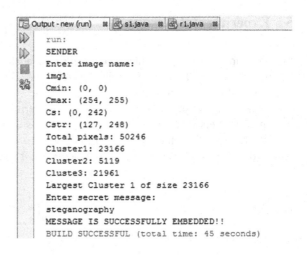

```
Output - new (run)    s1.java    r1.java
run:
SENDER
Enter image name:
img1
Cmin: (0, 0)
Cmax: (254, 255)
Cs: (0, 242)
Cstr: (127, 248)
Total pixels: 50246
Cluster1: 23166
Cluster2: 5119
Cluste3: 21961
Largest Cluster 1 of size 23166
Enter secret message:
steganography
MESSAGE IS SUCCESSFULLY EMBEDDED!!
BUILD SUCCESSFUL (total time: 45 seconds)
```

Fig. 10 Stego-image

Fig. 11 Receiver side
clustering output

```
RECEIVER
Enter image name:
img1
Cmin: (0, 0)
Cmax: (254, 255)
Cs: (0, 242)
Cstr: (127, 248)
Total pixels: 50246
Cluster 1: 0,38  0,39  0,40  0,41  0,42  0,43  0,45
Cluster 2: 16,111  16,112  16,115  17,108  17,110
Cluster 3: 0,0  0,1  0,2  0,3  0,4  0,5  0,6  0,7
Total no of pixels in Cluster1: 23166
 Total no of pixels in Cluster2: 5119
 Total no of pixels in Cluste3: 21961
Largest Cluster 1 of size 23166
```

receiver. For example, the message 'steganography' is sent as shown in Fig. 9. The stego-image which is obtained to be sent to the receiver is shown in Fig. 10.

At the receiver side, the same stego-image is received and clustering is performed. Largest cluster as shown in Fig. 11 is chosen to apply Kekre's Transform, and the message is extracted as shown in Fig. 15. The three clusters obtained and message received are shown in Figs. 12, 13, and 14.

Fig. 12 Cluster 1 (Receiver side)

Fig. 13 Cluster 2 (Receiver side)

Fig. 14 Cluster 3 (Receiver side)

Fig. 15 Receiver side
process snapshot of message
extraction

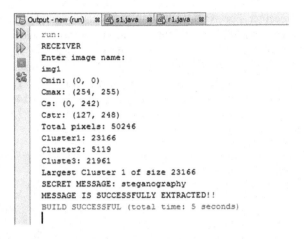

```
Output - new (run)    s1.java    r1.java

run:
RECEIVER
Enter image name:
img1
Cmin: (0, 0)
Cmax: (254, 255)
Cs: (0, 242)
Cstr: (127, 248)
Total pixels: 50246
Cluster1: 23166
Cluster2: 5119
Cluste3: 21961
Largest Cluster 1 of size 23166
SECRET MESSAGE: steganography
MESSAGE IS SUCCESSFULLY EXTRACTED!!
BUILD SUCCESSFUL (total time: 5 seconds)
```

6 Conclusion

The Fuzzy C Strange Points Clustering Algorithm followed by Kekre's Transform along with LSB technique was used to achieve steganography. The results gathered indicated that using the FCS algorithm gives better quality clusters. If we verify the original image with the stego-image, its quality is almost the same. Even if intruders doubt that some camouflaged communication is going on, it will be difficult for them to extract the hidden data as they would not be aware of the transform that has been employed. As a future scope, steganography can be achieved in a similar way by replacing the FCS points clustering technique with the redesigned FCS points clustering methodology [11].

References

1. Zaidoon Kh. AL-Ani, A.A. Zaidan, B.B. Zaidan, Hamdan. O. Alanazi (2010), "Overview: Main Fundamentals for Steganography" JOURNAL OF COMPUTING, VOLUME 2, ISSUE 3, MARCH 2010, ISSN 2151-9617
2. Sangeeta Dhall, Bharat Bhushan and Shailender Gupta (2015), "An In-depth Analysis of Various Steganography Techniques" International Journal of Security and Its Applications Vol. 9, No. 8 (2015), pp. 67–94
3. Jasleen Kour, Deepankar Verma, (2014), "Steganography Techniques-A Review Paper" International Journal of Emerging Research in Management and Technology, ISSN:2278-9359 Vol. 3, Issue-5
4. Terence Johnson, Santosh Kumar Singh, (2016), "Fuzzy C Strange Points Clustering Algorithm" IEEE International Conference On Information Communication And Embedded System (ICICES 2016) "ICICES 2016", February 25–26,2016 S A Engg College, Chennai, India. E-ISBN-978-1-5090-2552-7, PoD ISBN: 978-1-5090-2553-4, INSPEC Accession Number: 161596528, https://doi.org/10.1109/icices.2016.751829, ieee, pp. 1–5

5. Dr. H. B. Kekre, Archana Athawale and Dipali Sadavarti (2010), "Algorithm to Generate Kekre's Wavelet Transform from Kekre's Transform." International Journal of Engineering Science and Technology Vol. 2(5), 2010, 756–767
6. Terence Johnson, Dr. Santosh Kumar Singh, Valerie Menezes, Edrich Rocha, Shriyan Walke, Diksha Prabhu Khorjuvekar, Sana Pathan, (2016), "Improved Steganography using Enhanced K Strange Points Clustering" International Journal of Applied Engineering Research © Research India Publications, ISSN 0973-4562 Volume 11, Number 9 (2016) pp. 6881–6885
7. Terence Johnson, Dr. Santosh Kumar Singh (2015), "Enhanced K Strange Points Clustering Algorithm." Proceedings of the 2nd International Research Conference on Emerging Information Technology and Engineering Solutions. EITES 2015, 978-1-4799-1838-6/15, IEEE Computer Society Washington, DC USA© 2015 IEEE, https://doi.org/10.1109/eites.2015.14, indexed in ACM Digital Library, pp. 32–37
8. Terence Johnson, Santosh Kumar Singh (2016), "Quantitative Performance Analysis for the Family of Enhanced Strange Points Clustering Algorithms" International Journal of Applied Engineering Research © Research India Publications, ISSN 0973-4562 Volume 11, Number 9 (2016) pp. 6872–6880
9. T Morkel, JHP Eloff and MS Olivier, (2005), "An Overview of Image Steganography", in Proceedings of the Fifth Annual Information Security South Africa Conference (ISSA2005), Sandton, South Africa, June/July 2005
10. H.B. Kekre, Tanuja Sarode, Rachana Dhannawat (2012), "Implementation and Comparison of Different Transform Techniques using Kekre's Wavelet Transform for Image Fusion." International Journal of Computer Applications (0975 – 8887) Volume 44–No 10, April 2012
11. Terence Johnson, Santosh Kumar Singh et. al (2016), "The redesigned Fuzzy C Strange Points Clustering Algorithm" 2016 2nd International Conference on Contemporary Computing and Informatics (IC3I), Greater Noida, India, 2016, IEEE Xplore, pp. 789–793. https://doi.org/10.1109/ic3i.2016.7918790.

Anomaly Detection in Phonocardiogram Employing Deep Learning

V. G. Sujadevi, K. P. Soman, R. Vinayakumar and A. U. Prem Sankar

Abstract Phonocardiogram (PCG) is the recording of heart sounds and murmurs. PCG compliments electrocardiogram in detection of heart diseases especially in the initial screenings due to its simplicity and low cost. Detecting abnormal heart sounds by algorithms is important for remote health monitoring and other scenarios where having an experienced physician is not possible. While several studies exist, we explore the possibility of detecting anomalies in heart sounds and murmurs using Deep-learning algorithms on well-known Physionet Dataset. We performed the experiments by employing various algorithms such as RNN, LSTM, GRU, B-RNN, B-LSTM and CNN. We achieved 80% accuracy in CNN 3 layer Deep learning model on the raw signals without performing any preprocessing methods. To our knowledge this is the highest reported accuracy that employs analyzing the raw PCG data.

Keywords Phonocardiogram (PCG) · Machine learning · Deep learning

1 Introduction

When the electrical activities of the heart results in the contraction and subsequent relaxation, the process involves opening and closing of the heart values due to its hemodynamics [1]. The vibrations that results from this action produces heart sounds

V. G. Sujadevi (✉) · K. P. Soman · R. Vinayakumar
Center for Computational Engineering and Networking, Amrita School of Engineering,
Amrita Vishwa Vidyapeetham, Amrita University, Coimbatore 641112, Tamil Nadu, India
e-mail: sujapraba@gmail.com

K. P. Soman
e-mail: kp_soman@amrita.edu

R. Vinayakumar
e-mail: vinayakumarr77@gmail.com

A. U. Prem Sankar
Center for Cyber Security Systems and Networks, Amrita School of Engineering,
Amrita Vishwa Vidyapeetham, Amrita University, Amritapuri 690525, Kerala, India
e-mail: premsankar.u@gmail.com

© Springer Nature Singapore Pte Ltd. 2019
H. S. Behera et al. (eds.), *Computational Intelligence in Data Mining*,
Advances in Intelligent Systems and Computing 711,
https://doi.org/10.1007/978-981-10-8055-5_47

and murmurs that results from the pumping action of the heart can be an indicator for the health of the heart. The monitoring of this sound has traditionally been performed by Stethoscope [1]. While this is a time-tested method, yet this requires highly trained cardiologist who can differentiate the abnormal sounds from the normal ones. Phonocardiograph helps recording this heart sound and that can be further analyzed by computing algorithms [2]. Various approaches into analyzing the Phonocardiogram signals have been studied [2, 3]. Extracting of the murmur sounds and the heart sounds using decomposing methods such as ensemble mode decomposition has been performed [3]. Studies have been done with the aim of automatic feature extraction of Phonocardiogram data waveforms [4].

Noise is one of the issue in the phonocardiogram data as the sounds of the muscles and other artifacts could interfere with the heart sounds and murmur that are generated by hemodynamics [5]. Studies have been done to denoise the heart sounds to be able to accurately identify the heart sounds. One such denoising method uses signal processing that combines wavelet packet transform and singular value decomposition (SVD) with significant success [5]. Signal processing method like Short Time Fourier Transform (STFT) combined with Deep learning method gives good results in the accurate diagnosis of normal and abnormal heart sounds [6]. Various signal-processing methods like fractal decomposition and wavelet feature fusions are used for the classification of PCG signals [7, 8]. Image processing technology is used for the conversion of one-dimensional heart sound signals into two-dimensional PCG for the feature extraction and recognition of the heart sounds [9]. Recent studies for the classification of the heart sounds employs neural network approach for the classification of the heart sounds [10]. We propose to study the classification of heart sounds as normal and abnormal using various Deep learning techniques like CNN, RNN, LSTM, GRU including bidirectional RNN and LSTM.

2   Deep Learning Architectures

This section discusses the network architectures particularly recurrent neural network (RNN), long short-term memory (LSTM), bidirectional-recurrent neural network (B-RNN), bidirectional-long short-term memory (B-LSTM) and convolutional neural network (CNN) concisely, additionally idea behind to map on the task of classifying heart sound in to either normal or abnormal.

2.1   Recurrent Neural Network (RNN) and Its Variants

Recurrent neural network is an enhanced model of traditional feed-forward network [11]. This contains a self-recurrent connection in unit that helps to carry out time related information from one time-step to another. This nature of RNN network has achieved good performance in various long standing artificial intelligence (AI) tasks.

In general, RNN accepts input $x = (x_1, x_2, \ldots, x_{T-1}, x_T)$ (where $x_t \in R^d$) and maps to hidden input sequence $h = (h_1, h_2, \ldots, h_{T-1})$ and output sequence $o = (o_1, o_2, \ldots, o_{T-1}, o_T)$ from $t = 1$ to T by iterating the Eqs. 1 and 2 recursively.

$$h_t = SG(w_{xh}x_t + w_{hh}h_{t-1} + b_h) \tag{1}$$

$$o_t = sf(w_{ho}h_t + b_o) \tag{2}$$

where w represents weight matrices, b terms denotes bias vectors and f is element wise non-linear *sigmoid* activation function. The significant issue of RNN is that vanishing and exploding gradient [12].

To reduce the vanishing and exploding gradient issue, LSTM has introduced a special unit, typically termed as a memory block, as shown in Fig. 1a. Long short-term memory (LSTM) is one of significant method of Deep learning that focuses on learning long-range temporal dependencies in large sequences of arbitrary length [12–14]. LSTM have established as a promising approach for sequence data modeling by solving the various tasks in the field of image processing, natural language processing, speech recognition and many others [15]. It computes output by recursively iterating Eqs. 3–7.

$$i_t = \sigma(w_{xi}x_t + w_{hi}h_{t-1} + w_{ci}c_{t-1} + b_i) \tag{3}$$

$$f_t = \sigma(w_{xf}x_t + w_{hf}h_{t-1} + w_{cf}c_{t-1} + b_f) \tag{4}$$

$$c_t = f_t \odot c_{t-1} + i_t \odot \tanh(w_{xc}x_t + w_{hc}h_{t-1} + b_c) \tag{5}$$

$$o_t = \sigma(w_{xo}x_t + w_{ho}h_{t-1} + w_{co}c_t + b_o) \tag{6}$$

$$h_t = o_t \odot \tanh(c_t) \tag{7}$$

where i, f, o, c term denotes the input gate output gate, forget gate and a memory cell respectively. Conventional RNN only facilitate to learn the past dependencies. Capturing future context in sequences is important as well. Bidirectional recurrent neural network (B-RNN) achieves this by capturing the long range dependencies of temporal patterns in sequence data in both the forward and backward directions. As shown in Fig. 1b, given an input sequence $x = (x_1, x_2, \ldots, x_{T-1})$ B-RNN maps input sequence to forward hidden states and backward hidden states and forward hidden states and backward hidden states to output sequences from $t = 1$ to T by iterating the Eqs. 8–10 recursively.

$$hf_t = SG(w_{xh}x_t + w_{hh}h_{t-1} + b_h) \tag{8}$$

$$hb_t = SG(w_{xh}x_t + w_{hh}h_{t-1} + b_h) \tag{9}$$

$$o_t = SG(w_{hfo}hf_t + w_{hbo}hb_t + b_o) \tag{10}$$

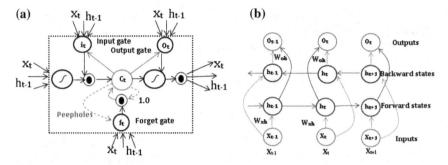

Fig. 1 **a** LSTM memory cell, **b** Bidirectional recurrent neural network

Convolutional neural networks Convolutional neural network (CNN) is an extension to traditional MLP in the context of inspiring the biological factors [15]. This was initially studied for image processing using Convolution 2D layers, pooling 2D layers and fully connected layer. Followed this, applied on natural language processing with Convolution 1D layer, pooling 1D layers and fully connected layer [16]. This infers that CNN takes image data in the form of 2D mesh and time series data in the form of 1D mesh in which the data are arranged in systematic time interval. Based on this, we used CNN, which is composed of Convolution 1D layer, pooling 1D layer fully connected layer and non-linear activation function as *ReLU*.

Convolution is a primary building block of CNN. Given a 1D signal of fixed length input vector $x = (x_1, x_2, \ldots, x_{n-1}, x_n, cl)$, (where $x_n \in R^d$ denotes features and $cl \in R$ denotes a class label), Convolution1D constructs a feature map *fm* by applying the convolution operation on the input data. A new feature map *fm* is obtained as

$$hl^i_{fm} = \tanh(w^{fm} x_{i:i+f-1} + b) \tag{11}$$

where b denotes a bias term. The filter hl is employed to raw input signal $\{x_{1:f}, x_{2:f}, \ldots, x_{n-f+1}\}$ as to generate a feature map as

$$hl = [hl_1, hl_2, \ldots, hl_{n-f+1}] \tag{12}$$

where $hl \in R^{n-f+1}$ and next we apply the max-pooling operation on each feature map as $\overrightarrow{hl} = \max\{hl\}$. This obtains the most significant features in which a feature with highest value is selected. However, multiple features obtain more than one features and those new features are fed to fully connected layer. A fully connected layer contains the sigmoid function that gives the values as binary such as 0 or 1. A fully connected layer is defined mathematically as

$$o_t = \sigma(w_{ho} hl + b_o) \tag{13}$$

3 Network Architecture

An intuitive overview of proposed Deep learning architecture for the task of classifying heart sound recordings in to either normal or abnormal is displayed in Fig. 2. This has not dependent on any of the feature engineering mechanism that was really considered as a daunting task. It directly accepts an input raw signals and itself learns the optimal feature representation by passing through many recurrent hidden layers. The newly formed feature representation are passed to fully-connected layer typically called as dense layer for classification, sigmoid for binary and softmax for multi-class classification.

3.1 Description of Dataset

As part of 2016 PhysioNet/CinC Challenge,[1] the database of heart recordings has been provided to participants and made available to the researchers for further enhancement in classifying the heart recordings either as normal or abnormal. The detailed description of the challenge data is discussed in detail by [17]. The heart

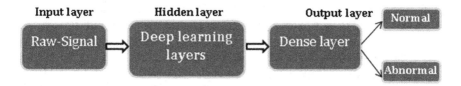

Fig. 2 Architecture of proposed system (inner units and their connections are not shown)

Table 1 Statistics of PCG training and validation database from 2016 PhysioNet/CinC challenge

Name of signal	Number of signals		Name of signal	Number of signals	
	Normal	Abnormal		Normal	Abnormal
Training-a	117	292	validation-a	40	40
Training-b	386	104	validation-b	49	49
Training-c	7	24	validation-c	4	3
Training-d	27	28	validation-d	5	5
Training-e	1958	183	validation-e	53	53
Training-f	80	34	validation-f	–	–
Total	2575	665	Total	151	150

[1]https://www.physionet.org/challenge/2016/.

sound recordings had been collected from various clinical or nonclinical environments such as in-home visits by several contributors around the world from both healthy subjects and patients. The details of the six datasets are given in Table 1.

4 Experiments and Results

TensorFlow (r0.11.0) [18] with Graphics processing unit (GPU) enabled Deep learning framework is used for all experiments in Ubuntu 14.04 operating system (OS). Deep learning algorithms contain several parameters and the performance of them is relying on the optimal parameters. To identify a suitable value for parameters such as learning rate, number of units, we conducted various trails of experiments. All experiments are trained using BPTT approach.

4.1 Hyper-parameter Tuning

To find suitable parameter for RNN, LSTM, GRU, B-RNN, B-LSTM, we used moderately sized architectures with one hidden layer, 32, 64 and 128 units in RNN and 32, 64 and 128 memory blocks in LSTM. 3 trails of experiment are done for each parameters related to units/memory blocks. Each experiment is run till 500 epochs. 64 units/memory blocks has shown highest accuracy in tenfold cross-validation for most of the Deep learning architectures. By considering the training cost, we decided to use 64 units/memory blocks for the rest of the experiments. For CNN, we run 2 trails of experiment with a CNN and pooling layer including number of filters 32, 64, 128 and filter length 2, 3 and 5. CNN with number of filters 64 and filter length 3 has attained highest accuracy in tenfold cross-validation. For the rest of the CNN experiments we decided to use these newly found parameters.

To find suitable learning rate, we run two trails of experiment for all Deep learning architectures till 500 epochs across varying learning rate in the range [0.01–0.5]. The average cross validation accuracy of these models are displayed in Fig. 3. Most of the Deep learning architecture has seen sudden peak in their accuracy at learning rate 0.1 and after that they followed fluctuations, finally accuracy decreased as well. B-LSTM has seen sudden decrease in accuracy at learning rate 0.2 and finally attained highest accuracy as well at 0.40, 0.45 and 0.45 in comparison to learning rate 0.1. This accuracy may be enhanced by running the experiments till 1000 epochs. as more complex architectures have showed less performance within 500 epochs. We decided to use 0.1 learning rate for the rest of the experiments after considering the training time and computational cost.

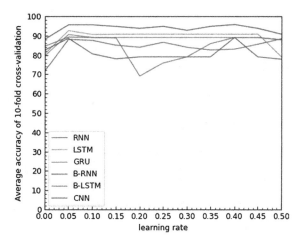

Fig. 3 Performance of various Deep learning architectures for heart sound classification across learning rate in the range [0.01–0.5]

4.2 Network Topologies

To find the optimal architecture, we used the following network topologies (1) RNN 1 layer (2) RNN 2 layer (3) RNN 3 layer (4) RNN 4 layer (5) LSTM 1 layer (6) LSTM 2 layer (7) LSTM 3 layer (8) LSTM 4 layer (9) GRU 1 layer (10) GRU 2 layer (11) GRU 3 layer (12) GRU 4 layer (13) B-RNN 1 layer (14) B-RNN 2 layer (15) B-RNN 3 layer (16) B-RNN 4 layer (17) B-LSTM 1 layer (18) B-LSTM 2 layer (19) B-LSTM 3 layer (16) B-LSTM 4 layer (17) CNN 1 layer (18) CNN 2 layer (19) CNN 3 layer We run 3 trails of experiments for all the network structures. Each trails of experiment are run till 500 epochs. Most of the Deep learning architectures have learned the normal category patterns within 250 epochs. For abnormal categories, each network required varied number epochs. The complex architecture in all networks has required large iterations to obtain highest accuracy. Finally, the best performed network topologies in each architectures are evaluated on the test data set and the detailed results is displayed in Table 2.

To alleviate the cost of training with CNN network, we apply fast fourier transform (FFT) on the raw signals. This facilitates to convert a signal into a frequency domain. To remove noise, high pass Butterworth filter is used. This filters out the noise above 240 beats per minute (4 Hz). Again, we apply FFT on the filtered signal to transform into to it's approximate frequency domain. Finally, we pass the fourier coefficients into CNN network.

To get an intuitive understanding related to the classification performance of signal as either normal or abnormal, we plotted Receiver operating characteristic (ROC) curves, as shown in Fig. 4a for RNN, LSTM, GRU, B-RNN, B-LSTM and CNN.

In general, Deep learning architecture passes the raw signals to more than one hidden layer to learn the optimal feature representation of them. In order to visualize and understand, the last layer values of the obtained feature representations of CNN

Table 2 Summary of test results

Algorithm	Accuracy	Precision	Recal	F-measure
RNN 1 layer	0.638	0.626	0.800	0.702
RNN 2 layer	0.653	0.672	0.683	0.678
RNN 3 layer	0.680	0.737	0.622	0.675
RNN 4 layer	0.688	0.674	0.806	0.734
LSTM 1 layer	0.688	0.736	0.606	0.665
LSTM 2 layer	0.685	0.710	0.694	0.702
LSTM 3 layer	0.700	0.734	0.689	0.711
LSTM 4 layer	0.715	0.747	0.706	0.726
GRU 1 layer	0.694	0.688	0.783	0.732
GRU 2 layer	0.691	0.713	0.706	0.709
GRU 3 layer	0.703	0.741	0.683	0.711
GRU 4 layer	0.712	0.743	0.706	0.724
B-RNN 1 layer	0.685	0.720	0.672	0.695
B-RNN 2 layer	0.680	0.728	0.639	0.680
B-RNN 3 layer	0.691	0.724	0.683	0.703
B-RNN 4 layer	0.715	0.744	0.711	0.727
B-LSTM 1 layer	0.724	0.740	0.744	0.742
B-LSTM 2 layer	0.709	0.716	0.756	0.735
B-LSTM 3 layer	0.700	0.698	0.772	0.734
B-LSTM 4 layer	0.736	0.741	0.778	0.759
CNN 1 layer	0.703	0.725	0.717	0.721
CNN 2 layer	0.773	0.771	0.773	0.770
CNN3layer	**0.798**	**0.815**	**0.806**	**0.810**

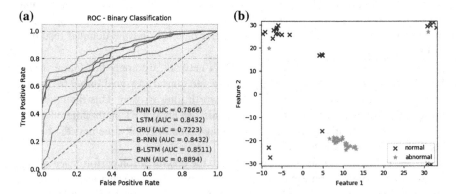

Fig. 4 **a** performance of various Deep learning architectures for heart sound classification across learning rate in the range [0.01–0.5], **b** 25 samples from Normal and Abnormal with their corresponding activation values of the last hidden layer neurons are represented using 2-dimensional linear projection (PCA)

3 layer architecture are redirected to t-SNE [19], as shown in Fig. 4b. It shows that signals with similar characteristics are clustered together.

5 Conclusion

In this paper, we evaluated the effectiveness of various Deep learning approaches towards classifying heart sounds as either normal or abnormal. Unlike the existing studies that largely depends on various feature engineering mechanisms our method did not require feature engineering and it just operated on raw input signals. In our experiment CNN has given the best results when compared to other methods. The primary reason is that, it used FFT with high pass Butterworth filter as it's feature engineering mechanism. This facilitates to separate noise from raw signals and to obtain information related to frequency domain. The reported results can be further enhanced by following hyper parameter mechanism for each deep network architecture. And to our knowledge this is the highest accuracy reported so far.

Acknowledgements We sincerely thank NVIDIA India for the K40 GPU card that was used in this study.

References

1. Barabasa, Constantin, Maria Jafari, and Mark D. Plumbley. "A robust method for S1/S2 heart sounds detection without ECG reference based on music beat tracking." Electronics and Telecommunications (ISETC), 2012 10th International Symposium on. IEEE, 2012.
2. Chen, Tien-En, et al. "S1 and S2 heart sound recognition using deep neural networks." IEEE Transactions on Biomedical Engineering 64.2 (2017): 372–380.
3. Jusak, Jusak, Ira Puspasari, and Pauladie Susanto. "Heart murmurs extraction using the complete Ensemble Empirical Mode Decomposition and the Pearson distance metric." Information Communication Technology and Systems (ICTS), 2016 International Conference on. IEEE, 2016.
4. Stainton, Scott, Charalampos Tsimenidis, and Alan Murray. "Characteristics of phonocardiography waveforms that influence automatic feature recognition." Computing in Cardiology Conference (CinC), 2016. IEEE, 2016.
5. Mondal, Ashok, et al. "A noise reduction technique based on nonlinear kernel function for heart sound analysis." IEEE Journal of Biomedical and Health Informatics (2017).
6. Faradisa, Irmalia Suryani, et al. "Identification of phonocardiogram signal based on STFT and Marquart Lavenberg Backpropagation." Intelligent Technology and Its Applications (ISITIA), 2016 International Seminar on. IEEE, 2016.
7. Chakir, Fatima, et al. "Phonocardiogram signals classification into normal heart sounds and heart murmur sounds." Intelligent Systems: Theories and Applications (SITA), 2016 11th International Conference on. IEEE, 2016.
8. Thomas, Rijil, et al. "Heart sound segmentation using fractal decomposition." Engineering in Medicine and Biology Society (EMBC), 2016 IEEE 38th Annual International Conference of the. IEEE, 2016.

9. Cheng, Xiefeng, et al. "Feature extraction and recognition methods based on phonocardiogram." Digital Information Processing, Data Mining, and Wireless Communications (DIPDMWC), 2016 3rd International Conference on. IEEE, 2016.

10. Grzegorczyk, Iga, et al. "PCG classification using a neural network approach." Computing in Cardiology Conference (CinC), 2016. IEEE, 2016.

11. Elman, Jeffrey L. "Finding structure in time." Cognitive science 14.2 (1990): 179–211.

12. Hochreiter, Sepp, and Jrgen Schmidhuber. "Long short-term memory." Neural computation 9.8 (1997): 1735–1780.

13. Graves, Alex, and Jrgen Schmidhuber. "Framewise phoneme classification with bidirectional LSTM and other neural network architectures." Neural Networks 18.5 (2005): 602–610.

14. Gers, Felix A., Nicol N. Schraudolph, and Jrgen Schmidhuber. "Learning precise timing with LSTM recurrent networks." Journal of machine learning research 3.Aug (2002): 115–143.

15. LeCun, Yann, Yoshua Bengio, and Geoffrey Hinton. "Deep learning." Nature 521.7553 (2015): 436–444.

16. Kim, Yoon. "Convolutional neural networks for sentence classification." arXiv preprint arXiv:1408.5882 (2014).

17. Liu, Chengyu, et al. "An open access database for the evaluation of heart sound algorithms." Physiological Measurement 37.12 (2016): 2181.

18. Abadi, Martn, et al. "TensorFlow: A system for large-scale machine learning." Proceedings of the 12th USENIX Symposium on Operating Systems Design and Implementation (OSDI). Savannah, Georgia, USA. 2016.

19. Maaten, Laurens van der, and Geoffrey Hinton. "Visualizing data using t-SNE." Journal of Machine Learning Research 9.Nov (2008): 2579–2605.

Secured Image Transmission Through Region-Based Steganography Using Chaotic Encryption

Shreela Dash, M. N. Das and Mamatarani Das

Abstract Information security is one of the challenging problems nowadays, and to solve this problem, a new algorithm which provides double layer security is proposed, by combining region-based steganography with chaotic encryption. Region-based steganography is the technique of protecting the hidden information in certain regions of interest, which is most efficient for data hiding. One of them is the edge region in image, which is most effective for data hiding. For encrypting data, we used CNN because of its random nature which is very challenging for hackers to know about the secret information. In the proposed work, the secret data are first encoded using CNN, and then, the scrambled message is embedded inside the edge region of cover image using matrix encoding scheme, which provides high security. The complete procedure is instigated in MATLAB, and the result analysis is given, which shows the strength of the technique.

Keywords Chaotic neural network (CNN) · Matrix encoding
Region-based steganography

1 Introduction

Because of the web advancement, a large number of people are sharing important information using digital media like pictures, audio or video through internet easily, but one of the real difficulties in sharing and transmitting any data through open

S. Dash (✉) · M. Das
Department of CSE, C. V. Raman College of Engineering,
Bhubaneswar 752054, Odisha, India
e-mail: shreelamamadash@gmail.com

M. Das
e-mail: mamataparida2005@gmail.com

M. N. Das
School of Computer Science, KIIT University, Patia, Bhubaneswar 751024, Odisha, India
e-mail: mndas_prof@kiit.ac.in

© Springer Nature Singapore Pte Ltd. 2019   535
H. S. Behera et al. (eds.), *Computational Intelligence in Data Mining*,
Advances in Intelligent Systems and Computing 711,
https://doi.org/10.1007/978-981-10-8055-5_48

channel is security. The mostly used two techniques for providing secure data communication are cryptography and steganography. Cryptography is the method which scrambles the secret information, so that even if the attacker knows that some kind of encryption is done, he/she cannot guess and decrypt/hack the message. It provides confidentiality, authenticity, and integrity of data. Steganography is the method of hiding the secret information inside any cover medium, and transmission through Internet is done in such a way that the unauthorized user is unable to know that any kind of hiding is done. The cover medium can be any digital image, digital audio, or digital video. Categorically, steganography methods can be either spatial domain or frequency domain based. In this paper, spatial domain method is used where we are directly dealing with the intensity of pixels. This technique is very simple and efficient than frequency domain. In the proposed method, the secret message is encrypted and then the encrypted message is embedded inside the cover media. This method combines both the concepts of cryptography and steganography, which provide high security to data. The rest of the paper is organized as follows: Sect. 2 presents related works; Sect. 3 proposed work; Sect. 4 encryption using CNN; Sect. 5 edge-based steganography; Sect. 6 simulation result; and Sect. 7 conclusion.

2 Related Work

Steganography techniques depend on either spatial domain or frequency domain. In spatial method, we straightforwardly manage the pixels of the picture to conceal the secret information. However, in frequency space, information is covered up in the frequency components by discovering DCT. Spatial domain methodology uses LSB-based steganography.

S. Dash [1] proposed a strategy which is effective to hide secret data without noteworthy alteration. The unethical persons cannot visualize any change. More security and high capacity are the key concepts of this method.

In LSB-based image steganography [2], it was suggested data to be hidden in bit color imaging of red, green, and blue, where more data can be stored.

Jarno Mielikainen described a steganography method in [3]. In the LSBM methodology, the random choice is given to increase or decrease one pixel from the cover image. The inserting is done utilizing a couple of pixels, where the very first pixel delivers a part of data, and a component of the two pixel values delivers the other part of data. This technique permits inserting an indistinguishable payload from LSB coordinating, however, with fewer changes to the cover picture. The evaluation of this technique demonstrates better conventional LSB coordinating as far as distortion and resistance against existing steganalysis.

Another steganography based on LSB is proposed in [4]. Here, covert key plays a role in overwriting the LSB of each byte of the cover picture. In this procedure, the XOR operation is performed between the binary bits of the covert key and least significant bit of each pixel in red component matrix. The result of the operation decides if the covert message will be embedded in blue or green matrix. If the result is 1, then 1 bit of secret message is kept in LSB of green pixel otherwise in blue pixel.

S. M. Masud Karim et al. [5] approached the issue in LSB matching revisited (LSBMR) algorithm, which selects suitable region in the cover image for embedding data. By hiding secret data in the specific region of the pixels, the detection becomes difficult. The maximum embedding capacity can be improved by modifying the parameters of neighbor pixels.

The region-based steganography schemes emphasize on pixel-value differencing [6]. Here, the secret bits are hidden in the most prominent edges.

In [7], chaos-based edge adaptive steganography is given. Most prominent edges are detected using Canny's edge detector [8]. Here encrypted message follows steganography. More secrecy is incorporated by doing XOR operation. This technique does not emphasize on capacity.

Another technique is proposed in [9], which provides better embedding capacity with more security. Here, the secrecy is maintained by keeping data in two different groups of pixels for more security.

Rig Das [10] planned a Huffman Encoding-based image steganography. This technique is more robust compared to others.

N. K. Kamila [11] planned a technique in which LSB steganography is combined with chaotic cryptography to make secure communication. The chaotic parameters need to be kept unchanged and hidden to provide more security. Due to the resemblance of both original and stego images, the cryptanalyst does not get any idea of secret communication. Therefore, it is less likely to be attacked.

3 Proposed Work

In the proposed model, the benefits of both the steganography and cryptography techniques are combined. Cryptography is the technique to transform the secure information through Internet in such a way that no one is able to understand it except the authorized users. Encryption is done by using public key, and for decryption, one needs the public key in the receiver side. Images and multimedia data which are transmitted over the insecure channel must also be encrypted due to security reason. But image encryption through traditional cryptosystem is not efficient due to the following two reasons.

1. Images are comparatively larger in size than text.
2. Time to process image is more than text data.

However, a chaotic sequence shows random behavior through its sensitive dependence on initial condition. It is non-converging and also non-periodic. Changing the initial value alone will generate a large number of deterministic and reproducible signals in a random manner. These random sequences can be converted into integer-valued sequences and can be used for image encryption efficiently. After encrypting the image, the encrypted secret image is embedded in the

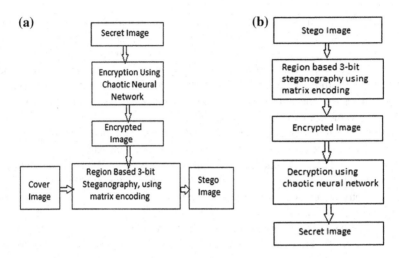

Fig. 1 a Chaotic encryption and embedding. **b** Chaotic decryption and extraction

cover image. For embedding, edge adaptive steganography with 3-bit LSB is used. In this paper, edge pixels are chosen for hiding as they are less sensitive to distortion.

The hidden message is then extracted using the reverse process of steganography. The extracted image is again passed through the chaotic neural network by initializing the same initial parameters μ and $\times 0$ to get the original message. The proposed model is given in Fig. 1a, b. This method provides a double layer security by using both cryptography and steganography. Although the attacker knows the technique, he/she cannot extract the secret message.

4 Chaotic Encryption

To solve many real-world problems, a powerful technique, called neural networks, is used. Neural networks have the ability to learn from experience. Researchers found a considerable measure of enthusiasm because of the way of confusion which is utilized as a part of different building disciplines, where cryptography must be a standout among the most potential applications. The properties of chaos like ergodicity, semi-arbitrariness, affectability reliance on introductory conditions, and system parameters have allowed confused progression as a promising option for the traditional cryptographic calculations. Chaos-based cryptography depends on perplexing elements of nonlinear frameworks or maps which are deterministic in nature, however straightforward. It can give a quick and secure means for information protection assurance, which is essential for digital media information transmission over the broadband web correspondence.

Suppose any image of size m × n is to be encoded using CNN, then the algorithm is:

Algorithm 1: Chaotic Encryption algorithm using neural network

Input: Secret Image IMG

Output: Encrypted secret image EIMG

Step 1: Convert the gray scale image IMG of size m x n into binary and keep it in the one dimensional array of size L.

Step 2: Initialize two parameters μ and X (1) to generate chaotic sequence.

Step 3 Determine X (1), X (2) ....X(L) by using the formula $X(i)=\mu*X(i-1)*(1-X(i-1))$;

Step 4: Create the binary representation of X(i) for i=0,1, ...L.

Step 5: Train the chaotic neural network by determining weight matrix.

For n=0 to L-1 do

    Let $g(n) = (d_0*2^0+d_1*2^1+d_2*2^2....+d_7*2^7)$

      For P=0 to 7 do

        For q=0 to 7 do

          Set $W_{pq} = 0$ if p is not equal to q

          Set $W_{pq} = 1$ if p is equal to q and b(8*n + p) = 0

          Set $W_{pq} = -1$ if p is equal to q and b(8*n + p) = 1

    if b(8*n + p) = 0

        $\Phi_p = -\frac{1}{2}$

    else

        $\Phi_p = \frac{1}{2}$

    end

    $d_i = f (\sum W_{pq} * d_i + \Phi_p)$ (where i=0,1...7) if (x >= 0)

        $f (x) = 1$

    else

        $f (x) = 0$

    end

    $g'(n)= \sum d_i'*2'$

end

---

5 Edge Adaptive LSB Steganography with Matrix Encoding

In this technique, the cover image of size M × N is split into two parts. The first part consists of only edge pixels, and second part consists of all the non-edge pixels. Canny's edge detection algorithm is used for edge detection. We choose edge pixels for embedding secret data as small changes done in edge pixels will not reflect the original image and will produce less distortion. The encrypted secret image is kept in the LSB of edge pixels using some XOR operations on the binary bits. If the size of the image is smaller, so less number of edge pixels will be detected. The edge pixels only may not be sufficient to hold all the secret bits. Therefore, remaining secret bits can be embedded into the non-edge pixels by using LSB technique. The following table is used for embedding.

Let p1, p2, p3 be the secret bits to hide in the Q_j which is the edge pixel and Q_{jR}, Q_{jG}, Q_{jB} be the red, green, and blue components of Q_j (Table 1).

Flowchart for Embedding and Extraction

The embedding algorithm uses region-based edge adaptive data hiding with matrix encoding (Fig. 2a) and we retrieve the secret message by using the extraction method (Fig. 2b).

6   Result and Analysis

The process for steganography has been implemented in MATLAB 14.0. After chaotic encryption of the following secret images Kid (50 × 50), Krishna (100 × 100), and Rabit (200 × 200), given in Fig. 3a, we get the encrypted image. The cover images we have taken are Brandy Rose (300 × 300), Monarch (500 × 500), and Lena (600 × 600), given in Fig. 3b.

The steps given in Fig. 4a describe the chaotic encryption and embedding technique. The steps given in Fig. 4b describe the extraction technique followed by the chaotic decryption. The embedding algorithm has been tested by calculating the PSNR, MSE, time taken to embed, and capacity of cover image to hold the amount of secret information. PSNR is of high importance for the evaluation of image quality. It depends on the pixel intensity change from original image and the stego image.

$$PSNR = 10 log_{10}(MAX^2/MSE) \text{ dB} \tag{1}$$

Table 1 Condition list

Condition	Action
$P_1 = Q_{jR7} \oplus Q_{ij8}$	No change
$P_1 \neq Q_{jR7} \oplus Q_{jR8}$	Invert Q_{jR8}
$P_2 = Q_{jG7} \oplus Q_{jG8}$	No change
$P_2 \neq Q_{jG7} \oplus Q_{jG8}$	Invert Q_{jG8}
$P_3 = Q_{jB7} \oplus Q_{jB8}$	No change
$P_3 \neq Q_{jB7} \oplus Q_{jB8}$	Invert Q_{jB8}

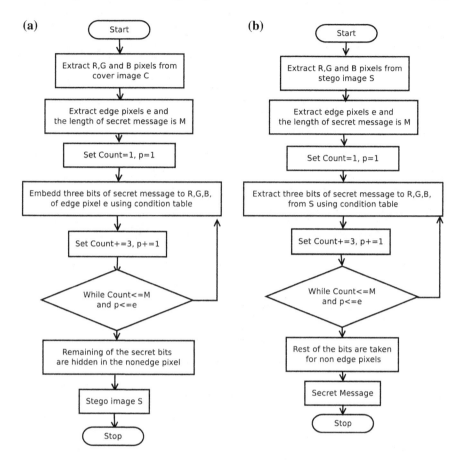

Fig. 2 **a** Embedding flowchart. **b** Extraction flowchart

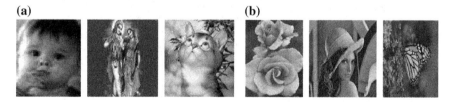

Fig. 3 **a** List of secret images. **b** List of cover images

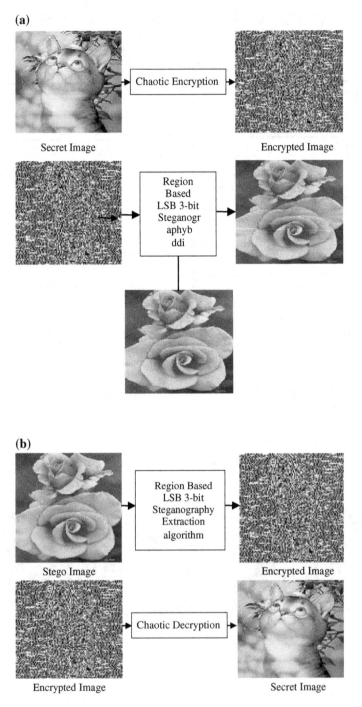

Fig. 4 **a** Chaotic encryption and steganography embedding technique. **b** Chaotic decryption and steganography extraction technique

Size of cover image	Size of secret image	PSNR
Brandy Rose (300 × 300)	Kid (50 × 50)	67.2
	Krishna (100 × 100)	61.2
	Rabit (200 × 200)	55.6
Monarch (500 × 500)	Kid (50 × 50)	71.7
	Krishna (100 × 100)	65.7
	Rabit (200 × 200)	59.6
Leena (600 × 600)	Kid (50 × 50)	73.3
	Krishna (100 × 100)	67.2
	Rabit (200 × 200)	55.9

Table 2 PSNR calculation for different cover and secret images

MAX = maximum possible pixel values in the image.

And MSE is defined as

$$MSE = 1/mn \sum_{i=0}^{m-1} \sum_{j=0}^{n-1} [I(i,j) - K(i,j)]^2 \tag{2}$$

where I is the cover image and K is the stego image of size m × n.

Table 2 shows the PSNR values for different sizes of cover and secret images.

7 Conclusion

Our proposed work provides two-level security to secret data. Here, a new system is given which combines the chaotic encryption with edge level LSB steganography with matrix encoding. In chaotic encryption, we analyzed and found that if the values of x and μ used for generating random numbers are 0.75 and 3.9, then it gives better result. These values should be kept secret and constant, so that even if cryptanalyst knows any presence of secret message, it is difficult for him/her to decrypt. These values are passed along with the embedded image and kept secretly in the cover image. Here, the steganography method uses XOR operation of last two LSB bits of the R, G, and B pixels, and then, three secret bits are hidden accordingly. The cover image is similar to the stego image so attacker cannot guess about secret communication. The PSNR value of the stego and original images is compared, and if it is high, the intruder cannot guess about secret hiding and break it. Through simulation we observed, secret message of size (200 × 200) can be embedded in (300 × 300) with PSNR more than 55. So our method provides more number of secret data that can be hidden in less size cover image. Thus, the combination of chaotic cryptography and LSB 3-bit steganography with matrix encoding is a powerful method to hide high payload secret data which enable people for secret communication. If the PSNR is high, we can achieve more by using some machine learning techniques. In the future, we can use some techniques to find the best location in the cover image for better hiding.

References

1. Shreela Dash, Mamatarani Das, Kartik Chandra Jena, "Region Based Data Hiding For High Payload", International Journal of Computer Science and Information Technologies, Vol. 6 (1), 2015, 913–919.
2. Abbas Cheddad, oan Condell, Kevin Curran, Paul McKevitt, F. Hartung, "Information hiding-a survey" Proceedings of the IEEE: Special Issue on Identification and Protection of Multimedia Content, Volume: 87 Issue:7, pp. I062–I078, July. 1999.
3. JarnoMielikainen, "LSB matching revisited," IEEE signal processing letters, vol. 13, no. 5, May 2006.
4. G. Karthigai Seivi, Leon Mariadhasan, K. L. Shunmuganathan "Steganography Using Edge Adaptive Image" International Conference on Computing, Electronics and Electrical Technologies [ICCEET} 2012, pp 1023–1027.
5. S. M. MasudKarim, Md. SaifurRahman, Md. Ismail Hossain "A New Approach for LSB Based Image Steganography using Secret Key" Proceedings of 14th International Conference on Computer and Information Technology (ICCIT) 2011.
6. Qinhua Huang and WeiminOuyang, "Protect Fragile Regions in Steganography LSB Embedding", 3rd International Symposium on Knowledge Acquisition and Modelling, 2010.
7. G. KarthigaiSeivi, Leon Mariadhasan, K. L. Shunmuganathan "Steganography Using Edge Adaptive Image" International Conference on Computing, Electronics and Electrical Technologies [ICCEET} 2012, pp 1023–1027.
8. Ratnakirti Roy, Anirban Sarkar, Suvamoy Changder "Chaos based Edge Adaptive Image Steganography" international Conference on Computational Intelligence (CIMTA), 2013, vol 10, pp. 138–146.
9. John Canny, "A Computational Approach to Edge Detection" IEEE Transactions On Pattern Analysis And Machine Intelligence, Vol. -8, No. 6, November 1986.
10. Rig Das, Themrichon Tuitthung, "A Novel Steganography Method For Image Based on Huffman Encoding", IEEE, 2012.
11. N.K Kamila, H. Rout, N. Dash, "Stego-Cryptography Using Chaotic Neural Network", American Journal Of Signal Processing, 2014.

Technical Analysis of CNN-Based Face Recognition System—A Study

S. Sharma, Ananya Kalyanam and Sameera Shaik

Abstract Face recognition is the essential security system, which is subjected to get more scrutiny in recent years especially in the field of research and also in industry. This study addresses the various approaches for recognizing face back on neural network by adopting convolutional neural network (CNN). The study has done on different techniques of face alignment, preprocessing techniques, and also in the size of the face images. This paper explains the computational analysis of face recognition system and emphasizes the accuracies and constraints of the images. The predominant face alignment approaches used are Dlib and constrained local model (CLM). For training, Tan-Triggs preprocessing technique is used in face image size of 96 × 96 and 64 × 64. The face recognition grand challenge (FRGC) dataset is used for the analysis, and it produced the accuracy of range from 90 to 98.30% on corresponding approaches.

Keywords Face recognition · CNN · Deep learning · Face alignment
Dlib · CLM · Tan-Triggs · FRGC

1 Introduction

The face recognition system is a computer-aided function that has a proper identification or verification of the pre-training of a particular label from a digital image of a label and therefore makes a bank of facial features. The face recognition system

S. Sharma (✉)
School of Electronics, Vignan's University, Guntur 522213
Andhra Pradesh, India
e-mail: sharma_ece@vignanuniversity.org

A. Kalyanam · S. Shaik
Department of Computer Science, Vignan's University, Guntur 522213
Andhra Pradesh, India
e-mail: ananyarao.kalyanam@gmail.com

S. Shaik
e-mail: sks_cse@vignanuniversity.org

© Springer Nature Singapore Pte Ltd. 2019
H. S. Behera et al. (eds.), *Computational Intelligence in Data Mining*,
Advances in Intelligent Systems and Computing 711,
https://doi.org/10.1007/978-981-10-8055-5_49

has culminated in the science of refined algorithms and mathematical modeling for appropriate features over the past decade. Although considerable attention has been paid to researchers and developers, it remains very difficult in real time [1].

Face detection systems are becoming the most compelling biometric affirmation module since they are not nosy or meddling. It is combined with another affirmation system to increase the accuracy of the privilege. Artificial intelligence and computer vision are also increasingly convoluted in facial recognition for out coming high precision [2]. A good facial recognition system is to have a high true acceptance rate (TAR) and much less false acceptance rate (FAR); the high FAR is 0.1%. The facial recognition procedure involves three main phases such as facial recognition, facial mark eradication, and facial avowal. There are many approaches to face the detection in the computer vision, but Dlib and CLM are mainly used here. The extracted facial features are trained with the fascinating neural network, and the characteristics are matched with the respective images of a person [3].

2 Technical Strategy

Technical strategy explains the overview of the approaches to be analyzed with face recognition systems. Figure 1 shows block diagram briefly and steps involved [4].

2.1 Face Alignment

Face alignment is a technology for computer vision to locate facial landmarks and also to identify the geometric shape of the face image. Face alignment is the essential and predominant task in the face recognition systems. Many recognition algorithms mainly depend on the face alignment to predict the persons. Face alignment algorithms operate by iteratively adjusting the position of the facial feature into canonical pose by which features relative to a fixed coordinate system can be identified and encode with the prior knowledge of face shape [5].

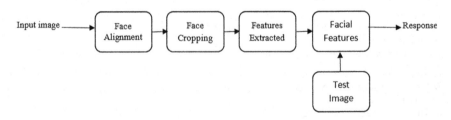

Fig. 1 Overview block diagram of face recognition

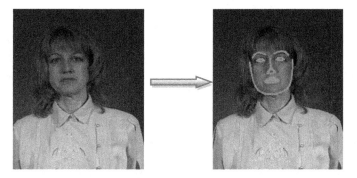

Fig. 2 Face alignment using Dlib

Fig. 3 Face alignment using CLM

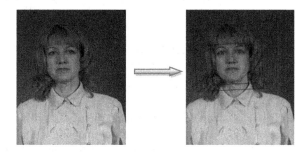

The positioning is done either by an un-supervised machine learning algorithms or by learning the deviation in appearance of a set of template regions surrounding individual feature of the image. In this paper, we exploit two methods for face alignment of images using Dlib [6] and constrained local model (CLM) [2]. We estimate the performance by comparing the recognition results using the images from detection with the aligned faces which are illustrated in Figs. 2 and 3 [7].

2.2 Face Cropping

The face cropping is the important process in face recognition or face verification to get high TAR or low FAR. The aligned images are from face alignment method, and only the faces are cropped out for facial feature extraction. Cropping the face image from the digital aligned images leads to have proper training in the neural networks. Since, the images that are to be trained are having only the features of the face of a person [2, 6, 5].

This paper highlights the result of the faces which are cropped into the size of 96 × 96 and 64 × 64 Figs. 4a, b and 5a, b shows.

Fig. 4 Face cropped after Dlib **a** 96 × 96 **b** 64 × 64

(a) **(b)**

Fig. 5 Face cropped after CLM **a** 96 × 96 **b** 64 × 64

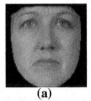

(a) **(b)**

3 Features Extraction

The extraction of features of aligned facial images is the most important function in any recognition of the face, which is eliminated by a deep neural method; it is a variation of deep learning. Neural networks of deep learning are the optimal solution for object detection, pattern recognition, and facial recognition [8].

The convolutional neural network [2] is one of the widespread deep learning techniques since it is well adapted to the invariance of the translation of images. These neural networks use three main principles for extracting features from facial images: local reception fields, common weights, and grouping.

3.1 Local Receptive Fields

In the convolutional neural network, the layer of hidden neurons has connections in small; localized pixel input image as opposed to other neural networks has a connection of each input pixel. To be more specific, each neuron in the first hidden layer would be connected to a small region of the input neurons. For hidden neurons in particular, the compound will be as follows.

That region in the input image is called the *local receptive field* for the hidden neuron which is a little window on the input pixels in Fig. 6a, b. Each link learns a weight, and the hidden neuron also learns a global bias. The receptive field slides from each pixel to have a link with each layer to hidden neurons [2, 6].

Fig. 6 **a** Input image pixel layer **b** hidden layer connection

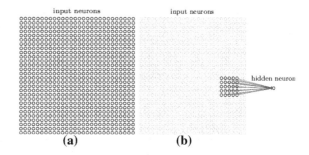

(a)   (b)

3.2 Shared Weights and Biases

The local receptive fields connected to the hidden layer with same stride length would share the same weight and bias throughout the hidden layer. For jth and kth hidden neurons, the output is

$$\sigma\left(b + \sum_{l=0}^{4} \sum_{m=0}^{4} w_{l,m}\, a_j + l, k + m\right)$$

where σ is neural activation function or sigmoid function, b is the shared bias, $w_{l,m}$ is the shared weights, and $a_{x,y}$ is to donate the input activation at position x, y.

The neurons in the first hidden layer detect exactly the same feature but at different locations in the input image. Hence, the map from input image hidden layer is called *feature map,* and the bias that meant for feature map is called *shared bias*. The shared weights and bias are called as *kernel or filter.* The face recognition many feature maps but the kernel reduces the parameters effectively [2, 6].

3.3 Pooling Layer

The pooling layer is chased by the convolution layer, which is used to provide information on the output of the convolution layer map in the condensed characteristics map. There are abounding pooling techniques for the convolutional neural network, though this survey exploits the maximum pooling for facial recognition systems [9].

Figure 7 shows the maximum pooling, recompensing the output of the maximum activation in the 2 × 2 input regions.

hidden neurons (output from feature map)

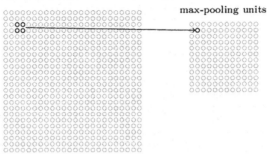

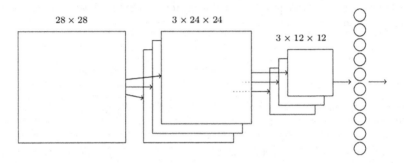

Fig. 7 Illustration of max pooling

Fig. 8 Architecture of CNN

3.4 Convolutional Network Architecture

This convolutional neural network architecture used for this survey is bringing together all the craftings described above which is following below in Fig. 8 [2, 6].

4 Datasets and Experimental Scenarios

Face recognition grand challenge (FRGC) dataset has been used for this analysis which has 11,284 2D frontal images of 286 individual labels. FRGC images are taken in the unconstrained environment with possible head orientations and varying distance between faces and camera [10].

The experimental scenario says the overview of the setup used for face recognition system for technical analysis; the maximum training images for a single person is 10 images. Figure 9 shows the examples of the dataset [6] (Table 1).

Fig. 9 An example of FRGC dataset

Table 1 Experimental scenarios

S. no	Criterion	Efficacy
1	Datasets	FRGC
2	Type	Grayscale
3	No. of images used in training per label	10
4	Number of tags	286
5	Algorithm used	CNN

5 Experimental Approaches

This paper has various technical analyses for face recognition system with different face alignment and cropping and preprocessing methods. Each approach has many advantages and disadvantages for FRGC datasets.

5.1 96 × 96 Face with Dlib Alignment Without Preprocessing

This approach uses Dlib face alignment for the input digital images of a person. The cropped faces in Fig. 10a, b are trained by CNN, and the features are extracted; thus, by testing against non-trained images, it gives the 96% of accuracy.

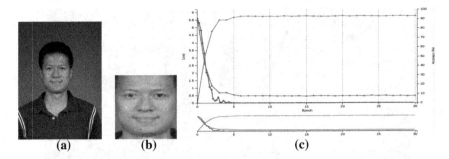

Fig. 10 **a** Input image **b** trained images **c** 96 × 96—Dlib alignment—without preprocessing

Figure 10c explains the various parameters like accuracy, training loss, and validation loss which showing that the output accuracy. Hence, the Dlib alignment for 96 × 96 face image can be used for face recognition systems which required high accuracy.

5.2 96 × 96 Face with Dlib Alignment with Preprocessing

The approach employed 96 × 96 facial images with Tan-Triggs preprocessing techniques; the facial images are extracted by using Dlib algorithm. This method gives the accuracy of 93% but has an advantage in false acceptance ratio of 0.01%. This approach can be deployed where the FAR is the primary concern in the face recognition or face verification systems. The following images show the training image and accuracy (Fig. 11).

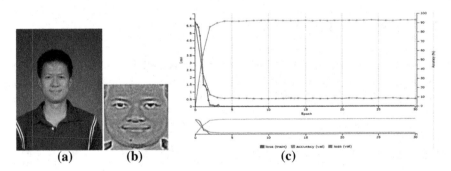

Fig. 11 **a** Input image **b** trained images **c** 96 × 96—Dlib alignment—with preprocessing

5.3   96 × 96 Face with CLM Alignment Without Preprocessing

This approach used constrained local model for face alignment and cropped the aligned faces into the size of 96 × 96 pixels. Thus, by trained using CNN algorithm, Fig. 12 shows the accuracy of 98.5% and FAR of 0.1%. The accuracy is little bit high that Dlib alignment approaches because of the extracted face is only the mask of the facial images. Hence the features are concentrated on the mask.

5.4   96 × 96 Face with CLM Alignment with Preprocessing

The CLM is used with the Tan-Triggs preprocessing technique on the facial mask and produces an accuracy of 95% with false acceptance rate of 0.01%. Tan-Triggs preprocessing techniques use Gaussian blurring and filters, then it normalizes the images with respect to their intensity values (Fig. 13).

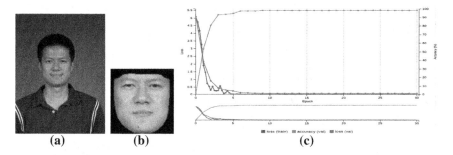

Fig. 12 **a** Input image **b** trained images **c** 96 × 96—CLM alignment—without preprocessing

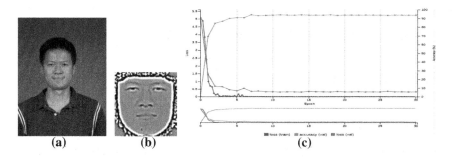

Fig. 13 **a** Input image **b** trained images **c** 96 × 96—CLM alignment—with preprocessing

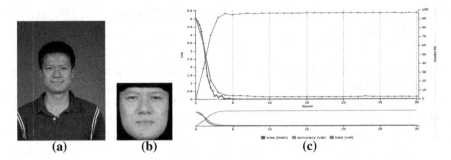

Fig. 14 **a** Input image **b** trained images **c** 64 × 64—CLM alignment—without preprocessing

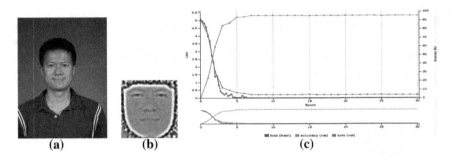

Fig. 15 **a** Input image **b** trained images **c** 64 × 64—CLM alignment—with preprocessing

5.5 *64 × 64 Face with CLM Alignment Without Preprocessing*

The size of the mask is reduced to 64 × 64 from digital images, and then the neural networks can be used to train the mask. The validation is done against the test images which results with the accuracy of 97% with very least FAR (Fig. 14).

5.6 *64 × 64 Face with CLM Alignment with Preprocessing*

The final approach of the face recognition system is processed with Tan-Triggs, and facial mask is resized to 64 × 64 pixels; the accuracy is higher of 96% (Fig. 15).

6 Results

See Table 2.

Table 2 Overall results

S. no	Approach	Accuracy (%)	FAR
1	96 × 96 face with Dlib alignment without preprocessing	96	0.1
2	96 × 96 face with Dlib alignment with preprocessing	93	0.01
3	96 × 96 face with CLM alignment without preprocessing	98.5	0.1
4	96 × 96 face with CLM alignment with preprocessing	95	0.01
5	64 × 64 face with CLM alignment without preprocessing	97	0.1
6	64 × 64 face with CLM alignment with preprocessing	96	0.01

7 Conclusion

This paper is a study of various technical analysis of convolutional neural network-based face recognition system. This analyzed eight different approaches by varying face alignment method, size of the facial image, and preprocessing technique. The result states that face alignment with CLM gives best accuracy for face recognition, and preprocessing technique provides the best solution for face verification. Dlib alignment also gives the good accuracy when long distance images are used for the training.

References

1. Florian Schrof, Dmitry Kalenichenko and James Philbin "*FaceNet: A Unified Embedding for Face Recognition and Clustering*" arXiv:1503.03832V3 [cs.CV], 17 June 2015.
2. K. Shanmugasundaram, S. Sharma and S. K. Ramasamy, "*Face recognition with CLNF for uncontrolled occlusion faces*," 2016 IEEE International Conference on Recent Trends in Electronics, Information & Communication Technology (RTEICT), Bangalore, 2016, pp. 1704–1708. https://doi.org/10.1109/rteict.2016.7808124.
3. Tim Rawlinson, Abhir Bhalerao and Li Wang "*Principles and Methods for Face Recognition and Face Modelling*" In Handbook of Research on Computational Forensics, Digital Crime and Investigation: Methods and Solutions. Ed. C-T. Li. Chapter 3, pages 53–78. 2010.
4. Rabia Jafri and Hamid R. Arabnia "*A Survey of Face Recognition Techniques*" Journal of Information Processing Systems, Vol. 5, No. 2, June 2009 pp. 41–68.
5. Vahid Kazemi and Josephine Sullivan "*One Millisecond Face Alignment with an Ensemble of Regression Trees*", CVPR 2014, computer vision foundation.
6. S. Sharma, K. Shanmugasundaram and S. K. Ramasamy, "FAREC — CNN based efficient face recognition technique using Dlib," 2016 International Conference on Advanced Communication Control and Computing Technologies (ICACCCT), Ramanathapuram, 2016, pp. 192–195. https://doi.org/10.1109/icaccct.2016.7831628.
7. V.V. Starovoitov, D.I Samal and D.V. Briliuk "*Three Approaches for Face Recognition*" 6-th International Conference on Pattern Recognition and Image Analysis October 21–26, 2002, Velikiy Novgorod, Russia, pp. 707–711.

8. D. Cristinacce and T. Cootes. *"A comparison of shape constrained facial feature detectors"*. In 6th International Conference on Automatic Face and Gesture Recognition 2004, Seoul, Korea, pages 375–380, 2004.

9. T. F. Cootes and C. J.Taylor. *"Active shape models"*. In 3rd British Machine Vision Conference 1992, pages 266–275, 1992.

10. T. F. Cootes, G. J. Edwards, and C. J. Taylor. *"Active appearance models"*. In 5th European Conference on Computer Vision 1998, Freiburg, Germany, volume 2, pages 484–498, 1998.

Application of Search Group Algorithm for Automatic Generation Control of Interconnected Power System

Dillip Khamari, Rabindra Kumar Sahu and Sidhartha Panda

Abstract A novel search group algorithm (SGA) technique with PID controller is proposed for an application toward multi-area interconnected power system-based automatic generation control (AGC). A reheat thermal power system over three unequal areas is considered. The system includes the nonlinearity parameters such as GRC and GDB. The supremacy of SGA tuned PID controller is projected with a comparative empirical result over recently published firefly algorithm (FA) optimized technique tuned PID controller for the similar multi-area power system which is interconnected. Simulation study confirms that the proposed SGA technique is better as compare to FA technique for the system.

Keywords Search Group Algorithm (SGA) · PID controller · Generation Rate Constraint (GRC) · Governor Dead Band (GDB) · Automatic Generation Control (AGC)

1 Introduction

In today's power system optimization arena, the automatic generation control (AGC) is a prominent factor for stable and secure power systems operation because of the growth in size of emerging power systems. Maintaining the steady frequency with the control of tie-line power flows by keeping real power balance in the restructured power system is the important function of AGC. The deviations are due to variation between the electrical load demand and generation which causes

D. Khamari · R. K. Sahu (✉) · S. Panda
Department of Electrical Engineering, Veer Surendra Sai University
of Technology (VSSUT), Burla 768018, Odisha, India
e-mail: rksahu123@gmail.com

D. Khamari
e-mail: dillip_apr9@yahoo.co.in

S. Panda
e-mail: panda_sidhartha@rediffmail.com

© Springer Nature Singapore Pte Ltd. 2019
H. S. Behera et al. (eds.), *Computational Intelligence in Data Mining*,
Advances in Intelligent Systems and Computing 711,
https://doi.org/10.1007/978-981-10-8055-5_50

undesirable effects. AGC in each area adjusts the generator set point automatically for the corresponding load change by calculating and driving area control error (ACE) to null. Deviation in net tie-line power interchange with the linear combination of frequency deviation in each area is defined as ACE in that area [1–4]. Different control schemes were proposed by researchers for AGC of power systems to accomplish better performance. For easy realization and simplicity of classical PID controller with its variant are more favorable to achieve zero ACE signals. In [5], the authors have studied the various AGC-based classical controllers in multi-area power system. E. S. Ali et al. reported the superiority of BFOA-tuned PI controller over genetic algorithm (GA) based two-area non-reheat thermal power system [6]. In [7], PID controller parameters were optimized by imperialist competitive algorithm (ICA) in a multi-area interconnected power system.

In [8], the authors were studied AGC of multi-area power system with 2-DOF PID controller using TLBO algorithm. U. K. Rout et al. were reported for two-area power system in AGC, using differential evolution (DE) method with PI controller [9]. They have studied the supremacy of DE approach over BFOA, GA, and conventional ZN-based PI controllers. In [10], authors reported the superiority of firefly algorithm (FA)-based PID controller in AGC over PSO, DE, GA, and BFOA-based optimization method for the similar interconnected multi-area power system. In recent past, the search group algorithm (SGA), a new meta-heuristic algorithm has been developed by Matheus Silva Goncalves et al. to deal with the optimization of truss structure, which is highly nonlinear, nonconvex optimization problem [11]. A new attempt has been proposed for optimal design of search group algorithm (SGA)-based PID controller in automatic generation control (AGC).

2 Material and Method

2.1 Power System Modeling

A three-area power system under investigation having GRC with GDB as nonlinearity parameter is depicted in Fig. 1. The power system for AGC of interconnected areas is also used in [10]. 2000 MW, 4000 MW, 8000 MW is the capacity of area-1, area-2, area-3, respectively. GRC is known as the rate of power generation in thermal power plant which is limited to a specified maximum rate. GRC around 3%/min is quite low for most of the reheat units [10]. GDB is termed as "the total magnitude of a sustained speed change within which there is no resulting measurable change in the position of the governor control valve." The speed GDB influences the power system dynamics. The typical width of governor's dead band is 0.06% (0.036 Hz) [10]. In Fig. 1 frequency bias parameters are B_1, B_2, and B_3, area control errors are ACE_1, ACE_2, and ACE_3, the controller outputs are u_1, u_2, and u_3, and the regulation parameters of speed governor are R_1, R_2, and R_3 in p.u. Hz; The time constants of speed governor are T_{G1}, T_{G2}, and T_{G3} in sec. The governor

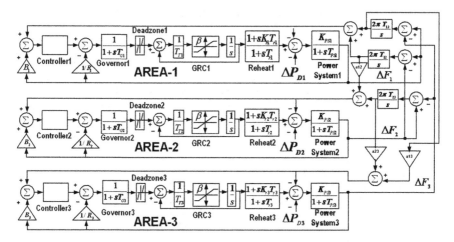

Fig. 1 Three-area power system with reheat, GRC and GDB are depicted as a block diagram

output commands are ΔP_{G1}, ΔP_{G2}, and ΔP_{G3} in p.u. The steam turbine time constants are T_{T1}, T_{T2}, and T_{T3} in sec. The steam turbine reheat time constants are T_{r1}, T_{r2}, and T_{r3} in sec. The reheat constants of steam turbine are K_{r1}, K_{r2}, and K_{r3}. The output power changes in turbine are ΔP_{T1}, ΔP_{T2}, and ΔP_{T3}. The load demand changes are ΔP_{D1}, ΔP_{D2}, and ΔP_{D3}. The tie-line power net incremental changes are ΔP_{Tie-1}, ΔP_{Tie-2}, and ΔP_{Tie-3} per unit of tie 1, 2, 3, respectively. The gains of power system are K_{PS1}, K_{PS2}, and K_{PS3}.

The time constants of power system are T_{PS1}, T_{PS2}, and T_{PS3} in sec. The synchronizing coefficients are T_{12}, T_{13}, and T_{23}. The system frequency deviations are ΔF_1, ΔF_2, and ΔF_2 in Hz. Appendix shows the relevant parameters and is taken from [10].

Equation (1) represents the relationship of the generator frequency ΔF with ACE as

$$ACE = B\Delta F + \Delta P_{Tie} \tag{1}$$

The reheat turbine transfer function is given in Eq. (2):

$$G_T(s) = \frac{\Delta P_T(s)}{\Delta P_V(s)} = \frac{1}{1+sT_T}\frac{1+sK_rT_r}{1+sT_r} \tag{2}$$

The governor transfer function is represented by Eq. (3):

$$G_G(s) = \frac{\Delta P_V(s)}{\Delta P_G(s)} = \frac{1}{1+sT_G} \tag{3}$$

The speed governing system transfer function is given by Eq. (4):

$$\Delta P_G(s) = \Delta P_{ref}(s) - \frac{1}{R}\Delta F(s) \tag{4}$$

The generator and load transfer function is represented by Eq. (5):

$$G_P(s) = \frac{K_{PS}}{1 + sT_{PS}} \tag{5}$$

where $K_{PS} = 1/D$ and $T_{PS} = 2H/fD$.

The generator and load system has two inputs $\Delta P_T(s)$ and $\Delta P_D(s)$ having one output as $\Delta F(s)$ given by Eq. (6) [2]:

$$\Delta F(s) = G_P(s)[\Delta P_T(s) - \Delta P_D(s)] \tag{6}$$

2.2 Control Structure and Objective Function

The PID controller block diagram is depicted in Fig. 2. Where the K_P is represented as proportional gain, K_I is represented as integral gain, and K_D is represented as derivative gains.

The controller's error inputs are the respective ACE are mentioned in Eqs. (7–9):

$$e_1(t) = ACE_1 = B_1\Delta F_1 + \Delta P_{Tie-1} \tag{7}$$

where $\Delta P_{Tie-1} = \Delta P_{Tie-1-2} + \Delta P_{Tie-1-3}$

$$e_2(t) = ACE_2 = B_2\Delta F_2 + \Delta P_{Tie-2} \tag{8}$$

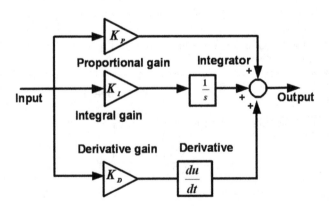

Fig. 2 Block diagram representation of PID controller structure

where $\Delta P_{Tie-2} = \Delta P_{Tie-2-3} - a_{12}\Delta P_{Tie-1-2}$

$$e_3(t) = ACE_3 = B_3\Delta F_3 + \Delta P_{Tie-3} \qquad (9)$$

where $\Delta P_{Tie-3} = \Delta P_{Tie-1-3} - a_{23}\Delta P_{Tie-2-3}$

$\Delta P_{Tie-1-2}$, $\Delta P_{Tie-1-3}$, $\Delta P_{Tie-2-3}$ are the tie-line power incremental changes which are connected between area-1 and area-2, area-1 and area-3, and area-2 & area-3, respectively. $a_{12} = \frac{P_{r1}}{P_{r2}}$, $a_{13} = \frac{P_{r1}}{P_{r3}}$, $a_{23} = \frac{P_{r2}}{P_{r3}}$ are used to considered in interconnected power system as loads inequalities [12]. The rated powers P_{r1} is for area-1, P_{r2} is for area-2, and P_{r3} is for area-3.

In the current context, integral time absolute error (ITAE) is treated as an objective function for tuning PID controller parameters. The objective function is given in Eq. (10).

$$J = ITAE = \int_{0}^{t_{sim}} (|\Delta F_i| + |\Delta P_{Tie-i}|) \cdot t \cdot dt \qquad (10)$$

The design problem is described as the given optimization problem by Eqs. (11–12):

$$\text{Minimize the value of } J \qquad (11)$$

subject to

$$K_{P\min} \leq K_P \leq K_{P\max}, K_{I\min} \leq K_I \leq K_{I\max}, K_{D\min} \leq K_D \leq K_{D\max} \qquad (12)$$

The ranges of the parameters of PID controller having lower bound −2 and upper bound as 2.

3 Proposed Method

3.1 Search Group Algorithm

M.S. Goncalves et al. developed a population-based optimization technique which is popularly known as search group algorithm (SGA) [11]. Key feature of SGA is comprised of five steps where each step is explained as follows.

3.1.1 Initial Population

Initial population P is created arbitrarily as per Eq. (13)

$$P_{ij} = X_j^{\min} + \left(X_j^{\max} - X_j^{\min} \right) U[0,1], \quad j = 1 \ to \ n, \ i = 1 \ to \ n_{pop} \tag{13}$$

The ith individual of the population P has jth design variable as P_{ij}. The sum of design variables is n. The population sum is n_{pop}. U[0,1] is an identical arbitrary variable that ranges between 0.0 to 1.0. $X_j^{\min}$ is the lower limit and $X_j^{\max}$ is the upper limit of the jth design variable.

3.1.2 Initial Search Group Selection

The objective function of each entity population is estimated after the formation of population. Then the objective functions are calculated, a standard tournament selection [11] is applied to construct the search group **R** by selecting n_g individuals from the population **P**. Every row of R is an individual symbolically. The ranking of the members of R is done after each iteration. If R_i symbolizes the ith row of R, then R_1 symbolizes the finest design and R_{ng} is the worst design in R, n_g is the number of members in search group.

3.1.3 Selection of Mutated Search Group

To amplify the global population search capability, n_{mut} individuals are substituted from R by generating new individuals according to Eq. (14)

$$X_j^{mut} = E[R_j] + t \, \varepsilon \sigma [R_j], \quad \text{for } j = 1, \ldots, n \tag{14}$$

Known mutated individual has the jth design variable denoted as X_j^{mut}, E is mean operator and σ is standard deviation operator, ε is a random variable. An "inverse tournament" selection is engaged in which the design with coarse objective function value is selected to be replaced.

3.1.4 Each Search Group Member Family's Generation

Every member of the search group generates set of individuals known as family by the perturbation described by Eq. (15)

$$X_j^{mut} = R_{ij} + \alpha \varepsilon \quad \text{for} \quad j = 1, \ldots, n \tag{15}$$

where the size of the perturbation is control by α. In the search process, after each iteration k the value of α reduces. The update of this parameter is given by the Eq. (16):

$$\alpha^{k+1} = b\,\alpha^k \tag{16}$$

where b is a parameter that describes the manner that α^k decreases in the successive iterations, the distance is controlled by α^k. In initial stage of iterations, α^k value is selected such a high value such that, it permits to explore the majority of the design phase. In the successive iteration, the value of α^k decreases and allows exploiting the design regions. The range of α^k is given by: $\alpha^0 \leq \alpha^k \leq \alpha_{\min}$ where $\alpha_{\min}$ is the smallest amount value of α^k, ascertaining a negligible new individuals movement including in the last iterations of the SGA also, α^0 is the value of α in the initial iteration. Let each family is represented by F_i, where i = 1 to n_g. The number of individuals created by every search group members solely depends on the value of its objective function. Better the quality of objective function, upper is the number of individuals it generates. The total number of designs calculated is kept constant for every iteration, i.e., equals to $(n_{pop} - n_g)$.

3.1.5 Selection of New Search Group

During the first $it^{\max}$ iterations, the fresh search group produced by the best global member of every family and this is known as global phase of algorithm. As the iteration number is greater than $it^{\max}_{global}$, the original search group is created by the best n_g individuals within all the families which is known as local phase of algorithm. $it^{\max}$ is max. number of iterations, $it^{\max}_{global}$ is the global phase max. number of iterations.

4 Implementation of Search Group Algorithm

The model is design and simulated in MATLAB with SIMULINK environment for the system under study. SGA program is written in MATLAB files (.m). Separate program is written and simulated for the model by (.m file through parameters of SGA and controller parameters) including a step load of 10% change in area-1 at 0 s (time as t). The parameters of SGA plays a vital role for the performance of SGA, and the range of these parameters are chosen as per the literature [11]. The SGA control parameter values used in the algorithm are shown in Table 1. SGA optimization method is implemented to optimize the PID controller parameters separately. PID controller having the unknown parameter is recorded at the end of $it^{\max}$. The optimized control parameters are mentioned in Table 2.

Table 1 SGA parameter values

Parameters	Value/expression
n_{pop}	100
α^0	2
α_{min}	0.01
it^{max}	50
it^{max}_{global}	$0.5 \times it^{max}$
n_g	$0.2 \times n_{pop}$
n_{mut}	$0.03 \times n_{pop}$
t	1, 2, 3 respectively for each muted individual

Table 2 SGA tuned PID controller parameters

Proportional gain			Integral gain			Derivative gain		
K_{P1}	K_{P2}	K_{P3}	K_{I1}	K_{I2}	K_{I3}	K_{D1}	K_{D2}	K_{D3}
+0.2107	−0.2203	+0.0742	−0.5772	−1.2715	+0.0502	−0.0287	−0.1047	−0.6537

4.1 Analysis of Results

A step load demand of 10% change is applied in area-1 at 0.0 s (time as t). The ITAE value and settling time are mentioned in Table 3. A comparative analogy of ITAE and settling time values has been done between proposed SGA tuned PID controller with recently published firefly algorithm tuned PID controller for the similar three-area power system model [10]. From Table 3, it is observed that, with the same structure of PID controller and ITAE objective function, a lesser ITAE value is generated with SGA tuned PID controller (ITAE = 10.1840) compared with FA tuned PID controller (ITAE = 30.9001). It can be seen from Table 3 that with SGA technique, the settling times in terms of F_1, ΔF_2, F_3, ΔP_{tie-12}, $\Delta P_{tie-13,}$ and ΔP_{tie-23} are improved compared to FA technique. This shows SGA performs better than FA. The superiority of SGA is demonstrated as compared to some other recently proposed optimization technique.

Table 3 Settling time and error

Controller/parameters		FA: PID controller [10]	SGA: PID controller
ITAE		30.9001	10.1840
Settling time (sec)	ΔF_1	26.56	16.26
	ΔF_2	26.72	15.74
	ΔF_3	26.55	15.50
	ΔP_{tie-12}	24.03	17.96
	ΔP_{tie-13}	19.66	11.04
	ΔP_{tie-23}	20.70	14.10

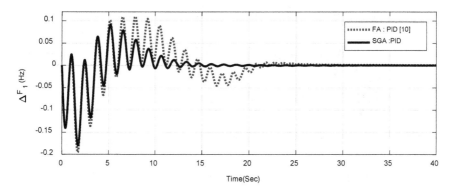

Fig. 3 Change in frequency of area-1 to 10% step load perturbation in area-1

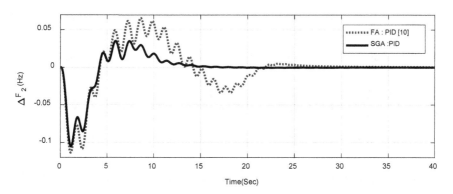

Fig. 4 Change in frequency of area-2 to 10% step load perturbation in area-1

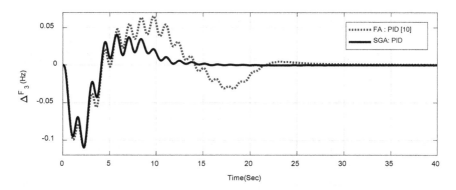

Fig. 5 Change in frequency of area-3 to 10% step load perturbation in area-1

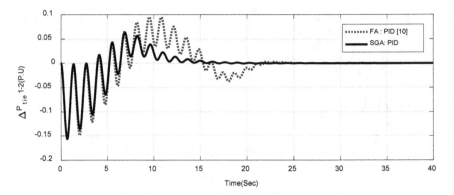

Fig. 6 Tie-line power deviation between area-1 and 2 to 10% step load perturbation in area-1

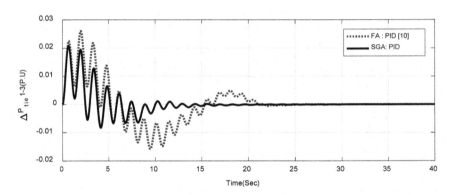

Fig. 7 Tie-line power deviation between area-1 and 3 to 10% step load perturbation in area-1

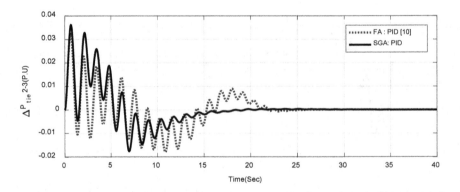

Fig. 8 Tie-line power deviation between area-3 and 3 to 10% step load perturbation in area-1

The system dynamic responses in terms of frequency and deviation in tie-line power are represented in Figs. 3, 4, 5, 6, 7, and 8. It has been observed from the Figs. 3, 4, 5, 6, 7, and 8 that robust dynamic performance is acquired by SGA tuned PID controller as compared to FA optimized PID controller [10].

5 Conclusion

In this work, SGA optimization technique has been proposed for tuning the parameters of PID controller for three interconnected unequal area thermal power system including GRC and GDB nonlinearities. The overall performance of the proposed method is demonstrated empirically. The dynamic responses are compared with recently published firefly algorithm-based optimized technique. Hence, it is concluded that the SGA tuned PID controller technique is very much effective, which provides better performance as compared to FA tuned PID controller optimization technique.

Appendix

System parameters [10]:

$f = 60$ (Hz); $B_1 = 0.3483$ (pu Hz), $B_2 = 0.3827$ (pu Hz), $B_3 = 0.3692$ (pu Hz); $D_1 = D_3 = 0.015$ (pu Hz), $D_2 = 0.016$ (pu Hz); $2H_1 = 0.1667$ (pu s), $2H_2 = 0.2017$ (pu s), $2H_3 = 0.1247$ (pu s); $R_1 = 3.0$ (Hz/pu), $R_2 = 2.73$ (Hz/pu), $R_3 = 2.82$ (Hz/pu); $T_{G1} = 0.08$ (s), $T_{G2} = 0.06$ (s), $T_{G3} = 0.07$ (s); $T_{T1} = 0.4$ (s), $T_{T2} = 0.44$ (s), $T_{T3} = 0.3$ (s); $K_{r1} = 0.5$, $K_{r2} = 0.5$, $K_{r3} = 0.5$; $T_{r1} = 10$ (s), $T_{r2} = 10$ (s), $T_{r3} = 10$ (s), $T_{12} = 0.2$ (pu/Hz), $T_{23} = 0.12$ (pu/Hz), $T_{31} = 0.25$ (pu/Hz), $P_{R1} = 2000$ MW, $P_{R2} = 4000$ MW, $P_{R3} = 8000$ MW.

References

1. Kundur P.: Power System Stability and Control, New Delhi: Tata McGraw Hill, (2009).
2. Elgerd, O.I.: Electric Energy Systems Theory—An Introduction, New Delhi: Tata McGraw Hill, (2000).
3. Hassan Bevrani.: Robust Power System Frequency Control, Springer, (2009).
4. Hassan Bervani, and Takashi Hiyama.: Intelligent AAGC, CRC Press, (2011).
5. Saikia, L.C., Nanda, J., Mishra, S.: Performance comparison of several classical controllers in AGC for multi-area interconnected thermal system. Int. J. Electr. Power Energy Syst. 33 (2011) 394–401.
6. Ali, E.S., Abd-Elazim, S.M.: Bacteria foraging optimization algorithm based load frequency controller for interconnected power system. Int. J. Electr. Power Energy Syst. 33 (2011) 633–638.

7. Shabani, H., Vahidi, B., Ebrahimpour, M.A.: Robust PID controller based on imperialist competitive algorithm for load-frequency control of power systems. ISA Trans. 52 (2012) 88–95.

8. Sahu, R.K., Panda, S., Rout, U.K., Sahoo, D.K.: Teaching learning based optimization algorithm for automatic generation control of power system using 2-DOF PID controller. Int. J. Electr. Power Energy Syst. 77(3) (2016) 287–301.

9. Rout, U.K., Sahu, R.K., Panda, S.: Design and analysis of differential evolution algorithm based automatic generation control for interconnected power system. Ain Sh. Eng. J. 4(3) (2013) 409–421.

10. Padhan, S., Sahu, R.K., Panda, S.: Application of firefly algorithm for load frequency control of multi-area interconnected power system. Elect. Power Comp. and Syst. 42(13) (2014) 1419–1430.

11. Gonçalves, M.S., Lopez, R.H., Fadel Miguel, L.F.: Search group algorithm: A new metaheuristic method for the optimization of truss structures design. Compt. Struct. 153(C) (2015) 165–184.

12. Golpira, H., Bevrani, H., Golpîra. H.: Application of GA optimization for automatic generation control design in an interconnected power system. Energy Conv. and Managt. 52 (2011) 2247–2255.

A Heuristic Comparison of Optimization Algorithms for the Trajectory Planning of a 4-axis SCARA Robot Manipulator

Pradip Kumar Sahu, Gunji Balamurali, Golak Bihari Mahanta
and Bibhuti Bhusan Biswal

Abstract This chapter presents four heuristic optimization techniques to optimize and compare the results inverse kinematics (IK) solutions, firefly, bat, particle swarm optimization (PSO) and teaching–learning-based optimization (TLBO) algorithm has been used as the optimization method for the intended purpose. In order to execute the algorithms, an objective function as the Euclidean distance between two points in space has been assigned. For each method, the convergence of the optimal position trajectory towards a set target point has been shown. The best cost plots for all algorithms have been presented. The error in positions of the obtained trajectory in X, Y and Z directions for different methods is depicted. In order to compare the outcomes of IK solutions obtained by executing the algorithms, a four-DOF SCARA robot manipulator has been considered for illustration purpose.

Keywords Inverse kinematics · Heuristic optimization methods
SCARA robot · Motion analysis

P. K. Sahu (✉) · G. Balamurali · G. B. Mahanta · B. B. Biswal
Department of Industrial Design, National Institute of Technology Rourkela, Rourkela
769008, Odisha, India
e-mail: pradipsahu2@gmail.com

G. Balamurali
e-mail: 515id1002@nitrkl.ac.in

G. B. Mahanta
e-mail: 516id1001@nitrkl.ac.in

B. B. Biswal
e-mail: bbbiswal@nitrkl.ac.in

© Springer Nature Singapore Pte Ltd. 2019   569
H. S. Behera et al. (eds.), *Computational Intelligence in Data Mining*,
Advances in Intelligent Systems and Computing 711,
https://doi.org/10.1007/978-981-10-8055-5_51

1 Introduction

The robotic systems are prominently being used for various industrial applications like assembly, machining, machine loading/unloading, welding and material handling. In order to minimize the manufacturing lead time in the aforementioned operations, the robotic systems must be processed through optimal path planning, trajectory planning and motion control. Besides, the precision and accuracy of the planning and control for the robotic manipulator motion must be taken care of. Therefore, the joint variables of the manipulator must be optimized to control the motion of the end-effector. In real-time application, motion and trajectory planning are performed in Cartesian space, whereas the control is articulated in joint space. The mathematical computation required in finding solutions to inverse kinematics problem of the manipulator is still a challenging task. The multiple solutions outcome of inverse kinematics (IK) problem raised the demand of achieving optimal solution for the intended task.

In the view of obtaining optimal solution, Rokbani et al. [1] proposed firefly algorithm as one of heuristic approach to IK problem for articulated robots. They intend to minimize the distance to a target point by establishing a fitness function between the forward position and the target position. A set of joint variables were attained by the firefly test for a three-link articulated manipulator. Aghajarian and Kiani [2] have established an adaptive neuro-fuzzy inference system (ANFIS) model for IK solution of PUMA 560 manipulator. They have used forward kinematics of the manipulator to get the training data for the planned ANFIS model. Ayyıldız and Cetinkaya [3] have presented a brief comparison of the results of IK solutions of four heuristic optimization algorithms, namely the gravitational search algorithm (GSA), the particle swarm optimization algorithm, genetic algorithm (GA) and the quantum particle swarm optimization (QPSO) algorithm for a developed four-DOF serial manipulator. After comparing the result of different algorithms for two scenarios, it was concluded that the QPSO algorithm happens to be the most effective approach for the defined task. Gigras et al. [4] have used artificial bee colony (ABC) algorithm, PSO for path planning in robotics. The comparison data reveal that ABC provides better result than PSO for the intended purpose. In order to optimize manipulator trajectory, Savsani et al. [5] have implemented seven meta-heuristic algorithms. They optimize the **joint Cartesian lengths, joint travelling distance and joint travelling time simultaneously** using those algorithms. The result shows that cuckoo search algorithm (CS), TLBO and ABC give better performance as compared to the other methods. Bayati [6] proposed cuckoo optimization algorithm (COA) and imperialist competitive algorithm (ICA) for an optimal solution to IK problem and compared the position errors with several other techniques. The simulation results of the work confirm that COA and ICA converge faster than the others. Amouri et al. [7] developed PSO and genetic algorithm (GA) to investigate the computation time and cost function value of distance for a continuum manipulator. The outcomes of numerical solution conclude that PSO method has less computation time as compared to GA approach in real-time implementations. However, the cost function was least in case of GA relative to PSO. Also, the comparison of results to several researches contributions shows that PSO and GA are better feasible methods than other methods. Huang et al. [8] developed a PSO algorithm and conducted simulation experiment to show its efficacy for solving IK problem of seven-DOF manipulator arm.

In the current research work, four different such as firefly, bat, TLBO and PSO algorithms are implemented so as to evaluate the optimal trajectory, best cost of the fitness function (length of the trajectory) optimal joint motion variables and positional errors. A brief comparison of the obtained results has been presented.

2 Heuristic Methodology

In this section, the four heuristic methodologies/algorithms used for the evaluation of trajectory and other parameters are comprehensively discussed.

2.1 Trajectory Planning Using Firefly Algorithm (FA)

In order to solve IK problem, the joint variables are treated as fireflies for the defined task. First, the firefly positions for the manipulator motion are taken in joint space. Then, the forward kinematics expression obtained in the Cartesian space is used to calculate the equivalent position of the end-effector. A fixed target point is defined to find the trajectory length as well as the error in position.

For a system having 'n' number of joints, a firefly position with iteration 'i' may be written as in Eq. (1). The firefly grows in swarms of 'n' individuals.

$$(q_j)_i = \left(\theta_1^j, \theta_2^j, \ldots\ldots\theta_n^j\right)_i \tag{1}$$

FA is an iterative optimization heuristic. The Cartesian position of end-effector, acquired with the IK solutions of firefly (j) at iteration (i), can be defined by Eq. (2).

$$X_{ji} = f\left(\theta_1^j, \theta_2^j, \ldots\ldots\theta_{1n}^j\right)_i \tag{2}$$

where f is the forward kinematic function of the manipulator.

The Euclidean distance among two fireflies i and j at x_i and x_j, respectively, is given by Eq. (3),

$$r_{ij} = \left\|x_i - x_j\right\| = \sqrt{\sum_{k=1}^{d} (x_{i,k} - x_{j,k})^2} \tag{3}$$

where $x_{i,k} = k$th component of the spatial coordinate x_i of ith firefly.

For spatial manipulator, the same is defined as in Eq. (4),

$$r_{ij} = \sqrt{(x_i - x_j)^2 + (y_i - y_j)^2 + (z_i - z_j)^2} \tag{4}$$

The movement of a firefly 'i' is attracted to a new brighter firefly 'j' can be defined as in Eq. (5),

$$x_i = x_i + \beta e^{-\gamma r_{ij}^2}(x_j - x_i) + \alpha \in \tag{5}$$

where

γ light absorption coefficient

β attraction coefficient base value

α mutation coefficient/randomization parameter

r   distance between two firefly

$\in$ (*rand*-1/2) = vector of random numbers drawn from a Gaussian distribution

rand   random number generator uniformly distributed in [0,1]

A fitness function can be assigned as the square root of the Euclidian distance from the end-effector position to the target position and can be defined as in Eqs. (3, 4). This fitness function after execution of the FA will result in optimal joint variables and that in turn effects in optimal trajectory plot.

Let us assume a case of 'n' number of fireflies with 'm' number of local optima for the IK problem. If n ≫ m, then the algorithm will converge. The 'n' fireflies are scattered evenly in the workspace. The algorithm converges to all the local optima for the iteration. The global optima are obtained by comparing the local optimal solutions.

2.2 Trajectory Planning Using Bat Algorithm (BA)

This BA is a meta-heuristic optimization algorithm. The echolocation activity of the bats with variable emission of the pulse rate and loudness is employed in the BA. This specific behaviour of the bats is implemented to get their target and classify kinds of insects. Bats produce loud sounds in order to locate potential prey. After hitting an object, this signals return back. They can easily read and understand the signals and predict the size of the object and the direction of movement. The BA assumes the principles that the whole bats use the echolocation to recognize distance. They have the idea about the difference among the target and circumstantial obstructions. The bats used to hover arbitrarily having velocity v_i at position x_i with a fixed frequency f_{min}, variable wavelength λ and loudness A_0 to hunt the target. The bats repeatedly alter the wavelength (or frequency) of produced pulsations to regulate the rate of pulse emission, $r \in [0, 1]$, according to object position. The variations of loudness lie in between a higher A_0 to a least A_{min}.

2.3 Trajectory Planning Using Teaching–Learning-Based Optimization (TLBO)

TLBO is a replication of a classical school learning systems, and no specific control parameters are needed for it. This algorithm needs common controlling parameters only such as population size and number of iterations for its working. TLBO consists of two phases named as teacher phase and learner phase. During the teacher phase, a teacher communicates knowledge directly to his/her students. Depending on the quality of the teacher, the learner obtains the knowledge. So, the teacher is commonly understood as a highly educated individual who trains learners/students to ensure superior results of the student in terms of grades or marks. The performance of a teacher's teaching effectiveness through the teacher phase is distributed under the Gaussian law. The rightmost part of the Gaussian distribution indicates that, out of all the students, some students know the entire resources taught by the teacher. The middle portion of the Gaussian distribution shows that maximum number of students may perceive the fresh learning properties. The leftmost part of the Gaussian distribution specifies that the teacher may not have any influence on students' understanding.

2.4 Trajectory Planning Using Particle Swarm Optimization (PSO)

PSO is efficient and reliable swarm intelligence-based method. The word 'swarm intelligence' refers to algorithms and distributed problem-solving technique motivated from the group behaviour of various animal societies and insect colonies search technique. It was observed by the researchers that the synchronous behaviour of flocking was followed by maintaining an optimal distance between an individual and their neighbouring members. Hence, velocity plays an important role in adjusting the optimal distance between particles. The individual particle determines the velocity for finding a food source on two factors, their own best experience from previous search and the best experience of all other members.

The particles in the population travel to their best position and towards the best particle position in the swarm. For evaluation of every particle in the swarm population, the permutation of jobs should be determined. The smallest position value is evaluated for each particle to obtain its corresponding permutation. After evaluation the algorithm will follow certain steps iteratively. For given position, velocity and fitness values, the particles will update their own best value (i.e. personal best value) if there is improvement in the fitness value. This is carried out for every particle, and the position and fitness values of the best particle in whole swarm population are updated as the global best value. Now the velocities of all particles are updated using their previous velocity, personal best value and the global best value. This provides with the new position of the particles. The smallest position value rule is used to determine the position so that computation of the fitness value for the particles can be performed again. To enhance the exploration of the search space, mutation or local search can be used for a certain group of particles in the population. A predetermined stopping criterion is set to terminate the iterations.

Same as the FA, a fitness function as the Euclidean distance has been assigned, and the same variable parameters are chosen for BA, TLBO and PSO.

3   Problem Definition and Objective Functions

The prime objective of the problem is to find an optimal trajectory (optimum trajectory length/best cost, optimal joint angle and joint distance, minimum error in positions). For illustration purpose, a pick and place operation through a four-DOF Quest SCARA manipulator as shown in Fig. 1 has been considered. It can be observed that the forward kinematics of the SCARA robot, the position vector does not have the fourth joint angle as its variable parameters. So that, joint angle has not taken into consideration for the said trajectory optimization task. In order to obtain, the trajectory for the pick and place operation between a source and a target point is defined. The Euclidean distance between those two points has been taken as the fitness function for the above four illustrated optimization algorithms. The best cost, optimal joint angle and joint distance, optimal trajectory and error in positions of different directions have been evaluated for a fixed number of iteration, and the outcomes are plotted to compare the effectiveness of the methods.

Figure 2 represents the kinematic model of the Quest SCARA robot. Table 1 shows the D-H kinematic parameters obtained for the manipulator.

The forward position kinematics of the SCARA manipulator end-effector is obtained as in Eq. (6),

$$\left. \begin{array}{l} P_x = a_1c_1 + a_2c_{12} \\ P_y = a_1s_1 + a_2s_{12} \\ P_z = -d_3 - d_4 \end{array} \right\} \tag{6}$$

Let us assume, the target point be (a, b, c). So the objective function as the Euclidean distance from a source point to the target point in 3D space is defined by Eq. (7),

Fig. 1 Quest SCARA robot model (MTAB India Pvt Ltd.)

Fig. 2 Kinematic model of
SCARA manipulator
(openrobotics.com)

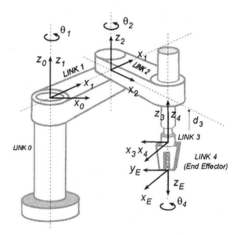

$$f = \sqrt{\left((p_x - a)^2 + (p_y - b)^2 + (p_z - c)^2 \right)} \qquad (7)$$

The sample population configuration to find the IK solution through heuristic optimization procedures for the objective function is presented in Eq. (7) and is given as,

$$S = \begin{bmatrix} \theta_{1,i} & \theta_{2,i} & d_{3,i} \\ \dots & \dots & \dots \\ \theta_{1,n} & \theta_{2,n} & d_{3,n} \end{bmatrix} \tag{8}$$

where

S the population configuration for the IK problem,
n population size (i = 1, 2, ...n).

4 Proposed Methodology

In order to evaluate the best solution for the aforesaid trajectory planning problem, a flowchart has been presented in Fig. 3. The process begins with the initialization of the manipulator kinematic parameters and optimization algorithm parameters. The feasibility of the start and end position is verified according to the defined parameter to find the forward kinematics. The optimization process begins for the given population producing midway positions. The forward kinematics has been used along with the obtained intermediate joint angles to find the end-effector intermediate positions. The distance objective fitness function for the given set of data is calculated till the end condition is attained. The optimal results are shown in the form of different performance parameters of the trajectory.

The kinematic parameters of upper bound and lower bound limits are presented in Table 1 in above section. Table 2 displays the parameters used to run all algorithms.

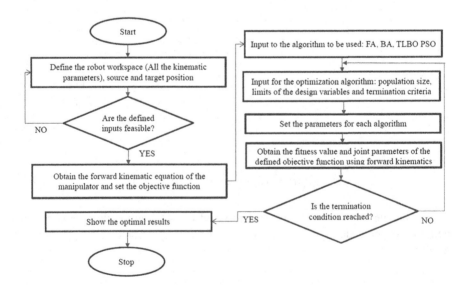

Fig. 3 Flowchart for manipulator trajectory optimization

Table 1 Kinematic parameters of Quest SCARA manipulator

Sl. no.	Joint angle θ_i (degree)	Joint distance d_i (mm)	Link length a_i (mm)	Twist angle α_i (degree)
1	$\theta_1 = \pm 135$	0	$a_1 = 200$	0
2	$\theta_2 = \pm 135$	0	$a_2 = 200$	180
3	0	$d_3 = 0\text{–}150$	0	0
4	θ_4	$d_4 = 150$	0	0

5 Results of FA, BA, TLBO and PSO

The performance of the trajectory planning of the manipulator with the heuristic optimization algorithms is evaluated by investigating the performance parameters. The planned optimal trajectories, error in motion in different directions and the convergence rate of the methods are chosen as the performance parameter. The modified algorithms (codes) of FA, BA, TLBO and PSO with parameters given in Table 2 are executed for 100 iterations for the same limits of the variable parameters in MATLAB R2015a software. Same target point locus has been chosen for all methods. The optimal position trajectory convergence of the four algorithms towards the target position is shown in Fig. 4. The results of best cost of the Euclidean distance objective function are presented as in Fig. 5. The errors in position in X, Y and Z directions (i.e. ex, ey and ez) for all the methods are represented in Fig. 6, Fig. 7 and Fig. 8, respectively. Figure 9 and Fig. 10 depict the variations of joint angle θ_1 and θ_2, respectively.

Table 2 Optimization parameters used for FA, BA, TLBO and PSO

Algorithms	Iteration	Population size	Ub and Lb of variable	γ	β	α	δ	α_{damp}	w	w_{damp}	C_1	C_2
FA	100	5	Same for all as given in Table 1	0.8	0.2	0.2	0.9	0.997	-	-	-	-
BA		500		-	-	-	-	-	-	-	-	-
TLBO		30		-	-	-	-	-	-	-	-	-
PSO		100		-	-	-	-	-	1	0.99	1.5	2

γ light absorption coefficient, β attraction coefficient base value, α mutation coefficient, α_{damp} mutation coefficient damping ratio, δ uniform mutation range, w inertia weight, w_{damp} inertia weight damping ratio, C_1 personal learning coefficient and C_2 global learning coefficient

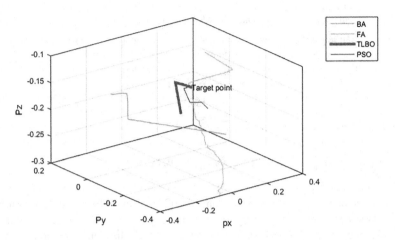

Fig. 4 Convergence of optimal position trajectories towards the target point

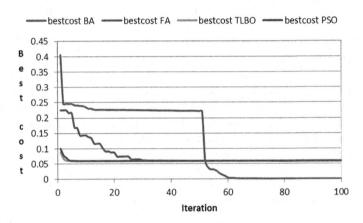

Fig. 5 Best cost plot with iteration

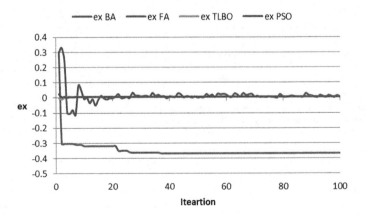

Fig. 6 Error in X-positions

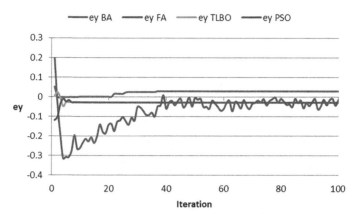

Fig. 7 Error in Y-positions

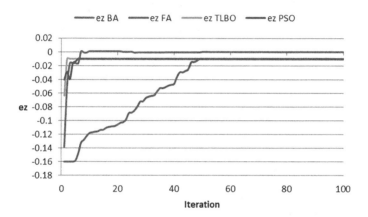

Fig. 8 Error in Z-positions

6 Comparison of Result of FA, BA, TLBO and PSO

It can be observed from Fig. 4 that the convergence of optimal position trajectory of TLBO (thicker magenta colour line) and PSO towards the target point is much faster, precise and accurate than that of FA and BA. Figure 5 reveals that the best cost convergence of TLBO (5 iterations) and PSO (6 iterations) faster as compared to FA (39 iterations) and BA did not after 100 iterations. Figures 6, 7 and 8 show that the error in positions in case of TLBO and PSO is much lower than that of FA and BA. That confirms the accuracy and preciseness of the TLBO and PSO. The variation of joint angles in case of results of TLBO and PSO is least relative to FA and BA as shown in Figs. 9 and 10. Since, PSO is swarm intelligence technique where velocities and acceleration of the particles are updated after every iteration. So the solution converges in less iteration compared to FA and BA. In case of

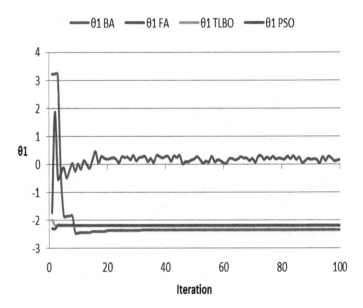

Fig. 9 Variation of joint angle θ_1

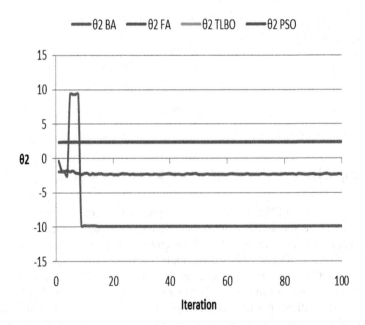

Fig. 10 Variation of joint angle θ_2

TLBO, there are no tuning parameters (for BA-position, velocity) which affect the solution accuracy. Therefore, the TLBO solution is better relative to the other three algorithms.

7 Conclusion

This research work presents an optimal trajectory planning approach using four different heuristic methods, viz. FA, BA, TLBO and PSO for Quest SCARA manipulator. The IK problem is evaluated for the objective function as Euclidean distance to a fixed target point. All the four stated algorithms were executed for 100 iterations, and the results of optimal trajectory, best cost and error in positions were obtained. The simulation results confirm that TLBO and PSO are producing optimal trajectory with least error in position. The variations in joint angles are also less. However, FA and BA are providing non-optimal trajectory with more positional error. The variation in joint angle is more. The convergence towards the target point is faster and precise for TLBO and PSO as compared to FA and BA. The above comparison shows that FA and BA are not suitable approaches for the optimal trajectory planning task. So this study reveals that a suitable optimal trajectory can be planned by implementing TLBO and PSO methods in less time.

References

1. Rokbani, N., Casals, A. and Alimi, A.M., 2015. Ik-fa, a new heuristic inverse kinematics solver using firefly algorithm. In Computational Intelligence Applications in Modeling and Control (pp. 369–395). Springer International Publishing.
2. Aghajarian, M. and Kiani, K., 2011, November. Inverse kinematics solution of PUMA 560 robot arm using ANFIS. In Ubiquitous Robots and Ambient Intelligence (URAI), 2011 8th International Conference on (pp. 574–578). IEEE.
3. Ayyıldız, M. and Çetinkaya, K., 2016. Comparison of four different heuristic optimization algorithms for the inverse kinematics solution of a real 4-DOF serial robot manipulator. Neural Computing and Applications, 27(4), pp. 825–836.
4. Gigras, Y., Jora, N. and Dhull, A., 2016, Comparison between Different Meta-Heuristic Algorithms for Path Planning in Robotics, International Journal of Computer Applications 142 (3):6–10.
5. Savsani, P., Jhala, R.L. and Savsani, V.J., 2016. Comparative study of different metaheuristics for the trajectory planning of a robotic arm. IEEE Systems Journal, 10(2), pp. 697–708.
6. Bayati, M., 2015. Using cuckoo optimization algorithm and imperialist competitive algorithm to solve inverse kinematics problem for numerical control of robotic manipulators. Proceedings of the Institution of Mechanical Engineers, Part I: Journal of Systems and Control Engineering, 229(5), pp. 375–387.
7. Amouri, A., Mahfoudi, C., Zaatri, A., Lakhal, O. and Merzouki, R., 2017. A metaheuristic approach to solve inverse kinematics of continuum manipulators. Proceedings of the Institution of Mechanical Engineers, Part I: Journal of Systems and Control Engineering, 231(5), pp. 380–394.

8. Huang, H.C., Chen, C.P. and Wang, P.R., 2012, October. Particle swarm optimization for solving the inverse kinematics of 7-DOF robotic manipulators. In Systems, Man, and Cybernetics (SMC), 2012 IEEE International Conference on (pp. 3105–3110). IEEE.

A Computer-Aided Diagnosis System for Breast Cancer Using Deep Convolutional Neural Networks

Nacer Eddine Benzebouchi, Nabiha Azizi and Khaled Ayadi

Abstract The computer-aided diagnosis for breast cancer is coming more and more sought due to the exponential increase of performing mammograms. Particularly, diagnosis and classification of the mammary masses are of significant importance today. For this reason, numerous studies have been carried out in this field and many techniques have been suggested. This paper proposes a convolutional neural network (CNN) approach for automatic detection of breast cancer using the segmented data from digital database for screening mammography (DDSM). We develop a network with CNN architecture that avoids the extracting traditional handcrafted feature phase by processing the extraction of features and classification at one time within the same network of neurons. Therefore, it provides an automatic diagnosis without the user admission. The proposed method offers better classification rates, which allows a more secure diagnosis of breast cancer.

Keywords Convolutional neural networks · CNN · Deep learning
Image classification · Breast cancer · Mammography · Diagnosis

1 Introduction

Cancer in general is a tumor related to the anarchic and indefinite proliferation of genetically modified cells. This proliferation is at the origin of the destruction of the base tissue and the extension of the tumor. In this case, the organism is not able to put it under control. The multiplication of tumor cells in one place constitutes a

N. E. Benzebouchi (✉) · N. Azizi (✉) · K. Ayadi
Labged Laboratory, Computer Science Department, Badji Mokhtar Annaba University,
PO BOX 12, 23000 Annaba, Algeria
e-mail: nasrobenz@hotmail.fr

N. Azizi
e-mail: azizi@labged.net

K. Ayadi
e-mail: ayadikhaled90@yahoo.fr

© Springer Nature Singapore Pte Ltd. 2019
H. S. Behera et al. (eds.), *Computational Intelligence in Data Mining*,
Advances in Intelligent Systems and Computing 711,
https://doi.org/10.1007/978-981-10-8055-5_52

583

malignant tumor or cancer. The propagation of cancer cells from the local tumor to other parts of the corps is a metastasis. In particular, breast cancer is the most repeated cause of death among women worldwide.

Breast cancer is among the most frequent and grievous cancers in the domain of public health, or so one in ten women is touched by this sickness during their lifetime [1]. However, the reduction of the mortality rate caused by this type of cancer as well as the promotion of the chances of recovery is possible only if the tumor has been taken care of in the first time its appearance. So as to ensure the early detection of such a tumor, radiologists have been led to increase the frequency of mammography, especially for the age group most concerned. In addition, each year, a large volume of mammography images must be analyzed, which requires intense work, a huge amount of time, and several interventions of different radiologists in order to help one another in decision-making. For it, several research studies have been directed toward the automation of mammography reading and decision-making [2].

The first work on automated mammography imaging systems is aimed at providing a second interpretation to radiologists to help them detect/diagnose at an early-stage malignant lesion regardless of their mass or microcalcifications. They are termed the computer-aided detection/diagnosis (CAD) systems.

The main purpose of automated system is to improve the diagnosis accuracy. In fact, CAD is used as a second opinion by the physicians to get the final diagnosis [3, 4], which can decrease human errors, and therefore to provide a uniform screening on a large scale and a better price.

The computers once trained can get much faster classifications, so this helps doctors in real-time classification. Machine learning for breast cancer diagnosis has achieved great development in recent years [5].

Deep learning is a branch of machine learning and can be applied to many problems such as image classification, voice recognition, and natural language processing. Convolutional neural networks (CNNs) are widespread, representing deep learning architectures, have encouraging results for image recognition applications, including medical imaging. CNNs were already used in the 1970s [6], have demonstrated an impressive record for difficult applications such as handwritten character recognition [7], and have improved the recognition rate for better computing approaches [8].

CNNs have demonstrated a qualitative and supreme evolution of technology to perform many complex image classification tasks, like the annual ImageNet challenges [9, 10].

The main functions of a computer-aided diagnosis system are defined as follows: segmentation, feature extraction, and classification which is the basic step in this process in order to obtain an end result.

This study introduces an image recognition system that utilizes a CNN so as to detect and classify abnormalities in mammograms. In general, a mammogram is classified as either normal, benign, or malignant (Fig. 1 shows an example).

The last two steps are merged into deep neural networks which perform both the automatic feature extraction and their classification. This paper is organized as

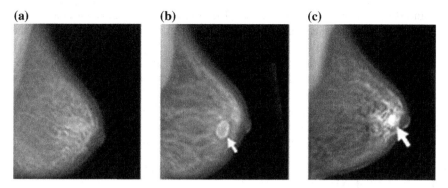

Fig. 1 Breast cancer mammogram images: **a** normal; **b** benign—not cancer; **c** cancer

follows. An overview of the related work presented in Sect. 2. Section 3 explains our method of preprocessing our data and represents our CNN architecture. In Sect. 4, we illustrate the results of our work. Finally, Sect. 5 presents a conclusion of this study as well as a general discussion of the results obtained.

2   Related Work

All women can be affected by breast cancer. Breast cancer is the leading cause of death in women. As such, several studies have been carried out to develop tools to diagnose this cancerous disease. Automated breast cancer detection has been achieved using different machine learning (ML) techniques.

In 2008, Verma [11] presented a new method for the classification of mammary masses for the diagnosis of breast cancer. The proposed methodology is based on the insertion of new neurons into the hidden layer. The classification rate is 94%.

In 2016, Sayd, A.M et al. [12] used magnetic resonance images in order to extract features for classifying mammography images into two classes: malignant, benign, used both KNN and LDA algorithms.

Also in 2016, Zhang, Qi et al. [13] used two unsupervised learning algorithms: the restricted Boltzmann machine (RBM) and the point-wise gated Boltzmann machine (PGBM) using deep learning to automatically extract the image features for the classification of breast cancer; the results showed an accuracy of 93.4%.

And also in 2016, Sidney ML of de Lima et al. [14] used two different types of data, images and texture and have extracted the features of each type of data using Zernike moments and multiresolution wavelets. The proposed approach combines the results of the both SVM and ELM algorithms for breast cancer classification with an accuracy of 94.11%.

In 2017, Alharbi, A. and Tchier, F. [15] proposed an automatic system for the detection of breast cancer based on the Saudi Arabian database by merging the results of the fuzzy and genetic algorithms.

These approaches mentioned above need a prior artistic step of extracting the characteristics of the images before the main recognition step, and they are not applicable in real time that a CNN. According to our knowledge, this is the first work using deep convolutional neural networks for computer-aided diagnosis system for breast cancers. This work is a continuation of our previous tasks [16–18].

3   Proposed Method

The creation of our network by the proposed approach, illustrated in Fig. 2, was obtained after several experiments and after a deep study of the literature for other tasks of pattern recognition. A preliminary step of manual segmentation on the database was carried out in order to extract the masses.

3.1   Segmentation

In our method, a manual segmentation of the mass of the learning base (our base includes mass images encircled by a red circle) is performed in order to extract the contour of the form to be analyzed using the ImageJ tool (Figure 3 shows an example).

3.2   CNN Architecture and Conception

We use 6 layers in our CNN architecture which are organized as follows: convolutional layer C1, subsampling layer S1, convolutional layer C2, subsampling

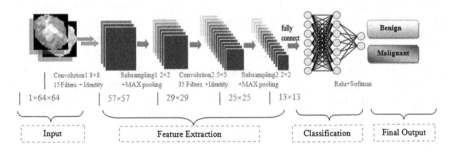

Fig. 2 Network architecture

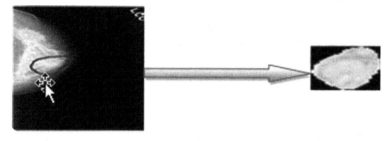

Fig. 3 Example of extraction of the mass [19]

Fig. 4 Model and training information

Model and Training Information

Model Type	MultiLayerNetwork
Layers	6
Total Parameters	431080
Start Time	
Total Runtime	
Last Update	2017-06-23 15:58:13
Total Parameter Updates	938
Updates/sec	2,38
Examples/sec	76,01

layer S2, dense layer D and finally output layer O (see Fig. 2). The main model parameters and training information of proposed CNN architecture are described in Fig. 4.

We develop a network with CNN architecture that avoids the phase of extracting traditional handcrafted features by processing the extraction of features and classification at one time within the same network of neurons and therefore provides an automatic diagnosis without the user admission.

The convolutional neural networks are currently the most powerful models for classifying images. They have two distinct parts. At the input, an image is provided in the form of a matrix of pixels. It has two dimensions of a grayscale image.

The first part of a CNN is the conventional part itself. It functions as an extractor of image characteristics. An image is passed through a succession of filters, or convoluted nuclei, creating new images called convolution maps. Some intermediate filters reduce the resolution of the image by a local maximum operation.

In the end, the convolution maps are combined in a feature vector, called a CNN code. This code CNN got out of it from the convolutive party is then connected in the entry of a second part, constituted by completely connected layers (multilayer perceptron). The role of this part is to combine the characteristics of the code CNN to classify the image. The output is a last layer with one neuron per category.

3.3 Training

Our neural network is performed after several performance tests. We start through the convolution blocks creation; a batch normalization step is applied after each convolutional layer to decrease the number of feature maps. A stochastic gradient descent is used with a momentum value of 0.9. L2 regularization method is also applied for weight and biases with a threshold equal to 0.0005. Finally, a low learning rate is fixed at 0.0001 to train our neural network.

We use two convolution layers of and two subsampling layers (see Fig. 2) using each one the identity activation function. A stride parameters are fixed at (1*1) and

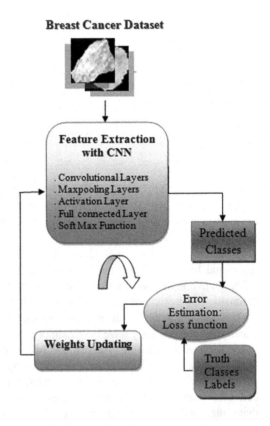

Fig. 5 Supervised training process of CNN classifier

(2*2) for convolutional and subsampling layers respectively. The MAX-pooling function is used with kernel size 2 × 2. For the dense layer, we use the widely used ReLU function; also the mean square error (MSE) function has been used to optimize the loss function. Finally, for the classification we use the function Softmax widely used.

Figure 5 shows the training process of CNN.

The proposed approach for breast cancer detection and classification uses a sample of 190 images from DDSM mammographic images database. Out of these, 95 images are benign and 95 images are malignant. In order to evaluate our approach, we use the cross-validation method of k-fold. After 13 epochs and 10 iterations, CNN was able to detect normal and abnormal classes of breast cancer with accuracy of **97.89%**.

4   Results

In our experiments, we use the digital database for screening mammography (DDSM) [20]; our database was developed, with 190 images, including 95 benign images and 95 malignant images. The implementation of the proposed work was done with Deeplearning4j. Deeplearning4j[1] is the first commercial-grade, open-source, distributed deep-learning library written for Java and Scala.

As shown in Table 1, **92** breast cancer images are correctly detected as malignant image by the proposed approach, and **94** non-cancer images were correctly classified as benign images.

In summary, 186 images were accurately labeled by the proposed method, resulting in **97.89% accuracy** with **sensitivity 98.9%, specificity 96.9%,** positive predictive value (**PPV**) **96.8%**, negative predictive value (**NPV**) **98,9%,** and **AUC 98.2%.** The sensitivity, specificity, PPV, NPV, and accuracy are defined in Eqs. 1, 2, 3, 4, and 5 [21]:

$$\text{Sensitivity} = \frac{\text{TP}}{\text{TP} + \text{FN}} \tag{1}$$

$$\text{Specificity} = \frac{\text{TN}}{\text{TN} + \text{FP}} \tag{2}$$

$$\text{PPV} = \frac{\text{TP}}{\text{TP} + \text{FP}} \tag{3}$$

$$\text{NPV} = \frac{\text{TN}}{\text{TN} + \text{FN}} \tag{4}$$

---

[1]https://deeplearning4j.org/

Table 1 Confusion matrix of CNN results

	Malignant	Benign
Malignant	92	1
Benign	3	94

Table 2 Obtained results of the proposed method

	Sensitivity	Specificity	PPV	NPV	AUC
Proposed CNN architecture	98.9	96.9	96.8	98.9	98.2

$$\text{Accuracy} = \frac{\text{TP} + \text{TN}}{\text{TP} + \text{TN} + \text{FP} + \text{FN}} \tag{5}$$

True positive (TP) describes the number of diseased individuals with a positive test, true negative (TN) describes the number of people not sick with a negative test, false positive (FP) is the number of non-diseased individuals with a positive test, and false negative (FN) is the number of diseased people with a negative test.

Table 2 summarizes the obtained results of the proposed CNN architecture.

The proposed approach was also evaluated according to the ROC curves [22], and the operational characteristic curve of the receiver (ROC) of our method has been plotted in Fig. 6. The area under the ROC (AUC) curve was **98.2,** and all the points of the curve are on the top half part of the ROC space; therefore, we can conclude that the proposed model has a good ROC curve.

Figure 7 shows model score versus iteration of our CNN; this is the value of the loss function on the current minibatch.

Fig. 6 ROC curve

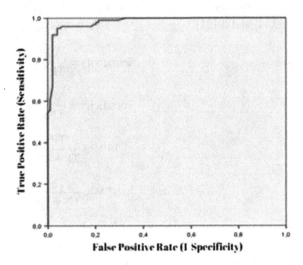

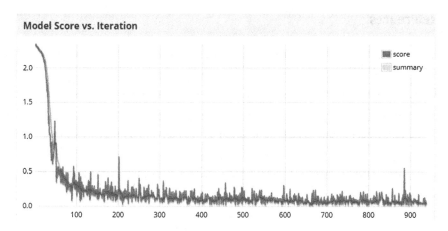

Fig. 7 Model score versus iteration

5 Discussion and Conclusion

Breast cancer is the leading cause of death in women. All women can be affected by this disease. Several studies in mammography imaging have been carried out to develop tools to diagnose this cancerous disease. Automated breast cancer detection has been achieved using different machine learning (ML) techniques.

In this study, we proposed a diagnostic aid system by classifying mammographic images in order to detect the nature of tumors (malignant/benign) using a convolutional neural network (CNN) as a binary classifier, with an image of the tumors (Malignant/Benign). The digital database for screening mammography (DDSM) is used to validate the robustness of our approach.

Proposed approach in this article avoids the design and/or the manual extraction of characteristics. Typical approaches are typically faced with difficult problems such as the conception of robust and easy methods and algorithms to calculate features, the assessment of these features and their relevance for the separation of classes, and the more general case of the selection of relevant features. Our approach based on convolutional neural networks circumvents these difficulties and avoids the delicate step associated with the features.

To conclude, proposed approach provides an automated breast cancer detection computer-aided system that enables the radiologists and the gynecologist in early diagnosis of breast cancer patients with high accuracy, comparing with state-of-the-art results using the same database (DDSM).

References

1. DeSantis, C., Siegel, R., Jemal, A.: Breast cancer facts and figures 2013–2014, Am. Cancer Soc. (2013) 1–38
2. Kourou, K., Exarchos, T.P., Exarchos, K.P., Karamouzis, M.V., Fotiadis D.I.: Machine learning applications in cancer prognosis and prediction Comput Struct Biotechnol J, 13 (2015) 8–17
3. Doi, K.: Computer-aided diagnosis in medical imaging: historical review, current status and future potential Comput Med Imaging Graph, 31 (2007) 198–211
4. Castellino, R.A.: Computer aided detection (CAD): an overview Cancer Imaging, 5 (2005) 17–19
5. Carneiro, G., Nascimento, J., Bradley, A.P.: Unregistered multiview mammogram analysis with pre-trained deep learning models. In International Conference on Medical Image Computing and Computer-Assisted Intervention, Springer International Publishing (2015) 652–660
6. Fukushima, K.: Neocognitron: A self-organizing neural network model for a mechanism of pattern recognition unaffected by shift in position. Biol Cybern, 36(4), (1980) 193–202
7. Cun, Y.L., Boser, B., Denker, J.S., Howard, R.E., Habbard, W., Jackel, L.D., and al.: Advances in neural information processing systems 2. Citeseer. ISBN 1-55860-100-; (1990) 396–404
8. Krizhevsky, A., Sutskever, I., Hinton, G.E.: ImageNet classification with deep convolutional neural networks. In: Pereira F, Burges CJC, Bottou L, Weinberger KQ Eds. Advances in neural information processing systems 25. USA: Curran Associates, Inc (2012) 1097–1105
9. Ioffe, S., Szegedy, C.: Batch normalization: Accelerating deep network training by reduc-ing internal covariate shift. (2015). URL: arXiv:1502.03167
10. He, K., Zhang, X., Ren, S., Sun, J.: Deep residual learning for image recognition (2015). arXiv:1512.03385
11. Verma, B.: Novel network architecture and learning algorithm for the classification of mass abnormalities in digitized mammograms. Artificial Intelligence in Medicine, 42, (2008) 67–79
12. Sayed, A.M., Zaghloul, E., Nassef, T. M.: Automatic Classification of Breast Tumors Using Features Extracted from Magnetic Resonance Images, Conference Organized by Missouri University of Science and Technology - Los Angeles, CA, Procedia Computer Science, 95 (2016) 392–398
13. Zhang, Qi., Xiao, Y., Dai, W., Suo, J., Wang, C., Shi, J., Zheng, H.: Deep learning based classification of breast tumors with shear-wave elastography, Ultrasonics, 72 (2016) 150–157
14. de Lima, S.M.L., da Silva-Filho, A.G., dos Santo, W.P.: Detection and classification of masses in mammographic images in a multi-kernel approach, Computer Methods and Programs in Biomedicine, 134 (2016) 11–29
15. Alharbi, A., Tchier, F.: Using a genetic-fuzzy algorithm as a computer aided diagnosis tool on Saudi Arabian breast cancer database, Mathematical Biosciences, 286 (2017) 39–48
16. Zemmal, N., Azizi, N., Dey, N., Sellami, M.: Adaptive semi supervised support vector machine semi supervised learning with features cooperation for breast cancer classification, Journal of Medical Imaging and Health Informatics, 6(2016) 53–62
17. Cheriguene, S., Azizi, N., Zemmal, N., Dey, N., Djellali, H., Farah, N.: Optimized Tumor Breast Cancer Classification Using Combining Random Subspace and Static Classifiers Selection Paradigms, Applications of Intelligent Optimization in Biology and Medicine, 96 (2015) 289–307
18. Zemmal, N., Azizi, N., Sellami, M.: CAD system for classification of mammographic abnormalities using transductive semi supervised learning algorithm and heterogeneous features, 12th International Symposium on Programming and Systems (ISPS), Algiers, Algeria, (2015)

19. Azizi, N., Tlili-Guiassa, Y., Zemmal, N.: A Computer-Aided Diagnosis System for Breast Cancer Combining Features Complementarily and New Scheme of SVM Classifiers Fusion, International Journal of Multimedia and Ubiquitous Engineering, 8(4) (2013) 45–58

20. Bowyer, K., Kopans, D., Kegelmeyer, W.R., Moore, Sallam, M., Chang, K., Woods, K.: The digital database for screening mammography. In Third international workshop on digital mammography, 58(1996) 27

21. Dey, N., Roy, A.B., Pal, M., Das, A.: FCM based blood vessel segmentation method for retinal images (2012). arXiv:1209.1181

22. Bradley, A.P.: The use of the area under the roc curve in the evaluation of machine learning algorithms. Pattern Recogn. 30(7) (1997) 1145–1159

Indian Stock Market Prediction Using Machine Learning and Sentiment Analysis

Ashish Pathak and Nisha P. Shetty

Abstract Stock market is a very volatile in-deterministic system with vast number of factors influencing the direction of trend on varying scales and multiple layers. Efficient Market Hypothesis (EMH) states that the market is unbeatable. This makes predicting the uptrend or downtrend a very challenging task. This research aims to combine multiple existing techniques into a much more robust prediction model which can handle various scenarios in which investment can be beneficial. Existing techniques like sentiment analysis or neural network techniques can be too narrow in their approach and can lead to erroneous outcomes for varying scenarios. By combing both techniques, this prediction model can provide more accurate and flexible recommendations. Embedding technical indicators will guide the investor to minimize the risk and reap better returns.

Keywords Machine learning · Sentiment analysis · Stock market
SVM

1 Introduction

This section describes the limitations of traditional approach in stock market analysis and lists the benefits of using machine learning and sentiment analysis.

1.1 Traditional Approach to Stock Market Analysis

Stock market is a very volatile in-deterministic system with vast number of factors influencing the direction of trend on varying scales and multiple layers. Efficient

A. Pathak (✉) · N. P. Shetty
Manipal Institute of Technology, Manipal University, Manipal 576104, India
e-mail: ashish.spathak33@gmail.com

N. P. Shetty
e-mail: pnishashetty@gmail.com

© Springer Nature Singapore Pte Ltd. 2019
H. S. Behera et al. (eds.), *Computational Intelligence in Data Mining*,
Advances in Intelligent Systems and Computing 711,
https://doi.org/10.1007/978-981-10-8055-5_53

Market Hypothesis (EMH) states that the market is self-correcting, i.e. current stock price reflects the most relevant cumulative price which is neither undervalued nor overvalued, and any new information is instantly depicted by the price change [1]. In layman's term, "The market is unbeatable," as you cannot gain any advantage over the market, but existing research proves otherwise. It is possible to predict the market trends by analysing the patterns of stock movement. Traditional approach applies the following models for this.

- Fundamental analysis
  This approach focuses mainly on a company's past performance and credibility. Performance measures like P/E ratios are utilized to filter stock which may incline towards a positive price surge. This approach is based on theory that profitable companies will continue to be so because of uptrend influenced by rewarding nature of the market.
- Technical analysis
  This approach is based on predicting the future prices by applying time series analysis on previous trends. Statistical techniques such as Bollinger Bands, simple moving averages are applied to predict the successive trends.

1.2 Modern Approach to Stock Market Analysis

Computer science provides us with cutting-edge tools for machine learning like SVM and EML which can analyse and perform knowledge discovery at large scales in short amount of time. Two approaches for prediction of stock market are proposed in this research.

- Qualitative Analysis
  News feeds regarding stock market highly affect the market trend and thus form a downhill movement in case of a negative news. Thus, the media/social network and stock market data are highly coupled and make the system more unpredictable. Existing research points out that in case of crisis, stocks mimic each other and lead to market crashes [1]. Nowadays, Twitter has come forth as the most reliable and fastest way of consuming media. With combined resources of news feed and Twitter feed, general population sentiment about a company can be highlighted. Text mining and sentiment analysis are useful tools for such a high-scale analysis.
- Quantitative Analysis
  Historical data is now readily available for most markets. Using this dataset, we can apply multiple machine learning models to give accurate results for future investments. These models can be trained for individual stocks with adjusted bias for most reflective features. These models can also be trained to work in different scenarios and overall market movement.

Traditional approach focuses on fundamental analysis and technical analysis to predict the market at a large scale which rarely translates to low-level individual Stock Prediction, but it can be clearly observed that individual stocks contribute to whole market movement rather than the other way around. Thus, focusing on individual stocks to predict market movement is a much more logical approach. With technology advancing at such a rapid pace and abundance of computing power, we can now easily strive towards a comprehensive system to accurately predict the market trend and reap beneficial financial returns. Existing research proves that modern approach outperforms traditional approach and can output the most accurate results [1].

2 Literature Survey

Mehak Usmani et al. in [2] proposed an intuitive idea of combining results from historical data, news and Twitter feed sentiment analysis. This dual approach predicts the stock market trend with high accuracy. It uses technical analyses like ARIMA and SMA to get an idea of the market trend. These models forecast the values based on proven mathematical models. This research considers other factors like depreciation and exchange rates. This research utilizes technical analysis for prediction which has been proven inferior to machine learning in terms of accuracy. Machine learning can handle noise and lack of information more efficiently. This approach has chances of inaccuracy for market scenarios not covered in training data.

The work proposed in [3] by Rodolfo C. Cavalcante et al. improves upon previously existing trading rules and produces results better than research proposed before. This research uses multiple proven market strategies to stimulate a real-time autonomous trader. This research focuses on short-term gains which is excellent for hands-off trading. Their model accumulates lot of revenue by trading in small time frames (minutes). Improvements can be made on choosing more features and making it more flexible.

In [4], Paul D. Yoo et al. investigated the success of machine learning models and event-driven models like sentiment analysis in predicting the stock market trends. It also illuminates the fact that macroeconomic conditions like international and political events affect market trends and need to be taken into consideration.

Alexander Porshnev et al. in [5] stated that addition of Twitter sentiment analysis does not add any valuable information to the prediction model and does not increase the accuracy. Thus, this research takes news feeds into consideration to add credibility to sentiment analysis.

The research was done by Dongning Rao et al. in [6], which provides great insight into proper implementation of sentiment analysis. They propose increasing the size of corpus (training data) with each test. This is done by adding non-polarizing words found in the test data not present in the corpus. The training data is refined by doing K-crossfold validation during each testing phase.

3 Methodology

The aim of this project is to build an application which outputs accurate recommendations in a quantifiable manner. For this purpose, three modules are implemented which are as follows:

- Machine learning module
- Sentiment analysis module
- Fuzzy logic module

These modules are integrated into a recommendation model in the following manner as shown in Fig. 1.

3.1 Machine Learning Module

The purpose of this module is to output Stock Prediction value. Stock Prediction value is the strength of difference in opening price and closing price. For this, we need to predict the closing price of the stock. This is achieved by applying machine learning on historical data of the stock. Research in [3] affirms that maximum number of features required to accurately predict a stock's closing price for a specific day are given as follows:

1. Opening price of prediction day
2. Lowest and highest prices of the prediction day
3. Simple moving average
4. Exponential moving average of opening and closing prices of the prediction day
5. Exponential moving average of lowest and highest prices of the prediction day

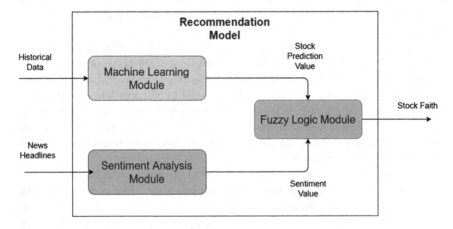

Fig. 1 Recommendation model for obtaining Stock Faith

6. Bollinger Bands of opening and closing prices of the prediction day
7. Bollinger Bands of lowest and highest prices of the prediction day

The training data is then fitted by a machine learning module and is used to predict the closing price of testing data through supervised learning. There are many regressors available **scikit-learn** library. Their accuracy was measured in terms of percentage error rate with accuracy calculated as shown in Eq. (1).

$$\left. \begin{array}{c} \frac{|Predicted\,Closing\,Price - Actual\,Closing\,Price|}{Actual\,Closing\,Price} \times 100 < \alpha \quad Accurate \\ else \qquad\qquad\qquad Inaccurate \end{array} \right\} \qquad (1)$$

where α is the acceptable error rate.

On finding accuracy for $\alpha = 2$ and 5, the accuracies observed are illustrated in Tables 1 and 2.

As it is obvious that Ridge Regressors give most accurate outcome for our dataset, it was selected to be used as the regressor for machine learning module to provide the Stock Prediction value.

The formula in Eq. (2) gives the Stock Prediction value.

$$\left(\frac{Actual\,Opening\,Price - Predicted\,Closing\,Price}{Actual\,Opening\,Price} \right) \times 100 + 50 \qquad (2)$$

3.2 Sentiment Analysis Module

The purpose of this module is to obtain the sentiment value of latest news headlines regarding each stock and output its average as sentiment value to fuzzy module.

The steps used in this module are as follows:

1. Data Collection
The data is collected by crawling through Indian financial news Web site www. moneycontrol.com. Minimum four news headlines are scraped for each stock and stored against the company symbol.

Table 1 Accuracy table for closing price prediction (error rate less than 2%)

Classifier	Accuracy (%)
Lasso	40.79
LassoLars	51.61
Elastic Net	40.79
Ridge Regressor	85.4
SVR (kernel = linear)	0.97
SVR (kernel = RBF)	0.97
Random forest	15.44
AdaBoost	3.99
Decision Tree	3.67

Table 2 Accuracy table for closing price prediction (error rate less than 5%)

Classifier	Accuracy (%)
Lasso	64.03
LassoLars	72.49
Elastic Net	64.03
Ridge Regressor	94.2
SVR (kernel = linear)	2.37
SVR (kernel = RBF)	2.37
Random forest	29.49
AdaBoost	7.49
Decision Tree	9.29

2. Tokenizing

Each news headline is broken down into sentences and then in turn broken down into words.

3. Lemmatizing

It is the process of reducing inflected (or sometimes derived) words to their word stem, base, or root form. For example, "the boy's cars are different colours" reduces to "the boy car be differ colour."

4. Finding Most Informative Features

Words that contribute most in adding polarity to a sentence are found.

The top ten most informative features that contributed most to the polarity are listed in Table 3.

5. Classifying features into positive and negative

These are then classified into positive and negative using nltk packages.

6. Adding these features to the sentiment analyser lexicon

These words are then added to the sentiment analyser wordlist with appropriate strength for positive and negative words.

7. Classifying the testing data into positive and negative sentiments using training set

Now, our sentiment analyser is ready for classifying financial news from our sources.

Now to feed our sentiment value to fuzzy logic module, it needs to be normalized on a scale of 0–100 as shown in Eq. (3).

$$\left(\frac{\sum_{i=1}^{n} \left(Polarity(News_i) \right)}{n} \right) \times 100 + 50 \tag{3}$$

where n is the number of news articles pertaining to each stock.

Table 3 Most informative features

Positive	Negative
Buy	Sell
Up	Down
Rise	Dip
Jump	Hold
Strong	Bear
Support	Impact
Grow	Decline
Fold	Fall
Double	Loss
Bag	Debt

3.3 Fuzzy Logic Module

The purpose of this module is to output Stock Faith which is the strength of recommendation.

The activation rules for this module are as follows:

- IF the News Sentiment was good or the Stock Prediction value was good, THEN the Stock Faith will be high.
- IF the Stock Prediction value was average, THEN the Stock Faith will be medium.
- IF the News Sentiment was poor and the Stock Prediction value was poor, THEN the Stock Faith will be low.

Complete operation is illustrated in Fig. 2.

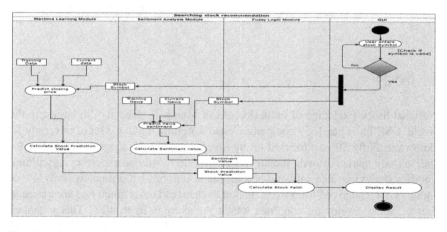

Fig. 2 Activity diagram

```
Symbol:                  ABBOTINDIA
opening price:                        4225.0          closing price:           4267.65
high price:                           4312.0          low price:               4225.0
predicted closing price:              4281.06
sentiment Value:                      85.0
closing prediction:                   51.33
Stock faith:                          62.47
news1:                   pos
news2:                   pos
news3:                   neu
news4:                   pos
                                               BUY
```

Fig. 3 Scenario for profit

```
Symbol:                  BAJAJCORP
opening price:                        377.75         closing price:            373.2
high price:                           377.75         low price:                371.5
predicted closing price:              373.07
sentiment Value:                      30.0
closing prediction:                   48.76
Stock faith:                          46.23
news1:                   neg
news2:                   neu
news3:                   neg
news4:                   neg
                                               SELL
```

Fig. 4 Scenario for loss

4 Result Analysis

Case 1: IF the News Sentiment was good or the Stock Prediction value was good, THEN the Stock Faith will be high as shown in Fig. 3.

Case 2: IF the News Sentiment was poor and the Stock Prediction value was poor, THEN the Stock Faith will be low as shown in Fig. 4.

5 Scope

National Stock Exchange of India (located in Mumbai) ranks at 12th largest in the world. NSE India has 1659 companies listed for public trading. Out of this, only 50 (known as Nifty 50) are focused on by investors. Nifty 50 acts as a barometer for Indian stock market growth. Indian economy relies mostly on exporting agricultural goods and services like software and technical support. Unfortunately, only 4% of India's GDP is derived from stock market exchange. This is much less compared to that of other developing countries which range from 20 to 40%. This untapped resource can be monetized more efficiently to contribute to the development of India.

6 Conclusion and Future Work

In this research, we propose that existing work [1–8] may integrate into a robust model to predict NSE stock market accurately. This model can be improved upon by defining refined fuzzy rules. Improving upon the training data's scale and time frame can result in better prediction. A trading model using the proposed methodology can be developed to compute total returns or investments in real time. This can prove the accuracy of the model. This model can successfully recommend the best stocks for investment.

References

1. Hellstrom, T. and Holmstrom, K. 1998, 'Predicting the Stock Market', Technical Report Series Ima-TOM-1997-07.
2. M. Usmani, S. H. Adil, K. Raza and S. S. A. Ali, "Stock market prediction using machine learning techniques," 2016 3rd International Conference on Computer and Information Sciences (ICCOINS), Kuala Lumpur, 2016, pp. 322–327.
3. R. C. Cavalcante and A. L. I. Oliveira, "An autonomous trader agent for the stock market based on online sequential extreme learning machine ensemble," 2014 International Joint Conference on Neural Networks (IJCNN), Beijing, 2014, pp. 1424–1431.
4. P. D. Yoo, M. H. Kim and T. Jan, "Financial Forecasting: Advanced Machine Learning Techniques in Stock Market Analysis," 2005 Pakistan Section Multitopic Conference, Karachi, 2005, pp. 1–7.
5. A. Porshnev, I. Redkin and A. Shevchenko, "Machine Learning in Prediction of Stock Market Indicators Based on Historical Data and Data from Twitter Sentiment Analysis," 2013 IEEE 13th International Conference on Data Mining Workshops, Dallas, TX, 2013, pp. 440–444.
6. D. Rao, F. Deng, Z. Jiang and G. Zhao, "Qualitative Stock Market Predicting with Common Knowledge Based Nature Language Processing: A Unified View and Procedure," 2015 7th International Conference on Intelligent Human-Machine Systems and Cybernetics, Hangzhou, 2015, pp. 381–384.
7. P. D. Yoo, M. H. Kim and T. Jan, "Machine Learning Techniques and Use of Event Information for Stock Market Prediction: A Survey and Evaluation," International Conference on Computational Intelligence for Modelling, Control and Automation and International Conference on Intelligent Agents, Web Technologies and Internet Commerce (CIMCA-IAWTIC'06), Vienna, 2005, pp. 835–841.
8. M. Qasem, R. Thulasiram and P. Thulasiram, "Twitter sentiment classification using machine learning techniques for stock markets," 2015 International Conference on Advances in Computing, Communications and Informatics (ICACCI), Kochi, 2015, pp. 834–840.

Exploring the Average Information Parameters over Lung Cancer for Analysis and Diagnosis

Vaishnaw G. Kale and Vandana B. Malode

Abstract Lung cancer seems to be a very common cause of death among the people all over the world. Hence, accurate detection of lung cancer increases the chance of survival of the people. The major problem with the treatment is the time constraint in several physical diagnoses that increases the death possibilities so basically this method is an approach to help the physicians to take more accurate decision in this regard. This paper comes up with a method which is based on average information statistical parameters using image processing for lung cancer analysis. The basic aim is to help the physicians to take decisions regarding possibilities of lung cancer. Image averaging is a digital image processing technique, which is mostly implemented to improve the quality of images that have been degraded by random noise. The average information parameters are among the statistical parameters that are implemented for lung cancer analysis, and hence, some of the parameters like Entropy, Standard Deviation, Mean, Variance, and MSE are considered in this paper. The selection of average information parameters is thoroughly based on the calculation of number of iterations carried over the lung images through the algorithm. This paper also successfully rejects null hypothesis test by implementing ANOVA. The images are microscopic lung images and the algorithm is implemented in MATLAB.

Keywords Average information · Statistical parameters · Lung cancer
ANN · ANOVA

V. G. Kale (✉)
Department of Electronics & Telecommunication, Dr. Vithalrao Vikhe Patil College
of Engineering, Ahmednagar 414111, Maharashtra, India
e-mail: vaishnaw25@rediffmail.com

V. B. Malode
Department of Electronics & Telecommunication, Jawaharlal Nehru Engineering College,
Aurangabad 431003, Maharashtra, India
e-mail: vandana_malode@yahoo.co.in

© Springer Nature Singapore Pte Ltd. 2019   605
H. S. Behera et al. (eds.), *Computational Intelligence in Data Mining*,
Advances in Intelligent Systems and Computing 711,
https://doi.org/10.1007/978-981-10-8055-5_54

1 Introduction

Lungs take care of proper functioning of the respiration of the human body. For normal growth, cells in the lungs divide and reproduce at a controlled rate to restore wounded tissues of the healthy body. Lung cancer [1, 2] develops, when cells inside the lungs multiply at an uncontrollable rate. These abnormal tissues of the lungs lead to cancer. Today there are many imaging techniques [1, 3] available with radiologists and physicians for the diagnoses of lung cancer such as X-ray, Computer Tomography (CT), High Resolution, Magnetic Resonance Imaging (MRI), and Positron Emission Tomography (PET). But each technique has some advantages with some shortcomings which do not give a complete assurance about the lung cancer, and also, the case history of the patient becomes important at the time of decision. Hence, there is a need of a method that could help the radiologists to reach a perfect result. Besides these medical imaging techniques, one more method that is implemented for lung cancer diagnosis is the lung biopsy [4]. Medical Imaging techniques are used to find out whether the cancer has spread over the lungs or not, but it lacks in accurate lung cancer diagnosis. A biopsy is a process in which small amount of lung tissue is taken for examination under electron microscope. Besides biopsies and surgical operations, imaging techniques are very important in the analysis of lung cancer. However, no test is ideal, and no scan can diagnose lung cancer, but biopsy can do that. But again, biopsy has some drawbacks that include difficulty in breathing, excessive bleeding, oozing out and also there is always a chance of spreading of cancer cells in the lungs as well as other parts of the body, due to the removal of small part of tissue and hence considered as the last option for the cancer diagnosis. It is often suggested when no other scan works.

The microscopic lung image is considered here for the statistical analysis which is obtained through biopsy taken through electron microscope [5], which is a powerful microscope that allows the researchers to view the specimen of the lung at nanoscale level. A small piece of lung tissue is taken, entrenched in paraffin, cut thin, placed on a glass slide, and then reagent is used in treating a specimen for microscopic examination. The resulting preparations are examined under microscope for lung cancer analysis. The images that are obtained through this process are called as microscopic lung images as seen in Fig. 1. The magnification of these images can be up to 400 times or even more which is very useful for the medical analysis. It is very difficult to visualize the microscopic images and take decisions as it may go wrong in number of cases, so it requires a robust method. Image processing with MATLAB is very useful in handling the microscopic lung images.

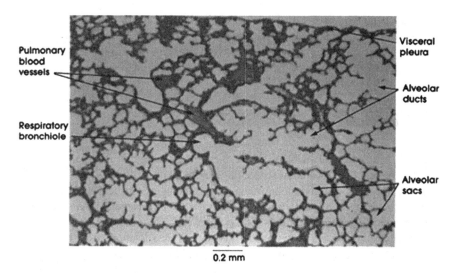

Fig. 1 Microscopic lung image

2 Methodology

The methodology used here is the extension of the algorithm used in [6], in which the statistical parameters used were Entropy, Standard Deviation, and texture factor for lung cancer analysis and diagnosis. These parameters were used to differentiate lung cancer from other lung diseases, as well as for lung cancer analysis. This method adds some more parameters into the analysis in order to improve the performance. This algorithm concentrates only on lung cancer analysis and diagnosis. In order to understand the methodology, the flow diagram of the algorithm in image processing needs to be understood and the parameters that are included in average information method. The selection of the parameters under this method is based on their average calculation principle used for the analysis. The parameter selection may vary method to method depending on the applications. Here the statistical parameters used are Entropy, Mean, Variance, Standard Deviation and Mean square Error. The input is the microscopic lung image, which is first normalized by resizing and then converting it into grayscale image. The quality of these images have been tested and verified. These images have been properly differentiated into cancerous and noncancerous microscopic lung images. The image of microscopic lungs is resized to 255 * 255 which is maintained throughout the implementation. Median filter is one of the best filters used to denoise such kind of medical images due to nonlinear nature of the noise. These images are having lots of variations in terms of pixel intensities and hence are not perfect for the processing, hence histogram equalization is applied for the image enhancement. Now the image is ready for the further processing, which involves implementation of average information method, finding out the similarities through correlation method and then finally the image classifier. The average information method is the statistical analysis that is carried out for lung cancer analysis. This analysis with image classifier together is used for the lung cancer diagnosis. The statistical analysis is used for lung cancer analysis, and averaging information method is one of the statistical methods used in this paper.

2.1 Statistical Analysis

Structural and statistical analysis is the subject of concerned for this method. As the image to be processed is a microscopic lung image, statistical analysis is considered, which can reveal the important information of the image. Statistical analysis [6, 7] is actually the analysis of random data. It does not try to understand the structure of image but provides their deterministic properties, which give the relationship between gray levels of an image. In this paper, the random data is nothing but the random pattern of the lung cancer. In order to analyze this random data, it is necessary to analyze its statistical properties. As no specific tools are available to process this random data, statistical analysis is one of the best solutions for the lung cancer analysis and diagnosis. They are demonstrated to feature a potential for the effective structure discrimination or disorder in the biomedical images. This type of analysis is done through statistical analysis of the microscopic lung images. There are some important statistical and mathematical parameters in this concerned, which are considered in this paper. These parameters can be analyzed against cancerous microscopic lung images to get an appropriate range for the lung cancer analysis. The identified range is obtained through number of iterations carried out for the specific image database only.

2.1.1 Average Information Method

The method is based on averaging of the intensity values for each pixel position in the image. Each scanned image has two components: One is constant signal component and the other is random noise component. In the averaging process, the signal component remains unchanged, but the noise component varies from frame to frame. Because the noise is random, it tends to cancel out while performing the summation. When the averaged image is computed, the image signal component has lot of influence over the summation as compared to the noise component. Based on the same principle, all the statistical parameters under this are selected. The study of these identified parameters helps us to analyze the cancerous as well as noncancerous lung images. The statistical parameters under average information considered are

(i) Entropy

It is an average information of the image. The lowest value of the Entropy means no uncertainty. It is zero if the event is sure or impossible, that is, $E = 0$ if $P = 0$ or 1. Entropy is supposed to be high throughout the image [6–8] and is calculated from Eq. (1)

$$E = -\sum_{x}^{m}\sum_{y}^{n} P[x, y] \log P[x, y] \qquad (1)$$

(ii) Mean

It calculates the mean of the gray levels in the image [6–8]. Mean is the most important and basic parameter of all statistical measures. The mathematical expression from Eq. (2) is used to calculate the mean of an image.

$$\mu = 1/N * M \sum_{x=0}^{M}\sum_{y=0}^{N} P[x, y] \qquad (2)$$

(iii) Variance

Variance [6–8] explains the distribution of gray levels over the image. The value of the Variance is expected to be high, if the gray levels of the image are spread out extensively. The formula for the variance used is shown in Eq. (3).

$$f(x, y) = \frac{1}{mn - 1} \sum_{(r,c)eW} [g(r, c) - \frac{1}{mn - 1} \sum_{(r,c)eW} g(r, c)]^2 \qquad (3)$$

(iv) Standard Deviation

Standard Deviation indicates a lot of variations that appears from the average value of the image which has the potential for measuring the variability in the image. The value of Standard Deviation is assigned to the center pixel of the image, which is calculated from Eq. (4). It is the square root of the variance [6–8].

$$f(x, y) = \sqrt{\frac{1}{mn - 1} \sum_{(r,c)eW} [g(r, c) - \frac{1}{mn - 1} \sum_{(r,c)eW} g(r, c)]^2} \qquad (4)$$

(v) Mean Square Error (MSE)

MSE represents the averaging of the squares of the errors between the two images [9]. The error is the amount by which the values of the reference image differs from the test image. It is actually the image quality measuring parameter. The mathematical expression for MSE is given in Eq. (5).

$$MSE = \sum_{0}^{m-1}\sum_{0}^{n-1} \|f(i, j) - g(i, j)\|^2 \qquad (5)$$

2.2   Correlation

Correlation is also a statistical technique, which shows how variables are robustly related with each other. It extracts the necessary information from an image. It is used to find the location in an image that is analogous to the reference image. Reference image is slid around the image to find the location, where the template overlaps the reference image to get aligned with similar values in the image. Correlation is a measure of gray level linear dependence between the pixels at the specified positions relative to each other [10].

$$correlation = \sum_{i=0}^{G-1} \sum_{i=0}^{G-1} \frac{\{i \times j\} \times P(i,j) - \{\mu_x \times \mu_y\}}{\sigma_x \times \sigma_y} \tag{6}$$

From Eq. (6), a correlation is calculated between the parameter values obtained by average information method and reference parameters of the noncancerous lung images, which is then given to image classifier for lung cancer diagnosis. An intelligent correlation analysis can help for better understanding of the image data as it finds the similarity between the two images.

2.3   Image Classifier

Neural Network [11, 12] is the method used as an image classifier in this paper for lung cancer diagnosis. The various values obtained for different statistical parameters under Average Information Method for cancerous and noncancerous microscopic lung images overlaps, which make it difficult to take a decision whether image is infected or not, hence Neural Network as an image classifier is used as a decision maker for the lung cancer diagnosis. Basically input–output pairs which in this case are the parameter values obtained through the algorithm and the desired output is the training data provided to ANN to build a network for generalization in order to diagnose new unseen cases of cancer, which is not present in the training data. Few parameter values for cancerous and noncancerous lung images goes beyond the specific calculated range, hence ANN is used to resolve this issue.

2.4   Standard Statistical Method

There are various standard statistical methods used in the image processing [13]. Analysis of variance (ANOVA) [14, 15] is used in this paper, which is a collection of statistical models used to analyze the differences among group means and their associated procedures (such as "variation" among and between groups), developed by statistician and evolutionary biologist, Ronald Fisher. In the ANOVA setting, the observed variance in a particular variable is partitioned into components attributable to different sources of variation. In its simplest form, ANOVA provides a statistical test of whether or not the means of several groups are equal and therefore generalizes the *t*-test to more than two groups. ANOVA is useful for comparing (testing) three or more means (groups or variables) for the statistical significance.

3   Proposed Method

The proposed method is implemented using image processing algorithm in MATLAB. The flow of the algorithm is as follows

1. The input or test image is a microscopic lung image.
2. The reference image is a healthy lung image.
3. The image database is pre-verified by the radiological experts as cancerous and noncancerous lung image.
4. Image is preprocessed by resizing it to 255 * 255, converted into grayscale image and then enhanced using a median filter.
5. The enhanced image is passed through the identified statistical parameters under average information method.
6. The statistical parameters of both test image and reference image are correlated using the correlation method.
7. The statistical parameters of both test image and reference image are correlated using the correlation method.
8. Similarities of both images are identified, but still there will be some values which falls beyond the calculated statistical parameter range.
9. Neural Network is used as a decision maker which classifies the test image into cancerous or noncancerous lung image.
10. ANN train and test images for lung cancer diagnosis.
11. Increased image database and input parameters has lead to an improved result.
12. Hypothesis test is also carried out by implementing the analysis of variance (ANOVA), which is one of the standard statistical methods.

4   Results and Discussion

The analysis of lung cancer is discussed in this section includes the actual results of the average information method and verified through a standard statistical method.

4.1 Results of Average Information Method

Statistics involves a discrete set of data that is characterized by Entropy, Mean, Variance, Standard Deviation, and MSE. The average information method is applied over predetermined cancer-infected microscopic lung images. With these calculations, specific range of each average information parameter has been identified for the analysis and diagnosis of lung cancer. The identified range of each parameter for cancerous lung is calculated as shown in Table 1.

The average range as observed in Table 1 is the statistical parameter value calculated for the current image. The image database is increased to 323 microscopic lung images including both cancerous as well as noncancerous. These images are already been pre-verified from radiological experts. Also, the parameters under average information are more in this paper for the analysis as compared to [6] in order to increase the accuracy of the method. The algorithm is tested when image database is increased to 323 microscopic lung images including both cancerous as well as noncancerous through which statistical parameter range is obtained that can be observed in Table 1.

In this paper, some more input parameters are added like Mean, MSE, and Variance as only few parameters are not enough to reach to any decision. Now the specific range is calculated for all the parameters when applied over the cancerous lung images. When new image is tested with these parameters and if the values of these parameters lie under the above-mentioned range as seen in Table 1, the decision regarding cancerous or noncancerous is taken. But suppose if some of the values overlap for the image, then the final decision is taken by ANN based on how many parameters lie in the range for the cancerous lung image.

It is also clear that the input parameter range increases as the number of iterations on the image increases with increase in image database. This increased range helps to improve the performance of the algorithm. Now the identified range is used for automatic run-time analysis and diagnosis of lung cancer, that is, without any manual interference.

Tables 2 and 3 show the calculations of parameter values for some of cancerous and noncancerous microscopic lung images. When these calculations are carried out

Table 1 Identified Range of average information parameters for cancerous lung image

Average information parameters	Minimum to maximum value for cancerous lung	Average range	Range for cancerous lung from graph	Range for noncancerous lung from graph
Mean	102–123	112.50	130.96	226.37
Standard deviation	50–59	54.50	58.60	49.64
Mean square error	55–140	97.50	155.085	232.947
Variance	$2.46 * 10^3$–$3.64 * 10^3$	$3.05 * 10^3$	$3.306 * 10^3$	$2.383 * 10^3$
Entropy	7.02–7.62	7.32	7.76	5.68

Table 2 Parameter values for noncancerous lung images

Average information parameters	Noncancer image 1	Noncancer image 2	Noncancer image 3	Noncancer image 4	Noncancer image 5
Mean	187.36	222.52	217.89	158.44	186.78
Standard deviation	40.08	30.05	36.75	40.03	45.91
Mean square error	235.92	244.05	176.21	228.7	199.01
Variance	$1.567 * 10^3$	866.20	$1.121 * 10^3$	$1.540 * 10^3$	$2.072 * 10^3$
Entropy	6.12	6.39	7.00	6.24	7.28

Table 3 Parameter values for cancerous lung images

Average information parameters	Cancer image 1	Cancer image 2	Cancer image 3	Cancer image 4	Cancer image 5
Mean	186.783	132.90	159.83	169.28	140.52
Standard deviation	32.68	48.554	52.813	51.21	43.69
Mean square error	220.78	118.89	162.88	179.72	136.92
Variance	$1.024 * 10^3$	$2.308 * 10^3$	$2.426 * 10^3$	$2.465 * 10^3$	$1.841 * 10^3$
Entropy	5.96	7.51	7.56	7.48	7.47

by applying the algorithm over the large image database, a range for all the parameters under average information method is obtained. This range helps to differentiate an image as cancerous or noncancerous lung image. Although it is not easy as it looks because some of the parameter values overlap and seems to be similar for both cancerous and noncancerous lung images, hence this confusion is eliminated by Artificial Neural Network which trains and tests the images for number of iterations.

The next thing is to calculate the accuracy of the algorithm for which 323 images are tested and the accuracy of the method is calculated as 68.42%. The accuracy of

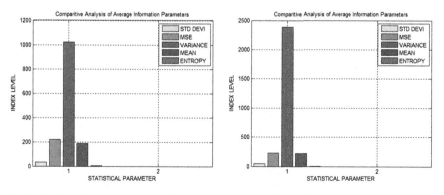

Fig. 2 Comparative analysis graph for **a** noncancerous and **b** cancerous lung

the algorithm is calculated based on how many images are correctly diagnosed as cancerous and noncancerous. As the image database is already been verified, it is compared with pre-diagnosis results. The average information method correctly diagnosis 221 images including cancerous and noncancerous out of 323 microscopic lung images.

Figure 2 shows the graph of statistical parameters versus index level. Run-time graph generation shows the impact of average information parameters on lung cancer diagnosis. The graphs as shown in Fig. 2a, b is a plot for the index level versus statistical parameters, which gives an idea about the variations in statistical parameter index level according to cancerous and noncancerous microscopic lung images. One can easily now differentiate the microscopic image as cancerous and noncancerous by observing the current graph. With subjective analysis, the graphs are having its own impact on lung cancer diagnosis.

4.2 Results of ANOVA

In the proposed system, 5 groups are considered according to the used parameters. ANOVA is applied over the proposed system and the important calculations found are as follows:

Total sum of squares (TSS) = 66050543

Sum of squares between the groups (SSB) = 5.68E + 07

Sum of squares within the groups (SSW) = 9.25E + 06

F ratio = SSB/SSW

$F_{(4, 45)}$ = 69.1, $p < 0.05$ (p = significance factor)

Critical value = **2.61** (approximately according to F-Distribution table for $F_{(4, 45)}$)

$F_{(4, 45)}$ is relative frequency

F test value > Critical value, which can be observed from Fig. 3, i.e., 69.1 > 2.61, hence the proposed method successfully rejects null hypothesis.

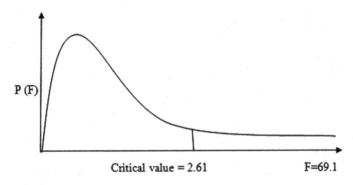

Fig. 3 Critical value calculation on F-distribution

5 Conclusions

Among numerous average information parameters, selective parameters are identified for lung cancer analysis and diagnosis. For the selection of statistical parameters, iteration method over predetermined lung cancer microscopic images is used. These statistical parameters under average information method have the ability to work effectively for lung cancer diagnosis. These parameters are tested and verified on microscopic lung images including cancerous and noncancerous lung images using image processing techniques with MATLAB. Out of 5 statistical parameters, Variance has shown good statistical response for cancerous lung images. Artificial Neural Network as an image classifier plays an important role in decision making, which decides whether the current image is cancerous or noncancerous and can be also observed through ANN performance graph. The result shows that accuracy improves with number of trained images, which shows that ANN works well as an image classifier for the proposed method. Also, the method is tested by one of the standard statistical method ANOVA, which successfully rejects null hypothesis. The accuracy of the method comes out to be 68.42% which is calculated on the basis of how many images are diagnosed correctly. This means that the proposed method is working satisfactorily, but still requires some more methods in addition, which could be a hybrid combination of mathematical, statistical and structural method or could be some new statistical or mathematical parameters that could fill up the gap that appears in this method inorder to improve the performance and accuracy of the algorithm. But surely this method is one of the major revolutionary steps toward the medical research field for lung cancer.

References

1. Joes Vilar, "Breathe Easy", How Radiologic Helps To Find and Fight Lung Diseases, European Society of Radiologic. Chapter 1.
2. J. B. Walter & D. M. Pryce, "The histology of lung cancer" PMC, US National Library of Medicine, National Institute of Health, pp. 107–116.
3. Kale Vaishnaw G., "Imaging Techniques for Lungs Analysis", International Journal of Scientific & Engineering Research (IJSER), Vol. 5, Issue 4, April 2014, pp. 1–4.
4. Muhammad Qurhanul Rizqie, Nurul Shafiqa Mohd Yusof, Rino Ferdian Surakusumah, Dyah Ekashanti Octorina Dewi, Eko Supriyanto and Khin Wee Lai, "Review on Image Guided Lung Biopsy", IJN-UTM Cardiovascular Engineering Center, Springer Science and Business Media Singapore 2015, pp. 41–50.
5. Vaishnaw Gorakhnath Kale, "An Overview of Microscopic Imaging Technique for Lung Cancer & Classification" International Journal of Innovation in Engineering, Research and Technology [IJIERT], ICITDCEME'15 Conference Proceedings, ISSNNo-2394-3696, pp. 1–4.
6. Kale Vaishnaw G., Vandana B. Malode, "New Approach of Statistical Analysis for Lung Disease Diagnosis using Microscopy images" IEEE-2016, pp. 378–383.

7. K. Punithavathy, M.M. Ramya, Sumathi Poobal, "Analysis of Statistical Texture Features for Automatic Lung Cancer Detection in PET/CT Images", International Conference on Robotics, Automation, Control and Embedded Systems–RACE2015.
8. Narain Ponraj, Lilly Saviour, Merlin Mercy, "Segmentation of thyroid nodules using watershed segmentation", Electronics and Communication Systems (ICECS), 2nd International Conference on, IEEE-2015.
9. Kale Vaishnaw G, "Lung Cancer Analysis by Quality Measures" International Journal of Modern Trends in Engineering and Research, Vol. 3, Issue 4, April 2016, Special Issue of ICRTET'2016, pp. 738–741.
10. David Jacobs, "Correlation and Convolution" Tutorial for CMSC 426, pp. 1–10.
11. Monica Bianchini and Franco Scarselli, "On the Complexity of Neural Network Classifiers: A Comparison between Shallow and Deep Architectures", IEEE Transactions on Neural Networks and Learning Systems, Vol. 25, No. 8, August 2014.
12. K. Balachandran, R. Anitha, "An Efficient Optimization Based Lung Cancer Pre-Diagnosis System with Aid of Feed Forward Back Propagation Neural Network (FFBNN)". Journal of Theoretical and Applied Information Technology 20 Oct 2013 Vol. 56 No. 2.
13. Jay L. Devore, Kenneth N. Berk, Modern Mathematical Statistics with Applications, © Springer Science+Business Media, LLC 2012.
14. K.elkourd, "Detect the Tumor with Numerical Analysis and With "ANOVA" Technique for MRI Image", International Journal of Engineering Issue 1, July 2013. ISSN: 2277-3754 ISO 9001:2008.
15. El. kourd Kaouther, Seif eddine Khelil, Saleh Hammoum, "Study With RK4 & ANOVA The Location Of The Tumor At The Smallest Time for Multi-Images" IEEE-2015.

A Hash-Based Approach for Document Retrieval by Utilizing Term Features

Rajeev Kumar Gupta, Durga Patel and Ankit Bramhe

Abstract Digital data increase on servers with time which resulted in different researchers focusing on this field. Various issues are arising on the server such as data handling, security, maintenance, etc. In this paper, an approach for the document retrieval is proposed which efficiently fetches the document according to the query which is given by user. Here hash-based indexing of the dataset document was done by utilizing term features. In order to provide privacy for the terms, each of this is identified by a unique number and each document has its hash index key for identification. Experiment was done on real and artificial dataset. Results show that NDCG, precision, and recall parameter of the work are better as compared to previous work on different size of datasets.

Keywords Information retrieval · Text feature · Text mining
Text ontology · Digital data · Feature selection

1 Introduction

With increasing the size of digital text data on the servers, text mining importance is increasing as this decreases lot of labor work for different use of text data. In this text mining research field, classification of information and retrieval of documentation is highly required. So combination of various data mining techniques is done

R. K. Gupta (✉) · A. Bramhe
Department of CSE, Sagar Institute of Science and Technology, Bhopal 462036,
Madhya Pradesh, India
e-mail: rajeevmanit12276@gmail.com

A. Bramhe
e-mail: ankitbramhe1112@gmail.com

D. Patel
Department of CSE, Rajiv Gandhi Proudyogiki Vishwavidyalaya, Bhopal 462036,
Madhya Pradesh, India
e-mail: durgapatel28@gmail.com

© Springer Nature Singapore Pte Ltd. 2019
H. S. Behera et al. (eds.), *Computational Intelligence in Data Mining*,
Advances in Intelligent Systems and Computing 711,
https://doi.org/10.1007/978-981-10-8055-5_55

while gathering information from the text document [1, 2]. Various researchers are working for improving accuracy of the work, but there is lot of improvement in the work for further increasing the parameters.

As text data are highly unorganized because it contains natural language, mining for retrieval of information from text data is crucial for researchers. Different pre-and post-processing steps are taken for improving the information quality [3]. While in case of text document information retrieval, it is found that most of the document data are open for all. Due to this, privacy of the text dataset is very low. So this work has focus on two issues: First is text information retrieval and second is privacy maintenance of the dataset. Ways to mine the text and cluster the documents for better processing are our concern.

Even any small activity of human produces electronic data. For example, when any person buys a ticket online, his details are stored in the database. As most of electronic or digital data available on servers are in the text form, these data are highly un-clustered or structureless but also suffers from the large amount of waste information. In this data, good quality of information is also available for the scientific and industry purpose. As most of the historic data are available in text which needs to be update but this required skilled labor or reader how have knowledge of the different terms for conversion. So considering all these facts, in 1960, Pittsburg University developed an system which performs these tasks efficiently. In mid-1960, University has developed a computer-enabled research assistant for performing the text reading [4]. In this computer programs, Boolean logics which were set with nearness expression in form of phrase were used.

These programs utilized full text query for retrieving document from the dataset. Here retrieval was done on the basis of content of the document not on the few set of keywords so this is termed as Full Text Retrieval. So if the user wants to fetch a document then this can be retrieved from the database based on the passed query. Here query is understood in form of terms and phrases. So this tends to find the FTR drawback where it does not understand the natural language of the user. But as FTR required less time for searching, so this fact is overcome in terms of time efficiency. So text mining algorithms are applied to develop a document retrieval system which takes less execution time with high relevant data and protection against intruder for query as well as database.

2   Related Work

A. Humad et al. [1], used text document clustering to group a set of documents based on the information it contains and to provide retrieval results when a user browses the Internet. In this work, results show that proposed work has retrieved the text document efficiently by prior classification of the text files in the document. Here work has focused on reducing the dimension of the dataset. So dimension reduction is done by two approaches: First is reducing noise or text which does not provide any information while second is removing of unwanted features from the document dataset.

Wild card approach used [5] has implemented the fuzzy approach where keyword set is used for matching the relevant text documents. Here comparison of these sets is done by the use of Edit Distance formula. So, collection of text features in form of these keyword set is done in this work. This reduces the calculation overhead for the work, while storage of the work was also handled. Here privacy of the data is needed to be maintained first for reducing the intruder activities.

N. Cao et al. [6] has introduced dual indexing of the text document retrieval. Here first index consists of document index where text files are ranked based on their similarity while in second index, words are indexed where list of similar group words are listed. Campus Net Search Engine (CNSE) is based on full text hunt engine, but it is not a complete text for indexing the document.

N. Cao et al. [7] searching the document by using multi-keyword technique. Here frequent words are arranged in tree data structure as per their IDF value. For each term and document, there is separate path, but number of path increases as number of documents is increasing. So for finding the new document, iterative steps are required for other related documents. Recursion requires time, and then comparison of word at each step also takes time, which increases as the document size in the dataset increases.

3 Proposed Methodology

The size of the data is increased day by day so data mining approaches are used to handle this huge amount of data. This work proposed a method for the document retrieval in the group without having any prior knowledge of the documents.

3.1 Preprocessing

Preprocessing is a method where document is converted into the feature vector [8]. This approach preprocesses the data to reduce the size of text document. In order to preprocess the data, stop word elimination strategy is used.

Data preprocessing reduces the size of the input text documents significantly. Stop words are functional words which occur regularly in the text (for example, a, the, an, of, etc., in English language), so classification of these words are useless [8–10]. Let D is document [India is a great country. It's a country of different religion and caste.], Stop word S is [a, are, an, and, am, for, is, its, when, where, etc.]. Then in preprocessing, subtraction operation is done on these sets. Here D-S = [India, great, country, country, different, religion, caste].

Pre_processing(D, S)

1. Loop 1:x // x number of words in D
2. Loop 1:y // y number of words in S
3. If Not(D[x]—S[y])
4. PD←D[xs]
5. Endif
6. EndLoop
7. EndLoop

3.2   Feature Term

The vector which holds the preprocessed data is used to find the characteristic of that document. To accomplish this, vector is compared with vector KEY (collection of keywords). The resultant vector will be considered as the feature vector for that document [11, 12]. To find the keyword for the vector Key, threshold is used. If the count of the word crosses the threshold limit, then it will be considered as the keyword or feature of that document. Equation 1 is used to find the features in the given document.

$$[\text{feature}] = \text{mini\_threshold}([\text{processed\_text}]) \tag{1}$$

3.3   Positive and Negative Feature Set

The positive and negative feature set is used to assign term ID. To identify each word uniquely, an ID is assigned to each term which is used in the dictionary of words. Here words coming from different document which are already present in the dictionary are not updated. So those terms which are not present in the dictionary are inserted in the dictionary with unique term. Figure 1 shows the block diagram of the learning model.

3.4   Document Hash Indexing

In this step, document index is decided based on the terms collected from the document. Here all the terms are arranged in decreasing order as per the frequency value of the terms in the document. So new order of the document term is 918465; this is based on the decreasing order of the term frequency. One hash number is generated for selected document in similar fashion other document in the dataset get collected. Now as per the index value, document is identified (Fig. 2).

Fig. 1 Block diagram of
proposed learning model

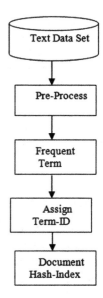

Fig. 2 Block diagram of
proposed searching model

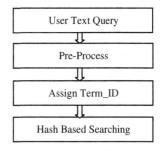

In this searching model, whole step of preprocessing and assigning term ID is same as done in previous steps, although term obtained after preprocessing is not filtered as per their frequency in query. So hash-based searching and retrieving related document is new for searching model. For convenience, words obtained from user text query are called as keywords.

3.5 Hash-Based Searching

In this step the keywords (terms) extract from the user text query have their own term ID whereas numbers are same as present in the dataset. Due to this term ID, privacy of the user query increases. Now all term ids that are present in the text query act as key for the hash function where each document set from the matched index is retrieved. Now apply intersection between those sets. So top most common elements are fetched which act as similar documents.

Let term ID be X which act as key and modulus function M is used for the hash-based indexing. So output of the function Y which is the index position of the insert key is given by the Eqs. 2 and 3.

$$M(X) \rightarrow Y \tag{2}$$

//Modulus function

$$M(X) = ||X, C|| \tag{3}$$

where C is fixed constant used for finding the index position of the key.

Proposed Algorithm
Input: DD //Document Dataset, Dict //Stopword-Dictionary
Output: Hash_Table

1. Loop 1:n // n number of document in dataset
2. PD←Pre_processing(DD[n], Dict)
   // Frequent Term
3. Loop 1:m // m number of keywords
4. C←Count(PD[m])
5. If C > T // T threshold
6. FT[n]←C
7. Endif
8. EndLoop
9. EndLoop
   // Assign Term ID
10. Final_Term←Unique(FT)
11. Loop 1:u // u number of unique terms
12. TID←Assign(Final_term, FT) // TID Term ID
13. EndLoop
    // Document Hash Table
14. Loop 1:n
15. SFT←Sort(FT[n]) // SFT Sorted frequent term
16. Hash_Table[SFT]←DD[n]
17. EndLoop

4 Experiment and Results

In order to implement above algorithm for document retrieval, MATLAB 2012a tool was used. Here same work can be implemented on other programming language as well. But as some of the function was inbuilt in the tool which help researcher to focus on the work. Experiment was done on real as well as on artificial dataset. Here different set of dataset was used for retrieving documents.

4.1 Evaluation Parameter

In order to evaluate the performance of the proposed approach, it is compared with the existing approach named "Verifiable Privacy-Preserving Multi-Keyword Text Search In The Cloud Supporting Similarity-Based Ranking" [13]. These approaches are compared in terms of precision, recall, and F_Score. Eqs. 4–6 are used to evaluate the performance of the proposed approach.

$$Precision = \frac{True\_Positive}{True\_Positive + False\_Positive} \tag{4}$$

$$Recall = \frac{True\_Positive}{True\_Positive + False\_Negative} \tag{5}$$

$$F\_Score = \frac{2 * Precision * Recall}{Precision + Recall} \tag{6}$$

Here let the query passed by the user is 'INDIA TAJ' then according to the pass query, some images will be pop up, now let for top five images shows in Fig. 3

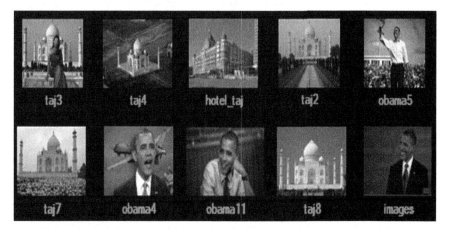

Fig. 3 Represent top five images for the query 'INDIA TAJ'

if Normalized Discounted Cumulated Gain is for this result. Then first it need that from the top five images how many images are relevant then other are consider as the irrelevant images.

Consider a vector L as the list of image represents the relevance by 1 and irrelevance by 0 so if the first image is relevant then first element in the vector is 1, if the second image is relevant then second element in the vector is 1, if the third image is irrelevant then third element in the vector is 0.

For above query let L = [1 1 0 1 0 1 0 0 1 0], then put this value in the Normalized Discounted Cumulated Gain formula where P = 5. Z_p is the total sum if all the values in the L vector are 1, i.e., [1 1 1 1 1 1 1 1 1 1], and i represents the position in the result such that i = {1, 2, 3, 4, 5}.

The NDCG measure is computed as given in Eq. 7.

$$NDCG \ @ \ P = Z_P \sum_{i=1}^{P} \frac{2^{l(i)} - 1}{\log(i+1)} \tag{7}$$

where P is the considered depth, $l(i)$ is the relevance level of the ith image, and ZP is a normalization constant that is chosen to let the optimal ranking's NDCG score to be 1.

4.2 Results

In order to evaluate the performance of the proposed approach, it is compared with the existing approach in terms of accuracy, precision value, recall value, and NDCG Values.

From Table 1, it is obtained that proposed work accuracy value is higher than previous work on different queries. As query set has good quality keywords, results of proposed work are also high. Figure 4 shows the graphical representation of Table 1.

From Table 2, it is obtained that proposed work precision value is higher than previous work on different queries. As query set has good quality keywords, results of proposed work are also high. Figure 5 shows the graphical representation of Table 2.

From Table 3, it is obtained that proposed work recall value is higher than previous work on different queries. As query set has good quality keywords, results of proposed work are also high. Figure 6 shows the graphical representation of Table 3.

Table 1 Comparison of accuracy value with previous work [13]

Comparison of Accuracy		
Query	Proposed work	Previous work [13]
Q1	1	0.4
Q2	1	0.6
Q3	0.8	0.4
Q4	0.8	0.4

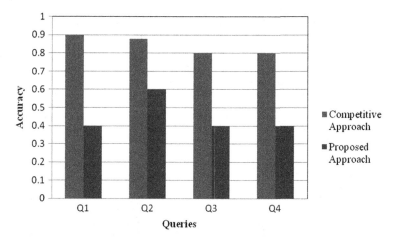

Fig. 4 Average accuracy comparison between proposed and previous work [13]

Table 2 Comparison of precision value with previous work [13]

Comparison of precision values		
Query	Proposed work	Previous work [13]
Q1	0.7857	0.7143
Q2	0.7143	0.7143
Q3	0.7143	0.7143
Q4	0.7857	0.7857

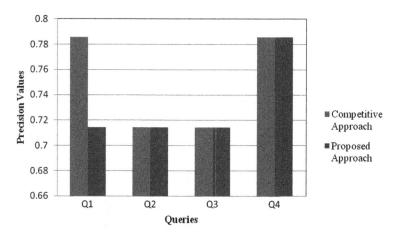

Fig. 5 Average precision value comparison between proposed and previous work

From Table 4, it is obtained that proposed work NDCG value is higher than previous work on different queries. As query set has good quality keywords, results of proposed work are also high. Figure 7 shows the graphical representation of Table 4.

Table 3 Comparison of recall value with previous work

Comparison of recall		
Query	Proposed work	Previous work [13]
Q1	0.8462	0.7143
Q2	0.7692	0.7143
Q3	0.7692	0.7692
Q4	0.8462	0.7857

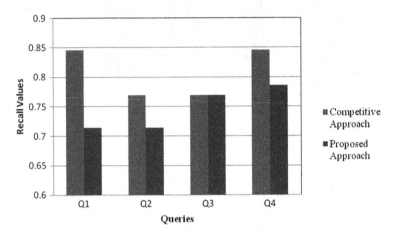

Fig. 6 Average recall value comparison between proposed and previous work

Table 4 Comparison of NDCG value with previous work

Comparison of NDCG values @5		
Query	Proposed work	Previous work [13]
Q1	1	0.3836
Q2	1	0.7227
Q3	0.7860	0.383
Q4	0.786	0.5531

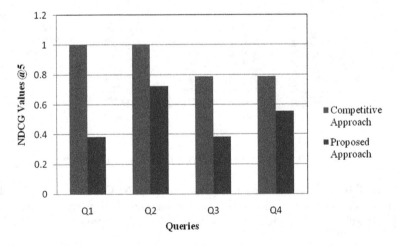

Fig. 7 Average NDCG values @5 comparison between proposed and previous work

5 Conclusions

In the past decade, size of digital data on the server increased exponentially, so to retrieve this stored data in a less time is a challenging task. To deal with this issue, lot of work has been done but they are mainly focused on the content classification where in our proposed work document are classified. Proposed work has increased the retrieval efficiency of the work in all different evaluation parameters. So use of hash-based indexing provides privacy with efficiency for document retrieval. As there is always some scope for the improvement because research is a never-ending process, same work can also we implemented in other environment.

References

1. Aparna Humad and Vikas Solanki, "A New Context Based Indexing in Search Engines Using Binary Search Tree", International Journal of Latest Trends in Engineering and Technology (IJLTET), Vol. 4, Issue 1 May 2014, pp. 26–30.
2. Souneil Park et al., "Disputant Relation-Based Classification for Contrasting Opposing Views of Contentious News Issues", IEEE Transactions on Knowledge and Data Engineering, Vol. 25, No. 12, December 2013, pp. 2740–2751.
3. K. Fragos, P. Belsis, and C. Skourlas, "Combining Probabilistic Classifiers for Text Classification", 3rd International Conference on Integrated Information Vol. 147 pp. 307–312, 2014.
4. Jian Ma, et al. "An Ontology-Based Text-Mining Method to Cluster Proposals for Research Project Selection". IEEE Transactions on Systems, Man, and Cybernetics—Part A: Systems and Humans, Vol. 42, No. 3, May 2012.
5. Jian Ma et al. "An Ontology-Based Text-Mining Method to Cluster Proposals for Research Project Selection". IEEE Transactions on Systems, Man, and Cybernetics—Part A: Systems and Humans, Vol. 42, No. 3, May 2012.
6. N. Cao et al., "Privacy-Preserving Multikeyword Ranked Search Over Encrypted Cloud Data," Proc. IEEE InfoCom, pp. 829–837, Apr 2014.
7. N. Cao, S. Yu, Z. Yang, W. Lou, and Y. Hou, "Lt Codes-Based Secure and Reliable Cloud Storage Service," Proc. IEEE InfoCom, pp. 693–701, 2012.
8. Dinesh Nepolean et al. "Privacy Preserving Ranked Keyword Search Over Encrypted Cloud Data", Vol. 4, No. 11, Nov 2013.
9. Fabrizio Silvestri, Raffaele Perego and Salvatore Orlando. Assigning Document Identifiers to Enhance Compressibility of Web Search Engines Indexes. In The Proceedings of SAC, 2004.
10. S. Ramasundaram, "Ngramssa Algorithm for Text Categorization", International Journal of Information Technology & Computer Science, Vol. 13, Issue No: 1, pp. 36–44, 2014.
11. Ning Cao et al., "Privacy-Preserving Multi-Keyword Ranked Search Over Encrypted Cloud Data", IEEE Transaction Parallel and Distributed Systems, Vol. 25, No. 1, Jan 2014.
12. Yuefeng Li et al., "Relevance Feature Discovery for Text Mining". IEEE Transactions on Knowledge.
13. Wenhai Sun et al., "Verifiable Privacy-Preserving Multi-Keyword Text Search in the Cloud Supporting Similarity-Based Ranking". IEEE Transactions on Parallel and Distributed Systems, Vol. 25, No. 11, November 2014.

Transform Domain Mammogram Classification Using Optimum Multiresolution Wavelet Decomposition and Optimized Association Rule Mining

Poonam Sonar and Udhav Bhosle

Abstract Author propose mammogram classification technique to classify the breast tissues as benign or malignant. Mammogram is segmented to obtain Region of Interest (ROI) and 2D DWT is obtained. GLCM feature matrix is generated for all the detailed coefficient of 2D DWT. Optimum Feature Decomposition Algorithm (OFDA) is used to discretize and optimize the features. Author propose Optimum Decomposition Selection Algorithm (ODSA) to select optimum decomposition from nine multiresolution wavelet decompositions of ROI using Euclidean distance between the feature matrixes. High-dimensional future space may degrade the performance of the classifier. Using propose algorithm, the size of feature matrix reduces to $[N \times F]^T$ from $[(N \times 9) \times F]^T$. Hence, dimension of search space reduces by approximately 90%. From the optimized feature vector and optimized decomposition, a signature feature vector matrix consisting of optimum decomposition and its optimum feature vector is generated to form transactional database. Association rules are generated using Apriori algorithm. These rules are optimized using multiobjective Genetic Algorithm with adaptive crossover and mutation. Mammogram is classified using Class Identification using Strength of Classification (CISCA) algorithm. The results are tested on two standard databases: MIAS and DDSM. It is noted that the propose scheme has advantage in terms of accuracy and computational complexity of the classifier.

Keywords Discrete wavelet transforms · GLCM feature extraction
Mammogram classification · Association rules · Genetic algorithm
MOGAACM and CISCA

P. Sonar (✉) · U. Bhosle
Department of Electronics and Telecommunication Engineering,
MCT's Rajiv Gandhi Institute of Technology, Mumbai 400058, India
e-mail: poonam.sonar@mctrgit.ac.in

U. Bhosle
e-mail: udhav.bhosle@mctrgit.ac.in

P. Sonar · U. Bhosle
University of Mumbai, Mumbai, India

© Springer Nature Singapore Pte Ltd. 2019   629
H. S. Behera et al. (eds.), *Computational Intelligence in Data Mining*,
Advances in Intelligent Systems and Computing 711,
https://doi.org/10.1007/978-981-10-8055-5_56

1 Introduction

Breast cancer is the second leading cause of cancer death in women. The American Cancer Society estimates that in 2017, around 252,710 new cases of invasive breast cancer will be diagnosed in USA [1]. Mammography is a breast screening method of detecting breast cancer at a very early stage to reduce the mortality rate. The mammogram technique is considered as the texture variations in breast region and displays changes most reliable for detecting breast cancer. Masses and microcalcifications are the two types of breast cancer indicators that can be seen on a mammogram [2]. Reading mammogram is very important task for radiologist. Thus, automatic classification using computer-aided diagnosis (CAD) is necessary. CAD helps a radiologist to interpret masses and microcalcifications and helps to mark 77% of the cancer that cannot be detected by radiologists [3].

Extensive work for mammogram feature extraction has been done by different researchers. Significant and most discriminative feature extraction improves the performance and precision of the classifier. Hence, feature selection and optimization is the key point in enhancing CAD efficiency [4]. Discrimination, reliability, independence, and optimality are the key points for selecting the features [5]. The features of digital image can be extracted in spatial domain or transform domain using different transforms such as DWT. Sheng Liu [6] presented novel multiresolution scheme for detection of speculated lesions of digital mammogram using 2D DWT multiresolution representation of mammograms and detection is performed from the coarse resolution to finest resolution using binary tree classifier. R. Mousa [7] proposed CAD for classification of abnormality in digital mammograms using new fuzzy classifier from wavelet coefficients. CAD system based on multiscale representation wavelet transform detection using wavelet coefficient thresholding for microcalcifications in digital mammogram is reported in [8]. It used Renyi's entropy for thresholding the wavelet coefficients. Ferria Borges [9] proposed to keep and use only biggest coefficients in the low frequency of the first level decomposition of wavelet to form signature vector for the corresponding mammogram. Rashed [10] used a multiresolution mammogram analysis to extract fractional amount of biggest wavelet coefficients using Daubechies 4, 8, 16 wavelets with fourth-level decomposition. Authors in [11] proposed mammogram classification technique for classifying breast tissue as normal, benign, and malignant classes using GLCM feature vector of 2D DWT of its ROI and selected the relevant feature satisfying t-test and f-test separately using BPNN classifier.

The literature survey highlights mammogram classification schemes based on wavelet decomposition. However, most of papers are not able to provide good accuracy as large dimension of feature space is key problem in mammogram images due to large variation in normal and abnormal tissue. High-dimensional future space may degrade the performance of the classifier. Significant and uncorrelated feature selection is important step in mammogram classification. In this paper, we proposed supervised mammogram classifier using discrete wavelet transform. The novelty of this method lies in the selection of most optimum multiresolution decomposition and associated optimum features for reduction in the search space. The optimized association rules are used for classification. The proposed technique reduces the complexity of the classifier and improves performance.

2 Proposed System

The proposed system is three stage optimization techniques. The first stage consists of feature optimization followed by selection of optimum decomposition from nine decompositions to reduce the feature search space. Finally, optimization of association rules is used for classification.

- Mammogram segmentation to obtain ROI, 2D wavelet decomposition up to third level and texture feature extraction using GLCM.
- Feature discretization and optimization for selection of discriminative features [12].
- Proposed Optimum multiresolution wavelet decomposition selection using OMDESA to select optimum decomposition.
- Use of optimized association rule mining using multiobjective Genetic Algorithm with adaptive mutation and crossover and classification of mammogram into benign and malignant class using Class Identification using Strength of Classification Algorithm (CISCA) [12].

2.1 Preprocessing and ROI Extraction

The preprocessing phase of digital mammograms generally involves suppression of noise, enhancement of mammograms intensity and contrast manipulation, background removal, edges sharpening, filtering, etc. It is necessary to improve the quality of mammogram images and make the feature extraction phase more reliable. Digitization noise introduced during image acquisition process is removed using median filtering. Image enhancement is achieved by performing contrast enhancement and thresholding on preprocessed mammogram.

Mammogram images consist of background noise and various artifacts such as wedges, labels, and unwanted pectoral muscles in the object area. Due to this, full mammogram images are not required for texture feature extraction and classification. Hence, the Region of Interest (ROI) is extracted from mammogram images. Removing such background information from the mammogram leads to considerable reduction in data size and further processing time. The extracted gray levels of the mammograms consisting of Region of Interest (ROI) are used for further analysis [13].

2.2   Multiresolution 2D DWT and Feature Extraction with Optimization

Multiresolution deals with hierarchical representation of information in the efficient manner. It is efficient framework for extracting, analyzing information at various levels of resolution, and for feature detection and extraction. Mammographic images can be represented in different resolutions by using 2D DWT. It preserves high- and low-frequency information. A 2D wavelet transform can be obtained by a combination of digital filter banks and down samplers. The concept of wavelet is elaborated in more detail by Daubechies [14]; he stated that wavelet functions are used as basis for representing other functions. The basis function is called as mother wavelet. A family can be generated by translation and dilations of the mother function [14]. Multiresolution scheme for wavelet decomposition developed by Mallat [15] is based on applying 2D DWT to the image. It produces four sets of different coefficients such as low-frequency coefficients (A), vertical high-frequency coefficients (V), horizontal high-frequency coefficients (H), and high-frequency coefficients in both directions (D) at each level of decomposition. Figure 1 shows 2D wavelet decomposition in three levels.

2D DWT is applied on all mammographic ROI to generate detail coefficient decomposition matrix for three resolution levels. GLCM and normalized GLCM matrices are calculated for the directions of 0^0, 45^0, 90^0, and 135^0. Every detailed decomposition subimage is expressed by its unique feature matrix of size [1 * 128], i.e., [32 features × 4 directions]. Then training feature vector is generated by concatenating all feature vector descriptor from all ROI as shown in Table 1. The extracted training feature vector along with class label is given as input to Optimum Feature Decomposition (OFD) algorithm [12]. It discretizes the continuous feature values into discrete feature intervals using cut points and generates optimized feature values with minimum inconsistency in its values. This optimized and discriminating feature vector is relevant and significant for forming transactional database for classification.

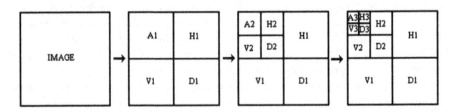

Fig. 1 Wavelet multiresolution decomposition

Table 1 Number of images for classification	Database	No. of training images		No. of testing images	
	MIAS	Benign	Malignant	Benign	Malignant
		35	35	13	11
	DDSM	78	78	40	40

3 Optimum Multiresolution Decomposition Selection Algorithm (OMDESA)

Multiresolution decomposition of mammogram results into number of decompositions. It is necessary to find how many and which decompositions have the capability for sufficient and satisfactory relevant information representation of original mammogram. Author proposed that, it is not necessary to keep and use all the decompositions for classification and proposed ODESA to keep only the significant decompositions based on Euclidean distance.

Proposed Algorithm

1. Training data set: Mammogram images are preprocessed, ROI extracted, and decomposed up to third decomposition level using 2D wavelet like Db3, Db4, and Bior3.7.
2. Consider the detailed coefficients (horizontal, vertical, and diagonal) of all the levels, thus you will get total nine decompositions per image.
3. Extract GLCM texture features of all the decompositions and normalize.
4. Discretize the features and select the significant feature using Optimum Feature Decomposition Algorithm (OFDA).
5. Now consider the first image, consider its first decomposition (suppose benign Fig. 1 LH2 of Db3). Calculate the Euclidean distance between the features of this decomposition with sets of all decompositions of all other images. Now take other decomposition (benign Fig. 1 HL of Db3) and carry out the same procedure for rest of the all images, so we will get a *decomposition distance matrix* of dimension $[9 * N \times ((9 * N) - 9)]$.
6. For every row of *decomposition distance matrix*, out of every nine consecutive values in a row, find out minimum Euclidean distance (this represents decomposition of every image), select that column, and neglect other values. This will reduce the dimension of decomposition distance matrix to $[9 * N \times ((9 * N) - 9)/9]$.
7. The most optimum decomposition selection which can significantly represent original mammogram image can be done as follows. Select the decomposition out of every 9 images which has occurred maximum times in a column. Repeat this process for all columns in decomposition distance matrix. Now we get optimized decomposition distance matrix, which will contain the final selected decompositions that will significantly represent the same information as in the original mammogram images.
8. Finally, optimized distance decomposition matrix and optimized feature vector are combined to form final *Signature Feature Vector*. This is used to form transactional database.

4 Optimized Association Rule Mining

Association rule mining is an important data mining technique for finding frequent patterns, associations, correlations among the set of items or features in huge database using user-specified minimum support and confidence thresholds [16]. Here, transactional database, which is generated from signature feature vector, is used to generate association rules using Apriori algorithm. It generates large number of association rules, since all the rules are not significant and relevant; rules are optimized using multiobjective Genetic Algorithm with adaptive crossover and mutation [12]. The multiobjective fitness function consists of comprehensibility (C (R)), predictive accuracy (P_{acc}), interestingness (Intt), and lift. All the rules are ranked as per their fitness function. For high rank rules, minimum crossover and mutation rate is assigned and for low rank rules maximum mutation and crossover rate is assigned. Antecedent part of association rule is used to separate the rules corresponding to benign and malignant class. From transaction database, Strength of Classification (SC) for benign and malignant class is calculated and compared to identify the class of the test transaction [12].

5 Experimental Results and Discussions

To validate the proposed mammogram classification technique, mammographic images from two standard databases, MIAS [17] and DDSM [18], are used. Table 1 shows the number of images taken for experimentation. Figure 2 shows few sample mammogram images from MIAS database. Figure 3 shows preprocessing and wavelet decomposition. Each mammographic ROI is decomposed into nine subimages using Db3, Db4, and biorthogonal3.7 wavelet.

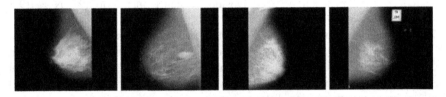

Fig. 2 Few sample mammogram images from MIAS database

(a)   (b)   (c)   (d)   (e)

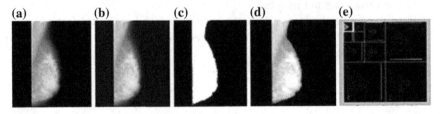

Fig. 3 Mammogram preprocessing: **a** original image **b** Median filtering **c** threshold image **d** Contrast enhancement and **e** wavelet decomposition view using third level

Table 2 Third level Boir3.7 wavelet decomposition features for MIAS database

| Image no. | Detail coefficient | Image feature vector | | | | | | |
		F_1	F_2	F_3	F_4	.. F_{25}	..	F_{128}
1	H1	0.2132	2.3453	1.8124	2.1234	18.2036		67.3456
	H2	0.0813	2.4561	1.8823	2.3243	17.2345		59.7867
	H3	0.0432	2.5666	1.9863	2.3556	17.1238		58.9876
	V1	0.1732	2.7890	1.7832	2.4892	16.1267		51.3482
	V2	0.0675	2.5672	1.4531	2.3987	16.9720		50.5631
	V3	0.0345	1.8975	1.3245	2.5892	16.5621		43.9812
	D1	0.2121	1.9762	1.0123	2.1567	16.7891		56.3215
	D2	0.1215	2.1230	1.0023	2.5678	15.2301		57.9321
	D3	1.1100	2.2339	0.9123	2.1345	15.1263		52.9021
2	H1	0.2421	2.5674	1.9765	2.5789	18.6732		55.6783
	H2	0.2152	2.5500	1.8876	2.4568	18.5432		55.6550
	H3	0.2021	2.4558	1.9178	2.2345	17.6789		56.9810
	V1	0.2366	2.4333	1.7654	2.4556	17.4321		54.5678
	V2	0.2298	2.0123	1.4300	2.2321	17.6782		52.3456
	V3	0.0876	1.0993	1.0842	2.0451	18.0987		47.9210
	D1	0.1462	2.0921	1.3452	2.0034	15.9876		43.0987
	D2	0.1342	0.9876	1.2398	1.4345	15.0876		44.9876
	D3	0.0432	0.0987	0.9342	1.0238	12.0097		41.0982
...								
70	H1	0.4237	2.6789	3.2453	2.8721	25.2987		87.7841
	H2	0.6783	2.9876	3.1208	2.6643	24.1222		67.9823
	H3	0.4321	2.5666	2.9064	2.9421	29.0067		50.5632
	V1	0.4122	2.4789	2.8743	2.7421	22.2567		70.5678
	V2	0.4007	1.9876	2.9990	2.7790	20.9887		45.8732
	V3	0.3556	0.2345	2.6543	2.4321	12.7890		43.9801
	D1	0.3212	2.1234	2.3211	2.5567	19.4567		64.3104
	D2	0.2987	2.3456	1.9805	2.5598	14.4567		53.0921
	D3	o.3211	0.02345	0.3432	2.2256	8.8899		52.9002

In this scheme, for every ROI image, third level wavelet decomposition generates nine subimages. For every mammogram ROI 1152 {(32) * No. of directions (4) *No of decompositions (9)} GLCM feature values are calculated. A feature vector matrix [N * F] corresponding to N training mammogram ROI and F number of features is shown for MIAS database using Bior3.7 decomposition (Table 2).

Since the dimension of the feature vector is very large. The significant features are selected using optimal feature decomposing algorithm [12] with parameters minperlevel = 10; mintofuse = 0.8. So, 13 significant features are selected from each decomposition. Table 3 shows optimized feature vector with the class label of image.

Table 3 Transform domain feature discretization and selection: optimized feature vector

Image label	F_1	F_2	F_3	F_4	F_5	..	F_{11}	F_{12}	F_{13}
1111	931	1008	2415	2439	2647		2745	2821	2934
1111	944	1175	2420	2444	2652		2750	2825	2935
1111	857	1104	2420	2444	2652		2752	2822	2939
1111	983	1087	2415	2439	2654		2744	2826	2939
1111	967	1168	2419	2443	2646		2751	2817	2944
...									
2222	942	1068	2429	2453	2658		2770	2828	2968
2222	938	1064	2426	2450	2665		2769	2825	2963
2222	940	1068	2422	2446	2654		2789	2823	2971

Table 4 Transactional database: signature feature vector along with class label

Optimized decomposition with class label	F_1	F_2	F_3	F_4	F_5	F_6	..	F_{11}	F_{12}	F_{13}
1111	678	931	1008	2415	2439	2647		2745	2821	2934
1111	708	944	1175	2420	2444	2652		2750	2825	2935
1111	643	857	1104	2420	2444	2652		2752	2822	2939
1111	667	983	1087	2415	2439	2654		2744	2826	2939
1111	784	967	1168	2419	2443	2646		2751	2817	2944
...										
2222	799	942	1068	2429	2453	2658		2770	2828	2968
2222	690	938	1064	2426	2450	2665		2769	2825	2963
2222	875	940	1068	2422	2446	2654		2789	2823	2971

Multiresolution approach results in large number of decomposition substructures with huge number of detailed coefficients. So, the significant decompositions contributing the most relevant information corresponding to original ROI are selected using proposed OMDESA. This optimum decomposition selected to generate signature feature vector to form final transactional database is shown in Table 4.

Here each mammogram ROI image is decomposed up to three levels to get nine decompositions per image. For each decomposition, 128 GLCM texture features are extracted. In first stage optimization, these 128 features are optimized to 13 most discriminative and highly correlated features. Hence, for N images, feature space is $[(N \times 9) \times 13]^T$. In second stage optimizations, one out of nine per image decompositions is selected. So, size of feature vector is reduced to $[N \times 13]^T$. This reduces the dimensionality of feature space by 90%. The transactional database containing significant decomposition and optimized feature vector forming signature feature vector is given as input to Apriori algorithm to generate association rules. Apriori algorithm generates large number of association rules. These rules are

Table 5 MOGAAMC results

Database	Association rules using Apriori algorithm		Optimized rules	
	Benign	Malignant	Benign	Malignant
MIAS	1700	2111	240	360
DDSM	2190	3042	530	711

Table 6 Mammogram classification results

Database	Wavelet	No. of optimized decompositions	Optimized features	Accuracy	
				Benign	Malignant
MIAS	Db3	70	13	91	90
	Db4	70	14	91	91
	Bior3.7	70	12	94	93
DDSM	Db3	156	25	96	92
	Db4	156	18	96	94
	Bior3.7	156	22	97	97

optimized to get significant and discriminative rules using MOGAAMC algorithm [12]. Table 5 shows Apriori algorithms rule and optimized rules.

CISCA [12] is used to classify each ROI into benign and malignant class. Mammogram classification results using CISCA are shown in Table 7. Similar procedure is carried out for DDSM database (Table 6).

6 Conclusion

Authors present wavelet transform domain mammogram classification scheme based on optimum selection of decomposition among nine decompositions per image. Proposed scheme involves three stage optimization techniques feature optimization, followed by optimum wavelet decomposition selection and optimized association rules for classification. Here complexity of total system has increased but dimensionality of features space is reduced by 90%. This low dimensionality of feature space narrows down the search space for the classification and reduces the complexity and enhances the speed of classifier. Results are tested on two standard databases, MIAS and DDSM, using three different wavelets Db3, Db4, and Bior3.7. Classification results show better performance as compared with the other techniques. Among three wavelets, Bior3.7 gives high performance on both MIAS and DDSM databases.

References

1. American Cancer Society, 2014. Detailed Guide: Breast Cancer. Available online at: http://www.cancer.org/cancer/breastcancer/detailedguide/, Accessed on 22 June 2017.
2. Michell, M., 2010. Breast Cancer: Contemporary Issues in Cancer Imaging, Cambridge University Press, UK.
3. Birdwell RL, Ikeda DM, O'Shaughnessy KF, Sickles EA. Mammographic characteristics of 115 missed cancers later detected with screening mammography and the potential utility of computer-aided detection. Radiology. 2001;219 1:192–202.
4. Aboul Ella Hassanien, Tarek Gabor, Machine Learning Applications in Breast Cancer Diagnosis In book: Handbook of Research on Machine Learning Innovations and Trends, Chapter: 20, Publisher: IGI Global, pp. 465–490.
5. Cheng, H., Shi, X., Min, R., Hu, L., Cai, X., & Du, H. (2006). Approaches for automated detection and classification of masses in mammograms. *Pattern Recognition*, (4), 646–668.
6. Liu, S., Babbs, C. F., & Delp, E. J., Multiresolution detection of speculated lesions in digital mammograms. IEEE Transactions on Image Processing, *10*(6), (2001), 874–884.
7. Mousa, R., Munib, Q., & Moussa, A., Breast cancer diagnosis system based on wavelet analysis and fuzzy-neural. Expert systems with Applications, *28*(4), (2005), 713–723.
8. Giuseppe Boccignone, Angelo Chianese and Antonio Picariellob, Computer aided detection of micro-calcifications in digital mammograms, Computers in Biology and Medicine, 30, (2000), 267–286.
9. Ferreira, C. B. R., & Borges, D. L., Analysis of mammogram classification using a wavelet transform decomposition. Pattern Recognition Letters, *24*(7), (2003), 973–982.
10. Rashed, E. A., Ismail, I. A., & Zaki, S. I., Multiresolution mammogram analysis in multilevel decomposition. Pattern Recognition Letters, *28*(2), (2007), 286–292.
11. Beura, Shradhananda et al. "Classification of Mammogram Using Two-Dimensional Discrete Orthonormal S-Transform for Breast Cancer Detection." Healthcare Technology Letters 2.2 (2015): 46–51. PMC. Web. 22 June 2017.
12. Poonam Sonar, Udhav Bhosle, Optimization of association rule mining for mammogram classification, International Journal of Image Processing, volume 11, issue 3, 67–85, June 2017.
13. Dengler J, Behrens S, Desaga JF. Segmentation of Macrocalcifications in Mammograms. IEEE Trans. On Medical Imaging. 1993;12(4):634–642.
14. Daubechies, I., 1992. Ten lectures on wavelets. SIAM CBMS-NSF Series on Applied Mathematics, vol. 61. SIAM.
15. S.G. Mallat, A theory for multiresolution signal decomposition: the wavelet representation, IEEE Transactions on Pattern Analysis and Machine Intelligence 7 (11) (1989) 674–693. on Letters 36 (2003) 2967–2991.
16. Agrawal R et al., "Mining associate on rules between sets of items in large databases", in proceedings of the ACM SIGMOD ICMD, Washington DC, 1993, pp. 207–216.
17. John Suckling, J Parker, D Dance, S Astley, I Hutt, C Boggis, I Ricketts, E Stamatakis, N Cerneaz, S Kok, et al. The mammographic image analysis society digital mammogram database. In Exerpta Medica. International Congress Series, volume 1069, pages 375–378, 1994.
18. TM Deserno, M Soiron, and JE de Oliveir. Texture patterns extracted from digitizes mammograms of different bi-rads classes. Image Retrieval in Medical Applications Project, release, 1, 2012. URL http://ganymed.imib.rwth-aachen.de/irma/datasetsen.php.

Noise Reduction in Electrocardiogram Signal Using Hybrid Methods of Empirical Mode Decomposition with Wavelet Transform and Non-local Means Algorithm

Sarmila Garnaik, Nikhilesh Chandra Rout and Kabiraj Sethi

Abstract Electrocardiogram (ECG) signal helps the physicians in the detection of cardiac-related diseases. Many noises like power line interference (PLI), baseline wander, electromyography (EMG) noise and burst noise are contaminated with the raw signal and corrupt the shape of the waveform which makes the detection faulty. So in recent years, many signal processing methods are proposed for removal of these noise artifacts effectively. In this paper, two hybrid methods, i.e., empirical mode decomposition (EMD) with wavelet transform filtering and EMD with non-local means (NLM) are proposed. The results are analyzed with performance parameters like signal to noise ratio (SNR), mean square error (MSE), and percent root mean square difference (PRD). The results exhibit better performance in hybrid method of EMD with NLM technique.

Keywords Wavelet · EMD · NLM · Hybrid method · SNR
MSE · PRD

1 Introduction

Signal processing tools are being used widely in biomedical applications due to its simplicity and better computational ability. Nowadays, most of the patients are dying due to cardiac-related diseases. ECG is a graphical illustration which

S. Garnaik (✉)
Department of EEE, VSS University of Technology, Burla 768018, Odisha, India
e-mail: sarmilagarnaik12@gmail.com

N. C. Rout
Department of EE, VSS University of Technology, Burla 768018, Odisha, India
e-mail: nikhilesh0007@gmail.com

K. Sethi
Department of ETC, VSS University of Technology, Burla 768018, Odisha, India
e-mail: kabirajsethi@yahoo.com

© Springer Nature Singapore Pte Ltd. 2019
H. S. Behera et al. (eds.), *Computational Intelligence in Data Mining*,
Advances in Intelligent Systems and Computing 711,
https://doi.org/10.1007/978-981-10-8055-5_57

639

analyzes the cardiac activity of heart and helps physicist to detect the cardiac-related diseases [1]. The frequency range of ECG signal generally lies between 0.5 and 100 Hz. But various noises like power line interference, baseline wander, EMG noise, burst noise are predominantly contaminated with the ECG signal and corrupt the waveform [2]. Hence, many important features of ECG signal are masked by the noise components. So extraction of raw signal without any disturbance in shape of waveform is an important issue. During recent years, various signal processing methods such as FIR filtering [3], adaptive algorithms [4], wavelet transforms [5], non-local mean algorithm [6], empirical mode decomposition [7]are being used to filter various noises from ECG signal.

FIR filtering window technique has disadvantage of fixed size of filter coefficients. Adaptive techniques overcome the disadvantage of time-invariant filters and are better suitable where stable characteristics are unknown [4]. Wavelet techniques transform the time domain analysis to frequency domain and show better performance of denoising the ECG signal. But wavelet techniques go through drawbacks like selection of mother wavelet and threshold function coefficients [6]. NLM filter with GPU algorithm exhibits better SNR value with its patch-based concept of self-similarity. EMD is an another technique [7], where the signal is decomposed into several intrinsic mode functions (IMFs) and different filtering methods are applied to these IMFs for noise reduction. In this paper, after application of EMD to the noisy ECG signal, two filtering methods, i.e., wavelet and NLM are applied separately to constitute two hybrid methods. First one is the EMD with wavelet filter, and the second one is the EMD with NLM filtering. Simulation of the above two methods is carried out using MATLAB 2016. The parameters like SNR, MSE, and PRD are evaluated and analyzed.

2   Methodology

2.1   Discrete Wavelet Filter

Wavelet filtering methods are widely used due to its better ability of noise reduction of ECG signals as compared to other adaptive and window technique methods [8]. Wavelet filter transforms the signal from its time domain to frequency domain using fundamental wavelet basis function known as mother wavelet function. The wavelet transform is processed through signal decomposition, where the signal is down sampled to its approximation and detail coefficients by passing the signals through low-pass filter and high-pass filter, respectively. The decomposition process of wavelet transform is shown in Fig. 1 with one level of decomposition. Wavelet filter has its form in both continuous and discrete nature. The basic difference between these two forms is that in discrete wavelet transform the dilation and translation factors are termed in terms of power of two and denoted as: $a = 2^j$, $b = k * 2^j$ [9].

Fig. 1 Wavelet transform decomposition process

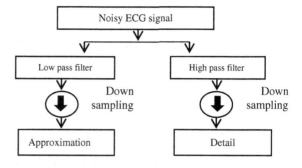

The mother wavelet function [5] is given by Eq. (1).

$$\Psi_{a,b}(t) = \frac{1}{\sqrt{|a|}} \psi\left(\frac{t-b}{a}\right) \tag{1}$$

where

$\Psi(t)$ Mother wavelet
a   Dilatation factor
b   Translation factor

For DWT [10], the wavelet function is written in Eq. (2).

$$\Psi_{a,b}(t) = \frac{1}{\sqrt{2^j}} \psi\left(\frac{t-k*2^j}{2^j}\right) \tag{2}$$

After decomposition, the raw signal can be reconstructed by up-sampling of decomposed signal. Recently, many thresholding techniques have been developed to get appropriate filter coefficients with different level of decomposition.

2.2 Non-local Means Method

Non-local means (NLM) filtering known as statistical neighborhood filtering was first introduced by Buades et al. in 2005. This method outcomes good results in denoising biomedical signals [11]. The problem found in local smoothing filter can be overcome by NLM filter which follows self-similarity concept. To reduce the computational complexity, the similarity between two points is not calculated in whole domain but in a defined nearby domain by performing the average of all weighted non-local values given in Eqs. (3) and (4). The weight vector represents the calculation of similarity between two points and is shown in Eq. (5).

Let a signal $s(n)$ be defined in a time domain of period T. The noisy signal $x(n)$ can be generated by adding noise to raw signal $s(n)$. A small duration called search

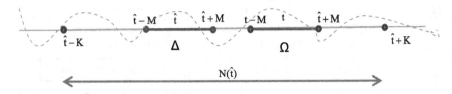

Fig. 2 Interpretation of non-local mean parameters

window $N(\hat{t})$ is chosen in Fig. 2 where $N(\hat{t}) = [\hat{t}- K, \hat{t} + K]$ having size $R = 2K + 1$. In the given time interval T, an approximation of the ECG signal $\hat{s}$ can be estimated for each point $\hat{t} \in T$, by means of a weighted sum of all point values $t \in T$. There are two patches called reference patch (Δ) and similarity window (Ω), where $\Delta = [\hat{t} - M, \hat{t} + M]$ and $\Omega = [t - M, t + M]$ having size $L = 2M + 1$.

For all value of $\hat{t} \in T$, computation of approximate ECG signal is given by

$$\hat{s}(\hat{t}) = \frac{1}{Z(\hat{t})} \sum_{t \in N(\hat{t})} \omega(\hat{t}, t) x(t) \qquad (3)$$

where

$$Z(\hat{t}) = \sum_{t \in N(\hat{t})} \omega(\hat{t}, t) \qquad (4)$$

and the weight vectors are represented as

$$\omega(\hat{t}, t) = e^{\dfrac{-\sum_{\delta \in \Delta} (x(\hat{t}+\delta) - x(t+\delta))^2}{2L\lambda^2}} \qquad (5)$$

Here λ denotes bandwidth parameter and controls the smoothness of denoising. Small value of λ causes noise fluctuation, and large value will make the waveform blurred. The value of M is to be chosen as half width of the R peak of QRS complex.

2.3 Empirical Mode Decomposition

Empirical mode decomposition (EMD) was proposed by Huang et al. [10] which is best suitable filter for a non-stationary and nonlinear signal. It is an iterative method where the signal is decomposed into a number of oscillatory parts known as intrinsic mode functions (IMFs) and the process is known as shifting process [7]. These functions satisfy following two criteria:

1. The difference between number of nearby extrema (including both minima and maxima) and the zero intersecting point ought to be zero or at most one amid a given timeframe.
2. Anytime in the interim, the mean estimation of upper envelope and lower envelope ought to be zero.

The decomposed IMFs include different bands of frequency starting from high to low. The lower-order IMFs contain high frequency, and higher-order IMFs contain comparatively low frequency. The final IMF is known as residue function which is monotonic in nature. Generally, decomposition process includes the first IMF which is of high-frequency and a low-frequency component. Again that low-frequency component is decomposed into second IMF function. This process continues till the final residue is of monotonic in nature. The whole algorithm is given below [12]

1. Find the upper envelope $x_u(t)$ and lower envelope $x_l(t)$ through cubic spline method.
2. Find mean value of above two as $m_l(t) = (x_u(t) + x_l(t)) / 2$ and then subtract it from the signal to get the difference $c_l(t) = x(t) - m_1(t)$.
3. Consider $c_l(t)$ as the new data and repeat the steps 1 and 2 until the obtained signal meets the two necessary conditions of an IMF. The first IMF is denoted as $I_1(t)$.
4. The residual signal $R_1(t)$ is given by $R_1(t) = x(t) - I_1(t)$.
5. Consider $R_1(t)$ as the new data and repeat the steps 1, 2, 3, and 4 for extracting all the IMFs. The moving methodology proceeds until the nth residue $R_n(t)$ becomes monotonic.

The summation of all IMFs and final residue gives raw signal which is given in Eq. (6).

$$x(t) = \sum_{i=1}^{n} I_n(t) + R_n(t) \tag{6}$$

2.4 Proposed Denoising Method

In this paper, EMD with wavelet and EMD with NLM methods are applied to the noisy ECG signal. The noises added to the raw signal are power line interference, baseline wander, EMG noise, and burst noise.

First of all, by the application of EMD algorithm, the noisy ECG signal is decomposed into a number of IMFs and a final residue. All the IMFs have different frequency value. S. Lahmiri proposed a hybrid system of EMD and wavelet technique which is investigated by taking all IMFs into consideration for denoising process [13]. But in this work, the IMFs which contain the noise frequency components are considered as contaminated IMFs and are selected for filtering process.

Then, wavelet and non-local means filters are applied separately to each selected IMFs. In case of wavelet denoising, discrete Meyer family is chosen due to its advantage of rapid decay and infinite differentiability with compact support in frequency domain [14]. Then, these filtered IMFs are added to the rest of the original IMFs in both cases for reconstruction of the raw signal. The whole process is shown in Fig. 3.

To compare the denoising level of different methods, the following parameters are chosen:

The mean square error (MSE) is given in Eq. (7).

$$\text{MSE} = \left(\frac{1}{N}\right) \sum_{n=0}^{N-1} [x(n) - y(n)]^2 \tag{7}$$

Another form of measurement is in terms of signal to noise ratio (SNR) and percent root mean square difference (PRD) parameter. It is defined in Eqs. (8) and (9).

$$\text{SNR(in dB)} = 10 \log \frac{\sum_{n=0}^{N-1} [x(n)]^2}{\sum_{n=0}^{N-1} [x(n) - y(n)]^2} \tag{8}$$

Fig. 3 Block diagram of ECG signal denoising using hybrid model

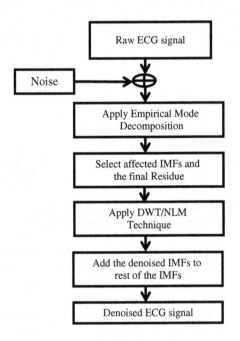

The percent root mean square difference (PRD) is defined by

$$PRD = 100 \sqrt{\frac{\sum_{n=0}^{N-1}[x(n) - y(n)]^2}{\sum_{n=0}^{N-1}[x(n)]^2}} \qquad (9)$$

where x(n) is the raw signal and y(n) is the output of the filter signal.

3 Results and Discussion

The raw ECG signal is obtained from physionet Massachusetts Institute of Technology—Beth Israel Hospital (MIT–BIH) with database number 100 [15]. To construct the noisy signal, PLI noise of 50 Hz, EMG noise of 150 Hz, baseline wander of 1 Hz, and burst noise are added to the raw signal. First, this noisy signal is applied directly to wavelet filter and non-local means filter separately. Wavelet family of discrete Meyer is chosen due to its advantage of rapid decay and high differentiability characteristics. Soft thresholding with decomposition level '2' is taken. For NLM method, the half width of the search window and the half width of the similarity window are taken as 1000 and 600, respectively. Figures 4 and 5 indicate the raw and noisy ECG signal, respectively. Figures 6 and 7 represent the denoised ECG signal by direct application of wavelet and NLM filters, respectively.

Secondly, using EMD, the generated ECG signal is decomposed into 12 numbers of IMFs and one final residue. The first three IMFs include the frequency components of noisy signal and are selected for denoising process. As baseline wander is of frequency 1 Hz, we need to filter out the final residue which has frequency less than 1 Hz. The selected IMFs and final residue are filtered through wavelet and NLM techniques separately. In both the cases, these filtered IMFs are again added to rest of the IMFs (IMF No. 4–12) for reconstruction of the raw signal.

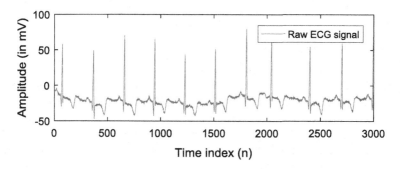

Fig. 4 Raw ECG signal

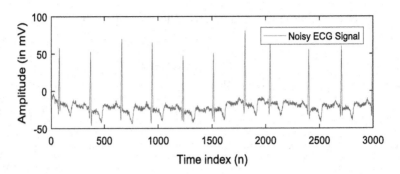

Fig. 5 Noisy ECG signal

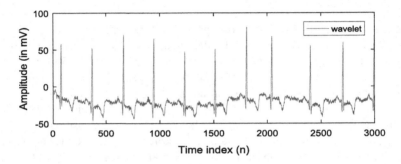

Fig. 6 Denoised ECG signal using wavelet transform

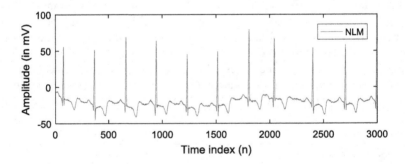

Fig. 7 Denoised ECG signal using NLM method

Figures 8 and 9 display the final denoised signal for both the hybrid methods, i.e., EMD with wavelet transform and EMD with NLM filter, respectively.

The performance analysis is portrayed in Table 1.

From the above table, it is noted that the proposed method of EMD with NLM filtering technique is more effective for denoising the artifacts from noisy ECG signal among all methods listed.

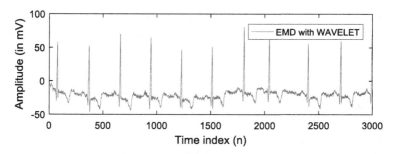

Fig. 8 Denoised ECG signal using EMD with wavelet transform

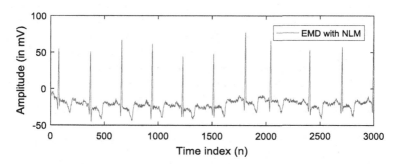

Fig. 9 Denoised ECG signal using EMD with NLM method

Table 1 Parameter comparison of different filtering methods

Types of filters	Name of filter	Parameter		
		SNR (in dB)	MSE	PRD
Wavelet	Discrete Meyer	26.1117	1.40	4.94
NLM	NLM	26.1927	1.383	4.9019
Hybrid model	EMD with wavelet	26.1939	1.382	4.9012
Hybrid model	EMD with NLM	26.1995	1.381	4.89

4 Conclusion

In the present work, two hybrid methods are proposed—one is EMD followed by wavelet-based filter, and the other is EMD followed by NLM filtering technique to remove the noises added to raw ECG signal. DWT is computed to level '2' using mother wavelet function of discrete wavelet type and scaling function. Discrete Meyer wavelet function removes noise from the ECG signal without any distortion of the ECG signal feature with the advantage of rapid decay and infinite differentiability with compact support in frequency domain. EMD with NLM filter exhibits better SNR value of 26.1995 dB, MSE of 1.381, and PRD of 4.89 than other hybrid method. Further, it overcomes the drawback of selection of mother wavelet and threshold function coefficients. Comparing all methods together, hybrid method of EMD with NLM technique is a more effective method for denoising of ECG signal.

References

1. Schamroth, L.: An Introduction to Electro cardiology. Blackwell Science, London. (1982)
2. Singh B, Singh P, Budhiraja S.: Various approaches to minimise noises in ECG signal: A survey. Fifth International Conference in Advanced Computing & Communication Technologies (ACCT), IEEE, (2015) 131–137
3. Mittal A, Rege A.: Design of digital FIR filter implemented with window techniques for reduction of power line interference from ECG signal. International Conference in Computer, Communication and Control (IC4), IEEE, (2015) 1–4
4. Sultana N, Kamatham Y, Kinnara B.: Performance analysis of adaptive filtering algorithms for denoising of ECG signals. International Conference in Advances in Computing, Communications and Informatics (ICACCI), IEEE (2015) 297–302
5. Patil, H. T., Holambe, R. S.: New approach of threshold estimation for denoising ECG signal using wavelet transform. In India Conference (INDICON), IEEE (2013) 1–4
6. Cuomo, S., De Michele, P., Galletti, A., Marcellino, L.: A GPU-Parallel Algorithm for ECG Signal Denoising Based on the NLM Method. 30th International Conference on Advanced Information Networking and Applications Workshops (WAINA), IEEE (2016) 35–39
7. Anapagamini, S. A., Rajavel, R.: Removal of artifacts in ECG using Empirical mode decomposition. International Conference on Communications and Signal Processing (ICCSP), IEEE (2013) 288–292
8. Biswas, U., Hasan, K. R., Sana, B., Maniruzzaman, M.: Denoising ECG signal using different wavelet families and comparison with other techniques. International Conference on Electrical Engineering and Information Communication Technology (ICEEICT), (2015) 1–6
9. El Hanine, M., Abdelmounim, E., Haddadi, R., Belaguid, A.: Electrocardiogram signal denoising using discrete wavelet transform. Computer Technology and Application, (2014). 5(2)
10. Huang, N. E., et al.: The empirical mode decomposition and the Hilbert spectrum for nonlinear and non-stationary time series analysis. In Proceedings of the Royal Society of London A: Mathematical, Physical and Engineering Sciences, Vol. 454, The Royal Society. (1998) 903–995
11. Buades, A., Coll, B., Morel, J. M.: A review of image denoising algorithms, with a new one. Multiscale Modeling & Simulation, (2005) 4(2) 490–530
12. Samadi, S., Shamsollahi, M. B.: ECG noise reduction using empirical mode decomposition based on combination of instantaneous half period and soft-thresholding. Middle East Conference on Biomedical Engineering (MECBME), IEEE (2014) 244–248
13. Lahmiri, S.: Comparative study of ECG signal denoising by wavelet thresholding in empirical and variational mode decomposition domains. Healthcare technology letters, 1(3), (2014) 104–109
14. Lu, J., Liu, H. P., Hsu, C. Y.: Discrete Meyer Wavelet Transform Features For online Hangul Script Recognition. Research Journal of Applied Sciences, Engineering and Technology, 4(20), (2012) 3905–3910
15. The MIT-BIH ECG Database: www.physionet.org/physiobank/database/mitdb/

A Path-Oriented Test Data Generation Approach Hybridizing Genetic Algorithm and Artificial Immune System

Gargi Bhattacharjee and Ashish Singh Saluja

Abstract Validating the correctness of software through a tool has started gaining a wide foothold in the business. A test data generator is one such tool which automatically generates the test data for software so as to attain maximum coverage. Researchers in the past have adopted different evolutionary algorithms to automatically generate a data set. One such often used procedure is Genetic Algorithm (GA). Due to certain flaws present in this approach, we have redefined the cause of concern for coverage in structural testing. In this paper, we have explored the properties of immune system along with GA. We have proposed a new hybrid algorithm—GeMune algorithm—inspired from these biological backdrops. Experimental results certify that the new algorithm has a better coverage compared to the use of only Genetic Algorithm for structural testing.

Keywords Test data generation · Genetic Algorithm · Artificial Immune System
Path coverage · White box testing · Software testing · Test cases
Evolutionary Computing · GeMune Algorithm

1 Introduction

Software testing [1, 2] uses numerous resources to ensure the quality of software. Structural testing, a sub-domain of testing, examines the internal structure of the code. While assessing the structure, the extent to which the source code of a program is executed when a set of test data is run is measured as its code coverage. This coverage becomes a guiding factor to determine the quality of software. The greater the coverage, lesser are the chances of failure. Certain criteria are used to

G. Bhattacharjee (✉) · A. S. Saluja
Department of IT, Veer Surendra Sai University of Technology,
Burla 768018, Odisha, India
e-mail: gbhattacharjee_it@vssut.ac.in

A. S. Saluja
e-mail: saluja.ashish01@gmail.com

© Springer Nature Singapore Pte Ltd. 2019   649
H. S. Behera et al. (eds.), *Computational Intelligence in Data Mining*,
Advances in Intelligent Systems and Computing 711,
https://doi.org/10.1007/978-981-10-8055-5_58

estimate the amount of coverage. One such criterion is path coverage. Path coverage illustrates the execution of all the possible paths at least once by the different set of input values. The set of test data creates the corresponding test cases. The prime objective of creating these test cases is to ensure that maximum number of test paths in the software under test is covered. Since the data set is large enough, we require a test data generator to perform this task. By automating the process of generating these values, the prime criteria of a tester get fulfilled. Thus, it reduces the human effort to be put into check the same. The usage of dynamic memory allocation makes the program highly unstable, thereby making it harder to anticipate the paths. This makes it nearly impossible for the tools to generate the data set. However in the past few decades, advances have been made in this research area. Several nature-inspired algorithms [3–7] have been used to bring out a solution. One such prominent algorithm to be worked in this domain is Genetic Algorithm (GA).

GA [8] is an evolutionary or nature-inspired technique which functions on the principles of Darwin's theory of **'Survival of the fittest'**. GA adopts the evolution paradigm of humans to produce offspring with better and fitter characteristics while propagating their genetic content. Here the data is represented as chromosomes. The following elaboration describes the operations of GA: the initial values (individuals) are selected either randomly or via seeding from a *mating pool*. The structure of an individual representing the chromosome depends upon the range of the problem domain. Each individual is evaluated by its corresponding fitness function to become a parent in the next generation. The process involved to choose the parents is known as *selection,* and the fitness value represents the quality of each individual with respect to the global optimal solution, i.e. how close the individual is related to the global optimum solution.

Once the mates to be combined together are decided, the two basic parameters of GA come into play—crossover and mutation. Crossover is the recombination operator addressing the random interchange and re-ordering of chromosomes from both the parents. Crossover is generally followed by mutation, in which, a bit from any position of the bit string is randomly flipped. Mutation helps to achieve diversity in the population. These operations lay foundation for fitter offspring. The entire procedure is again followed till the time it hits the stopping criterion.

But, in certain cases, it may happen that the quality of individuals may degrade. It may also lead to slow convergence rate resulting in a weak local optimal solution producing an uncertainty over determining the global maxima. A major flaw of the approach is that the number of iterations to get the optimal solution remains blur. These shortcomings may hinder the objective to be completely attained. These flaws [9] sparked the urge of infusing other properties to attain a better outcome.

The earliest work reported of generating test data through Genetic Algorithm was stated by Pargas et al. [10] and C.C. Michael et al. [11]. Girgis [12], Ghiduk et al. [13] and Andreou et al. [14] exercised GA to check for the data flow through the program. Researchers have focused on adapting GA to generate data set for satisfying varied coverage criteria. S. Khor and Peter. G [15] extended their concept to create data set for branch coverage, and [16] focused on conditional coverage, while some worked on attaining path coverage [17–20]. Likewise, several other works too have been noted in the literature [21–23]. The literature states that analysts have propounded a set of varied approaches inspired from nature and interfaced them with the concept of GA to attain complete coverage for structural testing. Our objective too has been inspired from the biological backdrop. We have redefined the cause of concern for coverage in structural testing in this paper by infusing the properties of Immune System (IS) along with GA.

Human body [24] becomes susceptible to illness when foreign particles (usually known as antigens) invade the system. These antigens stimulate certain responses which are recognized by our immune system (IS). The IS gets activated and starts producing complimentary pair (antibodies), thus acting as a barrier to prevent the widespread of the antigen. It ensures safety through striking the foreign elements by either recognizing the antigen from the memory cells or using cloning to produce new antibodies while undergoing hyper-mutation process and saves those antibodies for future references. The IS has a greater ability to set apart the self-body (antibodies) from the non-self-body (antigens). Farmer et al. [25] extended this basic mechanism to their problem, thereby getting the recognition of Artificial Immune System (AIS).

During the 90s, the AIS [26] emerged as a new path of Computational Intelligence (CI) mimicking the fundamentals of biological IS. It draws inspiration from the response of blood cells (B cells and T cells) towards any antigen and the process invoked to dismantle them. The properties of B cells and T cells have led to the development of three main algorithms: negative selection algorithm, clonal selection algorithm (CLONALG) and artificial immune network (AIN). Since our approach has been inspired by CLONALG, we present an elaborate description of the algorithm.

CLONALG originates from the biological principle of immunity put forward by Burnet [27]. The conversion of this principle into algorithms was proposed by Castro and Zuben [28]. The algorithm is based on the principles of selection and affinity. The process of evolution is based on the probability method. It executes its operation similar to that of Genetic Algorithm; the only difference being clonal immune algorithm results in a set of local optimal values, whereas the solution obtained from GA tries to find a global optimal solution. The initial population follows the under-mentioned series of operations.

Affinity Evaluation—It draws its analogy from the objective function of GA to select better antibodies to fight against antigens. The numeral value represents the antibody's fitness to be chosen for hyper-mutation process and to reach towards an optimal solution.

Density Calculation—It is used to calculate the similar type of antibodies present in the population.

Selection—It executes similarly to that of GA's selection operator, i.e. to select those antibodies which are fitter and which can undergo the hyper-mutation process in order to clone the antibodies.

Hyper-Mutation Process—It reflects the most important section of the AIS paradigm. It is where the cloning of antibodies is performed in order to fight against the antigens. It can be visualized as the body being subjected to a vaccine which will reproduce itself to protect the body. Similar to GA, the cloned antibodies go through mutation process and suppress those antibodies which are weak and replace them with the fitter cloned ones, so that the population of the antibodies after discarding the weaker ones is maintained in the population maintenance stage.

The algorithm stands as:

1. Antigens correspond to the problem that is required to be resolved by an instrumentally selected antibody to procure an optimal solution.
2. Initially, antibodies (population) are produced.
3. The affinity between antigen and antibodies is evaluated. It also helps in determining the quality of the antibodies.
4. Check it against the terminating condition. If the condition evaluates to a non-positive value, it is then followed by density calculation.
5. Immune operator governs the cloning and mutation of the healthy antibodies to be retained and discards the weaker antibodies, thus maintaining the population size.
6. Thus, the next generation population are the result of affinity, density and cloning.

The need to obtain an optimal result depends upon the appropriate selection of the immune operator.

Correlating the notion of test data generation for attaining a complete structural testing with AIS method can be described as: antigens become the targeted test nodes to be traversed automatically by the antibodies that behave as a set of test cases in order to attain maximum coverage of test paths, i.e. antibodies become the population depending upon which the path will be traversed. As stated earlier, affinity expresses how quickly the antibodies respond to the antigens. Here, affinity is the path between the antibodies and the antigen. Density estimates the percentage of population behaving similarly. This reflects the fact that the test set produced for the same sub-domain consists of values that will make the control flow through the same function and follow the same path. In addition to the above properties, diversity and vaccination also do form an integral part of the process. Diversity reflects to the various dimensions of the antibodies that are in the search space corresponding to a particular antigen. Vaccination remains the most vital of all; it acts as a function to build antibodies and helps in a greater coverage of antigens with diversified values. Artificial Immune System has not been much utilized in software testing. Yet, the recent advances suggest that immune system has been hybridized with Genetic Algorithm to generate data set [29–31].

The paper has been organized as follows—Sect. 2 discusses our proposed work followed by Sect. 3 which focuses on our case study. The next section analyses our results. Finally, the last section concludes our work along with highlighting the future extension of this topic.

2 Proposed Hybrid Approach—GeMune Algorithm

This research basically brings forth a hybrid approach to generate test data to attain maximum coverage for structural testing. The proposed approach infuses the concept of *clonal selection algorithm,* along with Genetic Algorithm. We have named our approach GeMune Algorithm since it is highly influenced by the properties of GA and AIS. The GeMune Algorithm is as follows:

> STEP 1: Initialize a random population of individuals
> STEP 2: *While (terminating condition not satisfied)*

Begin
> STEP 2.1: Evaluate the fitness value on the basis of GA's objective function.
> STEP 2.2: Randomly select the fit parents
> STEP 2.3: Perform crossover according to the vaccines (Regenerator operator)
> - Let any antigen (Ag) invade the system.
> - Antibodies (Ab) are cloned with the help of regenerator operator.
> - Apply the antibodies to the antigen.
> - Generate the affinity between the antibodies and the antigen.
>
> STEP 2.4: Apply mutation.
> *End*

End

The working principle of this algorithm can be summarized as—the preliminary initialization of the individuals is done to evaluate their respective fitness function. The parents for the operation are randomly selected from the pool. As suggested earlier, sometimes crossover may lead to certain degeneracy of individuals; therefore, to avert it, vaccination was applied alongside crossover. We have named it as the regenerator operator as it helps in producing multiple antibodies to be used against antigens. At this stage, the Euclidean distance between the positions of the parents is calculated. Multipoint crossover is then carried out with respect to the binary values of each parent in accordance with the binary value of Euclidean distance between them. This crossover is repeated considering the Euclidean distance in order to produce clones between the antibodies so as to generate a higher precision affinity towards the problem. Finally, mutation is carried out and the outputs become the individuals for the subsequent iteration. We have considered the number of iterations as the stopping condition. Figure 1 depicts the block diagram of our approach.

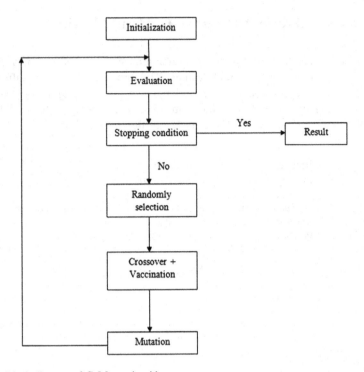

Fig. 1 Block diagram of GeMune algorithm

3 Case Study

We have implemented our approach using a program to find the greatest among three numbers. Figure 2 represents the CFG of the code portraying the possible set of paths that can be executed.

For every path to be traversed in the code, the value generated at that particular instance of the conditional statement makes the count for the next step, i.e. it leads to the level of attainment of path coverage. We apply our algorithm to dynamically generate the set of test cases that would cover all the possible test paths.

4 Result Analysis

The under-mentioned figures (Figs. 3, 4 and 5) show a comparative analysis of the result obtained on applying GA and our proposed GeMune approach. We have considered the stopping criterion for both the algorithms as a set of 20 iterations.

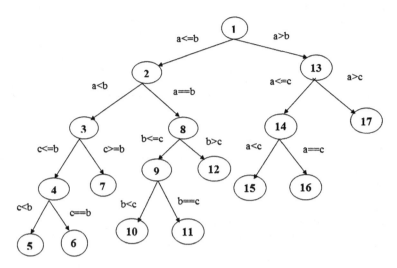

Fig. 2 Control flow graph for finding the greatest among three numbers

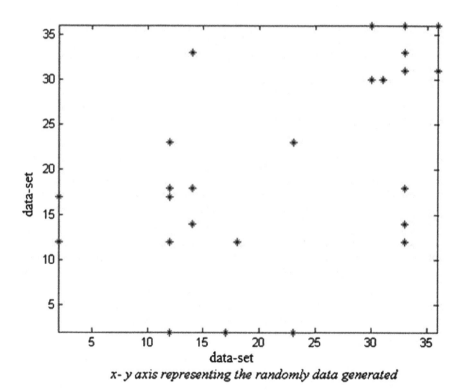

Fig. 3 Randomly generated data through GA

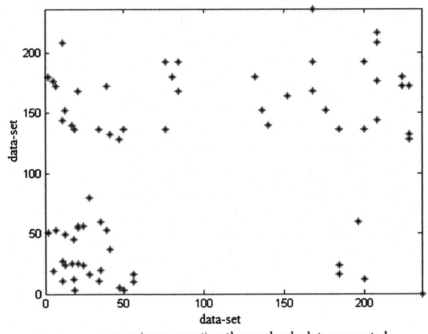

x- y axis representing the randomly data generated

Fig. 4 Randomly generated data through GeMune algorithm

5 Conclusion

The Artificial Immune System is adopted from the natural biological phenomenon to deal with foreign molecules by generating complimentary pairs. AIS has found its foothold in numerous applications of the computer science and has evolved as one of the reliable, efficient and effective techniques. Here, we have attempted to extend the concept of AIS merged with GA to the problem domain of dynamic test case generation. The process of dealing a problem with not just a single candidate but multiple candidates helps to achieve the target in less time and reduces the costs as well. The formulated hybrid process overcomes each other's drawbacks. Thus, the combination approach solves the problem with slightly higher rate. In future to come, the algorithm can be evolved with great optimizing result not only for the simple conditional statements but also for looping constructs.

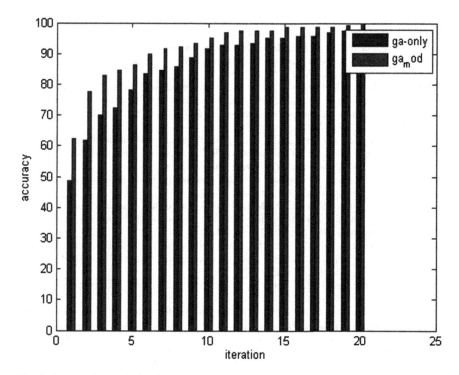

Fig. 5 Comparative analysis of average path coverage

References

1. Dijkstra, Edsger W. "The humble programmer." *Communications of the ACM* 15.10 (1972): 859–866.
2. Myers, Glenford J., Corey Sandler, and Tom Badgett. *The art of software testing*. John Wiley & Sons, 2011.
3. McMinn, Phil. "Search-based software test data generation: a survey." *Software testing, Verification and reliability* 14.2 (2004): 105–156.
4. Li, Huaizhong, and Chiou Peng Lam. "Software Test Data Generation using Ant Colony Optimization." *International Conference on Computational Intelligence*. 2004.
5. Nayak, Narmada, and Durga Prasad Mohapatra. "Automatic test data generation for data flow testing using particle swarm optimization." *Contemporary Computing* (2010): 1–12.
6. Srivastava, Praveen Ranjan, et al. "Automated test data generation using cuckoo search and tabu search (CSTS) algorithm." (2012): 195–224.
7. Srivatsava, Praveen Ranjan, B. Mallikarjun, and Xin-She Yang. "Optimal test sequence generation using firefly algorithm." *Swarm and Evolutionary Computation* 8 (2013): 44–53.
8. Mitchell, Melanie. *An introduction to genetic algorithms*. MIT press, 1998.
9. Goldberg, David E. *Genetic algorithms*. Pearson Education India, 2006.
10. Pargas, Roy P., Mary Jean Harrold, and Robert R. Peck. "Test-data generation using genetic algorithms." *Software Testing Verification and Reliability* 9.4 (1999): 263–282.
11. Michael, Christoph C., et al. "Genetic algorithms for dynamic test data generation." *Automated Software Engineering, 1997. Proceedings, 12th IEEE International Conference*. IEEE, 1997.

12. Girgis, Moheb R. "Automatic Test Data Generation for Data Flow Testing Using a Genetic Algorithm." *J. UCS* 11.6 (2005): 898–915.
13. Ghiduk, Ahmed S., Mary Jean Harrold, and Moheb R. Girgis. "Using genetic algorithms to aid test-data generation for data-flow coverage." *Software Engineering Conference, 2007. APSEC 2007. 14th Asia-Pacific.* IEEE, 2007.
14. Andreou, Andreas S., Kypros A. Economides, and Anastasis A. Sofokleous. "An automatic software test-data generation scheme based on data flow criteria and genetic algorithms." *Computer and Information Technology, 2007. CIT 2007. 7th IEEE International Conference on.* IEEE, 2007.
15. Khor, Susan, and Peter Grogono. "Using a genetic algorithm and formal concept analysis to generate branch coverage test data automatically." *Automated Software Engineering, 2004. Proceedings. 19th International Conference on.* IEEE, 2004.
16. Tracey, Nigel, et al. "Automated test-data generation for exception conditions." *Software-Practice and Experience* 30.1 (2000): 61–79.
17. Bueno, Paulo Marcos Siqueira, and Mario Jino. "Automatic test data generation for program paths using genetic algorithms." *International Journal of Software Engineering and Knowledge Engineering* 12.06 (2002): 691–709.
18. Chen, Yong, and Yong Zhong. "Automatic path-oriented test data generation using a multi-population genetic algorithm." *Natural Computation, 2008. ICNC'08. Fourth International Conference on.* Vol. 1. IEEE, 2008.
19. Lin, Jin-Cherng, and Pu-Lin Yeh. "Automatic test data generation for path testing using GAs." *Information Sciences* 131.1 (2001): 47–64.
20. Mansour, Nashat, and Miran Salame. "Data generation for path testing." *Software Quality Journal* 12.2 (2004): 121–136.
21. Srivastava, Praveen Ranjan, et al. "Use of genetic algorithm in generation of feasible test data." *ACM SIGSOFT Software Engineering Notes* 34.2 (2009): 1–4.
22. Hermadi, Irman, and Moataz A. Ahmed. "Genetic algorithm based test data generator." *Evolutionary Computation, 2003. CEC'03. The 2003 Congress on.* Vol. 1. IEEE, 2003.
23. Sofokleous, Anastasis A., and Andreas S. Andreou. "Automatic, evolutionary test data generation for dynamic software testing." *Journal of Systems and Software* 81.11 (2008): 1883–1898.
24. Janeway, Charles A., et al. *Immunobiology: the immune system in health and disease.* Vol. 1. Singapore: Current Biology, 1997.
25. Farmer, J. Doyne, Norman H. Packard, and Alan S. Perelson. "The immune system, adaptation, and machine learning." *Physica D: Nonlinear Phenomena* 22. 1–3 (1986): 187–204.
26. Dasgupta, Dipankar. "Advances in artificial immune systems." *IEEE computational intelligence magazine* 1. 4 (2006): 40–49.
27. Burnet, F. M. "Clonal selection and after." *Theoretical Immunology* 63 (1978): 85.
28. De Castro, L. Nunes, and Fernando J. Von Zuben. "The clonal selection algorithm with engineering applications." *Proceedings of GECCO.* Vol. 2000. 2000.
29. Bouchachia, Abdelhamid. "An immune genetic algorithm for software test data generation." *Hybrid Intelligent Systems, 2007. HIS 2007. 7th International Conference on.* IEEE, 2007.
30. Pachauri, Ankur. "Use of clonal selection algorithm as software test data generation technique." *Advanced Computing & Communication Technologies (ACCT), 2012 Second International Conference on.* IEEE, 2012.
31. Xu, Xiaofeng, et al. "A path-oriented test data generation approach for automatic software testing." *Anti-counterfeiting, Security and Identification, 2008. ASID 2008. 2nd International Conference on.* IEEE, 2008.

To Enhance Web Response Time Using Agglomerative Clustering Technique for Web Navigation Recommendation

Shraddha Tiwari, Rajeev Kumar Gupta and Ramgopal Kashyap

Abstract An organization needs to comprehend their customers and clients conduct, inclinations, and future needs which rely on their past conduct. Web usage mining is an intuitive research point in which customers and clients' session grouping is done to comprehend the exercises. This exploration examines the issue of mining and breaking down incessant example and particularly centered around lessening the quantity of standards utilizing shut example procedure and it likewise diminish filters the span of the database utilizing agglomerative grouping strategy. In the present work, a novel technique for design mining is introduced to tackle the issue through profile-based closed sequential pattern mining utilizing agglomerative clustering (PCSPAC). In this research, the proposed method is an improved version of Weblog mining techniques and to the online navigational pattern forecasting. In the proposed approach, first, we store the Web data which is accessed by the user and then find the pattern. Items with the same pattern are merged and then the closed frequent set of Web pages is found. Main advantage of our approach is that when the user next time demands for the same item then it will search only partial database, not in whole data. There is no need to take input as number of clusters. Experimental results illustrate that proposed approach reduces the search time with more accuracy.

Keywords Web usage mining · Closed sequential patterns · Sequence tree
User profile · Weight · Web database · Web services · Agglomerative
clustering

S. Tiwari · R. K. Gupta (✉) · R. Kashyap
Department of CSE, Sagar Institute of Science and Technology,
Bhopal 462036, Madhya Pradesh, India
e-mail: rajeevmanit12276@gmail.com

S. Tiwari
e-mail: shraddha@gmail.com

R. Kashyap
e-mail: ramgopalkashyap@sistec.ac.in

© Springer Nature Singapore Pte Ltd. 2019   659
H. S. Behera et al. (eds.), *Computational Intelligence in Data Mining*,
Advances in Intelligent Systems and Computing 711,
https://doi.org/10.1007/978-981-10-8055-5_59

1   Introduction

Today, associations rely upon programming for the most part for their online sites to relate with clients. Holding late clients and pulling in potential ones push these associations run over in striking approaches to make their sites more valuable and productive. The WWW [1] is an enormous wellspring of data that can come either from the Web content, depicted by the billions of HTML pages straightforwardly available, or from the Web use, spoke to by the enrollment data gathered day by day by every one of the servers around the globe. Web usage mining is the sort of data mining which manages the extraction of intriguing information of the WWW. Web use mining [2] has numerous applications, e.g., personalization of Web substance, support of the outline, suggestion frameworks, pre-bringing [3] and storing, and so forth. There are various advantages of Web utilization mining, primarily in Web-based business. Clients can be focused with fitting notice. Additionally, items which are identified with clients can be proposed progressively while perusing the site. As indicated by, the utilization mining process is partitioned into three sections. The initial segment begins with information cleaning and pre-preparing. Second part is the pre-handled information which is dug for some concealed and gainful data. Furthermore, the last piece of the Weblog mining process closes by breaking down the mining comes about.

Web utilization mining is the way toward applying information mining strategies to the disclosure of use designs from Weblogs information and to distinguish Web clients' conduct. In Web use mining, information can be gathered at the server side, customer side and intermediary servers.

Bunching [4] are exceptionally helpful and dynamic territories of machine learning research. It consolidates the things into a similar gathering which is having same characteristics [5]. It elevates to isolate gathering of information things (the informational collection) in bunches which things having closeness to each other and not at all like the things in different groups. It likewise overcomes with the issue of information over-burdening on the Internet.

2   Literature Survey

Yi Pan and HongYan Du [2] present a novel prefix chart-based calculation for mining shut successive itemsets. The new approach assembled a proficient prefix chart structure and utilized variable length bit vectors to show the relationship between the database and its things. Raut et al. [6] proposed fluffy various leveled grouping for making the bunches of online Web records in light of fluffy equality connection. This strategy is utilized as a part of developing groups with dubious limits. Azimpour-Kivi Mozhgan and Azmi Reza [7] proposed a Web session bunching in view of comparative patterns. This calculation bunched the site pages in view of comparability measure for Web sessions grouping utilizing an agglomerative bunching method with arrangement.

Varghese Nayana Mariya and John Jomina [8] proposed bunch enhancement procedure utilizing fluffy rationale. Web site page get to design is gathered from Weblog record as info and afterward dispense with unessential information things. Omar Zaarour et al. [4] proposed a change in the Weblog digging methodology for the expectation of online navigational example. Sheetal Sahu et al. [9] proposed a neural system instruments to find valuable data from the site information. It supports to site administration, business and bolster administrations, personalization, and system movement stream examination, and so on. Neural-based strategy is utilized to locate the present pattern which relies upon the execution of the grouping of the quantity of solicitations. The self-organizing map (SOM) is utilized to bunch the things in the types of gathering to recognize Web client's navigational examples.

Jerry Chun et al. [10] proposed the WAP-tree idea which can deal with the successive examples in powerful condition. It holds just the regular 1-arrangements from the value-based database. On the off chance that a few groupings are erased from the value-based database then it plays out some operation to partition the kept incessant 1-arrangements. Fan Muhan [5] proposed a technique to locate the incessant shut examples when new information arrives. It utilizes a sliding window to catch data conveniently and precisely. This data stream is divided into several basic windows. Here, all frequently closed patterns are mined as well store and update in each basic window. H. Ryang and U. Yun [11] proposed a novel algorithm which having the list structure. It is used to find the high-utility patterns over data streams based on a sliding window mode. It avoids the generation of candidate patterns so that it does not consume high computational resources. Finally, it efficiently works in complex dynamic systems.

3 Problem Description

Visit arrangement mining is a critical part identified with Web information and now a testing information mining work. The mining continuous succession has turned into a critical segment of numerous expectation or proposal frameworks. The online stores each time need clients' next thing expectation as pages prone to visit [12, 13]. It additionally likes to purchase together which items. The current calculations utilized for visit grouping mining could be ordered either as correct or rough calculations. Precise successive grouping mining calculations, as a rule, read the entire database a few times, and if the database is enormous, at that point, visit arrangement mining is not good with restricted accessibility to PC assets and continuous requirements. So the issues in the present situation are as follows:

(1) Web data partition is not used some conditional parameters just like profile constraint (Income, Age, and Experience) for partition of Web data.

(2) Numerous past consecutive mining calculations hints at no significance of pages where as each page has diverse significance. So the current techniques' perform reaction time is additionally moderate. Site required sensible rough techniques for breaking down information where the calculation speed is more critical than the accuracy. Every time the **whole database scans** for searching the frequent pattern not partial database. At the time of program execution, **number of clusters** is required as an input parameter.

4 Proposed Work

This exploration utilizes a novel approach profile-based closed sequential pattern mining utilizing agglomerative clustering (PCSPAC). It is utilized as a conduct investigation to distinguish client's navigational examples during the time spent Web usage mining. It relies upon the execution of the grouping of the measure of solicitations. Here, the proposed approach utilizing agglomerative grouping forgets to the halfway information of Web information. The information is gathered from the site www.getglobalindia.org as Weblog. Each Weblog has 13 parameters or data, i.e., Internet-Protocol Address, Browser, Version, Operating System, and so forth. So at preprocessing the Weblog is channel by chose characteristics and get just required information. After gathering the required data as indicated by client session-wise for finding the client conduct. In the following stage, the info bolster is assembled by client utilizing interface. On the off chance that the thing support is not exact and equivalent to the given help and then it creates the incessant thing utilizing pruning system of the thing.

4.1 Proposed PCSPAC Algorithm Framework

The following Fig. 1 shows that the process of PCSPAC algorithm generates useful closed sequential pattern using Web data.

Step 1: Collection of Web navigation history of Web site.

Step 2: Apply preprocessing techniques to remove noise from Weblog data and also convert into proper format.

Step 3: Now choose the attributes of profile for similarity matching and input the support value. So here it is matching the attribute similarity to other user navigation pattern. It also finds the frequent pattern using the given support threshold value in the model.

Fig. 1 Process of PCSPAC
algorithm

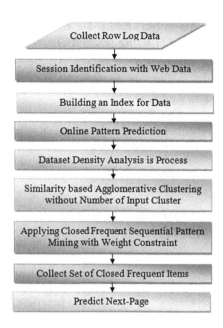

Step 4: Now generate rules using closed sequential pattern of Web data.

Step 5: For clustering the Web data, it requires to first find the smallest probability value and after that merges these two by taking smallest value in the form of cluster.

Step 6: Now select different size of Web data and generate the rules.

Step 7: So put those items in the cache which are having higher frequency.

Step 8: For the next item prediction put some items in the cache which are having higher frequency. Sometimes if next item is not available in the cache then it scans the related item from the cluster Web data, not the whole data.

4.2   Pseudocode of PCSPAC Algorithm

The profile-based closed sequential pattern mining using agglomerative clustering (PCSPAC) approach is applied to discover frequent sequential patterns by using agglomerative algorithm for producing the cluster of Web dataset. This cluster is used to access the partial Web dataset. By using closed sequential, it generates fewer candidates set for generating the rules so that response time increases. The merging method in realism is reconstructing a small Pattern tree. So at this time this tree having the Web pages of Web site in proportional sessions. The pseudocode of the proposed algorithm is

Algorithm (*Profile-based Closed Sequential Pattern Mining using Agglom-erative Clustering (PCSPAC)*)

Procedure: PCSPAC (Support, Attributes, Web Dataset)

```
{
  D = find the web data based on similarity of attributes
  int i=0;
  do {
     Compute the mean weight of prefix item x
     if (x.item.weight >= support)
     {   Output prefix; }
  i=i+1; }
  while(x.item.count>=i);

  // for finding closed sequential pattern in tree
  int j=0;
  do
  {    Generate the rules or pattern of given frequent item x
     if (sub-pattern <= super-pattern)
     {
             remove sub-pattern; }
  j=j+1;  }
  while(x.item.count>=j);

  //clustering of web data set
   int k=0;
  Generate the cluster of given frequent item
  int totCluster = CountNoOfItem();
  int sdv = search(smallest distance value);
  int nsdv = searchNearstItem(sdv);
  do {
      if (sdv <= nsdv)  {
         Merge-cluster(sdv);
         Generate-ClosedPatterTree();  }
   k=k+1;  }
   while(k<= totCluster);

SetCacheItem(ClosedPatterTree);
```

4.2.1   Closed Web Sequential Access Pattern

Closed sequential pattern is used to generate less and accurate rules. It will merge sub-pattern to super pattern if super pattern also having same weight. Figures 2, 3, 4, and 5 show the step-by-step process for the sequential access pattern.

The set FS consists of 17 frequent sequential patterns, the closed sequential patterns discovered are shown in Tables 1, 2, 3, and 4. The set of closed sequential patterns are as follows:

CS = {AA:2, ABB:2, ABC:4, CA:3, CABC:2, CB:3} as shows in Fig. 6.

Fig. 2 After first transaction

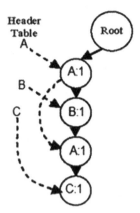

Fig. 3 After second transaction

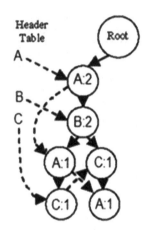

Fig. 4 After third transaction

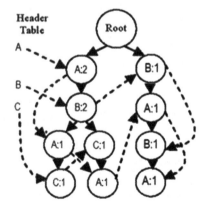

Fig. 5 After fourth
transaction

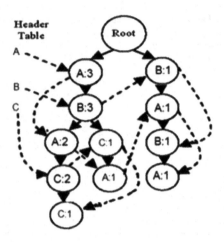

Table 1 Sequence database
of WAP-tree

TID	Web access sequence	Freq. subsequence
100	ABDAC	ABAC
200	EAEBCA	ABCA
300	BABFAE	BABA
400	ABACFC	ABACC

Here Minimum Support = 75% = 3

Table 2 Item with count for
pruning the items

Item	Count
A	4
B	4
C	3
D	1
E	2
F	2

Table 3 Sample sequential
database of Web access
sequence

User_Id	Web access sequence
1	C A A B C
2	A B C B
3	C A B C
4	A B B C A

Minimum Support is 2

Table 4 Closed sequential
pattern set

Frequent set (FS)		
A:4	B:4	C:4
AA:2	BB:2	CA:3
AB:4	BC:4	CAB:2
ABB:2		CAB:2
ABC:4		CAC:2
AC:4		CB:3
		CBC:3
		CC:2

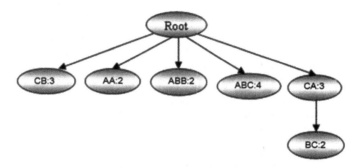

Fig. 6 Pattern tree constructed from the closed sequential Web access patterns

4.2.2 Agglomerative Clustering

Agglomerative clustering calculates minimum distance clustering as single linkage various leveled bunching or closest neighbor grouping. Separation between two bunches is characterized by the base separation between objects of the two groups.

5 Result Analysis

All experiments were conducted on a 2 GHz Intel Core2 Duo processor PC with 4 GB main memory running Microsoft Windows XP. The algorithms were implemented in Asp.Net with C# and were executed using 10% support value. In these experiments, a real dataset of www.getglobalindia.com is used, which is having click stream data from an e-commerce Web store and it has been used widely to assess the performance of frequent pattern mining. This dataset contains sequences of 9701 customers with a total of 56000 purchases in 47 distinct product categories.

This proposed approach presents utilitarian investigation on these different informational collections (e.g., 1010, 2020, 3030, 4040, and 5050 sessions) and furthermore with various help (e.g., 10, 20, 40, 60, 80, and 90%). The portrayal of results on the working of PCSPAC conversely with a recently created design mining calculation is the speediest example mining calculation. The fundamental reason for this exploration is to verify that the successive traversal designs with weight limitation can be produced by fusing a help and weight page with grouping is adequate. At first, demonstrating the quantity of consecutive examples can be managed through clients distribute weights, the adequacy in wording of runtime of the PCSPAC calculation, and the perfection of successive examples. Also, demonstrating the PCSPAC includes related things in the store. Third, it is utilizing Web administrations which naturally refresh weight of each page inconsistently. It additionally diminishes forward and backward time while finding next page from store since it has related pages as of now in reserve. The accompanying Table 5

Table 5 Running time (in Ms) with different size and different support

DB-Size	Sup-10%	Sup-20%	Sup-40%	Sup-60%	Sup-80%	Sup-90%
1010	6912	6723	6609	6707	6504	6099
2020	18607	18314	18453	18352	18372	18273
3030	19843	17857	18637	18743	18253	18284
4040	25324	24202	24498	25403	26462	24443
5050	87578	77983	84531	54315	40633	32373

demonstrates the running time (in ms) when the database record estimate is diverse with various backings.

Fig. 7 demonstrates the examination between WAP-Tree and PCSPAC algorithm with various help. Here utilizing record estimates 5050 in the database. It demonstrates that when record measure is 1010, then the proposed calculation enhances the effectiveness to 2.44%. So also when record estimate is 2020, 3030, 4040, and 5050, it enhances the effectiveness 7.41, 9.69, 10.15, 13.29, and 18.44%.

In information mining, a large portion of analyst utilize calculations to recognize already unrecognized examples and patterns covered up inside tremendous measures of information. These examples are utilized to make prescient models that attempt to gauge future conduct.

These models have various functional business applications—it is utilized by banks to choose which clients to affirm for advances and advertising group utilize them to figure out which prompts focus on crusades.

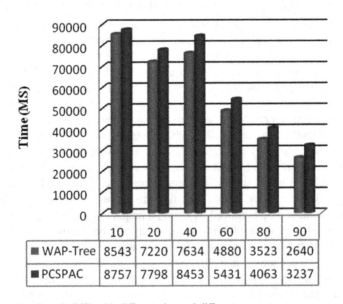

Fig. 7 Running time (in MS) with different size and different support

(a) **Accuracy**: Accuracy is an evaluation metrics on how a model performs.

$$ACCURACY = (TP + TN)/(TP + TN + FP + FN)$$

where TN is the number of true negative cases, FP is the number of false positive cases, FN is the number of false negative cases, and TP is the number of true positive cases

(b) **Precision**: Prediction is a calculation of positive predicted values precision, which is the fraction of retrieved documents that are relevant. The precision is calculated using the formula as:

$$Precision = (TP)/(TP + FP)$$

(c) **Recall:** Review in data recovery is the portion of the records that are pertinent to the inquiry and that are effectively recovered.

$$Precision = (TP)/(TP + FN)$$

The experiments are done with 1010, 2020, 3030, 4040, and 5050 records to calculate

Figure 8 shows the accuracy of the preprocessed Web dataset using different number of records. If records are 1010, 2020, 3030, 4040, and 5050, then accuracies (in %) are 46, 41, 53, 64, and 71%. So when number of records increases accuracy is also maintained.

Figure 9 shows the precision of the preprocessed Web dataset using different number of records. If records are 1010, 2020, 3030, 4040, and 5050, then precisions

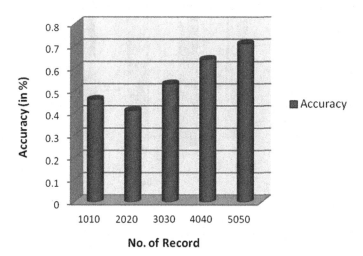

Fig. 8 Accuracy obtained for Web site using different no. of records

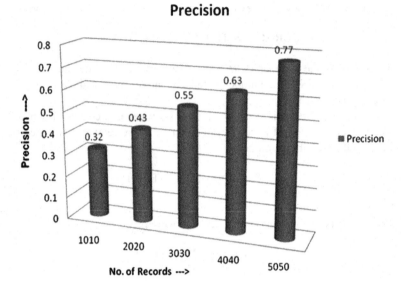

Fig. 9 Precision obtained for Web site using different no. of records

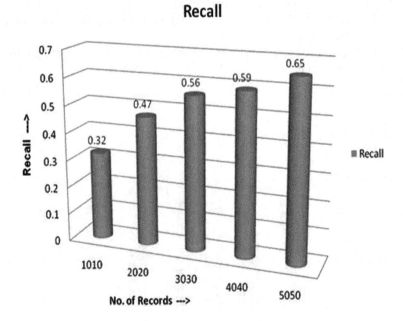

Fig. 10 Recall obtained for Web site using different no. of records

are 32, 43, 55, 63, and 77. So when number of records increases precision is also maintained.

Figure 10 shows the recall of the preprocessed Web dataset using different number of records. If records are 1010, 2020, 3030, 4040, and 5050, then recall are 32, 47, 56, 59, and 65. So when number of records increases recall is also maintained.

6 Conclusion

In this examination, a novel approach PCSPAC is proposed to find shut consecutive examples and output just incomplete Web information for next thing forecast. It separates expansive Web information by coordinating client profile closeness in light of a few properties. In order to filter the Web information, every information is arrange in groups and unions them inconclusive bunch. Shut regular pages are less and helpful guidelines as grouped by agglomerative bunching. The real downsides of the customary approach for mining designs are that size of each page is refreshed physically; yet by proposed strategy, it is refreshed consequently utilizing Web administrations. In the event that Web information measure is 5050 and applying support is 10, 20, 40, 60, 80, and 90%, then it enhances the proficiencies 2.44, 7.41, 9.69, 10.15, 13.29, and 18.44%. It performs quick reaction and extracts outcome because of this it is persuasive sufficient to complete tremendously figuring exorbitant operations in a generally short measure of time for finding next page expectation.

In future work, other information mining calculations can be actualized in cloud to proficiency deal with extensive Web information of numerous hospital sites in circulated condition for finding any basic infections. So there are numerous territories like parallel successive example, gathering of comparative sort of clients, in disseminated servers.

References

1. Etzioni, O., The world-wide Web: quagmire or gold mine? Communications of the ACM, 39 (11) (1996).
2. Yi Pan, HongYan Du – "A Novel Prefix Graph Based Closed Frequent Itemsets Mining Algorithm", IEEE International Conference on Computational Science and Engineering.
3. Kosala, R and Blockeel, H. Web mining research: a survey. SIGKDD Explorations, July, 2 (1) (2000).
4. Miss. Vrinda Khairnar and Miss. Sonal Patil, "Efficient clustering of data using improved K-means algorithm", Imperial Journal of Interdisciplinary Research (IJIR), Vol. 2, Issue-1, 2016.
5. Fan Muhan et al., "A Mining Algorithm for Frequent Closed Pattern on Data Stream based on Sub Structure Compressed in Prefix Tree", IEEE Proceedings of CCIS, pp. 434–439, 2016.
6. Raut Ms. Anjali B., and Bamnote Dr. G.R., "Clustering Method based on Fuzzy Equivalence Relation," International Conference on Computer & Communication Technology (ICCCT), pp. 666–671, 2011.

7. Azimpour-Kivi Mozhgan, and Azmi Reza, "A Webpage Similarity Measure for Web Sessions Clustering Using Sequence Alignment," International Symposium on Artificial Intelligence and Signal Processing (AISP), IEEE, pp. 20–24, Jun. 2011.
8. Varghese Nayana Mariya, John Jomina, "Cluster Optimization for Enhanced Web Usage Mining using Fuzzy Logic," World Congress on Information and Communication Technologies, IEEE, pp. 948–952, 2012.
9. Sheetal Sahu, Praneet Saurabh, Sandeep Rai, "An Enhancement in Clustering for Sequential Pattern Mining Through Neural Algorithm Using Web Logs". IEEE Proc. of the 2014 Sixth International Conference on Computational Intelligence and Communication Networks, pp 758–764, 2014.
10. Jerry Chun, Wensheng Gan, Tzung Pei Hong, "Efficiently Maintaining the Fast Updated Sequential Pattern Trees With Sequence Deletion", IEEE Access - The Journal for Rapid open access publishing, Vol. 2, pp. 1374–1383, 2014.
11. H. Ryang and U. Yun., "Efficient High Utility Pattern Mining for Establishing Manufacturing Plans with Sliding Window Control", IEEE - Expert Systems with Applications, Vol. 57, pp. 214–231, 2017.
12. R. Moriwal, and V. Prakash, "An efficient Algorithm for finding frequent Sequential traversal Patterns from Web Logs Based on Dynamic Weight Constraint", Proceedings of the Third International Conference on trends in Information, Telecommunication and Computing. vol. 150, (Springer Science + Business Media New York 2013) (2013).
13. Zaarour Omar et al., "Effective web log mining and online navigational pattern prediction," International Journal of Knowledge Based Systems, Elsevier, Vol. 49, pp. 50–62, Sept 2013.

Query-Optimized PageRank: A Novel Approach

Rajendra Kumar Roul, Jajati Keshari Sahoo and Kushagr Arora

Abstract This paper addresses a ranking model which uses the content of the documents along with their link structures to obtain an efficient ranking scheme. The proposed model combines the advantages of *TF-IDF* and PageRank algorithm. *TF-IDF* is a term-weighting scheme that is widely used to evaluate the importance of a term in a document by converting textual representation of information into a vector space model. The PageRank algorithm uses hyperlink (links between documents) to determine the importance of a Web document in the corpus. Combining the relevance of documents with their PageRanks will refine the retrieval results. The idea is to update the link structure based on the document similarity score with the user query. Results obtained from the experiment indicate that the performance of the proposed ranking technique is promising and thus can be considered as a new direction in ranking the documents.

Keywords Cosine-similarity · Pagerank · Query-optimized · Spearman's footrule · TF-IDF

1 Introduction

The dynamic Web which contains a huge volume of digital documents is growing in an exponential manner due to the popularity of the Internet. This makes difficult for the search engine to arrange the retrieved results from the Web according to the

R. K. Roul (✉) · K. Arora
Department of Computer Science, BITS-Pilani, K. K. Birla Goa Campus,
Zuarinagar 403726, Goa, India
e-mail: rkroul@goa.bits-pilani.ac.in

K. Arora
e-mail: kushagrarora786@gmail.com

J. K. Sahoo
Department of Mathematics, BITS-Pilani, K. K. Birla Goa Campus,
Zuarinagar 403726, Goa, India
e-mail: jksahoo@goa.bits-pilani.ac.in

© Springer Nature Singapore Pte Ltd. 2019
H. S. Behera et al. (eds.), *Computational Intelligence in Data Mining*,
Advances in Intelligent Systems and Computing 711,
https://doi.org/10.1007/978-981-10-8055-5_60

673

user's choice. *Ranking* which is one of the powerful technique in machine learning can shed light in this direction by arranging the retrieved results in a better manner. Many research works have been done in this field [1–7].

The existing PageRank algorithm [8, 9] has many short comings. The current paper tries to improve the existing PageRank algorithm. The problem with the traditional PageRank algorithm is that the link structure used in the calculation of PageRank assumes that a surfer will jump from a page to the other uniformly which may lead to topic drifting, i.e., suppose a user is looking for the documents on *computer science* and some documents may have outgoing links to *biological documents* (because many biological documents also related to computer science) then those documents also incorporate in the PageRank calculation. The proposed approach overcomes this by biasing the next jump of the user to only those documents which are relevant to the particular query he is searching for. The modified link structure which is input to the PageRank algorithm will contain those output links which are connected to a document having some standard relevance (which in our case is nonzero cosine-similarity with the user query). When these resulted documents from a normal ranking function (cosine-similarity in the proposed approach) with the updated link structure are supplied to the existing PageRank algorithm, the new ranks are obtained. These new ranks are query dependent and give a new direction to the existing PageRank algorithm that incorporates the needs of the user as well. Modifying the existing PageRank algorithm and considering it for restructuring the links based on query would be more beneficial while implementing it in search engine with datasets which have good link structure such as research journal databases that have a lot of citations and may not be linked to the topics as well. Hence, if their structure is modified, one may be able to get better results. Empirical results on different datasets show the effectiveness of the proposed approach.

The rest of the paper is organized on the following lines: Sect. 2 discusses the background of those techniques which are used in the proposed approach. In Sect. 3, we have discussed the proposed technique used for query-optimized PageRank. Section 4 discusses the experimental results of the proposed work. Finally, in Sect. 5, we concluded the work with some future enhancement.

2 Basic Preliminaries

2.1 Vector Space Model

An algebraic model called Vector Space Model (VSM) [10] aimed to facilitate information retrieval by modeling the documents as a set of terms. VSM transforms a full text version to a vector which has various patterns of occurrence. It represents document (D) as a vector of words, $D = (w_{1j}, w_{2j}, w_{3j}, w_{4j}, \ldots, w_{nj})$, where w_{ij} is weight of ith word in jth document.

2.2 Term Frequency and Inverse Document Frequency

Term Frequency (*TF*) measures how often a term *t* occurs in a document *D* whereas inverse document frequency (*IDF*) measures the importance of *t* in the entire corpus *P*. *TF-IDF* [11] is a technique which finds the importance of terms in a document based on how they appear in the corpus. If *t* appears in many documents, its importance goes down. Therefore, the common terms need to be filtered out. *TF-IDF* is calculated using Eq. 1.

$$TF\text{-}IDF_{t,D} = TF_{t,D} \times IDF_t \tag{1}$$

where

$$TF_{t,D} = \frac{number\ of\ occurance\ of\ t\ in\ D}{total\ length\ of\ D}$$

$$IDF_t = log\left(\frac{number\ of\ documents\ in\ P}{number\ of\ documents\ contain\ the\ term\ t}\right)$$

2.3 Cosine-Similarity

Cosine-similarity[1] is one among the standard similarity techniques which measures the similarity between two document vectors, say $\vec{D_1}$ and $\vec{D_2}$ and can be represented using Eq. 2:

$$cosine\text{-}similarity(\vec{D_1}, \vec{D_2}) = \frac{\vec{D_1} \cdot \vec{D_2}}{|\vec{D_1}| * |\vec{D_2}|} \tag{2}$$

If the angle between $\vec{D_1}$ and $\vec{D_2}$ is near to zero, then they share most of the common terms between them, and if the angle approaches toward 90°, then the dissimilarity between them increases. To retrieve the document relevant to a query, it is necessary to have a measure that computes the degree of similarity between the query and each of the document in the corpus. As in the vector space model we deal with vectors, the cosine-similarity between the vectors is used to obtain the similarity measure.

---

[1]https://radimrehurek.com/gensim/tutorial.html.

The cosine-similarity between the document $\overrightarrow{D_1}$ and the query $\overrightarrow{Q}$ is represented in Eq. 3.

$$cos\text{-}sim(\overrightarrow{D_1}, \overrightarrow{Q}) = \frac{\overrightarrow{D_1} \cdot \overrightarrow{Q}}{|\overrightarrow{D_1}| * |\overrightarrow{Q}|} \qquad (3)$$

3 Proposed Approach

This section discussed the query-optimized PageRank algorithm starting with a detail description of the PagaRank algorithm. The PageRank of a document combined with its *TF-IDF* vector will generate the desire ranking. Details are discussed in Sect. 3.2.

3.1 PageRank Algorithm

PageRank determines how important the Web site is by counting the number and quality of links to a page where the links are dynamic. It analyzes each link and thus assigns a weight to each element of a hyperlink set of documents. It basically measures the importance of a page compared to all other pages on the Web. A link to a page leads to an increase in the weight, while the page with no link can be neglected as it is less significant. The following steps are used to compute the PageRank:

1. Add an edge directed from node i to node j in the graph when Web site i references j. Since only the connections between the Web sites are important, any navigational links such as next and back buttons are ignored while computing their PageRanks.

2. In the proposed approach, all the pages will get equal importance that is linked by a single page. Thus, the importance of each page will be $\frac{1}{n}$, iff a node has n outgoing edges. Let us denote the transition matrix of the graph G by A and represented as

$$A = \begin{bmatrix} x_{11} & x_{12} & x_{13} & \cdots & x_{1n} \\ x_{21} & x_{22} & x_{23} & \cdots & x_{2n} \\ \vdots & \vdots & \vdots & \ddots & \vdots \\ x_{d1} & x_{d2} & x_{d3} & \cdots & x_{dn} \end{bmatrix}$$

3. Supposing that initially the importance is uniformly distributed among all the nodes then the initial rank vector v will have all the entries as $\frac{1}{k}$, if G has k nodes. Since the importance of a Web page is increased by each incoming link using step 1, the rank of each page will be updated by adding the importance of the incoming links to the current value. It is equivalent to multiplying A with v. Thus, after first iteration, the new importance vector is $v_1 = Av$. We can keep

iterating the above process to get the sequence $v, Av, A^2v, ..., A^kv$. This is called the PageRank of the Web graph G.

4. Since the dataset is so large, the graph G is not expected to be connected. Likewise, our dataset may have plain descriptive pages that do not have any outgoing links. Thus, for any directed Web graph, one require a non-ambiguous meaning of the rank of a page. To overcome such problems, damping factor (p) is used which is a positive constant ranging between 0 and 1. The typical value of damping factor is 0.15. The PageRank of G is defined in Eq. 4.

$$PageRank(G) = (1\text{-}p)A + pB \tag{4}$$

where

$$B = \frac{1}{n} \begin{bmatrix} 1 & 1 & 1 & \dots & 1 \\ 1 & 1 & 1 & \dots & 1 \\ \vdots & \vdots & \vdots & \ddots & \vdots \\ 1 & 1 & 1 & \dots & 1 \end{bmatrix}$$

3.2 Query-Optimized PageRank

It is a general thought to develop a technique which can combine the PageRank of a Web page (i.e., document) with its *TF-IDF* vector to get advantages of both these techniques. In this paper, a framework is drawn by using the content of Web page and the outlink information synchronously though primarily we have modified the popular "PageRank" algorithm. For a given user query Q, rank and return the documents which have nonzero cosine-similarity score with the query Q. In those return documents, we modify the link structure matrix to compute the new ranks. All the links which have a zero cosine-similarity with the query are removed as an outlink; therefore, for every page, one gets a smaller link structure. This idea basically emphasizes that a surfer who has started searching for a query will only jump to those pages which are related to the query in a certain way. Once the link structure is modified, adjustments are made by including the damping factor and then the new ranks are calculated using the steps mentioned below:

1. *Preprocessing of documents*:
   Consider a corpus P having C number of classes (where $C = \{C_1, C_2, ..., C_n\}$) and each class has m number of documents $D = \{D_1, D_2, ..., D_m\}$. Documents are preprocessed by removing the stop-words, and stemming is done for the entire corpus using porter stemming algorithm[2]. The stemmed vocabulary forms a dictionary of all the features (i.e., terms). Assume that the corpus dimension is $p \times r$ (p is the number of documents and r is the number of terms of the corpus P).

---

[2]http://tartarus.org/martin/PorterStemmer/.

Table 1 Term-document table

	D_1	D_2	D_3	...	D_p
t_1	t_{11}	t_{12}	t_{13}	...	t_{1p}
t_2	t_{21}	t_{22}	t_{23}	...	t_{2p}
t_3	t_{31}	t_{32}	t_{33}	...	t_{3p}
.	.	.	.	...	.
.	.	.	.	...	.
.	.	.	.	...	.
t_r	t_{r1}	t_{r2}	t_{r3}	...	t_{rp}

2. *Finding TF-IDF vectors*:
   From all the preprocessed documents and terms, the text data are later represented in numerical values. The formal VSM has been chosen for this purpose which converts every document into vector using *TF-IDF* values called *document-term* vectors (i.e., documents in the term space) as shown in Table 1.

3. *Finding cosine-similarity with the query vector:*
   Similar to the documents, the *TF-IDF* vector of the query is obtained. The cosine-similarity is calculated between the query vector and each document vector. The documents are arranged in the decreasing order of their cosine-similarity values, and all the documents that have zero cosine-similarity are removed from the corpus.

4. *Computing the ranks of the documents:*
   The ranks of the documents are calculated using the PageRank algorithm on the remaining documents in the corpus. Initially, same importance is given to all the documents. After that, the rank is updated by adding the importance of the incoming links to the current value of the document. The process is repeated for all the documents. The ranks are returned after incorporating the damping factor to the obtained rank matrix.

4 Experimental Work

To implement the PageRank algorithm with *TF-IDF*, a specific research dataset has been chosen called dbpedia[3] which includes both link structure and content. This dataset does not have a predefined set of relevant documents for any given query, so the measures such as accuracy, precision, recall, and F-measure could not be com-

---

[3]http://wiki.dbpedia.org/Datasets.

puted directly. This led to use a method called Spearman's footrule [12] to check accuracy of our ranking approach as compared to standard cosine-similarity ranks.

Spearman's footrule is applied to both the techniques (query-optimized and cosine-similarity) for obtaining the rankings. Assuming the size of the dataset to be N, this in turn implies that the rankings should range between 1 and N. As the rankings given by each of the two techniques being compared is basically a permutation of the other, hence there are no ties allowed. Let us say that the result of the rankings is permutations σ_1 for the proposed approach and σ_2 for the ranking based on the cosine-similarity score. This permutation is over S, the set of overlapping results when the top 'k' rankings of each model are considered. Spearman's footrule is computed using Eq. 5.

$$Fr^{|S|}(\sigma_1, \sigma_2) = \sum_{i=1}^{|S|} |(\sigma_1(i) - \sigma_2(i)|$$ (5)

$Fr^{|S|}$ is zero when the two lists are identical. When $|S|$ is even, $\frac{1}{2}S^2$ achieved its maximum value of $\frac{1}{2}S^2$ and when $|S|$ is odd, it achieved the maximum value of $\frac{1}{2}(|S| + 1)(|S| - 1)$. $Fr^{|S|}$ value will lie between 0 and 1, and this can be achieved by dividing the obtained result with its maximum value which is independent on the size of the overlap. The normalized Spearman's footrule (*NFr*) for $|S| > 1$ is computed using the Eq. 6.

$$NFr = \frac{Fr^{|S|}}{max\ Fr^{|S|}}$$ (6)

Thus, *NFr* will range between 0 and 1. Tables 2, 3, 4, 5, 6, and 7 show the rankings for different queries (both unigram and bi-gram) using the cosine-similarity, and the proposed query-optimized PageRank algorithm (*NFr* shows the difference between the *NFr* value obtained on cosine-similarity and *NFr* value obtained on the proposed approach). Only one query "Massachusetts" (Table 4) had three results, and Spearman coefficient turned out to be 0 for that since the retrieved documents are very less

Table 2 Query: agriculture (NFr = 0.333)

Cosine-similarity ranking	Query-optimized PageRanking
Agriculture	Agriculture
Algeria	Africa
Albania	Algeria
Almond	African_union
Accountancy	Albania
Africa	Almond
2005_Atlantichurricane_season	2005_Atlantichurricane_season
Aberdeen	Aberdeen

Table 3 Query: wikipedia (NFr = 0.821)

Cosine-similarity ranking	Query-optimized PageRanking
2005_Lake_TAnganyika_earthquake	African_Great_Lakes
African_Darter	2005_Lake_TAnganyika_earthquake
African_Black_Oystercatcher	African_Grey_Hornbill
African_Jacana	Abstract_art
African_dwarf_frog	06-02-2006
African_Great_Lakes	African_Brush_tailed_Porcupine
African_Grey_Hornbill	02-08-2000
Abstract_art	16_Cygni_Bb
06-02-2006	African_Buffalo
African_Brush_tailed_Porcupine	Almaty
02-08-2000	African_Darter
16_Cygni_Bb	African_Black_Oystercatcher
African_Buffalo	African_Jacana
Almaty	African_dwarf_frog

Table 4 Query: massachusetts (NFr = 0)

Cosine-similarity ranking	Query-optimized PageRanking
Abu_dhabi	Abu_dhabi
2004_Atlantichurricane_season	2004_Atlantichurricane_season
Alternative_rock	Alternative_rock

Table 5 Query: nobel (NFr = 0.635)

Cosine-similarity ranking	Query-optimized PageRanking
Alfred_Nobel	Albert_Einstein
Albert_Einstein	Alan_Turing
ABO_blood_group_system	Aberdeen
Alan_Turing	Alfred_Nobel
Action_potential	ABO_blood_group_system
Arican_Amrican_literature	Action_potential
Aberdeen	Arican_Amrican_literature

Table 6 Query: Roman Empire (NFr = 0.661)

Cosine-similarity ranking	Query-optimized PageRanking
1st_century_BC	14th_century
13th_century	13th_century
10th_century	Abacus
5th_century	9th_century
3rd_century	11th_century
11th_century	10th_century
6th_century	5th_century
Abacus	6th_century
14th_century	1st_century
1st_century	Akkadian_Empire
Akkadian_Empire	16th_century
9th_century	Aachen
16th_century	1st_century_BC
Aachen	3rd_century

Table 7 Query: general history (NFr = 0.714)

Cosine-similarity ranking	Query-optimized PageRanking
9th_century.txt	12th_century
1st_century_BC.txt	11th_century
12th_century.txt	9th_century
6th_century.txt	13th_century
African_slave_trade.txt	10th_century
Acceleration.txt	20th_century
13th_century.txt	4th_century
Adriaen_van_der_Donck.txt	1st_century_BC
Alfred_the_Great.txt	6th_century
10th_century.txt	African_slave_trade
Acts_of_Union_1707.txt	Acceleration
20th_century.txt	Adriaen_van_der_Donck
4th_century.txt	Alfred_the_Grea
11th_century.txt	Acts_of_Union_1707

and the link structure could not refine the ranks much based on the cosine-similarity of the documents with the query. A nonzero spearman's score says that the ranking given by the proposed approach that incorporates the cosine-similarity with PageRank is different, and puts forward a new direction of research for a modified PageRank which is query dependent.

5 Conclusion

This paper developed a ranking model that utilized the link structure and the cosine-similarity of the documents with the query, so as to improve the set of retrieved pages and the ranking. It has brought together the merits of ranking using *TF-IDF* weights and the PageRank algorithm. Unlike the *TF-IDF* weighting scheme, it does not just concentrate on the content of the documents nor like the PageRank algorithm rely solely on the link structure rather it ranks based on the relevance of the documents and their outlinks to the query. Our approach puts forward new unexplored ideas on PageRank algorithm which take cares the user's preferences. This work can be further extended as follows:

- To compare the ranking model of the proposed approach with other standard and recognized ranking models to understand the true efficiency of our approach.
- Modifications on the PageRanking based on query and subsetting the link structure accordingly may improve existing ranking systems such as on research journal databases, Wikipedia database (only on those links which are in Wikipedia), etc.
- Further, each query may be viewed as a mixture of various topics where each document is considered to have a set of topics that are assigned to it via. latent dirichlet allocation. This way PageRank matrix will receive more relevant and important outlinks.

References

1. A. J. Roa-Valverde and M.-A. Sicilia, "A survey of approaches for ranking on the web of data," *Information Retrieval*, vol. 17, no. 4, pp. 295–325, 2014.
2. H. Wu, Y. Hu, H. Li, and E. Chen, "A new approach to query segmentation for relevance ranking in web search," *Information Retrieval Journal*, vol. 18, no. 1, pp. 26–50, 2015.
3. J. B. Vuurens and A. P. de Vries, "Distance matters! cumulative proximity expansions for ranking documents," *Information Retrieval*, vol. 17, no. 4, pp. 380–406, 2014.
4. S. Gugnani and R. K. Roul, "Article: Triple indexing: An efficient technique for fast phrase query evaluation," *International Journal of Computer Applications*, vol. 87, no. 13, pp. 9–13, 2014.
5. Y. Wang, J. Lu, J. Chen, and Y. Li, "Crawling ranked deep web data sources," *World Wide Web*, vol. 20, no. 1, pp. 89–110, 2017.
6. P. Chahal, M. Singh, and S. Kumar, "An efficient web page ranking for semantic web," *Journal of The Institution of Engineers (India): Series B*, vol. 95, no. 1, pp. 15–21, 2014.

7. R. K. Roul and S. Sanjay, "Cluster labelling using chi-square-based keyword ranking and mutual information score: a hybrid approach," *International Journal of Intelligent Systems Design and Computing*, vol. 1, no. 2, pp. 145–167, 2017.
8. L. Page, S. Brin, R. Motwani, and T. Winograd, "The pagerank citation ranking: Bringing order to the web." Stanford InfoLab, Tech. Rep., 1999.
9. S. Gugnani, T. Bihany, and R. K. Roul, "A complete survey on web document ranking," *International Journal of Computer Applications*, vol. ICACEA, no. 2, pp. 1–7, 2014.
10. G. Salton, A. Wong, and C.-S. Yang, "A vector space model for automatic indexing," *Communications of the ACM*, vol. 18, no. 11, pp. 613–620, 1975.
11. K. Sparck Jones, "A statistical interpretation of term specificity and its application in retrieval," *Journal of documentation*, vol. 28, no. 1, pp. 11–21, 1972.
12. P. Diaconis and R. L. Graham, "Spearman's footrule as a measure of disarray," *Journal of the Royal Statistical Society. Series B (Methodological)*, pp. 262–268, 1977.

Effects of Social Media on Social, Mental, and Physical Health Traits of Youngsters

Gautami Tripathi and Mohd Abdul Ahad

Abstract The widespread use of social media has revolutionized the mode of communication in today's world. It has become an unparalleled mode of interaction around the globe. With such deep penetration of this technology in people's life, it has brought about many changes in their lifestyle including their health. The concept of health in today's world not only refers to the physical fitness but also encompasses the mental and social well-being. With the advent of technology, most individuals are exposed to the Internet and social networking and spend a major portion of their time using these for performing one or the other activities. The constant and excessive use of social media has affected human life in multiple ways. This paper lucidly examines the impacts of social media on the physical, mental, and social health aspects of youngsters.

Keywords Social media · Health · Well-being · Social behavior

1 Introduction

Social networking has emerged as one of the most sought-after modes of interaction among all age groups, especially among youngsters. According to [1], Facebook is the most widely used social network worldwide with around 1.94 billion monthly active users for March 2017 and 1.28 billion daily active users on average for March 2017. The total user base of Facebook in India is about 213 million [2]. Figure 1 shows a countrywise detailed statistics on the number of active users on Facebook in the year 2017.

G. Tripathi (✉) · M. A. Ahad
Department of Computer Science and Engineering, School of Engineering
Sciences and Technology, Jamia Hamdard (Hamdard University),
New Delhi 110062, India
e-mail: gautami1489@gmail.com

M. A. Ahad
e-mail: itsmeahad@gmail.com

H. S. Behera et al. (eds.), *Computational Intelligence in Data Mining*,
Advances in Intelligent Systems and Computing 711,
https://doi.org/10.1007/978-981-10-8055-5_61

685

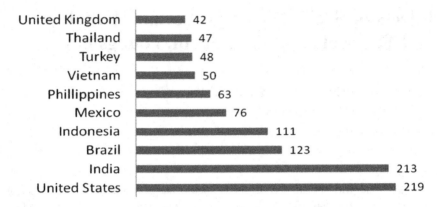

Fig. 1 Country wise number of Facebook users as of April 2017 (in millions) [2]

The debate on the effects of social networking is hard to be construed as it is a topic that divides viewpoints of people. Some people believe it to be a remarkable tool, but others are worried about the impact it has on people's lives. The pace at which social networking is changing the way people interact with each other might soon replace the conventional and long-established mode of interactions among people.

The dictionary [3] defines social networking as "The use of dedicated Web sites and applications to interact with other users or to find people with similar interests." Researches show that the choice of social networking sites depends on the demographics, but one fact that remains unchanged across all regions is that people are deeply engrossed in using these modes of interactions leading to addiction in some extreme cases. The growing popularity of social networking is the result of easy Internet availability and accessibility using smartphones and other handheld devices. Today, mobile social networking services have grown at a rapid rate resulting in the development of mobile apps for various social media. For instance, more than 80% of the users access Facebook via mobile devices [4]. Social networking sites allow you to connect and communicate with others, share photos, videos and information, organize events, chat, and play online games. The focus of this paper is to critically understand the various factors affecting the physical, mental, and social aspects of human health.

2 Related Work

The ubiquitous nature of social media has enticed numerous researches in the area. Some of the major research areas that have been covered to study the impact of social media and networking on human lives includes body image and disordered eating [5], security [6], customer value creation [7], social networking and Internet addiction, mobile social networking services [8], smartphones leading to social networking Web sites adoption, and health [9]. Grace Holland and Marika Tiggemann [5] have reviewed the impact of the use of social networking sites on body image and disordered eating outcomes. Body dissatisfaction occurs when views of the body are negative and involves a perceived discrepancy between a person's assessment of their actual and ideal body [10] which can lead to maladaptive consequences for both physical and mental health including depression, anxiety, low self-esteem, and eating disorders [11, 12, 13]. Since 1990, Internet usage has increased by approximately 50% each year [14]. In July 2015, it was estimated that the Internet was used by 43.4% of the world's population while in July 2016, it grew to 46.1%, whereas the world population changed by mere 1.13% [15]. As of November 2016, 77% of online adults were using social media in one or the other forms [16]. The authors of [17] showed that the appearance comparisons made through social and traditional media negatively impacts women's appearance satisfaction, thoughts of dieting, and their exercise behavior. Furthermore, the comparisons made through social media had the worst impact on women's mood causing harm to young women's mental and physical health. The researchers in [18] used a questionnaire to collect information on social media among Pharmacy students of Kenyatta University, Nairobi, Kenya. In their study, it was observed that 45% students use social media networking for communication with real and virtual friends. Primack et al. [19] showed that the use of multiple social media platforms leads to symptoms of depression and anxiety. The author of [20] talked about the purpose of Internet usage and learning by Internet. The researchers in [21] discussed the role of Facebook in higher education and the general perception of the social networking sites. They also discussed how the use of Facebook differs for faculty and students. In [22], the authors talked about the relationship between social media usage and depressive symptoms among children.

3 Social Media and Human Health

Social media allows people to connect and communicate with each other based on their similar interests, beliefs, and values. It has made it easier for people to find and communicate with individuals who share similar ideologies. The pervasive nature of social media has influenced every aspect of human life in multiple ways. Figure 2 shows the different facets of social networking. Facebook, being the most widely used social networking site with a user base of 213 million in India, is a more informal social platform than LinkedIn which is primarily a professional network [1, 2]. LinkedIn is stronger in identity and reputation, whereas Facebook is more inclined toward presence, conversations, and communications.

The exorbitant use of social media has confined the human thinking and representation in compliance to the site's database. When people present themselves

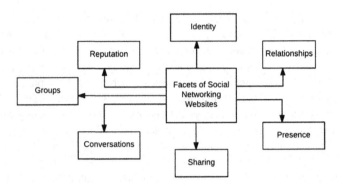

Fig. 2 Facets of social networking Web sites

using these networking sites, they tend to represent their likes, dislikes, interests, and habits by filling in the check marks conforming to a specified database, thus seeing and liking things they already agree to. With the passage of time, this might solidify individual opinions making them less open-minded and creative.

It is observed that this effect is primarily due to two reasons, firstly the friends with whom we share similar ideologies and secondly powerful algorithms used by social media that tends to show us things we have already shown interest in. Thus, an individual's representation of themselves is caged in the perspective of these sites inventing individual's taste, what they are and what they want to see. As an ill effect of social media, most people are trapped in an echo chamber where all they see and like is themselves, thus changing their perspective on various issues. This causes serious harm to their overall personality and health. Figure 3 shows different aspects of human health. The three major aspects of human health are mental, physical, and social health. Physical health is related to physical fitness and the physical capacity of the body to perform tasks referring to the efficient functioning of the body and its systems. Mental health defines a state of well-being in which an individual is capable of recognizing his abilities working in a productive and fruitful manner and coping up with the normal stresses of life. Social health involves being able to interact with others and participate in the community in both an independent and cooperative manner. If any of these gets affected, it has a negative effect on people's well-being.

The wellness program at University of California, Riverside, [23] defines seven dimensions of wellness that integrate all states of physical, mental, and social well-being.

Figure 4 shows the seven dimensions of well-being that contributes toward the overall social, physical, and mental health of humans. Thus, the effects of social networking on any of these dimensions of wellness have its direct impact on the three aspects of human health.

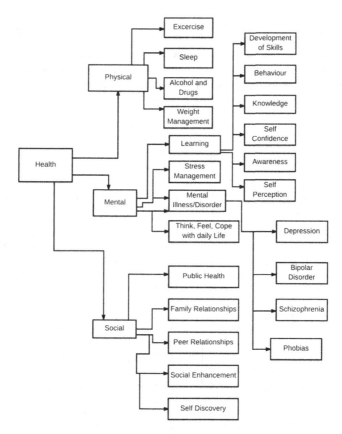

Fig. 3 Different aspects of human health

Fig. 4 Seven dimensions of wellness [23]

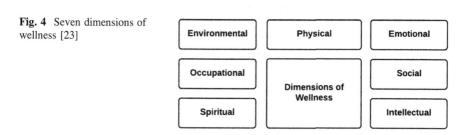

4 Procedure

The study was conducted in the National Capital Region in India. Data was collected through an online questionnaire which was given to 497 participants. The questionnaire was intended to cover almost all aspects of social media and networking with a focus on the impact of social media on the human health. The study was conducted in three phases. In the first phase, the demographics of the participants and their social media usage pattern were collected. The second phase focuses on the purpose of social media and networking. Lastly, the impacts of excessive usage of social media and networking on the various health aspects including the social, mental, and physical health of the participants were studied and critically analyzed.

5 Results and Observations

This section presents the various observations and analysis of the responses received from the participants. A total of 497 participants were involved in this cross-sectional study. Their responses were recorded online and are briefly presented below.

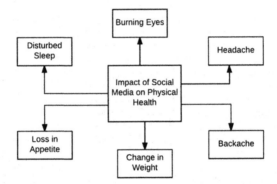

Fig. 5 Impact of social media on physical health

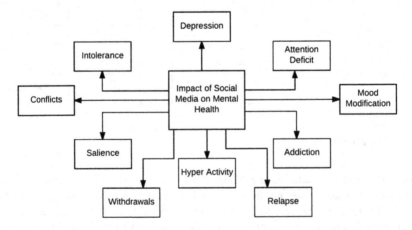

Fig. 6 Impact of social media on mental health

Fig. 7 Impact of social
media on social health

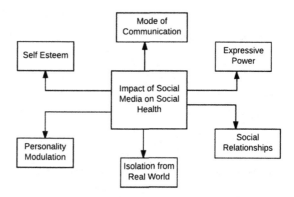

The responses obtained from the participants were analyzed to find out the effects
of social media and networking on the three aspects of human health. Figures 5, 6,
and 7 show the areas of impact of social media on the physical, mental, and social
aspects of human health, respectively. The impact of social media and networking on
the physical health was observed to be totally negative while for mental and social
health, the results were both positive as well as negative.

Tables 1 and 2 present the responses of the participants on demographics, social
media usage pattern, and health traits. It was observed that 97% of participants

Table 1 Demographics and Internet usage pattern of the participants

Total number of participants = 497		
Demographics/Internet usage pattern	Observations	Percentage
Gender	Male	63
	Female	37
Access Internet from	Home only	26
	School/University/College only	28
	Both home and School/College/University	43
	Other combinations	17
Daywise time spent using the Internet (in hours)	1–5	67
	6–10	18
	11–15	7
	16–20	3
	More than 20	1
Activities involved	Online shopping	11
	Entertainment	31
	Social networking/chatting	49
	Access information/studies/learning	9

Table 2 Responses of the participants on the various health traits

Questions on health traits	Responses of the participants (Percentage wise)		
	Yes	No	Don't know/don't want to disclose
Physical health			
When using Internet for long hours do you experience any kind of health issues?	66	31	3
Do you feel spending long hours communicating on social networking has affected your health in any way?	55	43	2
Do you feel that using social media and networking has reduced the amount time spent on physical activities like exercising, walking, hanging out with friends, playing outdoors?	62	36	2
Mental health			
Do you feel that you are addicted to Internet?	16	73	11
When offline, do you constantly think about getting online?	31	67	2
Does going offline makes you restless and you feel like going online immediately and you feel restless when not able to connect on social network?	41	56	3
Have you ever been involved in an argument with your family/teachers/elders for showing their disagreement toward your frequent use of social media?	73	18	9
Do you often update your pictures online on the social networking Web site?	49	41	10
Have you ever been involved in any kind of threat or other form of offensive/abusive content sent online to others?	4	78	18
Social health			
Do you prefer using social networking sites instead of spending time with family/relatives/doing household work?	63	21	6
Have you ever pretended to be someone else while using online chat or instant messaging?	21	76	3
Do you constantly check your social network profile while sitting in groups or physically attending any other gathering?	56	42	2
Have you ever used your smartphone or pretended to do so to avoid conversations with others?	43	51	6
Do you think you are isolating from society/family/relatives because of your involvement in online networking?	39	59	2
Do you feel emotionally connected to your online friends you have never met personally?	16	78	6
Since the time you started using social media to connect and communicate with others, was there any change in the number of your offline friends or your relationship with them?	23	54	23

accepted to be experiencing multiple health issues because of their prolonged use of social media using one or more electronic devices. Nearly 80% of the participants accepted to have frequent mood modifications and attention deficit. The other major problems were hyperactivity, addiction, conflicts, withdrawal, etc. The positive aspects of social networking on mental health were increased self-confidence, development of skills, awareness, etc. Few participants accepted that social media has helped them to expand their social circle (21%) and increased their expressive power (13%). Nearly 30% of respondents believed that it has provided a vast source of learning. But a majority of participants observed negative impacts on their social relationships with 36% accepting that social media has isolated them from the real world and had created a communication gap with their family and relatives. Nearly 29% participants accepted that they often prefer using social media over real-world communications and try to ignore the physical presence of their peers. More than 60% showed negative mood modifications and personality modulations and were often involved in arguments/fights with their elders when questioned on their prolonged social media usage.

6 Conclusions and Future Works

The results show that prolonged hours spent on social media, and networking has not only caused serious damage to the physical health of individuals but has also affected their social relationships and behavior. Moreover, it was also observed that reliance on social media and online communications has trapped individuals in an echo chamber rendering them inefficient of thinking beyond that leading to depression, hyperactivity, mood modifications, loss of reasoning and analysis, and attention deficit.

This study also has certain limitations that may create interesting opportunities for future research. Firstly, the study was a cross-sectional study which was limited to the national capital region in India, and only college students were employed as the research sample. While college students could present the largest segment of social media users, future research may generalize research findings using longitudinal study to other user populations in different environments (e.g., the workplace). Secondly, this paper focuses on generalized usage of social media and is not specific to a particular site. Future research may further examine the security aspects on specific social media. As social media encompasses a global environment, it will be interesting to look at cultural differences of social media usage in future studies. Statement on Ethical Approval

References

1. Facebook NewsRoom, Company Info, Statistics https://newsroom.fb.com/company-info/.
2. Countries with the most Facebook users as of April 2017, Statista 2017 https://www.statista.com/statistics/268136/top-15-countries-based-on-number-of-facebook-users/.
3. English Oxford Living Dictionaries https://en.oxforddictionaries.com/definition/social_network.
4. Tsay-Vogel, M (2016) Me versus them: Third-person effects among Facebook users. new media and society, 18(9):1956–1972.
5. Holland, Grace, and Marika Tiggemann (2016) A systematic review of the impact of the use of social networking sites on body image and disordered eating outcomes. Body image 17:100–110.
6. Hajli, Nick, and Xiaolin Lin (2017) Exploring the security of information sharing on social networking sites: The role of perceived control of information, Journal of Business Ethics 133.1:111–123.
7. Zhang, M., et al. (2016) Influence of customer engagement with company social networks on stickiness: Mediating effect of customer value creation, International Journal of Information Management. http://dx.doi.org/10.1016/j.ijinfomgt.2016.04.010.
8. Yang, Shuiqing, Bin Wang, and Yaobin Lu (2016) Exploring the dual outcomes of mobile social networking service enjoyment: The roles of social self-efficacy and habit, Computers in Human Behavior 64:486–496.
9. Ceglarek, Peter JD, and L. Monique Ward (2016) A tool for help or harm? How associations between social networking use, social support, and mental health differ for sexual minority and heterosexual youth, Computers in Human Behavior 65:201–209.
10. Grogan, S (2017) Body image: Understanding body dissatisfaction in men, women and children. Routledge, Tailor and Francis Group, 3rd Edition, London and New York.
11. Dittmar, H(2009) How do "body perfect" ideals in the media have a negative impact on body image and behaviors? Factors and processes related to self and identity. Journal of Social and Clinical Psychology, 28(1):1–8.
12. Grabe, S., Ward, L. M., and Hyde, J. S (2008) The role of the media in body image concerns among women: a meta-analysis of experimental and correlational studies. Psychological bulletin, 134(3):460–476.
13. Groesz, L. M., Levine, M. P., and Murnen, S. K. (2002) The effect of experimental presentation of thin media images on body satisfaction: A Meta analytic review. International Journal of eating disorders, 31(1):1–16.
14. INTERNET GROWTH STATISTICS, Today's road to e-Commerce and Global Trade Internet Technology Reports, By Internet World Stats http://www.internetworldstats.com/emarketing.htm.
15. Internet Live Stats http://www.internetlivestats.com/internet-users/.
16. Social Networking Use, In Pew Research Center http://www.pewresearch.org/data-trend/media-and-technology/social-networking-use/.
17. Fardouly, J., Pinkus, R. T., and Vartanian, L. R. (2017) The impact of appearance comparisons made through social media, traditional media, and in person in women's everyday lives. Body Image, 20:31–39.
18. Ogaji, I. J., Okoyeukwu, P. C., Wanjiku, I. W., Osiro, E. A., & Ogutu, D. A. Pattern of use of social media networking by Pharmacy students of Kenyatta university, Nairobi, Kenya. Computers in Human Behavior, 66, (2017):211–216.
19. Primack, B. A., Shensa, A., Escobar-Viera, C. G., Barrett, E. L., Sidani, J. E., Colditz, J. B., and James, A. E. (2017) Use of multiple social media platforms and symptoms of depression and anxiety: A nationally-representative study among US young adults. Computers in Human Behavior, 69:1–9.
20. Ruzgar, Nursel Selver (2005) A research on the purpose of Internet usage and learning via Internet. TOJET: The Turkish Online Journal of Educational Technology 4.4:27–32.

21. Roblyer, Margaret D., et al (2010) Findings on Facebook in higher education: A comparison of college faculty and student uses and perceptions of social networking sites. The Internet and higher education 13.3:134–140.
22. McCrae, N., Gettings, S. and Purssell, E. (2017) Social Media and Depressive Symptoms in Childhood and Adolescence: A Systematic Review Adolescent Res Rev, 1–16. https://doi.org/10.1007/s40894-017-0053-4.
23. Seven Dimensions of Wellness, University of California Riverside https://wellness.ucr.edu/seven_dimensions.html.

Document Labeling Using Source-LDA Combined with Correlation Matrix

Rajendra Kumar Roul and Jajati Keshari Sahoo

Abstract Topic modeling is one of the most applied and active research areas in the domain of information retrieval. Topic modeling has become increasingly important due to the large and varied amount of data produced every second. In this paper, we try to exploit two major drawbacks (topic independence and unsupervised learning) of latent Dirichlet allocation (LDA). To remove the first drawback, we use Wikipedia as a knowledge source to make a semi-supervised model (Source-LDA) for generating predefined topic-word distribution. The second drawback is removed using a correlation matrix containing cosine-similarity measure of all the topics. The reason for using a semi-supervised LDA instead of a supervised model is not to overfit the data for new labels. Experimental results show that the performance of Source-LDA combine with correlation matrix is better than the traditional LDA and Source-LDA.

Keywords Correlation · F-measure · Latent Dirichlet Allocation · Source-LDA · Topic modeling

1 Introduction

In a semantic information retrieval model when many words of a document are combined, a topic gets generated. Latent Dirichlet allocation (LDA) [1] is an unsupervised topic modeling and is very popular in the domain of text mining [2, 3], topic extraction, and image classification [4]. However, traditional LDA is not used supervised labels which adds expert knowledge into the learning procedure and also the topics in them are independent in nature. Hence, traditional LDA has limited utilities

R. K. Roul (✉)
Department of Computer Science, BITS-Pilani, K. K. Birla Goa Campus,
Zuarinagar 403726, Goa, India
e-mail: rkroul@goa.bits-pilani.ac.in

J. K. Sahoo
Department of Mathematics, BITS-Pilani, K. K. Birla Goa Campus,
Zuarinagar 403726, Goa, India
e-mail: jksahoo@goa.bits-pilani.ac.in

© Springer Nature Singapore Pte Ltd. 2019
H. S. Behera et al. (eds.), *Computational Intelligence in Data Mining*,
Advances in Intelligent Systems and Computing 711,
https://doi.org/10.1007/978-981-10-8055-5_62

for many real-world applications. Several variations of LDA have been developed by researchers in order to generate supervised learning, and many research works have done in document labeling [5–11].

In semi-supervised LDA (ssLDA), some of the documents in a corpus have associated labels and many of them do not have. Assignment of topics can be either approximate (at the document-level) or detailed (at the word-level). Compared to LDA, ssLDA has the following advantages:

i. Multiple-label constraints, i.e., multiple topic assignments are supported by ssLDA.
ii. It allows two level of labeling assignment, whereas traditional LDA has only one level.

- Document-level assignment of labels to a document.
- Word-level correspondence between words and labels within a document.

iii. Can able to process automatically a mixture of documents with and without tags. Because of semi-supervised in nature, it has a good control and flexibility over the training dataset used for classification.

In this paper, we developed a methodology where the ssLDA (Source-LDA is used for this purpose) is run on a corpus of documents. It generates a list of topics. The correlation between all the topics is computed which creates a correlation matrix. All these topics are appended to a list which generates a label for the document. The experiment is carried out on five well-known machine learning datasets, and results have shown the efficiency of the proposed approach.

The rest of the paper is organized on the following lines: In Sect. 2, we have discussed the background details of the techniques used in our approach. Section. 3 discusses the proposed methodology used for query optimized page rank. Section 4 discusses the experimental analysis of the proposed work. Finally, in Sect. 5, we concluded the work with some future enhancement.

2 Basic Preliminaries

2.1 Term Frequency and Inverse Document Frequency

Term frequency (TF) measures how often a term t occurs in a document D where as inverse document frequency (IDF) measures the importance of t in the entire corpus P. TF-IDF is a technique which finds the importance of terms in a document based on how they appear in the corpus. If t appears in many documents then its importance goes down. Therefore, the common terms need to be filtered out. TF-IDF is defined as

$$TF\text{-}IDF_{t,D} = TF_{t,D} \times IDF_t$$

where

$$TF_{t,D} = \frac{number\ of\ occurance\ of\ t\ in\ D}{total\ length\ of\ D}$$

$$IDF_t = \log \left(\frac{number\ of\ documents\ in\ P}{number\ of\ documents\ contain\ the\ term\ t} \right)$$

2.2 Correlation Matrix

The correlation (cr) [12] between two terms t_i and t_j is computed using Eq. 1

$$cr_{t_i t_j} = \frac{C_{t_i t_j}}{\sqrt{\left(V_{t_i} * V_{t_j} \right)}} \tag{1}$$

where $C_{t_i t_j}$ is the covariance between t_i and t_j, and V_{t_i} and V_{t_j} are their variance respectively as given below.

$$V_{t_i} = \frac{1}{b-1} \sum_{m=1}^{b} \left(X_{im} - \overline{X_i} \right)^2$$

$$V_{t_j} = \frac{1}{b-1} \sum_{m=1}^{b} \left(X_{jm} - \overline{X_j} \right)^2$$

where $\overline{X_i}$ and $\overline{X_j}$ represent the mean of b documents of t_i and t_j, respectively. Covariance is a measure of the joint variability of two terms t_i and t_j and is defined in Eq. 2.

$$C_{t_i t_j} = \frac{1}{b-1} \sum_{m=1}^{b} \left(X_{im} - \overline{X_i} \right) \left(X_{jm} - \overline{X_j} \right) \tag{2}$$

3 Methodology

3.1 Data Acquisition and Preprocessing of the Documents:

Given a corpus of documents $(D_1, D_2, \ldots, D_n)$ from different categories (or branch of engineering), all these documents are preprocessed by removing stop-words, stemming is done using porter stemming algorithm[1], and finally all possible monogram

---

[1]http://snowball.tartarus.org/algorithms/porter/stemmer.html.

and bi-gram keywords $(k_1, k_2, \ldots, k_m)$ are extracted which constitute the feature set. All these documents are converted into vectors using vector space model (VSM) where the *TF-IDF* weights of the keywords are used to constitute the vector.

3.2 Technique Used for Latent Dirichlet Allocation

LDA finds the mixture of topics in a document by observing all the words in a document where each topic is characterized by distribution over words (the distribution is multinomial), and it is shown in Fig. 1. It performs two steps of processing: generation and inference.

Generative process
For D documents, T topics, N words and using Dirichlet distribution on α (Dirichlet prior concentration parameter of per document-topic distribution) and β (Dirichlet prior concentration parameter of per topic-word distribution), we first generate multinomial distribution θ (topic distributed over document) and ϕ (word distributed over document).

1. Choose $\theta_i Dir(\alpha)$ where $i \in \{1, 2, \ldots, D\}$.
2. Choose $\phi_k Dir(\beta)$ where $k \in \{1, 2, \ldots, K\}$.
3. For each word positions i, j where $i \in \{1, 2, \ldots, D\}$ and $j \in \{1, 2, \ldots, N\}$.

   – Choose a topic $z_{i,j} Multinomial\left(\theta_i\right)$.
   – Choose a word $w_{i,j} Multinomial\left(\phi_{z_{i,j}}\right)$.

Inferences Process
Now we use Gibbs sampling to learn various distributions (a set of topics and the probabilities of the associated words, the particular topic mixture of each document and the topic of each word). Gibbs sampling treats all θ's are independent to each other. Similarly, all ϕ's are also treated independent. Mathematically, the inference

Fig. 1 Working methodology of LDA

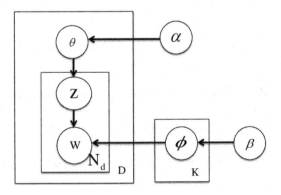

process is shown in Eq. 3.

$$P(W, Z, \theta, \phi; \alpha, \beta) = \prod_{i=1}^{K} P\left(\phi_i; \beta\right) \prod_{j=1}^{D} P\left(\theta_j; \beta\right) \prod_{t=1}^{N} P\left(Z_{j,t} \vee \theta_j\right) P\left(W_{j,t} \vee \phi_{z_{j,t}}\right)$$

(3)

Example Consider two type of biased dices: document and topic dices. D document dices have K faces denoting K topics and K topic dices contain V faces denoting V words (V vocabulary, i.e., total number of words). Since the dices are biased so in generative process, we chose a topic from multinomial document-topic distribution and a word from multinomial topic-word distribution. In the inference step, we use above results to find biases of both dices, thus finding probability of a topic in a document and a word in a topic. It is analogous to rolling a document dice and the list of topics (faces) which are more likely to appear. Similarly, for a topic dice when it is rolled the list of words (faces) which are more likely to appear, thus, obtaining all document-topic and word-topic distributions (biases).

3.3 Technique Used for Source-LDA

We first calculate $X_i = n_{wi} + \epsilon$ (where ϵ is any small positive number and n_{wi} is the number of times the word w_i from the corpus vocabulary appears in the knowledge source document).

Algorithm:
For each of the T topics ϕ_t :
      If $t \leq K$ then
           Choose $\phi_t \sim Dir(\beta)$
      Else
           Choose $\lambda_t \sim N(\mu, \sigma)$
           Choose $\delta_t \leftarrow \left[X_{t,1}^g(\lambda_t), X_{t,2}^g(\lambda_t), \dots, X_{t,V}^g(\lambda_t)\right]$
           Choose $\phi_t \sim Dir(\delta_t)$
      End If
End For
For each of the D documents d :
      Choose $N_d Poisson(\xi)$
      Choose $\theta_d Dir(\alpha)$
      For each of the N_d words $w_{n,d}$:
           Choose $Z_{n,d} \sim Multinomial(\theta)$
           Choose $w_{n,d} \sim Multinomial(\phi_{Z_{n,d}})$
      End For
End For

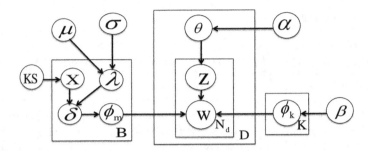

Fig. 2 Working methodology of Source-LDA

Here, δ is Dirichlet prior concentration parameter of per topic-word distribution from knowledge source, $Z_{i,j}$ is topic assignment for the word $w_{i,j}$ and W_{ij} is jth word of ith document. Here the new plate which works as an extension in Source-LDA (shown in Fig. 2) serves for the following purpose. Given a knowledge source KS, it forms a vector X_{mV} where m is the number of topics and V is the vocabulary of knowledge source. An element of $X_{ij} = n_{wj} + \epsilon$ (where n_{wj} is the number of times the word w_i from the corpus vocabulary appears in the knowledge source document, here each knowledge source document indicates a topic document). Here, λ serves as a damping factor, i.e., it controls the influence of knowledge source on δ_t and thus on ϕ. Most of the generative and inference process are similar to that of LDA. However, it has some differences in both which is explained below:

Generative Process

In Source-LDA, we generate multinomial distribution ϕ in two different ways. First one is $Dir(\beta)$ and second one is $Dir(\delta)$ where δ is obtained from knowledge source. Let T is the total number of topics, and it is defined as $T = M + K$, where M is the topics in knowledge source and K is an unlabeled topics (usually input by user according to the dataset). In LDA, only K word distributions over topics ϕ_k are generated, whereas in Source-LDA, T word distributions over topics are generated where M of them are inspired from knowledge source ϕ_m.

Inference process:

Inference process in Source-LDA is moreover similar to that LDA except the fact that it uses extra word over topic distributions generated from source extended model ϕ_m.

Example Taking the same dice example as discussed in Sect. 3.2, we have D document dices and K topic dices. Here, we have D document dices and $T = (M + K)$ topic dices. The difference in Source-LDA is we know the bias of M dices in beforehand, i.e., we know topic dice $t \in \{1, 2, \ldots, M\}$ which words (faces) are most likely to appear, whereas the bias of the remaining K dices is still unknown. This prior knowledge helps in finding topic dice for a word more effectively. A word W_{ij} looks for a topic dice where it is highly likely to appear, thus making use of topic dice (from knowledge source) to look for its bias. This helps in producing more accurate distributions, thus obtaining better results.

3.4 Generating Correlation Matrix

Before presenting the algorithm, we first focus on what we are trying to achieve with this correlation matrix and how including this in our approach will help us to improve the Source-LDA.

1. Consider the corpus (as mentioned in Sect. 3.1) having preprocessed documents. Run Source-LDA on the corpus. After running the Source-LDA, a list of topics for a document d gets generated and this happens for all the documents in the corpus.
2. Choose the top n topics (based on frequency of occurrence) from this list and denote it by a new list L. For example, let us say that this new list is [dog, religion].
3. Now find the correlation between each topic of the corpus using the Wikipedia dataset. Suppose *dog* is correlated to the topics *[animal, nature, pet, and mammal]* while *religion* is correlated to the topics *[god, atheism, and supernatural]*.
4. Now, append all correlations of these two topics to L and thus obtain a new list $L = [dog, animal, nature, pet, mammal, religion, god, atheism, supernatural]$ for the document d. Had the original label for 'd' been dog or religion using the correlation matrix would have made no difference. But if the original label said mammal or atheism then clearly using the correlation matrix to append the list of topics predicted leads to an increase in the number of true positives. It however also leads to an increase in the number of false positives. So there is a trade-off here. By specifying the correct threshold 't' (as mentioned in the steps below) for the matrix, we can gain a net increase in F-measure as can be seen in the experimental result.

- Compute the TF-IDF on all the Wikipedia articles and then compute the cosine-similarity between all articles in the Wikipedia.
- For each topic, make a new list by adding topics which have cosine-similarity greater than a threshold 't'.

For each Topic T_i in Result $[k]$:
  For each Topic T_j :
    If Cosine $-$ Similarity$(T_i, T_j) > t$:
      Append T_j to Result $[k]$
    End If
  End For
End For

4   Experimental Work

Five benchmark datasets are used for experimental work (DMOZ,[2] 20-Newsgroups,[3] Reuters,[4] Classic3[5], and WebKB[6]) and procured 41 more random articles from Wikipedia, i.e., 61 total topics from Wikipedia. For demonstration purpose, we have discussed here the experimental work carried out on 20-Newsgroups (20-NG) dataset. Similar experimental works have been carried out on all other four datasets, and the results are shown in Tables 1, 2, 3, 4, and 5. We run the Source-LDA algorithm on 20-NG dataset with Wikipedia articles as our knowledge base over 4000 iterations, which gave us topics for each word in the corpus. We then choose the top n topics (here $n = 6$) for each document which is based on the numerical frequency for topic related to each word and return it as the output for each document. The same testing dataset is used to get top topics from LDA. As we already know the topics of 20-NG dataset, the topics returned by LDA is compared with Source-LDA to get various comparative measures like weighted precision, recall, and F-measure. To compute the weighted precision, first we compute the precision of each class label, i.e., true positives and false positives of each label. This is done by matching the top topics returned by Source-LDA and topic list which already have (of known 20-NG) for all the documents. Weight of a particular class label of a document is computed using Eqs. 4 and 5.

$$Precision = \frac{\text{number of relevant topics returned by Source-LDA}}{\text{total number of topics in 20-NG}} \qquad (4)$$

$$Recall = \frac{\text{number of relevant topics returned by Source-LDA}}{\text{number of times the topics actually appear in 20-NG label list}} \qquad (5)$$

$$F - measure = 2 * \left(\frac{precision * recall}{precision + recall} \right)$$

Weights are calculated in similar way as in precision. The weighted precision and recall are used to compute F-measure. Figure 3 shows the accuracy of Source-LDA combined with correlation in comparison with LDA, Source-LDA, and LDA with correlation on different datasets. From the results, it can be observed that Source-LDA with correlation has better performance compared to LDA, Source-LDA, and LDA with correlation.

The reason why we are getting such high precision in Source-LDA is due to high number of true positives and very less false positives, as the Source-LDA model gives the correct labels quite accurately. However, it has low recall or comparable

[2]https://www.dmoz.org/.

[3]http://qwone.com/~jason/20Newsgroups/.

[4]http://www.daviddlewis.com/resources/testcollections/reuters21578/.

[5]http://www.dataminingresearch.com/index.php/2010/09/classic3-classic4-datasets/.

[6]http://www.cs.cmu.edu/afs/cs/project/theo-20/www/data/.

Table 1 Performance on 20-Newsgroups dataset

Technique	Precision	Recall	F-measure
LDA	75.972	82.110	78.921
Source-LDA	99.281	66.027	79.301
(LDA + Correlation)	74.876	84.335	79.325
(Source-LDA + Correlation)	70.967	98.130	82.366

Table 2 Performance on Reuters dataset

Technique	Precision	Recall	F-measure
LDA	70.836	75.225	72.964
Source-LDA	88.376	71.527	79.063
(LDA + Correlation)	72.765	76.045	74.369
(Source-LDA + Correlation)	81.967	88.675	85.189

Table 3 Performance on WebKB dataset

Technique	Precision	Recall	F-measure
LDA	72.654	79.110	75.744
Source-LDA	91.281	74.225	81.874
(LDA + Correlation)	74.849	80.980	77.794
(Source-LDA + Correlation)	85.348	89.745	87.491

Table 4 Performance on Dmoz dataset

Technique	Precision	Recall	F-measure
LDA	73.985	79.456	76.622
Source-LDA	89.595	80.337	84.713
(LDA + Correlation)	74.223	81.546	77.712
(Source-LDA + Correlation)	87.764	83.223	85.433

Table 5 Performance on Classic3 dataset

Technique	Precision	Recall	F-measure
LDA	76.881	71.458	74.070
Source-LDA	92.667	72.562	81.391
(LDA + Correlation)	77.432	75.256	76.328
(Source-LDA + Correlation)	84.629	82.445	83.522

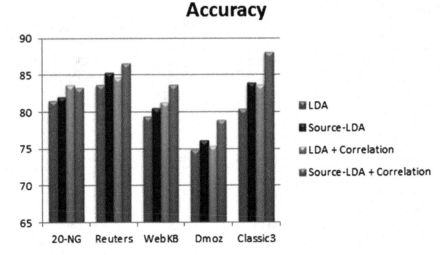

Fig. 3 Comparisons of accuracy on different datasets

with the recall of LDA, since it fails to give all the cases for a particular label. In Source-LDA with correlation, due to the addition of the correlated topics, the precision is lost because of increase in false positives in the similar topic list. But the advantage is regained with a drastic increase in recall, as almost all the topics related to a particular label are reflected in the correlation matrix, and thus the F-measure also increases.

5 Conclusion

In this paper, a topic labeling was proposed using semi-supervised LDA on a reliable knowledge source (such as Wikipedia) along with the correlation matrix containing cosine-similarity measure of all the topics. Five well-known machine learning datasets are used for experimental purpose. From the empirical results, it is concluded that Source-LDA combines with correlation matrix gave a better results compared to the traditional LDA. Experimentally, it is also observed that the performance of the proposed model increases with an increase in the number of topics to certain extent. This work can be further extended by exploring novel topic models for hierarchical labeled data. Also, this approach can be tested on Web document clustering and classification for better performance.

References

1. D. M. Blei, A. Y. Ng, and M. I. Jordan, "Latent dirichlet allocation," *Journal of machine Learning research*, vol. 3, no. Jan, pp. 993–1022, 2003.
2. S. Lacoste-Julien, F. Sha, and M. I. Jordan, "Disclda: Discriminative learning for dimensionality reduction and classification," in *Advances in neural information processing systems*, vol. 21, 2008, pp. 897–904.
3. J. Zhu, A. Ahmed, and E. P. Xing, "Medlda: maximum margin supervised topic models for regression and classification," in *Proceedings of the 26th annual international conference on machine learning*. ACM, 2009, pp. 1257–1264.
4. L. Du, L. Ren, L. Carin, and D. B. Dunson, "A bayesian model for simultaneous image clustering, annotation and object segmentation," in *Advances in neural information processing systems*, 2009, pp. 486–494.
5. R. K. Roul, S. R. Asthana, and G. Kumar, "Study on suitability and importance of multilayer extreme learning machine for classification of text data," *Soft Computing*, vol. 21, no. 15, pp. 4239–4256, 2017.
6. K. Toutanova and M. Johnson, "A bayesian lda-based model for semi-supervised part-of-speech tagging," in *Advances in neural information processing systems*, 2008, pp. 1521–1528.
7. A. Mukherjee and B. Liu, "Aspect extraction through semi-supervised modeling," in *Proceedings of the 50th Annual Meeting of the Association for Computational Linguistics: Long Papers-Volume 1*. Association for Computational Linguistics, 2012, pp. 339–348.
8. D. Wang, M. Thint, and A. Al-Rubaie, "Semi-supervised latent dirichlet allocation and its application for document classification," in *Proceedings of the the 2012 IEEE/WIC/ACM International Joint Conferences on Web Intelligence and Intelligent Agent Technology-Volume 03*. IEEE Computer Society, 2012, pp. 306–310.
9. S. Bodrunova, S. Koltsov, O. Koltsova, S. Nikolenko, and A. Shimorina, "Interval semi-supervised lda: Classifying needles in a haystack," in *Mexican International Conference on Artificial Intelligence*. Springer, 2013, pp. 265–274.
10. R. K. Roul and S. Sanjay, "Cluster labelling using chi-square-based keyword ranking and mutual information score: a hybrid approach," *International Journal of Intelligent Systems Design and Computing*, vol. 1, no. 2, pp. 145–167, 2017.
11. S. Jameel, W. Lam, and L. Bing, "Supervised topic models with word order structure for document classification and retrieval learning," *Information Retrieval Journal*, vol. 18, no. 4, pp. 283–330, 2015.
12. C. D. Manning, P. Raghavan, H. Schütze *et al.*, *Introduction to information retrieval*. Cambridge university press Cambridge, 2008, vol. 1, no. 1.

Diffusion Least Mean Square Algorithm for Identification of IIR System Present in Each Node of a Wireless Sensor Networks

Km Dimple, Dinesh Kumar Kotary and Satyasai Jagannath Nanda

Abstract Most of the real-world practical systems are inherently dynamic and their characteristics are represented by transfer functions which are IIR in nature. In literature-distributed estimation, algorithms have been developed for stable FIR system. In this paper, a distributed estimation technique is developed for identification of IIR system present at each node of a wireless sensor network. The distributed parameter estimation generally based on two modes of cooperation strategies: Incremental and Diffusion. In case of change in network topology, the diffusion mode of cooperation works well and shows robustness to link and node failure. Thus, an infinite impulse response diffusion least mean square (IIR DLMS) algorithm is introduced. In simulation, its performance is compared with the incremental version (infinite impulse response incremental least mean square algorithm (IIR ILMS)). Superior performance by the proposed approach is reported for parameter estimation of two IIR systems under various noisy environments.

Keywords Least mean square algorithm · Infinite impulse response system
Mean square error · Parameter estimation

1 Introduction

In recent years due to substantial use of wireless sensor network, the distributed estimation of parameters has been a focus area for many researchers, and wireless sensor network provides application of environment monitoring (temperature,

K. Dimple · D. K. Kotary · S. J. Nanda (✉)
Department of Electronics and Communication Engineering, Malaviya National
Institute of Technology, Jaipur 302017, Rajasthan, India
e-mail: nanda.satyasai@gmail.com

K. Dimple
e-mail: 2015pec5343@mnit.ac.in

D. K. Kotary
e-mail: 2015rec9510@mnit.ac.in

© Springer Nature Singapore Pte Ltd. 2019
H. S. Behera et al. (eds.), *Computational Intelligence in Data Mining*,
Advances in Intelligent Systems and Computing 711,
https://doi.org/10.1007/978-981-10-8055-5_63

sound, humidity, pollution vibration), battlefield surveillance, health care, home automation [1]. Sensor nodes in wireless sensor node (WSN) have less processing capability and are powered by small batteries. So, there is a requirement to develop a method which needs less amount of power and communication for processing of observed data [2]. A set of nodes are deployed in any geographical area which measures the raw observation, needs to estimate some parameter of interest from a noisy environment. The traditional method of parameter estimation is a centralized method which creates lots of communication overhead to the fusion center and decreases the lifetime of the entire network very soon.

Alternatively, to increase the lifetime of the WSNs and make it adaptive according to environment change distributed parameter estimations [3, 4] technique were proposed, here each and every sensor estimates the local parameters and shared these parameters to the neighbor's node to estimate the global parameter. Based on the cooperation among the nodes, two modes of cooperation strategies are used, i.e., Incremental and Diffusion [5]. In incremental strategy [6], each node updates their local weight with the help of data collected by the node and its immediate neighbors in a cyclic manner as shown in Fig. 1a. As per this strategy, a predefined incremental path is required for connecting each sensor available in that environment which is good for small area network and required less number of communication among the nodes. Diffusion strategy [7], shown in Fig. 1b, the weight of a particular node is estimated with the help of sum of estimated weight of subset of neighbors N_k and data $X_k(i)$ observed by that particular node. This estimated value is again send to the neighboring node. In diffusion mode, each node has access to more number of nodes. This strategy can work for changes in network topology and perform better for the large network as compared to the incremental mode of cooperation.

In this paper, the objective is to implement the Diffusion LMS for IIR system in order to estimate the parameters of each sensor node. Literature [8, 9] has reported the both algorithm Incremental LMS and Diffusion LMS for FIR system. An adaptive algorithm for IIR parameter estimation in non-distributed environment is reported in [10]. There are developments for parameter estimation with

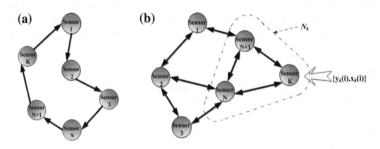

Fig. 1 Cooperation strategy among the sensor nodes **a** Incremental **b** Diffusion

nature-inspired optimization and adaptive techniques for incremental strategies [11–16]. But there is no literature that has significantly contributed to diffusion aspect. This motivates to develop the Diffusion LMS for IIR system.

2 Problem Formulation

Consider a sensor network with 'K' nodes which are deployed in any environment and measured data by kth node denoted by $y_k \in \mathfrak{R}^n w^0$, k = [1, 2, 3.........K]. Data collected by each sensor are added with noise $na_k(i)$ which is assumed to be uniformly distributed white Gaussian.

The input data vector $X_k(i)$ is uncorrelated with the noise $na_k(i)$ and here every sensor node is considered as an IIR plant [11] and output of kth IIR plant is described in Eq. (1):

$$y_k(i) = \sum_{q=0}^{Q} a_q * X_k(i-q) + \sum_{p=1}^{P} b_p * y_k(i-p) + na_k(i) \tag{1}$$

The main objective is to estimate the global parameters of interest (i.e., $[a_0 a_1 \ldots \ldots a_Q b_1 b_2 \ldots \ldots b_P]$) associated with measured data by the sensor node. w^0 can be calculated by centralized or distributed approaches. Zeroes and poles of the IIR plant are denoted by $a_q(0 \leq q \leq Q)$ and $b_p(1 \leq p \leq P)$.

In case of distributed system identification [12], the model shown in Fig. 2 is taken into consideration. In Fig. 2 at the kth node, the output of node can be modeled by following Eq. (2):

$$\hat{y}_k(i) = \left(\frac{\widehat{A}(i,z)}{1 - \widehat{B}(i,z)} \right) X_k(i) \tag{2}$$

where $\hat{A}_k(i,z)$ and $\hat{B}_k(i,z)$ are feed forward and feedback transfer function of kth instant, respectively, given in Eqs. (3)–(4):

$$\widehat{A}_k(i,z) = \sum_{q=0}^{Q} \hat{a}_{k,q}(i) * z^{-q} \tag{3}$$

$$\widehat{B}_k(i,z) = \sum_{p=1}^{P} \overset{\Lambda}{b}_{k,p}(i) * z^{-p} \tag{4}$$

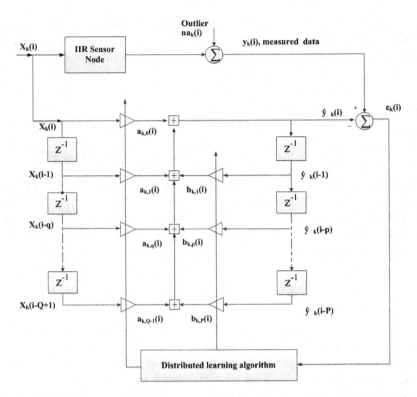

Fig. 2 Adaptive IIR system identification at a single sensor node of wireless sensor network

So, the output of model IIR system of sensor node can be expressed in Eq. (5) as

$$\widehat{y}_k(i) = \sum_{q=0}^{Q} \widehat{a}_q(i) * X_k(i-q) + \sum_{p=1}^{P} \widehat{b}_p(i) * \widehat{y}_k(i-p) \tag{5}$$

The error at kth sensor node is computed by subtracting desired filter output $y_k(i)$ to the model filter output $\widehat{y}_k(i)$ can be written in Eq. (6) as

$$\varepsilon_k(i) = y_k(i) - \widehat{y}_k(i) \tag{6}$$

The optimal weight vector which contains the forward and feedback path filter coefficients can be estimated by minimizing cost function in Eq. (7) as

$$J(w) = \arg \min \sum_{k=1}^{K} \left\| y_k(i) - X_k^T(i)w \right\| \tag{7}$$

3 Incremental LMS for IIR System Identification

The incremental mode of cooperation strategy requires a cyclic path to estimate the parameter of interest from nodes present in the environment [14]. Sensor nodes use their own collected data $\{x_k(i), y_k(i)\}$ and estimated parameters of immediate neighbors to determine parameters of interests. Main steps are given in Eqs. (8)–(10):

$$\varphi_0(i) = w(i-1) \tag{8}$$

$$\varphi_k(i) = \varphi_{k-1}(i) - c^* \widehat{\nabla}_k(i) \tag{9}$$

$$w(i) = \varphi_K(i) \tag{10}$$

Step1   Initialize $\varphi_0(i) = w(i-1)$ at node 1 that is current global estimate parameter for w^0 at each time iteration i. The c in Eq. (9) is convergence parameter.

Step2   Estimate the parameter of interest in incremental fashion [15] means $\varphi_{k-1}(i)$ is the estimated parameter of the previous node used to estimate a parameter of immediate forward node, i.e., $\varphi_k(i)$ as in Eq. (9) supported by Eqs. (11)–(14)

$$\widehat{\nabla}_k(i) = -2\,\varepsilon_k(i)^*[\alpha_{0k}(i)\alpha_{1k}(i)\ldots\ldots\,\alpha_{Qk}(i)\beta_{1k}(i)\beta_{2k}(i)\ldots\ldots\beta_{Pk}(i)]^T \tag{11}$$

$$\varepsilon_k(i) = y_k(i) - \widehat{y}_k(i) \tag{12}$$

$$\alpha_{qk}(i) = u_k(i-k) + \sum_{p=1}^{P} b_p(i)\,\alpha_{qk}(i-p) \quad 0 \leq q \leq Q \tag{13}$$

$$\beta_{pk}(i) = y_k(i-k) + \sum_{p=1}^{P} b_p(i)\beta_{pk}(i-p) \quad 1 \leq p \leq P \tag{14}$$

Step3   Estimation process is done in a cyclic manner so every node present in the network calculates their local estimate and at the end of procedure $w(i)$ takes the local estimate $\varphi_K(i)$ of node K. This algorithm repeats until and unless convergence criteria are fulfilled. After convergence, the global estimation is obtained from $w(i)$.

4   Diffusion LMS for IIR System Identification

Diffusion strategy does not require any cyclic path for connecting each sensor nodes to estimate the parameters. Proposed diffusion strategy for distributed estimation of IIR system identification performs in two steps as shown in Eqs. (15)–(16):

$$w_k(i-1) = \sum_{l \in N_{k,i-1}} m_{kl}\,\phi_l(i-1) \tag{15}$$

$$\phi_k(i) = w_k(i-1) - c^* \nabla_k^D \tag{16}$$

where N_k is the set of neighbor nodes linking that particular nodes, i.e., $(l = 1, 2 \ldots\ldots N_k)$,

Step1   represents that at any given time $i-1$ node k has access to a set of unbiased local estimates $\{\varphi_k(i-1)\}_{k \in N_k}$ from its neighborhood nodes N_k. This local estimation is combined at the node k which gives a gross estimate $w_k(i-1)$. m_{kl} is the merger coefficients [13] that carry the information about the sensor network topology. It is used to find which node $l \in N_k$ should allow giving their local estimates $\{\phi_k(i-1)\}$ to the nodes N_k if node k and node l are not connected then the value of merger coefficient is zero otherwise it is 1. The coefficients m_{kl} give rise to a merger matrix $M = [m_{kl}]$. The merger coefficient is defined in Eq. (17):

$$m_{kl} = \frac{m_{kl}}{\displaystyle\sum_{r \in N_{k,i-1}} m_{kl}} \tag{17}$$

There every node in N_k will have different neighborhood for connected sensor network [12] and another condition for merger coefficient is defined in Eq. (18):

$$\sum_l m_{kl} = 1 \quad l \in N_{k,i-1} \forall k \tag{18}$$

It helps to collect information from the nodes spread throughout the network [12]. So, we assume that M is a stochastic matrix.

Step2   Local estimate $\varphi_k(i)$ is the updated estimate at ith iteration with a gross estimate at previous iteration and gradient ∇_k^D by kth node at that iteration. c   is   the   convergence   parameter   matrix $c = diag[\mu \ldots\ldots\ldots \mu \nu_1 \nu_2 \ldots\ldots\ldots \nu_P]$ which contains same convergence coefficient for forward tap weight, i.e., μ and different convergence coefficients for feedback tap weight, i.e., $\nu_1 \nu_{2\ldots} \nu_P$. For calculating the gradient ∇_k^D of kth node, some equations are given below.

The model output for IIR plant is given in Eqs. (19)–(20):

$$\widehat{y}_k(i) = w_k^T(i)*U_k(i) \tag{19}$$

$$w_k = [a_{0k}a_{1k}\ldots\ldots\ldots a_{Qk}b_{1k}b_{2k}\ldots\ldots\ldots b_{Pk}]^T \tag{20}$$

where w_k is time-varying weight vector which possesses forward filter path coefficient and feedback filter path coefficient at kth sensor node in Eq. (21)

$$U_k = [x_k x_{k-1}\ldots\ldots\ldots x_{k-Q}\widehat{y}_{k-1}\widehat{y}_{k-2}\ldots\ldots\ldots \widehat{y}_{k-P}]^T \tag{21}$$

U_k is a signal vector which has information of input data vector and output data vector of the model at kth the node. The $y_k(i)$ is IIR plant output defined in Eq. (22)

$$y_k(i) = \sum_{q=0}^{Q} a_q(i)*X_k(i-q) + \sum_{p=1}^{P} b_p(i)*y_k(i-p) + na_k(i) \tag{22}$$

Error $\varepsilon_k(i)$ at node k is defined by Eq. (23) supported by Eq. (24)

$$\varepsilon_k(i) = y_k(i) - w_k(i)*U_k(i) \tag{23}$$

$$\nabla_k^D = -2*(y_k(i) - w_k(i)*U_k(i))*[\alpha_{0k}(i)\alpha_{1k}(i)\ldots\ldots\alpha_{Qk}(i)\beta_{1k}\beta_{2k}\ldots\ldots\beta_{Pk}] \tag{24}$$

where

$$\alpha_{qk}(i) = x_k(i-k) + \sum_{p=1}^{P} b_p(i)\,\alpha_{qk}(i-p); \quad 0 \leq q \leq Q$$

$$\beta_{pk}(i) = y_k(i-k) + \sum_{p=1}^{P} b_p(i)\beta_{pk}(i-p); \quad 1 \leq p \leq P$$

The complete operation of our algorithm can be summarized as the local estimates ϕ_k are given to all neighbors node where local estimates combined by merger coefficient. Gross estimates $w_k(i-1)$ at node k can find out by using Eq. (15). At each node, local estimates are updated with Eq. (16). This process repeated until the MSE is minimized. After convergence, the global estimates are obtained from w_k.

5 Simulation Studies

In this section, the performance of the proposed IIR DLMS algorithm is compared with the existing IIR ILMS for two IIR systems. The IIR system is present at each node of a sensor network and its structure is given in Fig. 2. The input is a zero mean white random signal with uniform distribution. The simulation is carried out for 5, 10, 20, and 30 dB SNR conditions on MATLAB environment. The noise is additive in nature and white random process uncorrelated with input. The number of sensor nodes is 20. As the training continues, the MSE at each node reduces continually and finally attains the lowest value after which no further reduction of MSE takes place. That stage leads to termination of the training process. The sum of squared error (SSE) is used here as performance measure during testing.

Example 1 A second-order IIR system [12] present at sensor node is given in Eq. (25)

$$H_p(z) = \left[\frac{0.05 - 0.4z^{-1}}{1 - 1.1314z^{-1} + 0.25z^{-2}} \right] \tag{25}$$

This can be modeled using second-order adaptive IIR filter as given in Eq. (26)

$$H_p(z) = \left[\frac{a_0 + a_1 z^{-1}}{1 - b_1 z^{-1} - b_2 z^{-2}} \right] \tag{26}$$

The convergence characteristics shown in Fig. 3 reveal that the minimum MSEs obtained using IIR DLMS is less than IIR ILMS. The simulation time of IIR ILMS and IIR DLMS is shown in Table 2. Table 3 shows the comparison of the sum of the square of error (SSE) during testing of IIR ILMS and IIR DLMS. It is observed from Table 3 that the SSE in the case of IIR DLMS is nearly similar to IIR ILMS (Table 1).

Example 2 The transfer function of a third-order IIR filter [12, 16] at sensor node is given in (27)

$$H_p(z) = \left[\frac{-0.2 - 0.4z^{-1} + 0.5z^{-2}}{1 - 0.6z^{-1} + 0.25z^{-2} - 0.2z^{-3}} \right] \tag{27}$$

Its model is represented in Eq. (28)

$$H_p(z) = \left[\frac{a_0 + a_1 z^{-1} + a_2 z^{-2}}{1 - b_1 z^{-1} - b_2 z^{-2} - b_3 z^{-3}} \right] \tag{28}$$

The convergence characteristics of Example 2 are shown in Fig. 4. It represents the minimum MSEs obtained using IIR DLMS same as the IIR ILMS. The estimated parameter values, simulation time and SSE values for the Example 2 are reported in Tables 4, 5 and 6 respectively.

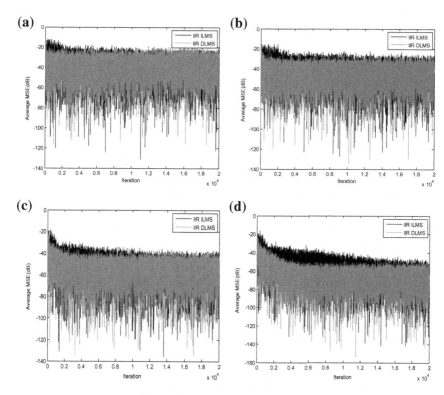

Fig. 3 Comparative results of convergence achieved by IIR ILMS and IIR DLMS for Example 2 under different noise conditions **a** SNR 5 dB, **b** SNR 10 dB, **c** SNR 20 dB, and **d** SNR 30 dB

Table 1 Estimated parameters obtained for Example 1 under different noise conditions using IIR ILMS and IIR DLMS during training

True	Estimated parameters using IIR ILMS				Estimated parameters using IIR DLMS			
	30 dB	20 dB	10 dB	5 dB	30 dB	20 dB	10 dB	5 dB
0.05	0.0520	0.0539	0.0683	0.0513	0.0528	0.0586	0.05269	0.0876
−0.4	−0.409	−0.401	−0.3884	−0.3626	−0.4420	−0.4527	−0.5294	−0.4279
1.1314	1.1104	1.1176	1.1523	1.2135	1.090	1.0806	0.9858	1.0916
−0.25	−0.230	−0.240	−0.2497	−0.3246	−0.2145	−0.2282	−0.2305	−0.2784

Table 2 Comparison of simulation time during training for Example 1

Noise (dB)	Time IIR ILMS (s)	Time IIR DLMS (s)
30	7.18	34.33
20	7.34	40.41
10	6.44	44.14
5	5.23	35.75

Table 3 Comparison of sum of square of error during testing for Example 1

Noise (dB)	SSE IIR ILMS	SSE IIR DLMS
30	0.000511	0.000602
20	0.0049	0.0078
10	0.0498	0.0975
5	0.1593	0.1816

Table 4 Estimated parameters obtained for Example 2 under different noise conditions using IIR LMS and IIR RLS during training

True	Estimated parameters using IIR ILMS				Estimated parameters using IIR DLMS			
	30 dB	20 dB	10 dB	5 dB	30 dB	20 dB	10 dB	5 dB
0.6	0.5980	0.5982	0.5959	0.5970	0.5960	0.5999	0.6066	0.56265
−0.25	−0.2611	−0.2604	−0.2599	−0.2627	−0.2664	−0.253	−0.2755	−0.2663
0.2	0.1984	0.1980	0.1977	0.1992	0.1946	0.1920	0.19596	0.16933
−0.2	−0.1826	−0.1826	−0.1862	−0.1784	−0.1709	−0.1696	−0.18076	−0.0357
−.4	−0.3857	−0.3878	−0.3885	−0.3957	−0.3749	−0.3884	−0.38310	−0.3331
0.5	0.5155	0.5160	0.5158	0.5257	0.52488	0.5216	0.51585	0.59467

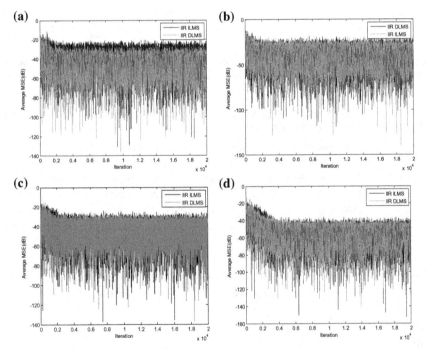

Fig. 4 Comparative results of convergence achieved by IIR ILMS and IIR IRLS for Example 1 under different noise conditions **a** SNR 5 dB, **b** SNR 10 dB, **c** SNR 20 dB, and **d** SNR 30 dB

Table 5 Comparison of simulation time during training for Example 2

Noise (dB)	Time IIR ILMS	Time IIR DLMS
30	27.39	216.81
20	26.32	226.05
10	24.57	197.17
5	25.25	198.34

Table 6 Comparison of sum of square of error during testing for Example 2

Noise (dB)	Sum of square of error IIR ILMS	Sum of square of error IIR DLMS
30	0.0119	0.000251
20	0.0140	0.0024
10	0.0351	0.0230
5	0.0855	0.0744

6 Conclusion

The paper introduces a Diffusion strategy of LMS algorithm for distributed parameter estimation of IIR systems present in each node of a sensor network. The simulation studies on two different order IIR systems reveal that the proposed IIR DLMS algorithm has less mean square error (MSE) than the IIR ILMS. The accuracy of filter weights in our algorithm is nearly same as IIR ILMS. It is also observed that as the noise strength is increased (decrease in SNR value) the parameter matching value also deviates. It reveals that the deterioration in the quality of effective identification under the presence of high noise. The overall performance of the proposed IIR DLMS makes it a suitable algorithm for parameter identification in distributed IIR systems. The proposed algorithm is suitable for large sensor network and robust to link failures.

References

1. Estrin, L. Girod, G. Pottie, M. Srivastava, Instrumenting the world with wireless sensor networks, in: Proceedings of the IEEE Inter- national Conference on Acoustics, Speech, Signal Processing (ICASSP), Salt Lake City, UT, vol. 4, May 2001, pp. 2033–2036.
2. I.F. Akyildiz, W. Su, Y. Sankarasubramaniam, E. Cayirci, A survey on sensor net-works, IEEE Commun. Mag. 40(8) (2002) 102–114.
3. Lopes, CassioG, and Ali H. Sayed. "Distributed processing over adaptive networks." *Proc. adaptive sensor array processing workshop*. 2006.
4. Sayed, Ali H., and Cassio G. Lopes. "Adaptive processing over distributed networks." *IEICE Transactions on Fundamentals of Electronics, Communications and Computer Sciences* 90.8 (2007): 1504–1510.

5. Majhi, B., Panda, G., & Mulgrew, B. Distributed identification of nonlinear processes using incremental and diffusion type PSO algorithms. In *IEEE Congress on Evolutionary Computation, 2009. CEC'09.* (pp. 2076–2082).
6. Lopes, Cassio G., and Ali H. Sayed. "Distributed adaptive incremental strategies: Formulation and performance analysis." *2006 IEEE International Conference on Acoustics, Speech and Signal Processing, 2006. ICASSP Proceedings.* Vol. 3. IEEE, 2006.
7. C.G. Lopes, A.H. Sayed, Diffusion least-mean squares over adaptive networks, in: Proceedings of IEEE International Conference on Acoustics, Speech, and Signal Processing (ICASSP), Honolulu, HI, April 2007, pp. 917–920.
8. C.G. Lopes, A.H. Sayed, Incremental adaptive strategies over distributed network, IEEE Trans. Signal Process. 55(August(8)) (2007) 4064–4077.
9. Cattivelli, F. S., and A. H. Sayed. "Diffusion LMS Strategies for Distributed Estimation." *IEEE Transactions on Signal Processing* 3.58 (2010): 1035–1048.
10. Turajlic, Emir, and Olja Bozanovic. "A novel adaptive IIR filter algorithm." *Telecommunications Forum (TELFOR), 2012 20th.* IEEE, 2012.
11. Shynk, John J. "Adaptive IIR filtering." *IEEE Assp Magazine* 6.2 (1989): 4–21.
12. Majhi, B. and Panda G. "Distributed and robust parameter estimation of IIR systems using incremental particle swarm optimization." *Digital Signal Processing* 23.4 (2013): 1303–1313.
13. Lopes, Cassio G., and Ali H. Sayed. "Diffusion least-mean squares over adaptive networks: Formulation and performance analysis." *IEEE Transactions on Signal Processing* 56.7 (2008): 3122–3136.
14. Li, Leilei, Yonggang Zhang, and Jonathon A. Chambers. "Variable length adaptive filtering within incremental learning algorithms for distributed networks." *Signals, Systems and Computers, 2008 42nd Asilomar Conference on.* IEEE, 2008.
15. Nedic, Angelia, and Dimitri P. Bertsekas. "Incremental subgradient methods for nondifferentiable optimization." *SIAM Journal on Optimization* 12, no. 1 (2001): 109–138.
16. Karaboga, Nurhan. "A new design method based on artificial bee colony algorithm for digital IIR filters." *Journal of the Franklin Institute* 346.4 (2009): 328–348.

Comparative Evaluation of Various Feature Weighting Methods on Movie Reviews

S. Sivakumar and R. Rajalakshmi

Abstract Sentiment analysis is a method of extracting subjective information from customer reviews. The analysis helps to reveal the consumer insights about the product, a theme, or a service. In the existing literature, various methods such as BoW and TF-IDF are employed for sentiment analysis and deep learning methods are not explored much. We made an attempt to apply Word2Vec feature weighting method for this problem. We carried out various experiments for sentiment analysis on a large dataset IMDB that contains movie review. We compared various feature weighting methods and analyzed using different classifiers, and the best combination was determined. From the experimental results, we conclude that Word2Vec with SGD is the best combination for sentiment classification problem on IMDB dataset. The result shown in the paper can be used as a base for future exploration of opinioned value on any textual data.

Keywords Sentiment classification · SVM · Stochastic gradient descent (SGD)

The original version of this chapter was revised: Post-publication corrections have been incorporated. The correction to this chapter is available at: https://doi.org/10.1007/978-981-10-8055-5_79

S. Sivakumar · R. Rajalakshmi (✉)
School of Computing Science and Engineering, VIT University,
Chennai 600127, Tamil Nadu, India
e-mail: rajalakshmi.r@vit.ac.in

S. Sivakumar
e-mail: ssivakumar.2016@vitstudent.ac.in

S. Sivakumar
Department of Computer Science, Dhanalakshmi College of Engineering,
Chennai 600127, Tamil Nadu, India

© Springer Nature Singapore Pte Ltd. 2019
H. S. Behera et al. (eds.), *Computational Intelligence in Data Mining*,
Advances in Intelligent Systems and Computing 711,
https://doi.org/10.1007/978-981-10-8055-5_64

721

1 Introduction

In microblog, millions of messages are posted daily. With the advance in the technology in Internet fields, more reviews are posted across the Internet. It is very difficult for the decision makers to interact and collect the opinions on the mass customers. Natural languages are used to write their opinions. Analyzing this sentimental opinion manually requires more resources, and it is a time-consuming process. Decision making is an important process in machine learning. As we have huge amount of online data, decision-making process can be made in a better way by analyzing the text with proper tools and expertise.

Sentiment is analyzed from a large corpus of review text. The polarity of text is determined as positive or negative. Sentiment analysis contributes in two ways. First, the business analyst can improve their product from the comments. They can sell their product in a better way. Second, the customer can view the opinion about the product, which helps to buy a right product. An automated NLP tool must be used for the extraction of the huge unstructured content. Later, analysis can be done on that content by using a text classifier. So, sentiment analysis gains attraction among the researchers.

We have conducted an experimental study on IMDB [1] dataset containing 25,000 movie reviews. Three feature weighting methods are applied on the dataset, viz. Bag of Words, TF-IDF, and Word2Vec. The performance of each feature weighting method is analyzed with various classifiers, viz. Random Forest, decision tree, support vector machine, Naïve Bayes, neural networks, K-nearest neighbor, and stochastic gradient descent classifier. We observed that performance of SGD with Word2Vec is good.

This paper is organized in the following way. The related work is presented in Sect. 2. The feature weighting methods and various classifiers are discussed in Sect. 3. The Section 4 includes the results and the comparative analysis of feature weighting methods. Finally, in Sect. 5, concluding remarks along with the future enhancement is presented.

2 Related Work

Andrew L. Maas [1] showed a model that mixes a blend of unsupervised and supervised strategies to learn word vectors catching semantic term–document data, in addition to that it captures rich sentiment message. The proposed model can use both consistent and multidimensional notion data. They proposed a model to use the record level assessment extremity explanations to exhibit in numerous online archives (e.g., star evaluations). They assessed the model utilizing little, broadly utilized estimation and subjectivity corpora and discover that it outplays of all previously presented strategies for sentiment categorization.

SidaWang et al. [2] demonstrated that: (i) The consideration of word bigram highlights gives steady picks up on assumption investigation errands; (ii) for short sentiment analysis, NB really shows improvement over SVMs (while for longer reports, the inverse outcome holds); (iii) a straightforward yet novel SVM variation utilizing NB log-count proportions including values reliably performs well crosswise over assignments and datasets. In light of these perceptions, they distinguish straightforward NB and SVM variations which beat most distributed outcomes on feeling examination datasets, once in a while giving another best in class execution level.

V. K. Singh et al. [3] proposed two methods of SentiWordNet by combining Adverb and Adjective Combined (SWN-AAC), SentiWordNet with Adverb, and Adjective alone and combining Adverb and Verb (SWN-AAAVC). The feature selection is done on document level and aspect level. They used 1000 instances present in the dataset. The document-level analysis SWN-AAAVC provides 82.9% positive which is higher than SWN-AAC 82% and Alchemy API 73.4%. They designed an aspect-level sentiment analysis algorithm on their own. This algorithm is simple and quick to implement. It produces results on the go and does not need any training.

Jiwei Li [4] have proposed a weight tuning method which filters out the useless information and concentrates only on the high-level information. The two models are weighted neural network (WNN) and binary-expectation neural network (BENN). They conducted experimental evaluation on sentence-level labels, document-level labels, and sentence representations for coherence evaluation. Recent popular variations of recursive models such as Matrix-Vector RNN (MV-RNN), Recursive Neural Tensor Network (RNTN), and Label-specific are implemented for sentence-level labels. The precision value for document-level analysis through WNN and BENN is better than state-of-the-art BoW model (unigram and digram). By weight tuning, the accuracy is increased to 0.936.

Raj K. Palkar et al. [5] used three standard datasets given by prestigious instructive organizations. Huge motion picture audits dataset v1.0 from Stanford University, dataset from Cornell university extremity dataset v2.0 and IMDB dataset recovered from University of California, Irvine are considered. Naïve Bayes, maximum entropy, support vector machine, classification and regression tree, and Random Forest are the classification algorithm; they have used for comparing the accuracy. The comparison is done on both unprocessed and preprocessed datasets. They do not involve with any handcrafted features during the training. The results given might be helpful in customizing the algorithm based on the task and domain.

Kamil Topal et al. [6] have proposed an emotion-based movie review analysis. Motion picture appraisals and audits at locales IMDB or Amazon are normally utilized by moviegoers to choose which motion picture to watch or purchase next. Right now, moviegoers make their decision to watch which motion picture based on the evaluation done on the picture at IMDB and Amazon. Based on the picture score and survey collected regarding the feeling content, the analysts generates an emotion map for the picture. This map helps to predict or anticipate about the motion picture. One can then choose to watch a picture which has some certain attractive designs in the emotion maps.

3 Materials and Methods

Sentiment analysis is a process of determining the customer attitude about the movie. For example, consider the following movie reviews.

Case 1: *The movie is economically planned by the producer.*
         Director has cleverly handled the actress in this film.
         Attractive locations are chosen to film this movie.
Case 2: *The dialogue is inconsistently delivered by the actors.*
         The theme of the movie is abused by the dialogue writer.
         Watching the movie with the family members will cause inconvenience.

The above two cases are the example for positive and negative reviews about a movie that was taken from the IMDB dataset. In case 1, the words 'economically,' 'cleverly,' and 'attractive' give positive opinion about the movie, whereas the words inconsistently, abused, and inconvenience convey negative opinion in the second case. To automate the process of classifying the sentiments based on opinions, we apply different feature weighting methods on various text classifiers and performed a comparative study by considering a publicly available dataset. Various methods that are used in our study are explained below.

3.1   Feature Weighting Methods

Feature weighting is a process of choosing a concerning features from the text based on the numerical value for further processing. We tried to explore the effect of feature weighting methods on the sentiment classification problem. Also, we tried to find out the impact of classification algorithm. Among the different weighting methods, we have chosen the following three methods. Different feature weighting techniques and their importance for URL classification were discussed in [7, 8].

Bag of Words (BoW). The BoW counts the number of times a word appears in the document. Using this, the document is compared and similarities are measured in various applications. Given a text T, a vector $V_T \varepsilon N^d$ is assigned to it, such that V_{Ti} is the number of times the ith word of the vocabulary has appeared in the text T. D is the vocabulary, which consists of all the words in the set of reviews.

TF-IDF. It means Term Frequency-Inverse Document Frequency. If reflects the importance of a word in a document. This is used as a weighting factor in various text mining applications. It increases the proportionally to the number of times a word appearing in the document, but offset by the frequency of the word in the corpus.

Word2Vec (W2V). The Word2Vec is used to produce word embeddings. This model is a two-layer neural networks that are trained to construct context of words. It takes a large corpus of text as input, and a word vector space is produced as an output. Each word in the corpus is assigned a vector in the space. The words sharing the common context in the corpus are located in close proximity in the vector space. The paragraphs are split into sentences, before processing of Word2Vec algorithm.

3.2 Text Classifiers

The purpose of the text classifiers in sentiment analysis problem is to determine the immanent value of the movie reviews. So, we apply the following methods to study the impact of the classifiers on the reviews.

Random Forest (RF): It is an ensemble approach. Ensemble methods are a divide and conquer approach used to improve the performance. A group of weak learner are combined together to form a strong learner. It starts with a supervised machine learning decision tree. In decision tree, the input is given at the top and the data gets bucketed into smaller sets, as we traverse down the tree.

Decision Tree (DT): Decision tree is a treelike graph structure. Each internal node corresponds to test on the attribute. Each branch on the node represents the outcome of the test. The leaf node represents the final outcome. It is suitable for any new possible scenario. It is simple to interpret and understand. Decision tree can be merged with other decision methods. If the values are uncertain or linked, the calculation gets complex.

Naïve Bayes (NB): It is based on Bayes theorem. It assumes the presence of a feature in a class is irrelevant to other feature. NB model is easy and useful for large datasets. The posterior probability shown in Eq. 1 is calculated using Bayes theorem.

$$P(c/x) = \frac{P(c/x) \cdot P(c)}{P(x)} \tag{1}$$

P(c/x)   is the posterior probability of class given predictor.
P(c)   is the prior probability of class.
P(x/c)   is the likelihood.
P(x)   is the prior probability of predictor.

Support Vector Machine (SVM): It is a supervised machine learning algorithm. It is commonly employed for classification and regression side as well. The idea is to divide the dataset into two classes to find the hyper plane. Margin is the distance between the data point and the hyper plane. The goal is to find a greatest possible margin for a chance of new data being classified correctly. Accuracy is better for smaller sets, while it takes more time in classification for larger datasets.

Neural Network (NN): They process one record at a time. It learns by comparing the classification of the record with known classification. The errors from initial classification are fed, to reduce the errors in further iteration. Backpropagation modifies the connection weigh using forward simulation, and it also done to train the network with known correct outputs. It provides a better accuracy for multi-dimensional data. They are hard to tune and train on larger dataset.

K-Nearest Neighbor (KNN): It uses all the available cases for classification. The new classification is done based on the Euclidean distance similarity measure. It is used in pattern recognition and statistical estimation. Euclidean distance is distance between two points defined as the square root of the sum of the squares of the differences between the corresponding coordinates of the points. The KNN complexity increases as the training dataset size increases.

Stochastic Gradient Descent (SGD): It is also known as incremental gradient descent. It performs updation on parameter for each training values. This method is useful for the parameters to be searched by an optimization algorithm. The update to the coefficients is performed for each training instance, rather than at the end of the batch of instances. Gradient descent algorithm makes prediction on each data instances, so it takes a longer time on very large datasets. Stochastic gradient descent gives a good result for a larger dataset in a quicker manner.

4 Results and Discussion

We performed comparative analysis of various feature weighting methods with different classification algorithms. For experimental setup, we used the system Intel Core i3 2.40 GHz, 6 GB RAM with 70 GB hard disk in Ubuntu 14.04 environment. We used IMDB dataset[1] from Stanford University. The dataset contain 25,000 movie reviews. Each instance contains three fields, namely id—unique identification, review—textual data about a movie, and sentiment—positive or negative. The dataset is divided into training set (80%—20,000 instances) and testing set (20%—5000 instances). All the experiments were performed using Python 2.7.6 version. For Word2Vec, we used gensim package and sklearn package is used for text classification.

We used precision, recall, accuracy, and F_1-measure as the performance metrics.
Accuracy. It is a proportion of correctly classified sentiments to total number of sentiments. Accuracy is shown in Eq. 2. It is used to determine the efficiency of the system.

$$\text{Accuracy} = \frac{\text{TP} + \text{TN}}{\text{TP} + \text{TN} + \text{FP} + \text{FN}} \tag{2}$$

where

TP   → number of positive sentiments that are correctly classified as positive,
FP   → positive sentiment, but classifier does not classify it as positive,
TN   → number of negative sentiments that are correctly classified as negative,
FN   → negative sentiment, but classifier does not classify it as negative.

Precision: It is the ratio of number of sentiments that is correctly judged as positive to total number of positively classified sentiments (P) which is shown in Eq. 3.

---

[1]http://ai.stanford.edu/~amaas/data/sentiment/

$$\text{Precision} = \frac{TP}{TP + FP} \tag{3}$$

Recall: It is the ratio of total number of positively predicted sentiments to total sentiments that are truly positive (R). Recall is given in Eq. 4.

$$\text{Recall} = \frac{TP}{TP + FN} \tag{4}$$

F_1: A combined measure that assesses the precision and recall trade-off is F-measure (Eq. 5), i.e., weighted harmonic mean. It is used to optimize the system toward precision or recall.

$$F_1 = \frac{2*P*R}{P + R} \tag{5}$$

In the first experiment, we combined BoW with various classifiers and analyzed the performance. The results are presented in Table 1. We obtained a high recall for SGD classifier as 86.7%, whereas BoW + SVM combination resulted in an accuracy of 84.4%, precision of 83.3%, and F_1 score of 85%.

In the second experiment, we combined TF-IDF with various classifiers and analyzed the performance. The results are presented in Table 2. We obtained a high accuracy and precision for SGD classifier as 75.1% and 78.1%, respectively, whereas TF-IDF + SVM combination produced a recall of 77.2% and F_1 score of 75.3%.

In the third experiment, we combined Word2Vec with various classifiers and analyzed the performance. The results are presented in Table 3. We obtained a high precision for RF classifier as 80.4%, whereas Word2Vec + SGD combination resulted in an accuracy of 82.2%, recall of 89.0%, and F_1 score of 83.6%.

The precision comparison of algorithm is shown in Fig. 1 and in Table 4. The precision of the SVM is high (83.3%) in BoW method, SGD is high (80.9%) in TF-IDF, and Random Forest is high (76.6%) in Word2Vec feature weighting. The F_1 score comparison of algorithm is shown in Fig. 2 and in Table 5. The F_1 score of the SVM is high in BoW method and TF-IDF method which is 85.0% and 75.3%, respectively, and SGD is high 83.6% in Word2Vec feature weighting.

Table 1 Performance of BoW method with different classifiers

	Accuracy	Precision	Recall	F_1 score
RF	0.821	0.818	0.829	0.822
DT	0.704	0.707	0.704	0.705
NB	0.787	0.809	0.756	0.792
SVM	**0.844**	**0.833**	0.864	**0.850**
NN	0.824	0.815	0.829	0.822
KNN	0.647	0.632	0.720	0.657
SGD	0.806	0.775	**0.867**	0.812

Table 2 Performance of TF-IDF feature weighting method with different classifiers

	Accuracy	Precision	Recall	F$_1$ score
RF	0.738	0.732	0.747	0.740
DT	0.624	0.620	0.633	0.626
NB	0.727	0.725	0.725	0.723
SVM	0.745	0.731	**0.772**	**0.753**
NN	**0.751**	0.738	0.761	0.749
KNN	0.666	0.659	0.682	0.668
SGD	**0.751**	**0.781**	0.693	0.751

Table 3 Performance of Word2Vec feature weighting method with different classifiers

	Accuracy	Precision	Recall	F$_1$ score
RF	0.809	**0.804**	0.817	0.826
DT	0.699	0.703	0.693	0.707
NB	0.748	0.763	0.723	0.755
SVM	0.725	0.746	0.684	0.735
NN	0.732	0.718	0.752	0.661
KNN	0.766	0.773	0.754	0.767
SGD	**0.822**	0.784	**0.890**	**0.836**

Fig. 1 Comparsion of text classifier with feature weighting methods based on precision

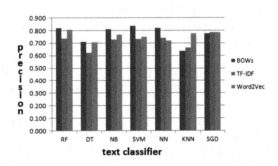

Table 4 Comparison of precision metrics on various machine learning algorithms

	BoW	TF-IDF	Word2Vec
RF	0.818	0.732	**0.804**
DT	0.707	0.620	0.703
NB	0.809	0.725	0.763
SVM	**0.833**	0.731	0.746
NN	0.815	0.738	0.718
KNN	0.632	0.659	0.773
SGD	0.775	**0.781**	0.784

Fig. 2 Comparsion of text classifier with feature weighting methods based on F_1 score

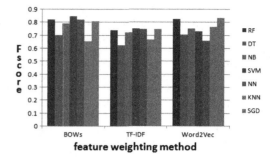

Table 5 Comparison of F_1 score metrics on various machine learning algorithms

	BOWs	TF-IDF	Word2Vec
RF	0.822	0.740	0.826
DT	0.705	0.626	0.707
NB	0.792	0.723	0.755
SVM	**0.850**	**0.753**	0.735
NN	0.822	0.749	0.661
KNN	0.657	0.668	0.767
SGD	0.812	0.751	**0.836**

5 Conclusion

The problem of sentiment analysis has been studied in this paper. Three feature weighting methods BoW, TF-IDF, and Word2Vec are combined with various classifiers such as Naive Bayes classifier, SVM, decision tree, Random Forest, stochastic gradient descent, and experiments were performed for sentiment analysis on a large dataset IMDB that contains movie review. From the experimental results, we conclude that Word2Vec with SGD is the best combination for sentiment classification problem on IMDB dataset. We achieved a high recall of 89% and an F1 of 83.6% for this method which is better than all the other combinations. This work can be extended in analyzing the online reviews from social media networks and by considering other advanced algorithms.

Acknowledgements We would like to thank Science and Engineering Research Board, Govt. of India, for funding this work (Award Number: ECR/2016/000484). We would also like to thank the management of VIT University, Chennai, for extending their support, where this research work was carried out.

References

1. Andrew L. Maas, Raymond E. Daly, Peter T. Pham, Dan Huang, Andrew Y. Ng, and Christopher Potts, "Learning Word Vectors for Sentiment Analysis", Proceedings of the 49th

Annual Meeting of the Association for Computational Linguistics: Human Language Technologies - Volume 1, pp. 142–150, June 19–24, 2011.

2. SidaWang and Christopher D. Manning, "Baselines and Bigrams: Simple, Good Sentiment and Topic Classification", Proceedings of the 50th Annual Meeting of the Association for Computational Linguistics: Short Papers - Volume 2, pp. 90–94, July 08–14, 2012.

3. V.K. Singh, R. Piryani, A. Uddin and P. Waila, "Sentiment Analysis of Movie Reviews: A new feature-based heuristic for aspect-level sentiment classification", International Mutli-Conference on Automation, Computing, Communication, Control and Compressed Sensing (iMac4 s), pp. 712–717, 22–23 March 2013.

4. Jiwei Li, "Feature Weight Tuning for Recursive Neural Networks", Neural and Evolutionary Computing, pp. 1–11, 13 Dec 2014.

5. Raj K. Palkar, Kewal D. Gala, Meet M. Shah and Jay N. Shah, "Comparative Evaluation of Supervised Learning Algorithms for Sentiment Analysis of Movie Reviews", International Journal of Computer Applications, Volume 142 – No.1, pp. 20–26, May 2016.

6. Kamil Topal and Gultekin Ozsoyoglu, "Movie Review Analysis: Emotion Analysis of IMDB Movie Reviews", IEEE/ACM International Conference on Advances in Social Networks Analysis and Mining (ASONAM), pp. 1170–1176, 8–21 Aug. 2016.

7. R. Rajalakshmi, 2015, "Identifying Health domain URLs using SVM", In: Proceedings of the Third International Symposium for Women in Computing and Informatics, WCI 15, pp. 203–208, ACM. https://doi.org/10.1145/2791405.2791441

8. R. Rajalakshmi and Sanju Xavier, 2017, "Experimental Study Of Feature Weighting Techniques For URL Based Webpage Classification", Procedia Computer Science, Volume 115, pp. 218–225, Elsevier. https://doi.org/10.1016/j.procs.2017.09.128

Dynamic ELD with Valve-Point Effects Using Biogeography-Based Optimization Algorithm

A. K. Barisal, Soudamini Behera and D. K. Lal

Abstract The paper focuses biogeography-based optimization (BBO) algorithm for solving the dynamic economic load dispatch problem (DELDP) of dispatchable units considering valve-point loading effects. BBO is a new biogeography-inspired optimization algorithm. Mathematical equations describe creation of species, movements from one habitat (island) to another and to get extinct. This algorithm follows two steps such as migration and mutation. The proposed method calculates economical schedule of units to satisfy load demand and ramp rate limits during operation to minimize total production cost. BBO search technique determines the global optimum dispatch solution. Various constraints like load balance constraints, operating limits, valve-point loading, ramp rate, and network loss coefficient are incorporated. The nonlinear nature of the generators in test system is considered to illustrate the effectiveness of the proposed method. The robustness of the methodology has been validated by demonstrating and comparing with previously developed techniques discussed in the literature.

Keywords Biogeography-based optimization · Dynamic economic load dispatch
Ramp rate · Valve-point effect

A. K. Barisal · D. K. Lal (✉)
Department of Electrical Engineering, Veer Surendra Sai University of Technology,
Burla 768018, Odisha, India
e-mail: laldeepak.sng@gmail.com

A. K. Barisal
e-mail: a_barisal@rediffmail.com

S. Behera
Department of Electrical Engineering, Government College of Engineering,
Bhawanipatna 76002, Odisha, India
e-mail: soudamini_behera@yahoo.co.in

© Springer Nature Singapore Pte Ltd. 2019
H. S. Behera et al. (eds.), *Computational Intelligence in Data Mining*,
Advances in Intelligent Systems and Computing 711,
https://doi.org/10.1007/978-981-10-8055-5_65

1 Introduction

Dynamic economic load dispatch (DELD) is one of the core tasks of power system operation and control. The purpose of the DELD is to minimize the cost of generation by fulfilling physical and operational constraints. Most of the literature covers static economic dispatch. In order to achieve true dispatch, the valve-point loadings on the generating units are to be included in DELD problem. The non-linearity of generator characteristics produces multiple local optimum solutions. Earlier, traditional optimization methods such as gradient projection method [1], Lagrangian relaxation (LR) [2], and dynamic programming (DP) [3] are used to solve the DELD problem. These methods find difficulties to give optimal solution due to their nonlinear and non-convex characteristic of generating units. The stochastic search algorithm such as simulated annealing (SA) [4, 5], genetic algorithm (GA) [6, 7], evolutionary programming (EP), and fuzzy optimization [8–11], and tabu search algorithm (TSA) [12], particle swarm optimization (PSO) [13, 14], and differential evolution (DE) [15] are successfully applied in solving nonlinear economic dispatch problems with discontinuous nature of the cost curves. In spite of the fact that these heuristic strategies do not generally ensure finding the global solution in finite time. They regularly give quick, sensible, and global optimal solutions. These techniques are based on heuristic rules to find and update positions of their candidates in the search space. The disadvantages of these strategies are necessity of huge memory. So, the blend of the techniques and deterministic strategies are required to take care of the optimization problems. The hybrid of EP and sequential quadrature programming (SQP) [16, 17] and PSO with SQP have been applied to solve DELD problem [18–20], since they can give close optimal solution. Artificial immune system (AIS) [21–28] has been implemented in various power system issues. It has many control variables and slow convergence nature that results unsuitability to address complex problems. The hybrid of bee colony optimization and SQP [20] gives attractive outcomes in tackling dynamic load dispatch issues.

Recently, another improvement idea, in light of biogeography, has been proposed by Simon [29]. BBO works in light of the two systems: relocation and mutation. BBO [29–32], nature-based calculations, has the attribute of sharing data between agents. Moreover, the calculation has certain kind of components which defeat few bad marks such as slow convergence, poor quality solutions, and tedious constraints satisfaction of the traditional strategies, for example, GA, EP, DE, PSO, and BFA [33] as specified in reference [32]. It has also less number of computational steps to yield optimal solution. In view of the above facts, BBO algorithm has been proposed for solving DELD problem. The technique has been applied to ten-unit test system to demonstrate its viability and relevance. The outcomes gotten from the preferred strategy are contrasted on those acquired from the literature.

2 Problem Formulation

The goal of the DELD is to plan the yields monetarily over a specific timeframe under different frameworks and operational limitations. The objective function is formulated as follows in Eq. (1).

$$MinF = \sum_{t=1}^{T} \sum_{i=1}^{N} F_{it}(Pg_{it})$$ (1)

where

F	Total cost over entire dispatch periods
T	Quantities of hour in the time horizon
N	Number of dispatchable units
$F_{it}(Pg_{it})$	The fuel cost in terms of its real power output Pg_{it} at a time t

Considering the valve-point impacts, the fuel cost capacity of the thermal unit is communicated as the aggregate of a quadratics and sinusoidal functions as given in Eq. (2),

$$F_{it}(Pg_{it}) = a_i Pg_{it}^2 + b_i Pg_{it} + c_i + |e_i(\sin(f_i(Pg_{it\,min} - Pg_{it})))|$$ (2)

$a_i, b_i, c_i, e_i,$ and f_i are constants of fuel cost function of unit i subject to the following real power balance constraint as presented in Eq. (3),

$$\sum_{i=1}^{N} Pg_{it} - P_{dt} - P_{lt} = 0, \quad t = 1, 2, \ldots T$$ (3)

where P_{dt} is the total assumed load demand at time t; P_{lt} the transmission loss at time t.

Real power operating limits is given in Eq. (4),

$$Pg_{it\,min} \leq Pg_{it} \leq Pg_{it\,max}, \quad i = 1, 2, \ldots T, \quad t = 1, 2, \ldots T$$ (4)

where $Pg_{i\,min}$ and $Pg_{i\,max}$ are the minimum and maximum real power outputs of ith generator, respectively.

Generator unit ramp rate limits are presented in Eq. (5)–(6).

$$Pg_{it} - Pg_{i(t-1)} \leq Ur_i \, i = 1, \ldots n$$ (5)

$$Pg_{i(t-1)} - Pg_{it} \leq Dr_i \, i = 1, \ldots n$$ (6)

Ur_i and Dr_i are ramp up and down rate limits of ith generator, respectively.

3 Biogeography-Based Optimization (BBO)

Biogeography depicts how species relocate starting with one island then onto the next, how new species emerge, and how species wind up plainly terminated. An environment is any island (range) that is topographically disconnected from different islands. The details of BBO are clearly explained in references [29–32].

4 System Under Study

The DELD issue of the ten-unit framework is settled by the proposed strategy to analyze the consequence of the proposed technique with hybrid strategies, such as hybrid DE, hybrid EP-SQP, modified EP-SQP, deterministically guided PSO, and PSO-SQP as reported in written works [16–20]. The load demand of the framework is partitioned by twenty-four interim hours. The framework data for the ten units is taken from reference [15]. Transmission losses have been overlooked and furthermore considered for examination comes about with those revealed in written works.

5 Simulation Results

The BBO method for the DELD issue depicted above has been linked to ten-unit test system to show the execution of the preferred technique. The ten-unit test system [15] with non-smooth cost capacities is utilized to confirm the achievability of the selected strategy. The software was created utilizing the MATLAB 7.01. After various trials of keep running with various estimations of BBO parameters tuning, the accompanying ideal estimations of BBO parameters have at last been found. Natural habitat measure = 30, habitat modification probability = 1, immigration probability limits = [0,1], step estimate for numerical integration = 1, mutation probability = 0.005, and maximum migration resettlement rate for every island = 0.5.

It is clear from Table 1 that the preferred strategy creates far better outcome contrasted with as of late announced outcomes in writing yet more opportunity for union contrasted with BCO-SQP [20]. The ideal scheduling of generating units for 24 h utilizing proposed strategy is presented in Table 2. Figure 1 delineates the meeting normal for ten-unit system without misfortunes to acquire ideal operating cost. The examination of results acquired by proposed strategy with misfortunes with other revealed brings about the writing, for example, EP and EP-SQP are given in Table 3. The normal time for the calculation of solution by proposed technique is observed to be 14.23 min. Besides, the proposed BBO technique gives prevalent union execution and quality solutions.

Table 1 Comparison of BBO results with other methods for ten-unit system without losses

Techniques	Production cost ($)	Computation time (minutes)
SQP [16]	1,05,11,163	1.19
EP [16]	10, 48,638	42.49
Hybrid EP [16]	10, 35,748	20.51
MHEP-SQP [17]	10, 28,924	21.23
DG PSO [18]	10, 28,835	15.39
Hybrid PSO-SQP [19]	10, 27,334	18.12
Shor's r- Algorithm [15]	10, 38,976	–
DE [15]	10, 33,958	–
Hybrid DE [15]	10, 31, 077	–
Hybrid BCO-SQP [20]	10, 32,200	2.15
Hybrid PSO-SQP [20]	10, 33,300	3.24
Hybrid PSO-SQP [20]	10, 34,100	3.33
BBO (proposed method)	10, 27,298	14.23

Table 2 Optimal scheduling of ten-unit system using proposed method

No. Hr	P_1 (MW)	P_2 (MW)	P_3 (MW)	P_4 (MW)	P_5 (MW)	P_6 (MW)	P_7 (MW)	P_8 (MW)	P_9 (MW)	P_{10} (MW)
1	152.81	140.09	94.738	121.47	124.47	147.34	93.102	85.549	21.153	55
2	226.32	219.43	75.565	87.031	123.78	123.38	93.356	85.612	20.526	55
3	228.33	223.27	153.42	120.73	125.37	122.96	122.68	85.633	20.597	55
4	226.48	302.85	196.69	125.94	172.55	126.09	94.402	85.575	20.42	55
5	303.56	310.92	239.18	119.72	124.13	124.71	97.57	85.144	20.059	55
6	303.3	310.21	302.74	120.34	173.2	130.6	127.16	85.136	20.312	55
7	302.44	389.58	304.32	120.45	172.55	122.67	128.92	85.705	20.363	55
8	378.7	397.93	296.14	118.81	172.65	122.67	128.7	84.996	20.41	55
9	456.4	458.87	300.34	121.86	172.86	122.75	129.7	85.851	20.367	55
10	456.6	458.15	316.48	170.66	221.33	159.32	128.71	85.328	20.422	55
11	457.16	459.77	338.26	219.03	224.42	157.77	129.34	84.802	20.452	55
12	459.4	459.64	338.55	249.76	235.11	158.88	129.31	114.2	20.157	55
13	456.39	396.84	322.34	245.77	224.54	135.95	129.45	85.364	20.345	55
14	379.71	396.77	298.86	212.89	222.18	123.75	129.22	85.313	20.302	55
15	379.08	396.94	264.35	181.06	172.77	122.14	99.866	84.741	20.049	55
16	304.06	317.73	198.23	184.02	172.23	122.95	93.525	85.765	20.497	55
17	303.9	309.32	219.25	145	125.67	123.97	92.341	84.899	20.653	55
18	303.55	380.47	281.45	118.59	169.14	122.22	92.956	84.171	20.456	55
19	380.09	397.12	298.46	121.04	172.99	123	122.26	85.888	20.162	55
20	456.44	459.57	322.76	168.46	222.24	152.99	128.71	85.615	20.209	55
21	381.02	396.22	309.69	181.35	221.69	144.27	129.26	85.101	20.395	55
22	302.83	316.92	288.04	132.29	173.66	123.79	129.79	85.39	20.285	55
23	225.84	309.08	210.99	111.93	124.5	87.495	101.13	85.571	20.453	55
24	226.3	230.55	165.04	63.08	122.97	122.9	92.83	85.241	20.091	55

Fig. 1 Convergence nature of ten-unit test system with losses

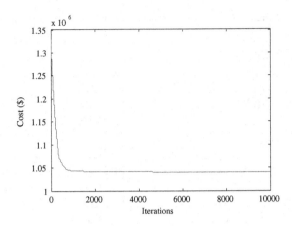

Table 3 Comparison of result of for ten-unit system with losses

Method	Costs ($/24 h)			CPU time
	Minimum	Average	Maximum	
EP [17]	1,054,685	1,057,323	NA	47.23
EP–SQP [17]	1,052,668	1,053,771	NA	27.53
BBO	1,040,465	142217	1,44,652	18.77

6 Conclusion

A productive calculation in light of geological dissemination of organic species is proposed in this paper to take care of ELD issue of generating units. The cost function displays the non-smooth and non-curved attributes, as the valve-point impacts are demonstrated and forced as rectified sinusoidal parts. The BBO method can locate better quality solutions and has better convergence attributes, computational productivity, and robustness. The viability of the proposed algorithm for the test system has been compared with recent publications for the similar system. The solution quality, dependability, and computational proficiency signify the prevalence of the proposed method.

References

1. Granelli, G.P., Marannino, P., Montagna, M., Silvestri, A.: Fast and efficient gradient projection algorithm for dynamic generation dispatching. IEE Proc. Generat.Transm. Distrib, vol. 136, no. 5, September 1989, pp. 295–302.
2. Hindi, K.S., Ab Ghani, M.R.: Dynamic economic dispatch for large-scale power systems: A Lagrangian relaxation approach. Elect. Power Energy Syst., vol. 13, no. 1, Feb. 1991, pp. 51–56.

3. Travers, D., Kaye, R.J.: Dynamic dispatch by constructive dynamic programming. IEEE Trans. on Power Syst., vol. 13, no. 1, 1998, pp. 72–78.
4. Panigrahi, C.K., Chattopadhyay, P.K., Chakrabarti, R., Basu, M.: Simulated annealing technique for dynamic economic dispatch. Elect. Power Compon. Syst., vol. 34, no. 5, May 2006, pp. 577–586.
5. Simopoulos, D.N., Kavatza, D., Vournas, C.D.: Unit commitment by an enhanced simulated annealing algorithm. IEEE Trans. Power Syst., vol. 21, no. 1, February 2006, pp. 68–76.
6. Li, F., Morgan, R., Williams, D.: Hybrid genetic approaches to ramping rate constrained dynamic economic dispatch. Elect. Power Syst. Res., vol. 43, no. 2, November 1997, pp. 97–103.
7. Walters, D.C., Sheble, G.B.: Genetic algorithm solution of economic dispatch with valve point loadings. IEEE Trans. Power Syst., vol. 8, no. 3, August 1993, pp. 1325–1331.
8. Yang, H.T., Yang, P.C., Huang, C.L.: Evolutionary programming based economic dispatch for units with non smooth incremental fuel cost functions. IEEE Trans. Power Syst., vol. 11, no. 1, February 1996, pp. 112–118.
9. Sinha, N., Chakrabarti, R., Chattopadhyay, P.K.: Evolutionary programming techniques for economic load dispatch. IEEE Trans. Evol. Computat. vol. &, no. 1, February 2003, pp. 83–94.
10. Han, X.S., Gooi, H.B., Kirchen, D.S.: Dynamic economic dispatch: Feasible and optimal solutions. IEEE Trans. Power syst. vol. 16, no.1, Feb. 2001, pp. 22–28.
11. Attaviriyanupap, P., Kita, H., Tanaka, E., Hasegawa, J. 'A fuzzy-optimization approach to dynamic economic dispatch considering uncertainties'. IEEE Trans. Power Syst., vol. 19, no. 3, August 2004, pp. 1299–1307.
12. Lin, M.M., Cheng, F.S., Tsay, M.T.: An improved tabu search for economic dispatch with multiple minima. IEEE Trans. Power Syst., vol. 17, no. 1, February 2002, pp. 108–112.
13. Gaing., Z.L.: Particle swarm optimization to solving the economic dispatch considering the generation constraints. IEEE Trans. Power Syst., vol. 18, no. 3, August 2003, pp. 1187–1195.
14. Chakrabarti, R., Chattopadhyay, P.K., Basu, M., Panigrahi, C.K.: Particle swarm optimization technique for dynamic economic dispatch. IE(India), vol. 87, 2006, pp. 48–54.
15. Yuan, X., Wang, L., Zang, Y., Yuan, Y.: A Hybrid differential method for dynamic economic dispatch with valve-point effects. Expert systems with applications, vol-36, 2009, pp. 4052–4048.
16. Attaviriyanupap, P., Kita, H., Tanaka, E., Hasegawa, J. 'A hybrid EP and SQP for dynamic economic dispatch with non smooth fuel cost function'. IEEE Trans. Power Syst., vol. 17, no. 2, May 2002, pp. 411–416.
17. Victoire, T.A.A., Jeyakumar, A.E.: Modified EP-SQP approach for dynamic dispatch with with valve point effects. Elect. Power and energy systems, vol. 17, 2005, pp. 594–601.
18. Victoire, T.A.A., Jeyakumar, A.E.: Deterministically guided PSO for dynamic dispatch considering valve-point effect. Elect. Power Syst. Res., vol. 73, no. 3, March 2005, pp. 313–322.
19. Victoire, T.A.A., Jeyakumar, A.E.: Reserved constrained dynamic dispatch of units with valve point effects'. IEEE. Trans. Power Syst., vol. 20, no. 3, August 2005, pp. 1273–1282.
20. Basu, M.: Hybridization of bee colony optimization and sequential quadratic programming for dynamic economic dispatch". Elect. Power and energ. syst., vol. 44, 2013, pp. 591–596.
21. de Castro, L.N., Timis, J.: Artificial Immune System: A Novel Paradigm to Pattern Recognition'. In Artificial Neural Networks in Pattern Recognition, SOCO-2002, University of Paiselely, U.K., 2002, pp. 67–84.
22. Dasgupta, D., Okline, N.A.: Immunity-Based Survey. In the Proceedings of IEEE international Conf. on System, Man and Cybernetics, vol. 1, Oct. 1997, pp. 369–374.
23. de Castro, L.N., Von Zuben, F.J.: Learning and Optimization using through the clonal Selection Principle. IEEE Trans. on Evolutionary Computation, vol. 6, June 2002, pp. 239–251.
24. de Castro, L.N., Von Zuben, F.J.: Artificial Immune System: Part 1-Basic Theory and Applications. Technical report, TR-DCA 01/99 Dec. 1999.
25. Burnet, F.M.: The Clonal Selection Theory of Acquired Immunity. Cambridge University Press Cambridge, U.K, 1959.

26. Panigrahi, B.K., Yadav, S.R., Agrawal, S., Tiwari, M.K.: A clonal algorithm to solve economic load dispatch. Elect. Power syst. Res., vol. 77, 2007, pp. 1381–1389.
27. Rahman, T.K.A., Suliman, S.I., Musurin, I.: Artificial Immune-Based Optimization Technique for Solving Economic Dispatch in Power System. Springer-Verlag, 2006, pp. 338–345.
28. Rahman, T.K.A., Yasin, Z.M., Abdullah, W.N.W.: Artificial Immune Based for Solving Economic Dispatch in Power system. IEEE National Power and Energy Conference Proceedings, Malaysia 2004, pp. 31–35.
29. Simon, D.: Biogeography-based optimization. IEEE Trans. Evolutionary Computation, vol. 12, no. 6, pp. 702–713, Dec. 2008.
30. D Simon, M Ergezer and D Du. 'Markov Analysis of Biogeography-Based Optimization'. [Online]. Available: http://academic.csuohio.edu/simond/bbo/markov/.
31. Simon, D., Ergezer, M., Du, D.: Population Distributions in Biogeography-Based Optimization with Elitism. IEEE International Conference on Systems, Man and Cybernetics, 2009.
32. Bhattacharya, A., Chattopadhyay, P.K.: Biogeography-based optimization for different economic load dispatch problems. IEEE Trans. Power Syst., vol. 25 no. 2, pp. 1064–77, May 2010.
33. Panigrahi, B.K., Pandi, V.R.: Bacterial foraging optimization: Nelder-Mead hybrid algorithm for economic load dispatch. IET Gen., Transm., Distrib., vol. 2, no. 4, pp. 556–565, 2008.

A Survey on Teaching–Learning-Based Optimization Algorithm: Short Journey from 2011 to 2017

Janmenjoy Nayak, Bighnaraj Naik, G. T. Chandrasekhar and H. S. Behera

Abstract Since the early days of optimization, basically there are two famous optimization methods such as evolutionary-based and swarm intelligence-based algorithms. These two algorithms are population-based metaheuristics and are used to solve many of the real-world complex computing problems. However, recent research of some of the multi-objective optimization algorithms reveals that those earlier developed metaheuristics are unable to solve the multi-dimensional problems due to their pitfalls such as adjustment of controlling parameters, probabilistic nature, own algorithmic-dependent parameters. Looking into such scenario, in 2011 a new population-based metaheuristic was developed by R.V. Rao called teaching–learning-based optimization (TLBO) algorithm. Since its inception, the applicability of TLBO has crossed many milestones as compared to other recently developed metaheuristics for its use in diversified problem domains of engineering. In this paper, a survey is conducted on TLBO and its variants along with the discussion on its range of applications from 2011 to 2017.

Keywords Teaching–learning-based algorithm · Population-based algorithm

J. Nayak (✉)
Department of Computer Science and Engineering, Sri Sivani College of Engineering
(SSCE), Srikakulam 532410, Andhra Pradesh, India
e-mail: mailforjnayak@gmail.com

B. Naik
Department of Computer Application, Veer Surendra Sai University of Technology,
Burla 768018, Odisha, India
e-mail: mailtobnaik@gmail.com

G. T. Chandrasekhar
Department of EEE, Sri Sivani College of Engineering,
Srikakulam 532410, Andhra Pradesh, India
e-mail: gtchsekhar@gmail.com

H. S. Behera
Department of IT, Veer Surendra Sai University of Technology,
Burla 768018, Odisha, India
e-mail: mailtohsbehera@gmail.com

© Springer Nature Singapore Pte Ltd. 2019
H. S. Behera et al. (eds.), *Computational Intelligence in Data Mining*,
Advances in Intelligent Systems and Computing 711,
https://doi.org/10.1007/978-981-10-8055-5_66

1 Introduction

Optimization is the process of choosing the best solution among the other available alternatives in the solution space. The process of optimization must be fully functional, efficient, and effective in choosing the solution for any particular problem. The process is to choose either maximization or minimization of a real function through the proper selection of value from any set of values. The goal of optimization is to satisfy the constraints with the adoption of maximized or minimized objective function values. Optimization algorithms can be categorized into different types based on their inspiration such as population algorithms (inspired by groups), evolutionary algorithms (inspired by biological evolution), swarm intelligence algorithms (inspired by behavior of different swarms), nature-inspired algorithms (inspired by phenomenon of nature), chemical algorithms (inspired by chemical processes), physical algorithms (inspired by physics). Among these optimization techniques, population-based metaheuristics have been greatly used in diversified range of problem domains due to their easy implementation, less complexity and capability of producing global optimal solutions. The algorithms maintain a set of candidate solutions, and in the population, every solution corresponds to an inimitable point in the search space. Moreover, the population-based algorithms perform much better as compared to other solo-search-based algorithms in solving real-life problems [1]. Since the development of the algorithms, these have been a key interest among a wide range of researchers and are successfully implemented in various application areas including mechanical design optimization, optimal power flow, multi-objective optimization, hydrothermal scheduling, power system, machining processes, scheduling of thermal power systems, power quality, and global optimization. Due to its easy implementation and less design issues, it attracts researchers to use it in various application areas.

Teaching learning-based optimization (TLBO) [2, 3] is a recently developed metaheuristic and is inspired by teaching and learning process among students and teachers in a typical real-life classroom scenario. Unlike other optimization processes, there is no requirement of any controlling parameter setting that makes it simple to implement for any problem. Being a population-based metaheuristic, it takes the advantage of maintaining a suitable balance between exploration and exploitation. Looking into its popularity, in this paper a brief survey is conducted on the applications of TLBO along with its different variants. Also, the criticism associated with the algorithm with the remedies and future scopes has been discussed. The remaining part of paper is sectioned as follows: Sect. 2 detailed the working principle of TLBO. Section 3 describes the variants of TLBO along with its hybridization of other algorithms, and the applications have been illustrated in Sect. 4. Section 5 discusses the criticism and the counterpart to criticism of the algorithm. The conclusion with the future scope has been indicated in Sect. 6.

2 Working Principle of TLBO

TLBO is a parameter-free population-based metaheuristic algorithm inspired by teacher and student/learner. The working principle of the algorithm is based on the outcome of a learner through the teaching process of a teacher. A teacher is an individual, possessing greater knowledge than his/her learners. The knowledge of a teacher is to be shared with the learners, and the outcome of the learners is the ultimate result from teaching. Moreover, the best effort from a teacher will be the ultimate indication toward getting more chance for a learner to secure good results. Apart from this, dissemination of knowledge is also possible among the friends of learners, which may also be a reason for increase in the knowledge level of learner. TLBO algorithm has two consequent phases: First one is teaching and other one is learning.

Phase-I (Teaching):

During this phase, the student's behavior is simulated through the teacher. A teacher always puts his best efforts in the class so as to bring his/her students next to his knowledge level. However, it is not practically feasible owing to the fact that there is a wide disparity among the knowledge levels of various students which can be classified in terms of average, good, and best. Hence, we need to consider the mean in order to calculate the overall knowledge level which is not only a random process but also dependent on various external factors.

Phase-II (Learning):

The behavior of the student based upon his discussion about his knowledge regarding a particular subject with his classmates or friends in that class is simulated by the learning phase. There may be cases where a student gains knowledge on a specific topic in due course of his/her discussion regarding the topic with his/her buddies in his class who tend to have a greater edge. The algorithm of TLBO may be realized in two phases as illustrated in Fig. 1.

In teaching phase, the level of student's knowledge is to be evaluated by the teacher, i.e., how much effort the teacher is giving in the class. Always, a teacher tries to give instructions to the students for enhancement in knowledge among the students. The output of a teacher can be simulated by the obtained results of the concerned students. For this, the mean results of all students in the class are considered. The population for learners (X) is chosen randomly and (X_{mean}) is the population mean. Subsequent to the computation of fitness for every student, the best solution is termed as $X_{teacher}$, having highest knowledge. Always, a teacher gives his/her superlative effort to teach in the class; however, the students may receive the things depending on the teaching quality. Also, most of the times, the student's quality (below average, average, or good) affects the result. With consideration to this, the quality factor of a teacher and the mean result of learners is represented as $rand(1)(X_{teacher} - T_F * X_{mean})$. Here, rand(1) lies in between a range of [0,1]. T_F is the teaching factor and is responsible for the modification of the mean

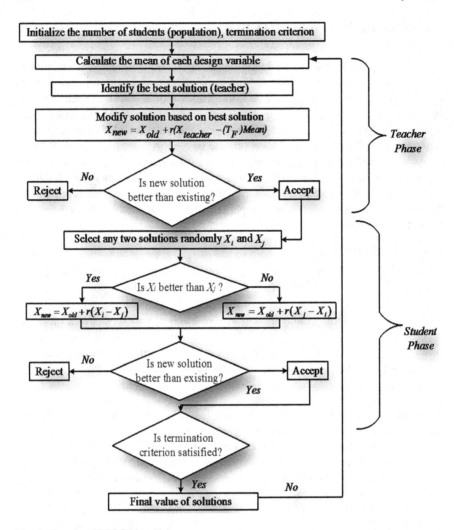

Fig. 1 Process of TLBO algorithm

value. The value of T_F is computed as $round(1 + rand(0, 1)(2 - 1))$, which means the generated value is either 1 or 2 (selected randomly).

3 Variants of TLBO

This section describes some variants of TLBO and their working principles in detail.

3.1 Elitist TLBO (E-TLBO)

In initial version of TLBO, two common algorithmic parameters, i.e., size of population and number of generation are considered. The concept of elitism is introduced in later version of TLBO called Elitist-TLBO (E-TLBO) [4], where in every generation; the worst solution in the population is replaced by elite solution. Also, Rajasekhar et al. [5] have developed OTLBO (opposition-based TLBO) with elitism concept. As a result, if duplicate solutions exist in the populations after replacing worst solutions with elite solution, then the duplicate solutions are modified by using mutation operation on those solutions, thereby avoiding trapping in the local optimal solutions. The concept of elitism is used in most evolutionary and swarm intelligence optimizations to avoid duplicate solutions in the population in every generation. The E-TLBO utilizes the concept of elitism by substituting the worst solutions in the population with best solution (elite solution) at the end of student phase.

3.2 Modified TLBO (MTLBO)

In MTLBO [6], the learner phase of the basic TLBO has been modified. Besides the teacher's teaching method in the classroom for the learners, there may be certain possibilities of enhancement of knowledge through some extra tutorial classes. To express this fact and by the inspiration of random scaling factor of differential evolution algorithm, Satapathy and Naik [6] have added an extra term to the basic equation of learner phase of TLBO algorithm. The range for such scaling factor is considered in between (0.5, 1) and hence, the equation is $0.5 * (1 + rand(0, 1))$. Here, "rand" indicates the randomly generated values of [0,1] with uniform distribution. The authors claim that by using MTLBO the global optimum solution can be easily found with the help of best fitness value in the population.

3.3 Weighted TLBO (WTLBO)

In WTLBO [31], a new weight factor has been added into the basic equations of TLBO. The objective to add such weight factor is to consider some parts of the last/ previous values of learners. During the computation of new learner's value [7], the value can be computed through a weight factor "w." Such inspiration is being considered from the concept that, while a teacher teaches any subject/topic, he/she expects that the learner has remembered some parts from the previously discussed classes. The weight factor is defined in Eq. (1).

$$w = \left(\frac{w_{\max} - w_{\min}}{\max.iteration}\right) \times i \tag{1}$$

Here, "$w_{\max}$ and $w_{\min}$" indicate the max and min values of "w," which are to be set as 109 and 0.1, respectively, "i" represents the current iteration number, and "max. iteration" is the maximum number of iteration. So, the equations of both teacher and learner phase may be updated as in Eqs. (2) and (3).

$$X_{new^g_{(i)}} = w \times X^g_{(i)} + rand \times (X^g_{Teacher} - T_F M^g) \tag{2}$$

$$X_{new^g_{(i)}} = \begin{cases} w * X^g_{(i)} + rand \times (X^g_{(i)} - X^g_{(r)}) \\ \qquad if f(X^g_{(i)}) < f(X^g_{(r)}) \\ w * X^g_{(i)} + rand \times (X^g_{(r)} - X^g_{(i)}) \quad otherwise \end{cases} \tag{3}$$

3.4 Quasi-oppositional TLBO (QOTLBO)

QOTLBO [8] is the combination of original TLBO and opposition-based learning [8]. This method is helpful for obtaining a global optimal solution as compared to random candidate solution. Also, it has been used to enhance the convergence speed.

Table 1 TLBO variants

Sl. no	Variant	Author	References	Year
1	Multi-objective TLBO (MOTLBO)	Niknam et al.	[9]	2012
2	Orthogonal TLBO (OTLBO)	Satapathy et al.	[10]	2013
3	Improved TLBO (I-TLBO)	Rao and Patel	[11]	2013
4	Simplified TLBO (STLBO)	Kai et al.	[12]	2013
5	Ameliorated TLBO (A-TLBO)	Li et al.	[13]	2013
6	Bare-bones teaching–learning-based optimization (BBTLBO)	Zou et al.	[14]	2014
7	Multi-class cooperative teaching–learning-based optimization algorithm with simulated annealing (SAMCCTLBO)	Chen et al.	[15]	2015
8	Discrete TLBO (DTLBO)	Li et al.	[16]	2015
9	Open MP TLBO	Umbarkar et al.	[17]	2015
10	Self-adaptive multi-objective TLBO (SA-MTLBO)	Yu et al.	[18]	2015
11	TLBO with learning experience (LETLBO)	Zou et al.	[19]	2015

Table 2 Hybridization of TLBO algorithm with other methods

Sl. no	Method/Algorithm	Author	References	Year
1	Local diversification strategy	Kundu et al.	[20]	2012
2	Differential evolution	Jiang and Zhou	[21]	2013
3	Brainstorming optimization algorithm	Krishnanand et al.	[22]	2013
4	Harmony search	Tuo et al.	[23]	2013
5	PSO	Zou et al.	[24]	2013
6	Double differential evolution	Ghasemi et al.	[25]	2014
7	Sequential quadratic programming	Krishnasamy and Nanjundappan	[26]	2014
8	PSO	Lim and Isa	[27]	2014
9	Fuzzy	Moghadam and Seifi	[28]	2014
10	Differential evolution + Chaotic perturbation	Yu et al.	[29]	2014
11	Dynamic group theory	Zou et al.	[30]	2014
12	Support vector machine	Cao and Luo	[31]	2015
13	Pareto tournament	Chaves-González et al.	[32]	2015
14	Simulated annealing	Chen et al.	[33]	2015
15	Support vector machine	Das and Padhy	[34]	2015
16	Genetic algorithm	Lahari et al.	[35]	2015
17	Radial basis function neural network	Omidvar et al.	[36]	2015
18	Differential evolution	Turgut et al.	[37]	2015
19	Bird-mating optimization	Zhang et al.	[38]	2015

In addition to these, some other variants are also developed and are listed in Table 1. Moreover, some researchers have hybridized TLBO with other methods to obtain better results as compared to TLBO that are listed in Table 2. This shows the hybridization of various techniques with TLBO algorithm for solving wide range of problems.

4 Applications of TLBO Algorithm

By searching through the keyword "TLBO algorithm" in Google Scholar, it is found that a total of 2,020 articles have been published, which are based on different diversified applications of engineering. From its inception in 2011, till date almost all range of researchers have applied the algorithm in various application domains. In this study, some similar applications are not being indicated and emphasis has been put forward on variety of applications (broadly classified) from all fields. The applications are classified into following categories and are sorted

Table 3 Applications of TLBO in electrical engineering

Author	Application	Year	References
Krishnanand et al.	Economic load dispatch	2011	[39]
Nayak et al.	Power system	2011	[40]
Nayak et al.	Power system	2011	[41]
Satapathy et al.	Power system	2012	[42]
Niknam et al.	Voltage regulation	2012	[43]
Ramanand et al.	Power dispatch	2012	[44]
Nayak et al.	Optimal power flow	2012	[45]
Niknam et al.	Micro grid operation	2012	[46]
Theja et al.	Power system	2012	[47]
Singh et al.	Directional current relay	2013	[48]
Khooban et al.	PID controller	2013	[49]
Sultana and Roy	Optimal capacitor placement	2014	[50]
Barisal	Load frequency	2015	[51]
Chatterjee et al.	Stability analysis	2016	[52]
Ghanavati	CMOS design	2017	[53]

Table 4 Applications of TLBO in data mining

Author	Application	Year	References
Satapathy and Anima Naik	Clustering	2011	[54]
Amiri	Clustering	2012	[55]
Naik et al.	Clustering	2012	[56]
Sahoo and Kumar	Clustering	2014	[57]
Safarinejadian et al.	Forecasting	2014	[58]
Wang et al.	Neural network	2014	[59]
Umbarkar et al.	Knapsack problems	2015	[60]
Davarpanah et al.	Cellular networks	2015	[61]
González-Álvarez et al.	Finding patterns in protein sequences	2015	[62]
Kanungo et al.	Clustering	2016	[63]
Nayak et al.	Classification	2016	[64]
Chen et al.	Neural network	2016	[65]

yearwise: electrical engineering (Table 3), data mining (Table 4), optimization (Table 5), and other applications (Table 6).

Besides the above applications, some more applications can also be found in Google Scholar. In brief to that, the popularity of algorithm is depicted from Fig. 2. Figure 2 indicates the total number of publications of TLBO algorithm till June 2017 in different reputed publishers/database such as IEEE, Springer, Science Direct, Taylor and Francis, World Scientific, Inderscience, and IGI Global. This search is made through the keyword search "TLBO algorithm."

Table 5 Applications of TLBO in optimization

Author	Application	Year	References
Rao et al.	Optimization of plate-fin heat exchanger	2011	[66]
Rao et al.	Mechanical design optimization	2011	[67]
Pare et al.	Optimization of cutting conditions in end milling process	2011	[68]
Venkata and Kalyankar	Parameter optimization of machining processes	2012	[69]
Venkata and Patel	Multi-objective optimization	2012	[70]
Rao and Patel	Constrained optimization	2012	[71]
Rajasekhar et al.	Global optimization	2012	[72]
Gordián-Rivera et al.	Constrained optimization	2012	[73]
Rao and Patel	Optimization of heat exchangers	2013	[74]
Rao and Patel	Optimization of thermoelectric cooler	2013	[75]
Yildiz	Optimization of multi-pass turning operations	2013	[76]
Baghlani and Makiabadi	Shape and size optimization of truss structures	2013	[77]
Dede and Ayvaz	Structural optimization	2013	[78]
Patel and Savsani	Optimization of a plate-fin heat exchanger design	2014	[79]
Chakravarthy et al.	Linear array optimization	2015	[80]
Rao and Waghmare	Design optimization of robot grippers	2015	[81]
Nama et al.	Optimization of geotechnical problem	2015	[82]
Qu et al.	Hypersonic re-entry trajectory optimization	2016	[83]
Cheng and Prayogo	Structural optimization	2017	[84]
Tiwari and Pradhan	Optimization of end milling process	2017	[85]
Toğan and Mortazavi	Optimization of skeletal structures	2017	[86]
Qu et al.	Aerodynamic shape optimization	2017	[87]

Apart from the above analysis, a yearwise analysis has also been represented to indicate the growth of research in TLBO algorithm since 2011. From 2011 to 2017, a detailed list of publications (yearwise) of TLBO algorithm is shown in Fig. 3. The list has been prepared using Google Scholar database. From Fig. 3, it is evident that the number of research publications in TLBO from 2011 has been increased substantially and indicates a potential growth in the research of TLBO algorithm.

Starting from 2011, twelve numbers of papers had been published, and in the year 2016, highest seven hundred and fifty-four numbers of papers are published in various journals and conferences. This indicates the popularity and usefulness of the algorithm.

Table 6 Applications of TLBO in other applications

Author	Application	Year	References
Knight et al.	Energy simulation	2011	[88]
Toğan	Design of steel frames	2012	[89]
González-Álvarez et al.	Motif finding	2012	[90]
Mohapatra et al.	Distribution network	2012	[91]
Rajasekhar and Das	Location of median line in 3D space	2012	[92]
Togan	Design of pin-jointed structures	2013	[93]
Naik et al.	Multi-cast routing	2013	[94]
Rao et al.	Image segmentation	2013	[95]
Singh and Dhillon	Design of LP and HP digital IIR filter	2013	[96]
Stephen et al.	Image enhancement	2013	[97]
Baykasoğlu et al.	Flow shop and job shop scheduling cases	2014	[98]
Akhlaghi and Nozhat	Designing plasmonic nanoparticles	2014	[99]
Khooban	Robotics	2014	[100]
Rao and More	Optimal tolerance design of machine elements	2014	[101]
Ji et al.	Demand forecasting	2014	[102]
Alneamy et al.	Disease diagnosis	2014	[103]
Bayram et al.	Oxygen concentration	2015	[104]
Crawford et al.	Solving set covering problems	2015	[105]
Agrawal et al.	Iris recognition system	2015	[106]
Cao and Luo	Fault diagnosis	2015	[107]
Balvasi et al.	Investigation of supper scattering plasmonic nanotubes	2016	[108]
Deng et al.	Mobile services	2016	[109]
Balasubramanian and Santhi	Unit commitment problem	2016	[110]
Zhang et al.	Resource allocation	2017	[111]
Ji et al.	Foundry industry	2017	[112]

5 Criticisms About TLBO

Although TLBO algorithm is successful in many applications, still in 2012 Čre-pinšek et al. [113] have commented some major points on the novelty and efficiency of the algorithm. They have experimented and discussed the misconceptions, such as "(a) TLBO is the best algorithm than other metaheuristics, (b) it does not require any algorithmic-specific parameter settings or parameter-less control". They have tested and verified the algorithm with some constrained and unconstrained benchmark functions. They criticized and raised some major points about the performance of TLBO algorithm. Later, Waghmare [114] commented on the points raised by C˘repinšek et al., and after the re-examination of the results of TLBO, he proved that the negative points pointed by C˘repinšek et al. about TLBO algorithm

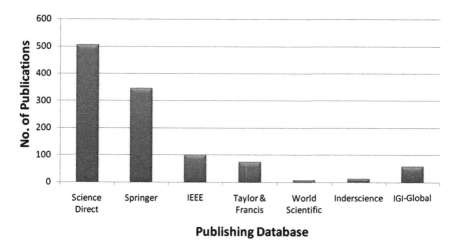

Fig. 2 Paper publications in TLBO in various reputed international publishers up to 2017

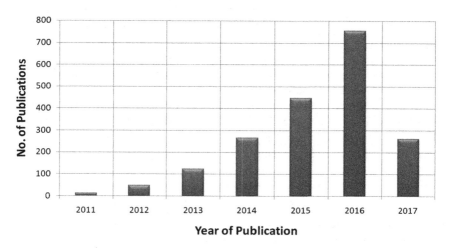

Fig. 3 Yearwise research publications of TLBO

are false. He clarifies that "TLBO only requires common controlling parameters as like other evolutionary and swarm-based algorithms. But like others, it is free from own algorithmic controlling parameter (such as mutation and crossover rate in GA, social and cognitive parameters in PSO) settings, which is a great advantage of TLBO with respect to complexity issues. All the benchmark functions (both constrained and unconstrained) considered by C˘repinšek et al. are again tested by Waghmare and proved that TLBO works correctly with some suitable contrary conclusions. However, our opinion is about the functionality and popularity of TLBO, is due to its simpler algorithmic structure and requirement of less dependent parameters. Moreover, the working procedure of TLBO in two different independent phases (teacher and learner) by suitably selecting the fittest candidate from one phase for others is the main functional aspect of the method.

6 Analysis

From the literatures published on TLBO, it is observed that the application area of the algorithm is quite diversified in a short span of time. Starting from optimization problems to diversified engineering problems, TLBO is successful as compared to other algorithms. A classified diagram of the application areas of TLBO can be visualized in Fig. 4. In this figure, the major application areas of the algorithm are covered under optimization and mechanical engineering problems. Besides those, some other areas of applications such as data mining and civil and electrical engineering problems are some of the important areas where researchers have used the algorithm substantially. The important advantage of this algorithm is its simplicity, easy to use and absence of any complex algorithmic tuning parameters. For solving large-scale problems, TLBO takes a special attention due to its requirement of less computational effort as compared to others. But for performance comparison and to maintain consistency among all the algorithms, the function evaluations should be remained same in case of all optimization algorithms. Moreover, Rao et al. [2] have emphasized on duplicacy removal from the population, which is a major aspect of any function or constraint optimization. Although some of the previously developed population algorithm like GA is concerned with repeated populations, Rao et al. have developed the algorithm for suitable removal of duplicate solutions. For more accurate extraction of duplicate values from the population, they further developed an improved version of TLBO called

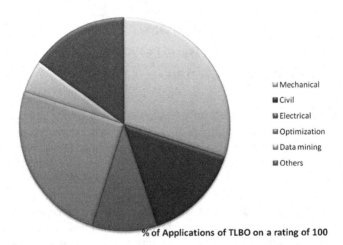

Fig. 4 Application areas of TLBO

Elitist TLBO (choosing the best from all good solutions). In fact, a number of other modified versions of TLBO have also been developed to show the effectiveness and potential growth of the algorithm.

7 Conclusions and Future Scope

TLBO is a population-based optimization algorithm, which has been widely acclaimed throughout the universe. Some interesting factors such as simplicity in its use, less parameter, finding global solutions make it more successful as compared to other algorithms. Also, several researchers have modified the algorithm with different strategies to make it more efficient. However, it cannot be claimed that after successful applications in various disciplines, TLBO is the best algorithm! This is because, according to "No Free Lunch Theorem," there does not exist such algorithm, which can be termed as the best among all and always there is a chance to improve than the previous one. Furthermore, a single optimization algorithm may not be fit to various problems of different nature.

Now-a-days, nature-inspired algorithms may be quite popular for their capability to find out the promising solutions for diversified applications, but that does not indicate the need of any urgent attention for solving any real-life problem. In reality, there is existence of many significant aspects those are still challenging. For an instance, many researchers have found that although several nature-inspired algorithms are successful to produce optimal solutions in reasonable time, still a significant gap lies in between theory and practice of the use of such algorithms. Issues such as rigorous mathematical analysis for the key factors of such algorithms, convergence analysis to get a optimal solution, balancing in between accuracy of the problem and required computational efficiency, proper tuning of algorithmic parameters are still to be resolved and need significant attention, which may be some important future aspects. In addition to that, TLBO may be found as an efficient algorithm and it is further to be used in many of such applications, where it has not been still used like regression analysis, financial forecasting, classification, healthcare applications.

Acknowledgements This work is supported by Technical Education Quality Improvement Programme, National Project Implementation Unit (a unit of MHRD, Govt. of India, for implementation of World Bank-assisted projects in technical education), under the research project grant (VSSUT/TEQIP/37/2016).

References

1. A., Prügel-Bennett, "Benefits of a population: five mechanisms that advantage population-based algorithms," *Evolutionary Computation, IEEE Transactions on*, Vol. 14, No. 4, pp. 500–517, 2010.
2. R. V., Rao, V. J., Savsani and D. P., Vakharia, "Teaching–learning-based optimization: an optimization method for continuous non-linear large scale problems," *Information Sciences*, Vol. 183, No. 1, pp. 1–15, 2012.
3. R. V., Rao and V. D., Kalyankar, "Parameter optimization of modern machining processes using teaching–learning-based optimization algorithm," *Engineering Applications of Artificial Intelligence*, Vol. 26, No. 1, pp. 524–531, 2013.
4. R. V. Rao, and V. Patel, "An elitist teaching-learning-based optimization algorithm for solving complex constrained optimization problems," International Journal of Industrial Engineering Computations, vol. 3, pp. 535–560, 2012.
5. A. Rajasekhar, R. Rani, K. Ramya, and A. Abraham,"Elitist teaching learning opposition based algorithm for global optimization," IEEE International Conference on Systems, Man, and Cybernetics SMC, pp. 1124–1129, 2012.
6. Satapathy, Suresh Chandra, and Anima Naik. "Modified Teaching–Learning-Based Optimization algorithm for global numerical optimization—A comparative study." Swarm and Evolutionary Computation 16 (2014): 28–37.
7. Roy, Provas Kumar, Aditi Sur, and Dinesh Kumar Pradhan. "Optimal short-term hydro-thermal scheduling using quasi-oppositional teaching learning based optimization." Engineering Applications of Artificial Intelligence 26.10 (2013): 2516–2524.
8. Roy Provas Kumar, Sur Aditi, Pradhan Dinesh Kumar (2013) Optimal short-term hydro-thermal scheduling using quasi-oppositional teaching learning based optimization, Engineering Applications of Artificial Intelligence 26, pp 2516–2524.
9. Niknam, T., Golestaneh, F., Sadeghi, M.S. (2012) θ -Multiobjective Teaching–Learning-Based Optimization for Dynamic Economic Emission Dispatch, IEEE Systems Journal, Volume: 6, Issue: 2, pp 341–352, https://doi.org/10.1109/JSYST.2012.2183276.
10. Satapathy Suresh Chandra, Naik Anima, Parvathi K. (2013). A teaching learning based optimization based on orthogonal design for solving global optimization problems, Springer Plus 2013, 2:130.
11. Rao, R. Venkata, and Vivek Patel. "An improved teaching-learning-based optimization algorithm for solving unconstrained optimization problems." Scientia Iranica 20.3 (2013): 710–720.
12. Kai Xia, Liang Gao, Lihui Wang, Weidong Li, Kuo-Ming Chao (2013) A Simplified Teaching-Learning-Based OptimizationAlgorithm for Disassembly Sequence Planning, IEEE 10th International Conference on e-Business Engineering (ICEBE), pp 393–398, https://doi.org/10.1109/ICEBE.2013.60.
13. Li, G., Niu, P., s, W., & Liu, Y. (2013). Model NOx emissions by least squares support vector machine with tuning based on ameliorated teaching–learning-based optimization. Chemometrics and Intelligent Laboratory Systems, 126, 11–20.
14. Zou, F., Wang,L., Hei,X., Chen,D., Jiang,Q., & Li, H. (2014b). Bare-bones teaching-learning-based optimization. The Scientific World Journal, https://doi.org/10. 1155/2014/136920.
15. Chen, Debao, Feng Zou, Jiangtao Wang, and Wujie Yuan. "SAMCCTLBO: a multi-class cooperative teaching–learning-based optimization algorithm with simulated annealing." Soft Computing 20, no. 5 (2016): 1921–1943.
16. Li, Jun-qing, Quan-ke Pan, and Kun Mao. "A discrete teaching-learning-based optimisation algorithm for realistic flowshop rescheduling problems." Engineering Applications of Artificial Intelligence 37 (2015): 279–292.

17. Umbarkar, A. J., N. M. Rothe, and A. S. Sathe. "OpenMP teaching-learning based optimization algorithm over multi-core system." International Journal of Intelligent Systems and Applications 7, no. 7 (2015): 57.

18. Yu, Kunjie, Xin Wang, and Zhenlei Wang. "Self-adaptive multi-objective teaching-learning-based optimization and its application in ethylene cracking furnace operation optimization." Chemometrics and Intelligent Laboratory Systems 146 (2015): 198–210.

19. Zou, Feng, Lei Wang, Xinhong Hei, and Debao Chen. "Teaching–learning-based optimization with learning experience of other learners and its application." Applied Soft Computing 37 (2015): 725–736.

20. Kundu, Souvik, et al. "A selective teaching-learning based niching technique with local diversification strategy." International Conference on Swarm, Evolutionary, and Memetic Computing. Springer Berlin Heidelberg, 2012.

21. Jiang, Xingwen, and Jianzhong Zhou. "Hybrid DE-TLBO algorithm for solving short term hydro-thermal optimal scheduling with incommensurable Objectives." In Control Conference (CCC), 2013 32nd Chinese, pp. 2474–2479. IEEE, 2013.

22. Krishnanand, K. R., Syed Muhammad Farzan Hasani, Bijaya Ketan Panigrahi, and Sanjib Kumar Panda. "Optimal power flow solution using self–evolving brain–storming inclusive teaching–learning–based algorithm." In International Conference in Swarm Intelligence, pp. 338–345. Springer Berlin Heidelberg, 2013.

23. Tuo, Shouheng, Longquan Yong, and Tao Zhou. "An improved harmony search based on teaching-learning strategy for unconstrained optimization problems." Mathematical Problems in Engineering 2013 (2013).

24. Zou, Feng, Lei Wang, Xinhong Hei, Qiaoyong Jiang, and Dongdong Yang. "Teaching-learning-based optimization algorithm in dynamic environments." In International Conference on Swarm, Evolutionary, and Memetic Computing, pp. 389–400. Springer International Publishing, 2013.

25. Ghasemi, Mojtaba, et al. "Modified teaching learning algorithm and double differential evolution algorithm for optimal reactive power dispatch problem: a comparative study." Information Sciences 278 (2014): 231–249.

26. Krishnasamy, Umamaheswari, and Devarajan Nanjundappan. "A refined teaching-learning based optimization algorithm for dynamic economic dispatch of integrated multiple fuel and wind power plants." Mathematical Problems in Engineering 2014 (2014).

27. Lim, Wei Hong, and Nor Ashidi Mat Isa. "Teaching and peer-learning particle swarm optimization." Applied Soft Computing 18 (2014): 39–58.

28. Moghadam, Ahmad, and Ali Reza Seifi. "Fuzzy-TLBO optimal reactive power control variables planning for energy loss minimization." Energy Conversion and Management 77 (2014): 208–215.

29. Yu, Kunjie, Xin Wang, and Zhenlei Wang. "An improved teaching-learning-based optimization algorithm for numerical and engineering optimization problems." Journal of Intelligent Manufacturing 27, no. 4 (2016): 831–843.

30. Zou, Feng, Lei Wang, Xinhong Hei, Debao Chen, and Dongdong Yang. "Teaching–learning-based optimization with dynamic group strategy for global optimization." Information Sciences 273 (2014): 112–131.

31. Cao, Junxiang, and Jianxu Luo. "A study on SVM based on the weighted elitist teaching-learning-based optimization and application in the fault diagnosis of chemical process." In MATEC Web of Conferences, vol. 22. EDP Sciences, 2015.

32. Chaves-González, José M., Miguel A. Pérez-Toledano, and Amparo Navasa. "Teaching learning based optimization with Pareto tournament for the multiobjective software requirements selection." Engineering Applications of Artificial Intelligence 43 (2015): 89–101.

33. Chen, Debao, Feng Zou, Jiangtao Wang, and Wujie Yuan. "SAMCCTLBO: a multi-class cooperative teaching–learning-based optimization algorithm with simulated annealing." Soft Computing 20, no. 5 (2016): 1921–1943.

34. Das, Shom Prasad, and Sudarsan Padhy. "A novel hybrid model using teaching–learning-based optimization and a support vector machine for commodity futures index forecasting." International Journal of Machine Learning and Cybernetics (2015): 1–15.

35. Lahari, Kannuri, M. Ramakrishna Murty, and Suresh C. Satapathy. "Partition based clustering using genetic algorithm and teaching learning based optimization: performance analysis." Emerging ICT for Bridging the Future-Proceedings of the 49th Annual Convention of the Computer Society of India CSI Volume 2. Springer International Publishing, 2015.

36. Omidvar, Mahyar, et al. "Selection of laser bending process parameters for maximal deformation angle through neural network and teaching–learning-based optimization algorithm." Soft Computing 19.3 (2015): 609–620.

37. Turgut, Oguz Emrah, and Mustafa Turhan Coban. "Optimal proton exchange membrane fuel cell modelling based on hybrid Teaching Learning Based Optimization–Differential Evolution algorithm." Ain Shams Engineering Journal 7.1 (2016): 347–360.

38. Zhang, Qingyang, Guolin Yu, and Hui Song. "A hybrid bird mating optimizer algorithm with teaching-learning-based optimization for global numerical optimization." Statistics, Optimization & Information Computing 3.1 (2015): 54–65.

39. Krishnanand, K. R., Bijaya Ketan Panigrahi, Pravat K. Rout, and Ankita Mohapatra. "Application of multi-objective teaching-learning-based algorithm to an economic load dispatch problem with incommensurable objectives." In International Conference on Swarm, Evolutionary, and Memetic Computing, pp. 697–705. Springer Berlin Heidelberg, 2011.

40. Nayak, Niranjan, Sangram Routray, and Pravat Rout. "A function based fuzzy controller for VSC-HVDC system to enhance transient stability of AC/DC power system." Swarm, Evolutionary, and Memetic Computing (2011): 441–451.

41. Nayak, N., S. K. Routray, and P. K. Rout. "A robust control strategies to improve transient stability in VSC-HVDC based interconnected power systems." Energy, Automation, and Signal (ICEAS), 2011 International Conference on. IEEE, 2011.

42. Satapathy, Suresh Chandra, Anima Naik, and K. Parvathi. "0–1 integer programming for generation maintenance scheduling in power systems based on teaching learning based optimization (TLBO)." In International Conference on Contemporary Computing, pp. 53–63. Springer Berlin Heidelberg, 2012.

43. Niknam, Taher, Rasoul Azizipanah-Abarghooee, and Mohammad Rasoul Narimani. "A new multi objective optimization approach based on TLBO for location of automatic voltage regulators in distribution systems." Engineering Applications of Artificial Intelligence 25, no. 8 (2012): 1577–1588.

44. Ramanand, K. R., K. R. Krishnanand, Bijaya Ketan Panigrahi, and Manas Kumar Mallick. "Brain storming incorporated teaching–learning–based algorithm with application to electric power dispatch." In International Conference on Swarm, Evolutionary, and Memetic Computing, pp. 476–483. Springer Berlin Heidelberg, 2012.

45. Nayak, M. R., C. K. Nayak, and P. K. Rout. "Application of multi-objective teaching learning based optimization algorithm to optimal power flow problem." Procedia Technology 6 (2012): 255–264.

46. Niknam, Taher, Rasoul Azizipanah-Abarghooee, and Mohammad Rasoul Narimani. "An efficient scenario-based stochastic programming framework for multi-objective optimal micro-grid operation." Applied Energy 99 (2012): 455–470.

47. Theja, Bagepalli Sreenivas, Anguluri Rajasekhar, and D. P. Kothari. "An intelligent coordinated design of UPFC based power system stabilizer for dynamic stability enhancement of SMIB power system." In Power Electronics, Drives and Energy Systems (PEDES), 2012 IEEE International Conference on, pp. 1–6. IEEE, 2012.

48. Singh, Manohar, B. K. Panigrahi, and A. R. Abhyankar. "Optimal coordination of directional over-current relays using Teaching Learning-Based Optimization (TLBO) algorithm." International Journal of Electrical Power & Energy Systems 50 (2013): 33–41.

49. Khooban, Mohammad Hassan, Alireza Alfi, and Davood Nazari Maryam Abadi. "Teaching–learning-based optimal interval type-2 fuzzy PID controller design: a nonholonomic wheeled mobile robots." Robotica 31.07 (2013): 1059–1071.

50. Sultana, Sneha, and Provas Kumar Roy. "Optimal capacitor placement in radial distribution systems using teaching learning based optimization." International Journal of Electrical Power & Energy Systems 54 (2014): 387–398.

51. Barisal, A. K. "Comparative performance analysis of teaching learning based optimization for automatic load frequency control of multi-source power systems." International Journal of Electrical Power & Energy Systems 66 (2015): 67–77.

52. Chatterjee, Shamik, Abishek Naithani, and V. Mukherjee. "Small-signal stability analysis of DFIG based wind power system using teaching learning based optimization." International Journal of Electrical Power & Energy Systems 78 (2016): 672–689.

53. Ghanavati, Behzad. "High-Accurate Low-Voltage Analog CMOS Current Divider Modify by Neural Network and TLBO Algorithm." Journal of Advances in Computer Research 8.27 (2017): 51–65.

54. Satapathy, Suresh, and Anima Naik. "Data clustering based on teaching-learning-based optimization." Swarm, evolutionary, and memetic computing (2011): 148–156.

55. Amiri, Babak. "Application of teaching-learning-based optimization algorithm on cluster analysis." Journal of Basic and Applied Scientific Research 2.11 (2012): 11795–11802.

56. Naik, Anima, Suresh Chandra Satapathy, and K. Parvathi. "Improvement of initial cluster center of c-means using teaching learning based optimization." Procedia Technology 6 (2012): 428–435.

57. Sahoo, Anoop J., and Yugal Kumar. "Modified teacher learning based optimization method for data clustering." Advances in Signal Processing and Intelligent Recognition Systems. Springer International Publishing, 2014. 429–437.

58. Safarinejadian, Behrooz, Masihollah Gharibzadeh, and Mohsen Rakhshan. "An optimized model of electricity price forecasting in the electricity market based on fuzzy timeseries." Systems Science & Control Engineering: An Open Access Journal 2.1 (2014): 677–683.

59. Wang, Lei, Feng Zou, Xinhong Hei, Dongdong Yang, Debao Chen, and Qiaoyong Jiang. "An improved teaching–learning-based optimization with neighborhood search for applications of ANN." Neurocomputing 143 (2014): 231–247.

60. Umbarkar, A. J., P. D. Sheth, and S. V. Babar. "Solving 0/1 Knapsack Problem Using Hybrid TLBO-GA Algorithm." In Proceedings of Fourth International Conference on Soft Computing for Problem Solving, pp. 1–10. Springer India, 2015.

61. Davarpanah, Danial, Mohammadreza Zamani, Mohsen Eslami, and Taher Niknam. "Joint successive base station switch off and user subcarrier allocation optimization for green multicarrier based cellular networks." In Electrical Engineering (ICEE), 2015 23rd Iranian Conference on, pp. 504–507. IEEE, 2015.

62. González-Álvarez, David L., Miguel A. Vega-Rodríguez, and Alvaro Rubio-Largo. "Finding patterns in protein sequences by using a hybrid multiobjective teaching learning based optimization algorithm." IEEE/ACM Transactions on Computational Biology and Bioinformatics (TCBB) 12.3 (2015): 656–666.

63. Kanungo, D. P., Janmenjoy Nayak, Bighnaraj Naik, and Himansu Sekhar Behera. "Hybrid Clustering using Elitist Teaching Learning-Based Optimization: An Improved Hybrid Approach of TLBO." International Journal of Rough Sets and Data Analysis (IJRSDA) 3, no. 1 (2016): 1–19.

64. Nayak, Janmenjoy, Bighnaraj Naik, and H. S. Behera. "Optimizing a higher order neural network through teaching learning based optimization algorithm." Computational Intelligence in Data Mining—Volume 1. Springer India, 2016. 57–71.

65. Chen, Debao, Renquan Lu, Feng Zou, and Suwen Li. "Teaching-learning-based optimization with variable-population scheme and its application for ANN and global optimization." Neurocomputing 173 (2016): 1096–1111.

66. Rao, R. Venkata, and Vivek Patel. "Thermodynamic optimization of plate-fin heat exchanger using teaching-learning-based optimization (TLBO) algorithm." optimization 10 (2011): 11.

67. Rao, Ravipudi V., Vimal J. Savsani, and D. P. Vakharia. "Teaching–learning-based optimization: a novel method for constrained mechanical design optimization problems." Computer-Aided Design 43.3 (2011): 303–315.

68. Pare, Vikas, Geeta Agnihotri, and C. M. Krishna. "Optimization of Cutting Conditions in End Milling Process with the Approach of Particle Swarm Optimization." International Journal of Mechanical and Industrial Engineering 1.2 (2011): 21–25.

69. Venkata Rao, R., and V. D. Kalyankar. "Parameter optimization of machining processes using a new optimization algorithm." Materials and Manufacturing Processes 27.9 (2012): 978–985.

70. Venkata Rao, R., and Vivek Patel. "Multi-objective optimization of combined Brayton and inverse Brayton cycles using advanced optimization algorithms." Engineering Optimization 44.8 (2012): 965–983.

71. Rao, R., and Vivek Patel. "An elitist teaching-learning-based optimization algorithm for solving complex constrained optimization problems." International Journal of Industrial Engineering Computations 3, no. 4 (2012): 535–560.

72. Rajasekhar, Anguluri, Rapol Rani, Kolli Ramya, and Ajith Abraham. "Elitist teaching learning opposition based algorithm for global optimization." In Systems, Man, and Cybernetics (SMC), 2012 IEEE International Conference on, pp. 1124–1129. IEEE, 2012.

73. Gordián-Rivera, Luis-Alfredo, and Efrén Mezura-Montes. "A combination of specialized differential evolution variants for constrained optimization." In Ibero-American Conference on Artificial Intelligence, pp. 261–270. Springer Berlin Heidelberg, 2012.

74. Rao, R. Venkata, and Vivek Patel. "Multi-objective optimization of heat exchangers using a modified teaching-learning-based optimization algorithm." Applied Mathematical Modelling 37.3 (2013): 1147–1162.

75. Rao, R. Venkata, and Vivek Patel. "Multi-objective optimization of two stage thermoelectric cooler using a modified teaching–learning-based optimization algorithm." Engineering Applications of Artificial Intelligence 26.1 (2013): 430–445.

76. Yildiz, Ali R. "Optimization of multi-pass turning operations using hybrid teaching learning-based approach." The International Journal of Advanced Manufacturing Technology 66.9–12 (2013): 1319–1326.

77. Baghlani, A., and M. H. Makiabadi. "Teaching-learning-based optimization algorithm for shape and size optimization of truss structures with dynamic frequency constraints." Iranian Journal of Science and Technology. Transactions of Civil Engineering 37.C (2013): 409.

78. Dede, Tayfun, and Yusuf Ayvaz. "Structural optimization with teaching-learning-based optimization algorithm." Structural Engineering and Mechanics 47.4 (2013): 495–511.

79. Patel, Vivek, and Vimal Savsani. "Optimization of a plate-fin heat exchanger design through an improved multi-objective teaching-learning based optimization (MO-ITLBO) algorithm." Chemical Engineering Research and Design 92.11 (2014): 2371–2382.

80. Chakravarthy, V. V. S. S. S., et al. "Linear array optimization using teaching learning based optimization." Emerging ICT for Bridging the Future-Proceedings of the 49th Annual Convention of the Computer Society of India CSI Volume 2. Springer International Publishing, 2015.

81. Rao, R. Venkata, and Gajanan Waghmare. "Design optimization of robot grippers using teaching-learning-based optimization algorithm." Advanced Robotics 29.6 (2015): 431–447.

82. Nama, Sukanta, Apu Kumar Saha, and Sima Ghosh. "Parameters Optimization of Geotechnical Problem Using Different Optimization Algorithm." Geotechnical and Geological Engineering 33.5 (2015): 1235–1253.

83. Qu, Xinghua, Huifeng Li, Ran Zhang, and Bo Liu. "An effective TLBO-based memetic algorithm for hypersonic reentry trajectory optimization." In Evolutionary Computation (CEC), 2016 IEEE Congress on, pp. 3178–3185. IEEE, 2016.

84. Cheng, Min-Yuan, and Doddy Prayogo. "A novel fuzzy adaptive teaching–learning-based optimization (FATLBO) for solving structural optimization problems." Engineering with Computers 33.1 (2017): 55–69.
85. Tiwari, Atul, and Mohan Kumar Pradhan. "Modelling and Optimization of End Milling Process Using TLBO and TOPSIS Algorithm: Modelling and Optimization." Handbook of Research on Manufacturing Process Modeling and Optimization Strategies (2017): 54.
86. Toğan, Vedat, and Ali Mortazavi. "Sizing optimization of skeletal structures using teaching-learning based optimization." An International Journal of Optimization and Control: Theories & Applications (IJOCTA) 7.2 (2017): 130–141.
87. Qu, Xinghua, Ran Zhang, Bo Liu, and Huifeng Li. "An improved TLBO based memetic algorithm for aerodynamic shape optimization." Engineering Applications of Artificial Intelligence 57 (2017): 1–15.
88. Knight, Jennifer L., and Charles L. Brooks III. "Multisite λ dynamics for simulated structure–activity relationship studies." Journal of chemical theory and computation 7.9 (2011): 2728–2739.
89. Toğan, Vedat. "Design of planar steel frames using teaching–learning based optimization." Engineering Structures 34 (2012): 225–232.
90. González-Álvarez, David L., Miguel A. Vega-Rodríguez, Juan A. Gómez-Pulido, and Juan M. Sánchez-Pérez. "Multiobjective Teaching-Learning-Based Optimization (MO-TLBO) for Motif Finding." In Computational Intelligence and Informatics (CINTI), 2012 IEEE 13th International Symposium on, pp. 141–146. IEEE, 2012.
91. Mohapatra, Ankita, Bijaya Ketan Panigrahi, Bhim Singh, and Ramesh Bansal. "Optimal placement of capacitors in distribution networks using a modified teaching-learning based algorithm." In International Conference on Swarm, Evolutionary, and Memetic Computing, pp. 398–405. Springer Berlin Heidelberg, 2012.
92. Rajasekhar, Anguluri, and Swagatam Das. "Teaching learning opposition based optimization for the location of median line in 3-d space." In International Conference on Swarm, Evolutionary, and Memetic Computing, pp. 331–338. Springer Berlin Heidelberg, 2012.
93. Togan, Vedat. "Design of pin jointed structures using teaching-learning based optimization." Structural Engineering and Mechanics 47.2 (2013): 209–225.
94. Naik, Anima, K. Parvathi, Suresh Chandra Satapathy, Ramanuja Nayak, and B. S. Panda. "QoS multicast routing using Teaching learning based Optimization." In Proceedings of International Conference on Advances in Computing, pp. 49–55. Springer India, 2013.
95. Rao, Chereddy Srinivasa, Kanadam Karteeka Pavan, and Allam Appa Rao. "An automatic medical image segmentation using teaching learning based optimization." In Proceedings of International Conference on Advances in Computer Science. 2013.
96. Singh, D. A. M. A. N. P. R. E. E. T., and J. S. Dhillon. "Teaching-Learning Based Optimization Technique for the Design of LP and HP Digital IIR Filter." Recent Advances in Electrical Engineering and Electronic Devices (2013): 203–208.
97. Stephen, M. James, and P. V. G. D. Reddy. "Image Enhancement for Fingerprints with Modified Teaching Learning based Optimization and New Transformation Function." Journal of Machine Learning Technologies 3, no. 1 (2013): 76.
98. Baykasoğlu, Adil, Alper Hamzadayi, and Simge Yelkenci Köse. "Testing the performance of teaching–learning based optimization (TLBO) algorithm on combinatorial problems: Flow shop and job shop scheduling cases." Information Sciences 276 (2014): 204–218.
99. Akhlaghi, Majid, Farzin Emami, and Najmeh Nozhat. "TLBO algorithm assisted for designing plasmonic nano particles based absorption coefficient." J. Optoelectron. Adv. Mater. Rapid Commun 8.9–10 (2014): 1–4.
100. Khooban, Mohammad Hassan. "Design an intelligent proportional-derivative (PD) feedback linearization control for nonholonomic-wheeled mobile robot." Journal of Intelligent & Fuzzy Systems 26.4 (2014): 1833–1843.
101. Rao, R. V., and K. C. More. "Advanced optimal tolerance design of machine elements using teaching-learning-based optimization algorithm." Production & Manufacturing Research 2.1 (2014): 71–94.

102. Ji, Gang, et al. "Urban water demand forecasting by LS-SVM with tuning based on elitist teaching-learning-based optimization." Control and Decision Conference (2014 CCDC), The 26th Chinese. IEEE, 2014.
103. Alneamy, Jamal Salahaldeen Majeed, and Rahma Abdulwahid Hameed Alnaish. "Heart disease diagnosis utilizing hybrid fuzzy wavelet neural network and teaching learning based optimization algorithm." Advances in Artificial Neural Systems 2014 (2014): 6.
104. Bayram, Adem, Ergun Uzlu, Murat Kankal, and Tayfun Dede. "Modeling stream dissolved oxygen concentration using teaching–learning based optimization algorithm." Environmental Earth Sciences 73, no. 10 (2015): 6565–6576.
105. Crawford, Broderick, Ricardo Soto, Felipe Aballay, Sanjay Misra, Franklin Johnson, and Fernando Paredes. "A teaching-learning-based optimization algorithm for solving set covering problems." In International Conference on Computational Science and Its Applications, pp. 421–430. Springer International Publishing, 2015.
106. Agrawal, Shikha, Shraddha Sharma, and Sanjay Silakari. "Teaching learning based optimization (TLBO) based improved iris recognition system." Progress in Systems Engineering. Springer International Publishing, 2015. 735–740.
107. Cao, Junxiang, and Jianxu Luo. "A study on SVM based on the weighted elitist teaching-learning-based optimization and application in the fault diagnosis of chemical process." In MATEC Web of Conferences, vol. 22. EDP Sciences, 2015.
108. Balvasi, Mohsen, Majid Akhlaghi, and Hossein Shahmirzaee. "Binary TLBO algorithm assisted to investigate the supper scattering plasmonic nano tubes." Superlattices and Microstructures 89 (2016): 26–33.
109. Deng, Shuiguang, Longtao Huang, Daning Hu, J. Leon Zhao, and Zhaohui Wu. "Mobility-enabled service selection for composite services." IEEE Transactions on Services Computing 9, no. 3 (2016): 394–407.
110. Balasubramanian, K. P., and R. K. Santhi. "Best compromised schedule for multi-objective unit commitment problems." Indian Journal of Science and Technology 9, no. 2 (2016).
111. Zhang, Wenyu, Shuai Zhang, Shanshan Guo, Yushu Yang, and Yong Chen. "Concurrent optimal allocation of distributed manufacturing resources using extended Teaching-Learning-Based Optimization." International Journal of Production Research 55, no. 3 (2017): 718–735.
112. Ji, Xiaoyuan, Hu Ye, Jianxin Zhou, Yajun Yin, and Xu Shen. "An improved teaching-learning-based optimization algorithm and its application to a combinatorial optimization problem in foundry industry." Applied Soft Computing (2017).
113. Črepinšek, Matej, Shih-Hsi Liu, and Luka Mernik. "A note on teaching–learning-based optimization algorithm." Information Sciences 212 (2012): 79–93.
114. Waghmare, Gajanan. "Comments on "A note on teaching–learning-based optimization algorithm." Information Sciences 229 (2013): 159–169.

Predicting Users' Preferences for Movie Recommender System Using Restricted Boltzmann Machine

Dayal Kumar Behera, Madhabananda Das and Subhra Swetanisha

Abstract Recommender system is one of the most important crucial parts for e-commerce domains, enabling them to produce correct recommendations to individual users. Collaborative filtering is considered as the successful technique for recommender system that takes rating scores to find most similar users/items for recommending items. In this work, in order to exploit user rating information, a model has been developed that uses Restricted Boltzmann Machine (RBM) to learn deeply and predict the ratings or preferences which are missed. The experiment is done on MovieLens benchmark dataset that compares with Pearson correlation and average prediction-type algorithms. Experimental result exhibits the performance of RBM to predict users' preferences.

Keywords Collaborative filtering · Restricted Boltzmann machine
Item-based filtering · Movie recommendation

1 Introduction

In the age of Web 3.0, the exponential growth of data and number of users of Internet result information overload problem. In movie domain, the number of movies and its number of viewers have experienced an enormous increase in the last decade. All this information may be particularly useful for certain users who plan to watch an unknown movie. However, the list of possibilities offered by the

D. K. Behera (✉) · M. Das
Department of CSE, KIIT University, Bhubaneswar, India
e-mail: dayalbehera@gmail.com

M. Das
e-mail: mndas_prof@kiit.ac.in

D. K. Behera · S. Swetanisha
Department of CSE, Trident Academy of Technology, Bhubaneswar, India
e-mail: sswetanisha@gmail.com

© Springer Nature Singapore Pte Ltd. 2019   759
H. S. Behera et al. (eds.), *Computational Intelligence in Data Mining*,
Advances in Intelligent Systems and Computing 711,
https://doi.org/10.1007/978-981-10-8055-5_67

Web search engine may be overwhelming. The evaluation of this long list search options is very time-consuming process so far as the user's sentiment is concerned.

Personalized recommender system aims to discover new item to the user from a large collection of data based on the past preferences and tastes. Popular recommendation techniques are content-based [1], collaborative filtering (CF)-based [2–4], knowledge-based [5], and hybrid recommendations [6, 7]. Content-based system recommends items that are similar to the user's past taste, whereas collaborative filtering-based recommendation is simple and recommends items that other users with similar tastes of the active user liked in the past.

However, recommendation system is considered to be a bigdata problem as the number of items and preferences of users is huge and unstructured. Most of the collaborative filtering algorithms are not suitable for very large datasets. Salakhutdinov et al. [8] proposed how Restricted Boltzmann Machine can be applied on large dataset and compare the result with SVD model. In this paper, we tried to employ Restricted Boltzmann Machine (RBM) to predict ratings of the user for which they have not given the rating and based on the predicted rating recommending movies to the user. We have used the MovieLens dataset for experiment and compare the result with Pearson correlation and weighted average prediction algorithms.

2 Related Works

Collaborative filtering [9] was introduced in year 1992 by Goldberg et al. to deal with information Tapestry. In the literature, different methods are proposed in order to increase predicted ratings' accuracy. To compute recommendations for a group of users, the popular method is KNN-based CF. In 2016, Ortega et al. [10] explained how group recommendations are performed using matrix factorization (MF), where proposed method is considering different size of the datasets [10]. Hernando et al. [11] also worked on factorization of rating data by representing that into two nonnegative matrices. In 2016, Koohi et al. [9] proposed a FCM-based collaborative filtering and compared the result with k-means and SOM for the MovieLens dataset. Silva et al. [4] proposed genetic algorithm-based recommendation for combining various feasible techniques. In [12], Ar et al. also used a genetic algorithm for optimizing the prediction process.

CF is also widely used in the field of movie recommendation. In 2015, Moreno et al. [13] proposed a framework to deal with cold-start problems in movies' recommendation. In 2016, Li et al. also proposed [14] a model to mine microblogs to deal with cold-start problems in movie and television RS. In the paper [15], Hou et al. proposed a model to predict the likeness of the movie trailer by the viewers. In 2016, Zhang et al. [16] proposed TWH model to incorporate text metadata into a low-dimensional semantic space and applied for content-based movie's recommendation. Wei et al. [6] come up with an approach called hybrid movie recommendation approach using tags and ratings. A hybrid model [17] using improved k-means, and GA for movie recommendation is proposed by Wang et al. to deal with the information overload problem.

3 Background

Let $U = \{u_1, ..., u_n\}$ denotes the set of users, $M = \{m_1, ..., m_p\}$ represents set of movies, and R as an n × p matrix of ratings $r_{i,j}$, with i ∈ 1, ..., n, j ∈ 1, ..., p. The users rating can be defined on a numerical scale from 1 to 5 (strongly like). If an user has not rated a movie, the corresponding matrix entry remains empty. A sample of users' rating for a movie is represented in Table 1. CF approaches are divided into two major categories: user-based and item-based approaches.

3.1 User-Based Approaches

User-based approaches measure the similarities between users. It recommends items that an user has not yet viewed but others similar users brought/viewed/rated that earlier. For example,

Alice likes "Star Wars" and "Empire Strikes Back." Bob likes "Star Wars." Though Bob's preferences are most likely preference of Alice, they can be treated as similar. Hence, Bob may be recommended with "Empire Strikes Back." The problems of user-based CF are that users are fickle and their taste changes over time, and it may also be affected by shilling attack.

3.2 Item-Based Approaches

It measures the pair-wise similarities between items. Based on the similarities between items, missing ratings can be predicted and items that are similar to user's preferred items can be identified. It recommends the top-k closest items.

Table 1 Rating database for user-item matrix

	Movie 1	Movie 2	Movie 3	Movie 4
User 1	5	3	–	5
User 2	–	5	3	–
User 3	5	–	4	3

3.3 Pearson (Correlation)-Based Similarity

To find the similarity among users, one popular method used in RS is Pearson's correlation coefficient. The $sim(A, B)$ in Eq. 1 [18] calculates similarity between two users A and B. The symbol $\bar{r}$ represents average ratings of the user, $r_{A,m}$ represents preference or rating of user A for movie m.

$$sim(A,B) = \frac{\sum_{m \in M}(r_{A,m} - \bar{r}_A)(r_{B,m} - \bar{r}_B)}{\sqrt{\sum_{m \in M}(r_{A,m} - \bar{r}_A)^2}\sqrt{\sum_{m \in M}(r_{B,m} - \bar{r}_B)^2}} \tag{1}$$

3.4 From Model to Predictions

When the model is developed using Pearson similarity measure mentioned in Eq. 1, the ratings can be predicted for any user-movie pair by using Pearson's rate prediction method as shown in Eq. 2 [18]. $pred$(A, m) represents predicted rating of user A for movie "m." N represents nearest neighbors having rating on movie "m" which has been calculated by using Eq. 1.

$$pred(A, m) = \bar{r}_A + \frac{\sum_{B \in N} sim(A, m) * (r_{B,m} - \bar{r}_B)}{\sum_{B \in N} sim(A, B)} \tag{2}$$

Another general approach to calculate user's preference is to average all the ratings of a movie by the neighbors of the active user. Prediction of rating of user A for movie "m" and number of neighbors N can be calculated as in Eq. 3. As a result, top-k movies can be recommended to the active user.

$$pred(A, m) = \frac{1}{N} \sum_{i=1}^{N} r_{i,m} \tag{3}$$

3.5 Restricted Boltzmann Machines (RBM)

Deep learning is an emerging field in the area of machine learning and also in the field of recommender system (both collaborative- and content-based approaches). RBM is a generative model that can slightly outperform matrix factorization and can be used as a deep model to learn. Through several forward and backward processes, RBM will be trained and is able to reveal which attributes are the most important in order to detect patterns.

RBM has two layers. The first layer is called as the visible (or input layer). The second layer is the hidden layer. There is no connection between the nodes in the same layer. Training RBM has two phases: (1) forward pass and (2) backward pass or reconstruction. In the forward pass, input data from all visible nodes are being passed to all hidden nodes. At the hidden layer's nodes, X (input data) is multiplied by a W (weight between the neurons) and added to hidden layer bias "h_bias." The resultant value is fed into the sigmoid function, which produces the node's output/ state. In the reconstruction phase, the samples from the hidden layer play the role of input. The produced output is a reconstruction, which is an approximation of the original input.

Stochastic gradient descent is used to find the optimal weight. It has again two phases, i.e., the positive phase increases the probability of training data and the negative phase decreases the probability of samples generated by the model. The negative phase is hard to compute; hence, contrastive divergence (CD) is used to approximate it. Gibbs sampling has been used during the calculation of CD.

Contrastive divergence is actually a matrix of values that are computed and used to adjust values of the W matrix. Changing W incrementally leads to training of W values. Then on each step (epoch), W is updated to a new value W' as shown in Eq. 4 where $\propto$ is the learning rate.

$$W' = W + \propto * CD \tag{4}$$

4 Experimental Design

The aim of this paper is to apply RBM to collaborative filtering. Users' rating dataset is normally sparse in nature and larger in size. So to speed up the execution and to predict users' preference, RBM has been used. The architectural design is depicted in Fig. 1.

Firstly, the dataset is divided into tenfold cross-validation subsets. In each case, 80% of the data are taken as training set and rest 20% as testing set. There are N users and P number of movies. The movies are rated from 1 to 5 by users. As few movies are rated by one user, most of the ratings on items are missing. The rating dataset is normalized as 0–1 where value closer to 1 represents higher rating.

Training phase of RBM involves two phases. In the forward pass, input data from all visible nodes are being passed to all hidden nodes. At the hidden layer's nodes, X (input data) is multiplied by a W and added to hidden layer bias h_bias. In the backward pass, the samples from the hidden layer are passed as input to the visible layer. The produced output is a reconstruction which is an approximation of the original input. To maximize the product of probabilities assigned to the training set V (a matrix, where each row of it is treated as a visible vector v), contrastive divergence is used. The weight matrix W is optimized as per Eq. 4 for predefined number of epochs.

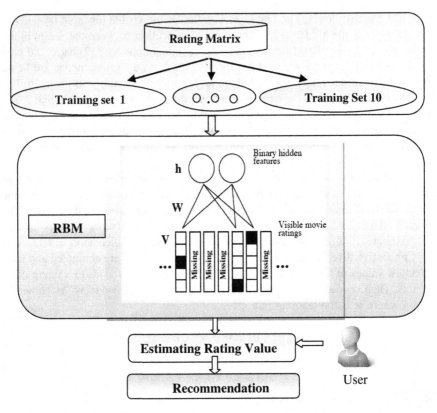

Fig. 1 Experimental model

Once the model is trained, it can be used to predict movies that an arbitrarily selected user from the testing set might like. This can be accomplished by feeding in the user's watched movie preferences (rating) into the RBM and then reconstructing the input.

The values that the RBM produced are the estimated value of the user's preferences for movies that the user has not watched based on the preferences of the users that the RBM was trained on. finally, top 20 best movies are chosen for recommendation.

5 Experimental Results

5.1 Acquiring Data

The dataset that we are using here is acquired from GroupLens that holds movies details, users, and movie ratings by these users. It contains one million ratings from 6000 users on 4000 movies. It is so sparse: about 7% of ratings are available and rest 93% of ratings are missing. Hence, it is challenging to predict missing rating in a too sparse dataset.

5.2 Importing Necessary Libraries

After downloading the data, necessary libraries are imported into the projects workspace. Libraries like "Tensorflow" and "Numpy" are taken to model and to initialize the RBM, "Pandas" to easily manipulate the datasets and "matplotlib" to plot the graph.

5.3 Loading and Formatting the Data

Load the dataset into the workspace. The input to the model contains x neurons, where x is the number of movies in the dataset. Each neuron contains a normalized rating value as 0–1. If user has not watched the movie, the value is 0 and for high rating the value is closer to 1. After loading the dataset, the rating is normalized as per the requirement. Unnecessary columns are deleted from the dataset.

5.4 Setting the Model Parameters

Sigmoid and Relu functions are taken as activation function. Mean absolute error (MAE) is taken to evaluate the recommendation. Each epoch uses certain batches with some size (e.g., each epoch uses 10 batches with size 100). Learning rate (alpha) is set to 0.9.

5.5 Training (Pseudocode)

The pseudocode of the proposed method is as below. Symbol used is explained in the earlier sections.

> *preprocesing( )*
> *For each epoch i do:*
> > *Calculate Contrastive Divergenc (CD) as:*
> > > *For each data point in batch do:*
> > > > *positive_gradient = 0, negative_gradient= 0 (matrices)*
> > > > *Pass data point through network, calculating vi and hi*
> > > > *hi=sigmoid(v * W+h_bais)*
> > > > *vi=sigmoid(h * transpose(W)+vb)*
> > > > *update positive_gradient = positive_gradient + X * hi-1*
> > > > *update negative_gradient = negative_gradient + vi * hi*
> > > *CD = (positive_gradient - negative_gradient) / no of datapoints.*
> > *Update weights and biases using eq. 4*
> > *Calculate error as difference between the data and its reconstruction*
> *Repeat for the next epoch for 100 times or until the error is marginal.*
> *Plot the error*

5.6 Recommendation (Pseudocode)

Pseudocode for recommendation of 20 best movies for a randomly selected user is mentioned below.

> *RScore - recommendation score*
> *currUser - current user selected from the testset*
> *prvw - previous weight*
> *prvhb - previous hidden layer bias*
> *prvvb - previous visible layer bias*
>
> *#Feeding in the user and reconstructing the input*
> *h0 = sigmoid(v0 * W + hb)*
> *v1 = sigmoid(h0 * W^T) + vb)*
> *feed = tf.Session.run(h0, feed_dict={ v0: currUser, W: prvw, hb: prvhb})*
> *rec = tf.Session.run(v1, feed_dict={ h0: feed, W: prvw, vb: prvvb})*
>
> *movies_df["RScore"] = rec[0]*
> *rec_movies_df=movies_df.sort_values(by="RScore",   ascending=False).head(20)*

Fig. 2 Learning set training
error

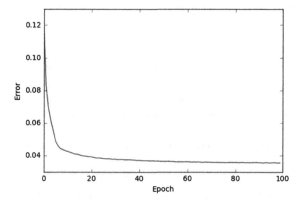

Fig. 3 Performance
comparison

5.7   Result Analysis

Figure 2 depicts the error rate during the training phase. In the testing phase after predicting the user's preferences, top 20 items are recommended. Then mean absolute error (MAE) is calculated by subtracting the recommendation score from the actual rating and divided by the number of movies considered. The MAE of the proposed method is 0.52. Whereas in case of average rating for k = 7 and neighborhood size 190, the mean absolute error is 0.74 and in case of Pearson correlation prediction, the mean absolute error is 0.58. MAE for different approaches is depicted in Fig. 3 (lower value indicates higher performance).

6   Conclusion

Recommender system helps the targeted users to save time in finding relevant items out of large collections as per their tastes. But users' preferences on the items are very less as compared to number of items available. So finding the most suitable similar users and estimating the missing ratings is a major challenge in a sparse dataset. In this work, a recommendation model is designed by using Restricted Boltzmann Machine to predict the user's preferences for movie that the user has not watched. The model is implemented in python language and some of the libraries used are tensorflow, numpy, and pandas. It is observed that the experimented model is able to handle sparse dataset. And also achieves a better result as compared to Pearson correlation prediction and average rating prediction.

Acknowledgements The authors would like to express thanks to all the reviewers for valuable comments and suggestions.

References

1. Salehi, M., Kamalabadi, I., Ghoushchi, M.: An Effective Recommendation Framework for Personal Learning Environments Using a Learner Preference Tree and a GA. IEEE Transactions on Learning Technologies. vol. 6. no. 4. (2013).
2. Bu, J., Shen, X., Xu, B., Chen, C., He, X., Cai, D.: Improving Collaborative Recommendation via User-Item Subgroups. IEEE Transactions on Knowledge and Data Engineering. vol. 28. no. 9, (2016).
3. Salah, A., Rogovschi, N., Nadif, M.: A dynamic collaborative filtering system via a weighted clustering approach. ScienceDirect. Neurocomputing 175 (2016) 206–215.
4. Silva, E Camilo-Junior, C., Pascoal, L., Rosa, T.: An evolutionary approach for combining results of recommender systems techniques based on collaborative filtering. ScienceDirect. Expert Systems With Applications 53 (2016) 204–218.
5. Mendoza, L., García, R., González, A., Hernández, G., Zapater, J: RecomMetz: A context-aware knowledge-based mobile recommender system for movie showtimes. ScienceDirect. Expert Systems with Applications 42 (2015) 1202–1222.
6. Wei, S., Zheng, X., Chen, D., Chen, C.: A hybrid approach for movie recommendation via tags and ratings. ScienceDirect. Electronic Commerce Research and Applications 18 (2016) 83–94.
7. Nilashi, M., Ibrahim, O., Ithnin, N.: Hybrid recommendation approaches for multi-criteria collaborative filtering. ScienceDirect. Expert Systems with Applications 41 (2014) 3879–3900.
8. Salakhutdinov, R., Mnih, A., Hinton, G: Restricted Boltzmann Machines for collaborative filtering. In Proceedings of the 24th International Conference on Machine Learning. (2007).
9. Koohi, H., Kiani, K.: User based Collaborative Filtering using fuzzy C-means. ScienceDirect. Measurement 91 (2016) 134–139.
10. Ortega, F., Hernando, A., Bobadilla, J., Hyung Kang, J.: Recommending items to group of users using Matrix Factorization based Collaborative Filtering. ScienceDirect. Information Sciences 345 (2016) 313–324.
11. Hernando, A., Bobadilla, J., Ortega, F.: A non negative matrix factorization for collaborative filtering recommender systems based on a Bayesian probabilistic model. ScienceDirect. Knowledge-Based Systems 97 (2016) 188–202.

12. Ar, Y., Bostanci, E.: A genetic algorithm solution to the collaborative filtering problem. ScienceDirect. Expert Systems With Applications 61 (2016) 122–128.
13. Moreno, M., Segrera, S., López, V., Muñoz, M.: Web mining based framework for solving usual problems in recommender systems. A case study for movies' recommendation. ScienceDirect. Neurocomputing 176 (2016) 72–80.
14. Li, H., Cui, J., Shen, B., Ma, J.: An intelligent movie recommendation system through group-level sentiment analysis in microblogs. ScienceDirect. Neurocomputing 210 (2016) 164–173.
15. Hou, Y., Xiao, T., Zhang, S., Jiang, X., Li, X., Hu, X., Han, J., Guo, L., Miller, L., Neupert, R., Liu T.: Predicting Movie Trailer Viewer's "Like/Dislike" via Learned Shot Editing Patterns. IEEE Transactions on Affective Computing, vol. 7, no. 1, January–March 2016.
16. Zhang, H., Ji, Y., Li, J., Ye, Y.: A Triple Wing Harmonium Model for Movie Recommendation. IEEE Transactions on Industrial Informatics, vol. 12, no. 1, February 2016.
17. Wang, Z., Yu, X., Feng, N., Wang, Z.: An improved collaborative movie recommendation system using computational intelligence. ScienceDirect. Journal of Visual Languages and Computing 25 (2014) 667–675.
18. Jannch, D., Zanker, M., Felfernig, A., Friedrich, G.: Recommender System- An Introduction, Cambridge University Press (2011).

Comparative Analysis of DTC Induction Motor Drives with Firefly, Harmony Search, and Ant Colony Algorithms

Naveen Goel, R. N. Patel and Saji Chacko

Abstract In industries, induction motor drives are more popular due to their brushless structure, low cost, low maintenance, and robust performance. Direct Torque Control (DTC) drive performance gained more importance due to its fast dynamic response and simple control structure. In order to enhance the performance of the speed control of DTC drives, this paper implements different optimization techniques to tune the speed PI controller. The simulation is carried out using MATLAB, and the results of genetic algorithm, ant colony, harmony search, and firefly optimization process are compared for different speeds of the DTC drive, with respect to peak overshoot and settling time.

Keywords Ant colony algorithm (ACO) · Direct torque control (DTC) Harmony search (HS) · Induction motor (IM) · Field-oriented control (FOC) Genetic algorithm (GA) · Firefly optimization (FFA)

1 Introduction

AC induction motor drives have advantages due to their simplicity, reliability, and low cost. As high accuracy is required for the Induction Motor (IM) drives used in industries, an efficient control problem has to be designed for DTC [1–3].

The features of IM drives depend on the applied control strategy. The control strategy and estimation of stator resistance of AC induction motor drives are more

N. Goel (✉) · R. N. Patel
Department of Electrical and Electronics Engineering, Shri Shankaracharya Technical Campus, Junwani, Bhilai 490020, Chattisgarh, India
e-mail: ngoel18@gmail.com

R. N. Patel
e-mail: ramnpatel@gmail.com

S. Chacko
Department of Electrical Engineering, Durg Polytechnic, Durg 490020, Chattisgarh, India
e-mail: chacosaji68@gmail.com

© Springer Nature Singapore Pte Ltd. 2019
H. S. Behera et al. (eds.), *Computational Intelligence in Data Mining*,
Advances in Intelligent Systems and Computing 711,
https://doi.org/10.1007/978-981-10-8055-5_68

complex, and this complexity affects substantially at the time of high performance. The main aim of the control method adopted is to provide the best possible performance of drive. Additionally, a very important requirement regarding control method is its simplicity (simple algorithm, simple tuning, and operation with small controller dimension).

Two high-performance control techniques for IM drives are field-oriented control (FOC) and direct torque control (DTC) [4–6]. DTC has the advantages of absence of coordinate transformation and minimum torque response time. In spite of the above advantages, it has some disadvantages like high torque ripple during starting, requirement of torque and flux estimator, change in stator resistance and high torque, and flux ripples.

Here, the DTC drive is modified which leads to the performance enhancement of drive. The hysteresis controllers [7] are replaced by Proportional Integral controllers. The conventional proportional integral (PI) is widely used in drive control system due to its simple control structure and easy to design, as the PI control method may not provide the good control if the PI parameters (k_p and k_i) do not adapt to the controller. This PI controller can be optimized using SA, PSO, genetic algorithm, ant colony, bacterial foraging (BG), harmony search (HS), firefly algorithm (FFA) [8].

The speed is controlled using closed-loop PI controller. Artificial intelligence (AI) techniques, GA, ACO, HS, and FFA, are implemented to tune this PI controller. In this paper, three optimization techniques are proposed. Due to the precise performance of the drive at different speeds and in different operating conditions, it needs low peak overshoot and less settling time. In this work, an integral time absolute error (ITAE) objective function is suggested for both conventional and AI tuning techniques. The objective function expression is given as in Eq. (1)

$$J = \int_0^T t \cdot |\Delta f| \tag{1}$$

2 Direct Torque Control

In the DTC drive, IM is fed by a VSI, which operates as a three-phase voltage source as shown in Fig. 1 [9–12]. The measured stator voltages v_{abc} and currents i_{abc} are transformed into $d - q$ axis voltages v_{ds}, v_{qs} and currents i_{ds}, i_{qs}. The speed of motor N_r is estimated with the N_r^* (reference speed), and the error is processed by the speed PI controller to produce a torque command T_e^* [13, 14].

The stator flux is calculated as in Eqs. (2) and (3):

$$\psi_{ds} = \int (V_{ds} - i_{ds}R_s)dt \tag{2}$$

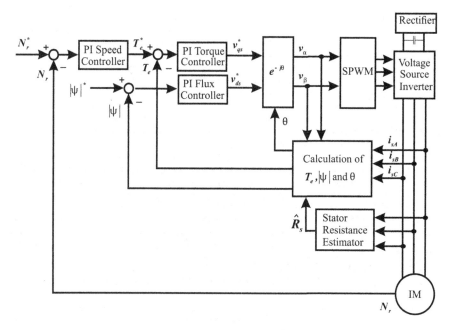

Fig. 1 Block diagram of modified DTC using PI controllers replaced hysteresis controllers

$$\psi_{qs} = \int (V_{qs} - i_{qs}R_s)dt \tag{3}$$

and flux linkage phasor magnitude is given by Eq. (4):

$$\psi_s = \sqrt{(\psi_{qs}^2 + \psi_{ds}^2)} \tag{4}$$

and the phase of stator flux linkage is expressed in Eq. (5):

$$\theta_s = \tan^{-1}(\psi_{qs}/\psi_{ds}) \tag{5}$$

and electromagnetic torque is given by Eq. (6):

$$T_e = (3/2)(P/2)(i_{qs}\psi_{ds} - i_{ds}\psi_{qs}) \tag{6}$$

Fig. 2 Block diagram of PI
controller

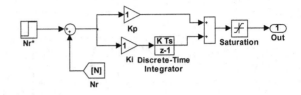

2.1 PI Controller

In the modified DTC, the limitation of hysteresis controllers can be overcome by
replacing the hysteresis controllers with PI controllers with SVM, which alleviates
the requirement of increasing the number of sectors. On the basis of errors between
the reference and estimated values of torque and flux it is possible to directly control
the inverter states in order to reduce the torque and flux errors within the prefixed
band limits. As the hysteresis controllers are having a specified bandwidth, its
switching selection is having a limited range for inverter. The output of the PI
controller used in industries for wide range in time domain is expressed as in Eq. (7)

$$v_c(t) = k_p(t) + k_i \int dt \tag{7}$$

To eliminate the steady-state error in control variable, the integral part is added
to the proportional controller. The speed error (e) is the input of PI controller, and
reference torque is the output as shown in Fig. 2.

3 Performance Enhancement of Speed of DTC Using Optimization Techniques

3.1 Genetic Algorithm

A genetic algorithm (GA) is a computational abstraction of biological evolution that
can be used to solve optimization problems. GA algorithm is population-oriented.
Consecutive populations of feasible solutions are generated in a stochastic manner
following laws similar to those of natural selection [15–17]. Three operators must
be specified to construct the complete structure of the GA procedure: selection,
crossover, and mutation operators.

3.2 Ant Colony Optimization

ACO technique was introduced by M. Dorigo in early 1900s as a novel nature inspired meta-heuristic for the solution of combinatorial optimization problem [18]. This algorithm is based on the real ant behavior in searching the source of food. It is evident that the shortest path has large pheromone concentrations, so that more ants tend to choose it to travel. There are three major phases in the ant colony algorithm: Initialization, constructing ant solution, updating pheromone.

The global updating rule is implemented in the ant system where all ants start their tours, and pheromone is deposited and updated on all edges based on Eq. (8):

$$\tau_{ij} \leftarrow (1 - \rho) \cdot \tau_{ij}(t) + \sum_{\substack{k \in colony\ that \\ used\ edge(i,j)}}^{m} \frac{Q}{L_k} \tag{8}$$

where ρ_{ij}: probability between the town i and j, Q: constant, L_k: length of the tour performed by K_{th} ant, ρ: evaporation rate.

3.3 Firefly Algorithm

Firefly algorithm is comparatively similar to particle swarm optimization technique, which is mimicked by the flashing pattern and the behavior of fireflies. The configuration of flashes is often matchless for an individual species which is used by them for catching prey. One of the unique characteristics of FF includes the attraction, which is first of its own features in any SI-based algorithm. In standard firefly algorithm, space is searched by moving the less brighter firefly moves towards the more brighter firefly. Firefly algorithm with mutation searches the search space by adding features to less brighter firefly from more brighter firefly. The extent of features to be added is decided by calculating the mutation probability of each firefly. Thus, FF deals with multimodal difficulties efficiently [19, 20]. To understand the standard FF, there are the following three idealized rules:

  i. Fireflies are fascinated to further fireflies of same sex as they are all unisex.
 ii. Fascination is proportional to brightness of the firefly, the distance will decrease, comparatively less bright one will get fascinated to the brighter one.
iii. The illumination of any firefly is given by the background of the objective function.

3.4 Harmony Search

HS was proposed by Zong Woo Geem in 2001 [21]. HS algorithm is inspired by the improvisation process of musicians, and it is a phenomenon-mimicking algorithm. The analogy between improvisation and optimization are as follows [22–24]:

(1) Each decision variable corresponds to each musician.
(2) The decision variable's value range corresponds to musical instrument's pitch range.
(3) To the solution, vector at certain iteration corresponds to musical harmony at a certain time.
(4) Audience's aesthetics corresponds to the objective function to be minimized or maximized.

The main steps of harmony search algorithm are as follows:

1. Initialize optimization problem and parameters.
2. Initialize harmony memory.
3. Improve a new harmony.
4. Update the new harmony.
5. Check for termination condition.

The parameters of HS algorithm are given in Table 1.

The Pseudocode for HS is:

Generate an initial solution of harmony memory size

Compute the fitness value of each harmony in harmony memory (HM)

for i = 1 to Number of iterations (or stop criteria)

Choose a value in random and check with HMCR and choose a solution

if it satisfies the condition with PAR

Update the solution with the bandwidth

else

Choose a new solution within the given bounds

Table 1 Parameters of harmony search algorithm

Harmony search property	Values of different parameters
Harmony memory size (population number)	20
Bandwidth (bw)	0.2
Harmony memory considering rate	0.95
Pitch adjustment rate	0.3
Max. no. of Iteration	100

Evaluate its fitness value

if the new fitness value is less than the worst one in HM, replace with the new harmony

end

4 Result and Discussion

The DTC drive with optimization techniques to tune speed PI controller is simulated in MATLAB/Simulink. In this section, the proposed GA, ACO, HS, and FFA are performed. The system performance is examined under various operating conditions. The figures and tables show the comparative performance analysis of DTC drive with the k_p and k_i of its speed PI controller tuned using GA, ACO, HS, and FFA optimization technique. The performance analysis is made with reference to overshoot and settling time for different cases as shown in Tables 2 and 3. Figures 3, 4, and 5 show the comparison of the settling time, peak overshoot, and %

Table 2 Speed reversal: comparison of four optimization techniques

Parameters	GA	ACO	FFA	HS
Kp	0.8398	0.1758	0.8196	1.1179
Ki	13.307	8.6587	14.9833	24.4201
Settling time (s)	0.1833	0.1446	0.1207	0.1062
Peak overshoot (rpm)	411.5698	451.0596	415.5191	413.3435
% Overshoot	2.9004	12.7459	3.8956	3.3437

Table 3 Variable load torque: comparison of four optimization techniques

Parameters	GA	ACO	FFA	HS
Kp	1.1	1.087	0.7339	2.0915
Ki	2.1	9.8741	15.0001	29.1512
Settling time (s)	0.2672	0.0553	0.0505	0.0577
Peak overshoot (rpm)	1050.1	928.2595	1016.1	925.6377
% Overshoot	16.6896	3.1453	12.9048	2.8514

Fig. 3 Graph of settling of drive

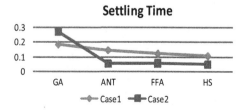

Fig. 4 Graph of peak
overshoot of speed

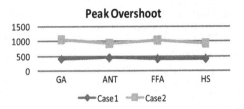

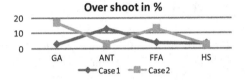

Fig. 5 Graph of % overshoot
of speed of drive

overshoot in speed using graph for the proposed optimization techniques, namely GA, ACO, FFA, and HS.

4.1 Case-I: Speed Variation

The drive is now subjected to a condition with speed profile as shown in Fig. 6 and at rated load torque of 5 Nm.

It is observed that the DTC drive tracks the reference speed smoothly. During speed reversal, it is seen that the settling time of the HS-PI speed controller is the lowest as compared to the GA-, ACO-, and FFA-tuned speed PI controller. The same observation regarding the performance of HS-tuned controller with regard to peak overshoot and % overshoot is made during drive starting condition and during change in speed from 400 to 1000 rpm at t = 4 s.

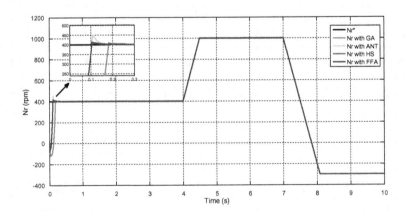

Fig. 6 Tracking of speed reversal at constant load torque

Fig. 7 Electromagnetic torque case-I

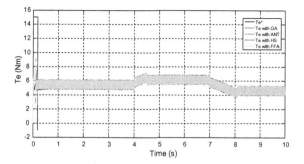

Fig. 8 Stator flux for case-I

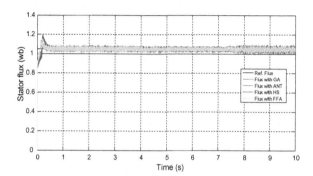

One of the major drawbacks of DTC drive is the torque ripples. Figure 7 shows the simulation results of the electromagnetic torque T_e developed. On comparison, it is observed that settling time and torque ripples of T_e for the HS-tuned speed controller are very less as compared to the other tuned controllers. Apart from that, it is seen during speed changes, the performance characteristics of stator flux response as shown in Fig. 8 for the HS-PI controller is optimum. In Fig. 9a, b, convergence of the fitness function using HSA for case-I and case-II is shown, respectively.

4.2 Case-II Variable Load Torque

In this case, the drive operates at constant speed of 900 rpm but with variable load torque. Initially, drive runs at no load till 4 s, after that the load torque increases to 2 Nm till 7 s. At 7 s, torque is again increased to 4 Nm till the end of the simulation. The simulation results for the speed response T_e developed and the stator flux are shown in Figs. 10, 11, and 12. Under this operating condition also, it is observed that the settling time, peak overshoot, and % overshoot of speed of the HS-PI controller are lowest among the other controllers. The other major observation is that the ripples in T_e during load transient for the HS-PI controller are the lowest.

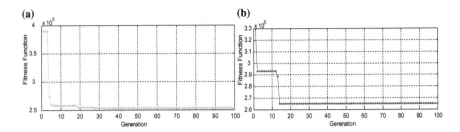

Fig. 9 **a** and **b** Fitness function convergence using HSA for case-I and case-II, respectively

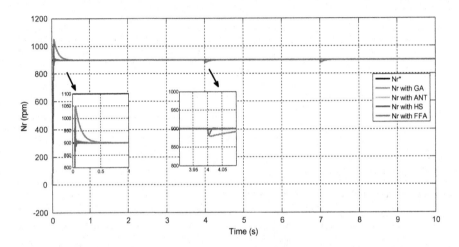

Fig. 10 Tracking of actual speed at constant speed and variable load torque

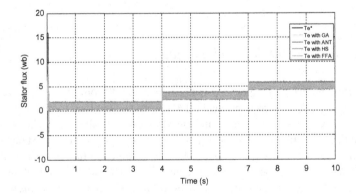

Fig. 11 Electromagnetic torque with constant speed and variable load torque

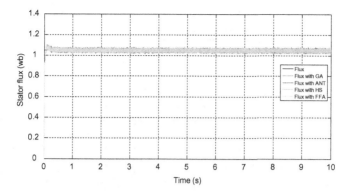

Fig. 12 Stator flux with constant speed and variable load torque

5 Conclusions

The paper presents a different approach for the tuning the gains of speed PI controller of a DTC drive using the four optimization techniques. The observed result shows optimum performance of the optimized tuned controller with reference to speed torque and flux response as compared to controller using GA-tuned method. The results also show that the tuning based on HS is simple, and it takes less time due to the less number of parameters to be initialized, and also, its performance index, i.e., the peak overshoot, settling time, and % overshoot, is better than the GA, ACO, and FFA-tuned speed controller.

References

1. Bose B.K. Bose: Modern Power Electronics and CA Drives. Prentice Hall (2002).
2. Finch John. W.: Controlled AC Electrical drives. IEEE Transaction on Industrial Electronics. 55(2) (2008) 481–491.
3. Kerkman R.J., Gary L.S., Schlegel D.W.: AC drives: Year 2000 (Y2K) and beyond. In: Applied Power Electronics Conference and Exposition (APEC'99) Fourteenth Annual, Vol. 1 (1999) 28–39.
4. Doncker De, Rik W., Donald W.: Novotny, The universal field oriented controller. IEEE Transactions on Industry applications. 30(1) (1994) 92–100.
5. Hung H. Le: Comparison of Field Oriented Control (FOC) and Direct Torque Control for Induction Motor Drives. In: Thirty-Fourth Industry Applications Annual Meeting, IEEE Conference, Vol. 2, Phoenix AZ (1999) 1245–1252.
6. Goel Naveen, Patel R.N., Chacko Saji: A Review of the DTC Controller and Estimation of Stator Resistance in IM Drives. International Journal of Power Electronics and Drive System. 6(3) (2015) 554–566.
7. Mohanty K.B.: A Direct Torque Controlled Induction Motor with Variable Hysteresis Band. In: 11th International Conference on Computer Modelling and Simulation, UK Sim (2009).
8. Ursem B.K., Vadstrup P.: Parameter identification of Induction motors using stochastic optimization. Journal of Applied Soft Computing. 4(1) (2004) 49–64.

9. Takahashi I., Ohmori Y.: High-performance direct torque control of an induction motor. IEEE Transactions on Industry Applications. 25(2) (1989) 257–264.
10. Depenbrock M.: Direct self-control (DSC) of inverter-fed induction machine. IEEE transactions on Power Electronics. 3(4) (1988) 420–429.
11. Abdul Wahab H.F., Sanusi H.: Simulink Model of Direct Torque Control of Induction Machine. American Journal of Applied Science. 5(8) (2008) 1083–1090.
12. Allirani S., Jagannathan V.: High Performance Direct Torque Control of Induction Motor Drives Using Space Vector Modulation, IJCSI. 7(6) 2010.
13. Lascu C., Ion B., Frede B: A modified direct torque control (DTC) for induction motor sensorless drive. IEEE Trans. on Industry Application. 36(1) (2000) 122–130.
14. Malhotra R., Singh N., Singh Y.: Genetic Algorithms: Concepts, Design for Optimization of Process Controllers. Computer and Information Science. 4(2) (2011) 39–54.
15. Bollam A., Phani KSV. K., Abhiram M., Chaluvaji S.: Optimization of Induction Motor Drives using DTC Technique with Modified Genetic Algorithm based PI Speed Controllers. Int. J. on Recent Trends in Engineering and Technology. 6(2) (2011) 252–257.
16. Chebre Md., Meroufel A., Bendaha Y.: Speed Control of Induction Motor Using Genetic Algorithm-based PI Controller. Acta Polytechnica Hungarica. 8(6) (2011) 141–153.
17. Goel Naveen, Patel R.N., Chacko Saji: Performance enhancement of Direct Torque Control IM drive using Genetic Algorithm. In: 2015 Annual IEEE India International Conference (INDICON), New Delhi, 2015 (2015) 1–6.
18. Dorigo M., Birattari M., Stutzle T.: Ant colony optimization. IEEE computational intelligence magazine, 1(4) (2006) 28–39.
19. Pal P., Dey R., Biswas K.R., Bhakta S.: Optimal PID Controller Design for Speed Control Of A Separately Excited DC Motor. A Firefly Based Optimization Approach. International Journal of Soft Computing, Mathematics and Control IJSCMC4.4 (2015) 39–48.
20. Ali E.S.: Firefly algorithm based speed control of DC series motor. International J. WSEAS Trans Syst Control. 10 (2015) 137–147.
21. Woo G.Z., Kim J.H., Loganathan G.V.: A new heuristic optimization algorithm: harmony search. Simulation. 76(2) (2001) 60–68.
22. Yang Xin-She: Harmony search as a metaheuristic algorithm. Music-inspired harmony search algorithm. Springer, Berlin Heidelberg (2009) 1–14.
23. Salem Md., Khelfi MD.F., Rochdi B.B.: Statistical Analysis of Harmony Search Algorithms in Tuning PID Controller. International Journal of Intelligent Engineering & Systems. 9(4) (2016) 98–106.
24. Hameedkalifullah A., Palani S.: Optimal tuning of PID power system stabilizer for multi machine power system using Harmony Search Algorithm. Journal of Theoretical and Applied Information Technology. 66(2) (2014) 513–520.

DNA Gene Expression Analysis on Diffuse Large B-Cell Lymphoma (DLBCL) Based on Filter Selection Method with Supervised Classification Method

Alok Kumar Shukla, Pradeep Singh and Manu Vardhan

Abstract The exponential growth of DNA dataset in the scientific repository has been encouraging interdisciplinary research on ecology, computer science, and bioinformatics. For better classification of cancer (DNA gene expression), many technologies are useful as demonstrated by a prior experimental study. The major challenging task of gene selection method is extracting informative genes contribution in the classification from the DNA microarray datasets at low computational cost. In this paper, amalgamation of Spearman's correlation (SC) and filter-based feature selection (FS) methods is proposed. We demonstrate the extensive comparison of the effect of Spearman's correlation with FS methods, i.e., Relief-F, Joint Mutual Information (JMI), and max-relevance and min-redundancy (MRMR). To measure the classification performance, four diverse supervised classifiers, i.e., K-nearest neighbor (K-NN), support vector machines (SVM), naïve Bayes (NB), and decision tree (DT), have been used on DLBCL dataset. The result demonstrates that Spearman's correlation in conglomeration with MRMR performs better than other combinations.

Keywords Diffuse large B-cell lymphoma · Feature selection · Relief-F
JMI · Max-relevance min-redundancy (MRMR) · K-NN · SVM
NB

A. K. Shukla (✉) · P. Singh · M. Vardhan
Department of Computer Science & Engineering, NIT, Raipur 492010,
Chhattisgarh (C.G), India
e-mail: akshukla.phd2015.cs@nitrr.ac.in

P. Singh
e-mail: psingh.cs@nitrr.ac.in

M. Vardhan
e-mail: mvardhan.cs@nitrr.ac.in

© Springer Nature Singapore Pte Ltd. 2019   783
H. S. Behera et al. (eds.), *Computational Intelligence in Data Mining*,
Advances in Intelligent Systems and Computing 711,
https://doi.org/10.1007/978-981-10-8055-5_69

1 Introduction

Recently, DNA microarray-based technology gained great attention due to its ability to measure expression levels for thousands of genes in single experiment. The discovery of biological patterns and class prediction is performed by the supervised, unsupervised, and semi-supervised feature selection techniques [1]. A typical microarray experiment involves the hybridization of an mRNA molecule to the DNA template. One of the exploration issues is to investigate and address the DNA microarray by the researchers, which is curse of dimensionality. It is not possible to solve the large number of attributes by using traditional methods. Therefore, researchers used effective feature selection (FS) technique by applying certain selection criteria to reduce the dimension of microarray dataset and provide relevant features to classification model for predicting tumors on the benchmark data. In this paper, we used the Spearman's correlation in conglomeration with feature selection method, namely MRMR [2], JMI [3], and Relief-F [4] for DLBCL classification. We investigated the effect of filter methods which depend on correlation between features (genes) and class (target). The rest of the paper is organized as follows. In Sect. 2, we explain the related work. Section 3 discusses the existing feature selection methods. The concept of proposed method is discussed in Sect. 4. Experimental results are shown in Sect. 5, finally, conclusion in Sect. 6.

2 Related Work

Many feature selection algorithms have been proposed for various applications including classification [5–7]. Most of the feature selection (FS) algorithms are based on statistical measures, i.e., chi-square (χ^2) method or population-based approach such as genetic algorithm (GA) and ant colony optimization (ACO). In general, estimation of FS is divided into four categories in the literature: filter method, wrapper methods, and ensemble and hybrid methods. In current trend on feature selection, hybrid and ensemble methods are the most powerful feature selection methods [8]. Hybrid method is designed by integrating two different methods as filter and wrapper. Ghareb et al. [9] used a hybrid search technique that combines the advantages of filter feature selection methods with an enhanced GA (EGA) in a wrapper approach to handle the high dimensionality of the feature space. Moradi and Gholampour [10] proposed HPSO-LS local search strategy which is embedded in the particle swarm optimization to select the less correlated and salient feature subset.

3 Feature Selection

Feature selection is an important step to reduce the high-dimensional datasets. It is generally used in the field of bioinformatics, machine learning, and data mining [11]. The aim of FS method is to choose a subset of the pertinent features of the original dataset of features based on some criteria such as correlation, redundancy, and inconsistency. In last five years, we found the commonly used FS techniques such as Laplacian scored, Relief-F, MRMR, JMI, and hybrid FS that work on huge number of features but small gene expression data size (ten to hundreds), when building classifiers. The informative gene is required when the data generally consist of several inappropriate, redundant, and noisy features. The relevant subset feature plays an important role to identify the early-stage tumor detection and cancer discovery. In addition, the exhaustive searches of all combinations of features are prerequisite for finding the optimal feature subsets for classifying of datasets. Here, we applied the existing filter feature selection methods on the DNA microarray datasets. The objective of FS is to choose the relevant features for classification model in the field of learning problems.

3.1 Filter Methods

Filter-based method selects a prominent feature subset (in high-dimensional datasets) without using a classification algorithm. To measure the evaluation performance, the wrapper-based methods are relatively slower than filter-based methods. In this paper, we choose the m among n most relevant attributes by filter methods which are related to coefficient correlation scores of each feature. Three feature selection methods are used in this study, namely Relief-F, minimum-redundancy maximum-relevancy (MRMR), and Joint Mutual Information (JMI). Relief-F is one of the powerful feature selection approach as comapre Relief [12]. Relief does not explicitly reduce the relevancy in selected genes. It helps us to focus of good feature sets and distinguish among instances of different classes which are better to predict a target. Relief-F algorithm is an improved version of Relief [13].

JMI method depends on mutual information (MI) as shown in Eq. 1. In order to measure the MI between variables, class label y and features X are well defined as:

$$I(y;X) = H(y) - H\left(\frac{y}{X}\right) \tag{1}$$

where $H(y)$ and $H\left(\frac{y}{x}\right)$ show the entropy and conditional entropy of the tangled variable. Here, we use Joint Mutual Information (JMI) to reduce the redundancy between data attributes and class y. The relevance of input attributes defined by the JMI is shown in Eq. 2

$$M_{JMI}(X) = \sum_{x_j \in S} I(x_k; x_j; y) \propto \sum_{x_j \in S} I(y; x_k/x_j) \qquad (2)$$

where $I(x_k; x_j; y)$ represents the MI between original features set x_k and selected features x_j with respect to target class y.

The last method is MRMR; researchers observed that minimization of max-dependency and maximal relevance on the feature sets is hard to understand. To overcome this issue, we introduced the minimum-redundancy and maximum-relevance (MRMR) feature selection framework. According to MI approach, the objective of FS is to choose a subset S with n features $[X = \{x_1, x_2, \ldots, x_n\}]$; they have the maximum dependency on the target class y called as max-dependency, as shown in Eq. 3

$$\max w(X, y) = I(y; x_1, x_2, \ldots, x_n) = H(y) - H\left(\frac{y}{x_1, x_2, \ldots, x_n}\right) \qquad (3)$$

The dependency among features X could be large. The randomly chosen two features from original set are highly correlated with each other. The corresponding class power would not change much if one of them were removed from subsets. Therefore, the minimal redundancy between features can be added to select optimal features as shown in Eqs. 4 and 5.

$$\min Z(X, y) = 1/|s^2| \sum_{x_j \in s} I(x_j; x_k) \qquad (4)$$

$$\text{Max } \emptyset(w, Z) = w - Z \qquad (5)$$

The criterion combining the above two constraints is called "minimum-redundancy maximum-relevance" (MRMR) as shown in Eq. 6

$$j_{MRMR}(\emptyset) = I(y; X) - 1/|s^2| \sum_{x_j \in s} I(x_j; x_k) \qquad (6)$$

where x_j is selected subset of features and x_k is original feature set.

3.2 Wrapper Methods

Wrapper method uses a classification algorithm to evaluate the accuracy produced by the use of the feature subset in the classification. In specific classifiers, wrapper methods can give a maximum classification performance, but generally they have high computational burden [14].

4 Proposed Methodology

The traditional feature selection method requires more computational cost to choose the prominent feature subset from the high-dimensional dataset due to expletive of dimension. The proposed framework is a combination of Spearman's correlation (SC) and filter-based feature selection method to select a subset of relevant features. This method considers both feature–class correlation and feature–feature correlation to determine a feature subset from datasets, and also enhances the classification accuracy and reduces the search complexity for generating feature subsets.

4.1 Proposed Feature Selection Strategy

The selection of feature subsets is based on evaluation function, which is toward highly correlated with the sample class and uncorrelated with features. The steps are as follows:

Step 1: According to first step, Spearman's correlation (SC) method [15] calculates the correlation of features ρ as mathematical expression shown in Eq. 7. Discover the top rank feature subsets by using SC and pass to second step for better classification.

$$\rho = \frac{\sum_{i=1}^{i=n}(x_i - \bar{x})((y_i - \bar{y}))}{\sqrt{\sum_{i=1}^{i=n}(x_i - \bar{x})\uparrow 2 * \sum_{i=1}^{i=n}(y_i - \bar{y})\uparrow 2}} \tag{7}$$

where ρ is a Spearman's rank correlation, x_i and y_i are the randomly selected attributes from n number of instances in sample different categories, and $\bar{x}$ and $\bar{y}$ is

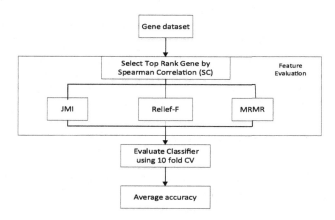

Fig. 1 Proposed evaluation framework

the mean of each random variable. Each feature is sorted in ascending order. The rank corresponds to their position in the sorted list.

Step 2: The aim of the proposed framework is to select the relevant feature subsets after by Spearman's rank correlation method with filter method to obtain the maximum score solution leading to the selection of the most relevant predictive subset of size n and the most relevant optimal subset of size m. The overall framework of the proposed method is shown in Fig. 1.

5 Classification

To measure the acceptability of feature subset for classification, different error estimation strategies have been suggested. We use tenfold CV method on the DNA microarray dataset, which randomly splits the dataset into training and testing data subgroups: Training dataset consists of 90% of the data samples and other testing subset consists of 10% of the data samples to estimate the performance based on confusion matrix. To evaluate the performance of the relevant subset diverse classifiers K-nearest neighbor (K-NN) [16], support vector machines (SVM) [17], naïve Bayes (NB) [18], and decision tree (DT) [19] have been used for classification.

5.1 Dataset

We present the effect of feature selection methods on diffuse large B-cell lymphoma (DLBCL) [20]. The DLBCL has 45 samples with expression values of 4026 genes. The classification performance of the proposed method on the DLBCL datasets is evaluated using the tenfold cross-validation approach as shown in Table 1. The dataset has been divided into a training set of 36 samples and a test set of 9 samples. We show the comparative analysis of effect of Spearman's correlation conglomeration with Relief-F, JMI, and MRMR by using four classifiers, i.e., support vector machine (SVM), naïve Bayes (NB), decision tree (DT), and K-nearest neighbor (K-NN). For performance evaluation, we have used accuracy as it is most used performance parameter [21]. To predict the percentage of correctly classified samples, the formulation of accuracy is shown in Eq. 8

$$\text{Accuracy}(\%) = \frac{\text{TP} + \text{TN}}{\text{TP} + \text{TN} + \text{FP} + \text{FN}} \tag{8}$$

Here, TP, TN, FP, and FN are true positive, true negative, false positive, and false negative in the independent datasets, respectively.

Table 1 Percentage of accuracies within diverse classification algorithms on DLBCL dataset for Relief-F, MRMR, and JMI methods using 10–100 selected gene

Classifiers	FS method	Number of gene selected					
		10	20	40	60	80	100
SVM	SC + Relief-F	92.50	93.50	97.50	95.00	95.50	98.00
	SC + MRMR	97.50	99.13	99.13	99.13	99.13	99.13
	SC + JMI	99.02	99.72	97.80	93.00	99.25	95.50
NB	SC + Relief-F	93.00	93.00	98.27	96.89	98.76	95.50
	SC + MRMR	97.50	**99.74**	98.00	97.50	97.81	97.50
	SC + JMI	95.50	98.14	97.82	97.32	96.23	95.50
K-NN	SC + Relief-F	89.00	93.50	95.50	96.00	95.48	93.00
	SC + MRMR	99.14	99.54	97.50	98.00	97.50	98.00
	SC + JMI	89.50	81.00	78.00	80.00	70.50	73.50
DT	SC + Relief-F	88.00	79.50	82.00	80.00	86.50	84.50
	SC + MRMR	75.00	82.50	81.00	83.50	81.50	82.50
	SC + JMI	84.00	85.00	84.50	88.50	89.00	86.50

5.2 Results and Discussion

The DLBCL dataset conatins 4026 genes and 45 instances. We select the top-ranked relevant genes by using Spearman's correlation methods and reduced selected correlated features again minimized by other feature selection methods, i.e., Relief-F, JMI, and max-relevance min-redundancy (MRMR). We have performed extensive experiments on DLBCL gene dataset by Relief-F, JMI, and MRMR methods and evaluated the performance with the help of the classifiers K-nearest neighbors, support vector machines, naïve Bayes, and decision tree. The demonstrations are shown in Table 1; average accuracy is computed as the average of the ten runs. Table 1 presents classification accuracies obtained for the above-mentioned gene expression dataset for three combinations of gene selection methods.

By using SVM classifier, SC + MRMR method gives the maximum accuracy of 99.13% in 20–100 genes and a minimum accuracy of 97.25% in 10 genes. SC + JMI method gives a maximum accuracy of 99.72% in 20 genes and a minimum accuracy of 93% in 60 genes. SC + Relief-F method gives a maximum accuracy of 98.00% in 100 genes and a minimum accuracy of 92.50% in 10 genes.

NB classifier with SC + Relief-F gives maximum accuracy of 98.76% in 80 genes and a minimum accuracy of 93% for 10, 20 genes. SC + MRMR method gives an accuracy of 99.74% in 20 genes, and minimum accuracy of 97.50% in 10, 60, and 100 genes. SC + JMI method gives maximum accuracy of 98.14% in 20 genes and a minimum accuracy of 95.50% in 10,100 gene.

SC + Relief-F gives a maximum accuracy of 96% in 60 genes and a minimum accuracy of 89.00% in 10 gene in case of K-NN classification. SC + MRMR method gives maximum accuracy of 99.54% in 20 genes and a minimum accuracy

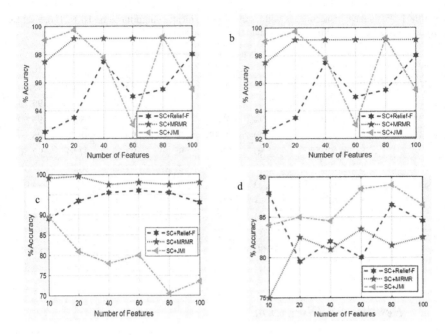

Fig. 2 Percentage of accuracies within **a** SVM, **b** NB, **c** K-NN, and **d** DT classification algorithms on DLBCL dataset

of 97.50% in 40, 80 genes. SC + JMI method gives accuracy of 89.50% in 10 genes and minimum accuracy of 70.50% in 80 genes by using K-NN.

SC + Relief-F method gives a maximum accuracy of 88.00% for 10 genes and a minimum accuracy of 79.50% for 20 genes when evaluated using DT. SC + MRMR method gives a maximum accuracy of 83.50% for 60 genes and a minimum accuracy of 75% for 10 genes using DT. SC + JMI method gives an accuracy of 89.00% for 80 genes and a minimum accuracy of 84% for 10 genes.

Figure 2a–d shows the comparison of three selection methods: SC + Relief-F, SC + JMI, and SC + MRMR method, and accuracy performance with respect to four classifiers on DLBCL datasets.

Figure 3a shows that SC + MRMR is outperforming with only 20 selected features when used in combination with SVM classifier. Figure 3b shows that SC + Relief-F method archives better accuracy in case of NB classification, and SC + JMI gives the better performance in NB and SVM classification as shown in Fig. 3c.

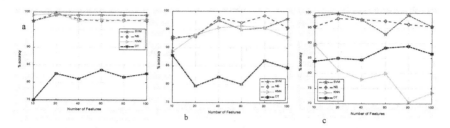

Fig. 3 Classification accuracy for DLBCL dataset based on **a** SC + MRMR, **b** SC + Relief-F, and **c** SC + JMI methods

6 Conclusion

To eliminate the low-ranked or irrelevant feature subsets, a combination of Spearman's correlation (SC) and filter-based feature selection (FS) is presented in this paper. Feature selection method aims to select a feature subset of the DLBCL dataset based on some criteria (redundancy and relevancy) and accurately classifying the instances of dataset into class labels. We have conducted experiments using four classifiers (K-NN, SVM, NB, and DT) and three FS (Relief-F, JMI, and MRMR) in conglomeration with SC method. The experimental result demonstrated that SC + MRMR method for 20 genes using NB classifier reaches the best performance in terms of accuracy on DLBCL dataset.

References

1. H. Liu, H. Motoda, R. Setiono, and Z. Zhao, "Feature Selection : An Ever Evolving Frontier in Data Mining," *J. Mach. Learn. Res. Work. Conf. Proc. 10 Fourth Work. Featur. Sel. Data Min.*, pp. 4–13, 2010.
2. H. Peng, F. Long, and C. Ding, "Feature selection based on mutual information: Criteria of Max-Dependency, Max-Relevance, and Min-Redundancy," *IEEE Trans. Pattern Anal. Mach. Intell.*, vol. 27, no. 8, pp. 1226–1238, 2005.
3. H. H. Yang and J. Moody, "Feature selection based on joint mutual information," *Proc. Int. ICSC Symp. Adv. Intell. Data Anal.*, pp. 22–25, 1999.
4. M. R.- Sikonja, "Theoretical and Empirical Analysis of ReliefF and RReliefF," *Mach. Learn. J.*, vol. 1, no. 53, pp. 23–69, 2003.
5. H. Lai, Y. Tang, H. Luo, and Y. Pan, "Greedy feature selection for ranking," *Proc. 2011 15th Int. Conf. Comput. Support. Coop. Work Des. CSCWD 2011*, pp. 42–46, 2011.
6. I. Guyon and A. Elisseeff, "An Introduction to Variable and Feature Selection," *J. Mach. Learn. Res.*, vol. 3, no. 3, pp. 1157–1182, 2003.
7. X. Liu, A. Krishnan, and A. Mondry, "An entropy-based gene selection method for cancer classification using microarray data.," *BMC Bioinformatics*, vol. 6, no. 1, p. 76, 2005.
8. H. Lu, J. Chen, K. Yan, Q. Jin, Y. Xue, and Z. Gao, "A Hybrid Feature Selection Algorithm for Gene Expression Data Classification," *Neurocomputing*, no. 2017, 2016.
9. A. S. Ghareb, A. A. Bakar, and A. R. Hamdan, "Hybrid feature selection based on enhanced genetic algorithm for text categorization," *Expert Syst. Appl.*, vol. 49, pp. 31–47, 2016.

10. P. Moradi and M. Gholampour, "A hybrid particle swarm optimization for feature subset selection by integrating a novel local search strategy," *Appl. Soft Comput. J.*, vol. 43, pp. 117–130, 2016.

11. S. A. Medjahed, T. A. Saadi, A. Benyettou, and M. Ouali, "Kernel-based learning and feature selection analysis for cancer diagnosis," *Appl. Soft Comput. J.*, vol. 51, pp. 39–48, 2017.

12. Y. Sun, "Iterative RELIEF for feature weighting: Algorithms, theories, and applications," *IEEE Trans. Pattern Anal. Mach. Intell.*, vol. 29, no. 6, pp. 1035–1051, 2007.

13. A. Arauzo-Azofra, J. Benitez, and J. Castro, "A feature set measure based on relief," *Proc. fifth Int. Conf. Recent Adv. Soft Comput.*, pp. 104–109, 2004.

14. R. Kohavi and H. John, "Wrappers for feature subset selection," *Artif. Intell.*, vol. 97, no. 97, pp. 273–324, 1997.

15. C. Spearman, "The Proof and Measurement of Association between Two Things," *Am. J. Psychol.*, vol. 15, no. 1, pp. 72–101, 2017.

16. K. Q. Weinberger and L. K. Saul, "Distance Metric Learning for Large Margin Nearest Neighbor Classification," *J. Mach. Learn. Res.*, vol. 10, pp. 207–244, 2009.

17. A. Ben-hur and J. Weston, "A user's guide to support vector machines," *Data Min. Tech. life Sci.*, pp. 223–39, 2010.

18. N. Friedman, D. Geiger, and M. Goldszmidt, "Bayesian Network Classifiers," *Mach. Learn.*, vol. 29, pp. 131–163, 1997.

19. W. Loh, "Classification and regression trees," *Data Min. Knowl. Discov.*, vol. 1, no. February, pp. 14–23, 2011.

20. "Diffuse Large B-cell Lymphoma Dataset." [Online]. Available: https://llmpp.nih.gov/lymphoma/data/clones.txt.

21. M. Sokolova, N. Japkowicz, and S. Szpakowicz, "Beyond accuracy, F-Score and ROC: A family of discriminant measures for performance evaluation," *Adv. Artif. Intell.*, vol. 4304, pp. 1015–1021, 2006.

Categorizing Text Data Using Deep Learning: A Novel Approach

Rajendra Kumar Roul and Sanjay Kumar Sahay

Abstract With large number of Internet users on the Web, there is a need to improve the working principle of text classification, which is an important and well-studied area of machine learning. Hence, in order to work with the text data and to increase the efficiency of the classifier, choice of quality features is of paramount importance. This study emphasizes two important aspects of text classification: proposes a new feature selection technique named Combined Cohesion Separation and Silhouette Coefficient (*CCSS*) to find the feature set which gathers the crux of the terms in the corpus without deteriorating the outcome in the construction process and then discusses the underlying architecture and importance of deep learning in text classification. To carry out the experimental work, four benchmark datasets are used. The empirical results of the proposed approach using deep learning are more promising compared to the other established classifiers.

Keywords Classification · Cohesion · Deep learning · Extreme learning machine · Multilayer ELM · Separation · Silhouette coefficient

1 Introduction

The number of Internet users and Web sites are increasing day by day. This in turn generates more digital documents on the Web and hence badly affects the classification of text data. Thus, selecting good features/terms out of a large volume of features is the need of the hour. In terms of information retrieval, there is no clear definition of a feature, but it is a distinctive attribute or characteristic of the data. The process of transforming raw data into features, which represent the model better, resulting in

R. K. Roul (✉) · S. K. Sahay
Department of Computer Science, BITS-Pilani, K. K. Birla Goa Campus,
Zuarinagar 403726, Goa, India
e-mail: rkroul@goa.bits-pilani.ac.in

S. K. Sahay
e-mail: ssahay@goa.bits-pilani.ac.in

© Springer Nature Singapore Pte Ltd. 2019
H. S. Behera et al. (eds.), *Computational Intelligence in Data Mining*,
Advances in Intelligent Systems and Computing 711,
https://doi.org/10.1007/978-981-10-8055-5_70

improved accuracy of the classification technique is known as feature engineering. This process can be categorized into three stages:

i. *Feature generation stage*: In this stage, candidate features are generated by predetermined kinds of sensing techniques from the training set.
ii. *Feature refinement stage*: Also known as dimensionality reduction stage, where refinement of features is done via feature extraction or feature selection.
iii. *Feature utilization stage*: After feature refinement stage is over, the refined features are used to represent the instances of the dataset. An appropriate classification model is selected to make use of these features.

Among the above three stages, feature refinement stage is important and becomes the main topic of discussion due to the following reasons:

- reduces the training time and storage size and improves the performance of the model by removing multicollinearity.
- data visualization is easier when the actual feature space is reduced to low dimensions such as dimensions of two or three.

In *feature selection*, the assumption is that the original feature set contains sufficient relevant features which can discriminate clearly between categories and thus irrelevant features are eliminated for better efficiency and accuracy. It selects a subset of informative features from the initial feature set and uses it for model construction. Further, the algorithms used for feature selection are classified into the following three categories:

i. *Filter methods* do not use any classifiers for feature selection; instead, features are selected on the basis of certain statistical properties. Hence, these methods are fast to compute and capture the usefulness of the feature set, which makes them more practical.
ii. *Wrapper methods* generate different subsets of features based on some algorithms and test each subset using a classifier. To find the score of feature subsets, wrapper methods use a predictive model, whereas filter methods use a proxy measure.
iii. *Embedded methods* combine the advantages of both the previous two methods, and their computational complexity lies between these two methods.

Choice of an efficient classifier is the next important step for text classification. Many traditional classifiers exist such as artificial neural network, support vector machine (SVM), Naive Bayes, decision trees, but they have their own limitations. Most of the classifier architectures are based on the approach of neural network, hence certain restrictions are imposed which stop them to solve many complex problems. Deep learning, the re-branding of neural network is a sub-branch of machine learning and is based on a set of network algorithms, completely data-driven, and progressively learning the data with high level of abstraction. The architecture of deep learning has been successfully applied in the domain of natural language processing, speech recognition, and image processing. The common restrictions found in conventional classifiers are not there in deep learning, hence it is able to solve any complex problem [1]. Recently developed Multilayer ELM [2] which relies on the architecture of deep learning can tackle such common problems and is capable of handling a large volume of data.

Many researchers have done work in text classification domain [3–7]. Thinking in the direction that *important features* and *efficient classifiers* in text classification can guarantee the best performance of the system, this paper proposes an innovative technique for selection of prominent features from a corpus to prepare an efficient reduced feature vector. This work is carried out in three stages as discussed below:

I. *Term clustering*:
   Initially, traditional k-means clustering technique is run on a given corpus of terms (extracted from the documents of all classes of the corpus) which generates k term-document clusters.

II. *Important feature selection based on the cohesion, separation, and silhouette coefficient score*:
   Next, three important well-known parameters (generally used to measure the efficiency of the cluster) called *cohesion, separation,* and *silhouette coefficient* scores are computed for all the features in each term-document cluster. The *total-score* of each term is calculated by adding the silhouette coefficient score to the ratio of cohesion and separation. All the terms in each term-document clusters are ranked based on their *total-score* and top m% terms are selected from each cluster. At the end, all these important terms are merged to generate an efficient reduced feature vector.

III. *Training and testing the classifiers*:
   The reduced feature vector along with the class labels are used to train the Multilayer ELM and other standard classifiers. The performance of each classifier is measured by comparing the classifier's prediction for a test document with the actual class label.

The experimental results using Multilayer ELM on four established datasets show the efficiency of our approach. We believe that our work may be one of the few research works where Multilayer ELM has been tested extensively in order to signify the importance and future applications of deep learning in the domain of text classification.

The rest of the paper is organized on the following lines: In Sect. 2, we briefly discuss the underlying architecture of ELM and Multilayer ELM and other details which are needed to implement the proposed approach. Section 3 describes the feature selection technique. The detail discussion along with the experimental works are covered in Sect. 4. Section 5 discusses the conclusion and future enhancement of the proposed work.

2 Basic Preliminaries

2.1 Extreme Learning Machine

Huang et al. suggested a feed-forward neural network having single layer and named it as Extreme Learning Machine (ELM) [8]. Many important characteristics of ELM such as no backpropagation, extremely fast learning speed, able to manage large dataset etc., make ELM more popular compared to other established classifiers.

ELM in Brief:
Consider N different examples (x_i, y_i), where $x_i = [x_{i1}, x_{i2}, ..., x_{in}]^T \in R^n$ and $y_i = [y_{i1}, y_{i2}, ..., y_{im}]^T \in R^m$, such that $(x_i, y_i) \in R^n \times R^m$, $i = 1, 2, ..., N$. ELM used an activation function $g(x)$ and L hidden layer nodes. Given input $\mathbf{x}$, ELM output function can be written as

$$y_j = \sum_{i=1}^{L} \beta_i g(w_i \cdot x_j + b_i) \tag{1}$$

where $j = 1, ..., N$, w_i and b_i are randomly generated hidden node parameters. $w_i = [w_{i1}, w_{i2}, w_{i3}, ..., w_{in}]^T$ is the weight vector that joins the 'n' input nodes to the ith hidden node. b_i is the bias of the ith hidden node. β is the weight vector that connects each hidden layer node to every output node and is represented as $\beta = [\beta_1, ..., \beta_L]^T$. $g(\mathbf{x})$ is responsible to map the input feature space of n-dimension to L-dimensional hidden layer space. Equation 1 in a reduced form can be represented as $H\beta = Y$, where Y and H are the output and hidden layer matrix respectively.

2.2 Multilayer ELM

In 2006, Hinton [9] proposed the deep belief network, which outperforms many SVMs, conventional multilayer neural network, and SLFNs. But the main problem remains is that it has slow learning speed. Aiming in the same direction, an artificial neural network having multiple hidden layers called ML-ELM is proposed by Kasun et al. [2]. The architecture of ML-ELM is shown in Fig. 1 and some important points related to the designing issue are listed below:

i. ML-ELM possesses all the properties of ELM since it combines ELM with ELM autoencoder (ELM-AE). Parameters that represent ML-ELM are trained layer-wise using ELM-AE in an unsupervised manner.
ii. During the training time, no iterations take place and hence the unsupervised training is very fast.

ELM Autoencoder: An autoencoder is a neural network that tries to reconstruct its input. The architectures of ELM-AE and ELM are almost similar except that ELM is supervised in nature, while ELM-AE is unsupervised and it can be stacked and

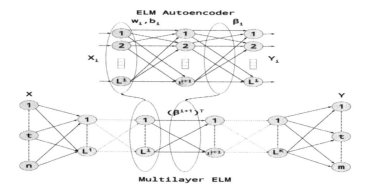

Fig. 1 Multilayer ELM and ELM autoencoder

trained in a progressive way. The stacked ELM-AEs will learn how to represent data; the first level will have a basic representation; the second level will combine that representation to create a higher-level representation and so on. Using Eq. 2, ML-ELM transfer the data between the hidden layers.

$$H_i = g((\beta_i)^T H_{i-1}) \tag{2}$$

where H_i is the ith hidden layer output matrix and β_i is the output weight vector of ELM-AE that placed before the ith hidden layer. $i = 0$ represents the input layer x. Regularized least square technique is used to calculate analytically the output weights of the connection between last hidden layer and nodes of the output layer.

2.3 Cohesion

Cohesion (compactness/tightness) determines the closeness of terms in a cluster by computing the distance of each term t from the centroid of the cluster. If the term t is highly cohesive, i.e., the distance between the term t and the centroid of the cluster is very small compared to other terms in the same cluster, then it defines the cluster very well. If the term t is present at the border of the cluster (far away from the centroid), then t is poorly cohesive to the cluster. Euclidean distance can be used to measure the distance of a term t from the centroid of the cluster c as follows: If $\vec{t} = (t_1, t_2)$ and $\vec{c} = (c_1, c_2)$, then $||\vec{c} - \vec{t}|| = \sqrt{(c_1 - t_1)^2 + (c_2 - t_2)^2}$, where $\vec{t}$ and $\vec{c}$ are term and centroid vectors respectively.

2.4 Separation

Separation (isolation) determines how distinct or well separated the term t is from other clusters. The separation score of t is the minimum among all the distances of the term t from the centroid of other clusters in which t is not a member. If the distance of t from the centroid of its neighboring cluster is high, then t is well separated from that cluster.

2.5 Silhouette Coefficient

Silhouette Coefficient [10] measures how much a term t is similar to its own cluster (i.e., cohesion) compared to other clusters (i.e., separation), and it can vary between -1 and 1. If the silhouette coefficient of the term t is high, then it is highly match to its own cluster and poorly match to other clusters. If most of the terms in a cluster have high coefficient values, then the cluster is well configured. Silhouette coefficient of the term t is defined in Eq. 3.

$$silhouette(t) = \frac{s(t) - c(t)}{max\big(c(t), s(t)\big)} \tag{3}$$

where $c(t)$ and $s(t)$ are the cohesion and separation score of the term t respectively. Silhouette coefficient needs to be positive, i.e., $c(t) < s(t)$, and $c(t)$ to be close to 0, i.e., the term t should be very close to its centroid as far as possible, since the coefficient obtains its maximum value of 1 when $c(t)$ is 0.

3 Proposed Approach

In this section, we put forward our proposed Combined Cohesion Separation and Silhouette Coefficient (*CCSS*) feature selection technique using the following steps.

1. *Preprocessing the corpus and converting document into vectors*:
   Given a corpus P having n classes, first, all the documents of each class are preprocessed using a suitable preprocessing algorithm and then the documents of all n classes are collected together which make the dimension of P as $b \times l$, where b and l are total number of terms and documents in the corpus P. In order to convert the documents into vector, the TF-IDF[1] weight of each term is calculated in the document space (t_{ij}) as shown in Table 1.
2. *Generating term-document clusters*:
   The corpus P is divided into k term-document clusters (term-doc), $td_i, i = 1$ to k

---

[1]https://radimrehurek.com/gensim/tutorial.html

Table 1 Term-document table

	d_1	d_2	d_3	$\ldots$	d_l
t_1	t_{11}	t_{12}	t_{13}	$\ldots$	t_{1l}
t_2	t_{21}	t_{22}	t_{23}	$\ldots$	t_{2l}
t_3	t_{31}	t_{32}	t_{33}	$\ldots$	t_{3l}
.	.	.	.	$\ldots$	.
.	.	.	.	$\ldots$	.
.	.	.	.	$\ldots$	.
t_b	t_{b1}	t_{b2}	t_{b3}	$\ldots$	t_{bl}

by running the k-means clustering algorithm [11] on it, where the dimension of each td_i now becomes $q \times l$. The reason to form clusters is not only to bring the related terms into the same group, but also, computation of the total-score for each term become more efficient using cohesion, separation, and silhouette coefficient because these techniques are unsupervised in nature.

3. *Selection of important features from each cluster*:
   Next aim is to select important features from each of the k clusters for maintaining uniformity without excluding any collection and it is done as follows:

   i. *Computation of centroid for each term-doc cluster*: From each term-doc cluster td_i, the centroid of all the term vectors $\vec{t_j}, j = 1, ..., q$ is computed by using Eq. 4.

   $$\vec{c_i} = \frac{\sum_{j=1}^{q} \vec{t_j}}{q} \tag{4}$$

   where $\vec{c_i}$ is the centroid vector (dimension of $1 \times l$) of ith term-doc cluster, td_i.

   ii. *Computation of cohesion score for each term*:
   To measure how cohesive is the term $\vec{t_j} \in td_i$ to the centroid $\vec{c_i} \in td_i$, the Euclidean distance is computed between $\vec{t_j}$ and $\vec{c_i}$ using Eq. 5.

   $$cohesion(\vec{t_j}) = (||\vec{c_i} - \vec{t_j}||) \tag{5}$$

   iii. *Computation of separation score for each term*:
   Equation 6 is used to measure how well separated a term $\vec{t_j} \in td_i$ from the centroid of other clusters $\vec{c_s}, \forall s \in [1, k]$ and $s \neq i$.

$$separation(\vec{t_j}) = \min\left(||\vec{c_s} - \vec{t_j}||\right) \qquad (6)$$

where $\vec{c_s}$ is the centroid of the sth cluster. In other words, it can be said that *separation* $(\vec{t_j})$ finds the minimum separation distance among all the distances computed between the term $\vec{t_j}$ and centroid of other clusters in which $\vec{t_j}$ is not a member.

iv. *Computation of silhouette coefficient for each term* :
The silhouette coefficient of the term $\vec{t_j} \in td_i$ is computed using Eq. 7.

$$silhouette(\vec{t_j}) = \frac{(separation(\vec{t_j}) - cohesion(\vec{t_j}))}{\max\left(cohesion(\vec{t_j}), separation(\vec{t_j})\right)} \qquad (7)$$

v. *Computation of total-score for each term of a term-doc cluster*:
Total-score of the term $\vec{t_j} \in td_i$ is computed using Eq. 8.

$$total\text{-}score(\vec{t_j}) = \frac{separation(\vec{t_j})}{cohesion(\vec{t_j})} + silhoutte(\vec{t_j}) \qquad (8)$$

The reason why we divided the separation score of a term with its cohesion score is that for any important term, the minimum separation value should be high, i.e., the term should be well separated from its nearest neighboring cluster and the cohesion value should be low, i.e., the term should tightly bound to the centroid of its own cluster, which can bring the *total-score* of that term high. But if it is considered in a reverse order, i.e., $\frac{cohesion}{separation}$, then the top (or important) terms will loose their score (or importance) in the cluster as the ratio will generate very low scores for them.

vi. *Selection of top m% terms from each term-doc cluster*:
By repeating Step 4(ii)–(v), terms of all k clusters will obtain their respective *total-scores*. Rank the terms in each term-doc cluster based on their *total-scores*. Select top m% terms from each term-doc cluster td_i and merge them into a list, which generates a reduced feature vector.

4. *Performance measurement of different classifiers*:
The reduced feature vector along with the class label of each document generates the training feature vector. Performance measurement of different classifiers can be done based on their test results.

4 Experimental Analysis

The experimental work of the proposed method is tested on four benchmark datasets (20-Newsgroups[2], Classic4[3], Reuters[4], and WebKB[5]). For all the datasets, k (used in k-means clustering) was set as 10 (decided empirically)[6]. The percentage of terms selected from each cluster was set as 1, 5, and 10%. The number of hidden layers was set as 5 for 20-Newsgroups and 3 for the remaining three datasets (decided empirically). The algorithm was tested on hidden layer nodes of different size both for ELM and ML-ELM, and the best results are obtained when the number of hidden layer nodes are larger than the input layer nodes as shown in Table 2. Figure 2 shows the performance comparison among different classifiers on different datasets using *CCSS* approach. It is evident from all the results that the performance of Multilayer ELM outperforms the other well-known classifiers.

Results show that the performance of *CCSS* are either comparable or better than the traditional classifiers. Table 3 shows the average F-measure comparisons of different established classifiers using *CCSS* technique. The bold result indicates the maximum F-measure obtained by a classifier using *CCSS* and it signifies that Multilayer ELM leaves behind other standard classifiers. The following points discuss some of the probable reasons of *"why Multilayer ELM outperforms other established classifiers ?"*:

 i. Multilayer ELM uses the ELM feature mapping mechanism to map the training feature vector into an extended feature space, hence it makes the features much simpler and linearly separable in the high-dimensional space which enhances its performance.
 ii. Impressed with an underlying design of the deep network.
 iii. Multiple nonlinear transformations of input data are possible by using multiple layers.
 iv. Higher-level abstraction of data is captured by Multilayer ELM using multiple layers, whereas the networks having a single layer fail to achieve it.
 v. By using representation learning, a different form of the input is learned at each layer within the network.

---

[2]http://qwone.com/~jason/20Newsgroups/.

[3]http://www.dataminingresearch.com/index.php/2010/09/classic3-classic4-datasets/.

[4]http://www.daviddlewis.com/resources/testcollections/reuters21578/.

[5]http://www.cs.cmu.edu/afs/cs/project/theo-20/www/data/.

[6]The code was run for different value of k and that value on which the approach obtained a better result is taken into consideration.

Table 2 Size of the input and hidden layer feature vectors

Dataset	No. of features	Input layer nodes for top-1%	Hidden layer nodes for top-1%	Input layer nodes for top-5%	Hidden layer nodes for top-5%	Input layer nodes for top-10%	Hidden layer nodes for top-10%
20-NG	16270	170	200	820	900	1630	1700
Classic4	15971	160	200	800	870	1600	1660
Reuters	13531	140	180	680	750	1360	1420
WebKB	7522	80	120	380	450	750	810

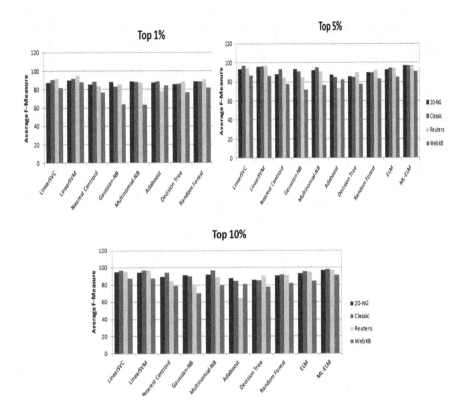

Fig. 2 F-measure comparison of different classifiers on *CCSS*

Table 3 Comparisons of ML-ELM with other established classifiers on CCSS approach

Classifier	20-NG (F-Measure-%)			Classic4 (F-Measure-%)			Reuters (F-Measure-%)			WebKB (F-Measure-%)		
	1%	5%	10%	1%	5%	10%	1%	5%	10%	1%	5%	10%
LinearSVC	87.515	93.454	94.921	90.146	95.659	96.550	91.524	94.076	95.477	81.836	86.492	87.571
Linear SVM	89.429	95.509	94.536	91.670	96.020	96.769	**94.926**	96.888	96.781	**87.785**	85.949	87.602
NC	85.225	87.908	89.582	88.345	93.279	93.992	83.451	83.986	84.519	76.456	77.728	78.611
GNB	87.783	93.315	91.138	83.022	90.986	89.721	85.148	84.877	79.145	63.930	71.888	70.298
MNB	88.662	92.190	91.938	88.169	94.668	96.917	87.526	90.668	88.693	63.170	76.448	79.623
Adaboost	87.334	87.182	87.688	88.394	84.586	84.586	77.928	73.314	64.800	84.302	82.282	80.883
DT	85.589	85.619	85.643	85.878	84.843	85.084	88.448	89.668	90.653	76.599	77.702	77.786
RF	88.761	89.718	90.571	88.719	89.555	91.673	90.925	91.753	91.076	81.502	83.076	81.987
ELM	88.421	92.334	92.886	90.185	94.576	95.632	93.327	94.452	94.556	80.536	84.675	84.243
ML-ELM	**93.803**	**96.743**	**96.926**	**95.707**	**96.540**	**98.078**	94.601	**96.938**	**96.958**	84.516	**90.910**	**91.278**

5 Conclusion

The paper proposed a novel feature selection technique *CCSS* which combined three important parameters (cohesion, separation, and silhouette coefficient) in order to generate an efficient reduced feature vector. Multilayer ELM as the classifier has been used for classifying the text data whose importance has been extensively measured and discussed in this paper. The proposed approach is summarized as follows:

- *term clustering and total-score calculation*: k-means clustering algorithm was run initially on a corpus to group the similar terms into k clusters. Cohesion, separation, and silhouette coefficient scores of every term in each cluster are computed, and finally with the help of these three scores, the *total-score* is generated.
- *final reduced feature vector preparation*: At the end, all the terms in each cluster are ranked together based on their total-score and top m% terms are selected as the important terms from a cluster. These important terms are merged into a list to finalize the reduced feature set.
- *training Multilayer ELM and other standard classifiers*: The reduced feature vector along with the labels of classes are used to train all conventional classifiers.

The encouraging results of the proposed approach show the stability and effectiveness of Multilayer ELM compared to different progressive classifiers in the domain of text classification. This justifies the stability, efficiency, and effectiveness of deep learning in the domain of text classification. This work can be extended by combining the Multilayer ELM feature space with other traditional classifiers which will further improve the classification results.

References

1. S. Ding, N. Zhang, X. Xu, L. Guo, and J. Zhang, "Deep extreme learning machine and its application in eeg classification," *Mathematical Problems in Engineering*, vol. 2015, 2015.
2. L. L. C. Kasun, H. Zhou, G.-B. Huang, and C. M. Vong, "Representational learning with extreme learning machine for big data," *IEEE Intelligent Systems*, vol. 28, no. 6, pp. 31–34, 2013.
3. R. K. Roul, A. Nanda, V. Patel, and S. K. Sahay, "Extreme learning machines in the field of text classification," in *Software Engineering, Artificial Intelligence, Networking and Parallel/Distributed Computing (SNPD), 2015 16th IEEE/ACIS International Conference on*. IEEE, 2015, pp. 1–7.
4. C. C. Aggarwal and C. Zhai, "A survey of text classification algorithms," in *Mining text data*. Springer, 2012, pp. 163–222.
5. X. Qiu, X. Huang, Z. Liu, and J. Zhou, "Hierarchical text classification with latent concepts," in *Proceedings of the 49th Annual Meeting of the Association for Computational Linguistics: Human Language Technologies: short papers-Volume 2*. Association for Computational Linguistics, 2011, pp. 598–602.
6. F. Sebastiani, "Machine learning in automated text categorization," *ACM computing surveys (CSUR)*, vol. 34, no. 1, pp. 1–47, 2002.

7. R. K. Roul, S. R. Asthana, and G. Kumar, "Study on suitability and importance of multilayer extreme learning machine for classification of text data," *Soft Computing*, vol. 21, no. 15, pp. 4239–4256, 2017.
8. G.-B. Huang, Q.-Y. Zhu, and C.-K. Siew, "Extreme learning machine: theory and applications," *Neurocomputing*, vol. 70, no. 1, pp. 489–501, 2006.
9. G. E. Hinton, S. Osindero, and Y.-W. Teh, "A fast learning algorithm for deep belief nets," *Neural computation*, vol. 18, no. 7, pp. 1527–1554, 2006.
10. P. J. Rousseeuw, "Silhouettes: a graphical aid to the interpretation and validation of cluster analysis," *Journal of computational and applied mathematics*, vol. 20, pp. 53–65, 1987.
11. A. K. Jain, M. N. Murty, and P. J. Flynn, "Data clustering: a review," *ACM computing surveys (CSUR)*, vol. 31, no. 3, pp. 264–323, 1999.

An Approach to Detect Patterns (Sub-graphs) with Edge Weight in Graph Using Graph Mining Techniques

Bapuji Rao and Sarojananda Mishra

Abstract The task of detecting pattern or sub-graph in a large graph has applications in large areas such as biology, computer vision, computer-aided design, electronics, intelligence analysis, and social networks. So work on graph-based pattern detection has a wide range of research fields. Since the characteristics and application requirements of graph vary, graph-based detection is not the only problem, but it is a set of graph-related problems. This paper proposes a new approach for detection of sub-graph or pattern from a weighted graph with edge weight detection method using graph mining techniques. The edge detection method is proposed since most of the graphs are weighted one. Hence this paper proposes an algorithm named EdWePat for detection of patterns or sub-graphs with edge weight detection rather node value.

Keywords Sub-graph · Pattern · Weight · Weighted graph

1 Introduction

Detection of pattern matching in graphs has a wide range of various research communities which include artificial intelligence, computer-aided design, computer vision, databases, biology, graph theory, information retrieval, electronics, and data mining. Graph-based pattern matching in semantic graphs has large numbers of

B. Rao (✉) · S. Mishra
Department of Computer Science Engineering and Applications, IGIT, Sarang,
Dehnkanal 759146, Odisha, India
e-mail: bapuji.research@gmail.com

S. Mishra
e-mail: sarose.mishra@gmail.com

© Springer Nature Singapore Pte Ltd. 2019
H. S. Behera et al. (eds.), *Computational Intelligence in Data Mining*,
Advances in Intelligent Systems and Computing 711,
https://doi.org/10.1007/978-981-10-8055-5_71

807

typed and attributed vertices as well as edges. Most of the graphs are neither typed nor attributed. Most of the graph matching algorithms perform matching strictly on the shape of the graph and does not bother about the semantics or meaning associated with the graph's vertices and edges [1]. Some algorithms matching based on vertex or edge types as well as structure by considering only semantics [2–5].

The basic idea of matching a pattern in a graph is to detect similarity of a specified pattern in a graph. A graph $G = (V, E)$ has a set of vertices V and a set of edges E. Each $e \in E$ is a pair of vertices (v_i, v_j) where $v_i, v_j \in V$. The vertices and edges of G could be typed and/or attributed. A pattern graph $P = (V_P, E_P)$ which defines the structural and semantic requirements of a sub-graph of G must match the pattern P to find the set M of sub-graphs of G. A graph $G' = (V', E')$ is said to be a sub-graph of G if and only if $V' \subseteq V$ and $E' \subseteq E$. So a match is a combination of (i) structural matching (isomorphism) or near structural matching and (ii) to find similarity between the vertices types and edges attribute values in P which must be in $m \in M$. Most of the graph matching approaches are able to find matches based on similarity of structure. A semantic graph is the representation of data in a graphical structure, where vertices refer as concepts. Semantic matching is a kind of matching graphs based on their vertex, edge types and attributes along with the graph structure. A graph matching algorithm returns a result based on the match of a specified pattern exactly or inexact matching. The actual aim of matching a graph is to detect the specific pattern's occurrences in a graph. The main aim of graph mining is to detect all those common patterns (sub-graphs) in the given graph [6].

The GRAMI and its extensions support directed and undirected graphs and can also be applied to both single and multiple labels (or weights) per node and edge. Further, the authors proposed CGRAMI, which supports the structural and semantic constraints, and AGRAMI, an approximate version which produces results with no false positives [7]. The authors Rao and Mishra [8] proposed an algorithm which compares two community graphs with node ID for similarity using graph mining techniques. The authors Rao et al. [9] proposed an algorithm which successfully detects a subcommunity graph from n-community graphs with node ID comparison using graph mining techniques.

2 Proposed Method

The authors have proposed an undirected weighted graph having 21 number of nodes, i.e., {1, 2, 3,..., 21} depicted in "Fig. 1." The edges with a specified weight and the edge have no specific directions. Hence it is called as undirected weighted graph. The authors want to detect two types of patterns, i.e., a loop pattern and a non-loop pattern. These patterns (sub-graphs) are to be detected based on its edge weight rather than node ID comparisons. The loop pattern with three nodes {1, 2, 3}

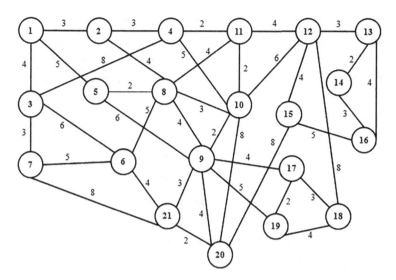

Fig. 1 Undirected weighted graph

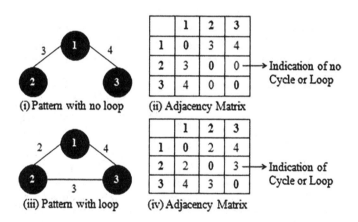

Fig. 2 Patterns to detect

having edges $\{\{(1, 2), (2, 1)\}, \{(1, 3), (3, 1)\}, \{(2, 3), (3, 2)\}\}$ and its corresponding weights are $\{2, 4, 3\}$. Similarly, a non-loop pattern with three nodes $\{1, 2, 3\}$ having edges $\{\{(1, 2), (2, 1)\}, \{(1, 3), (3, 1)\}\}$ and its corresponding weights are $\{3, 4\}$. These two patterns (sub-graphs) corresponding adjacency matrices are depicted in "Fig. 2." In case of loop pattern, the weight of the edge is shown with a nonzero

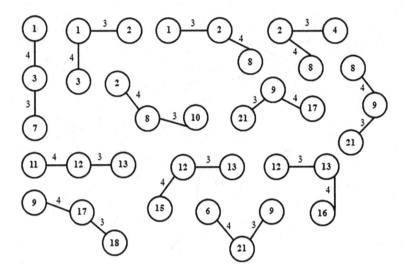

Fig. 3 Detected non-loop patterns

Fig. 4 Detected loop patterns

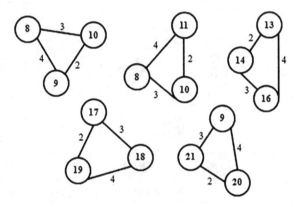

value in the (2, 3) cell of the adjacency matrix. To implement the proposed algorithm in the directed weighted graph, the condition needs to be modified.

The authors have successfully detected 12 numbers of non-loop patterns and 5 numbers of loop patterns depicted in "Fig. 3" and "Fig. 4," respectively. For this, the authors have proposed an algorithm, EdWePat, which efficiently detects those non-loop and loop patterns (sub-graphs) based on the edge weights comparison rather than node ID comparison using graph mining techniques.

3 Proposed Algorithm

```
Algorithm EdWePat(M, P)
M[tn][tn]: Weighted Adjacency Matrix of the graph.
P[n][n]: Weighted pattern to detect where n=3.
tn: Total number of nodes in the weighted graph.
edge: Assign 0 for no loop; otherwise the 'edge weight'
for loop in the pattern.
rows[]: Array to hold total counts of weights of the edge
of the pattern.
{ if(P[2][3]=0) then edge:=0;  // no loop
  else edge:=P[2][3];  // loop
  x:=0;
    for i:=1 to n-1 do
      for j:=i+1 to n do
      { x:=x+1;
        rows[x]:= Find(M, tn, P[i][j]);
      }
    tr := Create_Patterns(rows, node, M);
    Show_Patterns(Result, tr);
}

Procedure Find (M, tn, Pattern)
M[tn][tn]: Weighted Adjacency Matrix of the graph.
Pattern: Edge value of pair of nodes.
node[][3]: Matrix to hold node pair and its edge weight.
{ r:=0;
  if(Pattern!=0) then
  {   for i:=1 to tn do
        for j:=1 to tn do
        {
         if (Pattern=M[i][j]) then
         { r:=r+1;
           node[r][1]:=i;   // from node number
           node[r][2]:=j;   // to node number
           node[r][3]:=Pattern;  // edge weight
         }
        }
    }
      return(r);
  }

Procedure Create_Patterns(rows, node, M)
rows[x]: Array to hold total counts of weights of the
edge of the pattern.
node[][3]: Matrix to hold node pair and its edge weight.
M[tn][tn]: Weighted adjacency matrix of the graph.
Result[][3]: Matrix to hold detected patterns node pair
and its edge weight.
{ row1:=rows[1]; row2:=rows[2]; row:=1;
  for  i:=1 to row1 do
```

```
{
   for j:=(row1+1) to (row1+row2) do
   {
      if (node[i][1]=node[j][1]) then
      {//edge=0 means no loop, edge=non-zero means a loop
         if(M[node[i][2]][node[j][2]]=edge) then
         { Result[row][1]:=node[i][1];
           Result[row][2]:=node[i][2];
           Result[row][3]:=node[i][3];
           row:=row+1;
           Result[row][1]:=node[j][1];
           Result[row][2]:=node[j][2];
           Result[row][3]:=node[j][3];
           row:=row+1;
         }
      }
   }
}
   return(row); }

Procedure Show_Patterns(Result, tr)
{ for i:=1 to tr in steps of 2 do
    Show(Result[i], Result[i+1]);
}

Procedure Show(a, b)
Mat[4][4]: Matrix to assign the detected pattern or sub-
graph's adjacency matrix.
a[3],b[3]: Arrays to hold two rows from the matrix
Result[][3].
{ for i:=1 to 4 do
    for j:= 1 to 4 do
        Mat[i][j]:=0;
  Mat[1][2]:=Mat[2][1]:=a[1];
  Mat[1][3]:=Mat[3][1]:=a[2];
  Mat[1][4]:=Mat[4][1]:=b[2];
  Mat[2][3]:=Mat[3][2]:=a[3];
  Mat[2][4]:=Mat[4][2]:=b[3];

if(edge!=0) then Mat[3][4]:=Mat[4][3]:=edge;
for i:=1 to 4 do
  for j:=1 to 4 do
    display(Mat[i][j]);
}
```

The algorithm EdWePat has three procedures, Find(), Create_Patterns(), and Show(). It passes two matrices M[tn][tn] and P [3][3] as arguments, where M is the adjacency matrix of the given undirected weighted graph of order tn × tn and P is the adjacency matrix of the sub-graph or pattern of order 3 × 3 which is to be detected from the matrix M. If the matrix P contains the edge weight at P [3, 7], then it is considered as a loop in the pattern or sub-graph. In such scenario, the edge weight P [3, 7] is assigned to the variable "edge"; otherwise, 0 is assigned to the variable "edge." This "edge" value will be used while displaying all the detected patterns having loop from the given undirected weighted graph.

The first procedure Find(M, tn, P[i][j]) is called total "n (n − 1)/2" times, where "n" is the order of the pattern or sub-graph. In the algorithm, the order of P is 3, i.e., n = 3. So the procedure Find(M, tn, P[i][j]) is called total three times. During the calling of procedure Find(M, tn, P[i][j]), it counts all the three edges weights frequency in case of pattern having loop or all the two edges weights frequency in case of pattern having no loop. All the detected edge weights "from node index," "to node index," and "the edge weight," are assigned to the matrix node[][3]. Its corresponding edge weight frequencies are assigned to the array rows[]. After calling the procedure Find(M, tn, P[i][j]) three times, the array rows[] are assigned with three sets of frequencies of edge weights in case of pattern with loop and two sets of frequencies of edge weights in case of pattern with no loop.

Then the second procedure Create_Patterns(rows, node, M) which passes three arguments such as the edge weight frequency array rows[], the matrix node[][3] which contains the detected edge weight details such as "from node index," "to node index," and "the edge weight" and the weighted adjacency matrix of the given graph. It starts detecting the similar node pairs by comparing the first and second edge weights from the matrix node[][3] and assign to the matrix Result[][3].

Finally, the procedure Show_Patterns(Result, tr) which passes the matrix Result [][3] contains "tr" number of resultant node pair combinations. This procedure has a sub-procedure called Show(Result[i], Result[i + 1]) which is called "tr/2" times. For every call, it passes two consecutive rows from the matrix Result[][3]. In the sub-procedure Show(a, b), the detected pattern from first two rows is created and assigned to the matrix Mat [4][4]. Finally, it displays the resultant pattern or sub-graph's adjacency matrix using Mat [4][4]. This way, all the detected patterns are created and displayed accordingly.

4 Experimental Results

The dataset depicted in "Fig. 5" of the proposed undirected weighted graph depicted in "Fig. 1" indicates total number of nodes at first row and from second row onwards indicates "from node ID," "to node ID," and "the weight of the edge" between those pair of nodes.

Using the dataset depicted in "Fig. 5," the authors have successfully created its corresponding adjacency matrix M[][]. To detect the non-loop pattern {{0, 3, 4}, {3, 0, 0}, {4, 0, 0}} which is assigned to the matrix, P[][]. Now the authors have passed two matrices M[][] and P[][] to the algorithm EdWePat. It first checks for loop or non-loop in P[][]. If loop, then the variable "edge" will be assigned the edge weight value of P [2][3]; otherwise, assigned a zero value. Then the procedure Find (M, tn, P[i][j]) passes every edge weight of the given pattern for detection in the adjacency matrix and assigns its corresponding edge weight frequency to the array

Fig. 5 Dataset of Fig. 1

```
21
1        2        3
1        3        4
1        5        5
2        4        3
2        8        4
3        4        8
3        6        6
3        7        3
4        10       5
4        11       2
5        8        2
5        9        6
```

rows[]. Finally, the procedure Create_Patterns(rows, node, M) passes all the edge weights frequencies available in the array rows[], the matrix node[][3] which contains all the node pairs and its corresponding edge weights, and the adjacency matrix M[][] of the given graph. By comparing matrix node[][3] matrix with the

```
The Weighted Adjacency Matrix.....
    0   1   2   3   4   5   6   7   8   9  10  11  12  13  14  15  16  17  18  19  20  21
 1  0   3   4   0   5   0   0   0   0   0   0   0   0   0   0   0   0   0   0   0   0   0
 2  3   0   0   3   0   0   0   4   0   0   0   0   0   0   0   0   0   0   0   0   0   0
 3  4   0   0   8   0   6   3   0   0   0   0   0   0   0   0   0   0   0   0   0   0   0
 4  0   3   8   0   0   0   0   0   0   5   2   0   0   0   0   0   0   0   0   0   0   0
 5  5   0   0   0   0   0   0   2   6   0   0   0   0   0   0   0   0   0   0   0   0   0
 6  0   0   6   0   0   0   5   5   0   0   0   0   0   0   0   0   0   0   0   0   0   4
 7  0   0   3   0   0   5   0   0   0   0   0   0   0   0   0   0   0   0   0   0   0   8
 8  0   4   0   0   2   5   0   0   4   3   4   0   0   0   0   0   0   0   0   0   0   0
 9  0   0   0   0   6   0   0   4   0   2   0   0   0   0   0   0   4   0   5   4   3
10  0   0   0   5   0   0   0   3   2   0   2   6   0   0   0   0   0   0   0   8   0
11  0   0   0   2   0   0   0   4   0   2   0   4   0   0   0   0   0   0   0   0   0
12  0   0   0   0   0   0   0   0   0   6   4   0   3   0   4   0   0   8   0   0   0
13  0   0   0   0   0   0   0   0   0   0   0   3   0   2   0   4   0   0   0   0   0
14  0   0   0   0   0   0   0   0   0   0   0   0   2   0   0   3   0   0   0   0   0
15  0   0   0   0   0   0   0   0   0   0   0   4   0   0   0   5   0   0   0   8   0
16  0   0   0   0   0   0   0   0   0   0   0   0   4   3   5   0   0   0   0   0   0
17  0   0   0   0   0   0   0   0   4   0   0   0   0   0   0   0   3   2   0   0
18  0   0   0   0   0   0   0   0   0   0   0   8   0   0   0   0   3   0   4   0   0
19  0   0   0   0   0   0   0   0   5   0   0   0   0   0   0   0   2   4   0   0   0
20  0   0   0   0   0   0   0   0   4   8   0   0   0   0   8   0   0   0   0   0   2
21  0   0   0   0   0   4   8   0   3   0   0   0   0   0   0   0   0   0   0   0   2   0
```

Fig. 6 Weighted adjacency matrix of Fig. 1

Fig. 7 Adjacency matrix of
non-loop pattern

```
The Pattern to Detect.....
          1   2   3
     1    0   3   4
     2    3   0   0
     3    4   0   0
```

adjacency matrix M[][] to detect the edge weight and assigns the respective "from node ID," "to node ID," and "edge weight" to the matrix Result[][3]. Finally, the procedure Show_Patterns(Result, tr) is called for displaying of the detected pattern's adjacency matrix.

The adjacency matrix of "Fig. 1" and the adjacency matrix of the non-loop pattern P[][] = {{0, 3, 4}, {3, 0, 0}, {4, 0, 0}} are depicted in "Fig. 6" and "Fig. 7," respectively. The algorithm EdWePat has successfully detected twelve non-loop patterns whose adjacency matrices are depicted in "Fig. 8." Similarly, when the loop pattern {{0, 2, 4}, {2, 0, 3}, {4, 3, 0}} was input to the algorithm EdWePat, it has successfully detected five loop patterns. The loop pattern's adjacency matrix is depicted in "Fig. 9," and the detected loop pattern's adjacency matrices are depicted in "Fig. 10." Finally, the non-loop and loop patterns can be

```
Pattern-1              Pattern-5              Pattern-9
          1   2   3              8  10   2             12  13  15
     1    0   3   4         8    0   3   4        12    0   3   4
     2    3   0   0        10    3   0   0        13    3   0   0
     3    4   0   0         2    4   0   0        15    4   0   0

Pattern-2              Pattern-6              Pattern-10
          2   1   8              9  21   8             13  12  16
     2    0   3   4         9    0   3   4        13    0   3   4
     1    3   0   0        21    3   0   0        12    3   0   0
     8    4   0   0         8    4   0   0        16    4   0   0

Pattern-3              Pattern-7              Pattern-11
          2   4   8              9  21  17             17  18   9
     2    0   3   4         9    0   3   4        17    0   3   4
     4    3   0   0        21    3   0   0        18    3   0   0
     8    4   0   0        17    4   0   0         9    4   0   0

Pattern-4              Pattern-8              Pattern-12
          3   7   1             12  13  11             21   9   6
     3    0   3   4        12    0   3   4        21    0   3   4
     7    3   0   0        13    3   0   0         9    3   0   0
     1    4   0   0        11    4   0   0         6    4   0   0
```

Fig. 8 Detected non-loop patterns

Fig. 9 Adjacency matrix of loop pattern

The Pattern to Detect.....

	1	2	3
1	0	2	4
2	2	0	3
3	4	3	0

Fig. 10 Detected loop patterns

Pattern-1

	9	10	8
9	0	2	4
10	2	0	3
8	4	3	0

Pattern-2

	11	10	8
11	0	2	4
10	2	0	3
8	4	3	0

Pattern-3

	13	14	16
13	0	2	4
14	2	0	3
16	4	3	0

Pattern-4

	19	17	18
19	0	2	4
17	2	0	3
18	4	3	0

Pattern-5

	20	21	9
20	0	2	4
21	2	0	3
9	4	3	0

shown as sub-graphs from "Fig. 8" and "Fig. 10." The resultant patterns or sub-graphs are depicted in "Fig. 3" and "Fig. 4," respectively.

The algorithm was written in C++ programming language and compiled with Turbo C++. The experiment was run under MS Windows 7 operating system and observed satisfactory results.

5 Conclusions

The authors have proposed an algorithm called EdWePat for detection of patterns or sub-graphs with edge weight detection method rather than node value comparison using graph mining techniques, since most of the graphs are weighted one. The authors have detected successfully loop patterns and non-loop patterns with three numbers of nodes, i.e., {1, 2, 3}. The algorithm was run using C++ programming language with a prepared dataset for the undirected weighted graph depicted in "Fig. 1" and observed satisfactory results for both loop and non-loop patterns. The proposed algorithm with little modification of condition for loop can be implemented for a directed weighted graph too. The future work lies in the detection of loop and non-loop patterns with four numbers of nodes, i.e., $\{1, 2, 3, 4\}$.

References

1. Washio, T., and Motoda, H.: State of the Art of Graph-based Data Mining. SIGKDD Explorations Special Issue on Multi-Relational Data Mining, Volume 5, Issue 1 (2003).
2. Coffman, T., Greenblatt, S., and Marcus, S.: Graph-Based Technologies for Intelligence Analysis. Communications of the ACM, Special Issue on Emerging Technologies for Homeland Security, 45–47, Vol. 47, No. 3 (2004).
3. Darr, T., Greenblatt, S., and Strack, D.: A Multi-INT Level 2–3 Fusion Framework for Counter-Terrorism. Presented at Working Together: R&D Partnerships in Homeland Security, April 27–28, 2005.
4. Greenblatt, S., Marcus, S., and Darr, T.: TMODS - Integrated Fusion Dashboard - Applying Fusion of Fusion Systems to Counter-Terrorism. In: 2005 International Conference on Intelligence Analysis, May 2–6, 2005.
5. Wolverton, M., Berry, P., Harrison, I., Lowrance, J., Morley, D., Rodriguez, A., Ruspini, E., and Thomere, J.: LAW: A Workbench for Approximate Pattern Matching in Relational Data. In: Proceedings of the Fifteenth Innovative Applications of Artificial Intelligence Conference (IAAI-03), 2003.
6. Gallagher, B.: Matching Structure and Semantics: A Survey on Graph-Based Pattern Matching, American Association for Artificial Intelligence (www.aaai.org). (2006).
7. Elseidy, M., Abdelhamid, E., Skiadopoulos, S., and Kalnis, P.: GRAMI: Frequent Subgraph and Pattern Mining in a Single Large Graph. In: 40th International Conference on Very Large Data Bases, Hangzhou, China, September 2014.
8. Rao, B., and Mishra, S. N.: An Approach to Finding Similarity between Two Community Graphs Using Graph Mining Techniques, International Journal of Advanced Computer Science and Applications (IJACSA), 466–475, Vol. 7, No. 5 (2016).
9. Rao, B., Maharana, H. S., and Mishra, S. N.: An Approach to Detect Sub-Community Graph in n-Community Graphs Using Graph Mining Techniques. In: 2016 IEEE ICCIC, India, Dec 2016, https://doi.org/10.1109/iccic.2016.7919676 (available in IEEE Xplore).

Comparative Performance Analysis of Adaptive Tuned PID Controller for Multi-machine Power System Network

Mahesh Singh, Aparajita Agrawal, Shimpy Ralhan and Rajkumar Jhapte

Abstract In this paper, Genetic Algorithm (GA) and Bacterial Foraging Optimization (BFO) are two optimization algorithms used for tuning of Proportional–Integral–Derivative (PID) controller parameters in designing a multi-machine power system network. These are popular evolutionary algorithms which are generally used for tuning of PID controllers. The proposed approach is easy for implementation as well as it has superior features. The computational techniques enhance the performance of the system, and the convergence characteristics obtained are also stable. The system scheduling BFO-PID and GA-PID controller is modeled using MATLAB platform. When a comparison in the performance of optimal PIDs with conventional PID controller is carried out, the performance of BFO-PID and GA-PID controller is better as it improves the speed, loop response stability, and steady-state error; the rise time is also minimized. The results after simulation show that the controller developed using the BFO algorithm achieves a faster response as compared to GA.

Keywords Genetic algorithm · Bacterial foraging optimization
Proportional–integral–derivative · Optimization algorithm

M. Singh · A. Agrawal (✉) · S. Ralhan · R. Jhapte
Department of EEE, Shri Shankaracharya Technical Campus,
Bhilai 491001, Chhattisgarh, India
e-mail: aparajita.agrawal6@gmail.com

M. Singh
e-mail: singhs004@gmail.com

S. Ralhan
e-mail: shimpys@gmail.com

R. Jhapte
e-mail: jhapte02@gmail.com

© Springer Nature Singapore Pte Ltd. 2019 819
H. S. Behera et al. (eds.), *Computational Intelligence in Data Mining*,
Advances in Intelligent Systems and Computing 711,
https://doi.org/10.1007/978-981-10-8055-5_72

1 Introduction

The demand for electricity is increasing day by day which also increases the demand of continuous operation of the power system. But this is a difficult task as the electrical power system has to face disturbances of a wide variety that occurs commonly and it should stay intact by these disturbances. So here stability plays an important role in operation of power system. Power system stability can be dictated as the ability of the system to regain or maintain its synchronism when any disturbance or fault occurs [1]. Proportional–Integral–Derivative (PID) controllers are widely used nowadays in power systems not only to improve small-signal stability but also for enhancing system damping. For the first time in the year 1922, Minorsky introduced the classical PID controller. Still in 1911, Elmer Sperry had already represented the most properties of these controllers. PID controllers are also commonly used in plants because of its robust and simple structure and operation [2]. Power systems are often subjected to disturbances which make system unstable; so automatic tuning for this controller is needed. The gain of the controller is tuned by the optimization algorithms. Two algorithms are used here for tuning of PID: First is Bacterial Foraging Optimization (BFO), and second is Genetic Algorithm (GA).

2 Problem Identification

2.1 Proportional–Integral–Derivative (PID) Controller

PID controller is commonly used in industrial control systems [3]. An error value as the difference between a desired outcome and a measured process variable is calculated through PID [4]. The error is minimized by the controller as it adjusts the input value. PID controller has no knowledge of the correct output which takes the system to set point but it can change the output in a direction so that the process is moved toward set point. Figure 1 shows the structure of PID controller [5].

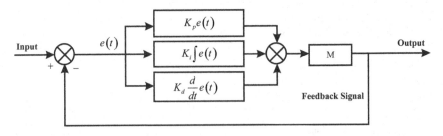

Fig. 1 Schematic of PID controller

In tuning of PID controller, three parameters are involved. The desired input u(t) is given by (1).

$$u(t) = K_p e(t) + K_i \int e(t) dt + K_d \frac{d}{dt} e(t) \tag{1}$$

where

K_p is proportional gain and has an effect on minimizing the rise time and K_i is integral gain and may result into poor transient response. It further has the effect on eradication of steady-state error, and K_d is derivative control and helps in reduction of overshoot and in improvement of the transient response.

2.2 *Error Minimization Technique*

The change in the value of error over the entire range of time has to be tracked by error minimization techniques [6] expressed by the following criterion given by (2–5).

A. Integral of absolute error (IAE)

$$IAE = \int_0^\infty |e(t)| \, dt \tag{2}$$

B. Integral of squared error (ISE)

$$ISE = \int_0^\infty e^2(t) \, dt \tag{3}$$

C. Integral of time multiplied by absolute error (ITAE)

$$ITAE = \int_0^\infty t \, |e(t)| \, dt \tag{4}$$

D. Integral of time multiplied by squared error (ITSE)

$$ITSE = \int_0^\infty t \, e^2(t) \, dt \tag{5}$$

Fig. 2 SMIB system

Generator   Transmission Line   Infinite Bus

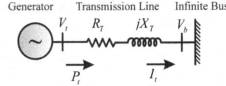

2.3 Single Machine Infinite Bus

The property of power system is that it is of higher order and complex with nonlinear components. The focus and analysis can be simplified on one machine by reducing the multi-machine power system to SMIB, i.e., Single Machine Infinite Bus [7]. Figure 2 shows a single line diagram of SMIB system.

During operation, many problems like fault, speed, or voltage deviation can occur which can be minimized as soon as possible by using adaptive tuning of controllers [8].

PID matrix for SMIB system is given by (6).

$$
\begin{bmatrix}
0 & \omega_0 & 0 & 0 & 0 \\
-\frac{K_1}{M} & 0 & -\frac{K_2}{M} & 0 & 0 \\
-\frac{K_4}{T'_{d0}} & 0 & -\frac{K_3}{T'_{d0}} & -\frac{1}{T'_{d0}} & 0 \\
-\frac{K_A K_5}{T_A} & 0 & -\frac{K_A K_6}{T_A} & -\frac{K_A}{T_A} & \frac{K_A}{T_A} \\
-\frac{K_1 K_p}{M} + \frac{K_2 K_4 K_d}{M T'_{d0}} & K_i - \frac{K_1 K_d \omega_0}{M} & -\frac{K_2 K_p}{M} + \frac{K_2 K_3 K_d}{M T'_{d0}} & \frac{K_2 K_d}{M T'_{d0}} & 0
\end{bmatrix}
\tag{6}
$$

2.4 Three-Machine Nine-Bus System

Another test system shown in Fig. 3 consists of a three-machine nine-bus system [9]. In this system, first unit, i.e., G_1, is taken as source and rest two units, i.e., G_2

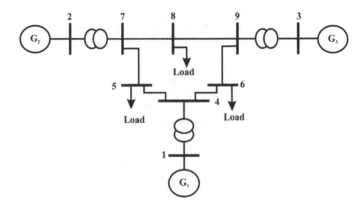

Fig. 3 Three-machine nine-bus system

and G$_3$, are considered for operation. Two conditions are considered for operation: First is under normal operating condition, and second is line outage in line 'a' between buses 5 and 7.

3 Methodology

3.1 Genetic Algorithm

In year 1989, as per the concept of fittest function's survival capability, a probabilistic search method was proposed by Goldberg. Genetic algorithm (GA) is a method by which global search and optimization is done by evolution and natural selection [10]. Figure 4 shows the flowchart of Genetic Algorithm [11]. Selection, crossover, and mutation are the steps of GA [12, 13].

3.2 Bacteria Foraging Optimization

Bacteria Foraging Optimization algorithm is new to the nature-inspired optimization algorithms, where health of bacteria is important [14]. Those species having better food searching ability have the greater chances of survival according to law of evolution [15]. The following are the stages of BFO [16] and is also shown in Fig. 5.

1. Chemotaxis: The process in which for search of food bacteria moves are called chemotaxis, which involves tumble and swim. Randomly, in any direction a unit walk is known as tumble, whereas if the unit walk is in the same direction, it is known as swim.

Fig. 4 Flowchart of Genetic Algorithm

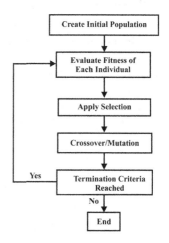

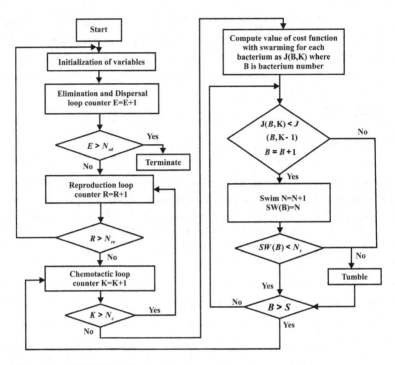

Fig. 5 Flowchart of BFO algorithm

2. Swarming: It is desired that when bacteria reach the nutrient-rich location (solution point), the other bacteria should attract by it, for them to converge at the nutrient-rich location (solution point) quickly than before.

3. Reproduction: In reproduction, health is to be considered, and according to health, all bacteria are arranged in reverse order. Here, only half the population survives, bacteria gets split into two from the middle and the least healthy bacteria die.

4. Elimination and dispersal: Elimination and dispersal are the part of the evolution. Events such as sudden rise in temperature of the local environment of the bacteria can kill them or disperse the bacteria that are currently in bacteria-rich location to entirely new location. Instead of eliminating these bacteria, these events may disperse them to near a food location. Over a long period of time, such events have spread the bacteria to all parts of our environment.

4 Results

4.1 Results with SMIB System

Table 1 shows the PID parameters with eigenvalues and damping ratio, which shows that the eigenvalues lie in the left half of s-plane indicating stability of system and damping ratio also improves indicating more stable system for BFO-PID.

The results of SMIB system under normal loading condition and light load condition are shown in Figs. 6 and 7, respectively. The speed deviation in BFO-PID reaches to a zero value in minimum time as compared to CPID and GA-PID, for both the cases.

4.2 Results of Three-Machine Nine-Bus System

Table 2 shows the PID parameters with eigenvalues and damping ratio, which shows that the eigenvalues lie in the left half of s-plane and damping ratio also improves indicating more stable system for BFO-PID.

The results of unit-2, i.e., G_2, for three-machine nine-bus system under normal operating condition and under line outage condition are shown in Figs. 8 and 9, respectively. The speed deviation in BFO-PID reaches to a zero value in minimum time as compared to CPID and GA-PID, for both the cases.

Table 3 shows the PID parameters with eigenvalues and damping ratio, which shows that the eigenvalues lie in the left half of s-plane and damping ratio also improves indicating more stable system for BFO-PID.

The results of unit-3, i.e., G_3, for three-machine nine-bus system under normal operating condition and under line outage condition are shown in Figs. 10 and 11, respectively. The speed deviation in BFO-PID reaches to a zero value in minimum time as compared to CPID and GA-PID, for both the cases.

Table 1 Tuned parameters of SMIB

	Normal operating condition			Light load condition		
	CPID	GA-PID	BFO-PID	CPID	GA-PID	BFO-PID
K_p	2.21	28.3880	16.137	2.21	28.3880	16.137
K_i	1.041	−0.1903	1.603	1.041	−0.1903	1.603
K_d	1.972	−0.9735	0.612	1.972	−0.9735	0.612
Eigenvalue	$-0.2 \pm 5.9i$	$-0.3 \pm 7.1i$	$-0.3 \pm 5.1i$	$-0.18 \pm 5.8i$	$-0.25 \pm 7i$	$-0.29 \pm 5.1i$
Damping ratio ξ	0.0338	0.0422	0.0587	0.0310	0.0357	0.0567

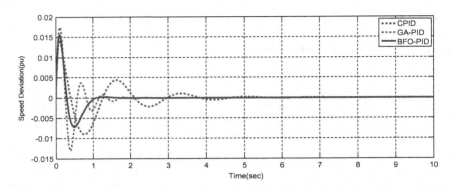

Fig. 6 Comparison of speed deviation of SMIB system

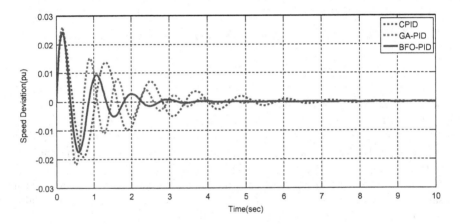

Fig. 7 Comparison of speed deviation of SMIB system with light load

Table 2 Damping ratio, Eigen values and controller paremers of unit -2

	Normal operating condition			Line outage condition		
	CPID	GA-PID	BFO-PID	CPID	GA-PID	BFO-PID
K_p	0.696	6.987	14.4304	0.696	6.987	14.4304
K_i	0.926	1.694	1.8189	0.926	1.694	1.8189
K_d	0.535	1.935	0.987	0.535	1.935	0.987
Eigenvalue	−0.039 ± 8.3085i	−0.0523 ± 8.3421i	−0.07 ± 8.3211i	−0.035 ± 8.536i	−0.0495 ± 8.425i	−0.066 ± 8.1112i
Damping ratio ξ	0.00469	0.00627	0.00841	0.00410	0.00587	0.00813

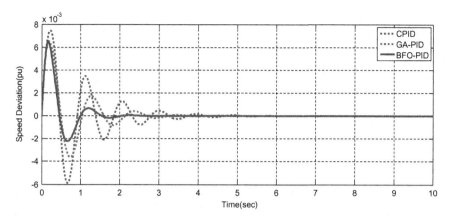

Fig. 8 Comparison of speed deviation of unit-2

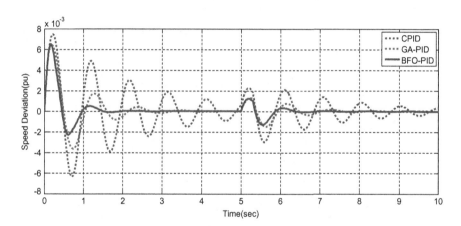

Fig. 9 Comparison of speed deviation of unit-2 with line outage

Table 3 Damping ratio, Eigen value and controller parameters of unit-3

	Normal operating condition			Line outage condition		
	CPID	GA-PID	BFO-PID	CPID	GA-PID	BFO-PID
K_p	0.696	19.573	45.8303	0.696	19.573	45.8303
K_i	0.926	0.197	0.4751	0.926	0.197	0.4751
K_d	0.535	0.5	0.3329	0.535	0.5	0.3329
Eigenvalue	−0.039 ± 8.3085i	−0.0442 ± 8.3563i	−0.0477 ± 8.4178i	−0.034 ± 7.803i	−0.0440 ± 9.251i	−0.0510 ± 9.716i
Damping ratio ξ	0.00469	0.00529	0.00566	0.00436	0.00475	0.00525

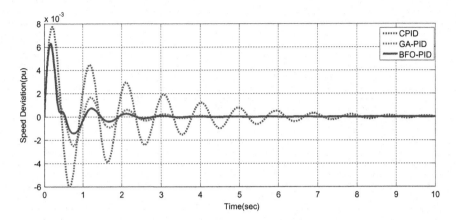

Fig. 10 Comparison of speed deviation of unit-3

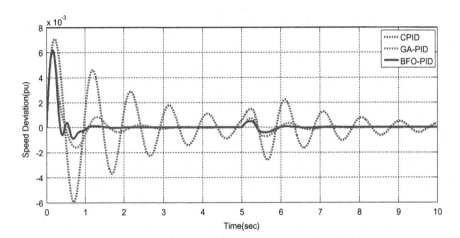

Fig. 11 Comparison of speed deviation of unit-3 with line outage

5 Conclusion

The performance of PID controller is analyzed for SMIB system as well as for three-machine nine-bus system by using optimal PID with different optimization algorithms. According to the results, it is clear that the performance of the BFO-PID controller is better as compared to GA-PID and conventional PID. The eigenvalues obtained by using BFO-PID show that the system is more stable in BFO-PID, and also, the overshoot is minimized. The deviation in speed reaches to zero value in minimum time. Hence, the BFO-PID provides better overall system stability and performances.

References

1. Kamdar, R., Kumar, M., Agnihotri, G.: Transient Stability Analysis and Enhancement of IEEE-9 Bus System. Electrical & Computer Engineering: An International Journal (ECIJ). Vol. 3. No. 2. (2014) 41–51.
2. Mo, H., Yin, Y.: Research on PID tuning of servo system based on Bacterial Foraging Algorithm. Seventh International Conference on Natural Computation. Vol. 3. (2011) 1758–1762.
3. Fan, X., Cao, J., Yang, H., Dong, X., Liu, C., Wu, Q., Gong, W.: Optimization of PID Parameters Based on Improved Particle-Swarm-Optimization. International Conference on Information Science and Cloud Computing Companion. (2013) 393–397.
4. Jalilvand, A., Vahedi, H., Bayat, A.: Optimal tuning of the PID controller for a buck converter using Bacterial Foraging Algorithm. International Conference on Intelligent and Advanced Systems. (2010) 1–5.
5. Lakshmi, K. V., Srinivas, P.: Optimal tuning of PID controller using Particle Swarm Optimization. International Conference on Electrical, Electronics, Signals, Communication and Optimization (EESCO). (2015) 1–5.
6. Singh, M., Patel, R. N., Jhapte, R.: Performance comparison of optimized controller tuning techniques for voltage stability. Control, Measurement and Instrumentation (CMI). IEEE First International Conference. IEEE (2016) 11–15.
7. Gandhi, P.R., Joshi, S.K.: Design of PID Power System Stabilizer using GA for SMIB System: Linear and Non-Linear Approach. International Conference on Recent Advancements in Electrical, Electronics and Control Engineering. (2010) 319–323.
8. Jebali, M., Kahouli, O., Abdallah, H. H.: Power System Stabilizer Parameters Optimization Using Genetic Algorithm. 5th International Conference on Systems and Control, Cadi Ayyad University, Marrakesh. (2016) 78–83.
9. Naresh, G., Raju, M. R., Ravindra, K., Narasimham, S.V.L.: Optimal Design of Multi-Machine Power System Stabilizer Using Genetic Algorithm. Innovative Systems Design and Engineering. Vol. 2. No. 4. (2011) 138–154.
10. Calistru, C. N., Timofte, D.: Symbolic to Genetic Tuning of PID Controllers. IEEE International Conference on E-Health and Bioengineering EHB. (2015) 19–21.
11. Sumanand, S.K., Giri, V. K.: Speed control of DC motor using optimization techniques based PID Controller. IEEE International Conference on Engineering and Technology (ICETECH). IEEE (2016) 581–587.
12. da Silva, W. G., Acarnley, P. P., Finch, J. W.: Application of Genetic Algorithms to the Online Tuning of Electric Drive Speed Controllers. IEEE Transactions on Industrial Electronics. IEEE (2000) 217–219.
13. Vishal, V., Kumar, V., Rana, K.P.S., Mishra, P.: Comparative Study of Some Optimization Techniques Applied to DC Motor Control. Advance Computing Conference (IACC), IEEE International. IEEE (2014) 1342–1347.
14. Precup, R.E., Borza, A. L., Radac, M. B., Petriu, E.M.: Performance Analysis of Torque Motor System with PID Controller Tuned by Bacterial Foraging Optimization Algorithm. IEEE international Conference on Computational Intelligence and Virtual Environments for Measurement Systems and Applications. IEEE (2014) 141–146.
15. Ali, A., Majhi, S.: Design of Optimum PID Controller by Bacterial Foraging Strategy. IEEE International Conference on Industrial Technology. IEEE (2006) 601–605.
16. Korani, W. M., Dorrah, H. T., Emara, H. M.: Bacterial Foraging Oriented by Particle Swarm Optimization Strategy for PID Tuning. IEEE International Symposium on Computational Intelligence in Robotics and Automation-(CIRA). IEEE (2009) 445–450.

A Competent Algorithm for Enhancing Low-Quality Finger Vein Images Using Fuzzy Theory

Rose Bindu Joseph and Devarasan Ezhilmaran

Abstract Soft computing methods and the fuzzy theoretic approaches, in particular, are widely known for their ability to tackle the uncertainties and vagueness that exist in image processing problems. This paper puts forward a distinctive enhancement algorithm for finger vein biometric images in which interval type-2 fuzzy sets are used. Finger vein biometrics is one of the latest reliable biometric systems that make use of the uniqueness of the finger vein patterns of individuals. Low contrast, blur, or noise often result in the lower quality of the captured finger vein images. For efficient enhancement of the finger vein images, interval type-2 fuzzy set is presented in this work and Einstein T-conorm is suggested for type reduction by combining the upper and lower membership functions. The performance assessment of the proposed algorithm is done by estimating the linear index of fuzziness and entropy. The experiments are performed using different vein pattern images, and the outcomes are analyzed by comparing with the existing methods. The performance evaluation visibly exhibits the efficiency of the recommended method in comparison with the existing methods.

Keywords Interval type-2 fuzzy sets · Einstein T-conorm · Finger vein recognition · Image enhancement

1 Introduction

Biometrics assist in identifying individual people based on their distinctive physiological or behavioral features. Finger vein recognition is a propitious biometric with commercial applications which is explored widely in the recent years. As a desirable biometric technique it has several matchless characteristics like hostility to

R. B. Joseph (✉) · D. Ezhilmaran
School of Advanced Sciences, VIT University, Vellore 632014, Tamil Nadu, India
e-mail: rosebindujoseph.p2013@vit.ac.in

D. Ezhilmaran
e-mail: ezhilmaran.d@vit.ac.in

© Springer Nature Singapore Pte Ltd. 2019
H. S. Behera et al. (eds.), *Computational Intelligence in Data Mining*,
Advances in Intelligent Systems and Computing 711,
https://doi.org/10.1007/978-981-10-8055-5_73

831

forgery, great precision, distinctiveness and stability, noninvasive and contactless image acquisition, faster verification, small and inexpensive capturing devices, and live body identification [1].

The vein patterns inside the fingers are captured using a CCD camera while exposing the finger to near infrared rays of 760–900 nm wavelength [2]. The light rays are absorbed by the hemoglobin in the vein which makes the vein pattern look darker than the other regions in the image [3]. Enhancement of the captured image is an inevitable stage in finger vein recognition as the captured images often come with low contrast, blurred, or noisy because of uneven illumination, scattering of light by the tissues, and the finger position misalignment. Once the difference in the contrast is upgraded between vein patterns and the neighboring tissues, further processes like feature extraction and verification become faster and easier.

Different techniques of enhancement have been implemented by various researchers to enhance the finger vein images. Various Gabor filters namely circular Gabor filters [4, 5], 2D Gabor filters [6], and adaptive Gabor filters [7] are used for finger vein enhancement. Curvelet transform [8] and ridgelet transform [9] are some of the transform-based methods that have been used for enhancing the vein patterns. Histogram equalization [10, 11] is another refined method which enhances the image contrast by efficiently spreading out the most recurrent intensity values. Despite the fact that these are well-established methods, these traditional approaches have many practical issues. Most of these techniques amplify noise and smoothening and noise reduction procedures in these methods consequently distort the edges which result in incorrect vein pattern. Also, the conventional techniques are time consuming as they involve several complex computations. The existing methods do not fully satisfy our requirement as the finger vein images usually have limited gray-level contrast and short dynamic range.

Fuzzy techniques are one of the prominent soft computing methods that have been successfully implemented in finger vein enhancement as they represent images as fuzzy sets which help in regulating the vagueness and imperfection in finger vein images. Cheng-Bo Yu [12] suggested an image enhancement algorithm based on fuzzy theory. Kwang Yong Shin [13] enhanced the image quality of finger vein patterns using an image fusion system with fuzzy technique. Oriented filters and fuzzy theory was the base of the enhancement technique proposed by Lin Yu [14] for finger vein images.

In 1975, Zadeh presented type-2 fuzzy set [15] by extending type-1 fuzzy set which he had introduced in 1965 [16]. Mizumoto and Tanaka [17] explained the set-theoretic operations of type-2 fuzzy sets, and Nieminen [18] gave additional details on the algebraic details of type-2 fuzzy sets. It was Karnik and Mendel [19, 20] who proposed the concept of type-2 fuzzy set centroid [21]. Liang and Mendel [22] developed fuzzy logic theory of interval type-2 systems. The major applications of type-2 fuzzy sets include preprocessing of data [23], decision making [24], and solution of fuzzy relation equations [25].

Coming to the image processing applications of type-2 fuzzy sets, Tizhoosh [26] used the concept for image thresholding. Castillo and Melin [27] proposed noise cancellation of images using type-2 fuzzy logic. Chaira [28] proposed enhancement of medical images with type-2 fuzzy sets.

This paper introduces an interval type-2 fuzzy-based image enhancement technique for finger vein images. Einstein T-conorm is used as an aggregate operator for upper and lower membership functions. This method works efficiently on low contrast finger vein images as well without any loss of information about finger vein features.

2 Proposed Method

A novel image enhancement method using interval type-2 fuzzy sets is presented here. Each image pixel is fuzzified using a suitable type-1 membership function. Appropriate fuzzy linguistic hedges are used to blur type-1 set for calculating the lower membership function (LMF) and the upper membership function (UMF). Einstein T-conorm is used for type reduction by combining the membership functions. This resultant image with the modified membership function is the enhanced image.

2.1 Assignment of Fuzzy Membership

Fuzzy sets are the sets on the universal set X which can accommodate degrees of membership which contrasts with the classical notion of crisp set for which any element is either a member or nonmember.

A fuzzy set S in a nonempty set X can be expressed as $S = \{(u, \mu_S(u)), u \in X\}$ where the membership function $\mu_S(u): X \rightarrow [0, 1]$. Clearly, the membership function $\mu_S(u)$ of an element u in S is a real number such that $0 \leq \mu_S(x) \leq 1$.

For a finger vein image A, let x_{ij} denote the intensity level at pixel (i, j) where the image has L gray levels. Let $M \times N$ be the image size. This means i varies from 0 to $M - 1$ and j from 0 to $N - 1$. The image is fuzzified by assigning a fuzzy membership value $\mu_A(x_{ij})$ to the gray level x_{ij} using Eq. (1).

$$\mu_A(x_{ij}) = \frac{x_{ij} - x_{\min}}{x_{\max} - x_{\min}} \tag{1}$$

Here, $x_{\min}$ and $x_{\max}$ are the smallest and the greatest values of gray levels of the pixels in the image. Clearly $x_{\min} \leq x_{ij} \leq x_{\max}$. Now the type-1 fuzzy set representing the image A is given by, $A = \{(x_{ij}, \mu_A(x_{ij})/0 \leq i \leq M - 1, 0 \leq j \leq N - 1)\}$.

2.2 Introduction to Interval Type-2 Membership

If the membership values of a fuzzy set are also fuzzy sets, then it is called a type-2 fuzzy set with secondary membership functions. The membership functions of a more generalized type-n fuzzy are type n−1 fuzzy sets. Type-2 fuzzy logic systems are beneficial in certain environments where it is challenging to decide a precise fuzzy set membership function. Linguistic as well as numerical uncertainties can be handled in a better way using these sets.

The membership function of a fuzzy set $\tilde{G}$ of type-2 is given by $\mu_{\tilde{G}}(x, v)$ which means $\tilde{G} = \int_{x \in X} \int_{v \in J_x} \mu_{\tilde{G}}(x, v)/(x, v), J_x \subseteq [0, 1]$. In the case of discrete sets, $\sum$ is used instead of $\int$. Here, $x \in X$ and $v \in J_x \subseteq [0, 1]$.

Footprint of uncertainty (FOU) is given by, $\text{FOU}(\tilde{G}) = \bigcup_{x \in X} J_x$. It is formed by combining all memberships which are primary.

When the secondary memberships in a type-2 fuzzy set all become unity, it turns into an interval type-2 fuzzy set. The upper and lower membership functions characterize these sets. Because of their reduced computational cost and adaptability, these sets are used extensively in many applications.

An interval type-2 fuzzy set $\tilde{G}$ is a type-2 fuzzy set in which the secondary membership function $\mu_{\tilde{G}}(x, v) = 1$, $\forall x \in X$, and $v \in J_x \subseteq [0, 1]$; i.e. $\tilde{G} = \{((x, v), 1)/ x \in X, \forall v \in J_x \subseteq [0, 1]\}$. $\mu_{\tilde{A}}^U(x)$ and $\mu_{\tilde{A}}^L(x)$ are the upper and lower membership values of the interval type-2 fuzzy sets, and they form the FOU.

The uncertainties in the image are more effectively modeled by the type-2 fuzzy sets because they possess fuzzy membership functions. For the membership function of each element, an interval is taken instead of a single value in the proposed interval type-2 fuzzy set. The type-1 membership function is blurred by assigning upper and lower membership values to each pixel for building the footprint of uncertainty and thus forming interval type-2 fuzzy set.

We obtain $\mu_{\tilde{A}}^U(x_{ij})$ and $\mu_{\tilde{A}}^L(x_{ij})$ of the interval type-2 fuzzy set by applying a fuzzy linguistic hedge along with the reciprocal on the type-1 membership values $\mu_A(x_{ij})$ of each pixel. They are given by Eqs. (2) and (3).

$$\mu_{\tilde{A}}^U(x_{ij}) = \left[\mu_A(x_{ij})\right]^{\alpha} \tag{2}$$

$$\mu_{\tilde{A}}^L(x_{ij}) = \left[\mu_A(x_{ij})\right]^{\frac{1}{\alpha}} \tag{3}$$

By experimenting different values for α, the most suitable value has been chosen as $\alpha = 0.5$ where the linguistic hedges are the *dilation* and *concentration* given by Eqs. (4) and (5).

$$\mu_{\tilde{A}}^U(x_{ij}) = \left[\mu_A(x_{ij})\right]^{0.5} \tag{4}$$

$$\mu_{\tilde{A}}^{L}(x_{ij}) = \left[\mu_{A}(x_{ij})\right]^{2} \tag{5}$$

Now the resultant interval type-2 fuzzy set is $\tilde{A} = \left\{ \left(x_{ij}, \mu_{\tilde{A}}^{U}(x_{ij}), \mu_{\tilde{A}}^{L}(x_{ij})\right) \forall x_{ij} \in A \right\}$. Clearly, $\mu_{\tilde{A}}^{U}(x_{ij}) \le \mu_{A}(x_{ij}) \le \mu_{\tilde{A}}^{L}(x_{ij}), \forall x_{ij} \in A$.

2.3 Type Reduction Using T-conorm to Obtain Enhanced Image

Triangular operators, often termed as T-operators, are vital components of fuzzy systems. They are the triangular norm (T-norm), the triangular conorm (T-conorm or S-norm), and negation functions which were introduced in order to evaluate the membership values of intersection, union, and complement of fuzzy sets.

Let $T: [0, 1] \times [0, 1] \rightarrow [0, 1]$. T is a T-norm, if and only if, for all $u, v, w \in [0, 1]$, $T(u, v) = T(v, u)$; $T(u, v) \le T(u, w)$, if $v \le w$; $T(u, T(v, w)) = T(T(u, v), w)$; $T(u, 1) = u$. A T-norm is Archimedean, if and only if, $T(u, v)$ is continuous and $T(u, u) < u, \forall u \in (0, 1)$.

Let $T^{*}: [0, 1] \times [0, 1] \rightarrow [0, 1]$. T^{*} is a T-conorm, if and only if, for all $u, v, w \in [0, 1]$, $T^{*}(u, v) = T^{*}(v, u)$; $T^{*}(u, v) \le T^{*}(u, w)$, if $v \le w$; $T^{*}(u, T^{*}(v, w)) = T^{*}(T^{*}(u, v), w)$; $T^{*}(u, 0) = u$. T-conorm is Archimedean, if and only if, T^{*} is continuous and $T^{*}(u, u) > u, \forall u \in (0, 1)$.

Einstein T-operators are the Einstein product (T-norm) and Einstein sum (T-conorm) which are strict Archimedean T-operators. They were named after Albert Einstein, as Einstein sum is analogous to Einstein's formulae for composition of velocities in his spatial relativity theory. Einstein T-operators have been extended as aggregation operators in interval-valued fuzzy logic systems. Einstein product (T-norm) and Einstein sum (T-conorm) are given by, $E(x, y) = \frac{x \cdot y}{1 + (1 - x) \cdot (1 - y)}$ (T $-$ norm) and $E^{*}(x, y) = \frac{x + y}{1 + x \cdot y}$ (T $-$ conorm) $\forall x, y \in (0, 1)$.

The upper and lower membership degrees that are obtained by Eqs. (4) and (5) are combined for type reduction using the Einstein T-conorm given by Eq. (6).

$$\mu_{A}^{E}(x_{ij}) = \frac{\mu_{\tilde{A}}^{U}(x_{ij}) + \mu_{\tilde{A}}^{L}(x_{ij})}{1 + \left[\mu_{\tilde{A}}^{U}(x_{ij}) \cdot \mu_{\tilde{A}}^{L}(x_{ij})\right]} \tag{6}$$

The resultant enhanced image after the modification of the membership function appears to be much clearer and brighter than the original image.

3   Experiment and Analysis

Performance of the recommended system is analyzed in this section. The proposed algorithm is applied on finger vein images and is compared with some existing algorithms, and the results are analyzed. Enhancement of the proposed algorithm is found to be better than the other methods. The experiments were conducted on a PC with 1.70 GHz CPU and 4.0 GB memory and executed using MATLAB 7:14:0.

Finger vein database (version 1.0) [29] from Hong Kong Polytechnic University has been used for this experiment and evaluation. Finger vein images are taken from both male and female volunteers. A contactless imaging device has been used for capturing the vein images. Finger vein images from 156 subjects are available in the current database. The images were captured in two different sessions having an average interval of 66.8 days between the sessions. Six image samples have been taken from the index finger and middle finger of each subject in one session.

Experiments were performed on several finger vein images from the database, and the results on four sample images are shown for reference. Figure 1 shows the enhanced output images for four vein images taken from the database. The image outputs with the proposed method are compared with a non-fuzzy method (histogram equalization), a type-1 fuzzy method (fuzzy intensifier operation), type-2 fuzzy method by Tizhoosh [30], type-2 fuzzy method with Hamacher T-conorm by Chaira [28], and the proposed interval type-2 fuzzy method with Einstein T-conorm. In [30], Tizhoosh has used Eq. (7) for modification of membership function.

$$\mu(x_{ij}) = \mu_A^L(x_{ij}) \cdot \beta + \mu_A^U(x_{ij}) \cdot (1 - \beta) \tag{7}$$

Here $\beta = \frac{mean(x_{ij})}{L}$. Here L is the number of gray levels.

In [28], Chaira has taken a version of Hamacher T-conorm given by Eq. (8).

Fig. 1 Enhanced output images for some sample finger vein images from the database

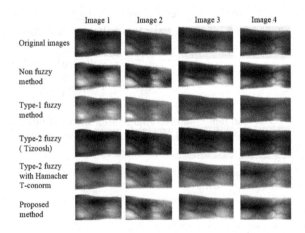

$$\mu^{enh}\left(g_{ij}\right) = \frac{\mu^{upper} + \mu^{lower} + (\beta - 2)\mu^{upper} \cdot \mu^{lower}}{1 - (1 - \beta)\mu^{upper} \cdot \mu^{lower}} \tag{8}$$

From Fig. 1, it is evident that the output images using proposed type-2 fuzzy method are much more enhanced than images from the other methods and they show clearer vein patterns. Even the low contrast images get better enhancement and clearer vein patterns using the proposed method.

The performance of the proposed method in comparison with various vein enhancement methods is thoroughly analyzed in terms of linear index of fuzziness and entropy. Linear index of fuzziness measures the vagueness in the image and is calculated using formula (9).

$$\gamma = \frac{2}{MN} \sum_{i=0}^{M-1} \sum_{j=0}^{N-1} \min\left(\mu_{ij}, 1 - \mu_{ij}\right) \tag{9}$$

Here, μ_{ij} is the membership value of the enhanced image of size $M \times N$. As the vagueness in the image decreases when it is enhanced, a decrease in the index of fuzziness shows better enhancement. Table 1 shows the comparison of linear index of fuzziness by the different methods where it indicates the lowest index value for the proposed method. This clearly characterizes the better enhancement by the proposed method.

Table 2 displays the entropy of the original images and the enhanced images using different approaches. Entropy is a statistical measure of randomness that can be used to describe the quality of the image. The results show the least entropy value for the proposed method, and this indicates that the enhancement by the interval type-2 fuzzy method with Einstein T-conorm is better compared to the existing methods. A graphical representation of comparison of entropies of different methods for different images is shown in Fig. 2. Entropies are plotted on the y-axis corresponding to the different sample images listed on the x-axis.

In the proposed algorithm, the upper and lower membership values are calculated using Eqs. (2) and (3). We have analyzed the results using different values of

Table 1 Linear index of fuzziness using different methods

	Type-1 fuzzy method	Type-2 fuzzy (Tizhoosh)	Type-2 fuzzy (Hamacher T-conorm)	Proposed method
Image 1	0.2518	0.2134	0.1554	0.1362
Image 2	0.2397	0.2063	0.1773	0.1582
Image 3	0.2413	0.2100	0.1009	0.0829
Image 4	0.2573	0.2187	0.1154	0.0963
Image 5	0.2607	0.2201	0.1968	0.1751
Image 6	0.2669	0.2556	0.2462	0.2213
Image 7	0.2501	0.2477	0.2230	0.2101
Image 8	0.2611	0.2278	0.1467	0.1249

Table 2 Entropy of output images using different methods

	Original image	Non-fuzzy method	Type-1 fuzzy method	Type-2 fuzzy (Tizhoosh)	Type-2 fuzzy (Hamacher T-conorm)	Proposed method
Image 1	6.0495	6.0382	6.0181	5.8550	5.4799	5.3297
Image 2	6.2119	6.2001	6.0917	5.9653	5.6389	5.5095
Image 3	5.9698	5.9113	5.9008	5.8061	5.0992	4.8511
Image 4	5.9631	5.9337	5.9189	5.8032	5.2317	5.0490
Image 5	6.0567	5.9949	5.9283	5.7841	5.5580	5.4908
Image 6	6.4412	6.3172	6.2681	6.1489	5.9988	5.9144
Image 7	6.4881	6.4117	6.3301	6.2560	6.0870	5.9451
Image 8	6.2120	6.2015	6.2001	6.0002	5.5362	5.3725

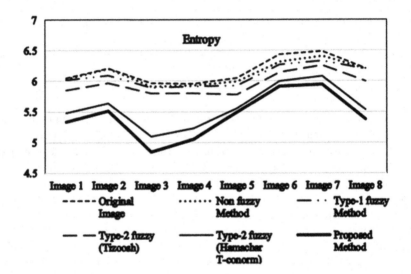

Fig. 2 Graphical representation of entropy of output images

Table 3 Performance evaluation using different values of α

		$\alpha = 0.5$	$\alpha = 0.75$	$\alpha = 0.9$
Image 1	Entropy	5.3297	5.3506	5.3566
Image 2	Fuzzy index	0.1362	0.1401	0.1402
Image 3	Entropy	5.5095	5.5268	5.5335
Image 4	Fuzzy index	0.1582	0.1625	0.1626
Image 5	Entropy	4.8511	4.8880	4.8910
Image 6	Fuzzy index	0.0829	0.0854	0.0856

α such as $\alpha = 0.5, \alpha = 0.75$, and $\alpha = 0.9$ in order to arrive at the best suitable value of α. The results are shown in Table 3. From the analysis, we understand that the enhancement is optimal when $\alpha = 0.5$.

4 Conclusions

This paper proposes a fuzzy theoretic algorithm based on interval type-2 fuzzy sets for enhancing the poor quality images of finger vein biometric database. Efficient enhancement of the finger vein database and query images is inevitable for faster and accurate feature extraction and matching in the case of finger vein biometrics. In the proposed algorithm, type-1 membership function is blurred using appropriate fuzzy linguistic hedges to form the upper and lower membership functions of the interval type-2 fuzzy set. This paper suggests Einstein T-conorm for type reduction by combining the upper and lower membership functions. The effectiveness of the algorithm is verified by comparison with existing methods using linear index of fuzziness and entropy. The performance evaluation clearly exhibits the advantage of the proposed method over the other existing methods. In future research, feature extraction of finger vein images using interval type-2 fuzzy systems can be experimented.

References

1. Hashimoto, J.: Finger vein authentication technology and its future. In: VLSI Circuits, Symposium on Digest of Technical Papers, pp. 5–8 (2006).
2. Miura, N., Nagasaka, A., Miyatake, T.: Feature extraction of finger-vein patterns based on repeated line tracking and its application to personal identification. Machine Vision and Applications. 15(4), 194–203 (2004).
3. Kumar, A., Zhou, Y.: Human identification using finger images. In: IEEE Transactions on image processing, pp. 2228–44 (2012).
4. Yang, J., Shi, Y., Yang, J., Jiang, L.: A novel finger-vein recognition method with feature combination. In: 16th IEEE International Conference on Image Processing, pp. 2709–2712 (2009).
5. Zhang, J., Yang, J.: Finger-vein image enhancement based on combination of gray-level grouping and circular gabor filter. In: International Conference on Information Engineering and Computer Science, pp. 1–4 (2009).
6. Kejun, W., Jingyu, L., Oluwatoyin, P.P., Weixing, F.: Finger vein identification based on 2-D gabor filter. In: 2nd International Conference on Industrial Mechatronics and Automation (ICIMA), Vol. 2, pp. 10–13 (2010).
7. Cho, S.R., Park, Y.H., Nam, G.P., Shin, K.Y., Lee, H.C., Park, K.R., Kim, S.M., Kim, H.C.: Enhancement of finger-vein image by vein line tracking and adaptive gabor filtering for finger-vein recognition. In: Applied Mechanics and Materials, Vol. 145, pp. 219–223 (2012).
8. Zhang, Z., Ma, S., Han, X.: Multiscale feature extraction of finger-vein patterns based on curvelets and local interconnection structure neural network. In: 18th International Conference on Pattern Recognition, Vol. 4, pp. 145–148 (2006).

9. Li, H.B., Yu, C.B., Zhang, D.M., Zhou, Z.M.: The Study on Finger Vein Image Enhancement Based on Ridgelet Transformation. In: 6th International Conference on Wireless Communications Networking and Mobile Computing (WiCOM), pp. 1–4 (2010).
10. Wen, X.B., Zhao, J.W., Liang, X.Z.: Image Enhancement of Finger-vein Patterns Based on Wavelet Denoising and Histogram Template Equalization. Journal of Jilin University (Science Edition), 2, p. 026 (2008).
11. Pi, W., Shin, J., Park, D.: An effective quality improvement approach for low quality finger vein image. In: International Conference on Electronics and Information Engineering (ICEIE), Vol. 1, pp. V1–424 (2010).
12. Yu, C.B., Zhang, D.M., Li, H.B., Zhang, F.F.: Finger-vein image enhancement based on muti-threshold fuzzy algorithm. In: 2nd International Congress on Image and Signal Processing, pp. 1–3 (2009).
13. Shin, K.Y., Park, Y.H., Nguyen, D.T., Park, K.R.: Finger-vein image enhancement using a fuzzy-based fusion method with gabor and retinex filtering. Sensors, 14(2), pp. 3095–3129 (2014).
14. You, L., Sun, L., Zhang, J.: Finger Vein Images Enhancement Method Based on Fuzzy Set and Oriented Filter. International Journal of Digital Content Technology and its Applications, 6(23), p. 271 (2012).
15. Zadeh, L.A.: The concept of a linguistic variable and its application to approximate reasoning-I. Information sciences, 8.3, 199–249 (1975).
16. Zadeh, L.A.: Fuzzy sets, Information and control, 8.3, 338–353 (1965).
17. Mizumoto, M., Tanaka, K.: Some properties of fuzzy sets of type 2. Information and control, 31(4), pp. 312–340 (1976).
18. Nieminen, J.: On the algebraic structure of fuzzy sets of type 2. Kybernetika, 13(4), pp. 261–273 (1977).
19. Karnik, N.N., Mendel, J.M., Liang, Q.: Type-2 fuzzy logic systems. IEEE transactions on Fuzzy Systems, 7(6), pp. 643–658 (1999).
20. Karnik, N.N., Mendel, J.M.: Operations on type-2 fuzzy sets. Fuzzy sets and systems, 122(2), pp. 327–348 (2001).
21. Karnik, N.N., Mendel, J.M.: Centroid of a type-2 fuzzy set. Information Sciences, 132(1), pp. 195–220 (2001).
22. Liang, Q., Mendel, J.M.: Interval type-2 fuzzy logic systems: theory and design, IEEE Transactions on Fuzzy systems, 8.5, pp. 535–550 (2000).
23. John, R.I., Innocent, P.R., Barnes, M.R.: Type 2 fuzzy sets and neuro-fuzzy clustering of radiographic tibia images. In: IEEE World Congress on Computational Intelligence Fuzzy Systems Proceedings, Vol. 2, pp. 1373–1376 (1998).
24. Yager, R.R.: Fuzzy subsets of type II in decisions. Cybernetics and System, 10(1–3), pp. 137–159 (1980).
25. Wagenknecht, M., Hartmann, K.: Application of fuzzy sets of type 2 to the solution of fuzzy equations systems. Fuzzy Sets and Systems, 25(2), pp. 183–190 (1988).
26. Tizhoosh, H.R.: Image thresholding using type II fuzzy sets. Pattern recognition, 38(12), pp. 2363–2372 (2005).
27. Melin, P., Castillo, O.: An intelligent hybrid approach for industrial quality control combining neural networks, fuzzy logic and fractal theory. Information Sciences, 177(7), pp. 1543–1557 (2007).
28. Chaira, T.: An improved medical image enhancement scheme using Type II fuzzy set. Applied soft computing, 25, pp. 293–308 (2014).
29. The Hong Kong Polytechnic University finger image database (version 1.0), http://www4.comp.polyu.edu.hk/~csajaykr/fvdatabase.htm.
30. Ensafi, P., Tizhoosh, H.: Type-2 fuzzy image enhancement. Image analysis and recognition, pp. 159–166 (2005).

An Adaptive Fuzzy Filter-Based Hybrid ARIMA-HONN Model for Time Series Forecasting

Sibarama Panigrahi and H. S. Behera

Abstract In this paper, linear ARIMA and nonlinear HONN especially pi-sigma neural network (PSNN) models are integrated to develop a hybrid ARIMA-HONN model for time series forecasting. Assuming the time series to be a sum of low-volatile and high-volatile components, the time series is first decomposed into constituent components by using an adaptive fuzzy filter. Then, the low-volatile and high-volatile components are modeled by using ARIMA and HONN models, respectively. The final prediction is obtained by combining the ARIMA predictions with HONN predictions. Using benchmark real-world datasets such as lynx, sunspot, temperature, passenger, and unemployment, the proposed ARIMA-HONN, ETS, ARIMA, ANN and two existing hybrid ARIMA-ANN models were simulated. Simulation results indicated the superiority of proposed model as compared to its counterparts for the datasets used.

1 Introduction

Time series analysis and forecasting is one of the hottest interdisciplinary research areas that have attracted researchers from diversified fields including finance, management, and engineering. Efficient time series forecasting (TSF) assists mankind to avoid financial crisis [1], in building next-generation power system [2], in providing better Internet service [3], and to be prepared for natural calamities like earthquake [4], rainfall, and drought. Conventionally, linear statistical models like exponential smoothing and autoregressive integrated moving average (ARIMA)

S. Panigrahi (✉)
Department of Computer Science and Engineering, Sambalpur University
Institute of Information Technology, Burla 768019, Odisha, India
e-mail: panigrahi.sibarama@gmail.com

H. S. Behera
Department of Computer Science and Engineering & Information Technology,
Veer Surendra Sai University of Technology, Burla 768018, Odisha, India
e-mail: hsbehera_india@yahoo.com

© Springer Nature Singapore Pte Ltd. 2019   841
H. S. Behera et al. (eds.), *Computational Intelligence in Data Mining*,
Advances in Intelligent Systems and Computing 711,
https://doi.org/10.1007/978-981-10-8055-5_74

were widely used for time series forecasting. These models assume the time series contains only linear patterns. However, most of the time series are nonlinear because of underlying physical data-generating process [5]. Therefore, nonlinear models like artificial neural network (ANN) have been gaining its popularity in time series forecasting. To improve the forecast accuracy, several evolutionary ANN models [6–8] for time series forecasting were developed. However, neither statistical models nor ANN models are alone sufficient to capture all patterns existing in a time series. This is because, most of the time series is neither linear nor nonlinear rather contains a combination of linear and/or nonlinear patterns, and linear models cannot handle nonlinear patterns equally well and vice versa. Therefore, recently several hybrid models by integrating statistical forecasting models with nonlinear soft computing models were reported. The hybrid models used ARIMA [9–13] and discrete wavelet transform (DWT) [14] to model linear components, whereas ANN [9–12, 14] and support vector machine (SVM) [13] to model the nonlinear component. Despite higher-order neural network (HONN), especially pi-sigma neural network have shown promising result in classification and regression, HONN models haven't been considered in these hybrid models. Motivated by this, a hybrid ARIMA-HONN model for time series forecasting is proposed.

The remaining of this paper is structured as follows. An overview on ARIMA and HONN especially pi-sigma network is presented in Sect. 2. The proposed hybrid methodology is explained in Sect. 3. Section 4 gives the experimental setup and simulation results. Finally, conclusions and future works are described in Sect. 5.

2 Preliminaries

2.1 ARIMA

ARIMA is one of the most popular statistical time series forecasting models. In ARIMA modeling procedure, first the time series is made stationary by differencing d times. Then, the consequential data are considered as a linear function of past p data values and q error sequences (as shown in Eq. 1).

$$y_t = \Theta_1 y_{t-1} + \Theta_2 y_{t-2} + \cdots + \Theta_p y_{t-p} + \varnothing_1 \varepsilon_{t-1} + \varnothing_2 \varepsilon_{t-2} + \cdots + \varnothing_q \varepsilon_{q-1} \quad (1)$$

where y_t represents actual value at time period t and ε_t represents the error sequence which is assumed to be white noise and is Gaussian distributed with a constant variance of σ^2. $\Theta_i (i = 1, 2, \ldots, p)$ are autoregressive (AR) coefficients, and $\varnothing_j (j = 1, 2, \ldots, q)$ are moving average (MA) coefficients. Considering all, the time series model is denoted as ARIMA (p, d, q). One of the most crucial parts in ARIMA modeling procedure is to identify the appropriate model order (p, q). The first practical approach to ARIMA model-building procedure was developed by Box and Jenkins [15]. In this procedure, first the model order and model coefficients are estimated. The model parameters p and q are identified using correlation analysis

[15]. Once a provisional model is specified, the model coefficients are calculated by using nonlinear optimization procedures like GMLE [16]. Akakine information criterion (AIC) is used for model validation, and the model with lowest value of AIC is chosen as the best model. Finally, the future values of the time series are predicted using selected ARIMA model, estimated model coefficients, and past data.

2.2 Pi-Sigma Neural Network (PSNN)

The pi-sigma neural network (PSNN) [17] (Fig. 1) is a fully connected feed-forward neural network. It has a single hidden layer. The number of hidden neurons is flexible and usually represents the order of PSNN. The weights connecting the inputs to the hidden neurons are trainable, whereas the weights connecting the hidden neurons with output neurons are nontrainable, i.e., set to unity. Hidden neurons have a summation unit and use a linear activation function. Output neurons have product units with nonlinear activation function. Thus, PSNN calculates the product of weighted summation of inputs and passes it through a nonlinear function. Shin and Ghosh [17] argued that PSNNs perform better, require less training time because of only one layer of trainable weights, require less memory, and have two times less computations as compared to multilayer perceptron. Traditionally, gradient-based training algorithms were used to train PSNN. However, in recent years, several evolutionary-based PSNN training algorithms [18–21] were reported.

Let input $x = (x_1, x_2, x_3, \ldots, x_i, \ldots, x_n)$ be the n-dimensional input vector for an m-order PSNN with bias vector at hidden layer $b = (b_1, b_2, b_3, \ldots, b_j, \ldots, b_m)$ and a vector of n weight vector $w = (w_1, w_2, w_3, \ldots, w_i, \ldots, w_n)$ with each weight vector $w_i = (w_{i1}, w_{i2}, \ldots, w_{ij}, \ldots, w_{im})$ and w_{ij} representing the weight between ith input and jth hidden neuron. Let the output vector of hidden layer

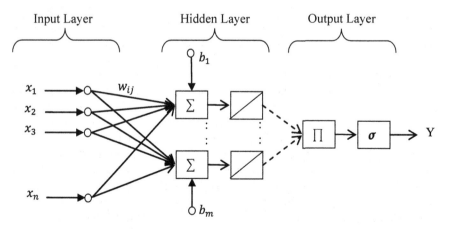

Fig. 1 Architecture of PSNN

$h = (h_1, h_2, h_3, \ldots, h_j, \ldots, h_m)$ with h_j be the output of jth hidden neuron, which can be computed by using Eq. 2. Then, the output is computed by using Eq. 3.

$$h_j = b_j + \sum_{i=1}^{n} (w_{ij}x_i) \tag{2}$$

$$Y = \sigma\left(\prod_{j=1}^{m} h_j\right) \tag{3}$$

3 Proposed Method

Over the past few years, hybrid forecasting models [11–16] by combining linear models with nonlinear models are becoming popular due to superior performance than individual models. Zhang [9] in 2003 for the first time proposed a hybrid ARIMA-ANN model for TSF. In his approach, the time series is considered to be a combination of linear and nonlinear components. ARIMA is applied directly to the original series to capture the linear patterns. Then, the residual error sequence obtained is considered as nonlinear and modeled by ANN. The final prediction is obtained by combining the ARIMA predictions with ANN predictions. Using the same approach, Faruk [10] proposed an optimized conjugated gradient-trained hybrid ARIMA-ANN model and used it for water quality time series prediction. To improve the forecast accuracy, Khashei and Bijari [11] considered the time series as a function of a linear and a nonlinear component. In this method, ARIMA is directly applied to the original series to obtain one forecast. Then, that one value and past error sequence are used by ANN to calculate the final forecast. Zhang [9], Faruk [10], and Khashei and Bijari [11] apply ARIMA directly on the original series considering the time series contains a linear pattern. However, ARIMA provides a better fit only when the time series is stationary and truly Gaussian. If any of the conditions is violated, ARIMA provides poor forecasts, and consequently, the resulting hybrid model will provide poor forecasts. To improve the forecast accuracy further, Babu and Reddy [12] and de Oliveira and Ludermir [13] first decompose the time series by exploring the nature of volatility of time series. These models have shown better forecast accuracy than the previous hybrid models. In these models, the time series is first decomposed into low-volatile and high-volatile components (as in Eq. 4) based on kurtosis (as in Eq. 5) value of the series. These models use moving average or exponential smoothing filter to decompose the series. Once the suitable length of filter is obtained, the series is decomposed into low-volatile and high-volatile series which are modeled by ARIMA and ANN models, respectively. Then, the final prediction (as in Eq. 6) is obtained by combining ARIMA predictions and ANN predictions. However, these models have a major drawback of losing more data points as the length of filter for which decomposition increases. Motivated by this need, in this paper a hybrid

ARIMA-HONN model is proposed which uses an adaptive fuzzy filter to decompose the time series. The use of the adaptive fuzzy filter avoids the loss of data values in the process of decomposition and thus can be applied to smaller time series also.

$$x_t = l_t + h_t \tag{4}$$

$$kurtosis(x) = \frac{E((x - E(x)))^4}{\left(E\left((x - E(x))^2\right)\right)^2} \tag{5}$$

$$\hat{x}_t = \hat{l}_t + \hat{h}_t \tag{6}$$

Algorithm 1: The proposed methodology

1: Given a time series $x = [x_1, x_2, \ldots, x_n]^T$.

2: Input the length of in-sample l_a and out-of-sample l_b data.

3: Decompose the series into low-volatile l_t and high-volatile components h_t using the adaptive fuzzy filter

    3.1: Define the universe of discourse $U = [U_l \ U_h]$ with $U_l = min_x - D$, $U_h = max_x + D$, $D = ((max_x - min_x) \times l_b)/l_a$

    3.2: for each length of interval $p = 1$ to l_a

        3.2.1: Partition the universe of discourse based on the length of interval p. $U = \{u_1, u_2, u_3 \ldots u_k\}$ with $k = (U_h - U_l)/p$ with each $u_i = [u_{il} \ u_{ih}], u_{1l} = U_l$ and $u_{kl} = U_h$ and $u_{ih} - u_{il} = p$.

        3.2.2: Calculate the defuzzified value m_i for each partition u_i i.e. $m_i = mean(T_j)$, with T_j is the jth data point that fall within the interval u_i. When no data point fall within the interval u_i, then $m_i = (u_{ih} + u_{il}) \div 2$

        3.2.3: for each time series data point y_j, if y_j belongs to the partition u_i then $l_{tj} = m_i$

        3.2.4: Calculate the fitness $F_p = |3 - kurtosis(l_t)|$

    3.3: Select the best length of interval P_{best} for which F has the minimum value.

    3.4: Using P_{best} as length of interval, calculate the low-volatile component l_t using step-3.2.3 and high-volatile component $h_t = x_t - l_t$.

4: Obtain the ARIMA predictions $\hat{l}_t$ on low-volatile component l_t.

5: Obtain predictions $\hat{h}_t$ on high-volatile component h_t using HONN.

6: Obtain Final predictions $\hat{y}_t$ by combining the ARIMA predictions $\hat{l}_t$ with HONN predictions $\hat{h}_t$.

Algorithm 1 presents the proposed methodology. This method assumes the time series to be a sum of a low-volatile component and a high-volatile component. The time series is first decomposed into two components based on nature of volatility. The decomposed component is said to be low-volatile, if the kurtosis of any of the decomposed series is 3. The residual component is considered as high-volatile. In order to decompose the series, an adaptive fuzzy filter concept is introduced. For this, the universe of discourse $U = [U_l U_h]$ for the time series is defined. Then, the universe of discourse $U = \{u_1, u_2, u_3, \ldots, u_k\}$ with $k = (U_h - U_l)/p$ is partitioned into k equal length intervals of length p. Then, the defuzzified values m_i for each partition $u_i = [u_{il} u_{ih}]$ are calculated. Then, for each time series data point y_j, the low-volatile component $l_{tj} = m_i$ is obtained by setting the defuzzified value m_i of the interval u_i within which y_j lies. The length of interval is varied between 1 to length of in-sample data, and corresponding kurtosis (l_t) is calculated. The length of interval for which kurtosis (l_t) is closest to 3 is selected, and the high-volatile component h_t is calculated by subtracting low-volatile component l_t from original series y_t. Then, the low-volatile and high-volatile components are modeled using ARIMA and HONN, respectively. The final prediction is obtained by combining ARIMA predictions with HONN predictions.

4   Experimental Setup and Results

In this study, three individual models such as ARIMA [22], ETS [22], and multi-layer perceptron (MLP) and two hybrid models such as Zhang's [9], Babu and Reddy's [12] models were considered for comparative performance analysis with the proposed model. The best-suited ARIMA and ETS models for time series are obtained from forecast package of R [22]. The rest of the simulations were implemented in MATLAB. Pi-sigma network was used as the HONN model and trained using backpropagation algorithm. The number of inputs is obtained by analyzing the autocorrelation function (ACF) of the time series. The order of PSNN was determined by simulating the hidden neurons between 1 and 20, and the results present the optimal number of hidden neurons for which it achieves smallest mean SMAPE over 50 executions. One neuron is present in the output layer.

Five datasets (described in Table 1) from the Time Series Data Library [23] were considered for comparative performance analysis. The time series is first divided into in-sample (80%) and out-of-sample (20%) sets. For all models excluding ARIMA and ETS, the in-sample data are again divided into train set (approximately 60% of original series) and validation set (approximately 20% of original series). The division of datasets, the number of significant lags, ARIMA and ETS models used in the simulations for all time series are presented in Table 2. Note that for the proposed model and Babu and Reddy's model the number of significant lags and ARIMA models used are different than that of Table 2. In all the methods, data are normalized (as in Eq. 7) using the min-max normalization technique. Hyndman and

Table 1 Datasets description

Time Series	Description
Lynx	Number of lynx trapped annually in Mackenzie River from 1821 to 1934
Sunspot	Wolf's sunspot numbers from 1700 to 1987
Passenger	Monthly international airline passenger from 1949 to 1960
Temperature	Mean monthly air temperature (Deg. F) Nottingham Castle from 1920 to 1939
Unemployment	Monthly Canadian total unemployment (thousands) from 1956 to 1975

Table 2 Datasets division, number of lags, ARIMA, and ETS models

Time series	Train	Validation	Test	Number of lags	ARIMA	ETS
Lynx	68	23	23	12	ARIMA (1, 1, 1)	ETS (A, N, N)
Sun spot	172	58	58	11	ARIMA (2, 0, 1)	ETS (A, N, N)
Passenger	86	29	29	12	ARIMA (3, 1, 3)	ETS (A, N, N)
Temperature	144	48	48	12	ARIMA (5, 0, 1)	ETS (A, N, N)
Unemployment	144	48	48	12	ARIMA (2, 0, 3)	ETS (A, Ad, N)

Koehler [24] conducted a comparative study on different measures of forecast accuracy. In this study, symmetric mean absolute percentage error (SMAPE) (as in Eq. 8) is considered.

$$x^{'} = \frac{y - min_x}{max_x - min_x} \tag{7}$$

$$SMAPE = \frac{1}{n}\sum_{j=1}^{n}\frac{\left|x_j - \hat{x}_j\right|}{\left(\left|x_j\right| + \left|\hat{x}_j\right|\right)/2} \tag{8}$$

In this study, the one-step-ahead prediction is calculated, and then, the models are evaluated. By making the above experimental setup, for each method on each dataset, 50 independent simulations were carried out. Table 3 represents the mean of SMAPE over best 30 out of 50 executions. It can be clearly observed that the proposed ARIMA-HONN model achieves best forecast accuracy in four datasets. Additionally, the proposed model achieved better SMAPE than ETS in 4 cases and all other models in all 5 cases.

Friedman and Nemenyi hypothesis test [25] was conducted to evaluate the models across all datasets. In order to rank the algorithms, Friedman's test was employed, and if the ranks are significantly different, a Nemenyi test was conducted

Table 3 Symmetric mean absolute percentage error (SMAPE) for all datasets (best values in bold)

	ARIMA [22]	ETS [22]	ANN	Babu and Reddy [12]	Zhang [9]	Proposed
Lynx	50.714	**48.781**	63.445	51.249	53.269	49.724
Sun spot	37.141	49.039	35.005	37.141	35.537	**34.228**
Passenger	6.8829	9.9966	5.3276	4.7027	5.0594	**4.504**
Temperature	4.8149	8.3699	4.1199	4.2225	4.4404	**4.054**
Unemployment	7.3117	6.5909	6.0006	6.2131	6.0949	**5.428**

Table 4 Ranks of all models considering SMAPE measure

Model	Rank SMAPE	Mean rank
Proposed	91.286	1
Babu and Reddy [12]	98.404	2
Zhang [9]	99.8236	3
ARIMA [22]	100.486	4
MLP	107.294	5
ETS [22]	115.822	6

[26]. Table 4 present the results with respect to SMAPE. It can be clearly observed that the proposed model obtained the best rank among all models considered in this study.

5   Conclusion

In this paper, linear ARIMA and nonlinear HONN models are integrated to devise a hybrid ARIMA-HONN model for time series forecasting. The proposed model assumes the time series to be a sum of low-volatile and high-volatile components. First, the time series is decomposed into high-volatile and low-volatile components by using an adaptive fuzzy filter. The low-volatile component is modeled using ARIMA, and high-volatile component is modeled using HONN. Then, ARIMA predictions are added with HONN predictions to obtain the final prediction. In order to evaluate the effectiveness of the proposed model, five datasets and four different models were considered. Simulation results demonstrated that no model is best to forecast all datasets. In some dataset, individual models perform better than hybrid models. However, the proposed hybrid model has shown statistically better performance than other models considered for the datasets used. In order to have a robust evaluation of results, Friedman and Nemenyi hypothesis tests were conducted. Results indicated the statistical superiority of proposed model for the datasets considered in this study.

References

1. Lin W.Y., Hu Y.H., Tsai C.F.: Machine Learning in Financial Crisis Prediction: A Survey. IEEE Transaction on Systems, Man, and Cybernetics—Part C: Applications and Reviews, 42 (2012) 421–436.
2. Raza M.Q., Khosravi A.: A review on artificial intelligence based load demand forecasting techniques for smart grid and buildings. Renewable and Sustainable Energy Reviews, 50 (2015) 1352–1372.
3. Meade N., Islam T.: Forecasting in telecommunications and ICT-A review. International Journal of Forecasting, 31(4) (2015) 1105–1126.
4. J. Reyes, A. Morales-Esteban, F. Martnez-lvarez, Neural networks to predict earthquakes in Chile, Applied Soft Computing, 13(2) (2013) 1314–1328.
5. Zhang G.P., Patuwo B.E., Hu M.Y.: Forecasting with artificial neural networks: The state of art, International Journal of Forecasting 14 (1988) 35–62.
6. Panigrahi S., Karali Y., Behera H. S.: Time Series Forecasting using Evolutionary Neural Network, International Journal of Computer Applications 75(10) (2013) 13–17.
7. Panigrahi S., Behera H. S.: Effect of Normalization Techniques on univariate time series forecasting using evolutionary Higher Order Neural Network, International Journal of Engineering and Advanced Technology 3(2) (2013) 280–285.
8. Panigrahi S., Karali Y., Behera H. S.: Normalize Time Series and Forecast using Evolutionary Neural Network, International Journal of Engineering Research & Technology, 2(9) (2013) 2518–2522.
9. Zhang G.: Time series forecasting using a hybrid ARIMA and neural network model, Neurocomputing 50(0) (2003) 159–175.
10. Faruk D. O.: A hybrid neural network and ARIMA model for water quality time series prediction, Engineering Applications of Artificial Intelligence 23 (2010) 586–594.
11. Khashei M., Bijari M.: A novel hybridization of artificial neural networks and ARIMA models for time series forecasting, Applied Soft Computing 11(2) (2011) 2664–2675.
12. Babu C. N., Reddy B. E.: A moving-average filter based hybrid ARIMA-ANN model for forecasting time series data, Applied Soft Computing 23 (2014) 27–38.
13. Oliveira J. F. de, Ludermir T. B.: A hybrid evolutionary decomposition system for time series forecasting, Neurocomputing 180 (2016) 27–34.
14. Khandelwal I., Adhikari R., Verma G.: Time Series Forecasting using Hybrid ARIMA and ANN Models based on DWT Decomposition, Procedia Computer Science 48 (2015) 173–179.
15. Box G.E.P., Jenkins G.: Time Series Analysis, Forecasting and Control, Holden-Day Incorporated, 1990.
16. Gaussian maximum likelihood estimation for ARMA models. I. Time series, Journal of Time Series Analysis 27(6) (2006) 857–875.
17. Shin Y., Ghosh J.: The pi–sigma network: An efficient higher-order neural network for pattern classification and function approximation, Neural Networks, IJCNN-91-Seattle International Joint Conference on Neural Networks, Seattle, WA, 1 (1991) 13–18.
18. Karali Y., Panigrahi S., Behera H. S.: A novel differential evolution based algorithm for higher order neural network training, Journal of Theoretical & Applied Information Technology, 56(3) (2013) 355–361.
19. Sahu K. K., Panigrahi S., Behera H. S.: A novel chemical reaction optimization algorithm for higher order neural network training, Journal of Theoretical & Applied Information Technology, 53(3) (2013) 402–409.
20. Panigrahi S., A Novel Hybrid Chemical Reaction Optimization Algorithm with Adaptive Differential Evolution Mutation Strategies for Higher Order Neural Network Training, International Arab Journal of Information Technology 14 (1) (2017) 18–25.

21. Panigrahi S., Bhoi A.K., Karali Y.: A modified Differential Evolution Algorithm trained Pi-Sigma Neural Network for Pattern Classification, International Journal of Soft Computing and Engineering 3 (5) (2013) 133–136.
22. Hyndman R. J., Khandakar Y.: Automatic Time Series Forecasting: The forecast Package for R, Journal of Statistical Software 27(3) (2008) 1–22.
23. Hyndman R. J.: Time Series Data Library, 2010. http://data.is/TSDLdemo.
24. Hyndman R. J., Koehler A. B.: Another look at measures of forecast accuracy, International Journal of Forecasting 22(4) (2006) 679–688.
25. Hollander M., Wolfe D.A., Chicken E.: Nonparametric statistical methods, Nonparametric Statistical Methods, John Wiley& Sons, Hoboken, NJ, 1999.
26. Demsar J.: Statistical comparisons of classifiers over multiple datasets, Journal of Machine Learning Research 7 (2006) 1–30.

An Evolutionary Algorithm-Based Text Categorization Technique

Ajit Kumar Das, Asit Kumar Das and Apurba Sarkar

Abstract In general, most of the organizations generate unstructured data from which extraction of meaningful information becomes a difficult task. Preprocessing of unstructured data before mining helps to improve the efficiency of the mining algorithms. In this paper, text data is initially preprocessed using tokenization, stop word removal, and stemming operations and a bag-of-words is identified to characterize the text dataset. Next, improved strength pareto evolutionary algorithm-based genetic algorithm is applied to determine the more compact set of informative words for clustering of text documents efficiently. It is a bi-objective genetic algorithm used to approximate the pareto-optimal front exploring the search space for optimal solution. The external clustering index and number of words described in the documents are considered as two objective functions of the algorithm, and based on these functions chromosomes in the population are evaluated and the best chromosome in non-dominated pareto front of final population gives the optimal set of words sufficient for categorizartion of text dataset.

Keywords Text mining · Feature selection · Text clustering · Cluster validation
Multi-objective evolutionary algorithm

1 Introduction

Unstructured data presents a high challenge in the field of text mining. Dimension reduction techniques in text documents select important features that give same distribution pattern of the dataset as obtained considering all features. In text mining,

A. K. Das · A. K. Das · A. Sarkar (✉)
Department of Computer Science and Technology, Indian Institute of Engineering
Science and Technology, Shibpur, Howrah 711103, West Bengal, India
e-mail: as.besu@gmail.com

A. K. Das
e-mail: writetoajit@yahoo.com

A. K. Das
e-mail: akdas@cs.iiests.ac.in

© Springer Nature Singapore Pte Ltd. 2019　　　　　　　　　　　　　　851
H. S. Behera et al. (eds.), *Computational Intelligence in Data Mining*,
Advances in Intelligent Systems and Computing 711,
https://doi.org/10.1007/978-981-10-8055-5_75

feature selection process initially goes through three basic steps composed of tokenization, stop word removal, and stemming. Tokenization is used to tokenize the documents into separate words, and thus a text document is considered as a collection of tokens. Stop word removal technique is applied on the tokenized documents to remove stop words from the documents to reduce its dimensionality. Stemming algorithms refer to the process of removing affixes from the words. Several stemming algorithms have been proposed over the years for optimal representation of the documents. The most widely cited stemming algorithm is the Porter stemming algorithm [11]. It uses a set of rules iteratively to remove suffixes from a word until no more suffix can be removed. The limitations of the method are: (i) It is not restricted to produce word stem, and as a result the word may be conflated which has different meaning from the stem. For example, the word "general" becomes "gener." (ii) Even if it produces a word stem, it is often overzealous. For example, "doing" becomes "doe" and "punish" becomes "pun." (iii) Like many other existing stemmers, it ignores prefixes completely; as a result, two opposite meaning words may remain as unrelated tokens. For example, "reliability" and "unreliability" are considered as two different tokens by this method. The Lovins stemmer [8] is similar in mechanism with Porter stemmer but has a larger set of suffixes, each of which may have multiple morphemes and does not apply its rules iteratively. Though it is more conservative than the Porter stemmer, still it suffers from over conflation and nonword stems. Krovetz [9] proves that meaning is essential during stemming specially for light English text. In Krovetz stemmer, the rule set obtained by Porter and Lovin methods is restructured to produce word stems using a non-iterative stripping mechanism. The main demerits of this algorithm are that it relies heavily on the integrity of the dictionary and tends to be too conservative. Thus, "prediction" is conflated to "predict" and "predictions" becomes "prediction," while it should be conflated more. The Xu and Croft stemmer algorithm [12] does not remove suffixes, but instead defines equivalence classes of words that should be conflated. Finding these equivalence classes can be computationally expensive and does not identify the stem and suffixes independently. The algorithm uses a normalized co-occurrence method to make inferences about whether word forms are of the same meaning. The stemming algorithm using Latent Semantic Analysis [5, 10], known as P stemmer, uses a set of prefix and suffix stripping rules in parallel generating multiple possible "parses" of the word in a computationally efficient way. For example, the word "prediction" would be parsed both as "pre diction" and "prediction." A dataset from the document set is prepared using term frequency-inverse document frequency (tf-idf) concept where number of documents is the objects and bag-of-words is the features representing the documents. As the bag-of-words in a text mining application is huge, so feature selection algorithm is very useful for selecting only the heavy weight words. Feature selection algorithms are of two types, namely filters and wrappers. Filter method selects important features using the knowledge from dataset itself. RELIEF [3] and its updated versions [6] are the popular filter-based feature selection methods. Nonparametric methods are also used to measure mutual information involving continuous features [7]. The wrapper methods [9] such as nearest neighbor classifier, a decision tree, a Naive Bayes method execute the learning algorithm to calculate the quality of feature subset. In spite of computationally hard, the wrapper methods have higher acceptance than filter methods for their better accuracy.

In the proposed work, Strength Pareto Evolutionary Algorithm (SPEA) [13] is used for selecting important words of the documents as well as for document clustering. First, the documents are partitioned into disjoint groups based on their class labels as available in the dataset. Also, the documents are partitioned into group of clusters using cosine similarity between every pair of documents. Using two groups of clusters, one fitness function of the SPEA is defined in terms of the external probability-based cluster validation index, calculated using the concept presented in the paper [4] and the other fitness function is considered as the number of words taken initially. The proposed SPEA algorithm tries to maximize the first objective function and the second one to be minimized. The algorithm runs iteratively, and after the convergence, the best solution gives the important features/words and the clusters of documents associated with it. The remaining work of the paper is organized as follows: The proposed SPEA-based important words selection and document clustering are discussed in Sect. 2. Section 3 demonstrates the experimental results and performance of the proposed method for a benchmark Newsgroup20 dataset. Finally, conclusions are made in Sect. 4.

2 Proposed Method

The multi-objective genetic algorithm namely SPEA is applied on the preprocessed document set to select important set of words for document clustering. The document set is a matrix where each row represents a document and each column represents the bag-of-words identified from these documents using Krovetz stemming method [9]. The matrix value gives the importance of the word in the document. Initially, the method selects a random solution in the search space. Later, applying objective function, the method searches for better solution by measuring the fitness of the solutions. Finding the optimal solution is an iterative process. Final optimal solution is achieved with the convergence of the algorithm. The method of finding the important words and the clusters of documents is described in subsections below.

2.1 Initial Population

Randomly, an initial population of binary strings is generated, where size of each string is equal to the number of words selected in preprocessing steps. Each string or chromosome gives the words by which the documents are characterized, where 0 and 1 in the chromosome represent presence and absence of the corresponding word, respectively. These chromosomes in initial population are the initial candidate solutions of the algorithm. If the population size is too small, then sufficient dimension is not available for searching effective solutions, and on the other hand, too large population size reduces the efficiency of the method. As all 1s in a chromosome give the words considered for representing the documents, so a vertical projection on the preprocessed document set with selected words is performed to obtain a sub-dataset associated to the chromosome. This sub-dataset is used to measure the fitness value of the chromosome.

2.2 Fitness Function

The proposed multi-objective genetic algorithm-based important word selection method uses two objective functions named as fitness function, based on which the quality of a candidate solutions is measured. The fitness functions are defined as: (i) number of words in consideration to represent the documents is minimized and (ii) external overlapped clustered probability index is maximized. The first fitness function is defined based on the number of 1s in the chromosome. The second fitness function is defined using the following procedure:

(a) In the training dataset, let us consider the number of document types or classes is s. The document set associated to the chromosome is first classified into s different groups based on the document types. Thus, a set of clusters $CL_1 = \{Cl_{11}, Cl_{12}, \ldots, Cl_{1s}\}$ are obtained. Next two steps give the second set of clusters CL_2.

(b) Cosine similarity is calculated for every pair of documents using Eq. 1, and thus a cosine similarity matrix is obtained. Two vectors with the same orientation have a cosine similarity of 1; higher the similarity value more similar the documents are and vice versa.

$$Sim(X, Y) = \frac{\sum_{i=1}^{n} X_i Y_i}{\sqrt{\sum_{i=1}^{n} X_i^2} \sqrt{\sum_{i=1}^{n} Y_i^2}} \tag{1}$$

where X_i and Y_i are ith components of documents X and Y, respectively. For each row of the similarity matrix, average similarity value is computed and the documents with similarity value greater than the average value are kept in a cluster. Thus, s number of clusters are generated, one for each document in the dataset.

(c) Obviously, all the s clusters are overlapped with each other and number of clusters is equal to the number of samples in the dataset. To obtain optimal set of overlapping clusters, cluster similarity index is computed between every pair of obtained clusters C_i and C_j using Eq. 2.

$$Cl(C_i, C_j) = \frac{|C_i \cap C_j|}{|C_i \cup C_j|} \tag{2}$$

Higher cluster similarity index implies more objects are overlapped between the clusters. The average similarity of each cluster with other clusters is computed. Then the cluster, say C_i, with the highest similarity index is considered, and the clusters with similarity greater than the average are merged with C_i. This process is repeated considering cluster with the next highest similarity index and continued until all objects are placed in at least one cluster. Thus, a set of t clusters $CL_2 = \{Cl_{21}, Cl_{22}, \ldots, Cl_{2t}\}$ are generated.

(d) Now, the overlapped external cluster validation index between two sets of clusters CL_1 and CL_2 is measured using the concept discussed in paper [4]. This validation index value ranges in [0–1]. Higher value indicates clusters obtained by both the clustering algorithms are more similar to each other.

2.3 External Population

In SPEA, to maintain the elitism property for allowing the best chromosome(s) from the current generation to carry over to the next generation directly, external population P' is also generated together with the internal population or simply population P. This guarantees that the quality of the solution obtained by GA will not be reduced from one generation to the other. Generally, the size of the external population is taken as one-fourth of the original population. The chromosomes of the external population are selected from the non-dominated pareto-optimal front obtained using fitness functions of the chromosomes in the current population. If the number of selected members for external population exceeds its size, then selected members are fed in an agglomerative hierarchical clustering algorithm and merging of clusters are done until number of clusters obtained is less or equal to the external population size. From each cluster, the chromosome closest to the mean of the cluster is chosen as member of P'.

2.4 Selection

During each successive generation, a portion of the existing population is selected to breed a new generation. Many popular selection methods such as roulette wheel selection, tournament selection, rank selection are widely used to select the best chromosomes during creation of mating pool. In the proposed algorithm, a binary tournament selection method is used, based on the strength assigned to the chromosomes as per the SPEA algorithm. The strength $S_1(ch_i)$ assigned to each chromosome ch_i

in external population is computed using Eq. 3.

$$S_1(ch_i) = \frac{n_i}{M+1} \tag{3}$$

where n_i is the number of current population members dominated by chromosome ch_i and M is the size of the population. So, the strength of a chromosome is directly proportional to the number of chromosomes dominated by it. The strength of each chromosome ch_j in internal population is defined using Eq. 4.

$$S_2(ch_j) = 1 + \sum_{i \in P' \wedge i \neq j} S_1(ch_i) \tag{4}$$

In Eq. 4, $i \neq j$ implies that the fitness of internal population member j is assigned as one more than the sum of the strength values of all external population members which weakly dominate j. So, according to SPEA algorithm the chromosome having less strength is considered as the best chromosome to be selected for the mating pool. This selection method is continued till the mating pool is filled.

2.5 Crossover

In the selection process, only the fittest chromosomes of the current population P are selected which gives some direction of searching solutions. For searching better solutions throughout the whole search space, offspring are created by crossover operation. Crossover is one of the basic operators of GA by which new chromosomes are generated to give the direction of searching solutions in the population toward local optima. As the main goal of GA is to make the population convergence, so crossover happens more frequently, generally in every generation. There are different types of crossover techniques, such as one-point crossover, uniform crossover, multi-point crossover, and in the proposed algorithm, single-point crossover with 0.8 crossover probability is considered for offspring generation.

2.6 Mutation

Mutation is a divergence operation used after crossover operation in GA. It is some-time necessary to break some chromosomes of a population out of a local optimum space and potentially discover a better optima space. As mutation is a divergence operator, it happens less frequently, so its effect is on less number of chromosomes in the population. Basically, mutation operation randomly selects a position in the chromosome and flips the bits in that position of the chromosome. The proposed algorithm performs mutation operation with mutation probability 0.02.

2.7 Elitism

There is a possibility in GA-based heuristic that some comparatively better solutions are removed from the population which may slower the process of finding the optimal solution, sometimes even make it quite difficult to achieve it. Elitism is the property used in GA to keep the copy of the better chromosomes and move them directly into the next generation. This can sometimes have a large dramatic impact on the performance of GA by ensuring that the GA does not waste time rediscovering previously discarded partial solutions. In our method, elitism is applied generating the external population set of chromosomes, where only the best non-dominated solutions are kept. The overall algorithm of the proposed method is presented in Algorithm 1.

Algorithm 1: TEXT CATEGORIZATION

Input: P = Population of size M; N = Length of each chromosome in P; *ArchiveSize* = External Population Size; G_{max} = Maximum number of generation; $P_{crossover}$ = Crossover probability; $P_{mutation}$ = Mutation probability;

Output: Important_words and clusters_of_documents

1 **repeat**
2   **foreach** $ch \in P$ **do**
3   Generate a set of clusters CL_1 and CL_2;
4   Compute *Cluster_index*;
5   *Objective_function_1* = Number of words in associated dataset ;
6   *Objective_function_2* = *Cluster_index*;
7   *Generation_Count* = 0;
8   Solve *Objective_function_1* and *Objective_function_2*;
9   S = set of non-dominated solutions;
10   **if** $|S| \leq$ *ArchiveSize* **then**
11   Move S to external population P';
12   **else**
13   Reduce S by SPEA algorithm and keep them in P';
14   Compute strength of each chromosome in P' by Eq. 3;
15   Compute strength of each chromosome in P by Eq. 4;
16   **repeat**
17   Apply tournament selection on P and P' to create P_{new};
18   **until** $|P| = |P_{new}|$;
19   **repeat**
20   Select chromosomes ch_1 and ch_2 randomly;
21   Apply single point crossover on ch_1 and ch_2 with $P_{crossover}$;
22   Generate two offspring of the parents;
23   Compute their *fitness values*;
24   Select the best two among parents and offspring;
25   **until** *crossover operation performs* $|P_{new}|/2$ *times*;
26   **foreach** $ch \in P_{new}$ **do**
27   Apply mutation on ch with probability $P_{mutation}$;
28   Compute *fitness values* of mutated offspring;
29   **if** *fitness values increases* **then**
30   Replace offspring by mutated one;
31 **until** *Generation_Count* = G_{max};
32 Find best chromosome ch_{best} of P';
33 *Important_words* = words associated to ch_{best};
34 *Clusters_of_documents* = clusters associated to ch_{best};
35 **return** *Important_words* and *Clusters_of_documents*;

Table 1 Parameters of the proposed algorithm

Parameter	Value
Population size (M)	300
Probability of crossover ($P_{Crossover}$)	0.8
Probability of mutation ($P_{Mutation}$)	0.02
Maximum number of successive generation (G_{Max})	1000

3 Experimental Results

The original dataset is copied, and a .csv file is created. In preprocessing step, stop words are removed from this .csv file. In the next step, stemming is done on this processed file. Term-document matrix is created with tf-idf method, which is finally used as input data for the proposed algorithm. These preprocessing and data preparation steps are implemented using R. The proposed algorithm is implemented in python and runs for 1000 generation on a desktop with Pentium core i5 processor. Experiments are carried out on benchmark Newsgroup20 dataset [1]. The dataset contains 1120 documents and 1500 features. The parameters used in the proposed algorithm are listed in Table 1. These parameters are selected after several test evaluation of the proposed algorithm on the dataset.

The proposed important words selection method (IWS) provides the optimal informative words, which can efficiently cluster the documents. To measure the effectiveness of the method, some existing feature selection methods like Correlated Subset Evaluation (CSE), Symmetrical Uncertain Attribute Evaluation (SUA), Relief Attribute Evaluation (REL), OneR Attribute Evaluation (ONE), Information Gain Attribute Evaluation (IGA), Gain Ratio Attribute Eval (GRA), and Correlation Attribute Evaluation (CoE) are used to reduce the dataset. The reduced datasets and the whole dataset (WAD) are used for various classifiers like (a) Naive Bayes classifiers: (i) Naive Bayes multinomial (NBMN), (ii) Naive Bayes multinomial text (NBMNT), (iii) Naive Bayes multinomial updateable (NBMNU); (b) function-based classifiers like (i) LOGISTIC classifier, (ii) SGDTEXT classifier, (iii) SMO classifier; (c) lazy classifiers like (i) LWL classifier, (ii) KSTAR classifier, (iii) IBK classifier; (d) meta-classifiers like (i) multi-class classifier (MULCLS), (ii) multi-scheme classifier (MULSCHM), (iii) voting classifier (VOTE); (e) tree-based classifiers like (i) DECISION STUMB, (ii) REEPTREE, (iii) LMT; and (f) rule-based classifiers like (i) ZEROR, (ii) JRIP, (iii) OneR to measure the classification accuracies using tenfold cross validation technique. All the feature selection techniques and classification methods are running using WEKA tool [2]. The results obtained by different methods are listed in Tables. 2, 3, 4, 5, 6, and 7. Tables show that, in most of the classifiers, proposed IWS algorithm gives better result than the other feature selection methods, which demonstrates the effectiveness of the method.

Table 2 Results for Bayes classifier

Classifier	NBMN (%)	NBMNT (%)	NBMNU (%)
IWS	87.76	64.75	87.09
WAD	87.10	64.52	87.09
CSE	87.09	64.52	87.09
SUA	87.09	64.52	87.09
REL	87.09	64.52	87.09
ONE	87.09	64.52	87.09
IGA	87.09	64.52	87.09
GRA	87.09	64.52	87.09
CoE	87.09	64.52	87.09

Table 3 Results for functions classifier

Classifier	LOGISTIC (%)	SGDTEXT (%)	SMO (%)
IWS	77.82	64.52	85.48
WAD	74.19	64.52	85.48
CSE	75.08	64.52	85.48
SUA	69.35	64.52	79.03
REL	70.96	64.52	80.64
ONE	72.58	64.52	83.87
IGA	69.35	64.52	79.03
GRA	69.35	64.52	79.03
CoE	74.19	64.52	83.87

Table 4 Results for Lazy classifier

Classifier	LWL (%)	KSTAR (%)	IBK (%)
IWS	81.04	78.47	83.87
WAD	74.19	64.52	77.42
CSE	80.64	80.64	83.87
SUA	74.19	64.52	80.64
REL	75.80	64.52	82.25
ONE	74.19	64.52	79.03
IGA	74.19	64.52	80.64
GRA	74.19	64.52	80.64
CoE	75.80	64.52	82.25

Table 5 Results for Meta classifier

Classifier	MULCLS (%)	MULSCHM (%)	VOTE (%)
IWS	77.41	64.52	64.52
WAD	74.19	64.52	64.52
CSE	75.80	64.52	64.52
SUA	69.35	64.52	64.52
REL	70.96	64.52	64.52
ONE	72.58	64.52	64.52
IGA	69.35	64.52	64.52
GRA	69.35	64.52	64.52
CoE	74.19	64.52	64.52

Table 6 Results for Tree classifier

Classifier	DECISION STUMB (%)	REEPTREE (%)	LMT (%)
IWS	79.41	79.03	87.09
WAD	77.41	69.35	77.41
CSE	77.41	79.03	87.09
SUA	77.41	74.19	79.03
REL	70.96	70.96	83.87
ONE	77.41	74.19	80.64
IGA	77.41	74.19	79.03
GRA	77.41	74.19	79.03
CoE	74.19	72.58	65.80

Table 7 Results for Rules classifier

Classifier	ZEROR (%)	JRIP (%)	OneR (%)
IWS	64.52	80.64	73.27
WAD	64.52	70.96	69.35
CSE	64.52	75.80	74.19
SUA	64.52	75.80	74.19
REL	64.52	74.19	69.35
ONE	64.52	70.96	74.19
IGA	64.52	70.96	74.19
GRA	64.52	80.64	74.19
CoE	64.52	80.64	69.35

4 Conclusion

The proposed SPEA-based IWS algorithm uses overlapped cluster validation index for defining one objective function for which a novel cosine similarity and cluster similarity-based overlapping clustering algorithm are applied in the paper. The experimental results show the usefulness of the method. As future work, the next version of SPEA algorithm, i.e., SPEA2 [14] will be used to remove various shortcoming of the SPEA algorithm. Along with the use of SPEA2 algorithm, we will try to use different external validity indices, such as RAND index and F-Measure for cluster evaluation. Finally, from the clusters of documents, the summerization of texts can be made.

References

1. Uci machine learning repository. https://archive.ics.uci.edu/ml/datasets/Twenty+Newsgroups
2. Using weka tool. http://www.cs.waikato.ac.nz/ml/weka/downloading.html
3. Arampatzis, A., der Weide, P.V., Koster, C., van Bommel, P.: Linguistically-motivated information retrieval. Encyclopedia of Library and Information Science (2000)
4. Campo, D., Stegmayer, G., Milone, D.: A new index for clustering validation with overlapped clusters. Expert Systems with Applications 64(1), 549–556 (2016)
5. Foltz, P.W.: Latent semantic analysis for text-based research. Behavior Research Methods, Instruments, & Computers 28(2), 197–202 (Jun 1996)
6. Harman, D.: How effective is suffixing? Journal of the American Society for Information Science 42(7), 7–15 (1991)
7. Hull, D.: Stemming algorithms: A case study for detailed evaluation. Journal of the American Society for Information Science 47(1), 70–84 (1996)
8. Jivani, A.G.: A comparative study of stemming algorithms. International Journal of Computer Technology and Applications 2(6), 1930–1938 (2011)
9. Krovetz, R.: Viewing morphology as an inference process. In: In Proceedings of the 16th Annual International ACM SIGIR Conference on Research and Development in Information Retrieval. pp. 191–202 (1993)
10. dos SantosEmail, J.C.A., Favero, E.L.: Practical use of a latent semantic analysis (lsa) model for automatic evaluation of written answers. Journal of the Brazilian Computer Society 21(21), 1–8 (November 2015)
11. Willett, P.: The porter stemming algorithm: then and now. Program 40(3), 219–223 (2006)
12. Xu, J., Croft, B.: Corpus based stemming using co-occurrence of word variants. ACM Transactions on Information Systems 16(1) (1998)
13. Zitzler, E., Thiele, L.: An evolutionary algorithm for multiobjective optimization: The strength pareto approach. Technical Report 43, Computer Engineering and Networks Laboratory (TIK), Swiss Federal Institute of Technology (ETH) Zurich, Gloriastrasse 35, CH-8092 Zurich, Switzerland. (May 1998)
14. Zitzler, E., Laumanns, M., Thiele, L.: SPEA2: Improving the strength pareto evolutionary algorithm. Tech. rep. (2001)

Short-Term Load Forecasting Using Genetic Algorithm

Papia Ray, Saroj Kumar Panda and Debani Prasad Mishra

Abstract Electrical power load forecasting has at all conditions been a basic subject in the energy trade. Load forecasting requires relative learning, reminiscent of neighborhood climate, and past load request information. The precision of load anticipating needs a huge impact on a power organization's system and making cost. Review load forecasting is along these lines essential, especially with the progressions happening inside the utility business in light of deregulation and dispute. A few outmoded approaches, for example, regression model, time approach model and pro framework have been proposed for without a moment's hesitation stack deciding by various levels of accomplishment. In this paper, ANN arranged through back development in the mix with the genetic algorithm is utilized. In back spread, the weights of neuron change as indicated by the edge plunge which may look out for close-by minima, so genetic algorithm is executed with backpropagation.

Keywords Artificial neural network (ANN) · Backpropagation (BP)
Short-term load forecasting (STLF) · Genetic algorithm (GA)

P. Ray
Department of Electrical Engineering, Veer Surender Sai University of Technology,
Burla 768018, Odisha, India
e-mail: papiavssut@gmail.com

S. K. Panda (✉)
Department of Electrical Engineering, Vignana Institute of Technology and Management,
Berhampur 761008, Odisha, India
e-mail: Saroj.panda89@gmail.com

D. P. Mishra
Department of Electrical Engineering, International Institute of Information Technology,
Bhubaneswar 751003, Odisha, India
e-mail: debani11@gmail.com

© Springer Nature Singapore Pte Ltd. 2019
H. S. Behera et al. (eds.), *Computational Intelligence in Data Mining*,
Advances in Intelligent Systems and Computing 711,
https://doi.org/10.1007/978-981-10-8055-5_76

863

1 Introduction

Forecasting is a wonder of comprehending what may happen to a framework in the following coming eras. In electrical power frameworks, there is an incredible requirement for precisely anticipating the load and vitality prerequisites since power eras and also circulation are an awesome budgetary risk to the state exchequer [1]. Precise load gauge furnishes framework dispatchers with convenient data to work the framework financially and dependably. It is additionally essential for accessibility of power is a standout among the most vital components of modern improvement, particularly for a growing nation like India. Care additionally must be taken that the vitality conjecture is neither excessively preservationist nor excessively idealistic [2]. On the off chance that the estimate is excessively preservationist, at that point it is likely that the creating limit may miss the mark regarding the real power request, bringing about confinements being forced on the power supply that might be negative to the monetary advancement of the nation. Then again, if the estimate is excessively hopeful, it might lead, making it impossible to the making of an overabundance producing limit, bringing about greater venture without getting any prompt returns [3, 4]. A creating nation like India cannot generally bear the cost of both of the two above conditions inferable from extensive weight on its constrained money-related assets. In this manner, there is an exceptionally solid requirement for creating electrical load estimating models utilizing the most recent accessions procedures.

Different determining models have been produced so far utilizing different graphical and measurable approaches, the auto backward and moving normal models being the most prevalent. Computerized reasoning procedures, for example, master frameworks and neural systems have demonstrated promising brings about numerous system. Recent progress in the applications of Artificial Neural Networks (ANN) technology to power systems in the areas of forecasting has made it possible to use this technology to overcome the limitations of the other methods used for electrical load forecasting [5, 6]. This is because of the way that as opposed to depending on unequivocal standards or numerical capacities between past load and climate change, neural systems draw a connection among information and yield information [7, 8]. In this manner, the neural systems that digress from depending on measurable models and expansive chronicled databases hold a decent guarantee with the end goal of load anticipating. Throughout the years, the utilization of artificial neural network (ANN) in control businesses has been developed in acknowledgment. This is an account of ANN is equipped for catching procedure data in a discovery way. Given sufficient information, yield information, the ANN can rough any nonstop capacity for subjective precision. This has been demonstrated in different fields, for example, plan recognition, calculation, framework recognizable authentication, forecast, standard train, fault detection and others.

The improvement of a decent ANN relies upon a few components. The principles considered are identified with the information being utilized. This is reliably with other discovery models where display qualities are unequivocally impacted by the nature of information utilized. The second component is the system engineering or model structure. The diverse system design brings about various estimation executions. Generally, multilayer discernment and its fluctuations are broadly utilized as a part of process estimation. The third component is the model size and many-sided quality. What is required is a precise model with best parameter settings. This is on account of a little system may not be ready to speak of the genuine circumstance because of its constrained capacity, while an expansive system may over-fit commotion in the preparation information and neglect to give great speculation capacity. At long last, the nature of a procedure shown is additionally firmly reliant on organizing preparing. This stage is basically a distinguishing proof of model parameters that fit the given information and is maybe an essential variable among all [9]. A few endeavors have been proposed by different analysts to reduce this preparation issue. These incorporate forcing limitations on the pursuit space, restarting preparing at numerous arbitrary focuses, altering preparing parameter, and rebuilding the ANN design [1]. Notwithstanding, some methodologies are issues particular and not all around to acknowledging and diverse specialists have a tendency to incline toward various approaches. Among these, one of the additionally encouraging strategies is by presenting adjustment of system, preparing utilizing genetic algorithm (GA). In this work, the GA-BPN demonstrate is utilized for removing the best weight networks for various layers of BPN along these lines estimating the future power request all the more precisely. Thus, this work gives advancement of association weights in ANN utilizing GA as methods for enhancing the versatility of the determining.

2 GA for Optimization

Optimization is the study of discovering choices that fulfill given imperatives and meet a particular objective at its ideal esteem. In building, imperatives may emerge from physical impediments and specialized particulars; in business, requirements are frequently identified with assets, including labor, gear, expenses, and time. The goal of worldwide advancement is to locate the "most ideal" arrangement in nonlinear choice models that much of the time have various problematic (nearby) arrangements. Without worldwide advancement apparatuses, architects and specialists are regularly compelled to agree to doable arrangements, frequently disregarding the ideal esteems. In down to earth terms, this infers sub-par outlines and operations, and related costs regarding unwavering quality, time, cash, and other assets. The established improvement systems experience issues in managing worldwide advancement issues. One of the principle reasons for their disappointment is that they can without much of a stretch be ensnared in neighborhood minima. Additionally, these strategies cannot produce or even utilize the worldwide data expected to locate the worldwide least for a capacity with different neighborhood minima [10].

The genetic algorithm solves optimization problems by mirroring the standards of organic development, over and over adjusting a populace of individual focuses utilizing rules displayed on quality blends in organic multiplication. Because of its irregular nature, the hereditary calculation enhances the odds of finding a worldwide arrangement. Along these lines, they demonstrate to be extremely productive and stable in hunting down worldwide ideal arrangements. It understands unconstrained, bound-obliged, and general advancement issues, and it does not require the capacities to be differentiable or nonstop [11].

The principal GA cycle made for this issue includes the going with strides:

1. Construct an initial population of chromosomes.
2. Select the mating pair of chromosomes.
3. Accomplish fitness scaling if vital.
4. Choice the mating sets of chromosomes.
5. Make new offspring through crossbreed and change operations.
6. Edge amasses for the general population to come.
7. On the occasion that procedure has focalized, restore the best chromosome as the arrangement, generally; go to step 2.

There are a couple of assortments of the basic genetic computation that rise up out of different modifications, which improve the execution on sensitive issues. Every now and again, a scaling of the execution measure is given to growing contrasting qualities in a mass. This methodology keeps up a vital separation from dominance by a super individual that would provoke to a joining at an adjacent ideal. The probabilistic idea of the assurance strategy gives an injection of creation even to the weakest individual from the populace. Correspondingly, potentially the best performing part (top of the line) will not be accessible for the general population to come, either in view of non-decision for mating or in view of the fundamental changes taking after crossbreed and transformation. In any interest, it is always charming to copy the tip best structure set up into the general population to come. In Appendix, Table 4 shows the GA-based backpropagation parameter.

3   The Proposed ANN Model

Artificial neural networks (ANNs) are a kind of computational knowledge propelled by the way the natural frameworks of people, for example, the mind procedure data. The human cerebrum is comprised of neurons, which are interconnected by dendrites and gather data by means of this association. ANNs are comprised of various straightforward and very interconnected handling components called neurons. Every one of the neurons in the mind works as one to ensure that all data got is prepared as effectively and precisely as could be expected under the circumstances. The manufactured neurons attempt to mimic this sort of conduct shown by the genuine neurons in the mind. ANNs learn by illustration and are designed for specific classes of issues or applications through a learning framework. There are various topologies that are used in ANNs, for example encourage forward systems, repetitive systems. The usually utilized ANNs are the nourish forward ANNs where the info signals are propagated from input layer through a concealed layer to the output layer. Recurrent Neural Networks (RNNs) or feedback ANNs have signals going in both ways inside the system [12].

Figure 1 shows a bolster forward neural system which is the most normally utilized neural system engineering for short-term load forecasting estimating. It involves an input vector which would for the most part contain inputs comprised of recorded load information, chronicled and determined climate parameters, day sorts, and additionally other load-influencing factors. It additionally contains a shrouded layer and after that a yield layer, and typically one output is adequate; however, this can be designed as required. In Appendix, Table 3 gives the conventional BP parameters.

The hybrid rate, change rate, and time opening were 0.9, 0, and 0.96, independently. The elitist method and grouped qualities heads were in like manner used. The fundamental masses made were fixed to 20, and the chromosome size was fixed to 16 bits. The sigmoid limit is presented in Eq (1) as given below.

$$F(X) = 1/1 + \exp(-x) \tag{1}$$

The sigmoid limit is applied as the start work between the data layer and the hid layer. The scaled load yield was procured by direct extension of the scaled yields and a slant regard from the disguised neurons. For the model, the architecture of the network was initialised to be a three-layer fully connected feed-forward network first and the weights between the layers were allowed to adjust according to the constraints given and the function to be optimised. The mean average percentage error (MAPE) was implemented as the fitness work that is updated by the genetic computation. The mathematical expression of the MAPE is given in Eq (2).

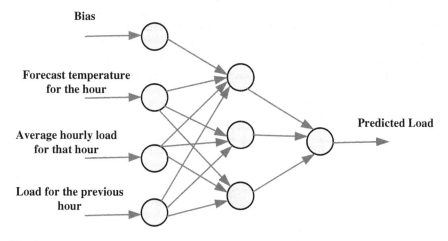

Fig. 1 ANN model

$$\text{MAPE} = 1/N(\sum NI = 1[|\text{ActualI} - \text{ForecastI}|/\text{ActualI} \times 100]) \qquad (2)$$

where N is the total number of hours, ActualI is the genuine load at the hour I, and ForecastI is the forecast value of the load at that hour. The weights between the layers were obliged to fluctuate in the vicinity of −4 and 4, and the anticipated yields were compelled to differ in the vicinity of 0 and 2. As the learning advanced, it was watched that a portion of the system associations was dispensed with by the hereditary calculation, as they were not found to have any significant effect on the yield. Each of the neural systems, hence advanced, was different from the underlying system appeared in Fig. 1.

The approximation of the parameters is known as the "preparation" of the system and is finished by the minimization of a misfortune work (for the most part a quadratic capacity of the yield mistake). Numerous enhancement techniques have been adjusted for this errand. The primary preparing calculation to be formulated was the back spread one, which practices a steepest drop strategy in light of the calculation of the angle of the misfortune work regarding the system parameters (i.e., the motivation behind why the initiation capacities must be differentiable). In load calculating system, this essential type of multi-layer feed forward network appeared above is as yet the most well known. In any case, there are countless plans, which may be appropriate for other applications. Since quantitative determining depends on removing designs from seeing past occasions and extrapolating them into the future, one ought to anticipate that NNs will be a great contender for this assignment. In fact, NNs are exceptionally appropriate for it, for no less than two reasons. First, it has been formally shown that NNs can inexact numerically any consistent capacity to the coveted precision. Secondly, NNs are information-driven strategies, as in it is a bit much for the specialist to propose speculative models and afterward assess their parameters. Given an example of information and yield vectors, the NNs can consequently delineate the connection between them. As these two qualities recommend, NNs should end up being especially valuable when one has a lot of information, however little from the earlier learning about the laws that oversee the framework that created the information. As far as hypothetical research in forecasting, NNs have advanced from figuring direct gauges toward processing both confidence in terms and contingent likelihood densities. As far as viable applications in forecasting, the achievement of NNs appears to rely upon the sort of issue under consideration.

4 Result and Discussion

The contextual investigation has been done for Xingtai Power Plant in Hebei territory, China, to estimate Xingtai's 24 h STLF in one day. The informational collection utilized for this review comprises of hourly load and climate circumstance information over the period June 10, 2006 to June 30, 2006. The gauging exactness of the customary BP model is analyzed in view of similar demonstrating periods. The information is isolated into three informational collections: the preparation informational index (11 days, from June 10 to June 20), the approval

Table 1 Training, validation, and testing data sets of GA-BP model

Data sets	Periods
Training data	6.10–6.20
Validation data	6.21–6.29
Testing data	6.30

Table 2 Actual and forecasted load

Time (hr)	Actual load in (MW)	GA-based BP forecasted load (MW)
1	943	942
2	914	912
3	907	880
4	875	880
5	873	870
6	872	880
7	931	930
8	976	990
9	1062	1050
10	1144	1139
11	1213	1162
12	1263	1219
13	1231	1210
14	1196	1162
15	1150	1140
16	1190	1190
17	1212	1239
18	1231	1240
19	1223	1240
20	1228	1202
21	1245	1219
22	1317	1279
23	1214	1202
24	1081	1090

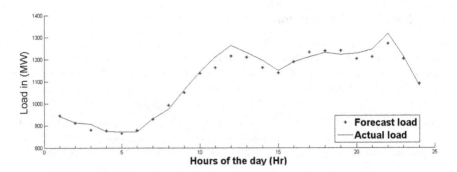

Fig. 2 Actual load and GA-BP model

informational collection (9 days, from June 21 to June 29), and the testing informational collection (1 day, June 30). The three informational indexes are recorded in Table 1.

Table 1 compresses the load information for this period. This examination uses Xingtai's load information to think about the gauging exhibitions of the GA-BP calculation with the customary BP show. The aggregate load values from June 10, 2006 to June 20, 2006 fill in as test information. Absolutely, 21 stack information for the Xingtai's load is accessible, as recorded in Table 2.

Before preparing the system, the example information ought to be standardized. Here, "MATLAB 7" is utilized to standardize the example information. Keeping in mind the end goal to spare the constrained space, this procedure is excluded.

The evaluation system contains 27 input parameters, so the input nodes $n1 = 27$ and output node $n3 = 24$, using the empirical formula, the hidden layer $n2 = \sqrt{(n1 \cdot n3)} + a$ where "a" is constant in the range 1–10. So, $n2 = 26$. Therefore, the BP network is set up in the order of 27-26-24.

5 Result Analysis

It is observed from Fig. 2 and Table 2 that the MAPE obtained by GA-based BP model is 0.05%. Thus, GA-based BP mode should be used for short-term load forecasting (STLF).

6 Conclusion

So, artificial neural network (ANN) technique for short-term load forecasting (STLF) is good. In this paper, STLF using a genetic algorithm (GA) is described. The data are taken from the Xingtai Power Plant in Hebei territory. The result

Table 3 Conventional BP parameters

Network type	MLPNN
Training algorithm	Backpropagation
Number of layers	3
Hidden layer neurons	60–80
Hidden layer activation function	Logsig, tansig
Output layer activation function	Purely
Training parameter goal	4 * 10 − 9
Performance function	MAPE
Epochs	10000
Learning rate	0.1

Table 4 GA-based backpropagation parameter

Population size	20
Elitism	0.96
Crossover	0.9
Mutation	0
Fitness function	F = 1 / (1 + MAPE)
Number of generation	20

shows that GA-based backpropagation model is good for STLF because the GA-based BP model is 0.05% as compared to other artificial techniques.

APPENDIX

See Tables 3 and 4.

References

1. Moghram, Rahman, S: Analysis and Evaluation of Five Short-Term Load Forecasting Techniques. IEEE Trans. PowerSystems, Vol. 4, No. 4, Oct. 1989, pp. 1484–1491.
2. Hayes, B; Madrid, Spain; Gruber, J; Prodanovic, M: Short-Term Load Forecasting at the local level using smart meter data. Power Tech, 2015 IEEE Eindhoven, June 29, 2015-July 2, 2015, pp: 1–6.
3. Papalexopoulos, A-D; T. Hesterberg, T-C: A Regression- Based Approach to Short-Term System Load Forecasting. IEEE Trans. Power Systems, Vol. 5, No. 4, Nov. 1990, pp. 1535–1547.
4. Hagan, MBehr, S-M: The Time Series Approach to Short-Term Load Forecasting. IEEE Trans. Power Systems, Vol. 2, No. 3, Aug. 1987, pp. 785–791.
5. Grzegorz, Dudek, G: Pattern-based local linear regression models for short-term load forecasting. Electric Power Systems Research, Volume 130, January 2016, Pages 139–147.

6. Yang; Tzer, H; Huang, C-M: A new short term load forecasting approach using self organizing fuzzy ARMAX models. IEEE Trans. Power Systems, Vol.13, Issue.1. 1998, pp. 217–225.

7. Pham, D,T; Karaboga, D: Intelligent Optimization Techniques, Genetic Algorithm, Tabu Search, Simulated Annealing and Neural Networks: Springer – Verlag, 2000.

8. Dash, P,K; Satpathy, H,P; Liew, A,C; Rahman, S: A Real time Short time Load Forecasting system using functional link Network. IEEE Power Trans. Systems, Vol.12, No.2, May 1997, pp. 675–680.

9. Baoa, Z,H,Y; Xinoga, T; Chiong, R: Hybrid filter – wrapper feature selection for short-term load forecasting. Engineering Applications of Artificial Intelligence, Volume-40, April 2015, pp. 17–27.

10. Ding, N ;RTE, D,E,S; Cedex, V; Benoit, F,C; Foggia, G; Besanger,Y: Neural Network Based Model Design for Short Term Load Forecast in Distribution System. IEEE Trans. On Power System. (Volume:31, Issue:1), Jan 2016, pp. 72–81.

11. Ray, P; Mishra, D: Signal Processing Technique based Fault Location of a Distribution Line, 2nd IEEE International Conference on Recent Trends in Information Systems (ReTIS), July 2015, pp. 440–445.

12. Ray, P; Mishra, P: Artificial Intelligence based Fault Location in a Distribution System, 13th International Conference on Information Technology (ICIT), Dec 2014, pp. 18–23.

A Dynamic Bottle Inspection Structure

Santosh Kumar Sahoo, M. Mahesh Sharma and B. B. Choudhury

Abstract In our market, most of the products are available in jars or bottles. So in view of maintaining proper specification of a particular bottle, the same should be properly investigated. The proposed bottle inspection has been concentrated through an artificial intelligent (AI) model and the performance of the said also evaluated. For this analysis, about 5000 bottle models are taken and their different properties have been considered for meeting large information to and from a data set, out of which they are categorized into two classes like defect-free and defective bottles. For analysis, an artificial intelligent scheme has been followed along with vision builder simulation tool which is carried out with a core i3 processor.

Keywords Artificial intelligent (AI) · Vision builder simulation tool
Machine vision (MV) system

1 Introduction

The vision processes are attentive toward identifying real items in an image and assigning properties to that objects. Researchers are adding vision structures at different applications basically industries to optimize cost, higher efficiency, and consumer gratification. All machine vision (MV) system comprises a mixture of

S. K. Sahoo (✉) · M. Mahesh Sharma
Department of Electronics and Instrumentation Engineering,
CVR College of Engineering, Vastunagar, Mangalpalli, Hydrabad 501510, Telengana, India
e-mail: Santosh.kr.sahoo@gmail.com

M. Mahesh Sharma
e-mail: mahesh95533@gmail.com

B. B. Choudhury
Department of Mechanical Engineering, Indira Gandhi Institute of Technology,
Sarang 759146, Odisha, India
e-mail: bbcigit@gmail.com

© Springer Nature Singapore Pte Ltd. 2019   873
H. S. Behera et al. (eds.), *Computational Intelligence in Data Mining*,
Advances in Intelligent Systems and Computing 711,
https://doi.org/10.1007/978-981-10-8055-5_77

hardware and software for acquiring and processing the captured image of an object.

The MV system includes different mechanisms for proper implementation of assigned task. Hereafter, the prime rudiments of a machine vision system are image sensor, illumination arrangement, processor with image processing software, image processing hardware and, final control element like robot or robotic arm. By considering the above factor, Tkayuki et al. [1] developed a communication robot used in shopping mall by using the machine vision techniques. In the developed structure, the robot has been used as a communication medium in a mall to interact with the customers in a pleasant manner. Mahdi et al. [2] suggested a unique method intended for sorting of raisins through machine vision system. Niko et al. [3] used a MV system for automatic quality control of welded rings by considering different image processing methods.

Victores et al. [4] verified a robot-aided tunnel inspection and maintenance system by MVI structure and proximity sensor integration. Cubero et al. [5] established advances in machine vision applications for automatic identification of fruits and vegetables in food industries using two different classifiers. Fuzzy K-nearest neighbor in addition with multi-layer perceptron neural network classifier is used to classify the radiographic images. Tobias et al. [6] described a MV system for estimating the size by weight limestone particles. This arrangement comprises of a PC, a charge-coupled device as camera and a machine-driven controller. An anticipated model permits fast, precise, reproducible, and robust 100% inspection of an items and also confirms high end of product quality. Li et al. [7] established an effective machine vision system for automatic classification of microcrack in eggshell by an application of different algorithms. The proposed scheme follows the image processing by considering the different aspects like color abnormality condition, broken caps, and size.

2 Methodology for Bottle Inspection System

In the first stage, bottle image to be inspected has captured by high-resolution smart camera. Noise in the captured image is reduced through image processing techniques and image quality has improved for further processing. The region of interest (defective parts of the bottle) is selected on the original image by cropping mechanism and then converted into grayscale image. The detailed methodology is shown in Fig. 1. In this research work, a more general noise model, where a noisy pixel is taken as an arbitrary value in the dynamic range according to some underlying probability distribution is introduced. Let $O(i, j)$ and $A(i, j)$ denote the intensity value of the original and the noisy image at position (i, j), respectively. Then for an impulse noise model with error probability 'P' is described as in Eq. (1) [6].

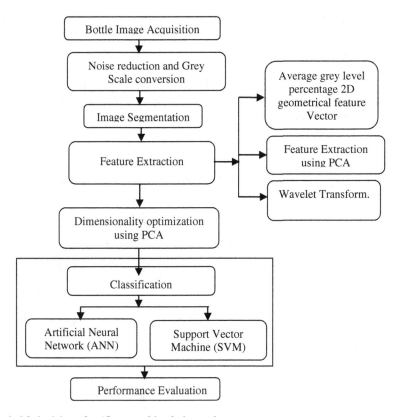

Fig. 1 Methodology for AI-centered bottle inspection system

$$A(x) = \begin{cases} O(i,j) & \text{with probability } 1-P \\ \eta(i,j) & \text{with probability } P \end{cases} \tag{1}$$

Here $\eta(i,j)$ is an identically distributed and independent random process with an arbitrary underlying probability density function. As histogram of original image O (i, j) is evaluated with a 'm' bins. So, the Probability 'P' is expressed [4] as $P = \frac{n_K}{\sum_{K=1}^{m} n_K}$, where n_K = Value at Kth bin, m = Total numbers of bins. This noise can be modeled with an independent, additive model, where the noise $\eta(i,j)$ has a μ mean Gaussian distribution described by its standard deviation, or variance.

The one-dimensional Gaussian distribution has the density function described as in Eq. (2)

$$G(x) = \frac{1}{\sqrt{2\pi}\sigma} e^{\frac{(x-\mu)^2}{2\sigma^2}} \tag{2}$$

where 'σ' is the standard deviation of the distribution, and 'μ' is the expectation of the distribution. From engineering point of view, the discrete wavelet analysis is a two-channel digital filter bank composed of the low-pass and the high-pass filters, iterated on the low-pass output. The low-pass filtering yields an approximation of a signal (at a given Scale), while the high-pass filtering yields the details that constitute the difference between the two successive approximations. Figure 2 shows a typical discrete wavelet two-level filter bank and Fig. 3 represents a typical two-level DWT for de-noising of a bottle sample.

Then the bottle image is segmented from the image background by an application of segmentation methods. After estimating all adaptive features by extracting the features, they are combined to form a dataset using mathematical concepts such as average grayscale two-dimensional feature vector, wavelet transform, and principal component analysis (PCA). These extracted features are considered as input variables and types of defective bottles are considered as output variables in the classification of defect-free bottle. In the classification stage, the intelligent techniques like artificial neural network (ANN) trained by back propagation algorithm, differential evaluation algorithm and support vector machine are used to classify the images as per predefined dimensions. All the three adaptive features and two training methods of ANN are employed one by one in both defective and defect-free bottle images. Feature numbers are reduced from 5000 to 2500 by

Fig. 2 A typical discrete wavelet filter bank

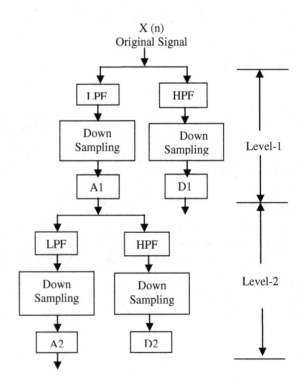

(a)   **(b)**

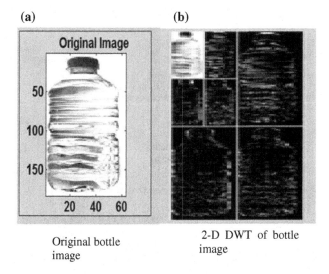

Original bottle image   2-D DWT of bottle image

Fig. 3 Typical two- level DWT for de-noising of a bottle sample

principal component analysis (PCA) for better computation. During the training and testing, the concentrated structures are applied as input to the classifier.

Again the classification is also performed through vision builder simulation window for validation purpose. This vision builder (VB) along with laboratory view window is used for automated inspection to solve visual inspection tasks including inspection, parts presence, and counting. Results of both the schemes are compared and final outcomes are conveyed. During the experiment, factors similar to illumination, focal length, and magnification factor of camera and workpiece positions are maintained constant throughout training and testing phase.

3 Experimental Setup for Based Bottle Inspection Model and Study of Its Outcome

Vision-centered scrutiny arrangement has a pair of innovative skills for contact-less measurement and inspection. These apparatus incorporates multitude methodologies including digital imaging, integrated circuit technology, embedded systems and software. Photographic view of machine vision inspection system is shown in Fig. 4. The inspection of bottle using the adaptive feature extractions with wavelet transform trained by support vector machine and adaptive model has been considered in this experimental work. The image of the bottle is grabbed by a high-resolution camera and sent to the personal computer through frame buffers.

The acquired capture image has been processed for noise optimization and feature extraction by using analytical tool like wavelet transforms. For analysis

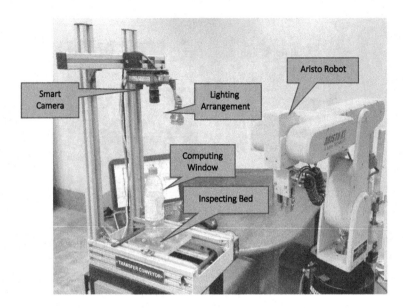

Fig. 4 Photographic view of AI-based bottle inspection system

point of view, the extracted features obtained from wavelet scheme have been trained again and classified for the desired response. Though there is not suitable standard for database available for bottle so 5000 images have been considered and resized in a standard dimension to form a big database.

During inspection interval, there is a chance for improper illumination and position changes of an object due to different factors, so that a sensory arrangement is implemented. Generally, the investigation is performed within the room at 60 W bulb lighting having illumination intensity of 50 lx and 500 lx, respectively. Figure 5 indicates the error variation at classification stage with the proposed arrangement.

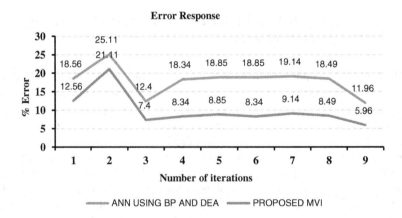

Fig. 5 Error response of MVIS at classification stage

The term classification here refers to the technique of categorizing an object depending on their attributes. This process needs a good quality of training data or information of an inspecting object which is obtained earlier. Considering the extracted features of bottle object, the proper classification related to defective and defect free is decided. In this classification stage, the rate of computational speed is compared by the different approach with the proposed sensor-centered MVI system. The comparison of computational speed is discussed here for proper evaluation. The capability of the proposed vision system is analyzed by comparing their performances in terms of three adaptive features. Single and multiple images are considered for the experiment. Table 1 indicates that the neural networks with sensor module in the three types of feature extraction method have a higher success ratio than the without sensor module method. In multiple zone images, wavelet-based feature extraction has produced the success ratio of 97.5% but the real-time inspection rate is 1.48 bottles/s. In average grayscale 2D feature extraction method has produced success ratio of 95% but the real-time inspection rate is 1.73 bottles/s.

Table 2 reveals that the neural networks using with sensor module in the three types of feature extraction method have higher success ratio than that of the without sensor module. With sensor module, wavelet-based feature extraction has produced the success ratio of 97.5% but the real-time inspection rate is 2.22 bottles/s. In average grayscale 2D feature extraction method has produced success ratio of 95% but real-time inspection rate is 2.5 bottles/s.

Figure 5 reveals that the proposed model provides better response than the other MVI system having without sensory arrangement.

Vision inspection system using machine vision with adaptive feature extraction is very much useful to inspect the quality level for inspection of bottles. It examines the defective bottles using artificial neural network and vision builder simulation platform. In this thesis, the defective and defect-free bottles have been inspected by considering with and without sensor module scheme, three feature extraction methods as well as the simulation platform.

3.1 Comparison of Overall Performance

The overall performance of the developed vision systems for with and without sensor module implementation is compared in terms of the three adaptive features. The complete comparison is shown in Table 3 and Fig. 6.

Ultimately, the performance comparisons between all methods provide information about the variations due to the selection of feature extraction method as well as training method of decision-making algorithm and types of image. From the above figure, it is concluded that sensor implementation over the MVI system has better efficiency than the other for estimated values of extracted features.

Table 1 Performance analysis of proposed vision inspection system by ANN using BP for defective bottle scrutiny

Sl no	Parameters	Without sensor module			With sensor module		
		AVG grayscale 2D feature vector	PCA-based features	Wavelet-based features	AVG grayscale 2D feature vector	PCA-based features	Wavelet-based features
1	Classification success ratio (%)	92.57	93.75	95	95	96.25	97.5
2	Failure rate (%)	6.256	5	3.75	3.75	2.25	2.5
3	False alarm rate (%)	1.254	1.25	1.25	1.25	1.25	0
4	Success ratio (%)	92.5	93.75	95	95	96.25	97.5
5	Real-time inspection rate (detected/s)	1.9	1.77	1.63	1.73	1.63	1.48

Table 2 Performance analysis of proposed vision inspection system by ANN using DEA for bottle inspection

Sl no	Parameters	Without sensor module			With sensor module		
		AVG grayscale 2D feature vector	Gaussian-based features	PCA-based features	AVG grayscale 2D feature vector	Gaussian-based features	PCA-based features
1	Classification success ratio (%)	92.57	93.75	95	95	95	97.5
2	Failure rate (%)	6.256	5	3.75	3.75	3.75	2.5
3	False alarm rate (%)	1.254	1.25	1.25	1.25	1.25	0
4	Success ratio (%)	92.5	93.75	95	95	95	97.5
5	Real-time inspection rate (detected/s)	2.85	2.58	2.5	2.5	2.42	2.22

Table 3 Comparison of overall performance

Sl no	Feature extraction method	Overall performance of VBIS	
		Without sensor (%)	With sensor (%)
1	2D feature vector	92.50	95.00
2	PCA features	93.25	96.00
3	Wavelet features	95	98

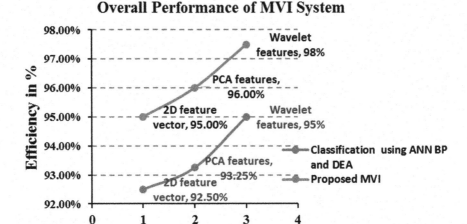

Fig. 6 Performance comparisons of proposed vision inspection system by ANN using BP and DEA for defective bottle inspection using with and without sensor implementation in MVI system

3.2 Comparison of Computational Time

The overall computational time of the developed vision-based inspection system model is compared to the three adaptive features. Sensor and without sensor-centered MVI models in both bottle images are considered for this evaluation along with the simulation platform like vision builder. The complete comparisons are shown in Table 4 and Fig. 7.

Table 4 Comparison of computational time of proposed vision inspection system for bottle inspection

Sl no	Feature extraction methods	Overall performance					
		BP		DEA		Gained computational time (s)	
		Without sensor	With sensor	Without sensor	With sensor	Without sensor	With sensor
1	2D feature vector	42	46	28	32	33.33	30.43
2	PCA features	45	49	31	33	31.11	28.65
3	Wavelet features	49	54	32	36	34.69	30.33
Average gained computational time (s)						33.04	29.80

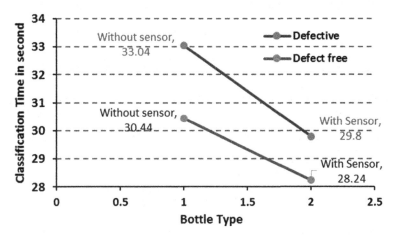

Fig. 7 Comparison of computational time of proposed vision inspection system for bottle inspection

4 Conclusions

AI-centered MVI system is one of the best suitable schemes for proper inspection of the defective bottles. This vision inspection system provides a good technology for inspecting the imperfections in bottles. Based on the comparison of computational times of all methods, it may be asserted that ANN using DEA computational time is comparatively less than that of ANN using BP. The percentage of average computational time gained in the model in bottle for bottle images without and with sensor implementations are 33.04 and 29.80, respectively. The computational time is more for bottle classification with sensor implementation than the without sensor implementation.

Again inspection of bottle using MVI system enables the user to examine the imperfections in the bottles in a natural manner which is very similar to standard inspection methods. It categorizes the different types of bottles as per standards. The MVI system is capable of classifying the bottle imperfections as per standard. The range of classification success ratio is 91.25–97.5%. Eventually, it concluded that the AI-based MVIS is most suitable for inspection of quality level for imperfections in bottles with an overall efficiency of 98% as shown in Fig. 6.

References

1. Takayuki Kanda, Masahiro Shiomi, Zenta Miyashita, Hiroshi Ishiguro: A communication robot in a shopping mall. IEEE transactions on Robotics (2010) Vol. 26(5) 897–913.
2. Mahdi Abbasgolipour, Mahmoud Omid, Alireza Keyhani.: Sorting Raisins by Machine vision system. Modern Applied Science (2010) Vol. 4(2) 49–60.
3. Niko Herakovic, Marko Simic, Francelj Trdic, Jure Skvarc.: A machine vision system for automated quality control of welded rings. Machine vision and Applications, Springer, (2010) 1–15.
4. J. G. Victores, S. Martinez, A. Jardon, and C. Balaguer.: Robot aided tunnel inspection and maintenance system by vision and proximity sensor integration. Automation in construction, Elsevier (2011) Vol. 20, 629–636.
5. Sergio Cubero, Nuria Aleixos, Enrique Molto.: Advances in machine vision applications for automatic inspection and quality evaluation of fruits and vegetables. Food Bioprocess Technology, Springer, (2011) Vol. 4, 487–504.
6. Tobias Andersson, Matthew J Thurley, Johan E. Carlson.: A machine vision system for estimation of size distributions by weight of limestone particles. Minerals Engineering, Elsevier, (2012) Vol. 25, 38–46.
7. Yongyu Li, Sagar Dhakal, Yankun Peng.: A machine vision system for identification of micro crack in egg shell. Journal of Food Engineering, Elsevier, (2012) Vol. 109, 127–134.

Feature Selection-Based Clustering on Micro-blogging Data

Soumi Dutta, Sujata Ghatak, Asit Kumar Das, Manan Gupta and Sayantika Dasgupta

Abstract The growing popularity of micro-blogging phenomena opens up a flexible platform for the public as communication media for the public. For any trending/non-trending topic, thousands of post are posted daily in micro-blogs. During any important event, such as natural calamity and election, and sports event, such as IPL and World Cup, a huge number of messages (micro-blogs) are posted. Due to fast and huge exchange of messages causes information overload, hence clustering or grouping similar messages is an effective way to reduce that. Less content and noisy nature of messages are challenging factor in micro-blog data clustering. Incremental huge data is another challenge to clustering. So, in this work, a novel clustering approach is proposed for micro-blogs combining feature selection technique. The proposed approach has been applied to several experimental dataset, and it is compared with several existing clustering techniques which results in better outcome than other methods.

Keywords Clustering · Feature selection · Micro-blogs

1 Introduction

In recent times, micro-blogging phenomena open up huge source of real-time information for the researchers. During any important event, such as natural calamity and election, and sports event, such as IPL and World Cup, thousands of messages are posted in micro-blogging sites. Due to fast information exchange nature of micro-blogging sites, it causes information to be overloaded. Micro-blogging post contains at most 140 characters. It also contains noisy and redundant data. For our experiment

S. Dutta (✉) · S. Ghatak · M. Gupta · S. Dasgupta
Institute of Engineering & Management, Kolkata 700091, India
e-mail: soumi.it@gmail.com

S. Dutta · A. K. Das
Indian Institute of Engineering Science and Technology Shibpur, Howrah 711103, India

© Springer Nature Singapore Pte Ltd. 2019
H. S. Behera et al. (eds.), *Computational Intelligence in Data Mining*,
Advances in Intelligent Systems and Computing 711,
https://doi.org/10.1007/978-981-10-8055-5_78

purpose, we have considered Twitter streaming dataset which are crawled using the Twitter API [14].

Twitter felicitates the searching technique by keywords or topic name to identify related tweets. It is not possible for any user to go through all the post/tweets who wants to know the outline of a topic. In such a scenario, an effective way to reduce the information load on the user is to group similar posts, so that the user might see only few messages in each cluster. Another challenge of micro-blogging data is large volumes, which intend more time to cluster the data into subgroups. So selected features can be identified from each cluster which represent the characteristics of the cluster. Feature selection focuses on reduction of overfitting. Feature selection employs dimension reduction for a given dataset where selected features are the important features which are sufficient to represent the dataset independently. This reduced dimension of the dataset can also reduce the clustering time effectively. Clustering also makes easier the data summarization task which is another well-established problem in information retrieval.

In the proposed clustering approach, dataset is preprocessed first. Latent Dirichlet Allocation (LDA) [1] topic modeler is a generative process that can be used to identify the features capable of identifying a topic in the dataset and returned list of reduced features which can be used to represent each topic in the entire dataset. Now, for cluster identification, hamming distance is measured between each topic-feature vector and a data tuple, and minimum distance value cluster is identified as destination cluster for that data tuple. Using this approach, entire dataset can be clustered into multiple groups.

The proposed clustering approach is applied to micro-blogs dataset related to four disaster events. Few classical clustering methods such as K-means and hierarchical are also applied to the same dataset. The performance of the different algorithms is evaluated using the standard clustering index measure. As a whole, the proposed approach achieves better performance than the classical methods for micro-blogging dataset.

The rest of the paper is organized as follows. A short literature survey on clustering micro-blogging dataset is presented in Sect. 2. Section 3 describes the proposed clustering approach. The micro-blog datasets used for the algorithm are described in Sect. 4, while Sect. 5 discusses the results of the comparison among the various clustering algorithms. The paper is concluded in Sect. 6 with some potential future research directions.

2   Related Work

Many prior works have been done on micro-blogging data clustering. Hill et al. [6] discuss how social network-based clusters can capture homophily along with the possibility that a network-based attribute approach might not only capture homophily but also can be used instead of demographic attributes to determine the similarity in user behavior, thus preserving privacy of the user base.

Cheong [2] attempts to detect intra-topic user and message clusters in Twitter, by incorporating an unsupervised self-organizing feature map (SOM) as an machine learning-based clustering tool. Thomas et al. [13] propose an efficient text classification scheme using clustering based on semi-supervised clustering as a complementary step to text classification. The method provides better accuracy than the similarity measure for text processing (SMTP) used for distance calculation.

Yang and Leskovec [16] have proposed a clustering method by using temporal patterns of propagation. Karypis et al. [8] propose a hierarchical clustering algorithm using dynamic modeling which takes the dynamic modes of clusters and the adaptive merging decision; that is, depending upon the difference in clustering model, it can discover natural clusters of various shapes and sizes. It supports a two-phase framework that has been built effectively using various graph representations suitable for various application domains.

Dueck et al. [3] propose an affinity propagation algorithm for clustering tweets. Dutta et al. [4] use a graph-based community detection algorithm for clustering tweets and later use the clustering output for summarization. Recently, Rangrej et al. [10] have conducted a comparative study on three clustering algorithms—K-means, affinity propagation, and singular value decomposition algorithm—and have compared their performance in clustering short text documents.

3 Micro-blog Clustering Algorithms

3.1 Data Preprocessing

Micro-blogging data often contains non-textual characters such as smileys, @usernames, exclamation/question marks, which acts as noise and degrades clustering performance. So dataset needs to be preprocessed. So, initially, stopwords, URLs, numerals, addressing, user mentions, e-mails, and special characters are removed from the dataset and stemmed.

This section describes the proposed methodology in detail. Each tuple in the dataset represents a single message or post or tweet(document) in Twitter. All the dataset is tokenized first, and a list of unique tokens are identified. Then, a document-term matrix is generated where rows(M) represent individual tweets and the columns represent distinct terms/tokens(Z). The entries in the matrix represent the presence(1) or absence(0) of a particular term/token in the post/tweet. Table 4 shows the corresponding matrix for the set of tweets in Table 1.

Next, dimension of document-term matrix is reduced using an Information Theoretic Approach. So, for each term/token conditional probability(p values) is evaluated using Bayes's Rule. The standard formula for Bayes's Rule is shown in Eq. 1:

$$P(H \mid E) = \frac{[P(E \mid H) * P(H)]}{P(E)} \tag{1}$$

Table 1 Document-term matrix for the toy dataset shown in Table 4

Tweet ID	$attr_1$	$attr_2$	$attr_3$	...	...	$attr_Z$
T_1	1	1	1	0	0	0
T_2	1	0	0	1	1	1
T_3	1	0	0	1	0	0
T_4	1	0	0	0	0	0
T_5	1	0	0	0	0	1
..	0	0	1	0	0	1
T_M	0	0	0	1	1	0

Here H represents occurrence of a token in entire dataset and E represents the occurrence of appearance of a token in all tweets.

Then, mean p value is computed which is compared with p value of each individual term/token. Terms/tokens are discarded from the document-term matrix whose p value is higher than mean p value. According to Shannon's theory of communication, the mean p value represents average information yield. The method derived by Shannon clearly states that the token with lower the p value yields the higher self-information content, shown in Eq. 2. Now, considering new subset of terms/tokens document-term matrix(M x G) is regenerated as shown in Table 2:

$$I(W_n) = f(P(W_n)) \tag{2}$$

The proposed algorithm aims to cluster the dataset. The methodology is briefly outlined in Algorithm 1. Before clustering we are considering a topic modeler approach to identify probable number of clusters. So, LDA (Latent Dirichlet Allocation) topic modeler is used here which is a generative process that can be used to identify the features capable of identifying a topic in the dataset. LDA or Latent Dirichlet Allocation is a statistical generative process that takes three inputs, n (the number of topics), alpha and theta (alpha and theta are hyperparameters, i.e., param-

Table 2 Reduced document-term matrix

Tweet ID	$attr_1$	$attr_2$	$attr_5$	...	$attr_G$
T_1	1	1	0	0	0
T_2	1	0	1	1	1
T_3	1	0	1	0	0
T_4	1	0	0	0	0
T_5	1	0	0	0	1
..	0	0	1	0	1
T_M	0	0	1	1	0

eters of prior distribution) and returns n-tuples, where each tuple represents a topic and its corresponding features are used to identify the topic. The overall runtime complexity [12] of LDA method is $O((NT)^t(N + t)^3)$.

The expression is a polynomial in nature when the total number of topics is constant. The inference function belongs to the NP-hard class of problems when the number of topics is large. The number of topics is relatively small in our experimental datasets, and thus, the LDA performs well. If the number of topics is large for any certain dataset, the performance of the LDA algorithm may decline. The inference function, thus, has to be augmented in a way, so that the LDA algorithm performs better even if the number of topics becomes significantly large for a dataset.

To apply LDA approach on the experimental dataset, we need to measure alpha and theta using Bayesian inference function, as shown in Eq. 3.

$$p(\tilde{x} \mid X, \alpha) = \int_\theta p(\tilde{x} \mid \theta)p(\theta \mid X, \alpha)d\theta \qquad (3)$$

With the help of Information Theoretic Approach (Shannon's approach), the optimal number of groups, into which a text dataset can be divided, is determined. Using the Shannon's proposed formula for calculating the information yield of a particular message, we have reduced the total number of features that are initially generated by extracting the corpus from the dataset, i.e., the number of unique features that can be used to represent the entire dataset. This number can be used as optimal number of topics to be detected from the dataset.

We compute this value as an inverse probability of zero elements. That is, if in the document-term matrix total number of elements is N and total number of zeros is K, the probability of occurrence of zeros can be calculated as $I = K/N$ as the event is an independent one. Thus, the probability of occurrence of nonzero elements would be $T = 1 - I$. The optimal number of topics is then derived as—ceiling [Length(Corpus)/T].

The LDA method not only identifies the topics but also returns a list of reduced features (term/token) that can be used to represent the entire dataset and also to reduce the size of data that can later be processed by the clustering algorithm. LDA returns the list of reduced feature K. Then, a topic-feature matrix (F) is generated considering T × K dimensions, where T is total number of topics and K is total number of features. If a feature f is present in a topic, the attribute is marked as 1; otherwise, it is marked as 0. Table 3 shows the corresponding matrix.

Algorithm 1 Micro-Blogging Data clustering algorithm based on feature selection

---

Input: L number of tweets
Output: Tweets partitioned into T number of clusters.
 Pre-process all tweets in the dataset by removing Stopwords, Special characters, User mentions, URLs, Emails;
 Stem all the tokens;
 Create the corpus C as the list of unique tokens that forms the entire dataset;
 Let N = distinct number of tokens in corpus;
 Compute DM = document-term matrix of order L x N;
 for each token in C **do**
   Compute p-value using conditional probability of occurrence of each term by Bayes rule;
 end for
 Calculate the mean p-value(mp);
 for all $C_i, i \in [1, L]$ **do**
   Remove token if p > mp
 end for
 Reform the document-term matrix(DM) based on the new reduced corpus;
 Use Bayesian inference to compute the parameters alpha and theta from DM;
 z= Number of zero elements in DM;
 N= Total number of elements in DM;
 P(z) = z/N;
 P(Nz) = 1-P(z) where Nz;
 Calculate number of topics T = Ceiling [Length(C) / P(Nz)];
 Run LDA(T, alpha, theta) store Feature/topic in list FL;
 Prepare document-feature matrix (D);
 Prepare F, a feature matrix that maps each feature per topic with respect to total features returned by LDA;
 i,j=0
 while data in D **do**
   for all $vector_i, i \in [1, F]$ **do**
     Calculate hamming distance between data and vector as d = Hamming (data, $vector_i$);
     Update the Distances[i][j++] vector with the hamming distance,d;
   end for
   for all $row_i, i \in [1, Distances]$ **do**
     Calculate and store the mean value of the present row in list row-mean
   end for
 end while
 while data in D **do**
   for all $x, x \in [0, distance]$ **do**
     for all $v, v \in [0, x]$ **do**
       if v <= row-mean[index(x)] **then**
         Mark and store corresponding tweet to cluster c = index(row-mean)
       else
         Continue
       end if
     end for
   end for
 end while

Table 3 Topic-feature matrix

Tweet ID	$attr_1$	$attr_2$	$attr_5$	...	$attr_K$
$Topic_1$	0	1	1	0	0
$Topic_2$	1	0	1	0	1
$Topic_3$	0	1	1	0	1
$Topic_K$	1	1	0	0	0

Considering LDA computed features, another smaller document-feature matrix(D) is generated to represent the tweets as binary vectors. The document-feature matrix is an L × K matrix, where L is total number of tweets and K is total number of features. The matrix is formed using the same principle as the document-feature matrix is prepared.

In the topic-feature matrix F, each row of the matrix represents a feature vector to describe a particular topic. To identify the tweet topic, co-relation mean hamming distance is measured for each topic considering the mean distance between all tweet vectors in document-feature matrix (D) and each topic vector in topic-feature matrix (F). Each tweet is assigned to that topic/cluster whose tweet-topic hamming distance is less than or equal to the mean distance value. The proposed clustering algorithm is based on greedy-based approach. All the tweets are assigned to distinct clusters, and overlapping clustering concept is not considered in the proposed approach.

4 Dataset for Clustering Algorithms

We have collected data from Twitter using API [14] with keyword-based matching. Keywords like "Uttarakhand" and "flood" are used to select tweets relevant to the Uttarakhand flood. Similarly, keywords 'Hyderabad,' 'bomb,' and 'blast' are used to extract tweets related to the HDBlast event, where the keywords such as 'Sandyhook' and 'shooting' are used to identify tweets related to the SHShoot event.

We have chosen tweets related to a specific event during natural disaster and socio-emergency. In this section, the datasets are described briefly which are used for clustering algorithms.

We considered tweets posted during the following emergency events.

1. **HDBlast**—two bomb blasts in the city of Hyderabad, India [7],
2. **SHShoot**—an assailant killed 20 children and 6 adults at the Sandy Hook elementary school in Connecticut, USA [11],
3. **THagupit**—a strong cyclone code-named Typhoon Hagupit hit Philippines [5],
4. **UFlood**—devastating floods and landslides in the Uttaranchal state of India [15].

Table 4 gives examples of tweets of Hagupit Typhoon.

Table 4 Examples of tweets related to the Hagupit Typhoon event

Tweet ID	Example tweets (extract from tweet text)
T_1	JTWC forecasts 22W (#Hagupit) to be near Yap by Thursday as 70-knot typhoon. Who'll bet it's more like 140-knots? http://t.co/lkGOJBcwOG
T_2	The maximum Storm surge height is 0.1 m in #Colonia, Micronesia. This height is estimated for 04 Dec 2014 03 HOURS UTC #HAGUPIT #22W
T_3	CURRENT INFRARED IMAGE OF TROPICAL DEPRESSION #22W (#HAGUPIT) http://t.co/tJLPGlC3wR
T_4	#CHINA (CMA) SAYS, #HAGUPIT TO HIT #PHILLIPPINES WITH MAXIMUM WIND SPEED OF 245 KM/H (CAT-4 HURRICANE) ON 06 DEC 2014 http://t.co/Q9AVnae896
T_5	#Hagupit Predicted Path/Track of #TyphoonHagupit http://t.co/ms4VpDwyTQ http://t.co/DMBJgdGIAO

5 Experimental Results

We have described the experimental results in this section. The proposed approach is compared here with several classical clustering methods such as K-means and hierarchical clustering algorithms. To compare the performance of the proposed clustering algorithm, we have used few clustering methods such as (1) K-means clustering, (2) hierarchical clustering on the tweet dataset. K-means is an unsupervised learning algorithm which follows a simple and easy method to classify a given dataset through a number of clusters. Hierarchical clustering involves creation of clusters that have ordering from top to bottom. Widely used toolkit, Weka is used for the above-mentioned methodologies.

We have used a set of standard metrics [9] to evaluate the quality of clustering produced by the different methodologies. The following metrics were used:

(1) Calinski–Harabasz (CH) index evaluates the average between and within-cluster sum of squares to identify the cluster validity measure.
(2) The DB (DB) index measures the average of similarity between each cluster and its most similar one.
(3) Dunn (D) index uses the maximum diameter among all clusters to estimate the intra-cluster compactness and the minimum pair-wise distance between objects in different clusters as the intercluster separation.
(4) I-Index (I) evaluates separation based on the maximum distance between cluster centers, and it also evaluates compactness based on the sum of distances between objects and their cluster center.

(5) Silhouette (S) index validates the clustering performance measuring the pairwise difference of between and within-cluster distances.

(6) Xie-Beni (XB) index defines the intra-cluster compactness as the mean square distance between each data object and its cluster center and the intercluster separation as the minimum square distance between cluster centers.

Among all the indices, few are maximization index and few are minimization index. Smaller values for DB index and Xie-Beni index indicate better clustering performance, while higher values for Dunn index, Silhouette index, Calinski-Harabasz index, and I-index indicate better clustering performance. The reader is referred to [9] for a detailed description of all these metrics.

Table 5 Selection of parameters for the proposed methodology. The best performance according to each metric is marked in bold in each case

Method	S	CH	DB	I	XB	D
Uttarakhand dataset						
Proposed method	**0.1925**	3.1057	**0.759**	0.0548	**2.0994**	**1.172**
K-means method	0.17	3.2381	1.5100	**2.7545**	67.5705	0.871
Hierarchical method	0.164	**5.8153**	0.898	1.0472	2.4204	0.858
HYDB dataset						
Proposed method	**0.2643**	3.336	**0.887**	0.111	**1.5077**	**1.201**
K-means method	0.037	3.6289	1.513	0.3264	70.1213	0.964
Hierarchical method	0.044	**5.3888**	0.967	**3.5806**	1.6293	0.797
Hagupit dataset						
Proposed method	**0.2708**	**43.4864**	**0.795**	0.1425	**0.1319**	**1.52**
K-means method	0.2065	0.2567	1.682	**0.8984**	508.5419	0.976
Hierarchical method	0.025	15.1149	1.026	0.001	1.439	0.766
Sandyhook dataset						
Proposed method	**0.2643**	11.1302	**0.651**	0.1119	**0.558**	1.057
K-means method	0.152	12.2388	1.566	**1.6661**	1616.9	0.731
Hierarchical method	0.102	**16.036**	0.936	0.0749	100.3326	**0.833**

5.1 Metrics for Evaluating Clustering

Table 5 compares the performance of the proposed methodology with that of the classical approaches, according to the various metrics. The best performance according to each metric is highlighted in boldface.

The proposed methodology performs better than all the baseline approaches according to all the metrics except I-index and CH index (for which hierarchical clustering achieves the best performance). These results indicate the superior performance of the proposed feature selection-based tweet clustering approach.

6 Conclusion

In the proposed work, we have presented a method to cluster a set of micro-blogging data or tweets by generating the topic model and then reducing the features from the generated topic model. This work introduces a simple and effective methodology for tweet clustering combining feature selection. The experimental results show that the proposed algorithm performs better than some standard clustering approaches. As future work, we would like to improve the clustering algorithm, so that it can apply for incremental dataset of micro-blogging Web sites. We can extend the work to generate a topic-oriented summary of tweets by using automated summarization techniques. Primarily, the clustering technique follows a greedy selection method, which can be later substituted by a suitable backtracking selection method to produce more coherent clusters and thus will improve the overall performance.

References

1. Blei, D.M., Ng, A.Y., Jordan, M.I.: Latent dirichlet allocation. J. Mach. Learn. Res. **3**, 993–1022 (2003). URL http://dl.acm.org/citation.cfm?id=944919.944937
2. Cheong, M., Lee, V.C.S.: A study on detecting patterns in twitter intra-topic user and message clustering. In: ICPR, pp. 3125–3128. IEEE Computer Society (2010). URL http://dblp.uni-trier.de/db/conf/icpr/icpr2010.html#CheongL10
3. Dueck, D.: Affinity propagation: clustering data by passing messages. Ph.D. thesis, Citeseer (2009)
4. Dutta, S., Ghatak, S., Roy, M., Ghosh, S., Das, A.K.: A graph based clustering technique for tweet summarization. In: 2015 4th International Conference on Reliability, Infocom Technologies and Optimization (ICRITO) (Trends and Future Directions), pp. 1–6 (2015). DOI https://doi.org/10.1109/ICRITO.2015.7359276
5. Typhoon Hagupit – Wikipedia (2014). http://en.wikipedia.org/wiki/Typhoon_Hagupit
6. Hill, S., Benton, A., Ungar, L., Macskassy, S., Chung, A., Holmes, J.H.: A cluster-based method for isolating influence on twitter (2016)
7. Hyderabad blasts – Wikipedia (2013). http://en.wikipedia.org/wiki/2013_Hyderabad_blasts
8. Karypis, G., Han, E.H.S., Kumar, V.: Chameleon: Hierarchical clustering using dynamic modeling. Computer **32**(8), 68–75 (1999). DOI https://doi.org/10.1109/2.781637. URL http://dx.doi.org/10.1109/2.781637

9. Liu, Y., Li, Z., Xiong, H., Gao, X., Wu, J.: Understanding of internal clustering validation measures. In: Proceedings of the 2010 IEEE International Conference on Data Mining,
   ICDM '10, pp. 911–916. IEEE Computer Society, Washington, DC, USA (2010). DOI https://
   doi.org/10.1109/ICDM.2010.35. URL http://dx.doi.org/10.1109/ICDM.2010.35
10. Rangrej, A., Kulkarni, S., Tendulkar, A.V.: Comparative study of clustering techniques for
    short text documents. In: Proceedings of the 20th international conference companion on World
    wide web, pp. 111–112. ACM (2011)
11. Sandy Hook Elementary School shooting – Wikipedia (2012). http://en.wikipedia.org/wiki/
    Sandy_Hook_Elementary_School_shooting
12. Sontag, D., Roy, D.: Complexity of inference in latent dirichlet allocation. In: J. Shawe-Taylor,
    R.S. Zemel, P.L. Bartlett, F. Pereira, K.Q. Weinberger (eds.) Advances in Neural Information
    Processing Systems 24, pp. 1008–1016. Curran Associates, Inc. (2011). URL http://papers.
    nips.cc/paper/4232-complexity-of-inference-in-latent-dirichlet-allocation.pdf
13. Thomas, A.M., Resmipriya, M.: An efficient text classification scheme using clustering. Pro-
    cedia Technology **24**, 1220–1225 (2016)
14. REST API Resources, Twitter Developers. https://dev.twitter.com/docs/api
15. North India floods – Wikipedia (2013). http://en.wikipedia.org/wiki/2013_North_India_floods
16. Yang, J., Leskovec, J.: Patterns of temporal variation in online media. In: Proceedings of the
    fourth ACM international conference on Web search and data mining, pp. 177–186. ACM
    (2011)

Correction to: Comparative Evaluation of Various Feature Weighting Methods on Movie Reviews

S. Sivakumar and R. Rajalakshmi

Correction to:
Chapter "Comparative Evaluation of Various Feature Weighting Methods on Movie Reviews" in:
H. S. Behera et al. (eds.), *Computational Intelligence in Data Mining*, Advances in Intelligent Systems and Computing 711, https://doi.org/10.1007/978-981-10-8055-5_64

In the original version of the book, the following post-publication corrections have been incorporated in the chapter "**Comparative Evaluation of Various Feature Weighting Methods on Movie Reviews**":

The second author, R. Rajalakshmi, has been changed as the corresponding author.

The second affiliation "**Department of Computer Science, Dhanalakshmi College of Engineering, Chennai 600127, Tamil Nadu, India**" of the corresponding author, R. Rajalakshmi, has been removed.

The correction chapter and the book have been updated with the changes.

The updated version of this chapter can be found at
https://doi.org/10.1007/978-981-10-8055-5_64

Author Index

© Springer Nature Singapore Pte Ltd. 2019
H. S. Behera et al. (eds.), *Computational Intelligence in Data Mining*,
Advances in Intelligent Systems and Computing 711,
https://doi.org/10.1007/978-981-10-8055-5

Printed in the United States
By Bookmasters